PROGRESS IN

Nucleic Acid Research and Molecular Biology

Volume 81

PROGRESS IN
Nucleic Acid Research and Molecular Biology

edited by

KIVIE MOLDAVE

Department of Molecular Biology and Biochemistry
University of California, Irvine
Irvine, California

Volume 81

AMSTERDAM • BOSTON • HEIDELBERG • LONDON
NEW YORK • OXFORD • PARIS • SAN DIEGO
SAN FRANCISCO • SINGAPORE • SYDNEY • TOKYO
Academic Press is an imprint of Elsevier

Academic Press is an imprint of Elsevier
525 B Street, Suite 1900, San Diego, California 92101-4495, USA
84 Theobald's Road, London WC1X 8RR, UK

This book is printed on acid-free paper. ∞

Copyright © 2006, Elsevier Inc. All Rights Reserved.

No part of this publication may be reproduced or transmitted in any form or by any means, electronic or mechanical, including photocopy, recording, or any information storage and retrieval system, without permission in writing from the Publisher.

The appearance of the code at the bottom of the first page of a chapter in this book indicates the Publisher's consent that copies of the chapter may be made for personal or internal use of specific clients. This consent is given on the condition, however, that the copier pay the stated per copy fee through the Copyright Clearance Center, Inc. (www.copyright.com), for copying beyond that permitted by Sections 107 or 108 of the U.S. Copyright Law. This consent does not extend to other kinds of copying, such as copying for general distribution, for advertising or promotional purposes, for creating new collective works, or for resale.
Copy fees for pre-2006 chapters are as shown on the title pages. If no fee code appears on the title page, the copy fee is the same as for current chapters.
0079-6603/2006 $35.00

Permissions may be sought directly from Elsevier's Science & Technology Rights Department in Oxford, UK: phone: (+44) 1865 843830, fax: (+44) 1865 853333, E-mail: permissions@elsevier.com. You may also complete your request on-line via the Elsevier homepage (http://elsevier.com), by selecting "Support & Contact" then "Copyright and Permission" and then "Obtaining Permissions."

For information on all Elsevier Academic Press publications
visit our Web site at www.books.elsevier.com

ISBN-13: 978-0-12-540081-7
ISBN-10: 0-12-540081-0

PRINTED IN THE UNITED STATES OF AMERICA
06 07 08 09   9 8 7 6 5 4 3 2 1

**Working together to grow
libraries in developing countries**

www.elsevier.com | www.bookaid.org | www.sabre.org

ELSEVIER   BOOK AID International   Sabre Foundation

Contents

Prokaryotic DNA Mismatch Repair 1
Nimesh Joseph, Viswanadham Duppatla, and Desirazu N. Rao

I. Introduction 1
II. DNA Mismatch Repair 3
III. Methyl-Directed DNA Mismatch Repair 7
IV. Molecular Structure of the DNA Mismatch Repair Proteins 15
V. Communication Between Mismatched Nucleotides and the Excision Machinery 22
VI. Effect of DNA-Damaging Agents on Mismatch Repair 27
VII. Methylation-Independent DNA Mismatch Repair 28
VIII. DNA Mismatch Repair vs Bacterial Virulence: An Ongoing Debate 31
IX. Applications of DNA Mismatch Repair Proteins and Processes 35
References 37

Pleiotropic Effects of Phosphatidylinositol 3-Kinase in Monocyte Cell Regulation 51
Sanaâ Noubir, Jimmy S. Lee, and Neil E. Reiner

I. Background 53
II. PI 3-Kinase and Lipopolysaccharide 54
III. PI 3-Kinase and $1\alpha,25$-Dihydroxyvitamin D_3 66
IV. PI 3-Kinase and Phagocytosis of Mycobacteria 75
V. Stable Gene Silencing of Isoforms of PI 3-Kinase and Perspectives 80
References 84

Selenocysteine Incorporation Machinery and the Role of Selenoproteins in Development and Health 97
Dolph L. Hatfield, Bradley A. Carlson, Xue-Ming Xu, Heiko Mix, and Vadim N. Gladyshev

I. UGA Codes for Sec 99
II. Novelty of Eukaryotic Sec tRNA$^{[Ser]Sec}$ 99
III. Sec Biosynthesis 109

IV. Sec Insertion into Protein... 112
V. Selenium and Selenoprotein Hierarchy.. 116
VI. Selenoprotein Identity, Function, and Targeted Removal 120
VII. Mouse Models for Elucidating the Role of Sec tRNA$^{[Ser]Sec}$ in Selenoprotein Biosynthesis and the Role of Selenoproteins in Development and Health.. 123
VIII. How Did UGA Evolve as the Sec Codon?.. 131
IX. Concluding Remarks 133
References 134

Indirect Readout of DNA Sequence by Proteins: The Roles of DNA Sequence-Dependent Intrinsic and Extrinsic Forces 143

Gerald B. Koudelka, Steven A. Mauro, and Mihai Ciubotaru

I. Introduction 144
II. Indirect Readout is a Common DNA Recognition Mechanism............. 145
III. Indirect Readout Strategies 146
IV. Structural and Physicochemical Basis for DNA Sequence-Dependent Structural Polymorphisms 148
V. Indirect Readout by Bacteriophage Repressor Proteins...................... 154
VI. *In Vivo* Studies of Cation-Dependent 434 Repressor Gene Regulatory Activity... 169
VII. Summary and Prospects.. 170
References 171

Repair of Topoisomerase I-Mediated DNA Damage ... 179

Yves Pommier, Juana M. Barcelo, V. Ashutosh Rao, Olivier Sordet, Andrew G. Jobson, Laurent Thibaut, Ze-Hong Miao, Jennifer A. Seiler, Hongliang Zhang, Christophe Marchand, Keli Agama, John L. Nitiss, and Christophe Redon

I. Introduction: Mammalian Topoisomerase Families, Top1 Functions, and Catalytic Mechanisms.. 180
II. Induction and Stabilization of Top1 Cleavage Complexes by Camptothecin and Anticancer Drugs and by Carcinogens and Endogenous DNA Lesions 184
III. Conversion of Top1 Cleavage Complexes into DNA Damage............... 188
IV. Repair of Top1-Associated DNA Damage.. 189
V. Checkpoint Response to Top1-Associated DNA Damage 210
VI. Conclusion and Perspective 213
References 214

Regulation of L-Histidine Decarboxylase and Its Role in Carcinogenesis ... 231

Wandong Ai, Shigeo Takaishi, Timothy C. Wang, and John V. Fleming

I. Introduction	232
II. Transcriptional Regulation of HDC	233
III. Posttranscriptional Regulation of HDC	239
IV. Potential Role of HDC in Cancer Development	251
V. Concluding Remarks	258
References	260

Eukaryotic Initiation Factor 2B and Its Role in Alterations in mRNA Translation That Occur Under a Number of Pathophysiological and Physiological Conditions ... 271

Neil Kubica, Leonard S. Jefferson, and Scot R. Kimball

I. Introduction	272
II. Regulation of eIF2B GEF Activity	272
III. eIF2B Subunit Composition and Function	275
IV. Role of eIF2B in Contributing to Alterations in mRNA Translation Under Various Pathophysiological and Physiological Conditions	277
V. Unanswered Questions	287
References	287

Role of Protein Tyrosine Phosphatases in Cancer ... 297

Tasneem Motiwala and Samson T. Jacob

I. Introduction	298
II. Genetic and Epigenetic Alterations of *PTP* Genes	299
III. Transformation-Related Phenotypes Attributable to PTPs	304
IV. PTPs as Drug Targets	311
V. Concluding Remarks	316
References	320

The Case for mRNA 5' and 3' End Cross Talk During Translation in a Eukaryotic Cell ... 331

Anastassia V. Komarova, Michèle Brocard, and Katherine M. Kean

I. Introduction	331
II. Classical Eukaryotic Cellular mRNAs	337

III. 5′-3′ RNA–RNA Interactions	339
IV. Examples of RNA–Protein Interactions	344
V. Protein–Protein Interaction Stories	354
VI. Concluding Remarks	358
References	358

Interferon Action and the Double-Stranded RNA-Dependent Enzymes ADAR1 Adenosine Deaminase and PKR Protein Kinase 369

Ann M. Toth, Ping Zhang, Sonali Das, Cyril X. George, and Charles E. Samuel

I. Introduction	369
II. Adenosine Deaminase Acting on RNA	370
III. Protein Kinase Regulated by RNA	390
IV. Possible Roles of ADAR1 and PKR in Human Genetic and Infectious Diseases	409
References	411

Establishment and Regulation of Chromatin Domains: Mechanistic Insights from Studies of Hemoglobin Synthesis 435

Emery H. Bresnick, Kirby D. Johnson, Shin-Il Kim, and Hogune Im

I. Introduction	436
II. β-Globin Locus as a Model System to Dissect Chromatin Domain Regulation	437
III. Protein Components of the Endogenous β-Globin Chromatin Domain	441
IV. Developmentally Dynamic Histone Modification Pattern of the Murine β-Globin Locus: Molecular Determinants and Functional Implications	449
V. Multistep Model of β-Globin Chromatin Domain Activation	450
VI. A Role for the LCR in Establishing and Regulating the Erythroid Cell-Specific β-Globin Chromatin Domain	454
VII. General Insights into Chromatin Domain Establishment and Regulation	455
VIII. Concluding Remarks	458
References	459

Detecting the Unusual: Natural Killer Cells 473

Armin Volz and Britta Radeloff

I. Introduction	474
II. Discovery and Characterization of Natural Killer Cells	474
III. Natural Killer Cell Receptors	479
IV. Expression of NK-Receptor Genes	501
V. Natural Killer Cells and Disease	514
VI. Concluding Remarks	518
References	519
Index	543

Prokaryotic DNA Mismatch Repair

Nimesh Joseph,
Viswanadham Duppatla, and
Desirazu N. Rao

*Department of Biochemistry,
Indian Institute of Science,
Bangalore 560 012, India*

I. Introduction	1
II. DNA Mismatch Repair	3
III. Methyl-Directed DNA Mismatch Repair	7
IV. Molecular Structure of the DNA Mismatch Repair Proteins	15
V. Communication Between Mismatched Nucleotides and the Excision Machinery	22
VI. Effect of DNA-Damaging Agents on Mismatch Repair	27
VII. Methylation-Independent DNA Mismatch Repair	28
VIII. DNA Mismatch Repair vs Bacterial Virulence: An Ongoing Debate	31
IX. Applications of DNA Mismatch Repair Proteins and Processes	35
References	37

Repair of base mismatches in *Escherichia coli* and related bacteria is performed by two molecular yet overlapping processes: the long-patch mismatch repair and very short patch mismatch repair pathways. DNA mismatch repair is inevitable to maintain genomic stability and is highly conserved from prokaryotes to eukaryotes. Availability of several completely sequenced bacterial genomes has helped in the identification of proteins, which are involved in the DNA mismatch repair process and their subsequent biochemical characterization. Comparative studies of the activities of these proteins have helped in elucidating molecular pathway involved in the complex process of DNA mismatch repair. The characteristic features of the prokaryotic DNA mismatch repair proteins and their biochemical activities are reviewed here.

I. Introduction

DNA is recognized as the informationally active chemical component of essentially all genetic material (with the exception of the RNA viruses). This macromolecule is assumed to be extraordinarily stable such that the high degree of fidelity required of a master blueprint can be maintained. But, the physicochemical constitution of genes does not guarantee lifelong stability or proper function. The primary structure of DNA is quite dynamic and subject to constant changes; from large-scale variations, such as transposition, to alterations in the chemistry or sequence of individual nucleotides. Many of these changes occur as a consequence of errors introduced during replication, recombination, and repair itself.

In general, DNA repair can be defined as a range of cellular responses associated with the restoration of the genetic instructions as provided by the normal primary DNA sequence and structure (1). In *Escherichia coli*, spontaneous mutation frequency in newly replicated DNA is expected to be of the order of 10^{-5}–10^{-6} when the involvement of any external factors is excluded (2). If DNA were copied badly, the living organisms would not have lived longer and if it were copied perfectly, there would have been no room for evolution. Genetic variation arising due to local sequence alterations, with appropriate care taken to limit such changes to tolerable levels, brings DNA repair genes under the category of evolution genes (3). The DNA repair system is a balanced mechanism, which preserves the replication fidelity together with unhindered evolution. In human copying system, DNA repair allows only an average of 3 base pair (bp) errors while replicating the 3 billion bps in the human genome (2).

The ability of cells to tolerate DNA damage is biologically as important as their ability to repair such alterations. It has been suggested that when DNA damage became an evolutionarily advanced problem for living cells, genetic outcrossing or sexual recombination evolved as an efficient means for exchanging "good" bits of DNA for "bad," without cells having to permanently retain a redundant copy of the entire genome. It is, therefore, important to understand the ability of bacterial cells to exchange genetic information by successfully tolerating DNA damage (4).

In view of the plethora of types of DNA lesions, no single repair process can cope with all kinds of DNA damage. There are wide spectra of responses existing in the living cells to maintain the genome integrity in the face of replication errors, environmental assaults, and the cumulative effect of age as far as human health is concerned. Sequencing the complete 1.2 Mb double-stranded DNA genome of Mimivirus, the largest known virus that grows in amoebae, has revealed that its 1262 putative open reading frames (ORFs), which include the coding sequences for components of all DNA repair pathways (5). This is a remarkable feature that is identified for the first time in a virus. Although terminally differentiated cells do not replicate their genomic DNA, there is increasing experimental evidence that some of the repair pathways are operating in these cells such as nucleotide excision repair and transcription-coupled repair. In neurons, the nontranscribed strand is also well repaired, which is designated as differentiation-associated repair (DAR) (6).

DNA MISMATCH REPAIR

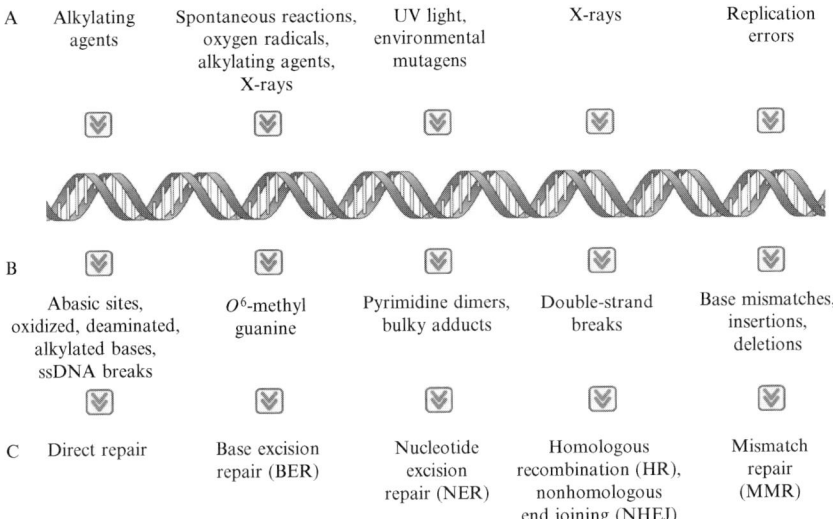

FIG. 1. DNA damage, repair mechanisms, and consequences. Represented here are the common DNA-damaging agents (A), examples of DNA lesions induced by these agents (B), and most relevant DNA repair mechanism responsible for the removal of the lesions (C).

DNA repair, in general, can occur by one of the two fundamental cellular events that involve either the direct reversal of DNA damage or its excision (2, 7). Some of the most common types of DNA damage and their sources are summarized in Fig. 1. As our understanding of the genetic, biochemical, and structural basis of DNA damage and repair has increased dramatically, it is not possible to discuss all the aspects of this complex process. There are several scientific reviews available, which deal with the multitude of molecular processes involved in repairing various DNA lesions. Here, only the prokaryotic DNA mismatch repair pathway will be discussed in detail.

II. DNA Mismatch Repair

Both prokaryotic and eukaryotic cells are capable of repairing mismatched base pairs in their DNA. Base mismatches in DNA can occur by several processes, such as replication errors, during homologous recombination as well as due to deamination of 5-methyl cytosine to thymine (2). Semiconservative synthesis of DNA accounts for the bulk of such base misincorporations.

In the absence of intervention by any cellular factors, the intrinsic error frequency of base mispairing during DNA synthesis is 10^{-1}–10^{-2}. The proofreading activity of DNA polymerase (associated $3' \rightarrow 5'$ exonuclease function) as well as the assistance of certain accessory proteins, such as single-strand DNA binding protein (SSB), reduces the error frequency to 10^{-5}–10^{-7}. The fidelity of DNA replication is further enhanced by a factor of 10^3 by the error surveillance activity of postreplicative mismatch correction thereby bringing down the cumulative error frequency as low as 10^{-10} (2, 8). For a mismatch repair event to contribute to the maintenance of the genetic fidelity, the correct base in the mispair must be distinguished from the incorrect one (9).

The most complex and well-understood bacterial mismatch repair pathways are the methyl-directed mismatch repair system of *E. coli* (9) and the related hex [*h*igh *e*fficiency, unknown (*x*)]-dependent mismatch repair system of *Streptococcus pneumoniae* (10, 11). Both systems have similar specificity for different mismatches, which they process in a strand-specific manner. Mutants deficient in mismatch repair have been well characterized in *E. coli*, *Salmonella typhimurium*, and *S. pneumoniae* (10–14). *E. coli* and *S. typhimurium* mutants of *mutH*, *mutS*, *mutL*, and *mutU* (same as *uvrD*, *uvrE*, and *recL*) and *S. pneumoniae* mutants of *hexA* and *hexB* all showed increased spontaneous mutation and recombination rates. In addition, the four *mut* mutants of *E. coli* showed the tex (*t*ransposon *ex*cision) phenotype, reflecting increased rate of spontaneous precise recombination between very short repeated sequences flanking integrated Tn5 and Tn10 transposons, thus leading to texs (15). These phenotypes implied that the functional MutHLSU and HexAB systems are inevitable to prevent spontaneous mutagenesis and recombination. In *E. coli*, inactivation of mismatch repair provides two sources of genetic variability: (i) it increases the frequency of base substitutions and small deletions/insertions by the lack of putative replicative DNA repair (16), and (ii) it eliminates a barrier against recombination between divergent (homologous) sequences (17, 18). The MutS protein plays an essential role in these two factors. It is the recognition by MutS of heterologies in the recombination intermediates, which improves the completion of recombination (19). On a biochemical level, MutS and MutL block RecA-mediated strand exchange between the fd and M13 genomes, which are 3% divergent, but not between M13 and M13 genomes (19). In the homeologous recombination, MutS has a greater effect than the MutL, which is effective only in combination with MutS. The close connection between the mutation avoidance and antirecombination was shown by the phenotype of *E. coli mutS* mutations, which are defective in both these processes (17). Study from Marinus group showed that C-terminal end of MutS is necessary for antirecombination but less significantly for mutation avoidance (20). Hong *et al.* (21) found that strains with a temperature sensitive *mutS* allele allow more frequent homeologus recombination at temperatures at which mutations occur at relatively low frequencies and show that the altered MutS protein is titrated by the mispairs encountered in interspecies mating.

Deficiency of mismatch repair system was found to activate intergeneric recombination between *E. coli* and *S. typhimurium* (22). Moreover, mismatch repair systems can conserve (antirecombination) and diversify (hyperrecombination) genetic information by virtue of the size of their repair patch. The mismatch repair system was found to be functional in the radio-resistant organism *Deinococcus radiodurans*, which was implicated in ensuring the fidelity of DNA replication and recombination. In contrast, cells devoid of MutS or MutL proteins were as resistant to γ-rays, mitomycin C, and UV-irradiation as wild-type bacteria, suggesting that the mismatch repair system is not essential for the reconstitution of a functional genome after DNA damage (23).

The mismatch repair systems are broadly classified into two, based on the length of the excision tract. The strand specificity of the repair is determined by secondary signals that can be located at considerable distance from the actual mismatch. The excision tract associated with these pathways can be large, 1000 bp or more, both *in vivo* (24, 25) and *in vitro* (26) and hence these systems are usually referred to as *long-patch mismatch DNA repair* (LPMMR) *systems*. This terminology does not imply that all the patches are necessarily long as the lower limit for the excision repair tract is not yet established (27).

Prokaryotic cells possess mismatch repair systems that are characterized by *short excision repair tracts* (typically 20 nucleotides or less) (28). They have a relatively restricted specificity with respect to the mismatches that they repair and are operational only within a limited sequence context (29). The *very short patch mismatch repair system* (VSPMMR) of *E. coli* is well characterized, which efficiently corrects the "T" in "GT" mismatches that occur in sequences resembling CC(A/T)GG. The internal "C" in this sequence is methylated by the deoxycytosine methylase (Dcm)-encoded methylase generating 5-methylcytosine (30). Dcm recognition sites are hot spots for mutation due to the spontaneous deamination of 5-methylcytosine to thymine yielding a "GT" mispair (31). Two of the genes necessary for LPMMR, *mutS* and *mutL*, are required for VSPMMR also (32, 33), although it remains unclear at which stage they are involved. MutS and MutL proteins are believed to regulate the very short patch repair (Vsr) strand-specific mismatch endonuclease (34) that recognizes "GT" mismatches in a CT(A/T)GG or NT(A/T)GG sequence context. Vsr endonuclease makes incisions 5′ to the underlined "T" and the mispaired "T" is removed by the 5′→3′ exonucleolytic activity of DNA polymerase I (35). MutL is shown to stimulate the binding of Vsr and MutS to heteroduplex DNA (36). The crystal structure of the *E. coli* Vsr endonuclease bound to a "GT" mismatch has been determined (37) and studies through structure-guided mutagenesis suggest that MutL causes a conformational change in the N-terminus of Vsr endonuclease, which enhances Vsr activity (38). *E. coli* strains which are deoxyadenosine methylase (Dam)-deficient are shown to be defective in VSPMMR, which is also associated with decreased levels of Vsr. Although the exact mechanism of the involvement of Dam is not clear, it is believed that the regulation of Vsr by Dam is probably posttranscriptional (39).

A second function of VSPMMR is to repair "UG" mismatches in Dcm sequences that arise by deamination of cytosine (40). Uracils arising by the deamination of cytosine are normally recognized by uracil-DNA glycosylase and repaired. The VSP repair of "UG" mismatches is not as efficient as the uracil-DNA glycosylase system, although it contributes to the maintenance of genetic fidelity at these sites (41). The "AG" and "AC" mispairs in *E. coli* are found to be corrected by a VSPMMR system that is independent of the *mutH, L,S* pathway and is methylation independent. This function is carried out by a protein MutY, which can also function as a DNA glycosylase with an associated 3'-AP (apurinic/apyrimidinic) lyase activity. MutY specifically removes "A" from "GA" or "CA" mispairs (42–44). Furthermore, there may also be other mismatch repair systems that have not yet been characterized. The requirements for VSPMMR are summarized in Table I.

TABLE I
COMMON DNA MISMATCH REPAIR PATHWAYS AND THE LIST OF FACTORS INVOLVED IN THE REPAIR PROCESS

Long-patch mismatch repair (LPMMR)	MutS
	MutL
	MutH
	Dam (N^6-adenine methylase at GATC)
	UvrD (DNA helicase II)
	RecJ (5' → 3' specific)
	ExoI (3' → 5' specific)
	ExoVII (5' → 3' specific)
	ExoX (3' → 5' specific)
	SSB (single-strand DNA binding protein)
	DNA polymerase III holoenzyme
	DNA ligase
	Cofactors : ATP, NAD^+, Mg^{2+}
Very short patch mismatch repair (VSPMMR)	MutS
	MutL
	Dcm (C^5-cytosine methylase at CCA/TGG)
	Vsr (very short patch repair endonuclease)
	DNA polymerase I
	DNA ligase
	Cofactors: ATP, NAD^+, Mg^{2+}

III. Methyl-Directed DNA Mismatch Repair

Of the several mismatch correction systems that have been identified in *E. coli*, the most interesting with respect to mechanism has been the methyl-directed LPMMR pathway. Direct proof for the existence of a mechanism for correction of mismatched base pairs was provided in the *E. coli* system by transfection with λ, ϕX174, and T7 heteroduplexes, marked genetically on the two DNA strands thereby allowing to score for the repair event (24, 25, 45–47). Such experiments had demonstrated that the incorrect base pairs were removed from the heteroduplexes prior to replication and, furthermore, implicated the products of *E. coli mutH*, *mutL*, *mutS*, and *uvrD* genes in this process (46–48). Since bacterial strains defective in these loci exhibited a mutator phenotype (49) (hence the terminology '*mut*' to represent genes encoded by these loci), it seemed likely that this set of genes directed the system involved in postreplication repair of DNA biosynthetic errors (50).

As pointed out by Wagner and Meselson (25), the function of such a system requires not only detection of base-pair mismatches but also a mechanism for discriminating the parental and newly synthesized strands as well. Since mismatches consist of normal Watson-Crick bases, the repair systems rely on secondary signals within the helix to identify the newly synthesized DNA strand containing the replication error. Several lines of evidence indicated that Dam methylation of d(GATC) sequences provide the necessary signal. Deficiency or overproduction of DNA methylase results in mutator phenotype (51–54). Dam methyltransferase (MT) of *E. coli* methylates adenines at N^6-position in the sequence 5′-GATC-3′ (55). As GATC modification occurs after DNA strand synthesis, newly synthesized DNA exists briefly in an unmethylated state and it is this transient absence of adenine modification that targets repair to the new DNA strand (56, 57). In addition, genetic analyses have suggested that Dam methylase participates in a pathway involving *mutH-*, *mutL-*, and *mutS*-encoded functions (58, 59). Lu *et al.* (26) had developed an *in vitro* assay that permits analysis of DNA mismatch repair activity in cell-free extracts of *E. coli*. The *in vitro* activity was found to be dependent on ATP, the state of methylation of mismatch heteroduplex, and the products of *mutH*, *mutL*, *mutS*, and *uvrD* loci (26, 60).

In *E. coli*, repair of the mismatched bases in DNA is initiated by the binding of a 97-kDa protein designated as MutS to the mismatch (61, 62). MutS is an ATPase that acts as a homodimer to bind base/base mismatches or small insertion/deletion loops that escape proofreading by the replicating polymerase (61). The results of DNA footprint analyses indicate that MutS protects a small (10–20 bp) region surrounding the position of each of the eight single base-pair mismatches but does not protect DNA devoid of mispairs (63). The *E. coli* mismatch repair system does not recognize and/or repair all mismatched base pairs with equal efficiency. The transition mismatches (GT and AC) are well repaired; whereas, the repair of some transversion mismatches (AG or CT) appear to depend on their position in the heteroduplex DNA (64, 65). It appears that the E. coli mismatch repair enzymes recognize and repair intrahelical mismatched bases but not the extrahelical bases in the looped-out structures (66).

MutS has variable affinity for different mismatches. In E. coli it forms strongest complexes with GT mismatches and single unpaired bases (67). To measure the interaction between E. coli MutS and mismatched base pairs, Brown et al. (68) carried out band shift analyses using synthetic DNA fragments containing all possible DNA mismatches as well as unpaired T (represented as ΔT). The order of affinity of MutS for heteroduplexes was found to be ΔT > GT > GG > AA ~ TT ~ TC > CA > GA > CC > GC. Toward gaining insight into the mechanism by which MutS discriminates between mismatch and homoduplex DNA, Wang et al. (69) have examined the conformations of specific and nonspecific MutS–DNA complexes by using atomic force microscopy. At homoduplex sites, MutS–DNA complexes exhibited a single population of conformations in which the DNA was bent. On the other hand, two populations of conformations were observed, bent and unbent, at mismatch sites. These results suggested that the specific recognition complex is one in which the DNA is unbent (69). Based on existing biochemical and crystallographic data it was proposed that (i) MutS binds to DNA nonspecifically and bends it in search of a mismatch; (ii) on specific recognition of a mismatch, it undergoes a conformational change to an initial recognition complex in which the DNA is kinked with interactions similar to those found in the crystal structures (70, 71); and (iii) MutS undergoes a further conformational change to the ultimate recognition complex in which the DNA is unbent (69).

Binding of MutS is followed by the loading of MutL (70 kDa) (72). The assembled MutS–MutL complex leads to the activation of the latent d(GATC) endonuclease associated with the monomeric MutH protein (25 kDa) (73, 74). The activated MutH incises the unmodified strand at the hemimethylated d(GATC) sequence (62). The observation that activated MutH protein can cleave a hemimethylated d(GATC) site located either 3' or 5' to the mismatch suggested that mismatch repair can be initiated by a single-strand break in the unmethylated strand that is located either 3' or 5' to the lesion. This inference was tested directly by examining the excision tracts that were generated by the methyl-directed mismatch repair system on a circular G.T heteroduplex that contained a single-hemimethylated d(GATC) site. Dideoxynucleotides were added to the reaction to terminate repair synthesis and allow visualization of the gaps by electron microscopy. Regardless of which strand was methylated, the gap was found to span the shortest path between the d(GATC) site and the mismatch. Analysis of the endpoints of the single-strand gaps indicated that each excision tract initiated at the d(GATC) and terminated within a 100-nucleotide region just beyond the mismatch. This clearly indicated that the excision must have occurred in a $3' \to 5'$ direction in one case and in a $5' \to 3'$ direction in the other (75). The excision reaction involves the removal of the unmodified strand spanning the d(GATC) site and the mismatched base. This depends on MutS, MutL, and cooperating action of DNA helicase II (UvrD or MutU) with an appropriate exonuclease [ExoI and ExoX with $3' \to 5'$ specificity whereas RecJ and ExoVII with $5' \to 3'$ specificity (76–79)]. Bidirectional excision implies that the methyl-directed repair system keeps track of the location of the strand signal with respect to the mispair (75, 80).

All members of the MutS and MutL family are shown to possess a conserved ATPase activity (81, 82). The weak ATPase activity of MutS is shown to be stimulated two- to fivefold by short hetero- and homoduplex DNAs (83). MutL is also shown to be a weak ATPase (84, 85) and is found to interact with the UvrD or helicase II (86, 87). Binding of MutL to DNA and a preference for single-stranded DNA was demonstrated by Bende and Grafstrom (88). The fact that the MutL ATPase activity is stimulated approximately threefold by double-stranded DNA and sevenfold by single-stranded DNA further supports the possible interaction between MutL and DNA (89). The ATPase active sites of MutS and MutL are essential for mismatch repair as indicated by sequence conservation, random mutagenic studies, and localization of a number of HNPCC mutations in humans (81, 84, 90, 91). Moreover, MutS and MutL homologues have been implicated in the repair of damaged DNA, such as transcription-coupled repair, and in apoptosis induced by DNA-damaging agents (92, 93).

Highly purified MutH introduces single-strand breaks immediately 5' to the dG residue of d(GATC) sequence if the DNA is hemimethylated or unmethylated at that site (74). Hemimethylated DNA molecules are incised on the unmodified strand. Single-strand incisions occur at a slower rate on unmethylated DNA whereas fully methylated DNA is not detectably cleaved. By incising unmethylated DNA strands, MutH activity provides a simple mechanism for strand discrimination based on methylation at d(GATC) sites. The rate of incision by the endonuclease activity of MutH is low (turnover rate is about 1 hr^{-1}) and the incision at this low rate is independent of the presence of the mismatch (63).

MutH homologues are only found in Gram-negative bacteria suggesting that different mechanisms are used for strand specificity in other organisms such as a free 3' end during DNA replication. In *E. coli*, the requirement for MutH can be alleviated if DNA substrate with a persistent strand break is used for mismatch repair. Transfection of *mutH* mutant *E. coli* with DNA containing a mismatch and a strand break was subjected to strand-specific rectification provided the bacterial strain also carried a ligase mutation (94).

This observation was further confirmed by (50) using the *in vitro* reconstituted mismatch repair system utilizing purified components. Claverys and Lacks (10) had suggested that undermethylation of d(GATC) sequences within newly replicated DNA might determine the strand specificity of only a small fraction of repair events, with the majority being determined by DNA termini such as those present at the ends of newly synthesized strands. The nature of the end-directed repair observed in the purified system (50) is consistent with this proposal, but biological data suggested that this reaction cannot be of major significance in the processing of biosynthetic errors. The reported mutabilities of *mutH-*, *mutL-*, and *mutS*-deficient strains of *E. coli* (58, 95) are of the same order and thus do not support this prediction. Another possibility is that termini-directed mismatch correction may be of significance in the processing of heteroduplex regions within recombination intermediates that contain exposed DNA ends.

Lahue *et al.* (96) had demonstrated the influence of different numbers of d(GATC) sequence on the efficiency of *in vitro* mismatch correction. Heteroduplexes containing four [$f1_{1-4}$], two [$f1h_{2,3}$], one [$f1h_1$] and [$f1h_2$], or no [$f1h_0$] d(GATC) sites in tandem were used for this study. The extent of *in vitro* mismatch repair varied over a range of atleast 20-fold, in the order $f1_{1-4} \geq f1h_1 > f1h_{2,3} > f1h_2 > f1h_0$, where the zero-site molecule was almost inert as a substrate. Reactivity decreased with the number of d(GATC) sites for substrate series $f1h_{2,3}$, $f1h_2$, and $f1h_0$, suggesting that the density of such sequences might affect the correction process. The high efficiency of repair of the $f1h_1$ heteroduplex was an exception to this observation (96). Radman and his colleagues (32) had reported similar observations in their transfection repair assays using ϕX174 containing zero, one, or two d(GATC) sites. These studies point to the fact that the sequence environment of a d(GATC) site or its placement relative to the mismatch can also contribute to the efficiency of mismatch repair (96).

MutL binds to the MutS-mismatch complex (72) and the weak endonuclease activity of MutH on umethylated d(GATC) sequence is greatly stimulated. The stimulation of MutH activity by MutL occurs in the absence of MutS and a mismatched base pair. This suggests that MutL is the component of the MutS–MutL complex responsible for activating MutH during mismatch repair *in vivo* and that the activation occurs via a direct physical interaction (97). The ability of MutL alone to activate MutH in an ATP-dependent manner provided the first line of evidence for regulation of protein–protein interactions through nucleotide binding to MutL. Studies using nonhydrolysable analogs of ATP indicated that ATP binding by MutL and not hydrolysis is essential for activating MutH. In addition, *MutL* mutants that retain ATP binding but are defective in ATP hydrolysis activate MutH better than the wild-type protein (84, 98). Activation of MutH does not require the presence of a γ-phosphate.

However, without a γ-phosphate, the MutL–ADP complex is less stable and falls apart more readily (*84*). The endonuclease activity of MutH is activated up to 50-fold in the presence of MutL, MutS, ATP, and a mismatched base (*62*). In addition, MutL is known to activate the unwinding activity of helicase II, which is another component of the mismatch repair process (*85*).

MutL serves in several ways to couple mismatch recognition and downstream mismatch repair events. First, a homodimer of MutL forms a complex with MutS and enhances ATP hydrolysis-dependent translocation presumably as part of the search for the strand discrimination signal. Second, MutL stimulates MutH endonuclease activity in an ATP-dependent manner (*81*). Third, MutL is required to load MutU (DNA helicase II or UvrD) at the site of the MutH-induced nick, facilitating DNA unwinding and subsequent exonucleolytic removal of the nascent strand (*99*) Helicase II translocates $3' \rightarrow 5'$ along a DNA strand in an ATP-driven reaction (*76*), unwinding DNA when it encounters regions of secondary structure. It is possible that loading of helicase II at the nicked d(GATC) site would require the helicase to be loaded on the unmethylated strand when the d(GATC) site is located $3'$ to the mismatch and on the methylated strand when the d(GATC) site is located $5'$ to the mismatch. Such a discrimination would require prior interaction between the mismatch and the d(GATC), mediated through MutS and MutL proteins along the contour of the DNA helix (*2*).

The excision step of mismatch repair requires any one of four single-stranded DNA specific exonucleases, ExoI, ExoVII, ExoX, or RecJ (*77–79*), as mismatch repair is abolished *in vitro* and *in vivo* only when all four exonucleases are lacking. Since all these exonucleases are highly specific for single-stranded DNA, the role of helicase II is inevitable in order to displace the incised strand thereby making it sensitive to attack by these single-stranded exonucleases (*29, 75*). Finally, single strand binding protein (SSB), DNA polymerase III holoenzyme, and DNA ligase are necessary for resynthesis and ligation (*9*). Paul Modrich and his colleagues had reconstituted the methyl-directed mismatch repair system *in vitro* using purified components. Thus, the overall mismatch repair process was found to be the coordinated activity of atleast 11 different proteins, such as MutS, MutL, MutH, UvrD, ExoI, ExoVII, ExoX, Rec J, SSB, DNA polymerase III holoenzyme, DNA ligase, and also cofactors, such as ATP, NAD^+, Mg^{2+}, and the four deoxyribonucleoside triphosphates (*50*) (Table I). A tentative pathway for DNA mismatch repair in *E. coli* is depicted in Fig. 2.

Phylogentic studies have divided *MutS homologues* (MSH) into two distinct families (*100*). MutS1 involved in mismatch repair and MutS2, which includes MSH4 and MSH5 group found in yeast and other eukaryotes and the prokaryotic MutS2 proteins. Large-scale genome sequencing efforts showed that MutS2 proteins include proteins from eubacterial, archaea, and plant sources.

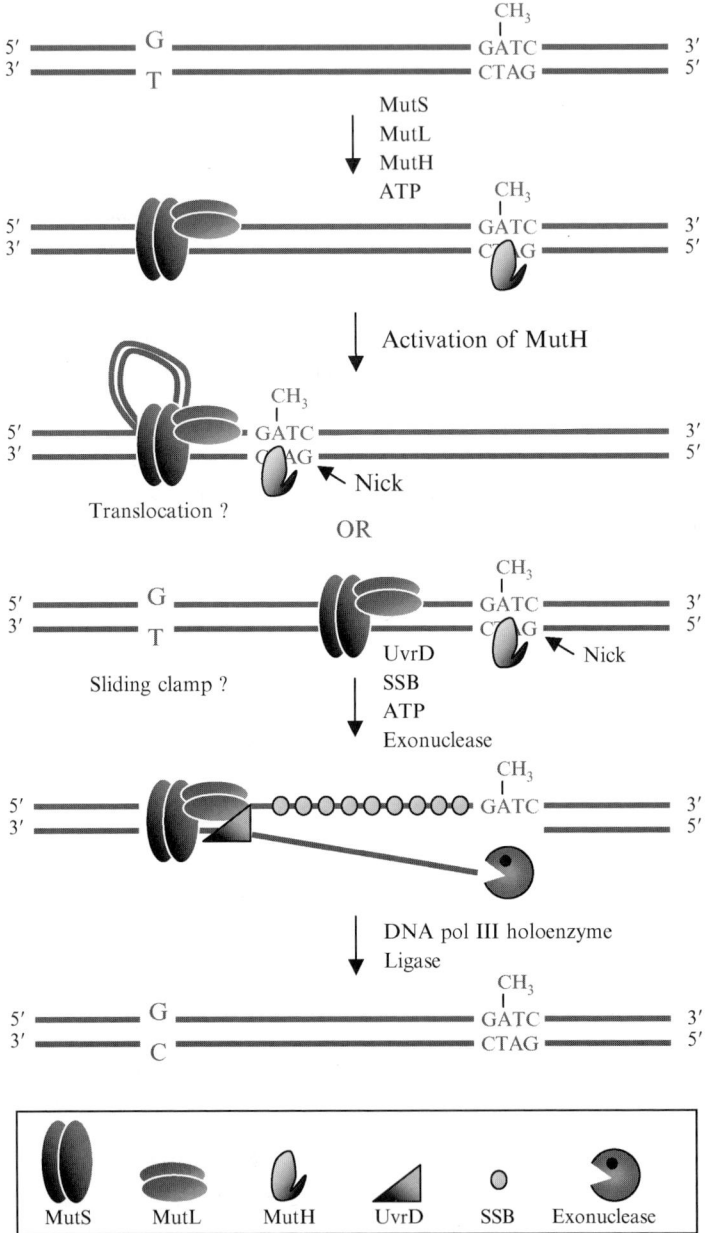

FIG. 2. Schematic representation of methyl-directed DNA mismatch repair in *E. coli*. Only the 5'-3' exonuclease activity is shown. (See Color Insert.)

The latter most likely derived from a cyanobacterial source and the result of a horizontal transfer from their plastids (*101*). Members of the MutS2 family have distinct structural features. They are slightly smaller than the MutS1 proteins; members of each family display a higher degree of sequence conservation. Members of the MutS2 family lack the 300 residue N-terminal region present in the members of the MutS1 family. By contrast, members of the MutS1 family lack a stretch of 250 amino acids that corresponds to the C-terminal domain present in the members of the MutS2 family (*102*). No function has been established for the members of the prokaryotic MutS2 group. A *mutS2*-encoding gene seems to always be present in the microorganisms that lack a *mutL* gene (*102*). It has been speculated that the MutS2 prokaryotic proteins are evolutionary precursors of the MSH4 and 5 proteins involved in meiotic recombination in eukaryotes.

In *Helicobacter pylori* (*H. pylori*), analysis of the complete genomic sequences (*103, 104*) failed to identify putative components of a mismatch repair (MMR) system other than ORF MPO621 encoding a protein with homology to MutS. It has been shown that HpMutS2 is not involved in MMR but inhibits homologous and homeologus recombination (*105*). Thus, it is possible that MutS2 proteins are candidates for controlling recombination and, therefore, genetic diversity in bacteria. In *H. pylori*, where there is a high level of genetic diversity, MutS2 could control the plasticity of genome by regulating both the integration of exogenous DNA and the reshuffling of sequences in their chromosomes (*100*). On the other hand, in *Bacillus subtilis*, a recent study showed that MutS2 does not have any recombination or mismatch repair activitiy (*106*).

Functional analysis a MutS2 homolog from *Pyrococcus furiosus*, a marine archae with an optimal growth of $100°C$, showed that it possess a thermostable ATPase activity and a thermostable DNA-binding activity. However, Pfu MutS2 does not have any detectable mismatch specific DNA-binding activity (*107*).

The extremely thermophilic bacteria *Thermus thermophilus* (*T. thermophilus*) HB8 is an aerobic, rod shaped, nonspirulatory Gram-negative bacterium which can grow at temperatures in excess of $75°C$. Takamatsou *et al.* (*108*) successfully demonstrated that *T. thermophilus mutS* gene rescued the hypermutability of the *E. coli mutS* mutant. It bound specifically with the G-T mismatch DNA even at $80°C$.

Inactivation of *Acinetobacter* sp. strain ADP1 MutS resulted in 3- to 17-fold increase in transformation efficiencies with DNA sequences that were 8% to 20% divergent relative to the strain ADP1. Strains lacking MutS exhibited increased spontaneous mutation frequencies and reversion assays demonstrated that the MutS preferentially recognized transition mismatches while having little effect on the repair of transversion mismatches. Comparison of the MutS amino acid sequences from the *Acinetobacter* strains with those from other Gram-negative bacteria clearly showed that *Acinetobacter* strain homolog represents a distinct evolutionary branch in this highly conserved protein family (*109*).

A *mutS* gene from *Pseudomonas putida* has been identified and found to encode a smaller MutS protein than do the genes of other bacteria. This gene is able to function in the *mutS* mutants of *E. coli* and *B. subtilis* (*110*). Van den Broek *et al.* (*111*) reported that MutS-dependent mismatch repair affects colony phase variation in *Pseudomonas* sp. strain PCL1171. Phase variation in *Pseudomonas* sp. is dependent on the accumulation and subsequent revival of mutations in the *gacA* and *gacS* genes—the two component regulatory system that affects secondary metabolites.

In *Vibrio cholerae*, a Gram-negative, noninvasive, enteric bacterium and the causative agent of the diarrhoeal disease cholera, several mutant genes involved in site-directed mismatch repair *mutL*, *mutS*, and *dam* have been identified (*112*, *113*). MutK encoding a 19-kDa protein has been identified, which is presumably involved in methyl-independent repair of DNA mismatch. MutK can reduce the spontaneous mutation frequency of *E. coli*, *mutS*, *mutL*, *mutU*, and *dam* mutants (*114*).

Natural transformation has been detected in many prokaryotic species. *Pseudomonas stutzeri* (*P. stutzeri*) is widely present in the environment and many members of this species are naturally transformable. A novel gene-acquisition mechanism, homology-facilitated illegitimate recombination (HFIR), during natural transformation was described for this strain. By using HFIR, the strain can integrate into its genome long stretches of fully heterologus DNA when they are linked on one side to a short homologous piece DNA that serves as a recombination anchor and thereby strongly facilitates illegitimate fusion of the heterologus parts of the molecules to resident DNA. Meier and Wackernagel (*115*) showed that upregulation of *P. stutzeri* MutS could enforce sexual isolation and downregulation could increase foreign DNA acquisition and that MutS affects mechanisms of HIFR.

The contribution of mismatch repair to limitation of interspecific recombination is different in different microorganisms. In *E. coli*, mismatch repair provides the main barrier and plays a much stronger role than sequence divergence, while mismatch repair contributes only a small part to sexual isolation in *B. subtilis* transformation (~16%). In *P. stutzeri*, the contribution of mismatch repair to sexual isolation is also small (~14–16%).

In *Hemophilus influenzae*, mismatch repair is more apparent in repairing DNA damage caused by oxidative compounds. Zaleski and Piekarowicz (*116*) showed that in an *H. influenzae dam* mutant treated with H_2O_2, mismatch repair is not targetted to newly replicated DNA strands and, therefore, mismatches are converted into single- and double-strand DNA breaks.

It was also shown that mismatch repair system of *E. coli* can prevent oxidative mutagenesis either by removing 7, 8dihydro-8-oxoguanine (8-oxoG) directly or by removing adenine misincorporated opposite 8-oxoG or both. (*117*).

IV. Molecular Structure of the DNA Mismatch Repair Proteins

The crystal structures of *E. coli* (*70*) and *Thermus aquaticus* (*71*) MutS, *E. coli* MutH (*118*), and the N-terminal (LN40) (*84*) and C-terminal (LC20) (*119*) domains of *E.coli* MutL have been solved (Fig. 3). The structure of MutH resembles a clamp with two "arms," each forming a subdomain, separated by a large cleft. The two arms share a hydrophobic interface and are connected by three polypeptide linkers (Fig. 3A). The interface between them is apparently flexible and allows the two subdomains to pivot relative to each other. A search of the protein sequence database found that MutH is homologous to the type II restriction endonuclease Sau3AI (*118, 120*). Both MutH and Sau3AI recognize the d(GATC) sequence and cleave 5′ to G. However, MutH cleaves an unmethylated strand either in a hemimethylated or unmodified duplex, whereas Sau3AI cleaves both strands regardless of their methylation state. When MutH makes a double-strand break, it cleaves each unmodified strand independently (*62, 118*).

A comparison of the crystal structure of MutH with those of cocrystal structures of several restriction endonucleases together with the multiple sequence alignment of MutH and related proteins suggested that F94, R184, and Y212 could be involved in the discrimination between a methylated or unmethylated adenine in the d(GATC) sequence. *In vitro*, *R184A* mutant displayed a strongly reduced endonuclease activity, whereas the Y212S variant had almost completely lost its preference for cleaving the unmethylated strand at hemimethylated d(GATC) sites. Furthermore, the Y212 variant was capable of cleaving fully methlyated d(GATC) sites at a comparable rate to unmethylated d(GATC) sites. This demonstrated that Y212 is important, if not the only one, in MutH for sensing the methylation status of the DNA (*121*).

A structural database search revealed that the tertiary structure of MutH is similar to those of type II endonucleases, PvuII and EcoRV (*118*), although these proteins do not share detectable sequence homology. Based on this structural similarity, a putative active site of MutH was identified comprising E56, D70, E77, K79, and K116, which when mutated to Ala abolished the endonuclease activity of MutH (*122–124*). Differences between MutH and PvuII-like restriction endonucleases provide some clues as to how a restriction enzymelike protein, MutH, evolved to be a regulated endonuclease that is specific for methyl-directed DNA mismatch repair. First of all, MutH is monomeric and makes a single-strand cleavage to initiate the mismatch repair

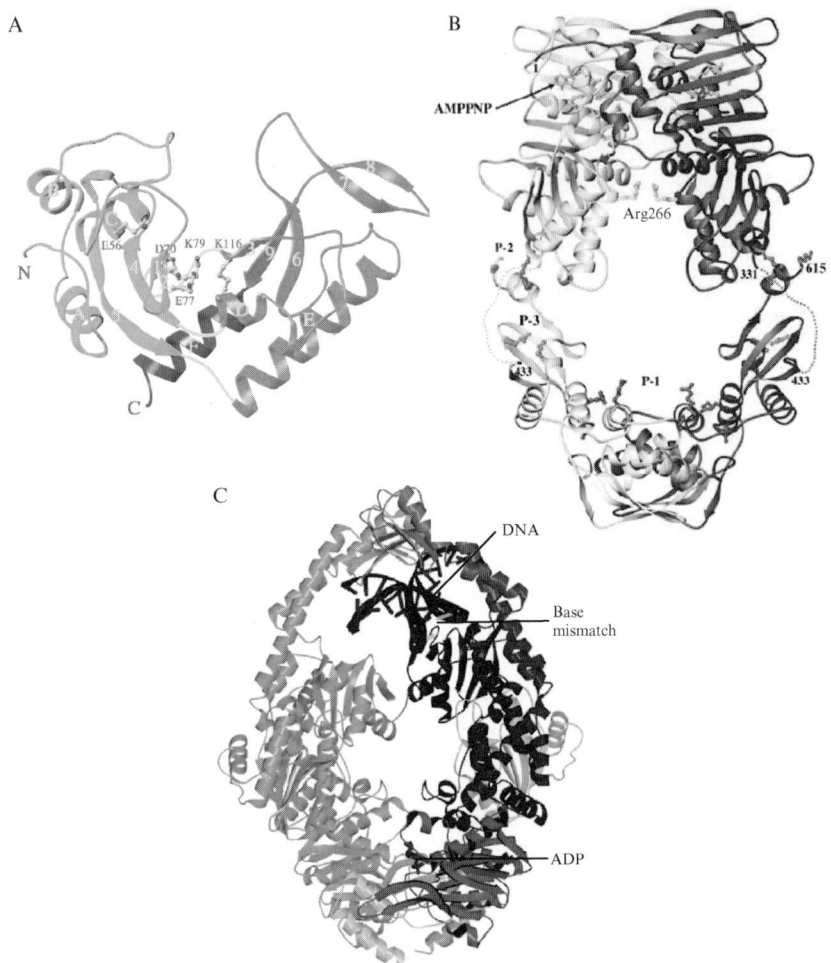

FIG. 3. The structures of *E. coli* MutH, MutL, and MutS proteins. (A) Ribbon diagram of MutH. The N-terminal subdomain is colored green and the C-terminal subdomain blue, the C-terminal "lever" red, and linker peptides between the two subdomains are yellow. The five active site residues are shown. (B) A ribbon diagram of full-length MutL (LN40 and LC20 placed to share a common dyad axis). The C-terminus of LN40 (331) and N-terminus of LC20 (433) are linked by a dotted line. AMPPNP (pink), Asn33 (Mg^{2+} chelation), Glu29 (ATP hydrolysis), and Arg266 (DNA binding) shown in yellow. Clusters of positively charged residues (DNA binding) are represented as P-1 (red), P-2 (blue), and P-3 (green) patches. (C) MutS structure with mismatch-binding monomer shown in domains: N-terminal mismatch-recognition domain, dark blue; connector domain, light blue; core domain, red; clamp, orange; ATPase domain, green with red ADP; and helix-turn-helix domain, yellow. The other monomer is shown in gray. DNA is shown in red, with a yellow mismatch. [Reprinted by permission from Macmilan Publishers *EMBO J.* **23,** 4134–4145 (2004); **17,** 1526–1534 (1998); *Curr. Opin. Struct. Biol.* **11,** 47–52 (2001).] (See Color Insert.)

while the active form of PvuII, like most of type II restriction enzymes, is dimeric and makes double-strand breaks. Another difference is that so far the active site of type II restriction enzymes has been found to be located in one structural domain (*125*) instead of two as observed in MutH. Finally, the DNA-binding groove in PvuII is formed between the two subunits, whereas it is between two subdomains in MutH. However, in both proteins, the DNA-binding groove exhibits "open and close" conformations. MutH has become a monomeric endonuclease, different from type II restriction enzymes, by enlarging the C-terminal half of the molecule such that it gains a second structural domain and develops a DNA-binding groove within a single subunit. MutH has also acquired additional residues to form an active site, such as Lys116, which has no counterpart in PvuII. Moreover, both the substrate-binding groove and the active site configuration are subject to changes. It is likely that through the C-terminal "lever," MutL and MutS help MutH to orient the two subdomains to receive DNA substrate and to configure its active site appropriately. The MutH-homologous restriction endonuclease Sau3AI, which is independently active, contains additional 270 residues at its C-terminus that may serve to alleviate the requirement of MutS and MutL for activation as MutH does (*118*).

There is evidence that MutH binds to the N-terminal domain (NTD) of MutL in an ATP-dependent manner, however, the interaction sites and the molecular mechanism of MutH activation have not yet been determined (*84, 99*). The protruding C-terminal "lever" of MutH is proposed to be a possible interaction site for MutL (*99*), which requires genetic and biochemical scrutiny. Toedt *et al.* (*126*) used a combination of site-directed mutagenesis and site-specific cross-linking to identify protein interaction sites between MutH and MutL. Unique cysteine residues were introduced in cysteine-free variants of MutH and MutL, which were differentially modified with 4-maleimidobenzophenone and were subjected to photochemical cross-linking. Four residues in MutH, which are present in helix E (Fig. 3A), Ser104, Val166, Leu167, and Arg172, were identified to be part of the potential interaction site for MutL (*126*).

Crystallographic (Fig. 3B) and biochemical studies have shown that MutL proteins contain an N-terminal ATPase region and a C-terminal dimerization domain (*84, 89, 119*), and it operates as a molecular switch by its interactions with MutH and MutS, regulated by ATP binding and hydrolysis (*99*). MutL and its homologues form a large family of proteins with members found in species ranging from archaebacteria to mammals (*81, 127*). Based on the crystal structures of the N- and C-terminal fragments of MutL and its ATP hydrolysis as well as DNA-binding activities, Guarne *et al.* (*119*) constructed a model of the full-length MutL protein. The N-terminal 300–400 residues of all MutL family members share extensive sequence homology. The C-terminal region is essential for homo- or heterodimerization of MutL and its homologues, but the amino acid sequence of this region is highly divergent among MutL homologues and no sequence or structural conservation has been described so far (*36, 84, 119, 128, 129*). However, BLAST analysis (www.ncbi.nlm.nih.gov/BLAST/) revealed that MutL homologues in Gram-positive bacteria, such as HexB, shared sequence similarity with human PMS2 and MLH3 in the C-terminal dimerization region (*119*).

Initially, MutL was not thought to possess any enzymatic activity (*81*). When the structure of an N-terminal 40-kDa fragment of MutL (LN40-residues 1–349), which encompasses all conserved residues in the MutL family, was determined, it was discovered that it is structurally similar to the ATPase fragment of DNA gyrase (*130*). When the residue in MutL, equivalent to the general base for ATP hydrolysis in DNA-gyrase—Glu29, was mutated to Ala, MutL lost the ATPase activity completely (*84, 89*). The ATPase region is found to be conserved among all MutL homologues and shares four sequence motifs with other GHKL (for *G*yrase, *H*sp90, Histidine *K*inase and Mut*L*) ATPase/kinase superfamily members (*131–133*). LN40 consisted of two domains; residues 20–200 formed the first domain and residues 207–331 made the second domain (*84*). In the absence of nucleotide ligand, the structure of LN40 was monomeric and partially unstructured. Solution studies indicated that binding of a nonhydrolyzable ATP analog, ADPPNP, transformed LN40 from monomeric to dimeric. In solution, the LN40–ADP complex dissociated to become monomeric and quickly released ADP (*84, 89*).

When the C-terminus of LN40 and the N-terminus of LC20 were placed adjacent to each other (the saddle-shaped LN40 opposite to the V-shaped LC20), the resulting MutL model contained a large central cavity (Fig. 3B). The linker region between the N-terminal ATPase and C-terminal dimerization domains shares no sequence similarity among MutL homologues. In *E. coli* MutL, it is found to be 100-residues long and dominated by Pro (20%), Ala (17%), Gln (12%), and charged residues (20%) (*119*). Bacterial MutL proteins are homodimers. Structurally, MutL can be dissected into a highly conserved N-terminal ATPase domain that is able to dimerize upon ATP binding, a flexible and poorly conserved linker, and a less conserved C-terminal domain that is involved in homodimerization (*36, 119*). Based on the crystal structure of the C-terminal *E.coli* MutL, dimerization was thought to be mediated by an internal subdomain comprising residues 475–569 (*119*). Based on computational analysis of all protein interfaces observed in the crystal structure and by mutational analysis, Kosinski *et al.* (*134*) suggest that the biological dimer interface is formed by a hydrophobic surface patch of the external subdomain (residues 432–474 and 570–615).

MutL was shown to bind both double- and single-stranded DNA with no sequence specificity (*87, 88, 135*). The DNA-binding activity of MutL was detected in the absence of a nucleotide cofactor but was enhanced in the presence of AMPPNP (89). The crystal structure of an LN40–AMPPNP complex revealed a positively charged groove inside the saddle-shaped LN40 dimer (Fig. 3B) (84). Mutation of Arg266 to Glu in the middle of the groove largely abolished the DNA-binding activity of the full-length MutL (124). Although LN40 alone binds DNA, the presence of the C-terminal dimerization region in the full-length MutL greatly enhanced DNA binding. Curiously, LC20 alone was not able to bind DNA (119).

To determine whether LC20 is directly involved in DNA binding of the full-length MutL, clusters of positively charged residues in LC20 were mutated to glutamate in three patches. Arg465, Arg468, Lys548, and Arg563 formed a large positively charged patch (P-1) on the concave surface of LC20 facing the central cavity; His606, Lys610, and Lys613 (P-2) at the C-terminus and Arg451 and Lys593 (P-3) on the convex surface of LC20 constituted two additional patches that are not facing the central cavity (Fig. 3B). All three patch-mutant proteins retained normal ATPase activity. The P-2 and P-3 mutant proteins showed reduced DNA-binding activity, but they behaved like wild-type protein in the mismatch repair assay, suggesting that DNA binding by the residues in P-2 and P-3 patches is functionally unimportant. On the other hand, the P-1 mutation reduced DNA-binding activity with concomitant increase in the mutation rate by nearly 100-fold in mismatch repair assay (119). Unlike the full-length MutL, LN40 and LC20 had little effect on UvrD helicase activity, despite the fact that they physically interact (84, 86, 119). Neither the N- nor C-terminal fragments of MutL retained any *in vivo* mismatch repair activity (91). Through protein cross-linking studies it was demonstrated that the interaction between LN40 and UvrD depended on the presence of both AMPPNP and DNA, whereas the interactions between LC20 and UvrD occurred in the absence of nucleotide or DNA (119).

Several studies have shown that MutL can interact with both MutH and MutS in the absence of DNA and that MutL can mediate binding of MutH to MutS thereby explaining the formation of a ternary complex (85, 97). Although the mechanism of MutH activation mediated by MutL is unknown, it has been suggested that a long-lived MutS-MutL-MutH endonuclease-competent intermediate exists (136). Giron-Monzon et al. (137) characterized the physical interactions between MutL and MutH by photochemical cross-linking and have suggested that (i) ATP-dependent dimerization of the MutL NTD is required for complex formation between MutL and MutH in the presence or absence of DNA; (ii) the MutL NTD is sufficient for physical interaction with MutH; (iii) the interaction site can be mapped to a region comprising residues Asn169, Ala251, Gln314, and Leu327 of MutL and residues Ser104, Glu156, and Arg172 of MutH; and (iv) a low-resolution molecular model of the MutL–MutH complex shows that both proteins can bind to the same DNA molecule.

Furthermore, the interaction site for MutH is proposed to be formed only in the closed, dimeric form of MutL, after ATP binding (137). The interface of MutL–MutS interaction and the residues involved are still elusive.

High-resolution crystal structures solved for the mismatch-binding protein MutS of *E. coli* (70) and *T. aquaticus* (71) homologues look very similar (Fig. 3C). These structures provide invaluable insight into how these proteins recognize such structurally diverse substrates as base–base mismatches and insertion–deletion loops. The structure of MutS can be better visualized as a pair of praying hands (and hence the nickname, *praying hands of fidelity*), with the thumbs folded inward and the DNA passing between the fingertips and the thumbs. Each subunit of MutS consists of five distinct domains. The amino terminal domain I, which forms the top segment of the thumb, contains the conserved amino acid motif GXFY(E), which is required for mismatch recognition. Domains II and III form the second and third thumb segments, domain IV forms the fingers, and the carboxy-terminal domain V represents heels of the palms. Domain V contains the ATP-binding site of MutS, which consists of the highly conserved Walker-type motifs (70, 71, 138). MutS forms a dimer in the absence of DNA largely because two α-helices of domain V of one subunit help in the formation of the ATP-binding pocket of the other subunit. But unlike the situation in the protein-DNA cocrystal in which the fingers and the top of the thumb can be seen to embrace the DNA, these domains are disordered in the structure of the protein alone (71).

One property of MutS proteins, recognized only in the last few years, further complicates analysis of its mechanism of action in DNA repair. The subunits of MutS dimer exhibit asymmetry in both their DNA binding and ATPase activities. The crystal structure of *E. coli* MutS dimer shows one Walker A site occupied by ADP, while the other site remains nucleotide free (70). Nucleotide-binding analysis of *E. coli* MutS revealed differential affinities of the two subunits for ATP, ADP, and nonhydrolyzable ATP analogues (139). *E. coli* MutS dimer was also shown to be capable of binding a nucleotide di- and triphosphate simultaneously (139). These results predicted asymmetry in the ATP hydrolysis activity of the two subunits, raising the possibility of up to nine different nucleotide-bound and nucleotide-free species occurring in the ATPase reaction. The asymmetry in the ATPase activity within the MutS dimer coincides with asymmetry in its DNA-binding activity. In both *T. aquaticus* and *E. coli* MutS-DNA crystal structures, only one subunit in the MutS dimer inserts a phenylalanine residue into DNA that stacks against the mismatched base; other hydrogen bonding and van der Waals contacts between the two subunits and DNA are also asymmetric (70, 71).

In an attempt to clarify the link between the asymmetric ATPase activity of the MutS dimer to its interaction with DNA during mismatch repair, Antony and Hingorani (140) showed that in *T. aquaticus* MutS dimer, one subunit

(S1) binds nucleotide with high affinity and the other (S2) with tenfold weaker affinity. In addition, their results showed that S1 hydrolyses ATP rapidly, while S2 hydrolyses ATP at a 30–50-fold slower rate. Furthermore, they showed that mismatched DNA binding to MutS inhibited ATP hydrolysis at S1 but slow hydrolysis continued at S2. The interaction between mismatched DNA and MutS is weakened when both subunits are occupied by ATP but remained stable when S1 is occupied by ATP and S2 by ADP. The various MutS species in the ATPase pathway, S1 ADP-S2 ATP and S1-ATP-S2 ADP, exhibit differences in interaction with mismatched DNA that are likely important for the mechanism of MutS action in DNA repair.

Although MutS proteins function as homodimers, the monomer subunits have different conformations. The formation of a structural heterodimer from identical protein subunits is particularly significant from an evolutionary perspective as the eukaryotic MutS complexes are found to exist only as heterodimers. Although the oligonucleotide substrates used in the crystallographic studies contained either a GT mismatch (70) or a single unpaired thymidine (71), the DNA was bent to a similar extent in both cases—about 60°—and the minor groove interactions in both structures involved the thymine and the highly conserved amino terminal motif GXFY(E) (70, 71). This motif is located in domain I, at the tip of the MutS thumb, where the phenylalanine (F) residue was shown to be essential for mismatch recognition in the MutS proteins of *E. coli* and *T. aquaticus* (141, 142). The crystal structure confirmed that the highly conserved helix-turn-helix domain at the C-terminus is important for dimerization of MutS. It also demonstrated that both the ATPase and DNA-binding sites are composite, utilizing domains from both subunits. Consistent with the composite nature of these sites, disruption of the helix-turn-helix domain of *T. aquaticus* MutS resulted in concomitant losses of dimerization, ATP hydrolysis, and mismatch binding (143).

Deletion analysis of MutS indicated that the dimerization domain is present at the C-terminal end of the protein, while the N-terminal end is important for binding mismatch-containing DNA (144). A P-loop motif for nucleoside triphosphate binding is located in the C-terminal half of MutS, and ATP binding or hydrolysis promotes dissociation of the MutS mismatch complex, a step that appears to be essential for the downstream steps in MMR. The crystal structures of MutS have failed to reveal the role of ATPase activity in the mismatch repair process. The composite ATPase domain of MutS is closely related to those found in DNA repair proteins UvrA and Rad50 (145) as well as those in "ABC" family of ATPases such as the cystic fibrosis gene product CFTR, the multidrug resistance protein Mdr, or the histidine permease HisP (146). The binding and hydrolysis of ATP is known to bring about major conformational changes in these proteins and the MSHs are no exception. Kato *et al.* (147) have shown that *T. thermophilus* MutS protein can have three different conformations based on direct observation made by small angle X-ray scattering. The conformation is drastically influenced by the presence of ADP and ATP. The ATP-bound form has the most compact conformation, the ADP-bound form has the most scattered conformation, and the nucleotide-free form has a conformation intermediate between the two. Their study clearly showed that the DNA-binding activity of MutS depends on the conformational changes triggered by both the binding and hydrolysis of ATP.

Indirect observations have been reported showing that adenine nucleotides modulate the conformation of MutS (*148–152*). While in the case of *E. coli* MutS partial proteolytic digestion in the presence of ADP, ATP, or ATPγS produced similar patterns, different from those obtained in the absence of nucleotides. For *T. aquaticus* MutS, no obvious differences could be observed between the proteolytic patterns obtained in the presence or absence of ATP (*152*).

Mutation of the conserved ATP-binding domain of MutS is associated with a dominant negative mutator phenotype *in vivo* (*90*). Addition of ATP to a mismatch-bound MutS complex was shown to trigger the formation of an α-loop structure with MutS at the base of the loop and the mismatch at the apex, as visualized by electron microscopy (*153*). Modrich and his colleagues (*81*) have proposed that this structure is generated by ATP hydrolysis-dependent bidirectional translocation of MutS away from the mismatch.

ATP binding after mismatch recognition by MutS is believed to serve as a switch that enables MutL binding and subsequent initiation of mismatch repair. The mechanism of this conformational switching by MutS is poorly understood. Crystallographic studies toward understanding the effects of ATP binding on the MutS structure using ATP-soaked crystals of MutS showed a trapped intermediate with ATP in the nucleotide-binding site. Local rearrangements of several residues around the nucleotide-binding site were observed suggesting a movement of the two ATPase domains of the MutS dimer toward each other. ATP binding increased affinity between the ATPase domains and the affinity was reduced in the presence of ADP (*154*).

V. Communication Between Mismatched Nucleotides and the Excision Machinery

The central issue about the mismatch repair mechanism is to understand the role of ATP hydrolysis and the mode of signal transduction by which mismatch recognition is coupled to site-specific initiation of excision. Strand discrimination in order to initiate the repair process requires prior interaction

DNA MISMATCH REPAIR

between the mismatch and the nearest hemimethylated d(GATC) mediated through MutS, MutL, and MutH proteins along the contour of the DNA helix. This process is further complicated as the DNA helicase II has to be loaded onto the correct strand of DNA in order to unwind the nicked strand toward the mismatched base, allowing it to be exonucleolytically cleaved. In order to explain the molecular mechanism underlying the communication between the mismatched base pair and the excision machinery, three models have been proposed for the mismatch repair process. The bidirectional excision capability of the methyl-directed pathway (*81*) requires that the repair system evaluate its location on the DNA helix with respect to that of the mispaired base to ensure loading of the proper excision system. The protein machinery responsible for determination of heteroduplex orientation operates over substantial helix contour length since a d(GATC) sequence can direct repair of a mismatch at a distance of 1 kb or more (*26, 56, 96, 155*).

Binding of MutS to the mismatched base was shown to protect the heteroduplex against restriction enzyme digestion, the recognition sequence of which is present very close to the mispair. This effect was mismatch dependent as MutS afforded no protection of the restriction site with an otherwise identical homoduplex DNA. The presence of ATP rendered the heteroduplex sensitive to restriction enzyme within 10 s of exposure to the endonuclease. Presence of ATPγS also resulted in some deprotection of the restriction site. These observations indicate that ATP binding by MutS might be sufficient to induce a large conformational transition within the MutS–DNA complex, or that the protein leaves the mismatch in the presence of the nucleotide (*153*). Paul Modrich and his colleagues (*153*) had used electron microscopy to examine the interactions of MutS and MutL with heteroduplex DNA to analyze the mechanism of methyl-directed correction and the structural nature of DNA–protein complexes involved in this reaction. It was shown that MutS mediates formation of ATP- and time-dependent α-shaped DNA loops. MutL, although not necessary for loop formation, could be a component of the loop structures and these DNA loops were suggested to be intermediates in the excision stage of the reaction (*153*). The *hydrolysis-dependent translocation model* (*27, 81*) describes the assembly of a MutS–MutL complex at the mismatch, which motors bidirectionally via ATP hydrolysis, creating an α-looped structure (Fig. 4A) (*153*). Such a DNA-tracking process was envisioned to link mismatch recognition by MutS with endonuclease activity of MutH at a nearby GATC site, which further leads to the loading of UvrD helicase and a single-strand DNA-specific exonuclease (*99*).

Studies on the human MSH proteins as well as *Escherichia coli* MutHLS proteins led to the proposal of a second mode of intermolecular interaction, which is termed as the *molecular switch model* (Fig. 4B) (*156–158*). This proposal was based on the observation that hMSH heterodimeric

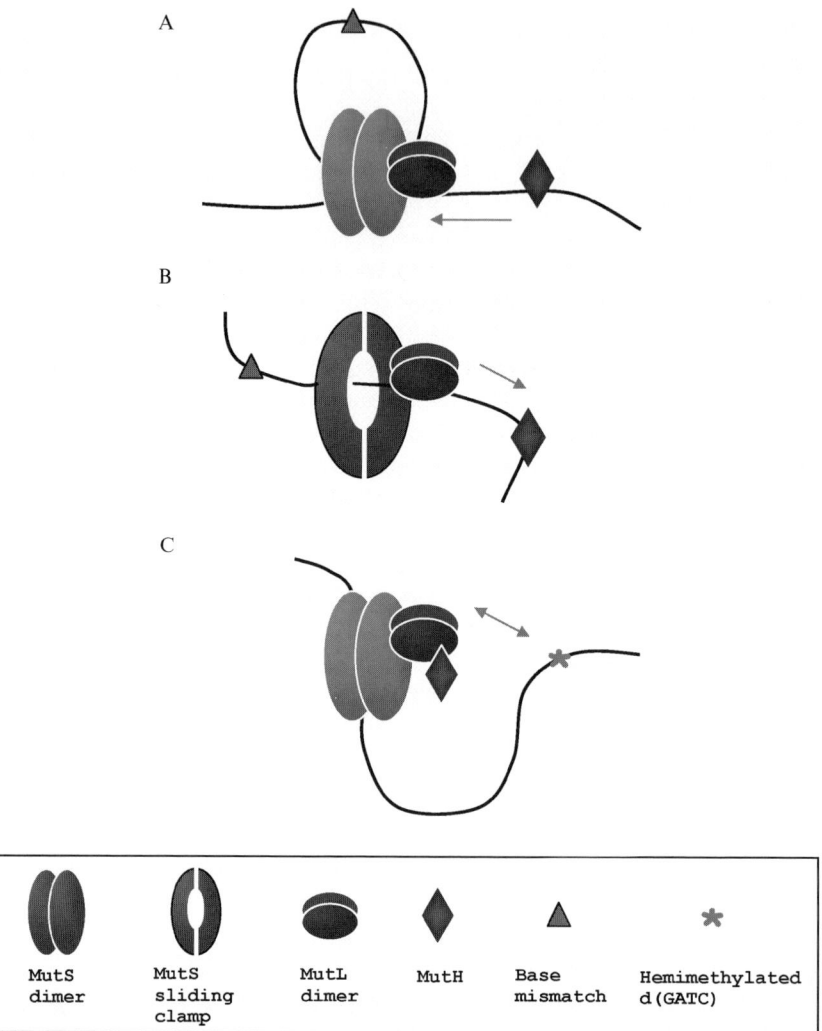

Fig. 4. Proposed models for the signal transduction between mismatched nucleotides and the excision machinery. (A) Hydrolysis-dependent translocation model. (B) Molecular switch model. (C) Static transactivation model.

complexes (hMSH2–hMSH6 or hMSH2–hMSH3) display significant mismatch-dependent ATPase activity (150, 156). Recognition of mismatched nucleotides by MutS was found to provoke ADP→ATP exchange that defines the protein as a molecular switch (159). Binding of ATP to *E. coli* MutS and hMSH proteins result in the formation of a DNA-sliding clamp capable of hydrolysis-independent diffusion for several thousand nucleotides. Repetitive rounds of mismatch-provoked ADP→ATP exchange results in the loading of multiple MutS hydrolysis-independent sliding clamps onto the adjoining DNA duplex (136, 148). H. influenzae MutS was reported to have reduced ATPase activity in presence of covalently closed circular GT heteroduplex DNA; the ATPase activity was increased upon linearization of the heteroduplex, suggesting the possible trapping of MutS molecules by circular heteroduplexes (160). It was proposed that a threshold number of localized ATP-bound MutS-sliding clamps are required to initiate mismatch repair (136). ATP hydrolysis was observed when the hMSH proteins dissociated from the DNA ends (148). This observation could account for the low-ATPase activity of hMSH proteins (149, 156, 161, 162). It also suggested that the rate-limiting step of hMSH ATP hydrolysis could be the exchange of bound ADP for ATP at the mismatch site rather than ATP hydrolysis during translocation. Based on these results, a modification of the hydrolysis-dependent translocation model was introduced, which incorporated a two-site ATPase, although ATP hydrolysis was still required to propel movement along the DNA (163). MutL is found to associate only with ATP-bound MutS-sliding clamps. Interaction of the MutS–MutL sliding clamp complex with MutH triggers ATP binding by MutL that enhances the endonuclease activity of MutH. Furthermore, MutL is shown to promote ATP-binding independent turnover of idle MutS-sliding clamps. Thus, the molecular switch model supports a mechanism of mismatch repair that relies on the activities of two dynamic and redundant ATP-regulated molecular switches. Multiple dynamic MutS–MutL sliding clamps contribute significant redundancy into the molecular switch model. The presence of multiple MutS–MutL sliding clamp complexes ensures that the mismatch repair reaction can be rapidly restarted from the last end point even if the protein components encounter catastrophic and/or induced dissociation. This process would appear completely iterative until the excision tract disrupts the mismatch that is responsible for the initial loading of MutS sliding clamps. The molecular switch model also satisfies the final and most significant requirement of mismatch repair reaction, the directionality of the excision tracts as it covers only the DNA region between d(GATC) incision to just past the mismatch. This model was proposed to be easily adaptable to the eukaryotic MSH and MLH homologues and ultimately to eukaryotic mismatch repair reaction (136).

Prediction of protein–protein interfaces of MutS, MutL, and MutH through *in silico* analyses (70, 71, 84, 118) led to the third mismatch repair mechanism that is termed as the *static transactivation model* (164). In this model, a DNA-scanning process concludes with identification of a mismatch that enhances the stability of ATP-bound MutS. This leads to the formation of a static MutS–MutL–MutH complex on or in the vicinity of the mismatched base. This heterotrimeric complex (bound to DNA) was suggested to collide in *trans* with a GATC site, provoking MutH endonuclease incision (Fig. 4C) (*164, 165*). It is not immediately evident how such a random collision in three-dimensional space could direct an excision tract from the Dam site toward the mismatch site.

Joshi and Rao (*151*) have put forth another mechanism to explain how the MutS–MutL complex communicates positional information of a mismatch to MutH. MutS exhibited a short DNase I footprint when bound to a mismatch in the absence of ATP and this footprint was dramatically expanded upon addition of ATP. High-resolution gel-shift analyses revealed that supershifted specific complexes, presumably containing multiple MutS homodimers on the same heteroduplex, were generated when ATP is hydrolyzed. Such complexes were suggested to be largely nonspecific in the absence of ATP or in presence of ATPγS. Specific MutS–MutL–heteroduplex ternary complexes were observed only upon ATP hydrolysis. Thus, it was suggested that loading of MutS onto a mismatch induces the formation of higher order complexes containing multiple MutS homodimers, presumably through an ATP hydrolysis-dependent "tread milling" action. Such a higher order MutS complex could productively interact with MutL under ATP-hydrolyzing conditions and generate a specific ternary complex, which might be proficient in communicating with MutH. This model neither depends on nor gives rise to the spooling of DNA. This suggestion was supported by the observation of footprint extension in ATP-hydrolyzing conditions, despite the heteroduplex ends being tethered to agarose beads that block helical rotations (*151*).

Peggy Hsieh, Wei Yang, and their coworkers had hypothesized that the MutS or MutSα proteins remain at the mismatched base as part of the recognition complex that contact excision-initiation signals by DNA bending (*164, 165*). To address this hypothesis, Wang and Hays (*166*) used internal barriers to protein sliding or DNA translocation by incorporating biotin–streptavidin blockades between the base mismatch and the preexisting single-strand breaks. All the existing models that explain the coupling of mismatch recognition to site-specific initiation of excision are based on studies using one or two purified proteins—MutS or MutSα, sometimes with MutL or MutLα—with mismatched DNA substrates for the analysis of protein-DNA interactions and ADP/ATP transactions under various conditions. Wang and Hays (*167, 168*) used human nuclear extracts, which were shown to correct mismatches in nicked plasmids with high efficiency and specificity. In HeLa nuclear extracts, base mismatch efficiently provoked the initiation of excision along the shorter nick-mispair path despite the intervening barriers. However, progress of excision and, therefore, mismatch correction were hampered. Thus, the cross talk between mismatch identification and excision initiation was proved to be through DNA bending and not by the sliding of the recognition complex along the DNA contour (*166*). This model only mimicked the initial generation of the *E. coli* strand-specific excision initiation signal and does not explain the efficiency or directionality of excision, which needs to be investigated further.

VI. Effect of DNA-Damaging Agents on Mismatch Repair

Cisplatin [*cis*-diaminodichloro platinum (II)] is a DNA-damaging drug that has shown spectacular success in the treatment of testicular ovarian and other tumors. Cisplatin forms DNA adducts that block replication and elicits a variety of cellular responses including nucleotide excision repair, recombination repair, and the triggering of apoptosis. The 1,2-interstrand cisplatin-DNA adducts induce significant distortions of the double helix and provide a structural signal for specific recognition by a variety of cellular proteins, including those involved in mismatch repair. It has been established that mismatch repair proteins mediate the cellular responses to cisplatin damage, but paradoxically they seem to sensitize rather than protect the cell. In *E. coli*, loss of mismatch repair confers cellular resistance to cisplatin cytotoxicity. Cisplatin analogues with a diaminocyclohexane (DACH) carrier ligand, such as oxaliplatin, do not elicit resistance in mismatch repair–deficient cells and therefore present promising therapeutic agents. *E. coli* MutS recognised the cisplatin-modified DNA with twofold higher affinity in comparison to the DACH-modified DNA. ADP stimulated the binding of MutS to cisplatin-modified DNA, whereas it had no effect on the MutS interaction with DNA modified by DACH adducts. Methylation deficient *E. coli*, *dam* mutants showed striking sensitivity to these compounds. The differential affinity of MutS for DNA modified with different platinum analogues could provide the molecular basis for the distinctive cellular responses to cisplatin and oxaliplatin (*169*).

Calmann and Marinus (*170*) used RecA-mediated strand exchange between homologous ϕX174 molecules, one that was platinated and other that was unmodified, and showed that strand transfer decreased in a dose-dependent manner. Addition of MutS to the reaction further decreased the rate suggesting that although mismatch repair was beneficial for mutation avoidance, its antirecombination activity on inappropriate substrates could be lethal to the cell.

Calmann *et al.* (*171*) showed that the C-terminal end of MutS is necessary for antirecombination and cisplatin sensitization and less significant for mutation avoidance. The inability of MutSΔ800 (C-terminal truncated MutS) to form tetramers may indicate that these are the active form of MutS.

Methylating agents, such as *N*-methyl-*N'*-nitro-*N*-nitrosoguanidine (MNNG), can react with DNA to create a variety of lesions acted upon by different DNA repair pathways. O6-methylguanine paired with cytosine or thymine is a substrate for the MutS protein of *E. coli*-DNA mismatch repair system (*172*). *E. coli dam* mutants are more susceptible to the cytotoxic action of MNNG than wild type (*173*). Mutations inactivating mismatch repair (*mutS, mutL*) in Dam background confer a level of resistance to MNNG similar to wild type indicating that mismatch repair can act on chemically modified substrates through MutS binding specifically to O6-methylguanine base pairs. Mismatch repair action at these base pairs may lead to the formation of nicks or gaps, which are converted to double-strand breaks requiring recombination to repair them. Calmann *et al.* (*171*) clearly show that the inhibition of recombinational repair by mismatch repair ensues because the homologous methylated DNA is perceived as homeologous DNA.

VII. Methylation-Independent DNA Mismatch Repair

Early evidence for the repair of mismatched bases in prokaryotes came from studies of transformation in a Gram-positive bacterium, *S. pneumoniae* (*10*). During the transformation of this organism, as in other Gram-positive bacteria, the DNA from the donor cell is converted to single-stranded segments upon entry into the recipient cell (*11*). By a process of recombination, these donor segments then replace homologous segments in the recipient chromosome to generate a heteroduplex region (*174, 175*). Genetic differences between the donor and the recipient result in the formation of DNA-containing mismatches. A striking feature of this phenomenon is the variation in the *integration efficiencies* of different genetic markers (*176, 177*). High-efficiency markers yield transformants with an efficiency approaching one transformant per genome equivalent of donor DNA entering the cell (e.g., marker *malM594* had a transformation efficiency of 0.98). On the other hand, the transformation efficiency of other markers studied varied from 0.05 to 0.5 [e.g., markers *malM567* (0.04) and *malM582* (0.5)] (*177*). The *mal* region of *S. pneumoniae* chromosome contains five genes involved in maltosaccharide utilization and exists as an operon consisting of *malR, malD, malX, malM*, and *malP*. The *malM* gene encodes for the enzyme amylomaltase (*178*). Mismatch repair was postulated to account for these differences, making the assumptions that the repair occurs preferentially on the donor strand and that its frequency depends on the identity of the mismatch. That is, the higher the repair efficiency, the lower the transformation efficiency (*179*).

Mismatch repair was invoked to account for another feature of pneumococcal transformation. When the donor DNA carried two closely linked genetic markers, one of low-integration efficiency and the other with high-integration efficiency, it was observed that the integration efficiency of the high-efficiency marker was lower than it would otherwise be. This effect of low-efficiency markers was explained by postulating that an excision repair process, which was activated by the low-efficiency marker, removed some or the entire donor strand including the high-efficiency marker. The distance dependence of this effect indicated that a donor strand segment of 1- to 2-kilobase (kb) long is eliminated together with the mismatched base. This length corresponds to the size of the average segment that is normally integrated (*177*).

S. pneumoniae mutants that no longer discriminate between high-efficiency and low-efficiency markers were isolated after mutagenic treatment of the wild-type cells (*180*). The mutations responsible for this phenotype were called *hex*, now redefined as *heteroduplex repair deficiency* (*10*). In such mutants, the low-efficiency markers are transformed with the same high efficiency that is characteristic of the high-efficiency markers, thus providing support for the concept of a positively acting mismatch repair system in *S. pneumoniae*. An interesting property of the *hex* mutants was their elevated levels of spontaneous mutation rates (*181*). This finding raised the possibility that the *hex*-dependent mismatch repair system played an important role in mutation avoidance.

Sequencing of the *hexA* gene of *S. pneumoniae* and comparing it with the *mutS* gene of *S. typhimurium* revealed that they were homologues (*182*), with the region around the ATP-binding site being particularly conserved. This provided compelling evidence that the mechanism of *hex*-dependent mismatch repair is related to the methyl-directed mismatch repair in *E. coli*. It also suggested that the HexA (95 kDa) protein is involved in the recognition of mismatched base pairs in *S. pneumoniae*. Although the *hexA* gene failed to complement an *E. coli mutS* mutation, its expression in the wild-type background resulted in a mutator phenotype (*183*). This observation suggested that either the HexA protein bound to mismatches but was unable to interact with the other Mut proteins or that it formed nonfunctional repair complexes. Ren et al. (*184*) had reported the identification of a *hexA* homologue in *Lactococcus lactis*, although its detailed functional characterization is lacking. *S. pneumoniae hexB* gene had been sequenced and was found to be a homologue of *E. coli mutL* and the *PMS1* gene of *Saccharomyces cerevisiae* (*185*).

In order to direct the mismatch repair to the donor strand, discrimination of the donor strand from the host strand in the heteroduplex is essential. Since *S. pneumoniae* lacks the methylation system for GATC sequences, it was proposed that transiently unligated ends flanking the mismatch in heteroduplex DNA might strand direct the repair system (*10*). Since the lagging strand of DNA is synthesized discontinuously, the breaks at the ends of Okazaki fragments (*186*) could serve to target the hex-dependent mismatch repair system to the daughter strand. This would permit correction of mistakes introduced as replication errors. Detailed analysis of the hex-dependent

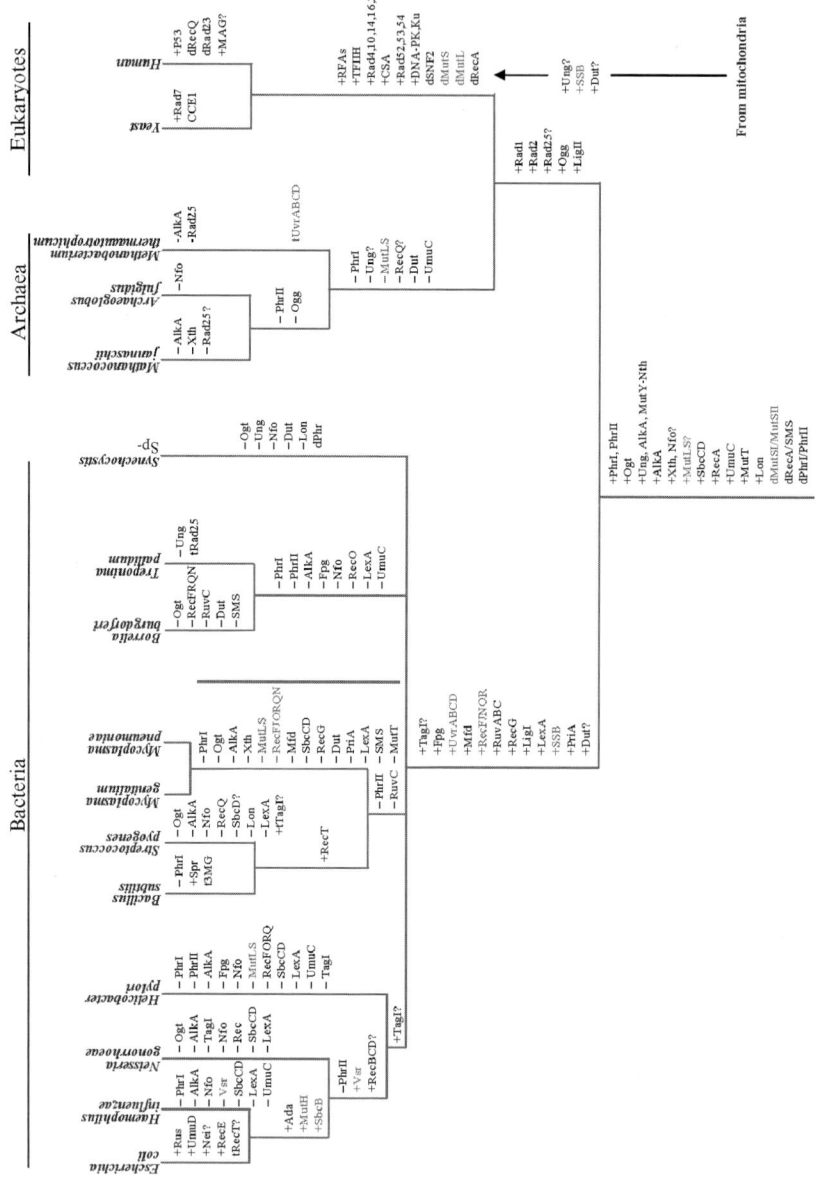

system will help in deciphering the molecular details of the mismatch repair process in the eukaryotes, especially in the relatively less understood area of strand discrimination, and role of strand breaks in directing the repair process to nascent DNA strand.

A global phylogenetic analysis of the prokaryotic DNA repair proteins to infer the evolutionary history of DNA repair pathways is shown in Fig. 5 (*187*). Aravind *et al.* (*188*) had put forth a theory stating that all eukaryotic MSH were transferred to the nucleus from the mitochondria. This is against the conclusion of Eisen and Hanawalt (*187*) who proposed that only the MSH1 is acquired from the mitochondria, which is consistent with the experiments that these genes function in mitochondrial mismatch repair. Culligan *et al.* (*189*) had done a more detailed analysis of the MutS evolutionary lineage and have suggested that *mutS* gene arose and evolved in the eubacteria and was transferred to eukaryotes through mitochondrial endosymbiotic events. This gene, now designated as *MSH1*, was then transferred to the nucleus and gave rise to all eukaryotic *MSH* genes. Whether MutS and MutL are ancient or not, their homologues are found in most bacteria. Thus, it is inferred that the absence of these genes in some species are due to gene loss. Multiple parallel losses of the *mutL* and *mutS* suggest that either these genes are particularly unstable and easily lost or that there is some advantage to the loss of these genes. The latter possibility is more likely in order to increase the mutation rate and hence allow a species or strain to evolve more readily in response to unstable changing environments (*187*).

VIII. DNA Mismatch Repair vs Bacterial Virulence: An Ongoing Debate

Gain or loss of the expression of factors that determine virulence at high frequency is an intrinsic feature of many bacterial pathogens. This phenomenon called phase variation, that is responsible for the reversible switching of surface antigens, was originally coined to describe the switching of *Salmonella* flagella antigens (*190*). Such hypermutable loci, represented as contingency loci, play an important role in facilitating bacterial adaptation to the dynamic

---

FIG. 5. Evolutionary gain and loss of DNA repair genes. The gain or loss of repair genes is traced onto an evolutionary tree of the species for which complete genome sequences were analyzed. Origins of repair genes (+) are indicated on the branches while loss of genes (−) is indicated along side the branches. Gene duplication events are indicated by a "d", while possible lateral transfers are indicated by a "t". [Reprinted by permission from Eisen & Hanawalt, Elsevier. *Mutat. Res.* **435**, 171–213 (1999).]

and unpredictable changes in the host environment (191). One of the commonest mechanisms of acquisition of hypermutation in contingency loci is mediated through changes in the number of repeats within simple sequence repeat tracts (microsatellites) located either in the promoter or ORF of a gene (191–193). Such simple sequence contingency loci are found in comparatively large numbers in the genomes of several pathogenic bacteria including *H. influenzae*, *Neisseria meningitides*, *H. pylori*, and *Campylobacter jejuni* (194–197). Hypermutable isolates of bacterial pathogens have been identified among natural populations of *E. coli* (198–200), *Salmonella enterica* (198), *Pseudomonas aeruginosa* (201, 202), *Neisseria meningitides* (203), and *Staphylococcus aureus* (204). The increased interest toward understanding the involvement of DNA mismatch repair genes in the phenomenon of phase variation has led to studies on the mismatch repair activities of certain pathogenic organisms such as *N. meningitides* (205–207), *P. aeruginosa* (202), and *S. pneumoniae* (208).

The majority of mutators among natural bacterial populations have been found to have defects in one or more components of the methyl-directed mismatch repair system particularly *mutS* (198, 199, 202, 209). Disruption of the *P. aeruginosa mutS* gene resulted in the generation of diverse colony morphologies in contrast with its parental wild-type strain that displayed monomorphic colonies. Clinical *P. aeruginosa* mutator isolates not only displayed an increase in the frequency of mutants with antibiotic resistance but also generated mutants whose antibiotic-resistance levels were higher than those measured for spontaneous resistant mutants derived from wild-type cells (210). On the contrary, *in silico* genome sequence analyses have suggested that the highly conserved *mutLS*-based postreplicative mismatch repair system is absent in mycobacteria (211). Springer *et al.* (212) have shown biological evidence for the lack of a classical mismatch repair function in mycobacteria that results in unusually high frameshifts in *Mycobacterium smegmatis* but not global mutation rates. However, despite the absence of mismatch repair machinery, *M. smegmatis* establishes a strong barrier to recombination between homeologous DNA sequences. Hemavathy and Nagaraja (213) had analyzed mycobacterial DNA for the presence of 6-methyladenine and 5-methylcytosine at Dam (GATC) and Dcm (CCA/TGG) sites using isoschizomer restriction enzymes. In all species of mycobacteria tested, Dam and Dcm recognition sequences were not methylated indicating the absence of these MTs. *In silico* analysis of the completely sequenced genomes of *Mycobacterium* species (http://www.tigr.org/tigr-scripts/CMR2/CMRHomePage.spl) failed to identify any putative *dam* and *dcm* genes, further confirming their observation. On the other hand, high-performance liquid chromatographic analysis of genomic DNA from *M. smegmatis* and *Mycobacterium tuberculosis* showed significant levels of 6-methyladenine and

5-methylcytosine, suggesting the presence of DNA MTs other than Dam and Dcm in these bacteria (*213*). Thus, the lack of mismatch correction and a high stringency of initiation of homologous recombination provide an adequate strategy for mycobacterial genome evolution, which occurs by gene duplication and divergent evolution (*212*).

H. influenzae, a common secondary invader after the infection by influenza virus, is the first free-living organism to have its entire chromosome sequenced, sneaking in just ahead of *E. coli* in that race (*214*). *H. influenzae* type b causes bacteremia and acute bacterial meningitis in infants and young children (less than 5 years of age). Nontypable *H. influenzae* causes ear infections (otitis media) and sinusitis in children and is associated with respiratory tract infections (pneumonia) in infants, children, and adults (*215*). The *mutH,L,S* gene products of *H. influenzae* are shown to complement a corresponding defect in *E. coli* and the purified proteins have been characterized functionally (*160, 216*) The pathogenic potential of *H. influenzae* critically depends on the phenomenon of phase variation of a number of surface-expressed molecules (*215*). Mono- and dinucleotide repeat tracts produce large numbers of DNA polymerase slippage events during replication that are corrected by the mismatch repair pathway. Inactivation of *mutS* in *H. influenzae* was found to destabilize dinucleotide repeat tracts of chromosomally located reporter constructs, implicating mismatch repair pathway in generating phase variation (*217*). Similar effect was seen in *N. meningitides* in which inactivation of *mutS* resulted in increased frequency of phase variation of genes presenting homopolymeric tracts of diverse length (*207*). On the contrary, abrogation of *mutS* function in two serotype b *H. influenzae* strains did not alter the phase variation rates of pilin, an important virulence factor (*192*). But Watson *et al.* (*218*) have reported hypermutable *mutS* mutants of *H. influenzae* in the sputum of cystic fibrosis patients. Unraveling the repair processes that control mutation rates of simple sequences in this bacterium is central to understanding the contribution of phase-variable genes in the host-pathogen relationship. *E. coli*, when grown in a glucose-limited environment, spontaneously and independently evolved greatly increased mutation rates and the mutations responsible for this effect was found to lie in the same region of the mismatch repair gene *mutL*. The position of the mutator mutations—in the region of MutL known as the ATP lid—suggested a possible deficiency in MutL ATPase activity as the cause of the mutator phenotype (*193*).

Dam methylase is found to have multiple cellular functions (*219*), which include directing postreplicative DNA mismatch repair to the correct strand (*56, 57*), guiding the temporal control of DNA replication (*220, 221*), and regulating the expression of multiple genes (including virulence factors) by differential promoter methylation (*222*). GATC sequences are unequally distributed within the chromosome and tend to cluster in promoter regions or within binding sequences for global regulators such as CRP, Fnr, and Integration Host Factor (IHF) (222, 223). The virulence of *dam* mutants has been studied most extensively with *S. typhimurium*. Heithoff et al. (224) found that *dam*-deficient *S. typhimurium* were proficient in colonization of the gastrointestinal tract but were unable to invade the mucosa; these avirulent mutants were effective as live vaccines against murine typhoid fever, providing cross-protective immunity to other *Salmonella* serovars.

Experimental data indicate that the lack of Dam methylation in *E. coli* influences the expression of a variety of genes. The impact of the expression on such a broad diversity of genes could be a reason for the loss of pathogenic properties in *dam* mutants of *S. typhimurium* (225) or lethality of such mutations in *V. cholerae* and *Yersinia pseudotuberculosis* (226). In *H. influenzae* it has been shown that the lack of Dam methylation influences the expression of a variety of genes as indicated by hypersensitivity to temperature, different antibodies and dyes (116).

Studies have shown that mutants of *S. enterica* serovar *typhimurium* lacking Dam methyltransferase (MTase) are highly attenuated for virulence in the mouse typhoid model. (224). DNA methylation is thus essential for virulence. Evidence suggests that the role of Dam methylation in *Salmonella* virulence is multifactorial. (227).

Dam MTase mutants of *S. enterica* secrete high amounts of proteins into the culture medium (227). Complementation with a plasmid borne *dam* gene restored the wild-type pattern of protein secretion, indicating that protein hypersecretion was solely caused by the *dam* mutation (227). This phenotype has been now reproduced in Dam$^-$ derivatives of other strains of *S. enterica* serovar *typhimurium*, such as the laboratory strain LY2 and the mouse virulent strain 14028S. Envelope instability thus might contribute to virulence alternation of *S. enterica dam*$^-$ mutants in the mouse typhoid model. Release of outer membrane proteins, such as OmpA, PAL, and Lpp, to the extracellular medium can occur in the presence of serum and this process has been proposed as a major phenomenon mediating host immune response. Therefore, overstimulation of host defences during infection might render *dam*$^-$ mutants unable to cause disease.

Reduction in the magnitude of the bacteraemia in the infant rat model with *H. influenzae* strain R3549 invasive nontypable-1(INT-1 *dam*$^-$) was an unexpected finding, but it was consistent with a report of reduced virulence of *S. typhimurium dam* mutants in the BALB/c mouse model (227). Watson et al. (228) have found a strain-dependent decrease in adherence, invasion, and intracellular replication in *H. influenzae dam* mutants, suggesting that Dam methylase activity is required for virulence. Mutants of *E. coli, S. typhimurium*, and *S. marcescens* lacking *dam* activity exhibited hypermutable phenotype (52, 225, 229). The *dam* mutants were hypersusceptible to the base analogue 2-aminopurine (2-AP) (*230*). 2-AP is an adenine base analogue that can occasionally pair with cytosine (*231*), a mismatch that initiates the methyl-directed mismatch repair system. This susceptibility to 2-AP was suppressed by either complementation *in cis* with a functional copy of *dam* or secondary mutations in *mutS* eliminating mismatch repair (*228*). *Serratia marcescens dam* mutants showed increased mutation frequency with UV radiation and are slightly more sensitive to inhibition of growth by UV irradiation than dam^+ strains (*229*). In contrast, *S. typhimurium dam* mutants did not show increased UV sensitivity (*225*). Surprisingly, *H. influenzae dam* mutants were not hypersusceptible to either UV radiation or hydrogen peroxide (*228*), the two DNA-damaging agents to which *E. coli dam* mutants were susceptible (*117, 230*). It is suggested that the DNA repair pathways in *H. influenzae* are less complex compared to those in the *E. coli* system, perhaps as a mode of reduction of "excess" genomic content. *H. influenzae* is adapted to its confined human environment where chemical and UV irradiation-induced DNA damages are not as common as in an environmental organism such as *E. coli* (*228*).

Genetic degeneration is intimately linked to all aspects of maintenance of DNA integrity and gene function. This is fuelled by continuous erosion of the genome by environmental and endogenous genotoxic agents. Knowledge acquired through studies on the mismatch repair pathways of prokaryotes can be extrapolated to the repair processes in eukaryotes, including humans. This will help in revealing the biological impact of genetic degeneration, including oncogenesis and age-related diseases, thereby uncovering new paradigms for prevention, diagnosis, and rational therapy. As famously remarked by the eminent French scientist and 1965 Nobel laureate Jacques Monod, "What's true for *E. coli* is true for an elephant."

IX. Applications of DNA Mismatch Repair Proteins and Processes

The ability of the *E. coli* MutHLSD system to identify and correct base mismatches has been exploited to develop novel methods in the identification and removal of mutations as well as in random mutagenesis. Mutation detection usually aims to accomplish one of two goals: to detect or exclude known mutations (*specific mutation testing*) or to scan known genes or exons for any mutation (*mutation scanning*). Smith and Modrich (*232*) had reported a method for removal of mutant sequences produced by polymerase that arise during DNA amplification using polymerase chain reaction (PCR). The mismatch binding of MutS has been exploited for mutation detection in several formats: solid-phase capture of mismatched heteroduplexes (*233*), mobility-shift assays (234), mismatch protection from exonuclease (MutEx), which also enables the mutation to be localized (235), and by utilization of an *in vitro* reconstituted MutHLS system (232). MutS mismatch binding has also been monitored by surface plasmon resonance (236), although this is likely to be a means of investigating MutS–mismatch interactions rather than as a mutation detection method. MutS also has the potential to enable whole genome screening using a technique called genome mismatch scanning (GMS) (237). Although technically quite demanding, its results toward identification of regions of genomes identical by descent are promising.

Wang and Liu (238) reported a procedure for directly fishing out subtle unknown mutations in bacterial genome with *T. thermophilus* MutS. Wild-type genomic and mutant DNA were mixed, digested with restriction enzymes, denatured, and reannealed. The heteroduplex DNA carrying mispaired bases were bound to MutS and recovered through Ni-NTA His-Bind® resin. The recovered DNA is cloned into plasmids, producing a mini library with inserts of the mutated regions. Further, DNA sequencing and genetic complementation demonstrated that this method was extremely efficient in fishing out the mutations from total genomic DNA (239). For easier detection of mutated regions, some researchers used MutS fused with biotinylated peptide (239) or MBP (maltose binding protein) (240) so that MutS could be more easily recovered through a simple affinity chromatography. Beaulieu *et al.* (241) have adapted the *E. coli* mismatch detection system, employing the factors MutS, MutL, and MutH, for use in PCR-based, automated, high-throughput genotyping and mutation detection of genomic DNA (designated as PCR candidate region mismatch scanning).

Lu and Hsu (242) described the use of *E. coli* MutY protein for the detection of mismatched GA in *p53* gene. A major limitation was that only GA mispairs were detected so that special mismatch-generating oligonucleotides containing A had to be used. Hsu *et al.* (243) reported the use of MutY in combination with thymine glycosylase for mismatch detection. In this method, DNA fragments amplified from normal and mutated genes by PCR were mixed, annealed, cleaved by mismatch repair enzymes, cleavage products separated by gel electrophoresis, and detected by autoradiography. Additionally, some accessory materials or equipments, such as fluorescent staining of MutS protected DNA (244), and a microelectronic MutS-DNA chip (245), and so on have been reported. Bellanne-Chantelot *et al.* (246) scanned 83 sequence-tagged sites (STSs) extracted from an 8 centimorgans (cM) human chromosome 21 region for polymorphism using immobilized MutS whereas Gotoh *et al.* (240) scanned bacterial genomes for unknown mutations by a combination of MutS and representational difference analysis (RDA) technique.

Stratagene has developed a highly efficient, rapid, and reproducible method for introducing random mutations in a cloned gene of interest. This method involves propagating the cloned gene in an *E. coli* strain called XL1-Red, which is deficient in three of the primary DNA repair pathways. The *mutS* (mismatch repair) (*81*), *mutD* ($3' \rightarrow 5'$ exonuclease of DNA polymerase III) (*247*), and *mutT* (hydrolyses 8-oxo dGTP) (*49*) genes were mutated in XL1-Red strain. The random mutation rate in this triple mutant was measured to be approximately 5000-fold higher than that of the wild type. This strain is particularly suitable for generating random mutations within a gene that has no selectable or screenable phenotype, and the method does not require extensive genetic or biochemical manipulations (*248, 249*). Thus, knowledge acquired in the field of DNA repair is gradually gaining momentum in getting translated into technological applications.

Acknowledgments

Arathi is acknowledged for her help in the preparation of this chapter. Work on the DNA mismatch repair in our laboratory was supported by the Department of Biotechnology through a research grant and the Proteomics program. Nimesh Joseph and Viswanadham Duppatla acknowledge Senior Research Fellowships from Council of Scientific and Industrial Research.

References

1. Yu, Z., Chen, J., Ford, B. N., Brackley, M. E., and Glickman, B. W. (1999). Human DNA repair systems: An overview. *Env. Mol. Genet.* **33**, 3–20.
2. Friedberg, E. C., Walker, G. C., and Siede, W. (1995). "DNA Repair and Mutagenesis." ASM Press, Washington DC.
3. Arber, W. (2000). Genetic variation: Molecular mechanisms and impact on microbial evolution. *FEMS Microbiol. Rev.* **24**, 1–7.
4. Papanicolaou, C., and Ripley, L. S. (1991). An *in vitro* approach to identifying specificity determinants of mutagenesis mediated by DNA misalignments. *J. Mol. Biol.* **221**, 805–821.
5. Raoult, D., Audic, S., Robert, C., Abergel, C., Renesto, P., Ogata, H., La Scola, B., Suzan, M., and Claverie, J. M. (2004). The 1.2-mb genome sequence of mimivirus. *Science* **306**, 1344–1350.
6. Nouspikel, T., and Hanawalt, P. C. (2002). DNA repair in terminally differentiated cells. *DNA Repair* **1**, 59–75.
7. Hoeijmakers, J. H. J. (2001). Genome maintenance mechanisms for preventing cancer. *Nature* **411**, 366–374.
8. Radman, M., Dohet, C., Bourgingnon, M.-F., Doubleday, O. P., and Lecomte, P. (1981). High fidelity devices in the reproduction of DNA. *In* "Chromosome Damage and Repair" (E. Seeberg and K. Kleppe, Eds.), pp. 431–445. Plenum Publishing Corp., New York.
9. Harfe, B. D., and Robertson, S. J. (2000). DNA mismatch repair and genetic instability. *Annu. Rev. Genet.* **11**, 359–399.
10. Claverys, J. P., and Lacks, S. A. (1986). Heteroduplex deoxyribonucleic acid base mismatch repair in bacteria. *Microbiol. Rev.* **50**, 133–165.

11. Lacks, S. A. (1988). Mechanisms of genetic recombination in Gram-positive bacteria. In "Genetic Recombination" (R. Kucherlapati and G. Smith, Eds.), pp. 43–85. American Society of Microbiology, Washington DC.
12. Radman, M., and Wagner, R. (1986). Mismatch repair in *E. coli*. *Annu. Rev. Genet.* **20**, 523–538.
13. Modrich, P. (1987). DNA mismatch correction. *Annu. Rev. Biochem.* **56**, 435–466.
14. Meselson, M. (1988). Methyl-directed repair of DNA mismatches. In "Methyl-Directed Repair of DNA Mismatches" (K. B. Low, Ed.), pp. 91–113. Academic Press, San Diego.
15. Lundblad, V., and Kleckner, N. (1985). Mismatch repair mutations in *Escherichia coli* K12 enhance transposon excision. *Genetics* **109**, 3–19.
16. Taddei, F., Radman, M., Maynard Smith, J., Toupance, B., Gugon, P. H., and Godeile, B. (1997). Role of mutator alleles in adaptive evolution. *Nature* **387**, 700–702.
17. Rayssiguier, C., Thaler, D. S., and Radman, M. (1989). The barrier to recombination between *Escherichia coli* and *Salmonella typhimurrium* is disrupted in mismatch repair mutants. *Nature* **342**, 396–401.
18. Vulic, M., Lenski, R. E., and Radman, M. (1999). Mutation, recombination and incipient speciation of bacteria in the laboratory. *Proc. Natl. Acad. Sci. USA* **96**, 7348–7351.
19. Worth, L., Jr., Clark, S., Radman, M., and Modrich, P. (1994). Mismatch repair proteins MutS and MutL inhibit RecA-catalysed strand transfer between diverged DNA. *Proc. Natl. Acad. Sci. USA* **91**, 3238–3241.
20. Calmann, M. A., Evans, J. E., and Marinus, M. G. (2005). MutS inhibits RecA-mediated strand transfer with methylated DNA substrates. *Nucleic Acids Res.* **33**, 3591–3597.
21. Hong, E. S., Yeung, H., Funchain, P., Slopska, M., and Miller, J. H. (2005). Mutants with temperature sensitive defects in *E. coli* mismatch repair system: Sensitivity to mispairs generated *in vivo*. *J. Bacteriol.* **187**, 840–846.
22. Radman, M. (1989). Mismatch repair and the fidelity of genetic recombination. *Genome* **31**, 68–73.
23. Mennecier, S., Coste, G., Servant, P., Bailone, A., and Sommer, S. (2004). Mismatch repair ensures fidelity of replication and recombination in the radioresistant organism *Deinococcus radiodurans*. *Mol. Genet. Genomics* **272**, 460–469.
24. Wildenberg, J., and Meselson, M. (1975). Mismatch repair in heteroduplex DNA. *Proc. Natl. Acad. Sci. USA* **72**, 2202–2206.
25. Wagner, R., and Meselson, M. (1976). Repair tracts in mismatched DNA heteroduplexes. *Proc. Natl. Acad. Sci. USA* **73**, 4135–4139.
26. Lu, A.-L., Clark, S., and Modrich, P. (1983). Methyl-directed repair of DNA base pair mismatches *in vitro*. *Proc. Natl. Acad. Sci. USA* **80**, 4639–4643.
27. Modrich, P. (1989). Methyl-directed DNA mismatch correction. *J. Biol. Chem.* **264**, 6597–6600.
28. Lieb, M. (1983). Specific mismatch correction in bacteriophage lambda crosses by very short patch repair. *Mol. Gen. Genet.* **191**, 118–125.
29. Modrich, P. (1991). Mechanisms and biological effects of mismatch repair. *Annu. Rev. Genet.* **25**, 229–253.
30. Marinus, M. G. (1984). Methylation of prokaryotic DNA. In "DNA Methylation" (A. Razin, H. Cedar, and A. D. Riggs, Eds.), pp. 81–109. Springer-Verlag, New York.
31. Duncan, B. K., and Miller, J. H. (1980). Mutagenic deamination of cytosine residues in DNA. *Nature* **287**, 560–561.
32. Jones, M., Wagner, R., and Radman, M. (1987). Mismatch repair of deaminated 5-methylcytosine. *J. Mol. Biol.* **194**, 155–159.
33. Lieb, M. (1987). Bacterial genes *mutL*, *mutS* and *dcm* participate in repair of mismatches at 5-methylcytosine sites. *J. Bacteriol.* **169**, 5241–5246.

34. Hennecke, F., Kolmar, H., Brundl, K., and Fritz, H. J. (1991). The *vsr* gene product of *E. coli* K-12 is a strand- and sequence-specific DNA mismatch endonuclease. *Nature* **353,** 776–778.
35. Dzidic, S., and Radman, M. (1989). Genetic requirements for hyper-recombination by very short patch mismatch repair: Involvement of *Escherichia coli* DNA polymerase I. *Mol. Gen. Genet.* **217,** 254–256.
36. Drotschmann, K., Aronshtam, A., Fritz, H.-J., and Marinus, H.-J. (1998). The *Escherichia coli* MutL protein stimulates binding of Vsr and MutS to heteroduplex DNA. *Nucleic Acids Res.* **26,** 948–953.
37. Tsutakawa, S. E., Jingami, H., and Morikawa, K. (1999). Recognition of a TG mismatch: The crystal structure of very short patch repair endonuclease in complex with a DNA duplex. *Cell* **99,** 615–623.
38. Monastiriakos, S. K., Doiron, K. M., Siponen, M. I., and Cupples, C. G. (2004). Functional interactions between the MutL and Vsr proteins of *Escherichia coli* are dependent on the N-terminus of Vsr. *DNA Repair* **3,** 639–647.
39. Bell, D. C., and Cupples, C. G. (2001). Very-short-patch repair in *Escherichia coli* requires the *dam* adenine methylase. *J. Bacteriol.* **183,** 3631–3635.
40. Gabbara, S., Wyszynski, M., and Bhagwat, A. S. (1994). A DNA repair process in *Escherichia coli* corrects U:G and T:G mismatches to C:G at sites of cytosine methylation. *Mol. Gen. Genet.* **243,** 244–248.
41. Duncan, B. K., and Weiss, B. (1982). Specific mutator effects of *ung* (uracil-DNA glycosylase) mutations in *Escherichia coli*. *J. Bacteriol.* **151,** 750–755.
42. Radicella, J. P., Clark, E. A., and Fox, M. S. (1988). Some mismatch repair activities in *Escherichia coli*. *Proc. Natl. Acad. Sci. USA* **85,** 9674–9678.
43. Au, K. G., Clark, S., Miller, J. H., and Modrich, P. (1989). *Escherichia coli mutY* gene encodes an adenine glycosylase active on G·A mispairs. *Proc. Natl. Acad. Sci. USA* **86,** 8877–8881.
44. Tsai-Wu, J. J., Liu, H. F., and Lu, A. L. (1992). *Escherichia coli* MutY protein has both N-glycosylase and apurinic/apyrimidinic endonuclease activities on A.C and A.G mispairs. *Proc. Natl. Acad. Sci. USA* **89,** 8779–8783.
45. Baas, P. D., and Jansz, H. S. (1972). Asymmetric information transfer during phi X174 DNA replication. *J. Mol. Biol.* **63,** 557–568.
46. Nevers, P., and Spatz, H. (1975). *Escherichia coli* mutants *uvr*D *uvr*E deficient in gene conversion of lambda heteroduplexes. *Mol. Gen. Genet.* **139,** 233–243.
47. Bauer, J., Krammer, G., and Knippers, R. (1981). Asymmetric repair of bacteriophage T7 heteroduplex DNA. *Mol. Gen. Genet.* **181,** 541–547.
48. Pukkila, P. J., Peterson, J., Herman, G., Modrich, P., and Meselson, M. (1983). Effects of high levels of DNA adenine methylation on methyl-directed mismatch repair in *Escherichia coli*. *Genetics* **104,** 571–582.
49. Cox, E. C. (1976). Bacterial mutator genes and the control of spontaneous mutation. *Annu. Rev. Genet.* **10,** 135–156.
50. Lahue, R. S., Au, K. G., and Modrich, P. (1989). DNA mismatch correction in a defined system. *Science* **245,** 160–164.
51. Marinus, M. G., and Morris, N. R. (1974). Biological function for 6-methyladenine residues in the DNA of *Escherichia coli* K12. *J. Mol. Biol.* **85,** 309–322.
52. Glickman, B. W. (1979). Spontaneous mutagenesis in *Escherichia coli* strains lacking 6-methyladenine residues in their DNA: An altered mutational spectrum in *dam*-mutants. *Mutat. Res.* **61,** 153–162.
53. Herman, G. E., and Modrich, P. (1981). *Escherichia coli* K-12 clones that overproduce *dam* methylase are hypermutable. *J. Bacteriol.* **145,** 644–646.
54. Marinus, M. G., Poteete, A., and Arraj, J. A. (1984). Correlation of DNA adenine methylase activity with spontaneous mutability in *Escherichia coli* K-12. *Gene* **28,** 123–125.

55. Palmer, B. R., and Marinus, M. G. (1994). The *dam* and *dcm* strains of *Escherichia coli*—a review. *Gene* **143**, 1–12.
56. Längle-Rouault, F., Maenhaut-Michel, G., and Radman, M. (1986). GATC sequence and mismatch repair in *Escherichia coli*. *EMBO J.* **5**, 2009–2013.
57. Modrich, P. (1994). Mismatch repair, genetic stability and cancer. *Science* **266**, 1959–1960.
58. Glickman, B. W., and Radman, M. (1980). *Escherichia coli* mutator mutants deficient in methylation-instructed DNA mismatch correction. *Proc. Natl. Acad. Sci. USA* **77**, 1063–1067.
59. McGraw, B. R., and Marinus, M. G. (1980). Isolation and characterization of Dam$^+$ revertants and suppressor mutations that modify secondary phenotypes of *dam*-3 strains of *Escherichia coli* K-12. *Mol. Gen. Genet.* **178**, 309–315.
60. Lu, A.-L., Welsh, K., Clark, S., Su, S.-S., and Modrich, P. (1984). Repair of DNA base-pair mismatches in extracts of *Escherichia coli*. *Cold Spring Harbor Symp. Quant. Biol.* **49**, 589–596.
61. Su, S., and Modrich, P. (1986). *Escherichia coli mutS*-encoded protein binds to mismatched DNA base pairs. *Proc. Natl. Acad. Sci. USA* **83**, 5057–5061.
62. Au, K. G., Welsh, K., and Modrich, P. (1992). Initiation of methyl-directed mismatch repair. *J. Biol. Chem.* **267**, 12142–12148.
63. Lahue, R. S., and Modrich, P. (1988). Methyl-directed DNA mismatch repair in *Escherichia coli*. *Mutat. Res.* **198**, 37–43.
64. Su, S.-S., Lahue, R. S., Au, K. G., and Modrich, P. (1988). Mispair specificity of methyl-directed DNA mismatch correction *in vitro*. *J. Biol. Chem.* **263**, 6829–6835.
65. Parker, B. O., and Marinus, M. G. (1992). Repair of DNA heteroduplexes containing small heterologous sequences in *Escherichia coli*. *Proc. Natl. Acad. Sci. USA* **89**, 1730–1734.
66. Fazakerley, G. V., Quignard, E., Woisard, A., Guschlbauer, W., van der Marel, G. A., van Boom, J. H., Jones, M., and Radman, M. (1986). Structures of mismatched base pairs in DNA and their recognition by the *Escherichia coli* mismatch repair system. *EMBO J.* **5**, 3697–3703.
67. Kramer, B., Kramer, W., and Fritz, H. J. (1984). Different base/base mismatches are corrected with different efficiencies by the methyl-directed DNA mismatch repair system of *E. coli*. *Cell* **38**, 879–887.
68. Brown, J., Brown, T., and Fox, K. R. (2001). Affinity of mismatch-binding protein MutS for heteroduplexes containing different mismatches. *Biochem. J.* **354**, 627–633.
69. Wang, H., Yang, Y., Schofield, M. J., Du, C., Fridman, Y., Lee, S. D., Larson, E. D., Drummond, J. T., Alani, E., Hsieh, P., and Erie, D. A. (2003). DNA bending and unbending by MutS govern mismatch recognition and specificity. *Proc. Natl. Acad. Sci. USA* **100**, 14822–14827.
70. Lamers, M. H., Perrakis, A., Enzlin, J. H., Winterwerp, H. H., de Wind, N., and Sixma, T. K. (2000). The crystal structure of DNA mismatch repair protein MutS binding to a G.T mismatch. *Nature* **407**, 711–717.
71. Obmolova, G., Ban, C., Hsieh, P., and Yang, W. (2000). Crystal structures of mismatch repair protein MutS and its complex with a substrate DNA. *Nature* **407**, 703–710.
72. Grilley, M., Welsh, K. M., Su, S. S., and Modrich, P. (1989). Isolation and characterization of the *Escherichia coli mutL* gene product. *J. Biol. Chem.* **264**, 1000–1004.
73. Grafström, R. H., and Hoess, R. H. (1983). Cloning of *mutH* and identification of the gene product. *Gene* **22**, 245–253.
74. Welsh, K. M., Lu, A. L., Clark, S., and Modrich, P. (1987). Isolation and characterization of the *Escherichia coli mutH* gene product. *J. Biol. Chem.* **262**, 15624–15629.
75. Grilley, M., Griffith, J., and Modrich, P. (1993). Bidirectional excision in methyl-directed mismatch repair. *J. Biol. Chem.* **268**, 11830–11837.

76. Matson, S. W. (1986). *Escherichia coli* helicase II (*uvrD* gene product) translocates unidirectionally in a 3' to 5' direction. *J. Biol. Chem.* **261**, 10169–10175.
77. Viswanathan, M., and Lovett, S. T. (1998). Single-strand DNA-specific exonucleases in *Escherichia coli*–roles in repair and mutation avoidance. *Genetics* **149**, 7–16.
78. Yamaguchi, M., Dao, V., and Modrich, P. (1998). MutS and MutL activate DNA helicase II in a mismatch-dependent manner. *J. Biol. Chem.* **273**, 9197–9201.
79. Burdett, V., Baitinger, C., Viswanathan, M., Lovett, S. T., and Modrich, P. (2001). In vivo requirement for RecJ, ExoVII, ExoI, and ExoX in methyl-directed mismatch repair. *Proc. Natl. Acad. Sci. USA* **98**, 6765–6770.
80. Cooper, D. L., Lahue, R. S., and Modrich, P. (1993). Methyl-directed mismatch repair is bidirectional. *J. Biol. Chem.* **268**, 11823–11829.
81. Modrich, P., and Lahue, R. (1996). Mismatch repair in replication fidelity, genetic recombination and cancer biology. *Annu. Rev. Biochem* **65**, 101–133.
82. Jiricny, J. (1998). Eukaryotic mismatch repair: An update. *Mutat. Res.* **409**, 107–121.
83. Bjornson, K. P., Allen, D. J., and Modrich, P. (2000). Modulation of MutS ATP hydrolysis by DNA cofactors. *Biochemistry* **39**, 3176–3183.
84. Ban, C., and Yang, W. (1998). Crystal structure and ATPase activity of MutL: Implications for DNA repair and mutagenesis. *Cell* **95**, 541–552.
85. Spampinato, C., and Modrich, P. (2000). The MutL ATPase is required for mismatch repair. *J. Biol. Chem.* **275**, 9863–9869.
86. Hall, M. C., Jordan, J. R., and Matson, S. W. (1998). Evidence for a physical interaction between the *Escherichia coli* methyl-directed mismatch repair proteins MutL and UvrD. *EMBO J.* **17**, 1535–1541.
87. Mechanic, L. E., Frankel, B. A., and Matson, S. W. (2000). *E. coli* MutL loads DNA helicase II onto DNA. *J. Biol. Chem.* **275**, 38337–38346.
88. Bende, S. M., and Grafström, R. H. (1991). The DNA binding properties of the MutL protein isolated from *Escherichia coli*. *Nucleic Acids Res.* **19**, 1549–1555.
89. Ban, C., Junop, M., and Yang, W. (1999). Transformation of MutL by ATP binding and hydrolysis: A switch in DNA mismatch repair. *Cell* **97**, 85–97.
90. Wu, T.-H., and Marinus, M. G. (1994). Dominant negative mutator mutations in the *mutS* gene of *Escherichia coli*. *J. Bacteriol.* **176**, 5393–5400.
91. Aronshtam, A., and Marinus, M. G. (1996). Dominant negative mutator mutations in the *mutL* gene of *Escherichia coli*. *Nucleic Acids Res.* **24**, 2498–2504.
92. Leadon, S. A. (1999). Transcription-coupled repair of DNA damage: Unanticipated players, unexpected complexities. *Am. J. Hum. Genet.* **64**, 1259–1263.
93. Wu, J., Gu, L., Wang, H., Geacintov, N. E., and Li, G. M. (1999). Mismatch repair processing of carcinogen-DNA adducts triggers apoptosis. *Mol. Cell. Biol.* **19**, 8292–8301.
94. Längle-Rouault, F., Maenhaut, M. G., and Radman, M. (1987). GATC sequences, DNA nicks and the MutH function in *Escherichia coli* mismatch repair. *EMBO J.* **6**, 1121–1127.
95. Schaaper, R. M., and Dunn, R. L. (1987). Spectra of spontaneous mutations in *Escherichia coli* strains defective in mismatch correction: The nature of *in vivo* DNA replication errors. *Proc. Natl. Acad. Sci. USA* **84**, 6220–6224.
96. Lahue, R. S., Su, S. S., and Modrich, P. (1987). Requirement for d(GATC) sequences in *Escherichia coli mutHLS* mismatch correction. *Proc. Natl. Acad. Sci. USA* **84**, 1482–1486.
97. Hall, M. C., and Matson, S. W. (1999). The *Escherichia coli* MutL protein physically interacts with MutH and stimulates the MutH-associated endonuclease activity. *J. Biol. Chem.* **274**, 1306–1312.
98. Yang, W. (2000). Structure and function of mismatch repair proteins. *Mutat. Res.* **460**, 245–256.

99. Dao, V., and Modrich, P. (1998). Mismatch-, MutS-, MutL-, and helicase II-dependent unwinding from the single-strand break of an incised heteroduplex. *J. Biol. Chem.* **273,** 9202–9207.
100. Eisen, J. A. (1998). A phylogenomic study of the MutS family of proteins. *Nucleic. Acid Res.* **26,** 4291–4300.
101. Malik, H. S., and Henikoff, S. (2000). Dual recognition-incision enzymes might be involved in mismatch repair and meiosis. *Trends Biochem. Sci.* **25,** 414–418.
102. Moreira, D., and Phillippe, H. (1999). Smr: A bacterial and eukaryotic homology of the C-terminal region of the MutS2 family. *Trends Biochem. Sci.* **24,** 298–300.
103. Alm, R. A., Ling, L. S., Moir, D. T., King, B. L., Brown, E. D., Doig, P. C., Smith, D. R., Noonan, B., Guild, B. C., de Jonge, B. L., Carmel, G., Tummino, P. J. *et al.* (1999). Genomic sequence comparison of two unrelated isolates of the human gastric pathogen *Helicobacter pylori*. *Nature* **397,** 176–180.
104. Tomb, J. F., White, O., Kerlavage, A. R., Clayton, R. A., Sutton, G. G., Fleischmann, R. D., Ketchum, K. A., Klenk, H. P., Gill, S., Dougherty, B. A., Nelson, K., Quackenbush, J. *et al.* (1997). The complete genome sequence of the gastric pathogen *Helicobacter pylori*. *Nature* **388,** 539–547.
105. Pinto, A. V., Mathieu, A., Marshin, S., Veaute, X., Ielpi, L., Labigne, A., and Radicella, J. P. (2005). Suppression of homologus and homeologus recombination by the bacterial MutS2 protein. *Mol. Cell* **17,** 113–120.
106. Rossolillo, P., and Albertini, A. M. (2001). Functional analysis of the *Bacillus subtilis yshD* gene, a *mutS* paralogue. *Mol. Gen. Genet.* **264,** 809–818.
107. Vijayvargia, R., and Biswas, I. (2002). MutS2 family protein from *Pyrococcus furiosus*. *Curr. Microb.* **44,** 224–228.
108. Takamatsu, S., Kato, R., and Kuramitsu, S. (1996). Mismatch DNA recognition protein from an extremely thermophilic bacterium, *Thermus thermophilus* HB8. *Nucleic Acid Res.* **24,** 640–647.
109. Young, D. M., and Ornston, L. N. (2001). Functions of the mismatch repair gene *mutS* from *Acinetobacter* sp. strain ADP1. *J. Bacteriol.* **183,** 6822–6831.
110. Kurusu, Y., Narita, T., Suzuki, M., and Watanable, T. (2000). Genetic analysis of an incomplete *mutS* gene from *Pseudomonas putida*. *J. Bacteriol.* **182,** 5278–5279.
111. van den Broek, D., Chin-A-Woeng, T. F., Bloemberg, G. V., and Lugtenberg, B. J. (2005). Role of Rpos and MutS in phase variation of *Pseudomonas* sp. PCL 1171. *Microbiology* **151,** 1403–1408.
112. Bera, T. K., Ghosh, S. K., and Das, J. (1989). Cloning and characterization of *mutL* and *mutS* genes of *Vibrio cholerae*: Nucleotide sequence of the *mutL* gene. *Nucleic Acid Res.* **17,** 6241–6251.
113. Bandyopadhyay, R., and Das, J. (1994). The DNA adenine MTase encoding gene (*dam*) of *Vibrio cholerae*. *Gene* **140,** 67–71.
114. Bhakat, K. K., Sharma, S., and Das, J. (1999). The *mutK* gene of *Vibrio cholerae*: A new gene involved in DNA mismatch repair. *J. Bacteriol.* **181,** 879–883.
115. Meier, P., and Wackernagel, W. (2005). Impact of mutS inactivation on foreign DNA acquisition by natural transformation in *Pseudomonas stutzeri*. *J. Bacteriol.* **187,** 143–154.
116. Zaleski, P., and Piekarowicz, A. (2004). Characterization of a *dam* mutant of *Haemophilus influenzae* Rd. *Microbiology* **150,** 3773–3781.
117. Wyrzykowski, J., and Volkert, M. R. (2003). The *Escherichia coli* methyl-directed mismatch repair system repairs base pairs containing oxidative lesions. *J. Bacteriol.* **185,** 1701–1704.
118. Ban, C., and Yang, W. (1998). Structural basis for MutH activation in *E. coli* mismatch repair and relationship of MutH to restriction endonucleases. *EMBO J.* **17,** 1526–1534.
119. Guarne, A., Ramon-Maiques, S., Wolff, E. M., Ghirlando, R., Hu, X., Miller, J. H., and Yang, W. (2004). Structure of the MutL C-terminal domain: A model of intact MutL and its roles in mismatch repair. *EMBO J.* **23,** 4134–4145.

120. Bujnicki, J. M. (2001). A model of structure and action of Sau3AI restriction endonuclease that comprises two MutH-like endonuclease domains within a single polypeptide. *Acta Microbiol. Pol.* **50**, 219–231.
121. Friedhoff, P., Thomas, E., and Pingoud, A. (2003). Tyr212: A key residue involved in strand discrimination by the DNA mismatch repair endonuclease MutH. *J. Mol. Biol.* **325**, 285–297.
122. Loh, T., Murphy, K. C., and Marinus, M. G. (2001). Mutational analysis of the MutH protein from *Escherichia coli*. *J. Biol. Chem.* **276**, 12113–12119.
123. Wu, T.-H., Loh, T., and Marinus, M. G. (2002). The function of Asp70, Glu77 and Lys79 in the *Escherichia coli* MutH protein. *Nucleic Acids Res.* **30**, 818–822.
124. Junop, M. S., Yang, W., Funchain, P., Clendenin, W., and Miller, J. H. (2003). *In vitro* and *in vivo* studies of MutS, MutL and MutH mutants: Correlation of mismatch repair and DNA recombination. *DNA Repair* **2**, 387–405.
125. Pingoud, A., and Jeltsch, A. (1997). Recognition and cleavage of DNA by type-II restriction endonucleases. *Eur. J. Biochem.* **246**, 1–22.
126. Toedt, G. H., Krishnan, R., and Friedhoff, P. (2003). Site-specific protein modification to identify the MutL interface of MutH. *Nucleic Acids Res.* **31**, 819–825.
127. Kramer, W., Kramer, B., and Williamson, M. S. (1989). Cloning and nucleotide sequence of DNA mismatch repair gene PMS1 from *Saccharomyces cerevisiae*: Homology of PMS1 to prokaryotic MutL and HexB. *J. Bacteriol.* **171**, 5339–5346.
128. Pang, Q., Prolla, T. A., and Liskay, R. M. (1997). Functional domains of the *Saccharomyces cerevisiae* MLH1p and PMS1p DNA mismatch repair proteins and there relevance to human hereditary nonpolyposis colorectal cancer-associated mutations. *Mol. Cell. Biol.* **17**, 4465–4473.
129. Wu, X., Platt, J. L., and Cascalho, M. (2003). Dimerization of MLH1 and PMS2 limits nuclear localization of MutLα. *Mol. Cell. Biol.* **23**, 3320–3328.
130. Wigley, D. B., Davies, G. J., Dodson, E. J., Maxwell, A., and Dodson, G. (1991). Crystal structure of an N-terminal fragment of the DNA gyrase B protein. *Nature* **351**, 624–629.
131. Bergerat, A., de Massy, B., Gadelle, D., Varoutas, P. C., Nicolas, A., and Forterre, P. (1997). An atypical topoisomerase II from archaea with implications for meiotic recombination. *Nature* **386**, 414–417.
132. Mushegian, A. R., Bassett, D. E., Jr., Boguski, M. S., Bork, P., and Koonin, E. V. (1997). Positionally cloned human disease genes: Patterns of evolutionary conservation and functional motifs. *Proc. Natl. Acad. Sci. USA* **94**, 5831–5836.
133. Dutta, R., and Inouye, M. (2000). GHKL, an emergent ATPase/kinase superfamily. *Trends Biochem. Sci.* **25**, 24–28.
134. Kosinski, J., Steindorf, I., Bujnicki, J. M., Giron-Monzon, L., and Friedhoff, P. (2005). Analysis of the quaternary structure of the MutL C-terminal domain. *J. Mol. Biol.* **351**, 895–909.
135. Drotschmann, K., Hall, M. C., Shcherbakova, P. V., Wang, H., Erie, D. A., Brownewell, F. R., Kool, E. T., and Kunkel, T. A. (2002). DNA binding properties of the yeast MSH2–MSH6 and MLH1–PMS1 heterodimers. *Biol. Chem.* **383**, 969–975.
136. Acharya, S., Foster, P. L., Brooks, P., and Fishel, R. (2003). The coordinated functions of E. coli MutS and MutL proteins in mismatch repair. *Mol. Cell* **12**, 233–246.
137. Giron-Monzon, L., Manelyte, L., Ahrends, R., Kirsch, D., Spengler, B., and Friedhoff, P. (2004). Mapping protein–protein interactions between MutL and MutH by cross-linking. *J. Biol. Chem.* **279**, 49338–49345.
138. Sixma, T. K. (2001). DNA mismatch repair: MutS structures bound to mismatches. *Curr. Opin. Struct. Biol.* **11**, 47–52.
139. Bjornson, K. P., and Modrich, P. (2003). Differential and simultaneous adenosine di- and triphosphate binding to MutS. *J. Biol. Chem.* **278**, 18557–18562.

140. Antony, E., and Hingorani, M. M. (2004). Asymmeteric ATP binding and hydrolysis activity of the *Thermus aquaticus* MutS dimer is key to modulation of its interactions with mismatched DNA. *Biochemistry* **43**, 13115–13128.
141. Malkov, V. A., Biswas, I., Camerini-Otero, R. D., and Hsieh, P. (1997). Photocross-linking of the NH_2–terminal region of Taq MutS protein to the major groove of a heteroduplex DNA. *J. Biol. Chem.* **272**, 23811–23817.
142. Yamamoto, A., Schofield, M. J., Biswas, I., and Hsieh, P. (2000). Requirement for Phe36 for DNA binding and mismatch repair by *Escherichia coli* MutS protein. *Nucleic Acid Res.* **28**, 3564–3569.
143. Biswas, I., Obmolova, G., Takahashi, M., Herr, A., Newman, M. A., Yang, W., and Hsieh, P. (2001). Disruption of the helix-u-turn-helix motif of MutS protein: Loss of subunit dimerization, mismatch binding and ATP hydrolysis. *J. Mol. Biol.* **305**, 805–816.
144. Wu, T.-H., and Marinus, M. G. (1999). Deletion mutation analysis of the *mutS* gene in *Escherichia coli*. *J. Biol. Chem.* **274**, 5948–5952.
145. Gorbalenya, A. E., and Koonin, E. V. (1990). Superfamily of UvrA-related NTP binding proteins. Implications for rational classification of recombination/repair systems. *J. Mol. Biol.* **213**, 583–591.
146. Hung, L. W., Wang, I. X., Nikaido, K., Liu, P. Q., Ames, G. F., and Kim, S. H. (1998). Crystal structure of the ATP-binding subunit of an ABC transporter. *Nature* **396**, 703–707.
147. Kato, R., Kataoka, M., Kamikubo, H., and Kuramitsu, S. (2001). Direct observation of three conformations of MutS protein regulated by adenine nucleotides. *J. Mol. Biol.* **309**, 227–238.
148. Gradia, S., Subramanian, D., Wilson, T., Acharya, S., Makhov, A., Griffith, J., and Fishel, R. (1999). hMSH2-hMSH6 forms a hydrolysis-independent sliding clamp on mismatched DNA. *Mol. Cell.* **3**, 255–261.
149. Gradia, S., Acharya, S., and Fishel, R. (2000). The role of mismatched nucleotides in activating the hMSH2-hMSH6 molecular switch. *J. Biol. Chem.* **275**, 3922–3930.
150. Wilson, T., Guerrette, S., and Fishel, R. (1999). Dissociation of mismatch recognition and ATPase activity by hMSH2-hMSH3. *J. Biol. Chem.* **274**, 21659–21664.
151. Joshi, A., and Rao, B. J. (2002). ATP hydrolysis induces expansion of MutS contacts on heteroduplex: A case for MutS treadmilling? *Biochemistry* **41**, 3654–3666.
152. Biswas, I., and Vijayvargia, R. (2000). Heteroduplex DNA and ATP induced conformational changes of a MutS mismatch repair protein from *Thermus aquaticus*. *Biochem. J.* **347**, 881–886.
153. Allen, D. J., Makhov, A., Grilley, M., Taylor, J., Thresher, R., Modrich, P., and Griffith, J. D. (1997). MutS mediates heteroduplex loop formation by a translocation mechanism. *EMBO J.* **16**, 4467–4476.
154. Lamers, M. H., Georgijevic, D., Lebbink, J. H., Winterwerp, H. H., Agianian, B., de Wind, N., and Sixma, T. K. (2004). ATP increases the affinity between MutS ATPase domains: Implications for ATP hydrolysis and conformational changes. *J. Biol. Chem.* **279**, 43879–43885.
155. Bruni, R., Martin, D., and Jiricny, J. (1988). d(GATC) sequences influence *Escherichia coli* mismatch repair in a distance-dependent manner from positions both upstream and downstream of the mismatch. *Nucleic Acids Res.* **16**, 4875–4890.
156. Gradia, S., Acharya, S., and Fishel, R. (1997). The human mismatch recognition complex hMSH2-hMSH6 functions as a novel molecular switch. *Cell* **91**, 995–1005.
157. Fishel, R., Acharya, S., Berardini, M., Bocker, T., Charbonneau, N., Cranston, A., Gradia, S., Guerrette, S., Heinen, C. D., Mazurek, A., Snowden, T., Schmutte, C. *et al.* (2000). Signaling mismatch repair: The mechanics of an adenosine-nucleotide molecular switch. *Cold Spring Harb. Symp. Quant. Biol.* **65**, 217–224.

158. Fishel, R. (1998). Mismatch repair, molecular switches, and signal transduction. *Genes Dev.* **12**, 2096–2101.
159. Joseph, N., Duppatla, V., and Rao, D. N. (2005). Functional characterization of the DNA mismatch binding protein MutS from *Haemophilus influenzae*. *Biochem. Biophys. Res. Commun.* **334**, 891–900.
160. Joseph, N., Sawarkar, R., and Rao, D. N. (2004). DNA mismatch correction in *Haemophilus influenzae*: Characterization of MutL, MutH and their interaction. *DNA Repair* **3**, 1561–1577.
161. Haber, L. T., and Walker, G. C. (1991). Altering the conserved nucleotide binding motif in the *Salmonella typhimurium* MutS mismatch repair protein affects both its ATPase and mismatch binding activities. *EMBO J.* **10**, 2707–2715.
162. Hess, M. T., Gupta, R. D., and Kolodner, R. D. (2002). Dominant *Saccharomyces cerevisiae* MSH6 mutations cause increased mispair binding and decreased dissociation from mispairs by MSH2-MSH6 in the presence of ATP. *J. Biol. Chem.* **277**, 25545–25553.
163. Blackwell, L. J., Martik, D., Bjornson, K. P., Bjornson, E. S., and Modrich, P. (1998). Nucleotide-promoted release of hMutSα from heteroduplex DNA is consistent with an ATP-dependent translocation mechanism. *J. Biol. Chem.* **273**, 32055–32062.
164. Junop, M. S., Obmolova, G., Rausch, K., Hsieh, P., and Yang, W. (2001). Composite active site of an ABC ATPase: MutS uses ATP to verify mismatch recognition and authorize DNA repair. *Mol. Cell.* **7**, 1–12.
165. Schofield, M. J., Nayak, S., Scott, T. H., Du, C., and Hsieh, P. (2001). Interaction of *Escherichia coli* MutS and MutL at a DNA mismatch. *J. Biol. Chem.* **276**, 28291–28299.
166. Wang, H., and Hays, J. B. (2004). Signaling from DNA mispairs to mismatch-repair excision sites despite intervening blockades. *EMBO J.* **23**, 2126–2133.
167. Wang, H., and Hays, J. B. (2002). Mismatch repair in human nuclear extracts. Quantitative analyses of excision of nicked circular mismatched DNA substrates, constructed by a new technique employing synthetic oligonucleotides. *J. Biol. Chem.* **277**, 26136–26142.
168. Wang, H., and Hays, J. B. (2002). Mismatch repair in human nuclear extracts. Time courses and ATP requirements for kinetically distinguishable steps leading to tightly controlled 5′ to 3′ and aphidicolin-sensitive 3′ to 5′ mispair-provoked excision. *J. Biol. Chem.* **277**, 26143–26148.
169. Zdraveski, Z., Mello, J. A., Farinelli, C. K., Essigmann, J. M., and Marinus, M. G. (2002). MutS preferentially recognizes cisplatin over oxaliplatin modified DNA. *J. Biol. Chem.* **277**, 1255–1260.
170. Calmann, M. A., and Marinus, M. G. (2004). MutS inhibits RecA-mediated strand exchange with platinated substrates. *Proc. Natl. Acad. Sci. USA* **101**, 14174–14179.
171. Calmann, M. A., Nowosielska, A., and Marinus, M. G. (2005). Separation of mutation avoidance and antirecombination functions in an *Escherichia coli mutS* mutant. *Nucleic Acid Res.* **33**, 1193–1200.
172. Rasmussen, L. J., and Samson, L. (1996). The *Escherichia coli* MutS DNA mismatch binding protein specifically binds O(6)-methylguanine DNA lesions. *Carcinogenesis* **17**, 2085–2088.
173. Jones, M., and Wagner, R. (1981). *N*-Methyl *N′*-nitro-*N*-nitrosoguanidine sensitivity of *Escherichia coli* mutants deficient in DNA methylation and mismatch repair. *Mol. Gen. Genet.* **184**, 562–563.
174. Fox, M. S., and Allen, M. K. (1964). On the mechanism of deoxyribonucleate integration in pneumococcal transformation. *Proc. Natl. Acad. Sci. USA* **52**, 412–419.
175. Lacks, S. (1962). Molecular fate of DNA in genetic transformation of *pneumococcus*. *J. Mol. Biol.* **5**, 119–131.

176. Ephrussi-Taylor, H., Sicard, A. M., and Kamen, R. (1965). Genetic recombination in DNA-induced transformation of pneumococcus. I. The problem of relative efficiency of transforming factors. *Genetics* **51**, 455–475.
177. Lacks, S. (1966). Integration efficiency and genetic recombination in pneumococcal transformation. *Genetics* **53**, 207–235.
178. Lacks, S. (1968). Genetic regulation of maltosaccharide utilization in pneumococcus. *Genetics* **60**, 685–706.
179. Ephrussi-Taylor, H., and Gray, T. C. (1966). Genetic studies of recombining DNA in pneumococcal transformation. *J. Gen. Physiol.* **49**, 211–231.
180. Lacks, S. (1970). Mutants of *Diplococcus pneumoniae* that lack deoxyribonucleases and other activities possibly pertinent to genetic transformation. *J. Bacteriol.* **101**, 373–383.
181. Tiraby, J. G., and Fox, M. S. (1973). Marker discrimination in transformation and mutation of pneumococcus. *Proc. Natl. Acad. Sci. USA* **70**, 3541–3545.
182. Priebe, S. D., Hadi, S. M., Greenberg, B., and Lacks, S. A. (1988). Nucleotide sequence of the *hexA* gene for DNA mismatch repair in *Streptococcus pneumoniae* and homology of *hexA* to *mutS* of *Escherichia coli* and *Salmonella typhimurium*. *J. Bacteriol.* **170**, 190–196.
183. Prudhomme, M., Mejean, V., Martin, B., and Claverys, J. P. (1991). Mismatch repair genes of *Streptococcus pneumoniae*: HexA confers a mutator phenotype in *Escherichia coli* by negative complementation. *J. Bacteriol.* **173**, 7196–7203.
184. Ren, J., Park, J. H., Dunn, N. W., and Kim, W. S. (2001). Sequence and stress-response analyses of the DNA mismatch repair gene *hexA* in *Lactococcus lactis*. *Curr. Microbiol.* **43**, 232–237.
185. Prudhomme, M., Martin, B., Mejean, V., and Claverys, J. P. (1989). Nucleotide sequence of the *Streptococcus pneumoniae hexB* mismatch repair gene: Homology of HexB to MutL of *Salmonella typhimurium* and to PMS1 of *Saccharomyces cerevisiae*. *J. Bacteriol.* **171**, 5332–5338.
186. Kornberg, A., and Baker, T. A. (1991). "DNA Replication." W. H. Freeman and Co., New York.
187. Eisen, J. A., and Hanawalt, P. C. (1999). A phylogenomic study of DNA repair genes, proteins, and processes. *Mutat. Res.* **435**, 171–213.
188. Aravind, L., Walker, D. R., and Koonin, E. V. (1999). Conserved domains in DNA repair proteins and evolution of repair systems. *Nucleic Acids Res.* **27**, 1223–1242.
189. Culligan, K. M., Meyer-Gauen, G., Lyons-Weiler, J., and Hays, J. B. (2000). Evolutionary origin, diversification and specialization of eukaryotic MutS homolog mismatch repair proteins. *Nucleic Acids Res.* **28**, 463–471.
190. Andrewes, F. W. (1922). Studies in group agglutination. *J. Pathol. Bacteriol.* **25**, 505.
191. Moxon, E. R., Rainey, P. B., Nowak, M. A., and Lenski, R. E. (1994). Adaptive evolution of highly mutable loci in pathogenic bacteria. *Curr. Biol.* **4**, 24–33.
192. Bayliss, C. D., Sweetman, W. A., and Moxon, E. R. (2004). Mutations in *Haemophilus influenzae* mismatch repair genes increase mutation rates of dinucleotide repeat tracts but not dinucleotide repeat-driven pilin phase variation rates. *J. Bacteriol.* **186**, 2928–2935.
193. Shaver, A. C., and Sniegowski, P. D. (2003). Spontaneously arising *mutL* mutators in evolving *Escherichia coli* populations are the result of changes in repeat length. *J. Bacteriol.* **185**, 6076–6082.
194. Hood, D. W., Deadman, M. E., Jennings, M. P., Bisercic, M., Fleischmann, R. D., Venter, J. C., and Moxon, E. R. (1996). DNA repeats identify novel virulence genes in *Haemophilus influenzae*. *Proc. Natl. Acad. Sci. USA* **93**, 11121–11125.
195. Saunders, N. J., Peden, J. F., Hood, D. W., and Moxon, E. R. (1998). Simple sequence repeats in the *Helicobacter pylori* genome. *Mol. Microbiol.* **27**, 1091–1098.

196. Saunders, N. J., Jeffries, A. C., Peden, J. F., Hood, D. W., Tettelin, H., Rappuoli, R., and Moxon, E. R. (2000). Repeat-associated phase variable genes in the complete genome sequence of *Neisseria meningitidis* strain MC58. *Mol. Microbiol.* **37**, 207–215.
197. Parkhill, J., Wren, B. W., Mungall, K., Ketley, J. M., Churcher, C., Basham, D., Chillingworth, T., Davies, R. M., Feltwell, T., Holroyd, S., Jagels, K., Karlyshev, A. V. *et al.* (2000). The genome sequence of the food-borne pathogen *Campylobacter jejuni* reveals hypervariable sequences. *Nature* **403**, 665–668.
198. LeClerc, J. E., Li, B., Payne, W. L., and Cebula, T. A. (1996). High mutation frequencies among *Escherichia coli* and *Salmonella* pathogens. *Science* **274**, 1208–1211.
199. Matic, I., Radman, M., Taddei, F., Picard, B., Doit, C., Bingen, E., Denamur, E., and Elion, J. (1997). Highly variable mutation rates in commensal and pathogenic *Escherichia coli*. *Science* **277**, 1833–1834.
200. Denamur, E., Bonacorsi, S., Giraud, A., Duriez, P., Hilali, F., Amorin, C., Bingen, E., Andremont, A., Picard, B., Taddei, F., and Matic, I. (2002). High frequency of mutator strains among human uropathogenic *Escherichia coli* isolates. *J. Bacteriol.* **184**, 605–609.
201. Oliver, A., Canton, R., Campo, P., Baquero, F., and Blazquez, J. (2000). High frequency of hypermutable *Pseudomonas aeruginosa* in cystic fibrosis lung infection. *Science* **288**, 1251–1254.
202. Oliver, A., Baquero, F., and Blazquez, J. (2002). The mismatch repair system (*mutS, mutL* and *uvrD* genes) in *Pseudomonas aeruginosa*: Molecular characterization of naturally occurring mutants. *Mol. Microbiol.* **43**, 1641–1650.
203. Richardson, A. R., Yu, Z., Popovic, T., and Stojiljkovic, I. (2002). Mutator clones of *Neisseria meningitidis* in epidemic serogroup A disease. *Proc. Natl. Acad. Sci. USA* **99**, 6103–6107.
204. Prunier, A. L., Malbruny, B., Laurans, M., Brouard, J., Duhamel, J. F., and Leclercq, R. (2003). High rate of macrolide resistance in *Staphylococcus aureus* strains from patients with cystic fibrosis reveals high proportions of hypermutable strains. *J. Infect. Dis.* **187**, 1709–1716.
205. Bucci, C., Lavitola, A., Salvatore, P., Del Giudice, L., Massardo, D. R., Bruni, C. B., and Alifano, P. (1999). Hypermutation in pathogenic bacteria: Frequent phase variation in meningococci is a phenotypic trait of a specialized mutator biotype. *Mol. Cell.* **3**, 435–445.
206. Richardson, A. R., and Stojiljkovic, I. (2001). Mismatch repair and the regulation of phase variation in *Neisseria meningitidis*. *Mol. Microbiol.* **40**, 645–655.
207. Martin, P., Sun, L., Hood, D. W., and Moxon, E. R. (2004). Involvement of genes of genome maintenance in the regulation of phase variation frequencies in *Neisseria meningitides*. *Microbiology* **150**, 3001–3012.
208. Lacks, S. A., Dunn, J. J., and Greenberg, B. (1982). Identification of base mismatches recognized by the heteroduplex-DNA-repair system of *Streptococcus pneumoniae*. *Cell* **31**, 327–336.
209. Li, B., Tsui, H. C., LeClerc, J. E., Dey, M., Winkler, M. E., and Cebula, T. A. (2003). Molecular analysis of *mutS* expression and mutation in natural isolates of pathogenic *Escherichia coli*. *Microbiology* **149**, 1323–1331.
210. Smania, A. M., Segura, I., Pezza, R. J., Becerra, C., Albesa, I., and Argarana, C. E. (2004). Emergence of phenotypic variants upon mismatch repair disruption in *Pseudomonas aeruginosa*. *Microbiology* **150**, 1327–1338.
211. Cole, S. T., Brosch, R., Parkhill, J., Garnier, T., Churcher, C., Harris, D., Gordon, S. V., Eiglmeier, K., Gas, S., Barry, C. E., Tekaia, F., Badcock, K. *et al.* (1998). Deciphering the biology of *Mycobacterium tuberculosis* from the complete genome sequence. *Nature* **393**, 537–544.

212. Springer, B., Sander, P., Sedlacek, L., Hardt, W. D., Mizrahi, V., Schar, P., and Bottger, E. C. (2004). Lack of mismatch correction facilitates genome evolution in *mycobacteria*. *Mol. Microbiol.* **53**, 1601–1609.
213. Hemavathy, K. C., and Nagaraja, V. (1995). DNA methylation in *mycobacteria*: Absence of methylation at GATC (Dam) and CCA/TGG (Dcm) sequences. *FEMS Immunol. Med. Microbiol.* **11**, 291–296.
214. Fleischmann, R. D., Adams, M. D., White, O., Clayton, R. A., Kirkness, E. F., Kerlavage, A. R., Bult, C. J., Tomb, J. F., Dougherty, B. A., and Merrick, J. M. (1995). Whole-genome random sequencing and assembly of *Haemophilus influenzae* Rd. *Science* **269**, 496–512.
215. Hood, D. W., and Moxon, E. R. (1999). Lipopolysaccharide phase variation in *Haemophilus* and *Neisseria*. In "Endotoxin in Health and Disease" (H. Brude, S. M. Opal, S. N. Vogel, and D. C. Morrison, Eds.), pp. 39–54. Marcel Dekker Inc, New York, NY.
216. Friedhoff, P., Sheybani, B., Thomas, E., Merz, C., and Pingoud, A. (2002). *Haemophilus influenzae* and *Vibrio cholerae* genes for *mutH* are able to fully complement a *mutH* defect in *Escherichia coli*. *FEMS Microbiol.* **208**, 123–128.
217. Bayliss, C. D., van de Ven, T., and Moxon, E. R. (2002). Mutations in *polI* but not *mutSLH* destabilize *Haemophilus influenzae* tetranucleotide repeats. *EMBO J.* **21**, 1465–1476.
218. Watson, M. E., Burns, J. L., and Smith, A. L. (2004). Hypermutable *Haemophilus influenzae* with mutations in *mutS* are found in cystic fibrosis sputum. *Microbiology* **150**, 2947–2958.
219. Marinus, M. G. (1996). Methylation of DNA. In "*Escherichia coli* and *Salmonella*: Cellular and Molecular Biology" (F. C. Neidhardt, J. L. Ingraham, B. Magasanik, K. B. Low, M. Schaechter, and H. E. Umbarger, Eds.), pp. 782–791. American Society for Microbiology Press, Washington, DC.
220. Messer, W., Bellekes, U., and Lothar, H. (1985). Effect of *dam* methylation on the activity of the *E. coli* replication origin *oriC*. *EMBO J.* **4**, 1327–1332.
221. Boye, E., and Lobner-Olesen, A. (1990). The role of *dam* methyltransferase in the control of DNA replication in *E. coli*. *Cell* **62**, 981–989.
222. Oshima, T., Wada, C., Kawagoe, Y., Ara, T., Maeda, M., Masuda, Y., Hiraga, S., and Mori, H. (2002). Genome-wide analysis of deoxyadenosine methyltransferase-mediated control of gene expression in *Escherichia coli*. *Mol. Microbiol.* **45**, 673–695.
223. Henaut, A., Rouxel, T., Gleizes, A., Moszer, I., and Danchin, A. (1996). Uneven distribution of GATC motifs in the *Escherichia coli* chromosome, its plasmids and its phages. *J. Mol. Biol.* **257**, 574–585.
224. Heithoff, D. M., Sinsheimer, R. L., Low, D. A., and Mahan, M. J. (1999). An essential role for DNA adenine methylation in bacterial virulence. *Science* **284**, 967–970.
225. Torreblanca, J., and Casadesus, J. (1996). DNA adenine methylase mutants of *Salmonella typhimurium* and a novel *dam*-regulated locus. *Genetics* **144**, 15–26.
226. Julio, S. M., Heithoff, D. M., Provenzano, D., Klose, K. E., Sinsheimer, R. L., Low, D. A., and Mahan, M. J. (2001). DNA adenine MTase is essential for viability and plays a role in the pathogenesis of *Yersinia pseudotuberculosis* and *Vibrio cholerae*. *Infect Immun.* **69**, 7610–7615.
227. Garcia-Del Portillo, F., Pucciarelli, M. G., and Casadesus, J. (1999). DNA adenine methylase mutants of *Salmonella typhimurium* show defects in protein secretion, cell invasion, and M cell cytotoxicity. *Proc. Natl. Acad. Sci. USA* **96**, 11578–11583.
228. Watson, M. E., Jr., Jarisch, J., and Smith, A. L. (2004). Inactivation of deoxyadenosine methyltransferase (*dam*) attenuates *Haemophilus influenzae* virulence. *Mol. Microbiol.* **53**, 651–664.
229. Ostendorf, T., Cherepanov, P., de Vries, J., and Wackernagel, W. (1999). Characterization of a *dam* mutant of *Serratia marcescens* and nucleotide sequence of the *dam* region. *J. Bacteriol.* **181**, 3880–3885.

230. Glickman, B., van den Elsen, P., and Radman, M. (1978). Induced mutagenesis in *dam⁻* mutants of *Escherichia coli*: A role for 6-methyladenine residues in mutation avoidance. *Mol. Gen. Genet.* **163,** 307–312.
231. Ronen, A. (1980). 2-Aminopurine. *Mutat. Res.* **75,** 1–47.
232. Smith, J., and Modrich, P. (1997). Removal of polymerase-produced mutant sequences from PCR products. *Proc. Natl. Acad. Sci. USA* **94,** 6847–6850.
233. Wagner, R., Debbie, P., and Radman, M. (1995). Mutation detection using immobilized mismatch binding-protein (MutS). *Nucleic Acids Res.* **23,** 3944–3948.
234. Lishanski, A., Ostrander, E. A., and Rine, J. (1994). Mutation detection by mismatch binding-protein, MutS, in amplified DNA—Application to the cystic-fibrosis gene. *Proc. Natl. Acad. Sci. USA* **91,** 2674–2678.
235. Ellis, L. A., Taylor, G. R., Banks, R., and Baumberg, S. (1994). MutS binding protects heteroduplex DNA from exonuclease digestion *in vitro*—a simple method for detecting mutations. *Nucleic Acids Res.* **22,** 2710–2711.
236. Babic, I., Andrew, S. E., and Jirik, F. R. (1996). MutS interaction with mismatch and alkylated base containing DNA molecules detected by optical biosensor. *Mut. Res.* **372,** 87–96.
237. Nelson, S. F., McCusker, J. H., Sander, M. A., Kee, Y., Modrich, P., and Brown, P. O. (1993). Genomic Mismatch Scanning—A new approach to genetic-linkage mapping. *Nature Genetics* **4,** 11–18.
238. Wang, J., and Liu, J. (2004). Directly fishing out subtle mutations in genomic DNA with histidine-tagged *Thermus thermophilus* MutS. *Mutat. Res.* **547,** 41–47.
239. Geschwind, D. H., Rhee, R., and Nelson, S. F. (1996). A biotinylated MutS fusion protein and its use in a rapid mutation screening technique. *Genet. Anal.* **13,** 105–111.
240. Gotoh, K., Hata, M., Miyajima, M., and Yokota, H. (2000). Genome-wide detection of unknown subtle mutations in bacteria by combination of MutS and RDA. *Biochem. Biophys. Res. Commun.* **268,** 535–540.
241. Beaulieu, M., Larson, G. P., Geller, L., Flanagan, S. D., and Krontiris, T. G. (2001). PCR candidate region mismatch scanning: Adaptation to quantitative, high-throughput genotyping. *Nucleic Acids Res.* **29,** 1114–1124.
242. Lu, A. L., and Hsu, I. C. (1992). Detection of single DNA-base mutations with mismatch repair enzymes. *Genomics* **14,** 249–255.
243. Hsu, I. C., Yang, Q., Kahng, M. W., and Xu, J. F. (1994). Detection of DNA point mutations with DNA mismatch repair enzymes. *Carcinogenesis* **15,** 1657–1662.
244. Sachadyn, P., Stanislawska, A., and Kur, J. (2000). One tube mutation detection using sensitive fluorescent dyeing of MutS protected DNA. *Nucleic Acids Res.* **28,** E36.
245. Behrensdorf, H. A., Pignot, M., Windhab, N., and Kappel, A. (2002). Rapid parallel mutation scanning of gene fragments using a microelectronic protein–DNA chip format. *Nucleic Acids Res.* **30,** E64.
246. Bellanne-Chantelot, C., Beaufils, S., Hourdel, V., Lesage, S., Morel, V., Dessinais, N., Le Gall, I., Cohen, D., and Dausset, J. (1997). Search for DNA sequence variations using a MutS-based technology. *Mutat. Res.* **382,** 35–43.
247. Scheuermann, R., Tam, S., Burgers, P. M., Lu, C., and Echols, H. (1983). Identification of the epsilon-subunit of *Escherichia coli* DNA polymerase III holoenzyme as the *dnaQ* gene product: A fidelity subunit for DNA replication. *Proc. Natl. Acad. Sci. USA* **80,** 7085–7089.
248. Bullock, W. O., Fernandez, J. M., and Short, J. M. (1987). XL1-Blue: A high efficiency plasmid transforming *recA Escherichia coli* strain with β-galactosidase selection. *Biotechniques* **5,** 376–379.
249. Greener, A., and Callahan, M. (1994). XL1-Red: A highly efficient random mutagenesis strain. *Strategies* **7,** 32–34.

Pleiotropic Effects of Phosphatidylinositol 3-Kinase in Monocyte Cell Regulation

SANAÂ NOUBIR,[*]
JIMMY S. LEE,[*] AND
NEIL E. REINER[*,†]

[*]Department of Medicine (Division of Infectious Diseases), University of British Columbia, Faculties of Medicine and Science, Vancouver Coastal Health Research Institute (VCHRI), Vancouver, British Columbia, Canada V5Z 3J5

[†]Department of Microbiology and Immunology, University of British Columbia, Faculties of Medicine and Science, Vancouver Coastal Health Research Institute (VCHRI), Vancouver, British Columbia, Canada V5Z 3J5

I. Background	53
A. Classification and Structure of PI 3-Kinases	53
B. Lipid Products as Downstream Mediators of PI 3-Kinase Signaling	53
II. PI 3-Kinase and Lipopolysaccharide	54
A. LPS Signaling Pathway and LPS Receptors	55
B. LPS-Induced Association and Coordinate Activation of p53/56^{lyn} and PI 3-Kinase	56
C. PI 3-Kinase-Dependent Activation of PKC-ζ in LPS-Treated Human Monocytes	58
D. Bimodal Regulation of IRAK-1 Dependent on CD14/TLR4, CR3, and PI 3-Kinase	60
E. LPS-Induced Adherence Is Regulated by Rho and PI 3-Kinase	61
III. PI 3-Kinase and 1α,25-Dihydroxyvitamin D$_3$	66
A. Vitamin D$_3$ Action and Signaling	67
B. A VDR·PI 3-Kinase Signaling Complex Regulates Myeloid Cell Differentiation	68
C. D$_3$-Induced Monocyte Antimycobacterial Activity Is Regulated by PI 3-Kinase	71
D. D$_3$-Induced Antimicrobial Activity Against *Salmonella enterica* Is PI 3-Kinase Dependent	74
IV. PI 3-Kinase and Phagocytosis of Mycobacteria	75
A. Phagocytosis of Mycobacteria: Receptors and Signaling	75
B. Cross-Talk Between CD14 and CR3: Regulation by PI 3-Kinase and Cytohesin-1	76

C. Rescue of Phagosome Maturation by Vitamin D_3 Is PI
3-Kinase Dependent 79
V. Stable Gene Silencing of Isoforms of PI 3-Kinase and Perspectives............ 80
A. Silencing of P110α Subunit of PI 3-Kinase: Differential Involvement
in Monocyte Adherence... 80
B. Perspectives.. 84
References 84

Mononuclear phagocytes are important regulators and effectors of both the innate and acquired immune responses, and they are also prominently involved in inflammation. Consequently, regulation of monocyte function is an intense area of research interest and recent studies have highlighted an important role for phosphoinositides in monocyte cell regulation. For nearly two decades now, phosphoinositides have been recognized to function as second messengers in signal transduction pathways initiated through cell-surface receptors. Class I phosphatidylinositol 3-kinase (PI 3-kinase or PI 3-K) is a lipid kinase that catalyzes the transfer of the γ-phosphate group of ATP to the 3′-position of the inositol ring of phosphatidylinositol (PI) and its derivatives. The phosphoinositide (P'tide) metabolites produced as a result are known to be involved in regulating a multitude of cellular events such as mitogenic responses, differentiation, apoptosis, cytoskeletal organization, membrane traffic along the exocytic and endocytic pathways (1), and various other aspects of monocyte function (2–5). This chapter will focus on those regulatory events involving PI 3-kinase that have been extensively studied in our laboratory and will emphasize the role of PI 3-kinase in three main areas. The first is activation of PI 3-kinase following lipopolysaccharide (LPS) stimulation, our efforts to identify downstream effectors, and a role for this activation in regulating changes such as monocyte adherence and endotoxin tolerance. The second area will focus on the role of PI 3-kinase in vitamin D3-induced monocyte maturation, the phagocyte oxidative burst, and antimicrobial activity during both infection with mycobacteria and *Salmonella*. Third, we will review aspects of the roles PI 3-kinase has been shown to play in regulating phagosome biology. Finally, the last part of this chapter will address important advances in a new approach to stable gene silencing in human monocytic cell lines using lentiviral-delivered small interfering RNA. The power of this strategy is illustrated by experiments that examined the specific role of the p110α subunit of PI 3-kinase in vitamin D3 but not LPS-induced monocyte adherence. To provide further perspective, several ongoing and promising experiments ultimately made possible by using stable and isoform-specific PI 3-kinase gene silencing are discussed.

I. Background

A. Classification and Structure of PI 3-Kinases

The phosphoinositide 3-kinases (PI 3-K) constitute a family of at least eight different lipid kinases that phosphorylate the hydroxyl group of the inositol ring of phosphoinositides at the 3' position. A considerable amount of research has led to the conclusion that an apparently vast diversity of cellular functions are differentially regulated by distinct PI 3-K family members and their corresponding isoforms (6–8). Based on their selective *in vitro* substrate specificity, the multiple PI 3-Ks have been divided into three classes (6). *In vitro*, class I PI 3-K isoforms phosphorylate phosphatidylinositol (PtdIns), PtdIns 4-phosphate, and PtdIns 4,5-bisphosphate. In intact cells, however, their preferred substrate appears to be PtdIns(4,5)P2. Class I PI 3-K is further divided into subclasses I_A and I_B, which signal downstream of tyrosine kinases and heterotrimeric G-protein–coupled receptors (GPCRs) respectively. All class I PI 3-K members are able to bind to Ras, but the role of this interaction in physiological PI 3-K signaling is not entirely clear.

Mammalian class I_A PI 3-Ks are heterodimers consisting of a regulatory subunit (p85α, p85β, p55, or other splice variants) and a p110 (α, β, or δ isoforms) catalytic subunit (9–11). Through their Src-homology 2 domain-containing p85 subunits, class I_A PI 3-Ks bind to phosphorylated tyrosine residues that are generated by activated tyrosine kinases in receptors and various receptor-associated adaptor proteins. The class I_Ap85 subunits prefer to bind to proteins containing the consensus phosphotyrosine motif Y(P)xxM in which x is any amino acid. Phosphotyrosine binding is thought to allow translocation of the cytosolic PI 3-Ks to the plasma membrane, where their lipid substrates and Ras reside. Nonphosphotyrosine-based recruitment mechanisms may also contribute to PI 3-K activation (12, 13). Some membrane proteins may even show a constitutive association with class I_A PI 3-Ks, as reported for CD2 (14). All mammalian cell types investigated express at least one class I_A PI 3-K isoform, and stimulation of almost every receptor that induces tyrosine kinase activity also leads to class I_A PI 3-K activation (15–17).

B. Lipid Products as Downstream Mediators of PI 3-Kinase Signaling

Certain types of ligand receptor interactions may trigger a rapid rise of cellular PtdIns(3,4,5)P3, and with some delay, PtdIns(3,4,)P2, the latter generated by the action of phosphoinositide 5-phosphatases. Several molecular targets for PtdIns(3,4,5)P3 and PtdIns(3,4,)P2 have been identified, which are translocated and activated on interaction with phosphoinositides. Two protein subdomains in particular, the FYVE domain (based on the first letters of four proteins containing this domain, *Fab*1, YOTB/ZK632.12, Vac1 and *EEA*-1) and the pleckstrin homology (PH) domain, have been defined whose structures confer specificity of lipid interaction at the molecular level. FYVE domains selectively bind PtdIns3P, whereas a subgroup of PH domains shows specificity for PtdIns(3,4)P2 and/or PtdIns(3,4,5)P3. PH domains that selectively bind PtdIns3P have also been identified (*18*). The potential for some PH domains to specifically interact with PtdIns(3,4)P2 and PtdIns(3,4,5)P3 correlates with *in vivo* data defining the same PH domain-containing proteins as PI 3-kinase effectors (*19, 20*). One of the most thoroughly studied PI 3-K effectors is the protein Ser/Thr kinase Akt/PKB (also called RAC), which translocates to the plasma membrane by binding PtdIns(3,4)P2. It has been conclusively demonstrated that the activity of type 1 phosphoinositide-dependent kinases (PDK), which phosphorylates and activates Akt/PKB on Thr^{308}, is also specifically controlled through the binding of PtdIns(3,4,5)P3, or PtdIns(3,4)P2 to its PH domain (*21–24*). Furthermore, PDK1s have been recently shown to phosphorylate and activate protein kinases C (PKC) ζ and δ directly (*25, 26*). Taken together, these data illustrate the complexity and fine tuning that has come to be recognized as characteristic of lipid-mediated signaling.

II. PI 3-Kinase and Lipopolysaccharide

Lipopolysaccharide (LPS), or endotoxin, is the major component of the outer surface of Gram-negative bacteria. LPS is an activator of both immune and inflammatory responses and has particularly potent effects on macrophages, monocytes, dendritic cells, endothelial cells, and others (*27, 28*). An important property of LPS is its ability to promote leukocyte adherence and the accumulation of leucocytes at inflammatory foci *in vitro* or *in vivo* (*29, 30*).

Our first indication that the monocyte response to LPS was regulated by PI 3-kinase derived from the finding that LPS treatment led to activation of this lipid kinase and this was required for the activation of monocyte PKC-ζ as well as the tyrosine kinase $p53/56^{lyn}$ (*4, 31, 32*). Based on these findings, further efforts were focused on understanding the role of the PI 3-kinase activation on monocyte cell regulation. Prior exposure of monocytes and macrophages to LPS induce a transient state of refractoriness to subsequent LPS restimulation, known as endotoxin tolerance (*33, 34*). The phenomenon of endotoxin tolerance has been widely investigated (*2, 33–35*), but to date, the molecular mechanisms of endotoxin tolerance remain to be resolved clearly. Recent research has identified distinct roles for PI 3-kinase in regulating monocyte adherence as well as endotoxin tolerance (*2, 36*). To put these findings in proper context, it is useful to review briefly key aspects of what is known about LPS signaling and different receptors for LPS.

A. LPS Signaling Pathway and LPS Receptors

It has become clear that cellular responses to LPS involve specific cell-surface receptors leading to the activation of pathways containing both tyrosine and serine/threonine protein kinases (32, 37, 38). Proximal events involved in at least one dominant LPS signaling pathway are dependant on the cell-surface molecule CD14 (39). The binding of LPS to mCD14 is enhanced by LPS-binding protein, a plasma protein (40). Unlike other more conventional receptors, CD14 is linked to the membrane via a glycosylphosphatidylinositol anchor, thereby lacking both transmembrane and cytoplasmic domains. The mechanisms, by which CD14 mediates stimulus–response coupling has been a central question in macrophages cell biology. Evidence has been presented to suggest that this may involve an auxiliary receptor subunit(s) that interacts with a ternary complex comprised of CD14, LPS, and LPS-binding protein. It has been shown that CD14 uses TLR-4 as a cell-surface coreceptor that mediates signal transmission. Genetic and other evidence has established that LPS signals predominantly through CD14 and TLR-4 (41, 42) and utilizes signaling elements in common with the interleukin-1 receptor including MyD88, IRAK, and TRAF6 to elicit cellular responses (43). TIRAP, also known as MAL, an adaptor protein in the TLR signaling pathway, has also been identified and shown to function downstream of TLR-4 (44, 45).

The β_2 integrin CD11/CD18 represents another putative LPS receptors. CD11b and CD11c, the alpha subunits of CR3 and CR4, respectively, are glycoproteins that associate noncovalently with the β_2 integrin subunit, CD18. Although there has been some ambiguity as to whether CR3 and CR4 are *bona fide* LPS signaling receptors, recent findings show that expression of CD11b/CD18 or CD11c/CD18 in Chinese hamster ovary cells confers a phenotype of LPS responsiveness to these cells which are normally CD14$^-$ (46–49). Of interest, CD11b/CD18 mutants, deficient in cytoplasmic domains thereby rendering them incapable of functioning to internalize Gram-negative bacteria, were still competent for LPS-induced cellular activation (47). This has led to the suggestion that CR3 like CD14 may function to initiate signal transduction via an as yet unidentified coreceptor (47).

To function as an LPS receptor, TLR4 must interact with a secreted protein, MD-2 (50–52). Site-directed mutagenesis has shown that one region of MD-2 is capable of binding TLR4 and may mediate its correct targeting to the cell membrane, whereas another region, rich in basic and aromatic residues, most likely functions to bind LPS (53). Incubation of naive cells with LPS triggers activation of IRAK-1 (IL-1R associated serine/threonine-specific protein kinase) and its association with MyD88, an adaptor protein linking IRAK-1 to the intracellular domain of TLR-4 (43, 54, 55). On stimulation, IRAK-1 rapidly translocates to the active receptor complex and interacts via its

N-terminal death domain with MyD88 (56–58). IRAK-1 then auto- and/or cross-phosphorylates itself at the active receptor complex resulting in phosphorylation in the kinase domain. Subsequently, IRAK-1 catalyzes multiple auto-phosphorylations in its ProST region, a domain rich in proline, serine, and threonine residues. IRAK-1 interacts with the downstream adapter TRAF6, which has polyubiquitinylated itself (59). TRAF6 then facilitates autophosphorylation of TAK1 (60, 61), which in turn phosphorylates I-κB-kinase (IKK) and MKK6 thus activating the NF-κB pathway as well as p38 MAP kinase. Rapid activation of nuclear factor-κB (NF-κB) in both monocytic (62, 63) and endothelial cells (64) is required for the production of proinflammatory cytokines, including interleukin-1 (IL-1), interleukin-6 (IL-6), and tumor necrosis factor (TNF) (65, 66). Tyrosine phosphorylation and activation of ERK1, ERK2, p38 MAP kinase, and c-Jun N-terminal kinase also appear to be important for cell activation by LPS (67).

B. LPS-Induced Association and Coordinate Activation of p53/56^{lyn} and PI 3-Kinase

Increased tyrosine phosphorylation of monocyte proteins (68, 69) and activation of the *src* family tyrosine kinases p53/56^{lyn}, p58^{hck}, and p59^{fgr} (37, 70, 71) have been reported to take place within minutes of exposure of macrophages to LPS. Moreover, we and others have shown that activation of serine-threonine protein kinases, such as p42 and p44 mitogen-activated protein kinases and PKC, also occurs following LPS stimulation (32, 38, 72, 73). Based on these findings and coupled with the knowledge that PI-3 kinase has been shown to be a downstream effector of tyrosine kinases (17), we investigated whether LPS brings about the activation of PI 3-kinase in human monocytes and whether this involves direct kinase–kinase interactions.

1. BACTERIAL LPS INDUCES CD14-DEPENDENT ACTIVATION OF PI 3-KINASE IN MONOCYTES

To examine whether LPS activates PI 3-kinase, inositol phospholipids were detected after extraction and separation of lipids by thin-layer chromatography (TLC). Levels of PtdIns 3,4,5-trisphosphate (PtdIns 3,4,5-P3) were elevated within minutes of exposure of human monocytes to LPS (Fig. 1). Analysis of ^{32}P-labeled lipid extracts of U937 cells by HPLC confirmed that levels of PtdIns 3,4,5-P3 increased rapidly following LPS treatment (31). Moreover, anti-PI 3-kinase immunoprecipitates prepared from unlabeled monocytes and assayed in an *in vitro* phosphorylation assay using PtdIns as substrate, showed higher enzymatic activity when these were prepared from lysates of LPS-treated cells as compared with control cells (31). These results suggest that increased levels of PtdIns 3,4,5-P in LPS-treated cells resulted from an

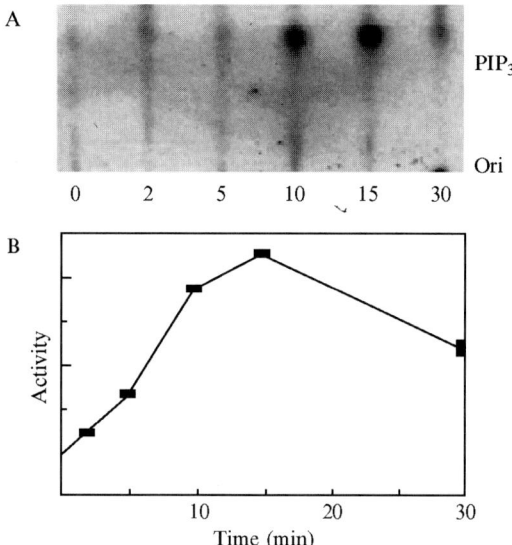

FIG. 1. PtdIns 3,4,5-P3 levels in monocytes after LPS treatment. (A) Human peripheral blood monocytes were labeled with Pi for 2 hr before incubation with LPS (100 ng/ml) for different periods. Lipids were extracted and separated by TLC using a solvent system. (B) Spots corresponding to PtdIns 3,4,5-P3 (PIP) were identified by comigration with standard. The spots were cut from the plate and analyzed by liquid scintillation counting. Activity is expressed relative to control. The visible spots at the origin reflect residual, water-soluble ^{32}P-labeled material in the samples, the amount of which is not relevant to the results. Adapted from the work of Herrera-Velit et al. (31), with permission.

increase in the specific activity of PI 3-kinase rather than decreased activity of phosphomonoesterases. PI 3-kinase activity was elevated as early as 2 min after LPS exposure and was observed with LPS concentrations in a range of 10–100 pg/ml, which are similar to those required to induce monocytic functional responses (31). Furthermore, activation of PI 3-kinase was shown to involve signaling through the monocyte cell-surface molecule CD14, since pretreatment of cells with Abs to CD14 abrogated LPS-induced increases in PtdIns 3,4,5-P3 (31).

2. LPS Activates $P^{53/56^{LYN}}$ and Promotes Its Association with PI 3-Kinase

Activation of several *src* family tyrosine kinases, including p53/56^{lyn}, p58^{fgr}, and p59^{hck} has been shown to occur in response to LPS, dependent on CD14. In the case of p53/56^{lyn}, a physical association of *lyn* with CD14 is involved (37). It has also been observed that PI 3-kinase may associate with and be activated by a variety of both receptor and nonreceptor tyrosine kinases (17, 74), including p53/56^{lyn} in B cells (75). Taken together, these findings suggested the possibility that activation of monocyte PI 3-kinase in response to LPS may involve its association with activated p53/56^{lyn}. This hypothesis in fact proved to be correct. Anti-*lyn* immunoprecipitates from LPS-treated THP-1 cells were examined and showed increased PI 3-kinase activity (31). Maximal *lyn*-associated PI 3-kinase activity was observed after 5–10 min of LPS treatment and was gone by 40 min. In addition, when parallel p53/56^{lyn} immunoprecipitates were examined, kinetics of *lyn* activation was maximal at 10 min, consistent with its maximal association with PI 3-kinase (31). Taken together, these results indicate that LPS brings about the coordinate activation of p53/56^{lyn} and PI 3-kinase and these enzymes become associated transiently for a period of time that corresponds to the duration of PI 3-kinase activation. Further evidence to suggest that activation of PI 3-kinase is dependent on its association with activated *lyn* was provided by the finding that inhibition of *lyn* with the tyrosine kinase inhibitor herbimycin abrogated activation of PI 3-kinase (31).

C. PI 3-Kinase-Dependent Activation of PKC-ζ in LPS-Treated Human Monocytes

1. LPS Induces Activation of PKC-ζ in Monocytes

In light of findings that LPS activated both PI 3-kinase and PKC in monocytes, an important objective was to determine whether these events were related and the isoform of PKC involved. Resolution of detergent-soluble lysates prepared from LPS-treated human blood monocytes using Mono Q chromatography revealed two principal peaks of myelin basic protein kinase activity. Immunoblotting and immunoprecipitation with different isoform-specific anti-PKC antibodies showed that the major and latest eluting peak was accounted for by PKC-ζ (4). Consistent with its identity as PKC-ζ, the LPS-activated monocyte kinase did not depend on the presence of lipids, Ca^{2+}, or diacylglycerol for activity. In addition, the kinase phosphorylated peptide epsilon and myelin basic protein with equal efficiency, but phosphorylated kemptide and protamine sulfate poorly, features all consistent with PKC-ζ. Translocation of PKC-ζ from the cytosolic to the particulate membrane fraction on exposure of monocytes to LPS provided further evidence for activation of the kinase (4).

2. PI 3-Kinase Is Required for LPS-Induced Activation of PKC-ζ

Previous work had suggested a role for PI 3-kinase metabolites in the activation of PKC. In particular, the PI 3-kinase metabolite PIP_3 was known to be an activator of the PKC isoforms ζ, ε, and δ (26, 76, 77). Based on this information and the observation from this laboratory and others that PKC activity was increased in LPS-treated monocytes (32, 38), we investigated the potential involvement of PI 3-kinase in LPS-induced activation of PKC-ζ. Human monocytes were incubated with two structurally unrelated PI 3-kinase inhibitors, wortmannin or LY294002, prior to the addition of LPS. Both inhibitors markedly attenuated activation of PKC-ζ (4). Other results showed that activation of PKC-ζ was abrogated in U937 cells transfected with a dominant negative mutant of the p85 subunit of PI 3-kinase (Fig. 2). These findings are consistent with a model in which PI 3-kinase is required for the activation of PKC-ζ in LPS-treated human moncytes. Further support for this argument was provided by experiments in which PKC-ζ activity was observed to be enhanced *in vitro* by the addition of phosphatidylinositol (3,4,5)P3 (4).

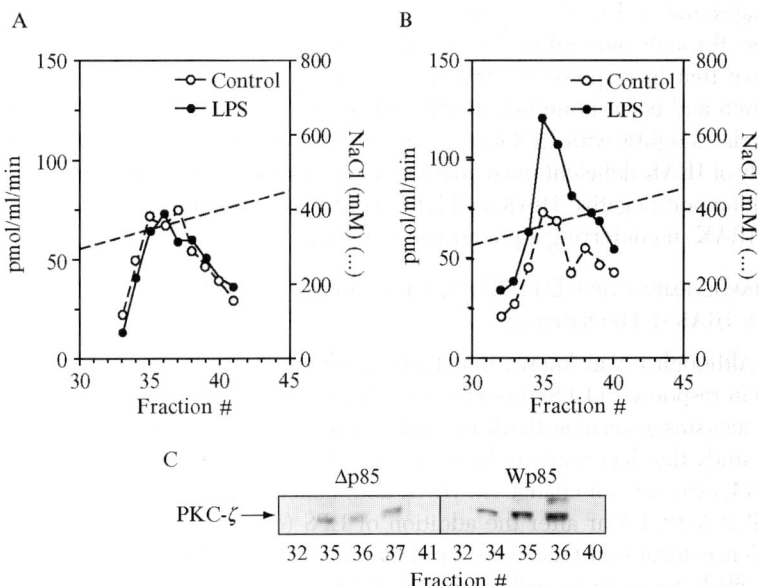

FIG. 2. Attenuation of LPS-induced activation of PKC-ζ by a dominant negative mutant of the p85 subunit (Δp85) of PI 3-kinase. Quiescent transfected U937 cells were treated either with LPS (1 μg/ml, 15 min) or medium alone and lysed in FPLC extraction buffer. Detergent lysates were fractionated by Mono Q chromatography. Aliquots (5 μl) of each fraction from cells transfected with Δp85 (A) or with wild-type p85 (Wp85) (B) were assayed for MBP kinase activity in triplicate. The standard deviations of each data point were <20% of the mean. (C) Aliquots from LPS-treated cells were analyzed by immunoblotting with anti-PKC-ζ antibodies. Adapted from the work of Herrera-Velit et al. (4).

D. Bimodal Regulation of IRAK-1 Dependent on CD14/TLR4, CR3, and PI 3-Kinase

1. IRAK-1 Degradation and Tolerance to LPS

Infection with Gram-negative bacteria often results in the development of a sepsis syndrome in which uncontrolled production of proinflammatory cytokines results in shock with multiorgan failure (78, 79). To prevent the morbidity and mortality that can develop under these conditions, a variety of intrinsic control mechanisms have evolved to exert negative feedback regulation of cellular responses to LPS. For example, macrophages and other phagocytic cells incubated with LPS become unresponsive or tolerant to a second challenge with endotoxin (33). One level at which this type of control may be exerted involves attenuation of LPS signaling. IRAK-1 plays a crucial role in facilitating formation of the multiprotein signaling complex or LPS signalosome. IRAK-1 links the active receptor complex to the central adapter and coactivator protein TRAF6. Thus, IRAK-1 functions as a switch or relay for the signalosome and is also capable of switching off or interrupting the signalosome through autoinduced removal by degradation. Recently, it has been shown that in response to prolonged LPS exposure, cellular levels of IRAK protein and correspondingly IRAK activity are markedly diminished, and this correlates tightly with LPS tolerance (80). Multiple lines of evidence including study of IRAK deficient mice, macrophages lacking IRAK, and over expression of dominant negative IRAK in THP-1 cells, have established an essential role for IRAK in conferring a phenotype of tolerance to LPS (43, 81, 82).

2. Involvement of CD14/TLR4, CR3, and PI 3-Kinase in IRAK-1 Degradation

Although it was known that IRAK undergoes rapid inactivation/degradation in response to LPS, providing negative feedback leading to LPS tolerance, mechanisms governing IRAK degradation were not fully understood. In previous study that had examined normal THP-1 cells, which expressed little or no CD14, evidence for changes in IRAK abundance and activity were not evident until at least 1.5 hr after the addition of LPS (80). However, several lines of evidence suggested that more rapid degradation of IRAK in response to LPS was likely to occur in cells that were $CD14^+$ (2). The prediction that rapid IRAK degradation was likely to involve CD14 proved to be correct. Based on a kinetic analysis comparing both $CD14^-$ and $CD14^+$ cells, we found that the expression of CD14 confers a phenotype of rapid IRAK degradation in response to LPS, a property not shared by cells expressing low or undetectable levels of CD14 (2). Whereas $CD14^-$ cells were also capable of degrading IRAK, this occurred with distinctly delayed kinetics, providing evidence for bimodal regulation of IRAK expression.

The discovery that IRAK degradation was in fact bimodal inferred that it was likely to be regulated by multiple mechanisms. This inference was confirmed with the use of blocking antibodies that established that only the rapid phase of IRAK degradation was TLR-4- and CD14-dependent (2). To address how the delayed phase was regulated, we considered other putative LPS receptors that might be involved. CR3 appeared to be a potential candidate as it had been shown to be capable of transducing LPS signals (47, 49). Blocking antibody experiments using THP-1rsv cells (CD14$^-$/CR3$^+$) clearly showed that CR3 was required for late phase IRAK degradation. Similar blocking experiments using both THP-1rsv (CD14$^-$) and THP-1wt (CD14$^+$) cells showed that CR3 functioned independently of TLR-4 in signaling delayed IRAK degradation (2).

The finding that dual receptors appeared to be involved in invoking independent pathways of IRAK degradation raised the important question of how these pathways were regulated. In examining this question, an important lead was provided by the findings that PKC activity was required for LPS-induced phosphorylation and degradation of IRAK and the demonstration that PKC-ζ associated with IRAK in LPS-treated THP-1 cells (83). These results, indicating a possible role for PKC-ζ in a pathway regulating IRAK degradation, led us to investigate whether this may be regulated by PI 3-kinase. As can be seen in Fig. 3, use of two distinct PI 3-kinase inhibitors, LY294002 and wortmannin, showed that PI 3-kinase was required for the early, but not the late phase of IRAK degradation. These findings suggested a model in which on exposure to LPS, PI 3-kinase becomes activated downstream of CD14/TLR4 and signals through a pathway leading to activation of PKC-ζ to bring about the rapid degradation of IRAK (Fig. 4). Investigation of other elements that may be involved in this pathway is presently underway. One potential candidate is phosphoinositide-dependent kinase type-1 because it is known to be activated by phosphatidylinositol 3,4,5-trisphosphate and 3,4-bisphosphate through its PH domain (84), and it has been shown to phosphorylate and activate PKC-ζ directly (25, 26).

E. LPS-Induced Adherence Is Regulated by Rho and PI 3-Kinase

1. Monocyte Adherence Induced by LPS Involves CD14 and LFA-1

Adherence of monocytes to endothelial cells is an essential requirement for the localization of these cells to sites of tissue inflammation (85–87). Several reports have shown that this process is dependent on the monocyte surface molecule lymphocyte function-associated antigen-1 (LFA-1) (CD11a/CD18; $\alpha_L\beta_2$) [(87–89) and reviewed in (90, 91)]. Intercellular adhesion

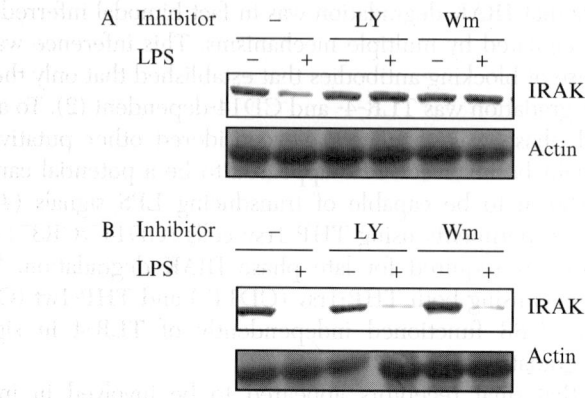

FIG. 3. PI 3-kinase is involved in early but not late phase IRAK-1 degradation. THP-1wt (CD14$^+$) cells were preincubated either with 20 μM LY294002 (LY), 50 nM wortmannin (Wm), or medium alone for 20 min at room temperature and then treated with 500 ng/ml LPS for 30 min (A) or 2 hr (B). Cells were then washed, lysed, and equal amounts of proteins were analyzed by SDS-PAGE and Western blotting with anti-IRAK-1 antibodies. Adapted from the work of Noubir et al. (2).

molecule-1 (ICAM-1) (CD54) has been identified as a high affinity counter-receptor for LFA-1 (92), and interactions between ICAM-1 and LFA-1 mediate several important functions in the immune system in addition to adherence (90). The affinity of LFA-1 under basal conditions for ICAM-1 or its other ligands is low, and LFA-1 must be activated to mediate stable adhesion (88, 89). Changes in the affinity of LFA-1 for ligand are known to be regulated by a process of "inside-out" signaling that converts LFA-1 into an activated form capable of mediating increased adhesion (91, 93). We examined signaling events required for LPS-induced monocyte adherence using a quantitative, microtiter adhesion assay, CD14 transfected THP1 cells, and immobilized sICAM-1. Experiments that examined competitive inhibition of LPS-induced adherence using mAbs to CD14, CD18, and CD11a provided direct evidence that LPS-induced adherence to sICAM-1 involved a CD14-mediated signal leading to activation of LFA-1. These findings were consistent with previous data showing that antibody cross-linking of cell surface CD14 induces LFA-1 activation (94). LPS effects on LFA-1 did not involve changes in the expression of CD18 or CD11a indicating that LPS-induced adhesion was rather related to increased affinity of LFA-1. Such a change in the properties of LFA-1 is presumably mediated by a specific pathway of inside-out signaling initiated through CD14.

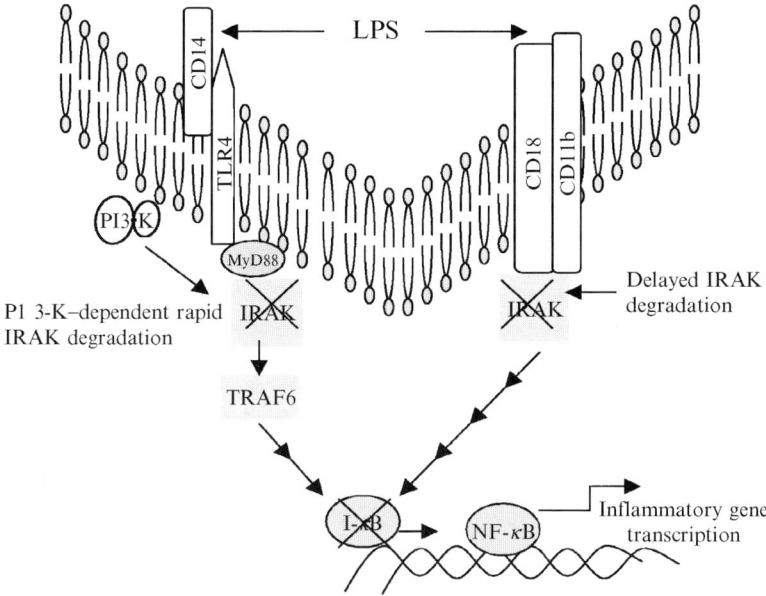

FIG. 4. Dual, independent pathways of IRAK-1 degradation are initiated separately through CD14/TLR-4 and CR3. Both CD14/TLR-4 and CR3 have the capacity to initiate the innate immune responses to LPS through IRAK-1–dependent pathways leading to activation of NF-κB–dependent gene transcription. Negative feedback is provided by dual, independent pathways also triggered through CD14/TLR-4 and CR3 leading to IRAK-1 degradation and attenuation of the inflammatory response. LPS signaling through CD14 and TLR-4 leads to rapid IRAK-1 degradation, and this is dependent on PI 3-kinase (PI3-K). In parallel, a delayed wave of IRAK-1 degradation initiated by CR3 occurs independently of both TLR-4 and PI 3-kinase. The latter pathway is likely to be of particular importance in cells that are CR3$^+$ and express little or no CD14. Adapted from the work of Noubir et al. (2).

2. Regulation by Rho and PI 3-Kinase and Requirement for Cytohesin

The signaling events that link cell stimulation to activation of LFA-1 are incompletely understood. The requirement for PI 3-kinase activity in a variety of leukocyte functions, together with its apparent role in the adhesion of platelets (95), lymphocytes (96), and neutrophils (97), made this enzyme an attractive candidate for mediating signaling through CD14 for monocyte adhesion. This hypothesis was supported by the fact that cytohesin-1 contains a PH domain that binds the PI 3-kinase metabolite, phosphatidylinositol 3,4, 5-triphosphate (PtdIns-3,4,5-P$_3$), leading to changes in properties of the protein (98). Cytohesin-1 is a regulatory or adaptor protein that has been shown to interact with the cytoplasmic tail of CD18 (99). Based on these findings, a potential role for PI 3-kinase in regulating the activity of LFA-1 was suggested.

Two different approaches provided evidence that PI 3-kinase regulated LPS-induced monocyte adherence. The first involved the use of wortmannin and LY294002 and the findings that both inhibitors abrogated LPS-induced adherence. The second approach demonstrated complete abrogation of LPS-induced adherence to sICAM-1 when a dominant negative mutant of PI 3-kinase ($\Delta p85$) was expressed in U937 cells (36). It has been shown previously that incubation of monocytes with LPS activates PI 3-kinase, leading to increased cellular levels of PtdIns-3,4,5-P_3 (31). Thus, the most likely mechanism for the attenuation of adherence by either wortmannin, LY294002, or $\Delta p85$ is inhibition of the formation of PtdIns-3,4,5-P_3.

Based on previous reports that the small G-protein Rho regulated PI 3-kinase activation in different cell systems (100–102), the possibility that LPS-induced adherence may be Rho-regulated and mediated by PI 3-kinase was investigated. A requirement for Rho in LPS-induced adherence was suggested by studies that used Clostridium difficile toxin B, which specifically inhibits Rho family proteins (103, 104). Pretreatment of THP-1wt ($CD14^+$) cells with toxin B attenuated LPS-induced adherence to sICAM, whereas PMA-induced adherence appeared to be mediated by a toxin B-insensitive pathway (36). The findings that both Rho and PI 3-kinase appeared to be essential for LPS-induced adherence raised the question as to whether they act independently or they are positioned together in a single signaling pathway. Toxin B prevented activation of PI 3-kinase in LPS-stimulated THP-1wt cells, suggesting that Rho regulates this LPS response in monocytes as well (36). This observation is consistent with previous reports showing involvement of Rho in PI 3-kinase activation in other systems (100–102). Although we cannot completely eliminate the possibility of a direct, PI 3-kinase–independent role for Rho in regulating monocyte adherence, the data suggest that LPS triggers Rho-mediated activation of PI 3-kinase, leading to downstream effects on LFA-1 and monocyte adherence.

An important question that arose from these observations was how PI 3-kinase activation modulates the properties of LFA-1. Recently, cytohesin-1 was shown to interact with the cytoplasmic tail of CD18 (99). Cytohesin-1 contains a domain homologous to the yeast Sec7 gene product and a PH domain. The Sec7 domain binds to and regulates LFA-1 (99), and this process is positively regulated by the binding of the PI 3-kinase metabolite PtdIns-3,4,5-P_3 to the cytohesin-1 PH domain (98). To directly address the role of cytohesin-1 in LPS-induced adherence, cytohesin-1–specific antisense oligonucleotides were used. The finding that antisense treatment of THP-1 cells, but not treatment with sense oligonucleotide, significantly attenuated LPS-induced adherence to ICAM-1 provided compelling evidence to suggest that cytohesin-1 plays an essential role in adherence induced by LPS (Fig. 5). Taken together, these results are consistent with a model (Fig. 6) in which LPS binding to CD14 switches on the small G-protein Rho leading to activation of PI 3-kinase. This is linked to increased monocyte adherence, dependent on changes in the adhesive properties of LFA-1. The latter appears to involve the interaction of PtdIns-3,4,5-P_3, cytohesin-1, and LFA-1.

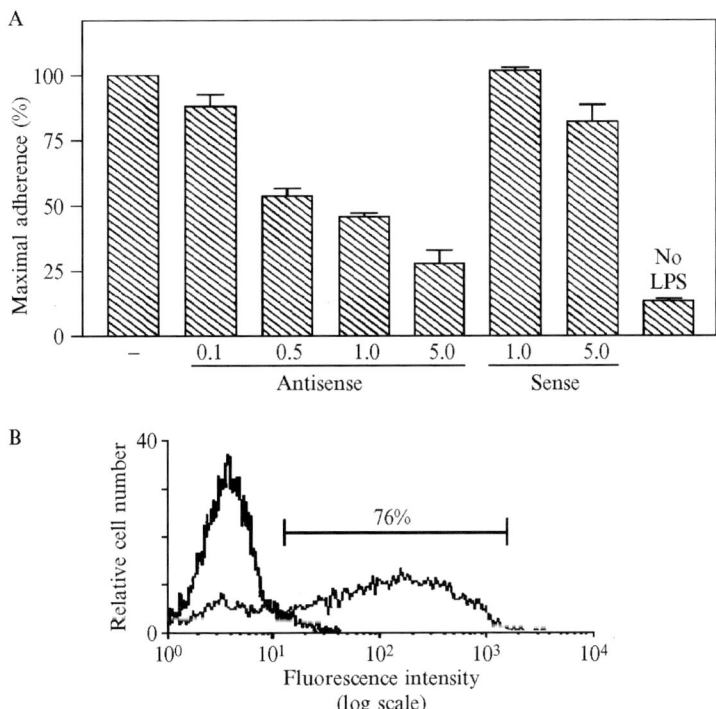

FIG. 5. Cytohesin-1 antisense S-oligos inhibit LPS-induced adherence to ICAM-1. (A) THP-1wt cells were incubated with various concentrations of S-oligos for 2 hr at 37°C in 250 µl of RPMI 1640 containing 2.5% lipofectamine. The medium was then adjusted to 1 ml and supplemented with 10% FCS, and culture was continued for 18 hr. Cells were then washed and tested in the adhesion assay (36). (B) THP-1wt cells were incubated with 5 µM fluoresceinated-5' antisense S-oligo using the same conditions as for unmodified S-oligo. Cells were washed and analyzed by FACS (36). The data shown in (A) are the means ± S.E. of values obtained in three independent experiments. Results in (B) represent one of two independent experiments with similar results. The tracing on the *left* in (B) represents cells incubated with medium alone (autofluorescence), and the tracing on the *right* corresponds to cells incubated with fluorescein-labeled, sense S-oligo. Adapted from the work of Hmama et al. (36).

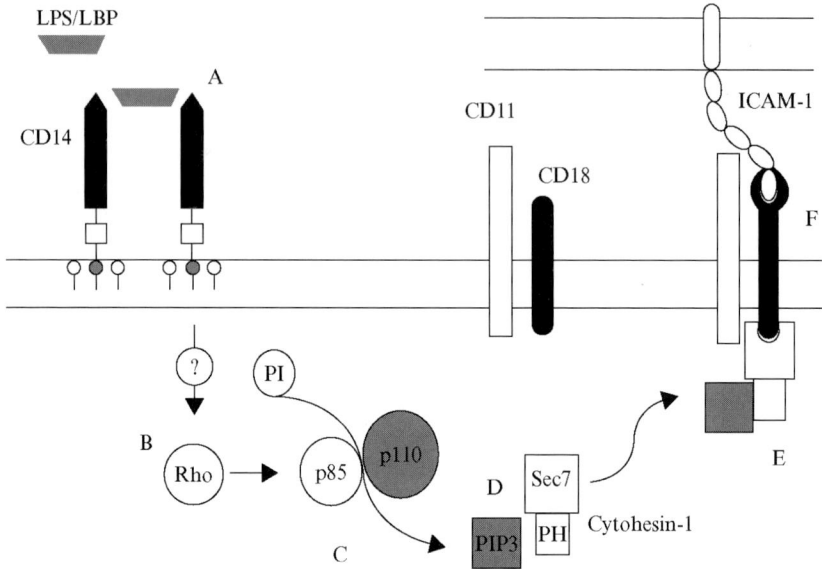

FIG. 6. LPS-induced monocyte adherence to ICAM-1. Monocyte adhesion to ICAM-1 involves activated LFA-1 and the adaptor molecule cytohesin-1. The latter is believed to be dependent on PtdIns-3,4,5-P_3 (PIP_3). LPS binding to CD14 (A) engages the small G-protein Rho (B) leading to activation of PI 3-kinase (C). PtdIns-3,4,5-P_3 (D) binds to the PH domain of cytohesin-1, thereby modifying its interaction, through its Sec7 domain, with the cytoplasmic tail of CD18 (E). This leads to altered properties of LFA-1 (F) and increased affinity for its counter receptor ICAM-1. Adapted from the work of Hmama et al. (36).

III. PI 3-Kinase and 1α,25-Dihydroxyvitamin D_3

1α,25-dihydroxycholecalciferol (Vit D_3 or D_3) is the biologically active form of vitamin D that plays an important role in numerous cellular and physiological processes. In addition to its well established importance in plasma calcium homeostasis (105, 106) and bone remodeling (107, 108), Vit D_3 also plays a functional role in the hematopoietic system. D_3 modulates the expression of several genes in promonocytic cell lines, thereby regulating their differentiation (109). Moreover, other physiological processes, such as cellular adherence and host resistance to infection, have also been attributed to Vit D_3 (110–114). Recent results from this laboratory showing that D_3 activates PI 3-kinase and PI 3-kinase activity is required for the D_3 induced myeloid cell differentiation as well as adherence and antimicrobial effector function are reviewed in the next section.

A. Vitamin D_3 Action and Signaling

The pleiotropic effects of D_3 are principally mediated through the intracellular vitamin D receptor (VDR). The VDR is a member of the superfamily of nuclear steroid, thyroid, and retinoic acid receptors. The classical mechanism of action of D_3 involves genomic signaling in which hormone binds to the VDR, a ligand-dependent transcription factor. The D_3·VDR complex then translocates to the nucleus, where it directly regulates transcription by binding to the vitamin D response element (VDRE) consensus sequence located upstream of many D_3-activated genes (*115*). This mode of signaling, referred to as "genomic action," is exemplified for instance in D_3-treated immature myeloid cells, which express the Cdk inhibitor p21 after hormone treatment. p21 has been shown to regulate gene expression and promote myeloid cell differentiation (*116*) and is transcriptionally induced by D_3 via a functional VDRE within the p21 promoter (*116*).

When immature myeloid leukemia cells, such as HL60, U937, THP-1, and M1, are incubated with D_3, they differentiate into cells expressing functional properties and differentiation markers of monocytes and/or macrophages, including CD14 and CD11b/CD18 (*117–120*). CD14 is undetectable on the surface of monocytic precursors and increases dramatically during differentiation into monocytes (*121–123*). This provides an excellent model in which to study mechanisms of myeloid cell maturation regulated by D_3. Induction of CD14 expression in response to D_3 occurs at the level of gene transcription and requires new protein synthesis (*123, 124*). However, a curious paradox of this response is that no VDRE consensus sequence has been identified in the promoter regions of CD14 or CD11b, and this also applies to certain other D_3-inducible genes (*123, 125*).

The region located at –128 to –70 bp upstream from the transcription start site of the CD14 promoter has been shown to be critical for CD14 expression in response to D_3 (*123*). Rather than a VDRE, this region contains two GC boxes that bind the nonreceptor transcription factor Sp1, and this is believed to be essential for CD14 expression (*123, 126*). Likewise, the integrin CD11b also possess a sequence at bp –60 from the transcription start site that binds Sp1 *in vitro* and *in vivo* and is essential to the promoter activity (*127*). In various attempts to define the molecular mechanisms regulating this type of cellular responses to D_3, efforts have been made to identify alternative signaling pathways to the classical mode of genomic action. Indeed, during the past decade, experimental evidence for "nongenomic" signaling has challenged the concept of exclusive VDR-mediated genomic action. For example, D_3 has been shown to stimulate the rapid formation of second messengers, including ceramides, cAMP, inositols, and Ca^{2+}, and to activate a variety of protein kinases including PKC, Raf, mitogen-activated protein (MAP) kinase, and Src family kinases (*125, 128–130*). MAP kinase activity was also shown to be activated by D_3 in keratinocytes, enterocytes, and in the promyelocytic cell lines HL60 and NB4 (*130–132*). The physiological complexity of nongenomic signaling itself and to what extent it acts cooperatively with VDR-mediated genomic signaling is unclear. Moreover, it has not been clear whether D_3 uses the classical VDR for nongenomic signaling or an alternative receptor is involved.

B. A VDR·PI 3-Kinase Signaling Complex Regulates Myeloid Cell Differentiation

1. D_3-INDUCED CD14 AND CD11B EXPRESSION IS PI 3-KINASE DEPENDENT

In light of the important roles played by PI 3-kinase in regulating the differentiation of a variety of cell types (*133–135*), the possibility that D_3-induced monocyte differentiation involves PI 3-kinase was investigated. Incubation of serum-starved THP-1 cells with 10 nM D_3 brought about a significant increase in PI 3-kinase activity, with a maximum response observed at 1 μM (*3*). Time course experiments using 100 nM D_3 detected PI 3-kinase activation as early as 5 min and a maximum level by 20 min (*3*). Subsequent experiments provided several lines of evidence to show that D_3-induced CD14 expression is PI 3-kinase dependent including: (a) D_3-induced CD14 expression was abrogated in THP-1 cells incubated with either wortmannin and LY294002 (*3*); (b) antisense targeting of p110α PI 3-kinase attenuated D_3-induced CD14 expression (Fig. 7); and (c) expression of a dominant negative mutant of PI 3-kinase ($\Delta p85$) in U937 cells also abrogated D_3-induced expression of CD14 (*3*). Similarly, CD11b induction in response to D_3 was also attenuated by wortmannin and LY294002 in both THP-1 cells and in human peripheral blood monocytes (*3*). Taken together, these results established that D_3-induced CD14 and CD11b expression are regulated by PI 3-kinase. Of particular note, D_3 induction of p21 expression in THP-1 cells was found to be resistant to both wortmannin and LY294002 (data not shown) (*3*).

2. INDUCTION OF CD14 IN RESPONSE TO D_3 INVOLVES THE ASSOCIATION OF PI 3-KINASE WITH THE VDR

Many responses to D_3, such as induction of expression of osteopontin, osteocalcin, calbindin, and 24-hydroxylase, are brought about by a mechanism involving binding of the VDR to a specific VDRE in the corresponding promoters (*136*). Since no VDRE has been identified within the CD14 gene promoter (*123*), this raised the question as to whether the VDR plays any role at all in regulating D_3-induced CD14 expression. Experiments using antisense S-oligo specific to VDR mRNA showed attenuation of D_3-induced CD14 expression (*3*).

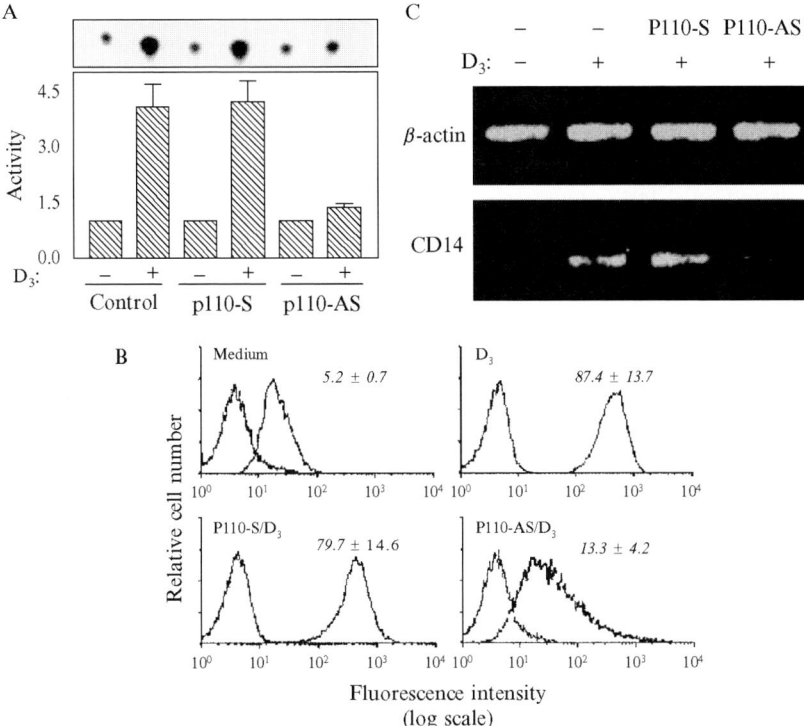

FIG. 7. Antisense S-oligo to the p110 subunit of PI 3-kinase inhibits D_3-induced CD14 expression. THP-1 cells were suspended in RPMI 1640 containing 2.5% Lipofectamine alone (Control), or the same solution containing 5 μM of either sense (S) or antisense (AS) S-oligo to p110. Cells were then incubated on a rotary shaker for 4 hr at 37°C. The incubation was brought up to 10 ml of medium, and cells were cultured for an additional 18 hr at 37°C and 5% CO_2. (A) THP-1 cells were stimulated with 100 nM D_3 for 20 min and assayed for PI 3-kinase assay (3). The upper rectangle shows PIP spots obtained in one representative experiment. The graphical data shown below represent PI 3-kinase activities (means ± SEM of values obtained in three separate experiments) (3). (B) Aliquots of control or S-oligo–treated cells (~1 × 10^6) were incubated in medium alone or with D_3 (100 nM, 24 hr), washed with staining buffer, and then labeled with anti-CD14 mAb or irrelevant mAb, followed by FITC-conjugated secondary Abs. Flow cytometric analyses were performed, and MFI indices were determined (3). Results are expressed as histograms of fluorescence intensity. Histograms displaced to the right represent cells stained with anti-CD14, and histograms on the left represent cells stained with irrelevant IgG2b. Bold, italicized values are means ± SEM ($n = 3$) of MFI indices (3). (C) Control or S-oligo–treated cells were incubated with D_3 (100 nM, 24 hr). Total RNA was extracted and RT-PCR was carried out for *CD14* and *β-actin* (3). Adapted from the work of Hmama *et al.* (3), with permission.

RT-PCR experiments showed that this was related to a marked reduction in CD14 mRNA accumulation in response to D_3 (3). Taken together, these findings indicate that the VDR is essential for D_3-induced CD14 gene expression.

Given the evidence that both PI 3-kinase and the VDR were found to be essential for induction of CD14 in response to D_3, it was important to know whether VDR was required for activation of PI 3-kinase. In fact, antisense targeting of VDR almost completely abrogated PI 3-kinase activation in response to D_3 (Fig. 8).

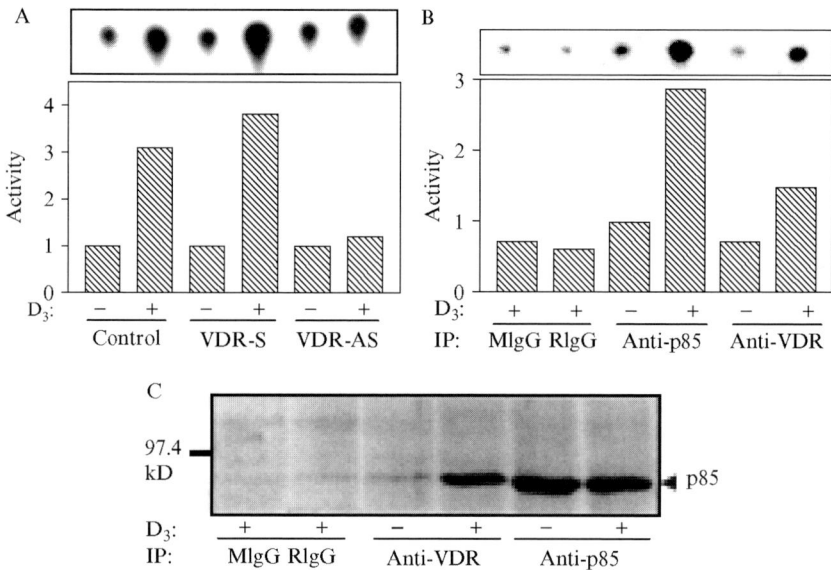

FIG. 8. The VDR is required for D_3-induced PI 3-kinase activation. (A) THP-1 cells were treated overnight with 5 µM of either sense (S) or antisense (AS) S-oligo to VDR (3). Cells were then stimulated for 20 min with 100 nM D_3, and lysates were prepared and assayed for PI 3-kinase activities (3). The upper rectangle shows PIP spots obtained in one representative experiment. The data shown below are PI 3-kinase activities (averages of values obtained in two separate experiments) (3). (B) Cells were stimulated with D_3 as in (A) and cell lysates were prepared and incubated overnight at 4°C either with 9A7 mAb (anti-VDR) or with mAb specific for the p85 subunit of PI 3-kinase. Immune complexes were adsorbed onto protein A and assayed for PI 3-kinase activities. The upper rectangle shows PIP spots obtained in one representative experiment. The data shown below are PI 3-kinase activities (average of two separate experiments). (C) Cell lysates prepared from either control cells or D_3-treated cells were immunoprecipitated with specific mAbs to either the VDR or the p85 subunit of PI 3-kinase. Immunoprecipitates were then analyzed by SDS-PAGE and immunoblotting with Abs to p85 (40). Immunoprecipitates with normal mouse IgG (MIgG) and normal rat IgG (RIgG) were used to control for the specificity of anti-p85 and anti-VDR Abs, respectively. Adapted from the work of Hmama et al. (3), with permission.

The finding that the D_3 receptor was required for both PI 3-kinase activation and CD14 expression in response to D_3 suggested the possibility that this might involve a signaling complex containing both the VDR and PI 3-kinase. Our results in fact indicated that immunoprecipitates of the VDR prepared from cells activated with D_3 contained PI 3-kinase activity and the p85 subunit (Fig. 8). Of particular interest was the finding that the association of p85 with the VDR was only observed in cells treated with hormone (Fig. 8). These results demonstrated that D_3 treatment induces the formation of a signaling complex containing the VDR and PI 3-kinase resulting in activation of the lipid kinase, which is required for induction of CD14 expression.

In summary, D_3-induced signaling in THP-1 cells initiated through the VDR appears to involve at least two distinct pathways. One is a genomic pathway involving p21, which regulates aspects of monocyte differentiation independently of PI 3-kinase. A second pathway is nongenomic and involves a signaling complex containing at least PI 3-kinase and VDR, which regulates the induction of CD14 and CD11b and perhaps other targets.

C. D_3-Induced Monocyte Antimycobacterial Activity Is Regulated by PI 3-Kinase

Several lines of evidence have indicated that D_3 regulates host resistance to *Mycobacterium tuberculosis*. D_3 deficiency and VDR receptor polymorphisms have been linked to increased susceptibility to infection with *M. tuberculosis* and *Mycobacterium leprae* (110–112). In this regard, D_3 production *in vivo* is promoted by exposure to ultraviolet light, providing a link between exposure to sunlight and antimycobacterial resistance (110, 113). In addition, D_3 has been shown directly to activate mononuclear phagocyte antimycobacterial activity (113, 114). However, the molecular basis through which D_3 regulates host resistance to *M. tuberculosis* has not been identified. To elucidate this further, we examined whether PI 3-K is involved in regulating D_3-induced antimycobacterial activity and investigated the mechanistic basis for this effector function.

1. D_3-Induced Monocyte Resistance to *M. tuberculosis* Is Regulated by PI 3-Kinase

To investigate the basis for the induction of monocyte antimycobacterial activity by D_3, THP-1 cells or human peripheral blood monocyte-derived macrophages were infected with *M. tuberculosis* H37Rv and then incubated with hormone. As expected, this resulted in decreased recovery of mycobacterial colony-forming units from D_3-treated cells compared with control cells. The potential role of PI 3-K in this response was studied by examining the growth of *M. tuberculosis* in cells treated with D_3 in the presence or absence of

LY294002 or wortmannin as well as in cells where the p110α subunit of PI 3-kinase was targeted by antisense. Treatment with either LY294002 or wortmannin inhibited PI 3-K activity in *M. tuberculosis*-infected, D_3-treated THP-1 cells and also significantly abrogated D_3-induced antimycobacterial activity (137). Essentially identical effects were observed in antisense-treated cells with reduced expression of the p110α subunit of PI 3-Kinase (137). Taken together, these findings indicated that activation of PI 3-K p85/p110α is required for the antimycobacterial action of D_3.

2. Phagocyte Oxidative Burst Is Required for the Antimycobacterial Action of D_3 and Is PI 3-Kinase–Dependent

The phagocyte oxidative burst is a major antimicrobial effector mechanism. To assess the role of macrophage reactive oxygen intermediates (ROI) in the antimycobacterial action of D_3, infected cells were incubated with hormone in the presence of either 4-hydroxy-TEMPO, PEG-SOD, or PEG-cat. These reagents act to eliminate ROI, either by scavenging (4-hydroxy-TEMPO) or enzymatically (PEG-cat or PEG-SOD), and each of them significantly attenuated the ability of D_3 to induce resistance to *M. tuberculosis* in THP-1 cells and monocyte-derived macrophages (137). Consistent with these observations, *M. tuberculosis*-infected macrophages secreted significant amounts of O_2 when treated with D_3. Of particular interest, this response was inhibited by LY294002 or wortmannin as well as by antisense targeting of p110α PI 3-K (137). These results suggested an interaction between *M. tuberculosis* infection and D_3 in triggering the phagocyte oxidative burst in human monocyte-derived macrophages as well as in THP-1 cells and that this process is regulated by PI 3-kinase.

3. Translocation of Phagocyte Oxidase Components to the Plasma Membrane

Phagosome oxidase activity requires recruitment of the cytosolic components, $p47^{phox}$ and $p67^{phox}$, to the membrane for assembly with flavocytochrome b_{558} (138, 139). The relative distribution of $p47^{phox}$ and $p67^{phox}$ in cytosolic and membrane fractions was examined. A marked translocation of both $p47^{phox}$ and $p67^{phox}$ to the membrane fraction was observed in *M. tuberculosis*-infected cells treated with D_3, whereas treatment of cells with either live or heat-killed *M. tuberculosis* alone did not bring about detectable changes (138, 139). We have shown that redistribution of $p47^{phox}$ and $p67^{phox}$ to the membrane fraction in response to infection with *M. tuberculosis* and treatment with D_3 was markedly reduced in cells pretreated with either of the PI 3-K inhibitors, LY294002 or wortmannin (Fig. 9). Consistent with this result, PI 3-kinase was shown to be required for triggering the NADPH

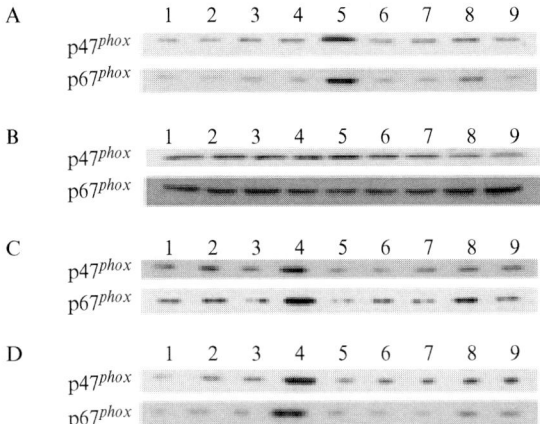

FIG. 9. THP-1 cell cytosol and membrane immunoblots probed with anti-p47^{phox} (upper panel) or anti-p67^{phox} (lower panel). In panels A and B, membrane and cytosolic fractions, respectively, were purified from resting cells (lane 1), cells differentiated with PMA (lane 2), differentiated cells coincubated with *M. tuberculosis* for 2 (lane 3) or 4 hr (lane 4), differentiated cells coincubated with *M. tuberculosis* for 4 hr followed by treatment with D$_3$ (lane 5), differentiated cells coincubated with heat-killed *M. tuberculosis* for 2 (lane 6) or 4 hr (lane 7), differentiated cells coincubated with heat-killed *M. tuberculosis* for 4 hr followed by treatment with D$_3$ (lane 8), and differentiated cells treated with D$_3$ only (lane 9). In panels C and D, membrane fractions were purified from differentiated cells treated in the presence or absence of the PI 3-K inhibitor, LY or Wm, respectively. Shown are PMA-differentiated cells (lane 1), cells coincubated with *M. tuberculosis* for 4 hr (lane 2), D$_3$-treated cells (lane 3), cells coincubated with *M. tuberculosis* followed by D$_3$ treatment (lane 4), PMA-differentiated cells treated with PI 3-K inhibitor (lane 5), cells treated with PI 3-K inhibitor followed by coincubation with *M. tuberculosis* for 4 hr (lane 6), cells treated with PI 3-K inhibitor followed by D$_3$ treatment (lane 7), cells treated with PI 3-K inhibitor followed by coincubation with *M. tuberculosis* for 4 hr and then D$_3$ treatment (lane 8), and cells coincubated with *M. tuberculosis* followed by treatment with PI 3-K inhibitor and then D$_3$ (lane 9). Adapted from the work of Sly *et al.* (137).

oxidase in response to particulate stimuli through activating F$_c\alpha$ and F$_c\gamma$ receptors in neutrophils (140). Similarly, based on effects of wortmannin, a role for PI 3-kinase in regulating the production of superoxide by human neutrophils in response to soluble stimuli, such as fMLP (141) and PMA (142), has also been suggested. In summary, a key regulatory pathway of D$_3$-induced host defense against tuberculosis has been identified and appears to represent another novel example of nongenomic signaling by vitamin D$_3$. PI 3-Kinase is required to trigger the oxidative burst in response to D$_3$ in *M. tuberculosis*-infected cells. However, activation of PI 3-kinase alone is insufficient to bring about full oxidase activation, since treatment of THP-1 cells with D$_3$ alone activates PI 3-kinase, but at best this elicits only modest superoxide production. Prior infection of cells with live *M. tuberculosis* was observed to prime cells for an enhanced D_3-induced oxidative burst (137). This suggests that live *M. tuberculosis* or some factor unique to the phagosome containing viable *M. tuberculosis* potentiates phagocytes to undergo a vigorous oxidative burst in response to D_3.

D. D_3-Induced Antimicrobial Activity Against *Salmonella enterica* Is PI 3-Kinase Dependent

Salmonella enterica serovar Typhimurium causes self-limiting gastroenteritis in immunocompetent adults and may also cause life-threatening extraintestinal infections (143). The phagocyte oxidative burst is a key host defense mechanism against *S. enterica* (144–146) and *Salmonella* SodCI and SodCII isoenzymes have been implicated in virulence (144, 146). In light of our new understanding of how D_3 induces macrophage resistance to *M. tuberculosis*, we examined the role of PI 3-kinase in D_3-induced killing of *S. enterica* as well as the potential roles for SodCI and SodCII in resisting this response (147).

Vitamin D_3 treatment had a pronounced antimicrobial effect against wild-type *S. enterica* serovar Typhimurium in THP-1 cells, reducing the number of viable bacteria by almost tenfold compared to the number of bacteria in untreated control cells. Incubation of monocytes with wortmannin or LY204002 prior to addition of D_3, abrogated this effect indicating that it was PI 3-K dependent (147). Vitamin D_3-induced killing of *S. enterica* was not dependent on nitric oxide (NO), as no nitrite was detectable in culture media by Griess assay. Furthermore, killing was not reversed by incubation of cells with NO inhibitor NG-monomethyl-L-arginine (L-MMA). Vitamin D_3-treated, *Salmonella*-infected THP-1 cells produced significantly more superoxide than did cells exposed to either infection alone or to D_3 alone (147). Furthermore, the amount of superoxide produced by vitamin D_3-treated, *Salmonella*-infected THP-1 cells was reduced to background levels by pretreatment with either LY294002 or wortmannin (147). These results indicated that D_3 induces bactericidal activity against intracellular *S. enterica* serovar Typhimurium in human monocyte-like cells that are PI 3-kinase dependent and suggest that the ulimate effector mechanism is the phagocyte oxidative burst. The latter argument is consistent with studies showing that control of *S. enterica* infection in the mouse model requires NADPH oxidase activity (144, 145, 148).

Further direct support for this model was provided by study of *SodC* mutants, which we found to be more susceptible to the effect of D_3 in THP-1 cells than were wild-type organisms. While each mutant strain (SodCI, SodCII, and the SodCI-SodCII double knockout) grew normally in untreated THP-1 cells, significant loss of viability was observed in D_3-treated cells. While this was observed for both individual SodCI and SodCII mutants, the effect was more pronounced in the SodCI-SodCII double mutant, which was nearly completely eliminated from D_3-treated cells (147). This result provided evidence that both SodC enzymes contribute to bacterial resistance to oxidative damage in a nonredundant fashion.

IV. PI 3-Kinase and Phagocytosis of Mycobacteria

A. Phagocytosis of Mycobacteria: Receptors and Signaling

Several receptors involved in the internalization of pathogenic mycobacteria by macrophages have been tentatively characterized, but not extensively studied. Binding of mycobacteria to macrophages is believed to occur in cholesterol-rich domains of the host cell plasma membrane (149), and involvement of β_2 integrins (complement receptor CR3 and CR4), together with other molecules including the mannose receptor, CR1 and CD14 (150), have been suggested. Other surface molecules, such as Toll-like receptors, are also involved in mycobacterial interactions with phagocytic cells (151). In monocytes and immature macrophages, which do not express the mannose receptor (152), CR3 and CD14 have been shown to be potential gates for entry of mycobacteria (153, 154). However, whether these two receptors act individually or cooperatively is not known.

Interest in signaling events that link cell stimulation to engagement of β_2 integrins has been considerable in recent years (155, 156). The mycobacterial cell surface glycolipid lipoarabinomannan (LAM) is known to enhance phagocytosis of mycobacteria (157). It has also been shown that LAM-regulated phagocytosis is dependent at least in part on CR3 (158). Studies by Kolanus et al. (99) led to the discovery that cytohesin-1 acts as an adaptor protein and interacts with the cytoplasmic tail of CD18 leading to changes in the functional properties of β_2 integrins. Cytohesin-1 contains a PH domain that binds the PI 3-kinase metabolite, phosphatidylinositol 3,4,5-trisphosphate (PtdIns-3,4,5-P3), leading to changes in the functional properties of the protein (98, 159). These findings have suggested a role for PI 3-kinase and cytohesin in regulating the activity of β_2 integrins (36, 160), including CR3. The receptors that cooperate to bring about optimal uptake of mycobacteria and how they are linked are important issues in innate immunity. In the next section, we have reviewed data from studies that focused on deciphering the mechanisms that lead to the internalization of *Mycobacterium bovis* bacillus Calmette-Guerin (BCG), the role of vitamin D_3 in regulating phagosome maturation as well as the important role of PI 3-kinase in regulating these events.

B. Cross-Talk Between CD14 and CR3: Regulation by PI 3-Kinase and Cytohesin-1

1. Role of CD14/CR3 in Phagocytosis of BCG and Involvement of TLR-2

We investigated the hypothesis that binding of mycobacteria to macrophages modulates the activity of cell-surface receptors so as to enhance bacterial internalization. Using THP-1 and CHO cells transfected to express individual receptors, evidence was obtained to show that CD14 played a major role in phagocytosis of *M. bovis* BCG and this required serum factors including lipopolysaccharide binding protein (LBP) (*161*). These results were consistent with a previous report showing that CD14 mediated uptake of mycobacteria by monocyte-derived microglial cells (*162*). Whereas our results showed that some degree of mycobacterial internalization occurred through CD14 alone and CR3 by itself functioned poorly as a phagocytic receptor for BCG uptake was significantly potentiated in the presence of both CD14 and CR3 (*161*).

Transfected CHO cells were also used to investigate the potential role of TLR-2 in regulating CD14-dependent phagocytosis. Exposure of CHO cells transfected with CD14 or TLR-2 alone to opsonized BCG led to a modest degree of bacterial ingestion. In contrast, coexpression of both TLR-2 and CD14 (CHO/CD14/TLR2) resulted in a four- to fivefold increase in the proportion of cells that ingested bacteria (*161*).

2. CD14-Dependent Phagocytosis of BCG Involves PI 3-Kinase and TLR-2

An important question raised by these findings was the mechanism of communication between CD14 and CR3. PI 3-kinase has been demonstrated to have a direct role in regulating a range of leukocyte functions dependent on rearrangement of the cytoskeleton such as adhesion (*163*), phagocytosis (*164, 165*), and phagosome biogenesis (*166–168*). These properties, together with the knowledge that the activities of β_2 integrin receptors can be modulated through PI 3-kinase (*36, 160*), identified this lipid kinase as an attractive candidate for regulating cross-talk between CD14 and CR3 in the context of the binding and uptake of mycobacteria. Incubation of THP-1wt cells with serum-opsonized BCG brought about a rapid and significant increase in PI 3-kinase activity with a maximum response observed within 10 min (~3.5-fold increase) (*161*). A threefold increase in PI 3-kinase activity was also observed in cells incubated with LAM in the presence of LBP. In contrast, THP-1wt cells incubated with BCG in the absence of opsonization with either serum or LAM had minimal effects on PI 3-kinase activity. PI 3-kinase inhibitors were used to examine further the potential involvement of the enzyme in CD14-induced phagocytosis. Preincubation of THP-1wt cells with either wortmannin or LY294002 inhibited phagocytosis in a dose-dependent manner with maximum inhibition of 78% and 75%, respectively. Taken together, these findings suggested that PI 3-kinase activation plays a central role in CD14-dependent phagocytosis of BCG (53). Furthermore, whether PI 3-kinase is involved in TLR-2–dependent ingestion of mycobacteria was investigated by examining the intracellular distribution of the PI 3-kinase product PtdIns-3,4,5-P_3 *in vivo*. CHO cells transfected with both CD14 and TLR-2 (CHO/CD14/TLR-2) were infected with BCG conjugated to the fluorescent tracer rhodamine B isothiocyanate (RITC) and then incubated with mAb RC6F8 and fluorescein isothiocynate (FITC)-conjugated secondary Ab. RC6F8 recognizes specifically PtdIns-3,4,5-P_3, with minimal cross-reactivity for either PtdIns-(4,5)-diphosphate or for the two monophosphates PtdIns-(3)-monophosphate and PtdIns-(4)-monophosphate (*169*). Marked colocalization of PtdIns-3,4,5-P_3 with RITC-BCG in CHO/CD14/TLR-2 cells, but not in CHO/CR3 or CHO/CD14 cells, was observed showing the important role of TLR-2 in the CD14-induced activation of PI 3-kinase (*161*). Finally, using the PI 3-kinase inhibitor LY294002, increased avidity of CR3 induced by LAM was also shown to be dependent on PI 3-kinase (*161*).

3. CR3-Mediated Phagocytosis of *M. Bovis* BCG Is Regulated by Cytohesin-1

An important question raised by these findings was the molecular basis by which activation of CD14/TLR-2 and PI 3-kinase–dependent signaling led to modulation of the properties of CR3. Cytohesin-1 is known to be a key regulator of leukocyte adhesion, and this is mediated by its interaction with the cytoplasmic tail of CD18 to bring about increased avidity of LFA-1 (CD18/CD11a) for ICAM-1 (*36, 99, 170*). The interaction of cytohesin-1 with β_2 integrins is positively regulated by the binding of the PI 3-kinase metabolite, PtdIns-3,4,5-P_3, to the cytohesin-1 PH domain (*98, 171*).

These findings suggested the hypothesis that CD14/PI 3-kinase–dependent enhancement of CR3 avidity brought about by mycobacterial cell wall fraction components may involve cytohesin-1. To investigate this possibility, cytohesin-1 expression was downregulated by incubation of THP-1wt cells in the presence of antisense S-oligos. Treatment of cells with antisense S-oligo to cytohesin-1 mRNA, but not with control sense S-oligo, eliminated the expression of this protein concomitant with significant attenuation of serum-dependent uptake of BCG (53% inhibition) (*161*).

Direct association between cytohesin-1 and CR3 was shown using flow cytometry based "pull down" experiments. Anticytohesin-1 Ab was coupled to latex beads and used to pull down cytohesin-1 and from lysates of activated cells. Beads were then stained with antibodies to either PtdIns-3,4,5-P_3 or CR3 and analyzed flow cytometry. LAM stimulation of THP-1wt cells increased the level of PtdIns-3,4,5-P_3 associated with cytohesin-1 (3.5-fold increase) concomitant with increased recruitment of CR3 (6.5-fold increase) (*161*). Pretreatment of cells with PI 3-kinase inhibitor or CD14 blocking mAb before LAM abrogated the recruitment of both PtdIns-3,4,5-P_3 and CR3 to cytohesin-1. Taken together, these results suggested a direct role for a CD14/PI3-kinase/cytohesin-1 signaling pathway resulting in enhanced CR3-dependent phagocytosis of BCG. In summary, several lines of evidence support a role for cytohesin-1 in CR3-dependent phagocytosis of BCG: (i) the uptake of serum- but not LBP-opsonized BCG was significantly attenuated in THP-1wt cells with reduced cytohesin-1, (ii) intracellular staining showed a CD14- and PI 3-kinase–dependent colocalization of cytohesin-1 with CR3 in LAM-stimulated cells, and (iii) pull-down experiments detected the physical association between cytohesin-1 and both PtdIns-3,4,5-P_3 and CR3 in LAM-stimulated cells (*161*).

These findings suggested a dynamic model regulating the uptake of mycobacteria by macrophages (Fig. 10). In this model, binding of BCG or LAM to CD14 triggers signaling through TLR-2 leading to activation of PI 3-kinase. This creates a point of bifurcation in which in one direction based on the effects of PI 3-kinase on the cytoskeleton alone, suboptimal levels of

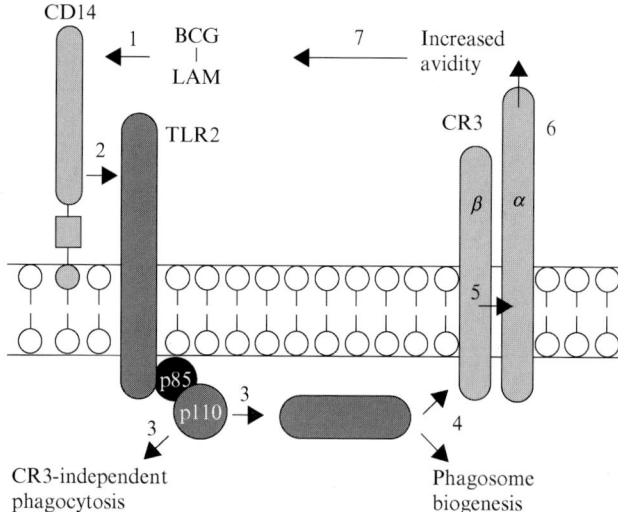

FIG. 10. Cooperativity between CD14 and CR3 mediates optimal uptake of BCG. Signaling through membrane-bound CD14 initiated by BCG or LAM leads to TLR-2–mediated activation of PI 3-K. The latter is responsible for triggering two pathways for bacterial internalization: a unique CD14 pathway that can function in the absence of CR3 and a cytohesin-1–regulated CR3-dependent pathway that together with CD14 results in optimal phagocytosis of BCG. Adapted from the work of Sendide *et al.* (*161*), with permission.

CD14-dependent bacterial ingestion can proceed independently of CR3. In the other direction, through the action of the PI 3-kinase product PtdIns-3,4,5-P_3, the molecular adaptor cytohesin-1 is recruited to the membrane where it binds to the β-chain of CR3 and converts it into an active receptor for BCG. Thus, CD14-dependent activation of PI 3-kinase through TLR-2 leads to recruitment of a second receptor that acts cooperatively to bring about optimal bacterial internalization.

C. Rescue of Phagosome Maturation by Vitamin D_3 Is PI 3-Kinase Dependent

To investigate processes regulating phagosome maturation, we established a technology to analyze phagosomes using flow cytometry (166). This involved labeling the plasma membrane of intact macrophages with the red-fluorescent marker PKH26 after which the cells were allowed to ingest latex beads resulting in the generation of phagosomes surrounded by a red-fluorescent membrane suitable for flow cytometry. Following cell disruption and partial purification of phagosomes, these vesicles were readily distinguished from both cell debris and free beads released from disrupted vacuoles. Using this system, we were able to show that phagosomes stained with specific mAbs showed progressive acquisition of both Rab7 and LAMP-1 consistent with their expected movement along the endocytic pathway. Alternatively, macrophages were preloaded with the lysosomal tracer FITC-dextran before membrane labeling with PKH and incubation with latex beads. Phagosome–lysosome fusion was then quantified on the basis of the colocalization of red and green signals [described in detailed (166)]. Using this system, we found that internalization of latex beads in the presence of lysates of *M. tuberculosis*, but not in the presence of lysates from the nonpathogenic organism *Mycobacterium smegmatis* markedly decreased phagosome acquisition of Rab7 and LAMP-1 and vesicle fusion with FITC-dextran–loaded lysosomes. Inhibition of phagolysosome fusion could be attributed, at least in part, to the mycobacterial cell wall glycolipid lipoarabinomannan. Further analysis showed complete rescue of phagosome maturation when cells were pretreated with Vit D_3 before exposure to lipoarabinomannan. To examine the role of PI 3-kinase in D_3-induced phagolysosome fusion, acquisition of FITC-DXT by phagosomes was studied in D_3-treated cells in the presence or absence of the PI 3-kinase inhibitor LY-294002. FITC-DXT–loaded THP-1 cells were treated with LY-294002 for 30 min and washed before D_3 stimulation. Twenty-four hours later, macrophages were labeled with PKH and exposed to latex beads in the presence or absence of LAM. The results indicated that in the absence of PI 3-kinase inhibitor, the ability of LAM to inhibit phagolysosome fusion was completely prevented by D_3. However, the ability of D_3 to prevent phagosome maturation induced by LAM was nearly completely abrogated by

LY-294002, suggesting that D_3 promotes phagolysosome fusion via a PI 3-kinase signaling pathway (*166*).

V. Stable Gene Silencing of Isoforms of PI 3-Kinase and Perspectives

Genetic approaches to assigning function to individual class I PI 3-kinase p110 isoforms have been limited since gene knockouts of p110α and p110β in mice were found to be embryonically lethal (*172, 173*). Consequently, it has not been possible to determine with precision the roles of these isoforms in immune cells. Recently, instead of gene knockouts, mutant p110δ mice were created and have been a useful alternative (*174*). Thus far, however, no phenotypes in monocytic cells from either knockout or mutant p110δ animals have been reported (*174–176*). Studying mononuclear phagocyte cell biology through genetic manipulation by nonviral transfection methods has been challenging due to the dual problems of low transfection efficiency and the difficulty in obtaining stable transfection. Methods involving cationic lipid and liposome-mediated delivery of DNA or physical methods, such as electroporation, result in low transfection efficiency in monocytic cells, loss of viability, and the difficulty of obtaining stable transfection (*177, 178*). An approach that has achieved greater success in monocytic cell lines is viral-mediated transduction. Although not all viruses can transduce monocytic cells efficiently, lentiviruses have been shown to do so at >90% efficiency (*179–181*). In what follows, we review our successful efforts to develop a strategy that allows for transduction of human monocytic cell lines using a lentiviral vector leading to stable silencing of a specific isoform of PI 3-kinase through RNA interference (RNAi) (*182*).

A. Silencing of P110α Subunit of PI 3-Kinase: Differential Involvement in Monocyte Adherence

1. STABLE SILENCING OF P110α PI 3-KINASE SUBUNIT IN HUMAN MONOCYTIC CELL LINES

RNAi is a sequence-specific posttranscriptional gene silencing mechanism initiated by the introduction of double-stranded RNA (dsRNA) into target cells (*183*). RNAi is a natural regulatory mechanism that occurs in many organisms, including plants, *Caenorhabditis elegans*, *Drosophila*, and mammalian cells [reviewed in (*183*)]. The RNAi pathway begins by processing dsRNA into short (<30 bp) dsRNA duplexes called small interference RNA (siRNA) by a host RNase Dicer. The siRNA then becomes incorporated into a multi-component nuclease complex called the RNA-induced silencing complex (RISC).

RISC then uses the siRNA sequence as a guide to recognize cognate mRNAs for degradation.

To produce recombinant lentiviral vectors for transduction of monocytes, we used the packaging cell line HEK 293T transiently cotransfected by the vector plasmid (pHR-U6-shRNA), helper plasmid (pCMVδR8.2), and envelope plasmid (pMD.G). This strategy [for detailed protocol see (182)] segregates the *trans*-acting sequences that encode for viral proteins (made by the helper plasmid) from the *cis*-acting sequences (regions recognized by viral proteins on the vector plasmid) involved in the transfer of vector sequences encoding a short hairpin RNA (shRNA) [reviewed in (184)]. This approach results in the production of progeny virions that do not contain any genetic elements that code for viral proteins. Therefore, transduced cells are not capable of producing any viral proteins. The vector plasmid contains a U6 promoter, which drives transcription of the shRNA coding sequence of interest as well as a CMV-driven enhanced green fluorescent protein (EGFP) reporter. These elements are flanked by long terminal repeats (LTRs) and contain *cis*-acting sequences allowing only the vector RNA to be packaged. Following transfection of HEK 293T cells, the plasmids pCMVΔR8.2 and pMD.G provide in *trans* the viral structural proteins required for the assembly of viral particles.

Using this approach, we harvested viral particles from conditioned medium and used them for transduction of monocytes. Transduction efficiency was then quantitated by GFP expression. The shRNA expressed from the integrated vector elements then silenced the target gene via the RNAi pathway. We designed a shRNA sequence specific to the PI 3-kinase p110α mRNA. THP-1 and U-937 cells were either mock transduced or transduced with viral vectors targeting either p110α, control shRNA, or with a U6 promoter control. Transduction of THP-1 and U-937 with the viral vectors targeting the p110α isoform resulted in nearly complete elimination of PI 3-kinase p110α isoform expression (Fig. 11). In contrast, transduction of cells either with the lentiviral vector expressing the control shRNA sequence, the U6-promoter control vector, or mock transduction did not affect p110α protein levels. This result was specific in that levels of other Class I_A PI 3-kinase catalytic subunits p110β and p110δ, or the p85α regulatory subunits were not affected. The stability of p110α silencing was confirmed by Western blotting of cells that were stored in liquid nitrogen for greater than 1 year, and cells that had been in continuous culture for more than 6 weeks.

2. Monocyte Adherence Induced by D_3, But not LPS Is Dependent on P110α

Previous results from this laboratory showed that LPS-induced adherence of THP-1 cells is dependent on PI 3-kinase (3). However, use of pharmacological inhibitors did not permit determination of whether this involved class I_A PI

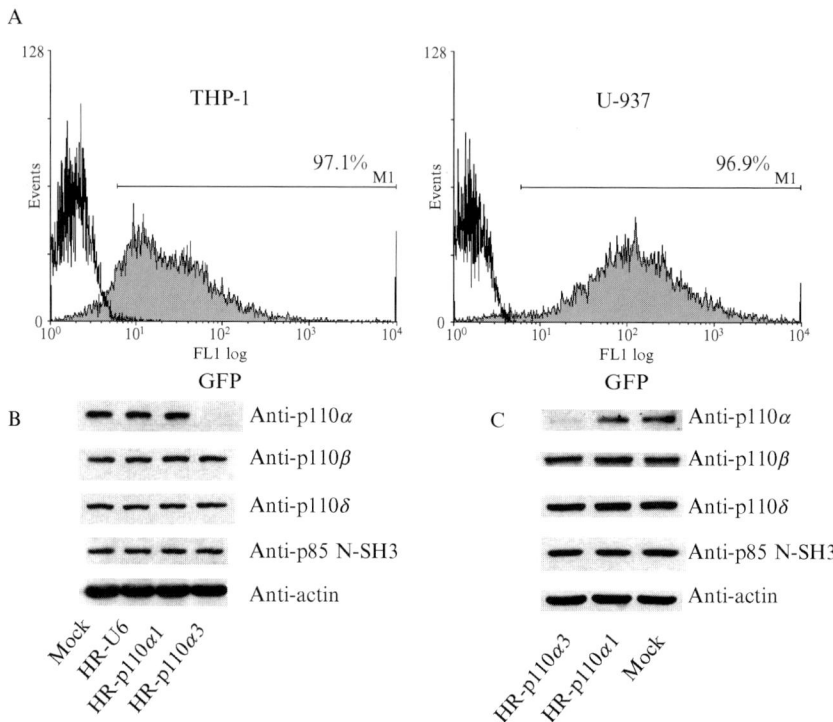

FIG. 11. Transduction of monocytic cell lines by lentiviral vectors is efficient and generates stable cell lines deficient in p110α. (A) Flow cytometry analysis of transduced (solid histogram) or mock transduced cells (clear histogram). Ten thousand cells were analyzed, and the GFP fluorescence intensity was measured on the FL1 channel. Approximately 97% of the THP-1 and U-937 cells were GFP-positive. The mean fluorescence intensity (MFI) for mock-infected cells was 8.5 for THP-1 and 6.6 for U-937. Transduced cells had MFI values of 38.3 for THP-1 and 132.3 for U-937. (B and C) Western blot analyses of class I_A PI 3-K p110 catalytic subunit isoforms (α, β, and δ), and p85 regulatory subunit in THP-1 cells and U-937 cells. Actin was used as a loading control. Adapted from the work of Lee et al. (182).

3-kinase and, if so, which p110 isoform (3). To address this question, LPS-induced adherence was examined in THP-1 cells rendered p110α-deficient using lentiviral transduction of shRNA. Consistent with previous findings that PMA-induced monocyte adherence was resistant to PI 3-kinase inhibitors (3), p110α-deficient cells showed normal PMA-induced adherence (182). Using PMA-treated cells as a control, LPS-induced adherence was determined as the percentage of PMA-induced adherence. Whereas PI 3-kinase inhibitor LY294002 significantly reduced LPS-induced adherence in all transduced cells, silencing of p110α did not affect LPS-induced adherence (182). Similar to LPS, LY294002 also inhibited D_3-induced adherence in all types of transduced cells. In contrast to LPS, however, monocyte adherence induced by D_3 was found to be attenuated in p110α-deficient cells (182). Moreover, this inhibitory effect was comparable to that observed in LY294002-treated cells.

Expression of the monocyte differentiation marker CD11b was also induced by D_3 in a PI 3-kinase–dependent manner, and gene silencing using shRNA showed that p110α was also required for this effect (Fig. 12). Taken together, these findings demonstrate that LPS and D_3 use distinct isoforms of PI 3-kinase to induce functional responses and lentiviral-mediated delivery of shRNA is a powerful approach to study the role of specific PI 3-kinase isoforms in monocytic cells.

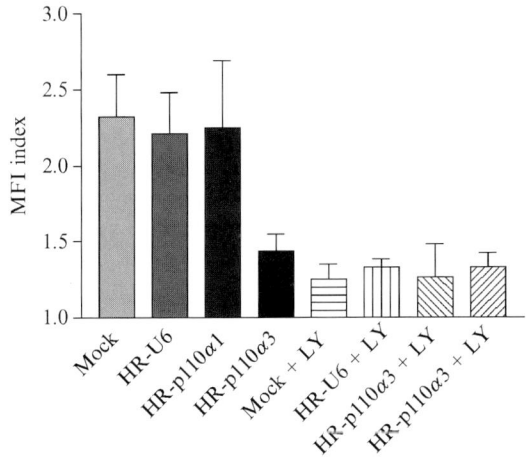

FIG. 12. CD11b induction by D_3 is attenuated in p110α-deficient THP-1 cells. THP-1 cells were either incubated or not with LY294002 (25 μM) for 30 min, followed by 100 nM D_3 for 72 hr at 37°C and 5% CO_2. Cells were then washed and stained with anti-CD11b RPE-conjugated antibody or isotype matched RPE-conjugated control antibody. After paraformaldehyde fixation, 10,000 cells were analyzed for RPE fluorescence. For transduced cells, the analysis was restricted to GFP-positive cells. Data were collected after compensation of GFP and RPE fluorochromes on FL1 and FL2 channels. MFI index is the ratio of (MFI of D_3-treated samples stained with anti-CD11b antibody–MFI of isotype-matched control antibody)/(MFI of anti-CD11b antibody-stained, untreated samples–MFI isotype-matched control antibody). Therefore, an MFI index of 1 indicates that there was no induction of CD11b by D_3, or a complete inhibition by LY294002. All of the LY294002 plus D_3-treated samples were significantly different from control cells treated with D_3 alone (post-ANOVA Tukey test, $p < 0.05$ for all pairs), but not from HR-p110α3 plus D_3 alone ($p > 0.05$). Error bars indicate S.D., $n = 3$. Adapted from the work of Lee et al. (182).

B. Perspectives

The ability of lentiviral vectors to transduce both dividing and nondividing cells and stable expression of the transgene make this a versatile strategy for gene silencing based on siRNA. Moreover, the finding that lentiviral-transduced cells expressing shRNA can be further manipulated by transfection with reporter plasmids, for instance luciferase, significantly expands the utility of this approach. One important application of this technique may be in gene therapy research, such as silencing genes in myeloid leukemia cells or other difficult to transfect cells to better understand their biology and to identify potential therapeutic targets. One area of investigation in our laboratory focuses on the construction of stable siRNA-silenced mutants of other isoforms of PI 3-kinase such as p110β and p110δ and studying their role in monocyte cell regulation.

Ongoing experiments also are taking advantage of discovery-based approaches such as gene array analysis. In modern genomics, traditional gene-by-gene approaches are not adequate to meet the demand of processing information generated from mapping the complex biology of the human genome. Microarray technology is rapidly defining broad patterns of gene regulation, insight into gene function, processes, and pathways. Substantial new knowledge is being generated about transcriptional regulation and biological control mechanisms. Using gene array technology coupled to specific siRNA silencing of individual PI 3-kinase isoforms we expect will yield substantial insight into the multiplicity of cellular events and cell regulation that are under the specific controls of these PI 3-kinase family members.

Acknowledgments

The authors thank the collaborators and trainees who contributed to the studies reviewed herein. This research was supported by Canadian Institutes of Health Research (CIHR) operating grant MOP-8633.

References

1. Vanhaesebroeck, B., Leevers, S. J., Ahmadi, K., Timms, J., Katso, R., Driscoll, P. C., Woscholski, R., Parker, P. J., and Waterfield, M. D. (2001). Synthesis and function of 3-phosphorylated inositol lipids. *Annu. Rev. Biochem.* **70,** 535–602.
2. Noubir, S., Hmama, Z., and Reiner, N. E. (2004). Dual receptors and distinct pathways mediate interleukin-1 receptor-associated kinase degradation in response to lipopolysaccharide. Involvement of CD14/TLR4, CR3, and phosphatidylinositol 3-kinase. *J. Biol. Chem.* **279,** 25189–25195.

3. Hmama, Z., Nandan, D., Sly, L., Knutson, K. L., Herrera-Velit, P., and Reiner, N. E. (1999). 1alpha,25-dihydroxyvitamin D(3)-induced myeloid cell differentiation is regulated by a vitamin D receptor-phosphatidylinositol 3-kinase signaling complex. *J. Exp. Med.* **190,** 1583–1594.
4. Herrera-Velit, P., Knutson, K. L., and Reiner, N. E. (1997). Phosphatidylinositol 3-kinase-dependent activation of protein kinase C-zeta in bacterial lipopolysaccharide-treated human monocytes. *J. Biol. Chem.* **272,** 16445–16452.
5. Petiot, A., Ogier-Denis, E., Blommaart, E. F., Meijer, A. J., and Codogno, P. (2000). Distinct classes of phosphatidylinositol 3′-kinases are involved in signaling pathways that control macroautophagy in HT-29 cells. *J. Biol. Chem.* **275,** 992–998.
6. Vanhaesebroeck, B., Jones, G. E., Allen, W. E., Zicha, D., Hooshmand-Rad, R., Sawyer, C., Wells, C., Waterfield, M. D., and Ridley, A. J. (1999). Distinct PI(3)Ks mediate mitogenic signalling and cell migration in macrophages. *Nat. Cell Biol.* **1,** 69–71.
7. Siddhanta, U., McIlroy, J., Shah, A., Zhang, Y., and Backer, J. M. (1998). Distinct roles for the p110alpha and hVPS34 phosphatidylinositol 3′-kinases in vesicular trafficking, regulation of the actin cytoskeleton, and mitogenesis. *J. Cell Biol.* **143,** 1647–1659.
8. Vieira, O. V., Botelho, R. J., Rameh, L., Brachmann, S. M., Matsuo, T., Davidson, H. W., Schreiber, A., Backer, J. M., Cantley, L. C., and Grinstein, S. (2001). Distinct roles of class I and class III phosphatidylinositol 3-kinases in phagosome formation and maturation. *J. Cell Biol.* **155,** 19–25.
9. Shepherd, P. R., Nave, B. T., Rincon, J., Nolte, L. A., Bevan, A. P., Siddle, K., Zierath, J. R., and Wallberg-Henriksson, H. (1997). Differential regulation of phosphoinositide 3-kinase adapter subunit variants by insulin in human skeletal muscle. *J. Biol. Chem.* **272,** 19000–19007.
10. Janssen, J. W., Schleithoff, L., Bartram, C. R., and Schulz, A. S. (1998). An oncogenic fusion product of the phosphatidylinositol 3-kinase p85beta subunit and HUMORF8, a putative deubiquitinating enzyme. *Oncogene* **16,** 1767–1772.
11. Dey, B. R., Furlanetto, R. W., and Nissley, S. P. (1998). Cloning of human p55 gamma, a regulatory subunit of phosphatidylinositol 3-kinase, by a yeast two-hybrid library screen with the insulin-like growth factor-I receptor. *Gene* **209,** 175–183.
12. Pleiman, C. M., Hertz, W. M., and Cambier, J. C. (1994). Activation of phosphatidylinositol-3′ kinase by Src-family kinase SH3 binding to the p85 subunit. *Science* **263,** 1609–1612.
13. Prasad, K. V., Janssen, O., Kapeller, R., Raab, M., Cantley, L. C., and Rudd, C. E. (1993). Src-homology 3 domain of protein kinase p59fyn mediates binding to phosphatidylinositol 3-kinase in T cells. *Proc. Natl. Acad. Sci. USA* **90,** 7366–7370.
14. Kivens, W. J., Hunt, S. W., III, Mobley, J. L., Zell, T., Dell, C. L., Bierer, B. E., and Shimizu, Y. (1998). Identification of a proline-rich sequence in the CD2 cytoplasmic domain critical for regulation of integrin-mediated adhesion and activation of phosphoinositide 3-kinase. *Mol. Cell Biol.* **18,** 5291–5307.
15. Stephens, L. R., Jackson, T. R., and Hawkins, P. T. (1993). Agonist-stimulated synthesis of phosphatidylinositol(3,4,5)-trisphosphate: A new intracellular signalling system? *Biochim. Biophys. Acta* **1179,** 27–75.
16. Wymann, M. P., and Pirola, L. (1998). Structure and function of phosphoinositide 3-kinases. *Biochim. Biophys. Acta* **1436,** 127–150.
17. Fry, M. J. (1994). Structure, regulation and function of phosphoinositide 3-kinases. *Biochim. Biophys. Acta* **1226,** 237–268.
18. Toker, A. (1998). The synthesis and cellular roles of phosphatidylinositol 4,5-bisphosphate. *Curr. Opin. Cell Biol.* **10,** 254–261.
19. Rebecchi, M. J., and Scarlata, S. (1998). Pleckstrin homology domains: A common fold with diverse functions. *Annu. Rev. Biophys. Biomol. Struct.* **27,** 503–528.

20. Shaw, G. (1996). The pleckstrin homology domain: An intriguing multifunctional protein module. *Bioessays* **18,** 35–46.
21. Alessi, D. R., James, S. R., Downes, C. P., Holmes, A. B., Gaffney, P. R., Reese, C. B., and Cohen, P. (1997). Characterization of a 3-phosphoinositide-dependent protein kinase which phosphorylates and activates protein kinase Balpha. *Curr. Biol.* **7,** 261–269.
22. Alessi, D. R., Deak, M., Casamayor, A., Caudwell, F. B., Morrice, N., Norman, D. G., Gaffney, P., Reese, C. B., MacDougall, C. N., Harbison, D., Ashworth, A., and Bownes, M. (1997). 3-Phosphoinositide-dependent protein kinase-1 (PDK1): Structural and functional homology with the *Drosophila* DSTPK61 kinase. *Curr. Biol.* **7,** 776–789.
23. Stokoe, D., Stephens, L. R., Copeland, T., Gaffney, P. R., Reese, C. B., Painter, G. F., Holmes, A. B., McCormick, F., and Hawkins, P. T. (1997). Dual role of phosphatidylinositol-3,4,5-trisphosphate in the activation of protein kinase B. *Science* **277,** 567–570.
24. Stephens, L., Anderson, K., Stokoe, D., Erdjument-Bromage, H., Painter, G. F., Holmes, A. B., Gaffney, P. R., Reese, C. B., McCormick, F., Tempst, P., Coadwell, J., and Hawkins, P. T. (1998). Protein kinase B kinases that mediate phosphatidylinositol 3,4,5-trisphosphate-dependent activation of protein kinase B. *Science* **279,** 710–714.
25. Le Good, J. A., Ziegler, W. H., Parekh, D. B., Alessi, D. R., Cohen, P., and Parker, P. J. (1998). Protein kinase C isotypes controlled by phosphoinositide 3-kinase through the protein kinase PDK1. *Science* **281,** 2042–2045.
26. Chou, M. M., Hou, W., Johnson, J., Graham, L. K., Lee, M. H., Chen, C. S., Newton, A. C., Schaffhausen, B. S., and Toker, A. (1998). Regulation of protein kinase C zeta by PI 3-kinase and PDK-1. *Curr. Biol.* **8,** 1069–1077.
27. Wenzel, R. P., Pinsky, M. R., Ulevitch, R. J., and Young, L. (1996). Current understanding of sepsis. *Clin. Infect. Dis.* **22,** 407–412.
28. Rietschel, E. T., Brade, H., Holst, O., Brade, L., Muller-Loennies, S., Mamat, U., Zahringer, U., Beckmann, F., Seydel, U., Brandenburg, K., Ulmer, A. J., Mattern, T. *et al.* (1996). Bacterial endotoxin: Chemical constitution, biological recognition, host response, and immunological detoxification. *Curr. Top. Microbiol. Immunol.* **216,** 39–81.
29. Pohlman, T. H., Stanness, K. A., Beatty, P. G., Ochs, H. D., and Harlan, J. M. (1986). An endothelial cell surface factor(s) induced *in vitro* by lipopolysaccharide, interleukin 1, and tumor necrosis factor-alpha increases neutrophil adherence by a CDw18-dependent mechanism. *J. Immunol.* **136,** 4548–4553.
30. Smedly, L. A., Tonnesen, M. G., Sandhaus, R. A., Haslett, C., Guthrie, L. A., Johnston, R. B., Jr., Henson, P. M., and Worthen, G. S. (1986). Neutrophil-mediated injury to endothelial cells. Enhancement by endotoxin and essential role of neutrophil elastase. *J. Clin. Invest.* **77,** 1233–1243.
31. Herrera-Velit, P., and Reiner, N. E. (1996). Bacterial lipopolysaccharide induces the association and coordinate activation of p53/56lyn and phosphatidylinositol 3-kinase in human monocytes. *J. Immunol.* **156,** 1157–1165.
32. Liu, M. K., Herrera-Velit, P., Brownsey, R. W., and Reiner, N. E. (1994). CD14-dependent activation of protein kinase C and mitogen-activated protein kinases (p42 and p44) in human monocytes treated with bacterial lipopolysaccharide. *J. Immunol.* **153,** 2642–2652.
33. McCall, C. E., Grosso-Wilmoth, L. M., LaRue, K., Guzman, R. N., and Cousart, S. L. (1993). Tolerance to endotoxin-induced expression of the interleukin-1 beta gene in blood neutrophils of humans with the sepsis syndrome. *J. Clin. Invest.* **91,** 853–861.
34. LaRue, K. E., and McCall, C. E. (1994). A labile transcriptional repressor modulates endotoxin tolerance. *J. Exp. Med.* **180,** 2269–2275.
35. Bone, R. C. (1996). Why sepsis trials fail. *JAMA* **276,** 565–566.
36. Hmama, Z., Knutson, K. L., Herrera-Velit, P., Nandan, D., and Reiner, N. E. (1999). Monocyte adherence induced by lipopolysaccharide involves CD14, LFA-1, and cytohesin-1. Regulation by Rho and phosphatidylinositol 3-kinase. *J. Biol. Chem.* **274,** 1050–1057.

37. Stefanova, I., Corcoran, M. L., Horak, E. M., Wahl, L. M., Bolen, J. B., and Horak, I. D. (1993). Lipopolysaccharide induces activation of CD14-associated protein tyrosine kinase p53/56lyn. *J. Biol. Chem.* **268**, 20725–20728.
38. Shapira, L., Takashiba, S., Champagne, C., Amar, S., and Van Dyke, T. E. (1994). Involvement of protein kinase C and protein tyrosine kinase in lipopolysaccharide-induced TNF-alpha and IL-1 beta production by human monocytes. *J. Immunol.* **153**, 1818–1824.
39. Wright, S. D., Ramos, R. A., Tobias, P. S., Ulevitch, R. J., and Mathison, J. C. (1990). CD14, a receptor for complexes of lipopolysaccharide (LPS) and LPS binding protein. *Science* **249**, 1431–1433.
40. Ulevitch, R. J., and Tobias, P. S. (1994). Recognition of endotoxin by cells leading to transmembrane signaling. *Curr. Opin. Immunol.* **6**, 125–130.
41. Qureshi, S. T., Gros, P., and Malo, D. (1999). Host resistance to infection: Genetic control of lipopolysaccharide responsiveness by TOLL-like receptor genes. *Trends Genet.* **15**, 291–294.
42. Poltorak, A., He, X., Smirnova, I., Liu, M. Y., Van Huffel, C., Du, X., Birdwell, D., Alejos, E., Silva, M., Galanos, C., Freudenberg, M., Ricciardi-Castagnoli, P. *et al.* (1998). Defective LPS signaling in C3H/HeJ and C57BL/10ScCr mice: Mutations in Tlr4 gene. *Science* **282**, 2085–2088.
43. Zhang, F. X., Kirschning, C. J., Mancinelli, R., Xu, X. P., Jin, Y., Faure, E., Mantovani, A., Rothe, M., Muzio, M., and Arditi, M. (1999). Bacterial lipopolysaccharide activates nuclear factor-kappaB through interleukin-1 signaling mediators in cultured human dermal endothelial cells and mononuclear phagocytes. *J. Biol. Chem.* **274**, 7611–7614.
44. Fitzgerald, K. A., Palsson-McDermott, E. M., Bowie, A. G., Jefferies, C. A., Mansell, A. S., Brady, G., Brint, E., Dunne, A., Gray, P., Harte, M. T., McMurray, D., Smith, D. E. *et al.* (2001). Mal (MyD88-adapter-like) is required for Toll-like receptor-4 signal transduction. *Nature* **413**, 78–83.
45. Horng, T., Barton, G. M., and Medzhitov, R. (2001). TIRAP: An adapter molecule in the Toll signaling pathway. *Nat. Immunol.* **2**, 835–841.
46. Flaherty, S. F., Golenbock, D. T., Milham, F. H., and Ingalls, R. R. (1997). CD11/CD18 leukocyte integrins: New signaling receptors for bacterial endotoxin. *J. Surg. Res.* **73**, 85–89.
47. Ingalls, R. R., Arnaout, M. A., and Golenbock, D. T. (1997). Outside-in signaling by lipopolysaccharide through a tailless integrin. *J. Immunol.* **159**, 433–438.
48. Ingalls, R. R., and Golenbock, D. T. (1995). CD11c/CD18, a transmembrane signaling receptor for lipopolysaccharide. *J. Exp. Med.* **181**, 1473–1479.
49. Medvedev, A. E., Flo, T., Ingalls, R. R., Golenbock, D. T., Teti, G., Vogel, S. N., and Espevik, T. (1998). Involvement of CD14 and complement receptors CR3 and CR4 in nuclear factor-kappaB activation and TNF production induced by lipopolysaccharide and group B streptococcal cell walls. *J. Immunol.* **160**, 4535–4542.
50. Nagai, Y., Akashi, S., Nagafuku, M., Ogata, M., Iwakura, Y., Akira, S., Kitamura, T., Kosugi, A., Kimoto, M., and Miyake, K. (2002). Essential role of MD-2 in LPS responsiveness and TLR4 distribution. *Nat. Immunol.* **3**, 667–672.
51. Schromm, A. B., Lien, E., Henneke, P., Chow, J. C., Yoshimura, A., Heine, H., Latz, E., Monks, B. G., Schwartz, D. A., Miyake, K., and Golenbock, D. T. (2001). Molecular genetic analysis of an endotoxin nonresponder mutant cell line: A point mutation in a conserved region of MD-2 abolishes endotoxin-induced signaling. *J. Exp. Med.* **194**, 79–88.
52. Shimazu, R., Akashi, S., Ogata, H., Nagai, Y., Fukudome, K., Miyake, K., and Kimoto, M. (1999). MD-2, a molecule that confers lipopolysaccharide responsiveness on Toll-like receptor 4. *J. Exp. Med.* **189**, 1777–1782.
53. Re, F., and Strominger, J. L. (2003). Separate functional domains of human MD-2 mediate Toll-like receptor 4-binding and lipopolysaccharide responsiveness. *J. Immunol.* **171**, 5272–5276.

54. Muzio, M., Natoli, G., Saccani, S., Levrero, M., and Mantovani, A. (1998). The human toll signaling pathway: Divergence of nuclear factor kappaB and JNK/SAPK activation upstream of tumor necrosis factor receptor-associated factor 6 (TRAF6). *J. Exp. Med.* **187,** 2097–2101.
55. Medzhitov, R., Preston-Hurlburt, P., Kopp, E., Stadlen, A., Chen, C., Ghosh, S., and Janeway, C. A., Jr. (1998). MyD88 is an adaptor protein in the hToll/IL-1 receptor family signaling pathways. *Mol. Cell* **2,** 253–258.
56. Burns, K., Martinon, F., Esslinger, C., Pahl, H., Schneider, P., Bodmer, J. L., Di Marco, F., French, L., and Tschopp, J. (1998). MyD88, an adapter protein involved in interleukin-1 signaling. *J. Biol. Chem.* **273,** 12203–12209.
57. Wesche, H., Henzel, W. J., Shillinglaw, W., Li, S., and Cao, Z. (1997). MyD88: An adapter that recruits IRAK to the IL-1 receptor complex. *Immunity* **7,** 837–847.
58. Muzio, M., Ni, J., Feng, P., and Dixit, V. M. (1997). IRAK (Pelle) family member IRAK-2 and MyD88 as proximal mediators of IL-1 signaling. *Science* **278,** 1612–1615.
59. Deng, L., Wang, C., Spencer, E., Yang, L., Braun, A., You, J., Slaughter, C., Pickart, C., and Chen, Z. J. (2000). Activation of the IkappaB kinase complex by TRAF6 requires a dimeric ubiquitin-conjugating enzyme complex and a unique polyubiquitin chain. *Cell* **103,** 351–361.
60. Lee, J., Mira-Arbibe, L., and Ulevitch, R. J. (2000). TAK1 regulates multiple protein kinase cascades activated by bacterial lipopolysaccharide. *J. Leukoc. Biol.* **68,** 909–915.
61. Kishimoto, K., Matsumoto, K., and Ninomiya-Tsuji, J. (2000). TAK1 mitogen-activated protein kinase is activated by autophosphorylation within its activation loop. *J. Biol. Chem.* **275,** 7359–7364.
62. Cordle, S. R., Donald, R., Read, M. A., and Hawiger, J. (1993). Lipopolysaccharide induces phosphorylation of MAD3 and activation of c-Rel and related NF-kappa B proteins in human monocytic THP-1 cells. *J. Biol. Chem.* **268,** 11803–11810.
63. Haeffner, A., Thieblemont, N., Deas, O., Marelli, O., Charpentier, B., Senik, A., Wright, S. D., Haeffner-Cavaillon, N., and Hirsch, F. (1997). Inhibitory effect of growth hormone on TNF-alpha secretion and nuclear factor-kappaB translocation in lipopolysaccharide-stimulated human monocytes. *J. Immunol.* **158,** 1310–1314.
64. Read, M. A., Cordle, S. R., Veach, R. A., Carlisle, C. D., and Hawiger, J. (1993). Cell-free pool of CD14 mediates activation of transcription factor NF-kappa B by lipopolysaccharide in human endothelial cells. *Proc. Natl. Acad. Sci. USA* **90,** 9887–9891.
65. Baeuerle, P. A. (1998). Pro-inflammatory signaling: Last pieces in the NF-kappaB puzzle? *Curr. Biol.* **8,** R19–R22.
66. Baeuerle, P. A., and Baltimore, D. (1996). NF-kappa B: Ten years after. *Cell* **87,** 13–20.
67. Guha, M., and Mackman, N. (2001). LPS induction of gene expression in human monocytes. *Cell Signal.* **13,** 85–94.
68. Weinstein, S. L., Gold, M. R., and DeFranco, A. L. (1991). Bacterial lipopolysaccharide stimulates protein tyrosine phosphorylation in macrophages. *Proc. Natl. Acad. Sci. USA* **88,** 4148–4152.
69. Weinstein, S. L., June, C. H., and DeFranco, A. L. (1993). Lipopolysaccharide-induced protein tyrosine phosphorylation in human macrophages is mediated by CD14. *J. Immunol.* **151,** 3829–3838.
70. English, B. K., Ihle, J. N., Myracle, A., and Yi, T. (1993). Hck tyrosine kinase activity modulates tumor necrosis factor production by murine macrophages. *J. Exp. Med.* **178,** 1017–1022.
71. Beaty, C. D., Franklin, T. L., Uehara, Y., and Wilson, C. B. (1994). Lipopolysaccharide-induced cytokine production in human monocytes: Role of tyrosine phosphorylation in transmembrane signal transduction. *Eur. J. Immunol.* **24,** 1278–1284.
72. Weinstein, S. L., Sanghera, J. S., Lemke, K., DeFranco, A. L., and Pelech, S. L. (1992). Bacterial lipopolysaccharide induces tyrosine phosphorylation and activation of mitogen-activated protein kinases in macrophages. *J. Biol. Chem.* **267,** 14955–14962.

73. Dong, Z., Qi, X., and Fidler, I. J. (1993). Tyrosine phosphorylation of mitogen-activated protein kinases is necessary for activation of murine macrophages by natural and synthetic bacterial products. *J. Exp. Med.* **177,** 1071–1077.
74. Cantley, L. C., Auger, K. R., Carpenter, C., Duckworth, B., Graziani, A., Kapeller, R., and Soltoff, S. (1991). Oncogenes and signal transduction. *Cell* **64,** 281–302.
75. Yamanashi, Y., Fukui, Y., Wongsasant, B., Kinoshita, Y., Ichimori, Y., Toyoshima, K., and Yamamoto, T. (1992). Activation of Src-like protein-tyrosine kinase Lyn and its association with phosphatidylinositol 3-kinase upon B-cell antigen receptor-mediated signaling. *Proc. Natl. Acad. Sci. USA* **89,** 1118–1122.
76. Nakanishi, H., Brewer, K. A., and Exton, J. H. (1993). Activation of the zeta isozyme of protein kinase C by phosphatidylinositol 3,4,5-trisphosphate. *J. Biol. Chem.* **268,** 13–16.
77. Toker, A., Meyer, M., Reddy, K. K., Falck, J. R., Aneja, R., Aneja, S., Parra, A., Burns, D. J., Ballas, L. M., and Cantley, L. C. (1994). Activation of protein kinase C family members by the novel polyphosphoinositides PtdIns-3,4-P2 and PtdIns-3,4,5-P3. *J. Biol. Chem.* **269,** 32358–32367.
78. Karima, R., Matsumoto, S., Higashi, H., and Matsushima, K. (1999). The molecular pathogenesis of endotoxic shock and organ failure. *Mol. Med. Today* **5,** 123–132.
79. Mayeux, P. R. (1997). Pathobiology of lipopolysaccharide. *J. Toxicol. Environ. Health* **51,** 415–435.
80. Li, L., Cousart, S., Hu, J., and McCall, C. E. (2000). Characterization of interleukin-1 receptor-associated kinase in normal and endotoxin-tolerant cells. *J. Biol. Chem.* **275,** 23340–23345.
81. Swantek, J. L., Tsen, M. F., Cobb, M. H., and Thomas, J. A. (2000). IL-1 receptor-associated kinase modulates host responsiveness to endotoxin. *J. Immunol.* **164,** 4301–4306.
82. Wesche, H., Gao, X., Li, X., Kirschning, C. J., Stark, G. R., and Cao, Z. (1999). IRAK-M is a novel member of the Pelle/interleukin-1 receptor-associated kinase (IRAK) family. *J. Biol. Chem.* **274,** 19403–19410.
83. Hu, J., Jacinto, R., McCall, C., and Li, L. (2002). Regulation of IL-1 receptor-associated kinases by lipopolysaccharide. *J. Immunol.* **168,** 3910–3914.
84. Currie, R. A., Walker, K. S., Gray, A., Deak, M., Casamayor, A., Downes, C. P., Cohen, P., Alessi, D. R., and Lucocq, J. (1999). Role of phosphatidylinositol 3,4,5-trisphosphate in regulating the activity and localization of 3-phosphoinositide-dependent protein kinase-1. *Biochem. J.* **337**(Pt. 3), 575–583.
85. Bevilacqua, M. P., Nelson, R. M., Mannori, G., and Cecconi, O. (1994). Endothelial-leukocyte adhesion molecules in human disease. *Annu. Rev. Med.* **45,** 361–378.
86. Bevilacqua, M. P., Pober, J. S., Wheeler, M. E., Cotran, R. S., and Gimbrone, M. A., Jr. (1985). Interleukin 1 acts on cultured human vascular endothelium to increase the adhesion of polymorphonuclear leukocytes, monocytes, and related leukocyte cell lines. *J. Clin. Invest.* **76,** 2003–2011.
87. Wallis, W. J., Beatty, P. G., Ochs, H. D., and Harlan, J. M. (1985). Human monocyte adherence to cultured vascular endothelium: Monoclonal antibody-defined mechanisms. *J. Immunol.* **135,** 2323–2330.
88. Dustin, M. L., and Springer, T. A. (1989). T-cell receptor cross-linking transiently stimulates adhesiveness through LFA-1. *Nature* **341,** 619–624.
89. van Kooyk, Y., van de Wiel-van, K., Weder, P., Kuijpers, T. W., and Figdor, C. G. (1989). Enhancement of LFA-1-mediated cell adhesion by triggering through CD2 or CD3 on T lymphocytes. *Nature* **342,** 811–813.
90. Springer, T. A. (1990). Adhesion receptors of the immune system. *Nature* **346,** 425–434.
91. Lub, M., van Kooyk, Y., and Figdor, C. G. (1995). Ins and outs of LFA-1. *Immunol. Today* **16,** 479–483.

92. Staunton, D. E., Marlin, S. D., Stratowa, C., Dustin, M. L., and Springer, T. A. (1988). Primary structure of ICAM-1 demonstrates interaction between members of the immunoglobulin and integrin supergene families. *Cell* **52,** 925–933.
93. Lollo, B. A., Chan, K. W., Hanson, E. M., Moy, V. T., and Brian, A. A. (1993). Direct evidence for two affinity states for lymphocyte function-associated antigen 1 on activated T cells. *J. Biol. Chem.* **268,** 21693–21700.
94. Beekhuizen, H., Blokland, I., and van Furth, R. (1993). Cross-linking of CD14 molecules on monocytes results in a CD11/. *J. Immunol.* **150,** 950–959.
95. Toker, A., Bachelot, C., Chen, C. S., Falck, J. R., Hartwig, J. H., Cantley, L. C., and Kovacsovics, T. J. (1995). Phosphorylation of the platelet p47 phosphoprotein is mediated by the lipid products of phosphoinositide 3-kinase. *J. Biol. Chem.* **270,** 29525–29531.
96. Shimizu, Y., Mobley, J. L., Finkelstein, L. D., and Chan, A. S. (1995). A role for phosphatidylinositol 3-kinase in the regulation of beta 1 integrin activity by the CD2 antigen. *J. Cell Biol.* **131,** 1867–1880.
97. Capodici, C., Hanft, S., Feoktistov, M., and Pillinger, M. H. (1998). Phosphatidylinositol 3-kinase mediates chemoattractant-stimulated, CD11b/CD18-dependent cell–cell adhesion of human neutrophils: Evidence for an ERK-independent pathway. *J. Immunol.* **160,** 1901–1909.
98. Klarlund, J. K., Guilherme, A., Holik, J. J., Virbasius, J. V., Chawla, A., and Czech, M. P. (1997). Signaling by phosphoinositide-3,4,5-trisphosphate through proteins containing pleckstrin and Sec7 homology domains. *Science* **275,** 1927–1930.
99. Kolanus, W., Nagel, W., Schiller, B., Zeitlmann, L., Godar, S., Stockinger, H., and Seed, B. (1996). Alpha L beta 2 integrin/LFA-1 binding to ICAM-1 induced by cytohesin-1, a cytoplasmic regulatory molecule. *Cell* **86,** 233–242.
100. Zhang, J., King, W. G., Dillon, S., Hall, A., Feig, L., and Rittenhouse, S. E. (1993). Activation of platelet phosphatidylinositide 3-kinase requires the small GTP-binding protein Rho. *J. Biol. Chem.* **268,** 22251–22254.
101. Gomez, J., Garcia, A., Borlado, L., Bonay, P., Martinez, A., Silva, A., Fresno, M., Carrera, A. C., Eicher-Streiber, C., and Rebollo, A. (1997). IL-2 signaling controls actin organization through Rho-like protein family, phosphatidylinositol 3-kinase, and protein kinase C-zeta. *J. Immunol.* **158,** 1516–1522.
102. Zheng, Y., Bagrodia, S., and Cerione, R. A. (1994). Activation of phosphoinositide 3-kinase activity by Cdc42Hs binding to p85. *J. Biol. Chem.* **269,** 18727–18730.
103. Just, I., Fritz, G., Aktories, K., Giry, M., Popoff, M. R., Boquet, P., Hegenbarth, S., and Eichel-Streiber, C. (1994). Clostridium difficile toxin B acts on the GTP-binding protein Rho. *J. Biol. Chem.* **269,** 10706–10712.
104. Just, I., Richter, H. P., Prepens, U., Eichel-Streiber, C., and Aktories, K. (1994). Probing the action of Clostridium difficile toxin B in *Xenopus laevis* oocytes. *J. Cell Sci.* **107**(Pt. 6), 1653–1659.
105. Boyle, I. T., Miravet, L., Gray, R. W., Holick, M. F., and Deluca, H. F. (1972). The response of intestinal calcium transport to 25-hydroxy and 1,25-dihydroxy vitamin D in nephrectomized rats. *Endocrinology* **90,** 605–608.
106. Nemere, I. (1995). Nongenomic effects of 1,25-dihydroxyvitamin D3: Potential relation of a plasmalemmal receptor to the acute enhancement of intestinal calcium transport in chick. *J. Nutr.* **125,** 1695S–1698S.
107. Holick, M. F., Garabedian, M., and Deluca, H. F. (1972). 1,25-dihydroxycholecalciferol: Metabolite of vitamin D3 active on bone in anephric rats. *Science* **176,** 1146–1147.
108. Raisz, L. G., Trummel, C. L., Holick, M. F., and Deluca, H. F. (1972). 1,25-dihydroxycholecalciferol: A potent stimulator of bone resorption in tissue culture. *Science* **175,** 768–769.
109. Takahashi, T., Nakamura, K., and Iho, S. (1997). Differentiation of myeloid cells and 1,25-dihydroxyvitamin D3. *Leuk. Lymphoma* **27,** 25–33.

110. Davies, P. D. (1985). A possible link between vitamin D deficiency and impaired host defence to Mycobacterium tuberculosis. *Tubercle.* **66,** 301–306.
111. Roy, S., Frodsham, A., Saha, B., Hazra, S. K., Mascie-Taylor, C. G., and Hill, A. V. (1999). Association of vitamin D receptor genotype with leprosy type. *J. Infect. Dis.* **179,** 187–191.
112. Wilkinson, R. J., Llewelyn, M., Toossi, Z., Patel, P., Pasvol, G., Lalvani, A., Wright, D., Latif, M., and Davidson, R. N. (2000). Influence of vitamin D deficiency and vitamin D receptor polymorphisms on tuberculosis among Gujarati Asians in west London: A case-control study. *Lancet* **355,** 618–621.
113. Crowle, A. J., Ross, E. J., and May, M. H. (1987). Inhibition by 1,25(OH)2-vitamin D3 of the multiplication of virulent tubercle bacilli in cultured human macrophages. *Infect. Immun.* **55,** 2945–2950.
114. Rook, G. A., Steele, J., Fraher, L., Barker, S., Karmali, R., O'Riordan, J., and Stanford, J. (1986). Vitamin D3, gamma interferon, and control of proliferation of Mycobacterium tuberculosis by human monocytes. *Immunology* **57,** 159–163.
115. Malloy, P. J., and Feldman, D. (1999). Vitamin D resistance. *Am. J. Med.* **106,** 355–370.
116. Liu, M., Lee, M. H., Cohen, M., Bommakanti, M., and Freedman, L. P. (1996). Transcriptional activation of the Cdk inhibitor p21 by vitamin D3 leads to the induced differentiation of the myelomonocytic cell line U937. *Genes Dev.* **10,** 142–153.
117. Brackman, D., Lund-Johansen, F., and Aarskog, D. (1995). Expression of leukocyte differentiation antigens during the differentiation of HL-60 cells induced by 1,25-dihydroxyvitamin D3: Comparison with the maturation of normal monocytic and granulocytic bone marrow cells. *J. Leukoc. Biol.* **58,** 547–555.
118. Oberg, F., Botling, J., and Nilsson, K. (1993). Functional antagonism between vitamin D3 and retinoic acid in the regulation of CD14 and CD23 expression during monocytic differentiation of U-937 cells. *J. Immunol.* **150,** 3487–3495.
119. Schwende, H., Fitzke, E., Ambs, P., and Dieter, P. (1996). Differences in the state of differentiation of THP-1 cells induced by phorbol ester and 1,25-dihydroxyvitamin D3. *J. Leukoc. Biol.* **59,** 555–561.
120. Abe, E., Miyaura, C., Sakagami, H., Takeda, M., Konno, K., Yamazaki, T., Yoshiki, S., and Suda, T. (1981). Differentiation of mouse myeloid leukemia cells induced by 1 alpha,25-dihydroxyvitamin D3. *Proc. Natl. Acad. Sci. USA* **78,** 4990–4994.
121. Ziegler-Heitbrock, H. W., and Ulevitch, R. J. (1993). CD14: Cell surface receptor and differentiation marker. *Immunol. Today* **14,** 121–125.
122. Antal-Szalmas, P., Strijp, J. A., Weersink, A. J., Verhoef, J., and Van Kessel, K. P. (1997). Quantitation of surface CD14 on human monocytes and neutrophils. *J. Leukoc. Biol.* **61,** 721–728.
123. Zhang, D. E., Hetherington, C. J., Gonzalez, D. A., Chen, H. M., and Tenen, D. G. (1994). Regulation of CD14 expression during monocytic differentiation induced with 1 alpha,25-dihydroxyvitamin D3. *J. Immunol.* **153,** 3276–3284.
124. Martin, T. R., Mongovin, S. M., Tobias, P. S., Mathison, J. C., Moriarty, A. M., Leturcq, D. J., and Ulevitch, R. J. (1994). The CD14 differentiation antigen mediates the development of endotoxin responsiveness during differentiation of mononuclear phagocytes. *J. Leukoc. Biol.* **56,** 1–9.
125. Bouillon, R., Okamura, W. H., and Norman, A. W. (1995). Structure-function relationships in the vitamin D endocrine system. *Endocr. Rev.* **16,** 200–257.
126. Zhang, D. E., Hetherington, C. J., Tan, S., Dziennis, S. E., Gonzalez, D. A., Chen, H. M., and Tenen, D. G. (1994). Sp1 is a critical factor for the monocytic specific expression of human CD14. *J. Biol. Chem.* **269,** 11425–11434.
127. Chen, H. M., Pahl, H. L., Scheibe, R. J., Zhang, D. E., and Tenen, D. G. (1993). The Sp1 transcription factor binds the CD11b promoter specifically in myeloid cells *in vivo* and is essential for myeloid-specific promoter activity. *J. Biol. Chem.* **268,** 8230–8239.

128. Gniadecki, R. (1998). Nongenomic signaling by vitamin D: A new face of Src. *Biochem. Pharmacol.* **56,** 1273–1277.
129. Kharbanda, S., Saleem, A., Emoto, Y., Stone, R., Rapp, U., and Kufe, D. (1994). Activation of Raf-1 and mitogen-activated protein kinases during monocytic differentiation of human myeloid leukemia cells. *J. Biol. Chem.* **269,** 872–878.
130. Marcinkowska, E., Wiedlocha, A., and Radzikowski, C. (1997). 1,25-Dihydroxyvitamin D3 induced activation and subsequent nuclear translocation of MAPK is upstream regulated by PKC in HL-60 cells. *Biochem. Biophys. Res. Commun.* **241,** 419–426.
131. Gniadecki, R. (1996). Activation of Raf-mitogen-activated protein kinase signaling pathway by 1,25-dihydroxyvitamin D3 in normal human keratinocytes. *J. Invest. Dermatol.* **106,** 1212–1217.
132. de Boland, A. R., and Norman, A. W. (1998). 1alpha,25(OH)2-vitamin D3 signaling in chick enterocytes: Enhancement of tyrosine phosphorylation and rapid stimulation of mitogen-activated protein (MAP) kinase. *J. Cell Biochem.* **69,** 470–482.
133. Kimura, K., Hattori, S., Kabuyama, Y., Shizawa, Y., Takayanagi, J., Nakamura, S., Toki, S., Matsuda, Y., Onodera, K., and Fukui, Y. (1994). Neurite outgrowth of PC12 cells is suppressed by wortmannin, a specific inhibitor of phosphatidylinositol 3-kinase. *J. Biol. Chem.* **269,** 18961–18967.
134. Kaliman, P., Vinals, F., Testar, X., Palacin, M., and Zorzano, A. (1996). Phosphatidylinositol 3-kinase inhibitors block differentiation of skeletal muscle cells. *J. Biol. Chem.* **271,** 19146–19151.
135. Rohrschneider, L. R., Bourette, R. P., Lioubin, M. N., Algate, P. A., Myles, G. M., and Carlberg, K. (1997). Growth and differentiation signals regulated by the M-CSF receptor. *Mol. Reprod. Dev.* **46,** 96–103.
136. Christakos, S., Raval-Pandya, M., Wernyj, R. P., and Yang, W. (1996). Genomic mechanisms involved in the pleiotropic actions of 1,25-dihydroxyvitamin D3. *Biochem. J.* **316**(Pt. 2), 361–371.
137. Sly, L. M., Lopez, M., Nauseef, W. M., and Reiner, N. E. (2001). 1alpha,25-Dihydroxyvitamin D3-induced monocyte antimycobacterial activity is regulated by phosphatidylinositol 3-kinase and mediated by the NADPH-dependent phagocyte oxidase. *J. Biol. Chem.* **276,** 35482–35493.
138. Clark, R. A., Volpp, B. D., Leidal, K. G., and Nauseef, W. M. (1990). Two cytosolic components of the human neutrophil respiratory burst oxidase translocate to the plasma membrane during cell activation. *J. Clin. Invest.* **85,** 714–721.
139. DeLeo, F. R., Allen, L. A., Apicella, M., and Nauseef, W. M. (1999). NADPH oxidase activation and assembly during phagocytosis. *J. Immunol.* **163,** 6732–6740.
140. Lang, M. L., and Kerr, M. A. (1997). Neutrophil Fc receptors signal through PI 3-kinase to trigger a respiratory burst. *Biochem. Soc. Trans.* **25,** S603.
141. Santoro, P., Cacciapuoti, C., Palumbo, A., Graziano, D., Annunziata, S., Capasso, L., Formisano, S., and Ciccimarra, F. (1998). Effects of wortmannin on human neutrophil respiratory burst and phagocytosis. *Ital. J. Biochem.* **47,** 13–18.
142. Yang, M., Wu, W., and Mirocha, C. J. (1996). Wortmannin inhibits the production of reactive oxygen and nitrogen intermediates and the killing of the *Saccharomyces cerevisiae* by isolated chicken macrophages. *Immunopharmacol. Immunotoxicol.* **18,** 597–608.
143. Fierer, J., and Guiney, D. G. (2001). Diverse virulence traits underlying different clinical outcomes of *Salmonella* infection. *J. Clin. Invest.* **107,** 775–780.
144. De Groote, M. A., Ochsner, U. A., Shiloh, M. U., Nathan, C., McCord, J. M., Dinauer, M. C., Libby, S. J., Vazquez-Torres, A., Xu, Y., and Fang, F. C. (1997). Periplasmic superoxide dismutase protects *Salmonella* from products of phagocyte NADPH-oxidase and nitric oxide synthase. *Proc. Natl. Acad. Sci. USA* **94,** 13997–14001.

145. Mastroeni, P., Vazquez-Torres, A., Fang, F. C., Xu, Y., Khan, S., Hormaeche, C. E., and Dougan, G. (2000). Antimicrobial actions of the NADPH phagocyte oxidase and inducible nitric oxide synthase in experimental salmonellosis. II. Effects on microbial proliferation and host survival *in vivo*. *J. Exp. Med.* **192**, 237–248.
146. Vazquez-Torres, A., Jones-Carson, J., Mastroeni, P., Ischiropoulos, H., and Fang, F. C. (2000). Antimicrobial actions of the NADPH phagocyte oxidase and inducible nitric oxide synthase in experimental salmonellosis. I. Effects on microbial killing by activated peritoneal macrophages *in vitro*. *J. Exp. Med.* **192**, 227–236.
147. Sly, L. M., Guiney, D. G., and Reiner, N. E. (2002). *Salmonella enterica* serovar Typhimurium periplasmic superoxide dismutases SodCI and SodCII are required for protection against the phagocyte oxidative burst. *Infect. Immun.* **70**, 5312–5315.
148. Farrant, J. L., Sansone, A., Canvin, J. R., Pallen, M. J., Langford, P. R., Wallis, T. S., Dougan, G., and Kroll, J. S. (1997). Bacterial copper- and zinc-cofactored superoxide dismutase contributes to the pathogenesis of systemic salmonellosis. *Mol. Microbiol.* **25**, 785–796.
149. Gatfield, J., and Pieters, J. (2000). Essential role for cholesterol in entry of mycobacteria into macrophages. *Science* **288**, 1647–1650.
150. Ernst, J. D. (1998). Macrophage receptors for *Mycobacterium tuberculosis*. *Infect. Immun.* **66**, 1277–1281.
151. Rook, G. A., Martinelli, R., and Brunet, L. R. (2003). Innate immune responses to mycobacteria and the downregulation of atopic responses. *Curr. Opin. Allergy Clin. Immunol.* **3**, 337–342.
152. Noorman, F., Braat, E. A., Barrett-Bergshoeff, M., Barbe, E., van Leeuwen, A., Lindeman, J., and Rijken, D. C. (1997). Monoclonal antibodies against the human mannose receptor as a specific marker in flow cytometry and immunohistochemistry for macrophages. *J. Leukoc. Biol.* **61**, 63–72.
153. Le, C. V., Carreno, S., Moisand, A., Bordier, C., and Maridonneau-Parini, I. (2002). Complement receptor 3 (CD11b/CD18) mediates type I and type II phagocytosis during nonopsonic and opsonic phagocytosis, respectively. *J. Immunol.* **169**, 2003–2009.
154. Khanna, K. V., Choi, C. S., Gekker, G., Peterson, P. K., and Molitor, T. W. (1996). Differential infection of porcine alveolar macrophage subpopulations by nonopsonized *Mycobacterium bovis* involves CD14 receptors. *J. Leukoc. Biol.* **60**, 214–220.
155. Dib, K. (2000). BETA 2 integrin signaling in leukocytes. *Front Biosci.* **5**, D438–D451.
156. Kucik, D. F. (2002). Rearrangement of integrins in avidity regulation by leukocytes. *Immunol. Res.* **26**, 199–206.
157. Strohmeier, G. R., and Fenton, M. J. (1999). Roles of lipoarabinomannan in the pathogenesis of tuberculosis. *Microbes. Infect.* **1**, 709–717.
158. Schlesinger, L. S., Hull, S. R., and Kaufman, T. M. (1994). Binding of the terminal mannosyl units of lipoarabinomannan from a virulent strain of *Mycobacterium tuberculosis* to human macrophages. *J. Immunol.* **152**, 4070–4079.
159. Dierks, H., Kolanus, J., and Kolanus, W. (2001). Actin cytoskeletal association of cytohesin-1 is regulated by specific phosphorylation of its carboxyl-terminal polybasic domain. *J. Biol. Chem.* **276**, 37472–37481.
160. Nagel, W., Zeitlmann, L., Schilcher, P., Geiger, C., Kolanus, J., and Kolanus, W. (1998). Phosphoinositide 3-OH kinase activates the beta2 integrin adhesion pathway and induces membrane recruitment of cytohesin-1. *J. Biol. Chem.* **273**, 14853–14861.
161. Sendide, K., Reiner, N. E., Lee, J. S., Bourgoin, S., Talal, A., and Hmama, Z. (2005). Crosstalk between CD14 and complement receptor 3 promotes phagocytosis of mycobacteria: Regulation by phosphatidylinositol 3-kinase and cytohesin-1. *J. Immunol.* **174**, 4210–4219.
162. Peterson, P. K., Gekker, G., Hu, S., Sheng, W. S., Anderson, W. R., Ulevitch, R. J., Tobias, P. S., Gustafson, K. V., Molitor, T. W., and Chao, C. C. (1995). CD14 receptor-mediated

uptake of nonopsonized *Mycobacterium tuberculosis* by human microglia. *Infect. Immun.* **63,** 1598–1602.

163. Navarro, A., Anand-Apte, B., Tanabe, Y., Feldman, G., and Larner, A. C. (2003). A PI-3 kinase-dependent, Stat1-independent signaling pathway regulates interferon-stimulated monocyte adhesion. *J. Leukoc. Biol.* **73,** 540–545.

164. Stephens, L., Ellson, C., and Hawkins, P. (2002). Roles of PI3Ks in leukocyte chemotaxis and phagocytosis. *Curr. Opin. Cell Biol.* **14,** 203–213.

165. Lutz, M. A., and Correll, P. H. (2003). Activation of CR3-mediated phagocytosis by MSP requires the RON receptor, tyrosine kinase activity, phosphatidylinositol 3-kinase, and protein kinase C zeta. *J. Leukoc. Biol.* **73,** 802–814.

166. Hmama, Z., Sendide, K., Talal, A., Garcia, R., Dobos, K., and Reiner, N. E. (2004). Quantitative analysis of phagolysosome fusion in intact cells: Inhibition by mycobacterial lipoarabinomannan and rescue by an 1alpha,25-dihydroxyvitamin D3-phosphoinositide 3-kinase pathway. *J. Cell Sci.* **117,** 2131–2140.

167. Fratti, R. A., Backer, J. M., Gruenberg, J., Corvera, S., and Deretic, V. (2001). Role of phosphatidylinositol 3-kinase and Rab5 effectors in phagosomal biogenesis and mycobacterial phagosome maturation arrest. *J. Cell Biol.* **154,** 631–644.

168. Rupper, A. C., Rodriguez-Paris, J. M., Grove, B. D., and Cardelli, J. A. (2001). p110-related PI 3-kinases regulate phagosome-phagosome fusion and phagosomal pH through a PKB/Akt dependent pathway in *Dictyostelium*. *J. Cell Sci.* **114,** 1283–1295.

169. Chen, R., Kang, V. H., Chen, J., Shope, J. C., Torabinejad, J., DeWald, D. B., and Prestwich, G. D. (2002). A monoclonal antibody to visualize PtdIns(3,4,5)P(3) in cells. *J. Histochem. Cytochem.* **50,** 697–708.

170. Geiger, C., Nagel, W., Boehm, T., van Kooyk, Y., Figdor, C. G., Kremmer, E., Hogg, N., Zeitlmann, L., Dierks, H., Weber, K. S., and Kolanus, W. (2000). Cytohesin-1 regulates beta-2 integrin-mediated adhesion through both ARF-GEF function and interaction with LFA-1. *EMBO J.* **19,** 2525–2536.

171. Nagel, W., Schilcher, P., Zeitlmann, L., and Kolanus, W. (1998). The PH domain and the polybasic c domain of cytohesin-1 cooperate specifically in plasma membrane association and cellular function. *Mol. Biol. Cell* **9,** 1981–1994.

172. Bi, L., Okabe, I., Bernard, D. J., Wynshaw-Boris, A., and Nussbaum, R. L. (1999). Proliferative defect and embryonic lethality in mice homozygous for a deletion in the p110alpha subunit of phosphoinositide 3-kinase. *J. Biol. Chem.* **274,** 10963–10968.

173. Bi, L., Okabe, I., Bernard, D. J., and Nussbaum, R. L. (2002). Early embryonic lethality in mice deficient in the p110beta catalytic subunit of PI 3-kinase. *Mamm. Genome* **13,** 169–172.

174. Okkenhaug, K., Bilancio, A., Farjot, G., Priddle, H., Sancho, S., Peskett, E., Pearce, W., Meek, S. E., Salpekar, A., Waterfield, M. D., Smith, A. J., and Vanhaesebroeck, B. (2002). Impaired B and T cell antigen receptor signaling in p110delta PI 3-kinase mutant mice. *Science* **297,** 1031–1034.

175. Jou, S. T., Carpino, N., Takahashi, Y., Piekorz, R., Chao, J. R., Carpino, N., Wang, D., and Ihle, J. N. (2002). Essential, nonredundant role for the phosphoinositide 3-kinase p110delta in signaling by the B-cell receptor complex. *Mol. Cell Biol.* **22,** 8580–8591.

176. Clayton, E., Bardi, G., Bell, S. E., Chantry, D., Downes, C. P., Gray, A., Humphries, L. A., Rawlings, D., Reynolds, H., Vigorito, E., and Turner, M. (2002). A crucial role for the p110delta subunit of phosphatidylinositol 3-kinase in B cell development and activation. *J. Exp. Med.* **196,** 753–763.

177. Kusumawati, A., Commes, T., Liautard, J. P., and Widada, J. S. (1999). Transfection of myelomonocytic cell lines: Cellular response to a lipid-based reagent and electroporation. *Anal. Biochem.* **269,** 219–221.

178. Liao, H. S., Kodama, T., Doi, T., Emi, M., Asaoka, H., Itakura, H., and Matsumoto, A. (1997). Novel elements located at −504 to −399 bp of the promoter region regulated the expression of the human macrophage scavenger receptor gene in murine macrophages. *J. Lipid Res.* **38**, 1433–1444.
179. Stripecke, R., Cardoso, A. A., Pepper, K. A., Skelton, D. C., Yu, X. J., Mascarenhas, L., Weinberg, K. I., Nadler, L. M., and Kohn, D. B. (2000). Lentiviral vectors for efficient delivery of CD80 and granulocyte-macrophage-colony-stimulating factor in human acute lymphoblastic leukemia and acute myeloid leukemia cells to induce antileukemic immune responses. *Blood* **96**, 1317–1326.
180. Bambacioni, F., Casati, C., Serafini, M., Manganini, M., Golay, J., and Introna, M. (2001). Lentiviral vectors show dramatically increased efficiency of transduction of human leukemic cell lines. *Haematologica* **86**, 1095–1096.
181. Introna, M., Barbui, A. M., Golay, J., Bambacioni, F., Schiro, R., Bernasconi, S., Breviario, F., Erba, E., Borleri, G., Barbui, T., Biondi, A., and Rambaldi, A. (1998). Rapid retroviral infection of human haemopoietic cells of different lineages: Efficient transfer in fresh T cells. *Br. J. Haematol.* **103**, 449–461.
182. Lee, J. S., Hmama, Z., Mui, A., and Reiner, N. E. (2004). Stable gene silencing in human monocytic cell lines using lentiviral-delivered small interference RNA. Silencing of the p110alpha isoform of phosphoinositide 3-kinase reveals differential regulation of adherence induced by 1alpha,25-dihydroxycholecalciferol and bacterial lipopolysaccharide. *J. Biol. Chem.* **279**, 9379–9388.
183. Shi, Y. (2003). Mammalian RNAi for the masses. *Trends Genet.* **19**, 9–12.
184. Buchschacher, G. L., Jr., and Wong-Staal, F. (2000). Development of lentiviral vectors for gene therapy for human diseases. *Blood* **95**, 2499–2504.

Selenocysteine Incorporation Machinery and the Role of Selenoproteins in Development and Health

Dolph L. Hatfield,[*]
Bradley A. Carlson,[*]
Xue-Ming Xu,[*] Heiko Mix,[†]
and Vadim N. Gladyshev[†]

[*]Molecular Biology of Selenium Section, Laboratory of Cancer Prevention, Center for Cancer Research, National Cancer Institute, National Institutes of Health, Bethesda, Maryland 20892

[†]Department of Biochemistry, University of Nebraska, Lincoln, Nebraska 68588

I. UGA Codes for Sec	99
II. Novelty of Eukaryotic Sec tRNA[Ser]Sec	99
A. tRNA and Gene Structures	99
B. Um34 Synthesis, a Highly Specialized Event in Sec tRNA[Ser]Sec Maturation	102
C. Um34 Sec tRNA[Ser]Sec Methylase	103
D. Amounts and Distributions of Sec tRNA[Ser]Sec Isoforms	104
E. Evolution of Sec tRNA[Ser]Sec and Occurrence of Sec Insertion Machinery	104
III. Sec Biosynthesis	109
A. Seryl-tRNA Synthetase	110
B. Phosphoseryl-tRNA[Ser]Sec Kinase and Sec Synthase	110
C. Selenophosphate Synthetase	111
IV. Sec Insertion into Protein	112
A. SECIS Elements	112
B. EFsec, SBP2, the L30 Ribosomal Protein, and Other Factors	114
C. UGA: To Stop or Not to Stop	116
V. Selenium and Selenoprotein Hierarchy	116
VI. Selenoprotein Identity, Function, and Targeted Removal	120
A. Identity of Sec UGA Codons	120
B. Selenoproteome	122
C. Targeted Removal	122
VII. Mouse Models for Elucidating the Role of Sec tRNA[Ser]Sec in Selenoprotein Biosynthesis and the Role of Selenoproteins in Development and Health	123
A. Transgenic Mouse Models	124
B. Conditional Knockout Mouse Models	127
C. Transgenic/Knockout Mouse Models	128
D. Transgenic/Conditional Knockout Mouse Models	129

E. Roles of Sec tRNA$^{[Ser]Sec}$mcm^{5}U and mcm^{5}Um
　　　in Selenoprotein Biosynthesis 130
VIII. How Did UGA Evolve as the Sec Codon? 131
　IX. Concluding Remarks 133
　　　References 134

　　The selenium field has grown dramatically over the last 10–15 years as many novel and exciting features about this trace element have been elucidated (1–11). One of the major areas of emphasis has been understanding the role of selenium in health (12). It has been known for many years that selenium is an essential micronutrient in the diet of mammals and that this element has numerous health benefits. It has roles in cancer (13) and heart disease prevention (14), inhibiting viral expression (15, 16), and delaying the progression of AIDS in HIV positive patients (17). In addition, selenium has been reported to have roles in immune function (18), male reproduction (19), mammalian development (20, 21), and slowing the aging process (18). We are beginning to better understand how selenium exerts many of these health benefits at the molecular level and the role that selenium-containing proteins, designated selenoproteins, play in this process. There are 25 known selenoproteins in humans and 24 in rodents (22). Selenium is inserted into selenoproteins in the form of the amino acid selenocysteine (Sec), which represents the 21st amino acid in the genetic code (1–3). Our interests in selenium research have centered around four main areas during the last several years: (i) the means by which Sec is biosynthesized and incorporated into protein; (ii) the generation of mouse models to elucidate the role of selenoproteins in development and health; (iii) identity and functions of selenoproteins; and (iv) the distribution and evolution of the Sec insertion machinery among eukaryotes. This review summarizes the current state of knowledge in these four areas. As discussed later, Sec is biosynthesized, unlike the common biosynthetic pathways of the other 20 protein amino acids, on its tRNA. Sec tRNA has many unique properties that distinguish it from all other naturally occurring tRNAs (Section II). In addition, the machinery for inserting Sec into protein is novel and unique to this amino acid. It is apparent that tremendous "effort" has been expended in evolution for inserting selenium into protein in the form of Sec as discussed in Section II.E. An important, unanswered question in the selenium field is whether selenoproteins (23) or low-molecular-weight selenocompounds (24) are responsible for the many health benefits exhibited by this trace element. Whereas selenocompounds have known cancer prevention effects, genetic studies involving polymorphic forms of selenoproteins [see (25) and references therein] and studies involving individual selenoproteins (26) suggest that selenium-containing proteins may also play a role in cancer prevention. Furthermore, studies involving the generation of transgenic mice encoding wild-type and mutant Sec tRNA transgenes ($trsp^t$) (27), Sec tRNA gene ($trsp$) conditional knockout mice (21), and $trsp$ knockout/$trsp$ transgenic mice (28, 28a) have provided mouse models for assessing the role of selenoproteins in health and development. The use of such mouse models as a tool to demonstrate the role of selenoproteins in health and development is discussed in detail in Section VII.

I. UGA Codes for Sec

Almost 20 years ago, the gene sequences of glutathione peroxidase 1 (GPx1) from mammals (29) and formate dehydrogenase from *Escherichia coli* (30) showed that both genes had an in-frame TGA codon in their open reading frames. Alignment of the genes with the amino acid sequences of the corresponding proteins demonstrated that the TGA codon aligned with Sec in the protein. Although it was tempting to conclude that TGA coded for Sec, this conclusion at that time had a fallacy. That is, Sec, unlike other amino acids, was biosynthesized on its tRNA. Thus, it was possible that an intermediate in the biosynthesis of Sec could be inserted into protein (e.g., phosphoserine [see (31) and references therein]) and then posttranslationally be modified to Sec. The intermediate, and not Sec, would then be the 21st amino acid in the genetic code. It was subsequently shown that Sec was biosynthesized on its tRNA in both bacterial (32) and mammalian cells (31) which provided conclusive evidence that Sec was the 21st amino acid. The expanded genetic code including Sec is shown in Fig. 1.

II. Novelty of Eukaryotic Sec tRNA$^{[Ser]Sec}$

Sec tRNA$^{[Ser]Sec}$ has been described as the key molecule (33) and central component (34) in the biosynthesis of selenoproteins. The reason for describing Sec tRNA in this manner is that it is the only known tRNA that governs the expression of only one class of proteins, the selenoproteins. The biosynthesis of Sec occurs on its tRNA, and the Sec moiety is synthesized from serine (Ser); thus, the tRNA molecule is designated as Sec tRNA$^{[Ser]Sec}$ (34). The structure of Sec tRNA$^{[Ser]Sec}$ and its evolutionary changes among those eukaryotes in which this tRNA and/or its gene have been sequenced are described in this section.

A. tRNA and Gene Structures

The structure of eukaryotic Sec tRNA$^{[Ser]Sec}$ has many novel features that distinguish it from all other tRNAs. For example, at 90–93 nucleotides in length, depending on the organism, it is the longest known tRNA in

5' Base	Middle base U	C	A	G	Middle base 3' Base
U	Phe	Ser	Tyr	Cys	U
	Phe	Ser	Tyr	Cys	C
	Leu	Ser	Stop	**Sec** / Stop	A
	Leu	Ser	Stop	Trp	G
C	Leu	Pro	His	Arg	U
	Leu	Pro	His	Arg	C
	Leu	Pro	Gln	Arg	A
	Leu	Pro	Gln	Arg	G
A	Ile	Thr	Asn	Ser	U
	Ile	Thr	Asn	Ser	C
	Ile	Thr	Lys	Arg	A
	Met / Initiator	Thr	Lys	Arg	G
G	Val	Ala	Asp	Gly	U
	Val	Ala	Asp	Gly	C
	Val	Ala	Glu	Gly	A
	Val	Ala	Glu	Gly	G

FIG. 1. The genetic code. Sec is highlighted in bold letters, indicating that it is the 21st amino acid. The dual function of UGA as Sec and termination and AUG as Met and the initiation codon are also shown.

eukaryotes. The genes for Sec tRNAs are three nucleotides shorter than the corresponding gene products as the CCA terminus is added posttranscriptionally. All known animal Sec tRNAs, as determined from sequencing the tRNA (bovine and rat) or the gene (human, mouse, rat, Chinese hamster, chicken, frog, zebra fish, fruit flies, and nematodes with the posttranscriptionally added CCA terminus) are 90 nucleotides long [reviewed in (1)], while the corresponding tRNA from *Dictyostelium* is 91 nucleotides (35) and those from various species of *Plasmodium* are 93 nucleotides (36, 37). The extra base in *Dictyostelium* Sec tRNA$^{[Ser]Sec}$ occurs in the D-stem and the extra three bases in *Plasmodium* occur in the long extra arm. The Sec tRNAs from *Chlamydomonas* (38) and *Tetrahymena* (35) are 90 nucleotides long and the gene for *Toxoplasma gondii* Sec tRNA (36) has also been identified and is 87 nucleotides in length.

The secondary structure of mammalian Sec tRNA$^{[Ser]Sec}$ is shown in a cloverleaf model in Fig. 2. It most likely exists in a 9/4 cloverleaf form (i.e., nine

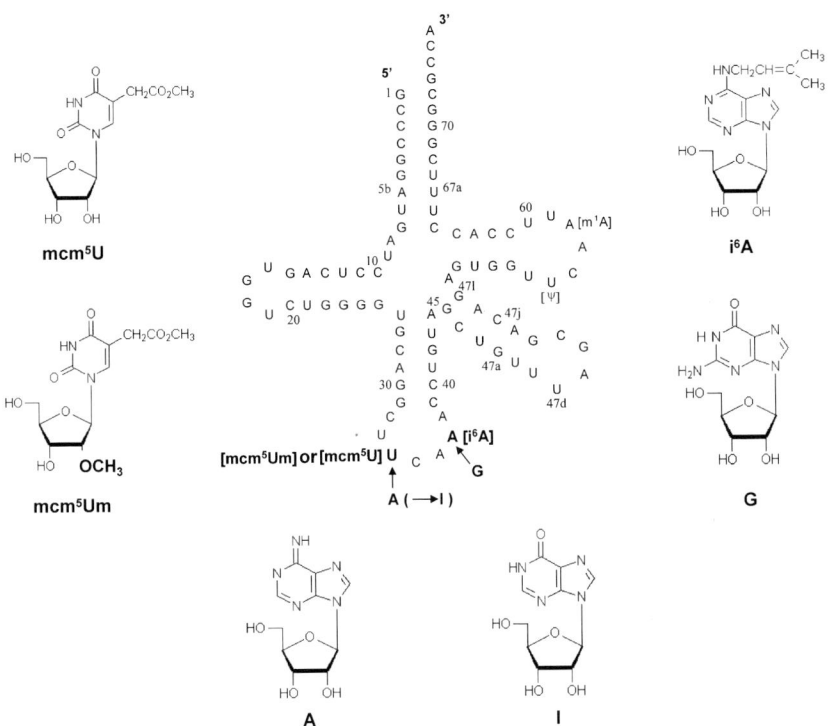

FIG. 2. Mammalian Sec tRNA[Ser]Sec. The primary structure of bovine Sec tRNA[Ser]Sec is shown in a 9/4 cloverleaf model (see text). The structures of the modified nucleosides at positions 34, mcm⁵U and mcm⁵Um, and 37, i⁶A, are also shown and the 2′-methylribose is highlighted in larger, bold letters. The A base change at position 34 that is converted intracellularly to I (indicated by an arrow in parenthesis) and the G base change at position 37 are also shown (see text).

paired bases in the acceptor stem and four in the TψC stem) (33, 39) which is another distinguishing feature of Sec tRNA[Ser]Sec. All other eukaryotic tRNAs exist in a 7/5 cloverleaf form (i.e., seven paired bases in the acceptor stem and five in the TψC stem). The extra bases in Sec tRNA[Ser]Sec that make it longer than other eukaryotic tRNAs are due to the presence of 13 nucleotides in the acceptor and TψC stem helices compared to 12 normally found in all other tRNAs and the atypically long variable arm (1). Sec tRNA[Ser]Sec has relatively few modified nucleosides as shown in Fig. 2, while other tRNAs may have as many as 15–17 modified nucleosides. The modified bases in Sec tRNA[Ser]Sec are 1-methyladenosine (m^1A) at position 58, pseudouridine (ψU) at position 55, N^6-isopentenyladenosine (i^6A) at position 37, and either 5-methoxycarbonylmethyluridine (mcm^5U) or 5-methoxycarbonylmethyl-2′-O-methyluridine (mcm^5Um) at position 34, which is the wobble position of tRNA. m^1A and ψU syntheses mark the initial nucleoside modifications to Sec tRNA$^{[Ser]Sec}$ and they likely occur in the nucleus of *Xenopus* oocytes (*40, 41*). i^{6}A and mcm^{5}U syntheses then occur, and they are likely carried out in the cytoplasm. The last step in the maturation process of Sec tRNA$^{[Ser]Sec}$ is the addition of Um34 (*40, 42*). Its synthesis is dependent on an intact primary and tertiary structure of Sec tRNA$^{[Ser]Sec}$, including the prior synthesis of the four modified bases at positions 34, 37, 55, and 58 (*42*). The syntheses of m^{1}A, ψU, i^{6}A, and mcm^{5}U are not as stringently connected to the primary and tertiary structures as that of Um34. In fact, Um34 addition is a highly specialized event as further discussed in Section II.B.

There are other features of Sec tRNA$^{[Ser]Sec}$ that further distinguish it as a novel tRNA. For example, the D-stem of Sec tRNA$^{[Ser]Sec}$ may contain six base pairs compared to three to four found in other tRNAs (*1*). The Sec tRNA$^{[Ser]Sec}$ gene is transcribed, unlike any other known tRNA gene, in that transcription begins at the first nucleotide within the coding sequence (*43*), while all other tRNAs are transcribed with a 5'-leader sequence that must be processed. The mature form of tRNA$^{[Ser]Sec}$ has a 5'-triphosphate moiety, and the possible role of this moiety in the function of this tRNA is not known. Numerous other novel features of tRNA$^{[Ser]Sec}$ transcription have been reported and these have been reviewed in detail elsewhere (*44*).

The Sec tRNA$^{[Ser]Sec}$ gene, designated *trsp*, has been examined in the genomes of many mammals, including humans, mice, rats, rabbits, cows, and Chinese hamsters [see (*1*) and references therein] and it occurs in single copy. *trsp* has also been sequenced in several higher and lower animals, including chickens, frogs, zebra fish, fruit flies, and nematodes (*1*). It also occurs in single copy in each of these animals with the exception of zebra fish where *trsp* exists in two gene copies (*1*). The genomes of humans and rabbits also contain a pseudogene, while that of Chinese hamsters has three pseudogenes (*1*).

B. Um34 Synthesis, a Highly Specialized Event in Sec tRNA$^{[Ser]Sec}$ Maturation

The Sec tRNA$^{[Ser]Sec}$ population in mammalian cells consists of two major isoforms, mcm^{5}U and mcm^{5}Um (Fig. 2). Their synthesis was discussed in Section II.A, and it was further noted that Um34 addition to Sec tRNA$^{[Ser]Sec}$ is a highly specialized event. In addition to being the last step in the maturation of this tRNA (*40*) and having a stringent dependence on an intact primary and tertiary structure for its addition (*42*), the methylation step is influenced by selenium status (*1*). Under conditions of selenium deficiency, the level of mcm^{5}U in rodent liver is enriched while that of mcm^{5}Um is reduced, and under the conditions of selenium sufficiency, the ratio of the two isoforms is reversed [reviewed in (*1*)]. These observations provided the first evidence that the two isoforms may have different roles in protein synthesis (45). Furthermore, several studies suggested that the enrichment of mcm^{5}Um correlated with the expression of GPx1 (27, 45), whereas the enrichment of mcm^{5}U correlated with the expression of TR3 (27). We have provided strong evidence that the presence of Sec tRNA$^{[Ser]Sec}$ Um34 is correlated with the expression of several selenoproteins and these proteins appear to have roles in stress-related phenomena (28, 28a). The same selenoproteins that are poorly expressed in the absence of Um34 are sensitive to selenium status and are downregulated under conditions of selenium deficiency (28). These findings are discussed in detail in Section VII.

It has also been known for sometime that the presence of Um34 in Sec tRNA$^{[Ser]Sec}$ causes dramatic changes in secondary and tertiary structures (46). This observation provided some of the early indications that the addition of Um34 is a highly specialized event and that the two isoforms may have different roles in protein synthesis.

C. Um34 Sec tRNA$^{[Ser]Sec}$ Methylase

The methylase that catalyzes the formation of Um34 on Sec tRNA$^{[Ser]Sec}$ has not been characterized, but a protein described by Ding and Grabowski (47), and designated SECp43, may have a role in Um34 synthesis. These investigators demonstrated by a protein overlay analysis in HeLa cell extracts that SECp43 existed in a complex with Sec tRNA$^{[Ser]Sec}$ and a 48-kDa protein. The requirements for the complex formation, the identity of the 48-kDa protein, and the role of SECp43 were not elucidated (47). Another protein, designated as the soluble liver antigen (SLA), had been previously identified as a 48-kDa protein that was targeted by antibodies in patients with an autoimmune chronic hepatitis (48). SLA precipitated Sec tRNA$^{[Ser]Sec}$ from human cell extracts. We have shown that the 48-kDa protein that forms a complex with SECp43 and Sec tRNA is most likely SLA (49, 50). Using RNA interference (RNAi) technology, we found that knockdown of SECp43 in mammalian cells dramatically reduced the formation of Um34, suggesting that this protein is involved in Um34 synthesis. In addition, we observed that SECp43 formed a complex with Sec tRNA$^{[Ser]Sec}$ and SLA. The targeted removal of either SECp43 or SLA affected the binding of the other to Sec tRNA$^{[Ser]Sec}$, even though tRNA$^{[Ser]Sec}$ attachment was most affected when SLA was the targeted member of the complex. Cellular location analysis using the green fluorescent protein (GFP) fused to either SECp43 or SLA demonstrated that the former protein is located primarily in the nucleus, while the latter one is found in the cytoplasm. Cotransfection of both proteins resulted in the nuclear translocation of SLA, suggesting that SECp43 may also serve as a chaperone for shuttling SLA and Sec tRNA$^{[Ser]Sec}$ between different cellular compartments. There is no sufficient data to designate SECp43 as the Um34 methylase, but this protein certainly has a role in the methylation process. The possible specific role of SLA is discussed in Section III.B.

D. Amounts and Distributions of Sec tRNA$^{[Ser]Sec}$ Isoforms

Several lines of evidence suggest that the levels of Sec tRNA$^{[Ser]Sec}$ in mammalian cells and tissues are not limiting. For example, reducing the Sec tRNA$^{[Ser]Sec}$ population by approximately ½ does not appear to affect the overall amount of selenoproteins expressed in mammalian cells grown in culture (51) or in those tissues and organs examined in mice (20, 21). Furthermore, an enrichment of the Sec tRNA$^{[Ser]Sec}$ population does not appear to affect selenoprotein expression in mammalian cells (52) or in tissues and organs of mice (21), although some minor differences have been observed (28, 53).

The levels and distributions of the Sec tRNA$^{[Ser]Sec}$ isoforms have been measured in mammalian cells grown in culture (54) and in a number of organs and tissues in mice (1, 27, 28) as summarized in Table I. Although the total amount of the Sec tRNA$^{[Ser]Sec}$ population varies considerably in different cell types and organs (see column 3 in Table IA and B, respectively), its levels decrease under conditions of selenium deficiency compared to selenium sufficiency. Furthermore, the amount of mcm^{5}Um is reduced under conditions of selenium deficiency, while that of the mcm^{5}U is enriched or remains approximately the same (compare columns 4 and 5 to 6 and 7 in Table IA and B, respectively). Thus, the level of mcm^{5}Um is far more sensitive to selenium status than mcm^{5}U, suggesting that Um34 methylase activity is responsive to increasing levels of selenium. Many of the stress-related selenoproteins are also sensitive to selenium status, and the Um34-containing species is directly involved in their biosynthesis (28, 28a).

E. Evolution of Sec tRNA$^{[Ser]Sec}$ and Occurrence of Sec Insertion Machinery

Transfer RNAs that decoded UGA and were aminoacylated with Ser, and the Ser moiety was converted to Sec, were observed simultaneously in 1989 in *E. coli* (32) and mammalian cells (31). The gene for Sec tRNA$^{[Ser]Sec}$ (*trsp*) was subsequently shown to be universal in the animal kingdom (55) and found in numerous species of prokaryotes (56). Although the occurrence of selenoproteins and the Sec insertion machinery was subsequently reported in archaea [see (22) and references therein], it was only recently that the Sec insertion machinery (57) and Sec tRNA$^{[Ser]Sec}$ (35–38) were observed in any eukaryotic organism outside the animal kingdom. *Chlamydomonas*, which is a green alga and a member of the plant kingdom, was found to contain the Sec insertion machinery and at least 10 selenoproteins (57). A relatively simple procedure was developed for partially purifying and then sequencing Sec tRNA$^{[Ser]Sec}$ by reverse transcription-polymerase chain reaction (RT-PCR) (38). A computational Sec tRNA genomic analysis program was also devised for detecting *trsp* in partially or completely sequenced genomes (57a), and this procedure in combination with that of isolating and sequencing the gene product by RT-PCR has provided an easy and simple method for determining whether an organism has the ability to insert Sec into protein. Using these two procedures, Sec tRNA$^{[Ser]Sec}$ was identified in two model organisms, *Dictyostelium discoideum* and *Tetrahymena thermophila* (35). The identification of UGA as a Sec codon in *T. thermophila* demonstrates that the sole termination codon in this organism [see (35) and references therein] has a shared function. It should also be noted that *T. thermophila* does utilize its Sec tRNA for making selenoproteins as its genome encodes selenoprotein genes (A. V. Lobanov and V. N. Gladyshev, unpublished data).

A phylogenetic tree consisting of the known Sec tRNA and/or *trsp* sequences was assembled to assess their phylogeny (Fig. 3). *T. thermophila* Sec tRNA exists on a branch, albeit distantly, with those of other eukaryotes, *T. gondii* and *Plasmodium*. *D. discoideum*, however, appears to be the most divergent eukaryotic Sec tRNA sequenced to date.

Phylogenetic analyses also showed that all *trsp* sequences in bacteria, archaea, and eukaryotes have a common origin. Eukaryotic *trsp* genes form a separate subgroup. However, identification of *trsp* sequences in sequence databases remains challenging due to distinct features of these molecules as compared to other tRNAs (see in an earlier section). Additional computational tools for prediction of *trsp* genes should be developed and applied to scan various genomic databases. These searches will not only enrich our understanding of these molecules, but will also provide a better understanding of the distribution of the Sec insertion machinery in eukaryotes, and will have the potential to identify other novel tRNAs.

The machinery involved in the insertion of Sec into protein is a highly sophisticated system that evolved for the synthesis and incorporation of a single amino acid, Sec, into protein. The driving force for evolving the Sec insertion machinery most likely was to take advantage of the unique chemical properties of the selenium atom in Sec. The pK_a of Sec is below physiological pH and Sec is therefore almost completely ionized intracellularly. Nature has clearly taken advantage of the fact that Sec is a better participant in certain redox reactions than Cys, which is the major amino acid involved in redox reactions. The pK_a of Cys is 8.3 which means that it is poorly ionized at physiological pH.

TABLE I
Amounts and Distributions of Sec tRNA$^{[Ser]Sec}$ Isoforms in Mammalian Cells and Organs

A. Cells in culture[a]

Cell line	Selenium supplementation[b]	% of total[c]	mcm^{5}U[d]		mcm^{5}Um[d]		mcm^{5}Um/mcm^{5}U[f]
			%	% of total[e]	%	% of total[e]	
HL-60	+(CDM)	9.6	38.5	3.70	61.5	5.90	1.60
HL-60	−(CDM)	7.5	61.3	4.60	38.7	2.90	0.63
HL-60	+(FCS)	9.4	55.3	5.20	44.7	4.20	0.81
HL-60	−(FCS)	7.4	77.0	5.70	23.0	1.70	0.30
RMT	+(CDM)	1.7	11.8	0.20	88.2	1.50	7.47
RMT	−(CDM)	1.4	35.7	0.50	64.3	0.90	1.80
CHO	+(FCS)	1.01	45.1	0.46	54.9	0.55	1.22
CHO	−(FCS)	0.86	56.2	0.48	43.8	0.38	0.78

B. Organs[g]

Organ	Selenium supplementation[b]	% of total[c]	mcm⁵U		mcm⁵Um		mcm⁵Um/mcm⁵U[f]
			%[d]	% of total[e]	%	% of total[e]	
Liver	+	4.5	33.3	1.50	66.7	3.00	2.00
	−	2.8	57.7	1.62	42.3	1.18	0.73
Kidney	+	7.5	33.7	2.52	66.3	4.97	1.97
	−	3.7	59.2	2.19	40.8	1.51	0.69
Heart	+	4.3	38.1	1.64	61.9	2.66	1.62
	−	3.2	66.4	2.12	33.6	1.08	0.51
Muscle	+	1.9	38.6	0.73	61.4	1.17	1.59
	−	1.5	73.3	1.10	26.7	0.40	0.35

[a] Taken from (54).
[b] CDM, chemically defined media; FCS, fetal calf serum.
[c] Percentage of tRNA[Ser]Sec population within the total seryl-tRNA population.
[d] Percentages of mcm⁵U and mcm⁵Um within the total tRNA[Ser]Sec population.
[e] Percentages of mcm⁵U or mcm⁵Um within the total seryl-tRNA population.
[f] Amount of mcm⁵Um divided by the amount of mcm⁵U.
[g] Taken from (45).

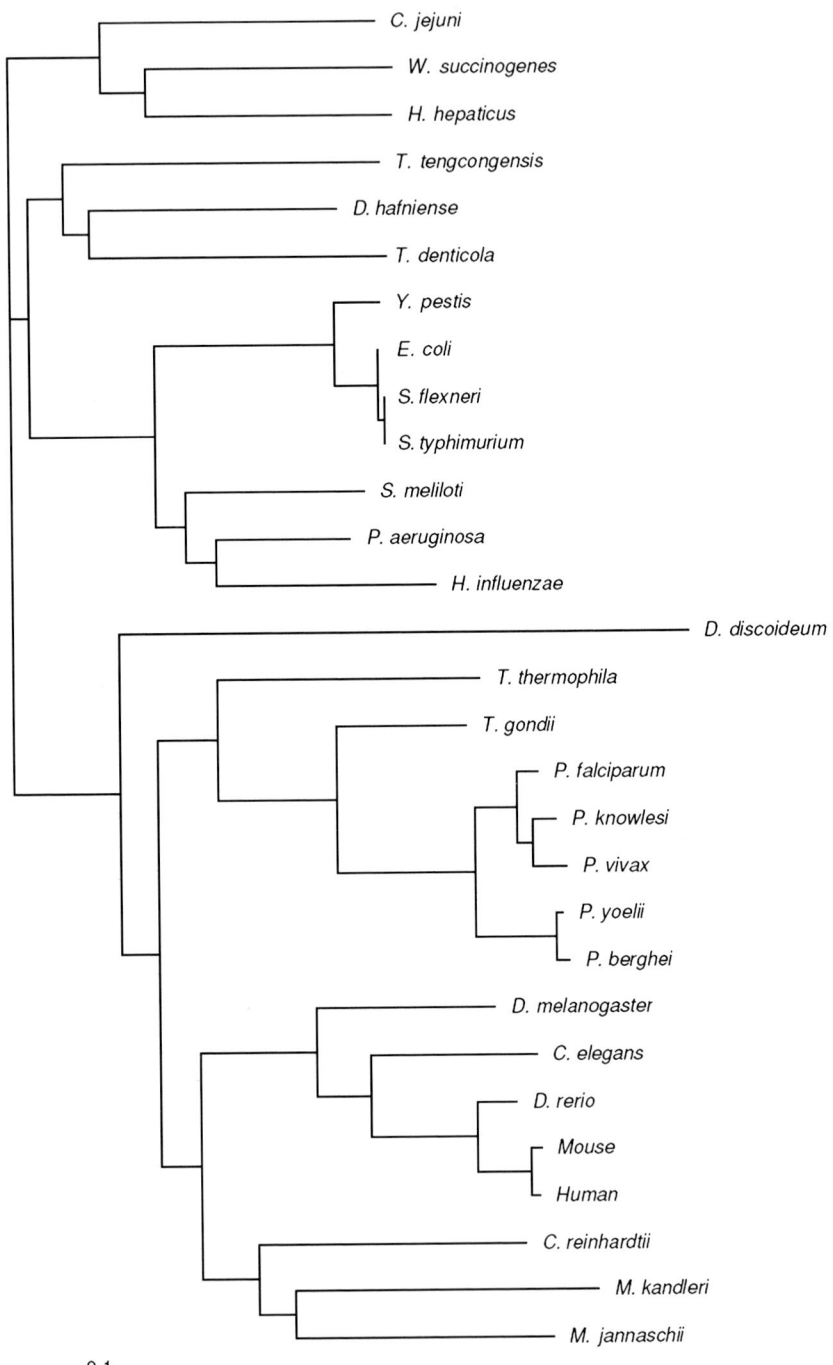

The Sec insertion machinery is present in representatives of the three life kingdoms, eubacteria, archaea, and eukaryotes. The resulting products of the Sec insertion machinery, selenoproteins, are essential for the survival of some organisms, for example, higher vertebrates (1–3), while others do not use this machinery at all, for example, yeasts and some higher plants [(22) and references therein]. Of the approximately 300 bacterial genomes sequenced and of the more than two dozen archaeal genomes sequenced, only about one-fourth use the Sec incorporating machinery to generate selenoproteins [(22) and references therein]. This observation raises an important question as to why some organisms, but not all organisms, have taken advantage of the possible benefits of selenoproteins in cellular metabolism, or alternatively, why some organisms have lost the ability to make selenoproteins?

Since Sec tRNA is the key molecule (33) and central component (34) in the Sec insertion machinery, and since our computational (36) and/or our isolation and sequencing techniques are so readily applicable to identifying Sec tRNAs (35, 38), focusing on the occurrence of *trsp* and/or Sec tRNA as a means of establishing the presence of the Sec insertion machinery should be a fruitful avenue for assessing eukaryotes that utilize this system. Sec tRNA identification provides a direct and easy means of evaluating whether an organism has the ability to insert Sec into protein as the distribution of Sec insertion machinery will most likely be determined only on an organism by organism basis.

III. Sec Biosynthesis

As discussed earlier, Sec is distinctive from the other 20 amino acids in protein in that its biosynthesis occurs on its tRNA (1–3). The other amino acids are first synthesized or taken up intracellularly and then attached to their

---

FIG. 3. Phylogenetic analysis of Sec tRNA$^{[Ser]Sec}$. An alignment of Sec tRNA sequences was generated using ClustalW multiple sequence alignment program. The phylogenetic tree was constructed using the Neighbor Joining method based on calculated distances between each pair of sequences. TreeView was used to visualize the tree. Scale bar in the bottom left corner displays a scale of 0.1 nucleotide substitutions per site. Accession numbers for the sequences are as follows: *D. rerio* (AF135237.1), *H. influenzae* (U32753.1), *P. aeruginosa* (AE004893.1), *M. jannaschii* (U67469.1), *M. kandleri* (AE010318.1), *S. meliloti* (AE007196.1), *S. typhimurium* (AE008874.1), *T. tengcongensis* (AE013138.1), *H. hepaticus* (AF134212.1), *W. succinogenes* (BX571659), *T. denticola* (AE017249.1), *Y. pestis* (AE017140.1), *C. jejuni* (AF486635.1), *D. hafniense* (NZ_AAAW00000000), *D. discoideum* (DDB0185230), *C. reinhardtii* (AY268554.1), *B. taurus* (X74110.1), *M. musculus* (L22019.1), *H. sapiens* (K02923), *C. elegans* (M34508.1), and *D. melanogaster* (NR_001709.1). *T. gondii* was from ToxoDB (www.toxodb.org) (TGG_995334). *Plasmodium* sequences were extracted from PlasmoDB: *P. falciparum* (103674902), *P. knowlesi* (Pk_614e10p1c), *P. vivax* (Pv_6576.phat_654), *P. yoelii* (98128547), and *P. berghei* (102857153).

cognate tRNA by an aminoacyl-tRNA synthetase that is specific for each amino acid and the cognate tRNA. Although the biosynthesis of asparagine and glutamine can occur on aspartate and glutamate tRNAs, respectively (58), this means of synthesizing these amino acids is restricted to certain life forms and is not universal in nature. As far as is known, Sec is synthesized on its tRNA in all organisms that encode the Sec tRNA insertion machinery and the codon for this amino acid is UGA.

A. Seryl-tRNA Synthetase

The fact that Ser is attached to Sec tRNA$^{[Ser]Sec}$ by seryl-tRNA synthetase indicates that the identity elements in this tRNA are for Ser and not Sec. The identity elements in Sec tRNA$^{[Ser]Sec}$ are located in the discriminator base and the long extra arm, both of which are essential to aminoacylation (59, 60). The acceptor, TψC, and D-stems also have a role in the identity process (61). Following the aminoacylation of Sec tRNA$^{[Ser]Sec}$ with Ser, the Ser moiety then serves as the backbone for the synthesis of Sec on its tRNA (31, 32, 62).

B. Phosphoseryl-tRNA$^{[Ser]Sec}$ Kinase and Sec Synthase

In bacteria, the enzyme that carries out the synthesis of Sec utilizing Ser on seryl-tRNA$^{[Ser]Sec}$ as the backbone for Sec formation has been thoroughly characterized (63). Sec synthase is a pyridoxal phosphate (PLP)-dependent protein. It catalyzes the removal of the hydoxyl group from Ser to form an aminoacrylyl intermediate that in turn serves as the acceptor for activated selenium. Once activated selenium is donated, selenocystenyl-tRNA$^{[Ser]Sec}$ is formed and is ready for insertion into the protein.

In mammals, the biosynthesis of Sec on its tRNA is not fully understood. In 1970, a kinase activity that phosphorylated a minor species of seryl-tRNA to form phosphoseryl-tRNA was reported in rooster liver (64) and a minor seryl-tRNA that decoded the nonsense UGA was found in bovine liver (65). The phosphoseryl-tRNA and the minor UGA decoding seryl-tRNA were subsequently identified as Sec tRNA$^{[Ser]Sec}$ (66), but the kinase activity remained elusive. We detected a candidate gene for mammalian phosphoseryl-tRNA$^{[Ser]Sec}$ kinase (*pstk*) using a comparative genomics approach that searched completely sequenced archaeal genomes for a kinase-like protein with the pattern of occurrence similar to that of components of the Sec insertion machinery (67). Mouse *pstk* was cloned, and the gene product (PSTK) expressed and characterized. PSTK specifically phosphorylated the seryl moiety on seryl-tRNA$^{[Ser]Sec}$ and in addition had a requirement for ATP and Mg^{2+}. The reaction was reversible, albeit poorly. Proteins with homology to mammalian

PSTK occur in *Drosophila*, *Caenorhabditis elegans*, *Methanopyrus kandleri*, and *Methanococcus jannaschii*, suggesting a conservation of its function across archaea and eukaryotes that synthesize selenoproteins, and the absence of this function in bacteria, plants, and yeast. The fact that PSTK has been highly conserved in evolution suggests that it plays an important role in selenoprotein biosynthesis and/or regulation (67). Although the roles of the kinase and phosphoseryl-tRNA$^{[Ser]Sec}$ in the biosynthesis of Sec have not been fully characterized (67, 67a), it is important to note that the formation of phosphoserine is consistent with a subsequent Sec synthase-catalyzed reaction as phosphorylated Ser would have a better leaving group than Ser in the PLP-dependent Sec biosynthetic pathway. The fact that some archaea carry out cysteine biosynthesis on tRNA^{Cys} by first aminoacylating the tRNA with phosphoserine and subsequently converting the phosphoserine moiety to cysteine by a PLP-containing Cys synthase (68) provides an excellent model that phosphoseryl-tRNA$^{[Ser]Sec}$ is a candidate intermediate in the biosynthesis of Sec [see also (67a)].

Therefore, is there a candidate Sec synthase? Computational analysis of sequenced genomes in archaea and eukaryotes known to utilize the Sec insertion machinery for homologs to the bacterial Sec synthase failed to reveal possible candidates. However, one protein that was linked to the Sec insertion machinery fits the characteristics of the candidate Sec synthase. This protein was identified several years earlier as SLA (48) and we have further characterized it as described in Section II.C. It is an attractive possibility that phosphoseryl-tRNA$^{[Ser]Sec}$ would serve as a substrate for the candidate Sec synthase (SLA) and the active donor of selenium would then displace the phosphate moiety to yield Sec.

C. Selenophosphate Synthetase

The active form of selenium that is donated to the intermediate in the biosynthesis of Sec has been identified in prokaryotes. The donor was characterized as monoselenophosphate, which is synthesized from selenide and ATP by selenophosphate synthetase (69). The active selenium donor has not been characterized in eukaryotes, but it may be the same selenium form (70–72). Two selenophosphate synthetase genes, designated *Sps1* and *Sps2*, have been identified in mammals (70–72). The gene product of *Sps2* is a selenoprotein, designated SPS2. The fact that SPS2 is a selenoprotein raises a question as to whether it might be involved in the autoregulation of its own biosynthesis (70). Once the activated form of selenium is donated to the intermediate, which is likely phosphoseryl-tRNA$^{[Ser]Sec}$, the biosynthesis of Sec on tRNA$^{[Ser]Sec}$ is completed.

IV. Sec Insertion into Protein

Since UGA can serve as both a stop codon and a Sec codon (Fig. 1), there are cellular mechanisms that distinguish between these two functions in addition to the Sec tRNA$^{[Ser]Sec}$ that decodes an in-frame UGA. Several factors have been identified that are required in the recoding of UGA as Sec and the insertion of this amino acid into protein. One novel feature of selenoprotein mRNAs is the occurrence of a *cis*-stem-loop structure known as the *SEC Insertion Sequence* (SECIS) element or elements in the 3'-untranslated region (3'-UTR) of selenoprotein mRNAs (73). SECIS elements are responsible for recoding the UGA codeword as Sec and bypassing stop. In addition to these two *cis*-acting factors (i.e., UGA and SECIS elements), there are several *trans*-acting factors involved in the insertion of this amino acid into protein which include: (i) the selenocystenyl-tRNA$^{[Ser]Sec}$-specific elongation factor, EFsec (74, 75); (ii) the SECIS-binding protein, SBP2 (76); and (iii) the L30 ribosomal protein, rpL30 (77). Each of these factors is discussed later.

A. SECIS Elements

SECIS elements are *cis*-acting RNA structures located in the 3'-UTR of all known eukaryotic selenoprotein genes and they serve as the signaling factors for recoding UGA as Sec (73). Eukaryotic SECIS elements are composed of: (i) two helices separated by an internal loop; (ii) a SECIS core structure consisting of a Quartet located at the base of helix 2; and (iii) an apical loop or bulge (Fig. 4). The Quartet is made up of four non-Watson-Crick interacting base pairs and appears to be the main functional site of the stem-loop structure as this is the site at which SBP2 and the L30 protein bind. In many SECIS elements, the apical loop is quite large and an additional ministem is formed that likely has a role in stabilizing the distal end of the element. The occurrence of this ministem was used to classify SECIS elements into form 1 and form 2 structures wherein form 1 lacks, and form 2 contains, the ministem (78). These two forms can be interconverted by mutations that either extend or shorten the apical loop and can naturally evolve from each other. Conservation within the primary structures of eukaryotic SECIS elements is virtually nonexistent with the highest degree of conservation occurring in the TGA of the 5' portion and the GA of the 3' portion of the Quartet. The adenine immediately preceding the Quartet and the two adenines in the apical loop occur in most of the selenoprotein genes sequenced thus far. However, the nucleotide preceding the Quartet could also be G as is the case in all known nematode SECIS elements. Less frequently, both U and C can be found at this position, especially in selenoprotein genes from certain protists.

Likewise, the AA tandem in the apical loop can be replaced with various combinations of nucleotides, but why certain pairs support Sec insertion, whereas

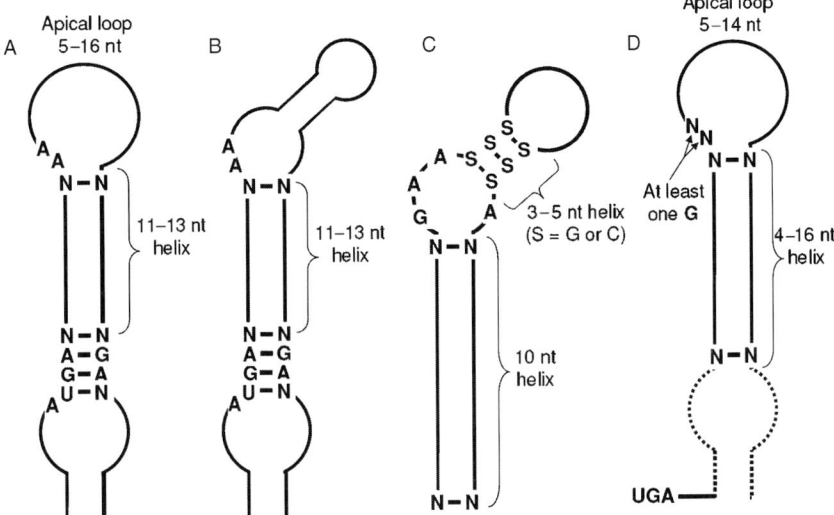

FIG. 4. Consensus structures of SECIS elements in eukaryotes, archaea, and bacteria. (A) Eukaryotic type I SECIS element, (B) eukaryotic type II SECIS element, (C) archaeal SECIS element, (D) bacterial SECIS element. Conserved nucleotides are indicated, along with some structural constraints. The indicated features represent the majority of SECIS elements in the three domains of life, but some additional deviations may occur.

others do not, is not fully understood. Mammalian SelM and SelO SECIS elements have CC in place of AA and these structures are functional (79).

There is a spatial requirement regarding the distances between in-frame Sec UGA codons and 3'-UTR SECIS elements. The minimal distance was measured to be between 51 and 111 nucleotides (73), suggesting that SECIS elements are both necessary and sufficient for Sec insertion, provided that mRNA bearing an in-frame UGA and a 3'-UTR SECIS element have access to the selenoprotein, ribosome-based protein synthesis machinery, and Sec-specific translation factors. The novel feature of a *cis*-SECIS element in the 3'-UTR directing the insertion of Sec in response to an in-frame UGA codon suggests that UGA-SECIS pairs can be designed in nucleotide sequences for targeting Sec insertion into protein.

It should also be noted that in archaea, SECIS elements are typically located in the 3'-UTR as in eukaryotes, but their structures differ from their eukaryotic counterparts (80). At least one archaeal selenoprotein gene encoding formate dehydrogenase has a predicted SECIS element in the 5'-UTR. The function of this structure has not been verified experimentally, but it fits the overall archaeal SECIS consensus. Bacterial SECIS elements are different from those that occur in both eukaryotes and archaea and they are located within the coding sequences of selenoprotein genes immediately downstream of Sec-encoding UGA codons (63). The bacterial SECIS elements are even less conserved than the archaeal and eukaryotic structures. It appears that only a single nucleotide, a guanosine, is highly conserved in the small apical loop of bacterial SECIS elements. This feature, as well as the distance between the Sec-encoding UGA codon and the SECIS structure could be used to predict novel selenoprotein genes in bacteria (81).

B. EFsec, SBP2, the L30 Ribosomal Protein, and Other Factors

Since the reviews published in *Progress in Nucleic Acid Research and Molecular Biology* are more focused on a personal perspective of research and since thorough reviews on the insertion of Sec into protein that involves both *cis*-acting and *trans*-acting elements have been published (8–10), these protein factors will only be discussed briefly here.

EFsec, which is also called eSelB, recruits selenocystenyl-tRNA and acts jointly with SBP2 to insert Sec into nascent polypeptides in response to UGA codons (74, 75). EFsec is specific for Sec and does not bind seryl-tRNA$^{[Ser]Sec}$ (74) or phosphoseryl-tRNA$^{[Ser]Sec}$ (67). It is different from the canonical elongation factor, EF-1α, which is involved in incorporation of the other 20 amino acids, although these two proteins have strong similarity except that EFsec contains a C-terminal extension. The C-terminal extension interacts with SBP2 but only when it is expressed as an isolated fragment (82). EFsec and SBP2 jointly constitute the functional equivalent of the single SELB factor in bacteria (63). The occurrence of SBP2 and EFsec as separate proteins in eukaryotes suggests a mechanism for rapid exchange of the Sec tRNA-EFsec complex (from empty to aminoacyl-tRNA bound) following Sec insertion.

The available evidence suggests that EFsec binds directly to the SECIS element, but further work is necessary to clarify the manner in which the interaction takes place and whether the C-terminal extension plays a role in the binding (9, 83). EFsec forms a complex with SBP2 and this complex formation is stimulated by the presence of selenocystenyl-tRNA$^{[Ser]Sec}$, but additional studies are required to fully elucidate its role in the recoding process outside of being an elongation factor (8–10).

In addition to SECIS elements and Sec tRNA$^{[Ser]Sec}$, the most studied of the factors involved in the insertion of Sec into protein is SBP2 (3, 3a, 9). This factor appears to have three essential functions that include binding to the SECIS core, binding to the ribosome, and the insertion of Sec into selenoprotein. SBP2 is recruited by the SECIS element to form a complex by binding to the SECIS Quartet and sequences preceding the Quartet (3, 3a, 8–10). Within the SECIS binding domain of SBP2 is an L7Ae RNA binding motif. The L7Ae binding motif occurs in a small class of proteins that seem for the most part to be associated with translation at the ribosome level or more specifically in ribosome interactions. SBP2 binds to the ribosome at the 28S RNA likely by binding with one or more kink turn structures (83). Although this factor remains quantitatively associated with ribosomes, it cannot simultaneously interact with the ribosome and the SECIS element. The role of the L7Ae domain in SBP2 may be to bind the kink turn in the SECIS element and similar structures in the 28S ribosome. rpL30 also has an L7Ae binding domain and this factor likely competes with SBP2 for binding to the SECIS element (9, 84). rpL30 also enhances the incorporation of Sec into protein (84). A model to accommodate the participation of both components in binding to the SECIS element has been proposed wherein SBP2 and rpL30 carry out different functions in the recoding of UGA and the SECIS element acts as a molecular switch upon protein binding (84). A somewhat different view of the role of rpL30 has been suggested wherein this element may serve in tethering the recoding machinery to the ribosome for translation (85). These observations have led Copeland and collaborators to propose a new and more detailed model for Sec insertion into protein wherein SBP2 is continually bound to the ribosome except at the time Sec is delivered to the A-site for decoding (9, 83). Thus, SBP2 can serve to select a subset of ribosomes and program them for Sec insertion competence by interacting with the SECIS element at the moment of Sec insertion. The role of rpL30 would then be to compete SBP2 off the SECIS element and back onto the ribosome, rendering the ribosome competent for another round of recoding (9).

Whether the incorporation of Sec into protein is an efficient or inefficient process *in vivo* is not fully understood, but both the positive (insertion) and negative (termination) aspects of decoding UGA have been thoroughly reviewed (3). However, *in vitro* assays showing efficient incorporation have not yet been achieved, suggesting that at least in part all factors involved in the Sec insertion machinery have not been identified. An example of this possibility is the report by Howard *et al.* (86) wherein they sought to uncover other elements that might be involved in the Sec insertion recoding process. These investigators found a stem-loop structure located a few nucleotides downstream of the UGA Sec codon in SelN that increased its translation in the absence of the 3'-UTR SECIS element. The recoding efficiency doubled in the presence of the normal SelN 3'-UTR SECIS and such "potential enhancing translational efficiency elements" were found in several, by not all, selenoprotein genes (86).

C. UGA: To Stop or Not to Stop

Some UGA Sec codons support both readthrough and termination during translation. For example, SelP mRNA from rat plasma contains 10 UGA Sec codons and the products from its translation occur in four isoforms (87). The shorter isoforms arise from termination at the second, third, and seventh UGA codons. These UGA codons dictate a cessation in protein synthesis as well as a continuation in protein production.

Another *cis*-feature that has a role in determining the efficiency of Sec insertion vs truncated selenoprotein release is the nucleotide context of the UGA Sec codon (73, 88–90, and see Table II). A purine immediately downstream of the UGA Sec codon in mammals favors termination, while a pyrimidine in this position favors readthrough (73, 88). An A at this site appears to favor termination slightly better than G, while U appears to favor insertion slightly better than C. The bases located in the second position within the downstream codon and the first codon (90) or the first two codons (89) located immediately upstream of UGA also have a role in the interplay of Sec insertion-UGA termination. It is important to emphasize that different Sec codons manifest different insertion efficiencies (87–90) and that nucleotide context is only one parameter influencing the fate of any specific UGA Sec codon. The effect of codon context on the efficiency of Sec insertion has also been reviewed elsewhere (1–3).

In addition, *trans*-acting factors have a major role in regulating Sec incorporation-translation termination interplay. SBP2, rpL30, EFsec, Sec tRNA$^{[Ser]Sec}$, the termination factor eRF1, and the eRF1- and ribosome-dependent GTPase eRF3 are all candidates that likely influence the fate of UGA Sec codons (89–93). As the influence of these elements on fate of UGA Sec codons have been discussed elsewhere (1, 6, 9, 10, 89–93), their possible interplay in the Sec insertion-translation termination process will not be further considered herein.

V. Selenium and Selenoprotein Hierarchy

Selenium-deficient mammals manifest a hierarchy with respect to the retention of selenium in different organs and to the expression of individual selenoproteins (1–3, 94, 95 for reviews). Laboratory animals maintained on diets low in selenium have substantially reduced amounts of this element in liver and kidney, while brain and testes retain most of their selenium (96–98). Since the levels of the Sec tRNA$^{[Ser]Sec}$mcm^{5}Um isoform (see Table I, and 45, 46, 54) and the efficiency of selenoprotein synthesis (90, 91, and references therein) are responsive to selenium status, these factors are more affected in

TABLE II

Possible Sites Involved in the Regulation of Selenoprotein Expression by Specific Sec tRNA[Ser]Sec Isoforms

Selenoprotein class[a]	cDNA length[b]	Exon number[c]	TCA exon position[d]	TCA to stop[e] (nt)	Stop to SECIS ATGA[f] (nt)	SECIS ATGA to poly A[g] (nt)	TGA context[h]	Molecular weight[i]	Subject to NMD[j]
Class I									
GPx1	923	2	Exon 1	462	57	219	TCTCTCTGAGGCACC	22.4	+++
GPx3	1415	5	Exon 2	459	544	189	AGCTACTGAGGTCTG	25.5	+++
SelR	863	5	Exon 3	64	76	391	AGATTCTGAATATTT	12.8	++
SelT	1606	6	Exon 2	438	67	846	GTATCCTGAGGGTAC	19.0	+
SelW	654	6	Exon 2	225	55	289	GGAGCTTGAGGCTAT	9.7	+++
Class II									
GPx2	1015	2	Exon 1	459	213	94	TCACTCTGAGGAACA	22.1	++
GPx4	958	7	Exon 3	372	75	141	TCGCAATGAGGCAAA	19.6	–
SelP	2072	4	Exon 1	963	284	509	GCCAGCTGATACTTG	42.4	–
			Exon 4	363	663	130	GAGAGCTGAGACACC		
			Exon 4	309			AAGCTCTGACGAAGG		
			Exon 4	186			ATTGCTTGACAGTGT		
			Exon 4	150			TGTAGCTGACAGGGG		
			Exon 4	84			GCTGCCTGACAAAAT		
			Exon 4	42			CCCAACTGAAGCTGA		
			Exon 4	36			TGAAGCTGAGATAAT		
			Exon 4	15			AGGAAGTGAAAATGA		
			Exon 4	9			TGAAAATGACATTCA		
Sep15	1463	5	Exon 3	208	564	152	TGCCGATGAAAATTG	17.8	–

(Continues)

TABLE II (*Continued*)

Selenoprotein class[a]	cDNA length[b]	Exon number[c]	TGA exon position[d]	Distances			TGA context[h]	Molecular weight[i]	Subject to NMD[j]
				TGA to stop[e] (nt)	Stop to SECIS ATGA[f] (nt)	SECIS ATGA to poly A[g] (nt)			
Class III									
TR1	3325	15	Exon 15	3	243	1420	GGCTGCTGAGGTTAA	54.6	–
TR3	1866	18	Exon 18	3	200	83	GGTTGCTGAGGCTAA	56.7	ND
Class IV									
DI1	1668	4	Exon 2	393	1172	90	TGCACCTGACCTTCA	29.7	–
DI2	5813	2	Exon 2	9	4556	298	GCCACCTGACCACCT	29.9	ND
DI3	1863	1	Exon 1	402	546	399	TGTACCTGACCACCG	31.7	ND
SPS2	2173	1	Exon 1	1167	520	153	AAGGGCTGAGGCTGC	47.9	–
TR2	2656	16	Exon 16	3	179	460	GGCTGCTGAGGCTGA	67.6	–
SelH	656	4	Exon 2	234	59	143	ACGAGCTGACGCGTG	13.0	++
SelI	6712	10	Exon 10	30	1235	4189	TCAGATTGACTAGGA	22.7	ND
SelK	1047	5	Exon 4	6	132	459	GGTGGATGAGGAAGG	10.7	–
SelM	687	5	Exon 2	294	71	133	GGAGGATGACAGTTG	16.4	–

	Total length[b]	Exons[c]	Exon (TGA)[d]	nt TGA–stop[e]	nt stop–ATGA[f]	nt ATGA–polyA[g]	Context[h]	MW[i]	NMD[j]
SelN	3428	12	Exon 9	487	918	796	GGAGGATGACAGTTG	62.6	ND
SelO	2360	9	Exon 9	6	96	254	GTAACATGATCCTCA	74.2	ND
SelV	1271	6	Exon 2	219	72	95	GGCCTCTGAAGCTAT	35.0	+
SelS	1175	6	Exon 6	3	335	208	GGCGGCTGAAACTAA	21.6	–

[a]Selenoproteins were placed into classes based on whether they were not rescued (Class I), partially rescued (Class II), rescued (Class III), or unknown (Class IV).
[b]Total length of cDNA coding region.
[c]Number of exons within the gene.
[d]Exon (within the gene) wherein the TGA resides.
[e]Number of nucleotides between the TGA Sec codon and the stop codon.
[f]Number of nucleotides between the stop codon and the highly conserved ATGA sequence within the SECIS element.
[g]Number of nucleotides between the highly conserved ATGA sequence within the SECIS element and the downstream poly A.
[h]Nucleotide context of the TGA Sec codon.
[i]Molecular weight of selenoproteins.
[j]mRNAs that were subject to NMD were assessed by measuring their recovery as determined in tissues by northern analysis and the levels assessed by PhosphoImager analysis (28). mRNAs that were very sensitive (>90% degradation), moderately sensitive (40–90% degradation), and weakly sensitive (10–40% degradation) to NMD are indicated by +++, ++, and +, respectively, while mRNAs that were not subject to NMD (<10% degradation) are indicated by –. ND, not determined.

liver and kidney than in brain and testes by changes in selenium status (1, 45, 46, 54). As an example of changes in selenium status affecting selenoprotein levels in varying degrees depending on the protein and organ, GPx1 activity was reduced in selenium-deficient rats to 1% in liver and to about 4–9% in heart, kidney, and lungs, while GPx4 activity was reduced to 25–50% in these tissues and unaffected in testes in selenium deficient compared to deficient rats, respectively (96, 97). Transgenic mice expressing an i^6A-deficient Sec tRNA$^{[Ser]Sec}$ isoform had reduced levels of selenium in their tissues and a similar hierarchy in the reduction of selenoprotein activities like that observed in selenium-deficient rodents, and the amounts of selenoprotein reduction varied in a tissue- and protein-specific manner (27).

An increased turnover of GPx1 mRNA has been attributed to the greater sensitivity of GPx1 activity to selenium deficiency (99–101). Under selenium-deficient conditions, the enhanced degradation of GPx1 mRNA is thought to occur by a surveillance pathway designated as nonsense-mediated decay (NMD), wherein the UGA Sec codon is recognized as nonsense (102–104). The position of the UGA Sec codon relative to the sole, downstream intron in GPx1 mRNA determines whether the mRNA is subject to NMD (105). However, during selenium deprivation, other selenoprotein mRNAs, such as D1, GPx4, and SelP, are not as sensitive to NMD as GPx1 despite the presence of introns downstream of their UGA codons (100, 104). Whether SBP2 may play a role in mRNA degradation of different selenoproteins due to preferential recognition of the SECIS element remains an open question (92, 93) and clearly more work is needed to elucidate our understanding of the varying degrees of sensitivity of different selenoprotein mRNAs to NMD.

VI. Selenoprotein Identity, Function, and Targeted Removal

A. Identity of Sec UGA Codons

Most selenoprotein genes are incorrectly annotated in genome sequence databases. The major reason for misannotation is incorrect interpretation of Sec-encoding UGA codons as signals that terminate protein synthesis. Since the translation stop signal is the most commonly used function of UGA codons, the misannotation of selenoprotein genes is a serious problem. In spite of the success in identification of selenoprotein genes in completely sequenced genomes (22, 106–110), including the human genome (22), these genes remain misinterpreted in major genome viewers and specialized sequence databases. There are three major types of errors. First, Sec-encoding UGA codons, which are present in C-terminal regions of selenoproteins are often recognized by the available annotation programs as stop signals. In this case, the coding sequences downstream of UGA codons are lost in annotation. Second, in selenoprotein genes in which Sec is in the N-terminal region, open reading frames (ORFs) are often interpreted to begin from an AUG downstream of Sec. In this case, the N-terminal regions are lost in annotated sequences. Third, exons containing sequences coding for Sec are not predicted. This results in ORFs lacking internal sequences.

Recent years have seen significant progress in the development of tools for identification of selenoprotein genes. Three principal methods have been developed that employ unique features of selenoprotein genes, including (i) presence of SECIS element (22, 81, 106–112), (ii) occurrence of conserved Cys-containing homologs (22, 81, 106, 111–113), and (iii) significant coding potential of sequences downstream of Sec-encoding UGA codons (106, 109, 111). These selenoprotein features are independent from each other and therefore allowed development of independent methods of selenoprotein gene identification.

The first method that was developed searched for SECIS elements (107, 108) and was initially sufficient for identification of these structures in EST databases and small genomes only. Although SECIS elements have low primary sequence conservation, their secondary structures are quite conserved, which allowed the development and computational description of SECIS consensus structures (81, 112, 114–116). In addition, calculation of the free energy of SECIS elements as a measure of their stability helped in computational description of eukaryotic structures. However, as bacterial, archaeal, and eukaryotic SECIS elements are different in sequence and structure, three different programs had to be developed for the identification of these structures (22, 81, 112, 114–116).

The SECISearch method has since been dramatically improved with the key advance being the ability to search for pairs of SECIS elements that are conserved in closely related genomes. As a result, SECISsearch was the principal method in identifying selenoprotein genes in the human genome in parallel with the searches of mouse and rat genomes (22). The same method was extended to predict selenoprotein genes in nematode (111) and Plasmodium (36) genomes.

A second tool involved in selenoprotein searches tested the coding potential of UGA codons (109). It was initially applied to fruit fly genomes and was extended to search for pyrrolysine codons in methanogenic archaea (117). Pyrrolysine is the 22nd amino acid to be included in the genetic code (118).

Finally, the third method takes advantage of the observation that most selenoproteins have homologs, typically in other organisms, in which Cys is present in place of Sec and the Sec/Cys pair is flanked by conserved regions. This method was particularly useful in characterizing selenoproteomes of prokaryotes (112, 113).

B. Selenoproteome

Application of combinations of the three bioinformatics methods/tools to various genomes and sequence databases resulted in identification of a large number of selenoprotein genes and often provided information on selenoproteomes, which are entire sets of selenoproteins in an organism. As discussed above, we now know that humans have 25 selenoprotein genes, whereas the genomes of rats and mice encode 24 selenoproteins (22). This difference is due to GPx6, a novel glutathione peroxide homolog of plasma GPx3, which occurs in the selenoprotein form in primates and some other mammals, such as pigs and cows, but is a Cys-containing protein in mice and rats. Evolutionary analyses revealed that it is the replacement of Sec with Cys in rodents that is responsible for this change in the active site residue. This conclusion is supported by the finding of fossil SECIS-like structure in a rodent GPx6 gene (22).

The number of selenoprotein genes is much lower in fruit flies and nematodes than in mammals. For example, *Drosophila melanogaster* has three selenoprotein genes (109, 110), whereas only a single selenoprotein, thioredoxin reductase, was identified in *Caenorhabditis elegans* and *Caenorhabditis briggsae* (111). In the case of nematodes, it was proposed that the entire system for Sec insertion is used to code for a single UGA codon in these animal genomes. Since thioredoxin reductase can also occur as a Cys-containing protein, it will not be surprising if a member of the animal kingdom is identified that lost altogether the ability to utilize Sec. This prediction is also consistent with the findings that some lower and higher eukaryotes, such as yeast and higher plants, respectively, do not utilize Sec.

It is also clear that unique selenoprotein genes evolved in some eukaryotes. This is illustrated by the finding of MsrA in *Chlamydomonas* (57) and SelU, SelJ, and paralogs of several other selenoproteins in fish (106, 119, 120). Moreover, analysis has identified four selenoprotein genes in *Plasmodium falciparum* and other Plasmodia (36, 37). These proteins do not have significant sequence homology to any other proteins outside of Plasmodia.

Finally, application of selenoprotein gene-identifying programs has revealed a remarkable variety of selenoprotein genes in bacteria, including some proteins that were previously thought to be limited to animals. Particularly revealing was the analysis of the Sargasso Sea environmental genome survey (113). Most of the novel selenoproteins were redox proteins in which catalytic Cys was replaced with Sec.

C. Targeted Removal

As noted above, there are 24 selenoproteins in rodents and 25 in humans (22). Targeting the removal of specific members of this protein class by genetic engineering in mice has shown that some are essential in development whereas others appear to be nonessential (*11*). GPx4 (*121*), TR1 (*122*), and TR3 (*123*) are all essential selenoproteins as their loss is embryonic lethal. The loss of GPx1 (*124*) or GPx2 (*125*) appears to have little or no effect on the animal's phenotype, but other studies provide insights into the roles of these selenoproteins as their targeted removal result in the loss of protective mechanisms against certain environmental stresses [see (*125*) and references therein]. There are other selenoproteins whose removal or mutation results in dramatic effects on health. Knockout of selenoprotein P (SelP) causes neurological problems that can be at least partially corrected by enriching the diet of affected animals with selenium (*126, 127*). Knockout of type 2 iodothyronine deiodinase results in a variety of defects, including an impaired adaptive thermogenesis and hypothermia in cold-exposed mice [see (*128*) and references therein], retarded cochlear development and hearing loss (*129*), and a pituitary resistance to T_4 (*130*). In addition, mutations in selenoprotein N (SelN) in humans result in several muscle disorders (*131, 132*). Mutations in SBP2 found within the human population result in affected individuals having reduced levels of a variety of selenoproteins and thyroid hormone dysfunction (*133*).

A number of embryonic lethal selenoproteins have now been conditionally knocked out using *loxP-Cre* technology, wherein embryonic lethal selenoprotein genes are targeted for removal in specific tissues and organs (*122, 123, 134*). These studies have examined the roles of essential selenoproteins in development and health and have further elucidated their roles. For example, TR1 was found to have major roles in embryogenesis of numerous tissues and organs, except heart (*122*), and TR3 in hematopoiesis and in heart development and function (*123*). On the other hand, the targeted removal of the nuclear form of GPx4 (designated snGPx4) has shown that the animals are viable and completely fertile but this form of GPx4 contributes to the overall structural stability of sperm chromatin (*134*). Knockout of SelP in liver by targeting the knockout of the Sec tRNA$^{[Ser]Sec}$ gene (*trsp*) has provided evidence of a transport function of this liver-derived protein in plasma, while this study substantiated SelP's essential role in brain (*135*).

VII. Mouse Models for Elucidating the Role of Sec tRNA$^{[Ser]Sec}$ in Selenoprotein Biosynthesis and the Role of Selenoproteins in Development and Health

Selenoproteins, unlike most other classes of proteins, are dependent on a single tRNA, Sec tRNA$^{[Ser]Sec}$, for their expression. Thus, by altering the expression of Sec tRNA$^{[Ser]Sec}$, selenoprotein synthesis can be modulated or perturbed providing a means of elucidating the roles of this class of proteins in development and health. Taking advantage of the fact that selenoprotein synthesis is dependent on the presence of Sec tRNA$^{[Ser]Sec}$, we generated several mouse models for assessing the roles of selenoproteins in development and health. The mouse models involve: (i) transgenic animals carrying wild-type or one of two mutant Sec tRNA$^{[Ser]Sec}$ transgenes (27 and Carlson et al., submitted), (ii) conditional knockout animals carrying a floxed Sec tRNA$^{[Ser]Sec}$ gene (trsp) that can be specifically targeted for removal using loxP-Cre technology (21), (iii) transgenic/standard knockout animals carrying wild-type or mutant trsp transgenes and a knockout of trsp in which the survival of the animal is dependent on the transgene (28, 28a), and (iv) double transgenic/conditional knockout animals carrying a Cre transgene under the control of a promoter targeted for a specific organ or tissue, and wild-type or mutant trsp transgenes and floxed trsp (B. A. Carlson et al., submitted). The mutant transgenes contain a mutation at either position 37 in Sec tRNA$^{[Ser]Sec}$ (27) or at position 34. As shown in Fig. 2, the base at position 37 is normally a highly modified base, i^{6}A, and that at position 34 is also a highly modified base, mcm^{5}U. We changed the base at position 37 to G and that at position 34 to A. Neither of these two mutant tRNAs contain Um34, which is the methyl group located on the 2′-hydroxylribose of mcm^{5}U (42). A in the wobble position of the anticodon is normally converted to I in tRNA and about 65% of the A34 mutant Sec tRNA$^{[Ser]Sec}$ is converted to I34 in mouse liver (B. A. Carlson et al., submitted).

These mouse models, their genotypes, the reason they were generated, and some of their uses are summarized in Table III. The main purposes for preparing these mouse models are to define the roles of the two Sec tRNA$^{[Ser]Sec}$ isoforms in selenoprotein synthesis and to assess the roles of selenoproteins in health and development. The mouse models and the roles of the mcm^{5}U and mcm^{5}Um isoforms are discussed in greater detail later.

A. Transgenic Mouse Models

Three separate constructs encoding wild-type or mutant trsp were prepared, which were used to generate transgenic mice (27). The two mutant trsp constructs consisted of a T → A mutation at position 34 and an A → G mutation at position 37 (see also Fig. 2). The construct encoding each trsp and used in generating trsp^t transgenic mice was 2.1 kb (27). This construct was used to prepare several transgenic founder mice containing different gene copy numbers and the wild-type, G37 and A34 transgenes are designated trsp^t, G37trsp^t, and A34trsp^t, respectively.

The resulting transgenic mice encoding multiple copies of trsp^t and G37trsp^t were initially studied and this investigation provided the first example of transgenic mice engineered to contain functional tRNA transgenes (27). Overexpression of wild-type Sec tRNA$^{[Ser]Sec}$ did not appear to affect

TABLE III
Mouse Models, Their Genotype Designations, the Reason for Generating the Mouse Line and Their Uses

Type	Genotype[a]	Purpose[b]	Uses[c]
Transgenic	$trsp^t$	Overexpress $trsp$	Determine if Sec tRNA$^{[Ser]Sec}$ is limiting (27)
			SP^d rescue in standard KO^e (28)
			SP^d replacement in targeted tissues and organs[f]
Transgenic	$G37trsp^t$	Role of Um34 modification in SP^d biosynthesis	Rescue of SPs^d in KO^e mice
			Replacement of SPs in conditional KO^e mice
			Roles in: Muscle growth (146)
			Prostate cancer[f]
			Mammary cancer[f]
Transgenic	$A34trsp^t$	Role of Um34 modification in SP^d biosynthesis	SP^d replacement only
Standard knockout	$\Delta trsp$	Mating with transgenic mice	SP^d rescue with $trsp^t$ and $trsp^tG37$
Conditional knockout	$\Delta trsp^{fl}$	Target $trsp$ removal using $loxP$-Cre technology	KO^e $trsp$ in various tissues and organs using promoters that are tissue and organ specific
Transgenic/standard KO^e	$trsp^t/\Delta trsp$	SP^d rescue	SP^d rescue in KO^e mice (28)
	$G37trsp^t/\Delta trsp$	SP^d rescue	Partial SP^d rescue in KO^e mice (28)
	$trsp^t/\Delta rsp^{fl}$	SP^d replacement	SP^d replacement in liver[g]
Transgenic/conditional KO^e (liver)	$G37trsp^t/\Delta trsp^{fl}$	SP^d replacement	SP^d replacement in liver[g]
	$A34trsp^t/\Delta trsp^{fl}$	SP^d replacement	SP^d replacement in liver[g]

[a] Genotype designation used throughout this report (see text).
[b] Reason the mouse models and lines were generated (see text).
[c] Uses—the various uses of the mouse models with accompanying references (see text).
[d] SP, selenoprotein(s).
[e] KO, knockout.
[f] A. M. Diamond, personal communication and collaborative study with J. Green (NCI, NIH).
[g] B. A. Carlson et al., submitted.

selenoprotein synthesis in the organs examined, providing further evidence that the Sec tRNA$^{[Ser]Sec}$ population was not limiting in protein synthesis [see also (20, 21, 52, and 51)]. The levels of numerous selenoproteins decreased, however, in mice carrying G37trsp^t in a protein and tissue-specific manner. GPx1 and TR1 were the most and least affected selenoproteins, and selenoprotein synthesis was most and least affected in liver and testes. Expression of the G37trsp^t product lacking the i^{6}A base modification altered the distribution of the two endogenous Sec tRNA$^{[Ser]Sec}$ isoforms, whereby the addition of Um34 was repressed and the level of the tRNA$^{[Ser]Sec}$mcm^{5}U was enriched. The total amount of the Sec tRNA$^{[Ser]Sec}$ population remained the same and the variations only occurred in the levels of the two isoforms. However, as the amount of the i^{6}A minus Sec tRNA$^{[Ser]Sec}$ increased, the amount of the Um34 isoform decreased and the amount of certain selenoproteins, for example, GPx1, also decreased. This observation led us to propose that the Um34 modification governed the translation of several selenoproteins that are involved in the lower echelon of selenoprotein hierarchy expression (28, 28a). That is, selenoenzymes, like GPx1, are sensitive to selenium status and are poorly expressed under conditions of limiting selenium, whereas other selenoenzymes, like TR1, are expressed even under conditions of limiting selenium [reviewed in (1)]. In subsequent studies, we realized that many of the selenoproteins responsive to selenium status serve largely stress-related functions, while those less sensitive to selenium status serve largely housekeeping functions and the stress-related members are the ones dependent on the Sec tRNA$^{[Ser]Sec}$ Um34 modification for their expression (see later).

As noted earlier, we also prepared transgenic mice encoding an A34 mutation in the Sec tRNA$^{[Ser]Sec}$ transgene (i.e., T-34 → A34 in which the T in the wobble position of the anticodon that is normally converted to mcm^{5}U in the tRNA gene product was changed to an A). Thus, the A34trsp^t product lacked mcm^{5}U and Um34, while the previously generated G37trsp^t product lacked i^{6}A and Um34. Similar effects on selenoprotein synthesis were observed with the A34 mutant transgenic mice as with the G37 mutant transgenic mouse (B. A. Carlson et al., submitted). Since the two mutant Sec tRNA$^{[Ser]Sec}$ isoforms lack a single, common modification, Um34, most certainly the similar effects of the mutant tRNAs on the selective expression of selenoproteins are due to the absence of this methyl group. In addition, the fact that the distributions of the two Sec tRNA$^{[Ser]Sec}$ isoforms are also shifted towards less mcm^{5}Um and more mcm^{5}U in selenium-deficient mice, and selenoproteins are reduced in a similar pattern in both selenium-deficient mice and the two mutant transgenic mice further substantiate our conclusion that Um34 is specifically involved in the synthesis of stress-related selenoproteins (see also in a later section). Several uses of G37trsp^t and A34trsp^t mice are summarized in Table III.

B. Conditional Knockout Mouse Models

Since the removal of *trsp* from the mouse genome is embryonic lethal (20, 21), the role of selenoproteins in the development and function of specific organs and tissues could not be studied. We therefore conditionally targeted the removal of *trsp* using *loxP-Cre* technology (21). Embryonic stem cells encoding *loxP-neo-loxP-trsp-loxP* in place of *trsp* were prepared and by transfecting this cell line with a IIb*Cre* transgene, we were able to select for a floxed *trsp* (*trsp^{fl}*) cell line. The latter cell line was used to prepare the conditional knockout mouse that encoded *trsp^{fl}* by standard electroporation procedures (21). We initially investigated the role of *trsp* in mouse mammary epithelium by deleting *trsp* using transgenic mice carrying the *Cre* recombinase gene under the control of the mouse mammary tumor virus long terminal repeat promoter or the whey acidic protein promoter. *trsp* was not completely removed in mouse mammary tissue, but the Sec tRNA$^{[Ser]Sec}$ population was reduced substantially to disrupt the pattern of selenoprotein expression in a protein-specific manner (21).

We also examined the role of selenoproteins in liver function (53). To address the role of selenoproteins in this organ, we mated homozygous floxed *trsp* (*trsp$^{fl/fl}$*) mice with transgenic mice carrying the *Cre* recombinase under the control of the albumin promoter that expressed the recombinase specifically in liver. Recombination was nearly complete in mice 3 weeks of age, which correlated with the loss of Sec tRNA$^{[Ser]Sec}$ expression and activities of selenoproteins. Total liver selenium was dramatically decreased that correlated with selenoprotein loss providing further evidence that selenoproteins account for most of the selenium in this organ. Levels of low-molecular-weight selenocompounds were virtually unaffected in liver. Plasma SelP levels were reduced approximately 75%, providing evidence that SelP is primarily exported from the liver. On the other hand, glutathione-*S*-transferase levels were elevated in selenoprotein-deficient liver, suggesting a compensatory activation of this detoxification program. Selenoprotein liver knockout mice appeared normal until about 24 hr before death and death appeared to be due to severe hepatocellular degeneration and necrosis with concomitant necrosis of peritoneal and retroperitoneal fat (53). These studies revealed an essential role of selenoproteins in liver function. It should also be noted that most animals died between 1 and 3 months of age, but it appears that these animals may be kept alive for extended periods of time on a diet enriched in other nutrients (U. Schweizer, L. Schomburg, and J. Kohrle, personal communication). The fact that these mice live much longer on a different diet provides an excellent model for assessing the role of selenoproteins in liver function and promoting health and preventing cancer in this organ.

We have also targeted the selective removal of *trsp* in either endothelial cells or myocytes in skeletal and heart muscle to elucidate the role of selenoproteins in cardiovascular disease (R. Shrimali et al., submitted). The loss of selenoprotein expression in endothelial cells was found to be embryonic lethal. We examined a 14.5-day-old embryo and it had numerous abnormalities, including necrosis of the central nervous system, subcutaneous hemorrhage, and erythrocyte immaturity. On the other hand, loss of selenoprotein expression in myocytes manifested no apparent phenotype until about day 12 after birth. On about day 12, affected mice suddenly developed decreased mobility and an increased respiratory rate, followed by death within approximately 5 hr. Pathological analysis revealed that mice lacking *trsp* in their myocytes had moderate to severe myocarditis with inflammation extending into the mediastinum. There was no evidence of inflammation of the skeletal muscle. These studies demonstrated an essential role of selenoproteins in endothelial cell development and an essential role of selenoproteins in proper function of cardiac muscle. They also provided a direct connection between the loss of selenoprotein expression in these cell types and cardiovascular disease (R. Shrimali et al., submitted).

As selenium has also been implicated in immune function (18), we have also initiated studies on the role of selenoproteins in immune function. The specific knockout of *trsp* in T and macrophage cells has been accomplished and we are examining the effects of selenoprotein loss on the adaptive and innate immune systems.

The conditional knockout of *trsp* provides an important tool for elucidating the roles of selenoproteins in health, development, and/or function in any tissue or organ for which there is a specific promoter. By using promoters regulating Cre expression that are switched on early in development of a specific tissue or organ, we can assess whether selenoproteins play a role in the development of that tissue or organ. By using promoters regulating Cre expression that are switched on after the tissue or organ is developed, we can assess whether selenoproteins play a role in the proper function of that organ or tissue. Finally, by using promoters that are switched on either early or late in the development of a specific tissue or organ, and provided the animal survives for a long period of time following the knockout of *trsp*, we can examine the animal's ability to handle various forms of stress [e.g., viral or bacterial infection, carcinogen(s), cancer driver gene(s), etc.]. Therefore, the role of selenoproteins (and selenium) in protecting the animal against environmental stresses (i.e., in overall health) can be assessed.

C. Transgenic/Knockout Mouse Models

Generation of transgenic mice carrying wild-type or mutant $trsp^t$ transgenes (27) has provided an excellent tool for rescuing mice carrying the standard knockout of *trsp*, which is embryonic lethal, with a transgene (28, 28a). We developed a strain of knockout-transgenic mice wherein the removed *trsp* (Δ*trsp*) was replaced initially with wild-type *trsp*^{t}. Mice rescued with *trsp*^{t} carried 20 copies of the wild-type transgene and thus the Sec tRNA$^{[Ser]Sec}$ population was enriched several fold. Little or no effect, however, was observed by the increased levels of Sec tRNA$^{[Ser]Sec}$ in the tissues examined, providing further evidence that this tRNA is not limiting in selenoprotein biosynthesis (20, 21, 27, 51, 52). Δ*trsp* mice were also rescued with the mutant transgene, G37*trsp*^{t} (28). The advantage of rescuing Δ*trsp* knockout mice with a mutant transgene is that there is no background of host Sec tRNA$^{[Ser]Sec}$ and selenoprotein expression is totally dependent on the mutant tRNA. As noted earlier, the *G37* mutant transgene yielded a tRNA that lacked two base modifications, i^{6}A37 and Um34. In mice rescued with G37*trsp*^{t}, several selenoproteins, including GPx1, GPx3, SelR, and SelT, were not detected. In other avenues pursued in this study, for example, in those involving northern blot analysis, some selenoproteins, such as SelW, were also found to be poorly expressed. Additional selenoproteins, however, including TR1 and TR3 and GPx4 were expressed in normal or reduced levels. The novel regulation of protein expression observed in this study occurred at the level of translation. Furthermore, in this initial rescue study, the data suggested that the Um34 modification had greater influence than the i^{6}A37 modification in regulating the expression of various mammalian selenoproteins and Um34 was required for synthesis of several members of this protein class (28, 28a). In a subsequent study, involving a second mutant *trsp*^{t} transgene, A34*trsp*^{t}, that also lacked Um34, and replaced selenoprotein synthesis in liver (B. A. Carlson *et al.*, submitted), we observed a similar pattern of an effect on selenoprotein expression confirming that Um34 was the base modification in Sec tRNA$^{[Ser]Sec}$ that was responsible for regulating the expression of some members within this protein class.

Many proteins that were poorly rescued in the G37*trsp*^{t} study appeared to be involved in responses to stress and their expression was also highly dependent on selenium in the diet. This led us to propose that Um34 was involved in synthesis of stress-related selenoproteins (28, 28a). Furthermore, the mRNA levels of the affected selenoproteins are regulated by selenium and are subject to NMD (Table II). The G37 rescue study described a novel mechanism of regulation of protein expression by tRNA modification that is in turn regulated by selenium status (28).

D. Transgenic/Conditional Knockout Mouse Models

Rescue of mice encoding a knockout of *trsp* with the *G37* mutant *trsp*^{t} as a model for studying the role of housekeeping and stress-related selenoproteins (28, 28a), for example, in health issues permitted us to examine their role in the whole animal, but did not focus on individual organs and tissues. By generating *trsp* conditional knockout mice carrying a specific promoter *Cre* and a mutant *trsp^t*, the role of housekeeping and stress-related selenoproteins in the development, health and/or proper function in specific organs or tissues can be assessed (B. A. Carlson *et al.*, submitted). The advantage of generating such mice is of course that these models provide important tools for examining individually housekeeping and stress-related selenoprotein function in development, tissue or organ function, and/or a variety of health issues in tissues and organs that selenium has been shown to play a role. Importantly, this mouse model will most certainly provide insights into one of the central questions in the selenium field which is "What are the contributions of selenoproteins vs low-molecular-weight selenocompounds in the cancer chemopreventive effects of selenium and other health benefits of this trace element?" We have targeted selenoprotein removal in liver and have replaced selenoprotein expression with either G37*trsp^t* or A34*trsp^t* and have found that the pattern of replacing selenoprotein expression is similar with both the G37 and A34 mutant transgenes (B. A. Carlson *et al.*, submitted). Both these mutant transgenes replaced housekeeping selenoprotein expression but not stress-related selenoprotein expression. This study, which shows that the two mutant Sec tRNA$^{[Ser]Sec}$ isoforms govern selenoprotein expression in virtually an identical manner without the influence of host wild-type Sec tRNA$^{[Ser]Sec}$, demonstrates that Um34 is responsible for the synthesis of stress-related selenoproteins.

E. Roles of Sec tRNA$^{[Ser]Sec}$mcm^{5}U and mcm^{5}Um in Selenoprotein Biosynthesis

The ability of a single methyl group on the ribosyl moiety in the wobble position of tRNA to regulate the expression of an entire subgroup of proteins has not been previously reported. The fact that Um34 on Sec tRNA$^{[Ser]Sec}$ can govern the expression of stress-related selenoproteins is a novel observation in the area of protein synthesis and raises questions as to how a single modification difference between mcm^{5}U and mcm^{5}Um can result in such dramatic effects on protein synthesis. Various characteristics of the 24 known selenoproteins in rodents (22) are summarized in Table II. These include length of the cDNA, the number of exons, the exon location of the Sec UGA codon, the number of bases between UGA and the termination codon, the number of bases between the termination codon and the first base of the SECIS element, the number of bases between the 3' end of the SECIS element and the beginning of the poly A sequence, the nucleotide context of UGA, the selenoprotein molecular weight, and whether the selenoprotein is subject to NMD. The selenoproteins are placed into one of four classes based on whether the individual member was rescued, partially rescued, not rescued, or its fate unknown in Δ*trsp* mice carrying *G37* mutant transgenes (*28, 28a*). Selenoprotein recoveries were based on western blot analyses, although the mRNA levels that were assessed by northern blot analyses were also suggestive of the efficiency of protein rescue. GPx1, GPx3, SelR, SelT, and SelW were poorly rescued as evidenced by the corresponding protein levels, and in addition, their mRNAs were subject to NMD. SelT mRNA, however, was more stable than the other, poorly rescued selenoprotein mRNAs. SelH and SelV mRNA levels were also regulated, but antibodies were not available for these selenoproteins and thus their ability to be rescued was not determined. Importantly, the fact that some of the poorly rescued selenoproteins have adequate mRNA levels strongly suggests that the Um34 effect on selenoprotein synthesis occurs at the translation level and is independent of NMD (*102, 103*).

A careful survey of each of the selenoprotein characteristics listed in Table II does not reveal any distinct parameters within Class I or II members that would account for their dependence on Um34 for expression. It may be that an unidentified *cis*-acting or *trans*-acting element is involved. In this regard, and as noted earlier, an additional stem-loop structure in several selenoprotein mRNAs that influences the level of translation and is located only a few bases downstream of the UGA codon was observed (*86*). However, this novel stem-loop structure occurred in SelN, SPS2, SelH, SelO, and SelT and does not appear to have any effect on translation of Class I and II members shown in Table II as a whole. Perhaps a more fruitful approach to unraveling the underlying mechanism(s) involved in the role of Um34 in stress-related selenoprotein synthesis is to assess whether different SECIS-binding proteins (SBPs) may be involved. A new SBP that appears to bind preferentially to different selenoprotein mRNAs has been observed (D. Driscoll, personal communication).

VIII. How Did UGA Evolve as the Sec Codon?

Two major ideas have been suggested to account for the means by which Sec entered the genetic code (*63, 136*). Initially, it was proposed that Sec was a component of the primordial genetic code and thus UGA dictated Sec in primitive anaerobic organisms (*63*). In this proposal, as the atmosphere became increasingly rich in oxygen due to photosynthetic organisms, the use of Sec was counterselected because of the sensitivity of this amino acid to oxidation. A subsequent hypothesis predicted that Sec evolved after all codons had assigned functions, and the number of selenoproteins accumulated rather than decreased in evolution (*136*). This latter proposal, which was in contrast to the idea that a declining use of Sec occurred in evolution, suggested that many eukaryotic selenoproteins, serving as antioxidant and redox proteins, were employed by aerobic organisms to function in antioxidant systems.

The above proposals do not tell us the means, however, by which UGA became the codon for Sec. Clearly, UGA is one of the most fascinating codons within the genetic code, as it likely has served more functions than any other codeword in evolution. In current genetic language, UGA serves as: (i) a termination codon (137); (ii) a Sec codon (31, 32); (iii) a cysteine codon in *Euplotes octocarinatus* (138); (iv) a tryptophan codon in mitochondria (139), *Mycoplasma,* and *Sprioplasma* (139, 140); (v) an inefficiently read tryptophan codon in *Bacillus subtilis* (141); and (vi) an inefficiently read codon in *E. coli* that is presumably decoded by tryptophan tRNA (142). In mammals, the UGA termination codon in rabbit β-globin mRNA has been shown to serve as many as eight functions (143), including a stop codon, a suppressor codon that supports partial read through for Arg- Cys-, Trp- and Ser-tRNAs [the latter tRNA is Sec tRNA$^{[Ser]Sec}$ that is aminoacylated with Ser (143)], and a translation reading gap codon with the abyss consisting of one, two, or three codons. Since globin mRNAs in other mammals terminate in UAA or UAG, but do not appear to serve as suppressor codons or to promote translation reading gaps, these functions in rabbit β-globin mRNA are most certainly associated solely with UGA. It would seem that if Sec entered the genetic code later rather than early in the code's development, then it is not too surprising that UGA was a good candidate to take on the additional function of serving to dictate Sec. The 22nd amino acid in the genetic code, pyrrolysine (118), most certainly evolved at a stage after the 20 other amino acids had their assigned codewords and pyrrolysine, like Sec, is encoded by a stop codon.

Furthermore, since other stop or infrequently read codons can code for Sec, provided the anticodon in Sec tRNA is complementary to the corresponding codon used in place of TGA (144, 145), it would seem that any of a number of codons could have evolved for Sec. However, the variety of functions of UGA suggests that this codon has been loosely programmed in evolution and therefore might be the most likely codeword to have evolved for the infrequently used Sec. This possibility would seem to be even more plausible if the inclusion of Sec in the genetic code occurred in evolution after the code had evolved rather than the code having evolved to accommodate Sec.

There are other compelling reasons for suggesting that Sec entered the genetic code late rather than early. That is, Sec is dramatically different from any other of the 20 protein amino acids in the mode of its basic biosynthetic steps and incorporation into protein. Unlike the other 20 amino acids, it requires a structural element in mRNA for proper insertion into protein in addition to the information specified by the genetic code. The Sec biosynthetic machinery is strikingly different from that of other amino acids and employs numerous additional Sec-specific components. These unique features of Sec biosynthesis and insertion certainly favor the view that Sec was added to the already existing genetic code to take advantage of the unique chemistry of selenium (136).

IX. Concluding Remarks

Since the genetic code was deciphered in the mid-1960s (137), there have been two new amino acid additions to the code. Sec, which is decoded by UGA, was the first addition. Since the mid-1980s, enormous progress has been made in understanding the mechanism of how Sec is synthesized and inserted into nascent selenopeptides in both bacteria and mammals. The bacterial Sec synthesis and protein insertion systems are virtually completely understood due to the investigations of Bock and collaborators [see (33, 63) for reviews]. In mammals, this machinery is far more complex, but many discoveries of how specific 3'-UTR mRNA structures, designated SECIS elements, function in recruiting the SBP(s), the Sec-specific elongation factor and selenocystenyl-tRNA$^{[Ser]Sec}$ into a large Sec insertion complex, the selenosome, have been accomplished. Sec tRNA$^{[Ser]Sec}$ is the central component in this unique amino acid system as it is used both as the site for Sec biosynthesis and its incorporation into protein. The resulting selenoproteins are a distinctive class of proteins in that their expression is dependent on this tRNA and on the presence of SECIS elements that dictate UGA as Sec and not stop. SECIS elements have been used as a focus in computational studies to identify numerous new selenoprotein genes. Their subsequent characterizations have led to, and will continue to lead to, elucidating many biological and health-related properties of selenium. Several novel mouse models, including transgenic, knockout, conditional knockout, transgenic/knockout, and transgenic/conditional knockout mouse lines, have been generated using the Sec tRNA$^{[Ser]Sec}$ gene (*trsp*) as a tool to study the roles of the two Sec tRNA$^{[Ser]Sec}$ isoforms in selenoprotein expression and the roles of selenoproteins, both housekeeping and stress-related members of this class, in development and health.

Acknowledgments

The authors express their sincere appreciation to Dr. Alexey Lobanov for his assistance with the figures. This research was supported by the Intramural Research Program of the National Institutes of Health, National Cancer Institute, Center for Cancer Research, and by National Institutes of Health grants awarded to Vadim N. Gladyshev.

References

1. Hatfield, D. L., and Gladyshev, V. N. (2002). How selenium has altered our understanding of the genetic code. *Mol. Cell. Biol.* **22**, 3565–3576.
2. Birringer, M., Pilawa, S., and Flohe, L. (2002). Trends in selenium biochemistry. *Nat. Prod. Rep.* **19**, 693–718.
3. Driscoll, D. M., and Copeland, P. R. (2003). Mechanism and regulation of selenoprotein synthesis. *Annu. Rev. Nutr.* **23**, 17–40.
3a. Copeland, P. R. (2003). Regulation of gene expression by stop codon recoding: Selenocysteine. *Gene* **312**, 17–25.
4. Schomburg, L., Schweizer, U., and Kohrle, J. (2004). Selenium and selenoproteins in mammals: Extraordinary, essential, enigmatic. *Cell Mol. Life Sci.* **61**, 1988–1995.
5. Schwiezer, U., Brauer, A. U., Kohrle, J., Nitsch, R., and Savaskan, N. E. (2004). Selenium and brain function: A poorly recognized liaison. *Brain Res. Rev.* **45**, 164–178.
6. Copeland, P. R. (2005). Making sense of nonsense: The evolution of selenocysteine usage in proteins. *Genome Biol.* **6**, 222.
7. Davis, C. D., and Irons, R. (2005). Are selenoproteins important for the cancer protective effects of selenium? *Curr. Nutr. Food Sci.* **1**, 201–214.
8. Small-Howard, A. L., and Berry, M. J. (2005). Unique features of selenocysteine incorporation function within the context of general eukaryotic translational processes. *Biochem. Soc. Trans.* **33**, 1493–1497.
9. Caban, K., and Copeland, P. R. (2006). Size matters: A view of selenocysteine incorporation from the ribosome. *Cell. Mol. Life Sci.* **63**, 73–81.
10. Hoffmann, P. R., and Berry, M. (2005). Selenoprotein synthesis: A unique translational mechanism used by a diverse family of proteins. *Thyroid* **15**, 769–775.
11. Schweizer, U., and Schomburg, L. (2005). New insights into the physiological actions of selenoproteins from genetically modified mice. *IUBMB Life* **57**, 737–744.
12. Hatfield, D. L., Berry, M. J., and Gladyshev, V. N., Eds. (2006). *In* "Selenium: Its Molecular Biology and Role in Human Health." Springer Science+Business Media, LLC, New York, NY, USA (In press).
13. Combs, G. F., and Lu, L. (2001). Selenium as a cancer preventive agent. *In* "Selenium: Its Molecular Biology and Role in Human Health" (D. L. Hatfield, Ed.), Chap 17, pp. 205–217. Kluwer Academic Publishers, Norwell, MA, USA.
14. Coppinger, R. J., and Diamond, A. M. (2001). Selenium deficiency and human disease. *In* "Selenium: Its Molecular Biology and Role in Human Health" (D. L. Hatfield, Ed.), Chap 18, pp. 219–233. Kluwer Academic Publishers, Norwell, MA, USA.
15. Beck, M. A. (2001). Selenium as an antiviral agent. *In* "Selenium: Its Molecular Biology and Role in Human Health" (D. L. Hatfield, Ed.), Chap 19, pp. 235–245. Kluwer Academic Publishers, Norwell, MA, USA.
16. Beck, M. A., Levander, O. A., and Handy, J. (2003). Selenium deficiency and viral infection. *J. Nutr.* **133**, 1463–1467.
17. Baum, M. K., Campa, A., Miguez-Burbano, M. J., Burbano, X., and Shor-Posner, G. (2001). Role of selenium in HIV/AIDS. *In* "Selenium: Its Molecular Biology and Role in Human Health" (D. L. Hatfield, Ed.), Chap 20, pp. 247–255. Kluwer Academic Publishers, Norwell, MA, USA.
18. McKenzie, R. C., Rafferty, T. S., Beckett, G. J., and Arthur, J. R. (2001). Effects of selenium on immunity and aging. *In* "Selenium: Its Molecular Biology and Role in Human Health" (D. L. Hatfield, Ed.), Chap 21, pp. 257–272. Kluwer Academic Publishers, Norwell, MA, USA.

19. Flohe, L., Brigelius-Flohe, R., Maiorino, M., Roveri, A., Wissing, J., and Ursini, F. (2001). Selenium and male reproduction. In "Selenium: Its Molecular Biology and Role in Human Health" (D. L. Hatfield, Ed.), Chap 22, pp. 273–281. Kluwer Academic Publishers, Norwell, MA, USA.
20. Bosl, M. R., Takaku, K., Oshima, M., Nishimura, S., and Taketo, M. M. (1997). Early embryonic lethality caused by targeted disruption of the mouse selenocysteine tRNA gene (Trsp). *Proc. Natl. Acad. Sci. USA* **94**, 5531–5534.
21. Kumaraswamy, E., Carlson, B. A., Morgan, F., Miyoshi, K., Robinson, G. W., Su, D., Wang, S., Southon, E., Tessarollo, L., Lee, B. J., Gladyshev, V. N., Hennighausen, L. et al. (2003). Selective removal of the selenocysteine tRNA [Ser]Sec gene (Trsp) in mouse mammary epithelium. *Mol. Cell. Biol.* **23**, 1477–1488.
22. Kryukov, G. V., Castellano, S., Novoselov, S. V., Lobanov, A. V., Zehtab, O., Guigo, R., and Gladyshev, V. N. (2003). Characterization of mammalian selenoproteomes. *Science* **300**, 1439–1443.
23. Diwadkar-Navsariwala, V., and Diamond, A. M. (2004). The link between selenium and chemoprevention: A case for selenoproteins. *J. Nutr.* **134**, 2899–2902.
24. Thompson, H. J. (2001). Role of low molecular weight, selenium-containing compounds in human health. In "Selenium: Its Molecular Biology and Role in Human Health" (D. L. Hatfield, Ed.), Chap 23, pp. 283–297. Kluwer Academic Publishers, Norwell, MA, USA.
25. Hu, Y. J., Dolan, M. E., Bae, R., Yee, H., Roy, M., Glickman, R., Kiremidjian-Schumacher, L., and Diamond, A. M. (2004). Allelic loss at the GPx-1 locus in cancer of the head and neck. *Biol. Trace Elem. Res.* **101**, 97–106.
26. Diwadkar-Navsariwala, V., and Diamond, A. M. (2004). The link between selenium and chemoprevention: A case for selenoproteins. *J. Nutr.* **134**, 2899–2902.
27. Moustafa, M. E., Carlson, B. A., El-Saadani, M. A., Kryukov, G. V., Sun, Q. A., Harney, J. W., Hill, K. E., Combs, G. F., Feigenbaum, L., Mansur, D. B., Burk, R. F., Berry, M. J. et al. (2001). Selective inhibition of selenocysteine tRNA maturation and selenoprotein synthesis in transgenic mice expressing isopentenyladenosine-deficient selenocysteine tRNA. *Mol. Cell. Biol.* **21**, 3840–3852.
28. Carlson, B. A., Xu, X. M., Gladyshev, V. N., and Hatfield, D. L. (2005). Selective rescue of selenoprotein expression in mice lacking a highly specialized methyl group in selenocysteine tRNA. *J. Biol. Chem.* **280**, 5542–5548.
28a. Carlson, B. A., Xu, X.-M., Gladyshev, V. N., and Hatfield, D. L. (2005). Um34 in selenocysteine tRNA is required for the expression of stress-related selenoproteins in mammals. In "Topics in Current Genetics" (H. Grosjean, Ed.), Vol. 12, Chap. 16, pp. 431–438. Springer-Verlag, Berlin-Heidelberg, Germany.
29. Chambers, I., Frampton, J., Goldfarb, P., Affara, N., McBain, W., and Harrison, P. R. (1986). The structure of the mouse glutathione peroxidase gene: The selenocysteine in the active site is encoded by the "termination" codon, TGA. *EMBO J.* **5**, 1221–1227.
30. Zinoni, F., Birkmann, A., Stadtman, T. C., and Bock, A. (1986). Nucleotide sequence and expression of the selenocysteine-containing polypeptide of formate dehydrogenase (formate-hydrogen-lyase-linked) from *Escherichia coli*. *Proc. Natl. Acad. Sci. USA* **83**, 4650–4654.
31. Lee, B. J., Worland, P. J., Davis, J. N., Stadtman, T. C., and Hatfield, D. L. (1989). Identification of a selenocysteyl-tRNA(Ser) in mammalian cells that recognizes the nonsense codon, UGA. *J. Biol. Chem.* **264**, 9724–9727.
32. Leinfelder, W., Stadtman, T. C., and Bock, A. (1989). Occurrence *in vivo* of selenocysteyl-tRNA(SERUCA) in *Escherichia coli*. Effect of sel mutations. *J. Biol. Chem.* **264**, 9720–9723.
33. Bock, A., Forchhammer, K., Heider, J., and Baron, C. (1991). Selenoprotein synthesis: An expansion of the genetic code. *Trends Biochem. Sci.* **16**, 463–467.

34. Hatfield, D. L., Choi, I. S., Ohama, T., Jung, J.-E., and Diamond, A. M. (1994). Selenocysteine tRNA(Ser)sec isoacceptors as central components in selenoprotein biosynthesis in eukaryotes. In "Selenium in Biology and Human Health" (R. F. Burk, Ed.), Chap 2, pp. 25–44. Springer-Verlag, New York, NY, USA.
35. Shrimali, R. K., Lobanov, A. V., Xu, X. M., Rao, M., Carlson, B. A., Mahadeo, D. C., Parent, C. A., Gladyshev, V. N., and Hatfield, D. L. (2005). Selenocysteine tRNA identification in the model organisms *Dictyostelium discoideum* and *Tetrahymena thermophila*. *Biochem. Biophys. Res. Commun.* **329**, 147–151.
36. Lobanov, A. V., Delgado, C., Rahlfs, S., Novoselov, S. V., Kryukov, G. V., Gromer, S., Hatfield, D. L., Becker, K., and Gladyshev, V. N. (2006). The *Plasmodium* selenoproteome. *Nucl. Acids Res.* **34**, 496–505.
37. Mourier, T., Pain, A., Barrell, B., and Griffiths-Jones, S. (2005). A selenocysteine tRNA and SECIS element in *Plasmodium falciparum*. *RNA* **11**, 119–122.
38. Rao, M., Carlson, B. A., Novoselov, S. V., Weeks, D. P., Gladyshev, V. N., and Hatfield, D. L. (2003). *Chlamydomonas reinhardtii* selenocysteine tRNA[Ser]Sec. *RNA* **9**, 923–930.
39. Hubert, N., Sturchler, C., Westhof, E., Carbon, P., and Krol, A. (1998). The 9/4 secondary structure of eukaryotic selenocysteine tRNA: More pieces of evidence. *RNA* **4**, 1029–1033.
40. Choi, I. S., Diamond, A. M., Crain, P. F., Kolker, J. D., McCloskey, J. A., and Hatfield, D. L. (1994). Reconstitution of the biosynthetic pathway of selenocysteine tRNAs in *Xenopus* oocytes. *Biochemistry* **33**, 601–605.
41. Sturchler, C., Lescure, A., Keith, G., Carbon, P., and Krol, A. (1994). Base modification pattern at the wobble position of *Xenopus* selenocysteine tRNA(Sec). *Nucleic Acids Res.* **22**, 1354–1358.
42. Kim, L. K., Matsufuji, T., Matsufuji, S., Carlson, B. A., Kim, S. S., Hatfield, D. L., and Lee, B. J. (2000). Methylation of the ribosyl moiety at position 34 of selenocysteine tRNA[Ser]Sec is governed by both primary and tertiary structure. *RNA* **6**, 1306–1315.
43. Lee, B. J., de la Pena, P., Tobian, J. A., Zasloff, M., and Hatfield, D. (1987). Unique pathway of expression of an opal suppressor phosphoserine tRNA. *Proc. Natl. Acad. Sci. USA* **84**, 6384–6388.
44. Hatfield, D. L., Gladyshev, V. N., Park, J. M., Park, S. I., Chittum, H. S., Huh, J. H., Carlson, B. A., Kim, M., Moustafa, M. E., and Lee, B. J. (1999). Biosynthesis of selenocysteine and its incorporation into protein as the 21st amino acid. In "Comprehensive Natural Products Chemistry, Vol. 4". (J. W. Kelly, Ed.), pp. 353–380. Elsevier Science, Oxford, England.
45. Chittum, H. S., Hill, K. E., Carlson, B. A., Lee, B. J., Burk, R. F., and Hatfield, D. L. (1997). Replenishment of selenium deficient rats with selenium results in redistribution of the selenocysteine tRNA population in a tissue specific manner. *Biochim. Biophys. Acta* **1359**, 25–34.
46. Diamond, A. M., Choi, I. S., Crain, P. F., Hashizume, T., Pomerantz, S. C., Cruz, R., Steer, C. J., Hill, K. E., Burk, R. F., McCloskey, J. A., and Hatfield, D. L. (1993). Dietary selenium affects methylation of the wobble nucleoside in the anticodon of selenocysteine tRNA([Ser]Sec). *J. Biol. Chem.* **268**, 14215–14223.
47. Ding, F., and Grabowski, P. J. (1999). Identification of a protein component of a mammalian tRNA(Sec) complex implicated in the decoding of UGA as selenocysteine. *RNA* **5**, 1561–1569.
48. Gelpi, C., Sontheimer, E. J., and Rodriguez-Sanchez, J. L. (1992). Autoantibodies against a serine tRNA-protein complex implicated in cotranslational selenocysteine insertion. *Proc. Natl. Acad. Sci. USA* **89**, 9739–9743.
49. Xu, X.-M., Mix, H., Carlson, B. A., Grabowski, P. J., Gladyshev, V. N., Berry, M. J., and Hatfield, D. L. (2005). Evidence for direct roles of two additional factors, Secp43 and SLA, in the selenoprotein synthesis machinery. *J. Biol. Chem.* **280**, 41568–41575.

50. Small-Howard, A., Morozova, N., Stoytcheva, Z., Forry, E. P., Mansell, J. B., Harney, J. W., Carlson, B. A., Xu, X.-M., Hatfield, D. L., and Berry, M. J. (2006). A supramolecular complex mediates selenocysteine incorporation *in vivo*. *Mol. Cell. Biol.* **26**, 2337–2346.
51. Chittum, H. S., Baek, H. J., Diamond, A. M., Fernandez-Salguero, P., Gonzalez, F., Ohama, T., Hatfield, D. L., Kuehn, M., and Lee, B. J. (1997). Selenocysteine tRNA[Ser]Sec levels and selenium-dependent glutathione peroxidase activity in mouse embryonic stem cells heterozygous for a targeted mutation in the tRNA[Ser]Sec gene. *Biochemistry* **36**, 8634–8639.
52. Moustafa, M. E., El-Saadani, M. A., Kandeel, K. M., Mansur, D. B., Lee, B. J., Hatfield, D. L., and Diamond, A. M. (1998). Overproduction of selenocysteine tRNA in Chinese hamster ovary cells following transfection of the mouse tRNA[Ser]Sec gene. *RNA* **4**, 1436–1443.
53. Carlson, B. A., Novoselov, S. V., Kumaraswamy, E., Lee, B. J., Anver, M. R., Gladyshev, V. N., and Hatfield, D. L. (2004). Specific excision of the selenocysteine tRNA[Ser]Sec (Trsp) gene in mouse liver demonstrates an essential role of selenoproteins in liver function. *J. Biol. Chem.* **279**, 8011–8017.
54. Hatfield, D., Lee, B. J., Hampton, L., and Diamond, A. M. (1991). Selenium induces changes in the selenocysteine tRNA[Ser]Sec population in mammalian cells. *Nucleic Acids Res.* **19**, 939–943.
55. Lee, B. J., Rajagopalan, M., Kim, Y. S., You, K. H., Jacobson, K. B., and Hatfield, D. (1990). Selenocysteine tRNA[Ser]Sec gene is ubiquitous within the animal kingdom. *Mol. Cell. Biol.* **10**, 1940–1949.
56. Heider, J., and Bock, A. (1993). Selenium metabolism in micro-organisms. *Adv. Microb. Physiol.* **35**, 71–109.
57. Novoselov, S. V., Rao, M., Onoshko, N. V., Zhi, H., Kryukov, G. V., Xiang, Y., Weeks, D. P., Hatfield, D. L., and Gladyshev, V. N. (2002). Selenoproteins and selenocysteine insertion system in the model plant cell system, *Chlamydomonas reinhardtii*. *EMBO J.* **21**, 3681–3693.
57a. Labanov, A. V., Kryukov, G. V., Hatfield, D. L., and Gladyshev, V. N. (2006). Is there a 23rd amino acid in the genetic code? *TIG* (in press).
58. Tumbula, D. L., Becker, H. D., Chang, W. Z., and Soll, D. (2000). Domain-specific recruitment of amide amino acids for protein synthesis. *Nature* **407**, 106–110.
59. Wu, X. Q., and Gross, H. J. (1993). The long extra arms of human tRNA((Ser)Sec) and tRNA (Ser) function as major identify elements for serylation in an orientation-dependent, but not sequence-specific manner. *Nucleic Acids Res.* **21**, 5589–5594.
60. Ohama, T., Yang, D. C., and Hatfield, D. L. (1994). Selenocysteine tRNA and serine tRNA are aminoacylated by the same synthetase, but may manifest different identities with respect to the long extra arm. *Arch. Biochem. Biophys.* **315**, 293–301.
61. Amberg, R., Mizutani, T., Wu, X. Q., and Gross, H. J. (1996). Selenocysteine synthesis in mammalia: An identity switch from tRNA(Ser) to tRNA(Sec). *J. Mol. Biol.* **263**, 8–19.
62. Sunde, R. A., and Evenson, J. K. (1987). Serine incorporation into the selenocysteine moiety of glutathione peroxidase. *J. Biol. Chem.* **262**, 933–937.
63. Bock, A. (2001). Selenium metabolism in bacteria. *In* "Selenium: Its Molecular Biology and Role in Human Health" (D. L. Hatfield, Ed.), Chap 2, pp. 7–22. Kluwer Academic Publishers, Norwell, MA, USA.
64. Maenpaa, P. H., and Bernfield, M. R. (1970). A specific hepatic transfer RNA for phosphoserine. *Proc. Natl. Acad. Sci. USA* **67**, 688–695.
65. Hatfield, D., and Portugal, F. H. (1970). Seryl-tRNA in mammalian tissues: Chromatographic differences in brain and liver and a specific response to the codon, UGA. *Proc. Natl. Acad. Sci. USA* **67**, 1200–1206.
66. Hatfield, D., Diamond, A., and Dudock, B. (1982). Opal suppressor serine tRNAs from bovine liver form phosphoseryl-tRNA. *Proc. Natl. Acad. Sci. USA* **79**, 6215–6219.

67. Carlson, B. A., Xu, X. M., Kryukov, G. V., Rao, M., Berry, M. J., Gladyshev, V. N., and Hatfield, D. L. (2004). Identification and characterization of phosphoseryl-tRNA[Ser]Sec kinase. *Proc. Natl. Acad. Sci. USA* **101**, 12848–12853.
67a. Diamond, A. M. (2004). On the road to selenocysteine. *Proc. Natl. Acad. Sci. USA* **101**, 12848–12853.
68. Sauerwald, A., Zhu, W., Major, T. A., Roy, H., Palioura, S., Jahn, D., Whitman, W. B., Yates, J. R., III, Ibba, M., and Soll, D. (2005). RNA-dependent cysteine biosynthesis in archaea. *Science* **307**, 1969–1972.
69. Glass, R. S., Singh, W. P., Jung, W., Veres, Z., Scholz, T. D., and Stadtman, T. C. (1993). Monoselenophosphate: Synthesis, characterization, and identity with the prokaryotic biological selenium donor, compound SePX. *Biochemistry* **32**, 12555–12559.
70. Guimaraes, M. J., Peterson, D., Vicari, A., Cocks, B. G., Copeland, N. G., Gilbert, D. J., Jenkins, N. A., Ferrick, D. A., Kastelein, R. A., Bazan, J. F., and Zlotnik, A. (1996). Identification of a novel selD homolog from eukaryotes, bacteria, and archaea: Is there an autoregulatory mechanism in selenocysteine metabolism? *Proc. Natl. Acad. Sci. USA* **93**, 15086–15091.
71. Kim, I. Y., and Stadtman, T. C. (1995). Selenophosphate synthetase: Detection in extracts of rat tissues by immunoblot assay and partial purification of the enzyme from the archaean *Methanococcus vannielii*. *Proc. Natl. Acad. Sci. USA* **92**, 7710–7713.
72. Low, S. C., Harney, J. W., and Berry, M. J. (1995). Cloning and functional characterization of human selenophosphate synthetase, an essential component of selenoprotein synthesis. *J. Biol. Chem.* **270**, 21659–21664.
73. Low, S. C., and Berry, M. J. (1996). Knowing when not to stop: Selenocysteine incorporation in eukaryotes. *Trends Biochem. Sci.* **21**, 203–208.
74. Tujebajeva, R. M., Copeland, P. R., Xu, X. M., Carlson, B. A., Harney, J. W., Driscoll, D. M., Hatfield, D. L., and Berry, M. J. (2000). Decoding apparatus for eukaryotic selenocysteine insertion. *EMBO Rep.* **1**, 158–163.
75. Fagegaltier, D., Hubert, N., Yamada, K., Mizutani, T., Carbon, P., and Krol, A. (2000). Characterization of mSelB, a novel mammalian elongation factor for selenoprotein translation. *EMBO J.* **19**, 4796–4805.
76. Copeland, P. R., Fletcher, J. E., Carlson, B. A., Hatfield, D. L., and Driscoll, D. M. (2000). A novel RNA binding protein, SBP2, is required for the translation of mammalian selenoprotein mRNAs. *EMBO J.* **19**, 306–314.
77. Chavatte, L., Brown, B. A., and Driscoll, D. M. (2005). Ribosomal protein L30 is a component of the UGA-selenocysteine recoding machinery in eukaryotes. *Nat. Struct. Mol. Biol.* **12**, 408–416.
78. Grundner-Culemann, E., Martin, G. W., III, Harney, W., and Berry, M. J. (1999). Two distinct SECIS structures capable of directing selenocysteine incorporation in eukaryotes. *RNA* **5**, 625–635.
79. Korotkov, K. V., Novoselov, S. V., Hatfield, D. L., and Gladyshev, V. N. (2002). Mammalian selenoprotein in which selenocysteine (Sec) incorporation is supported by a new form of Sec insertion sequence element. *Mol. Cell. Biol.* **22**, 1402–1411.
80. Rother, M., Resch, A., Gardner, W. L., Whitman, W. B., and Bock, A. (2001). Heterologous expression of archaeal selenoprotein genes directed by the SECIS element located in the 3′ non-translated region. *Mol. Microbiol.* **40**, 900–908.
81. Zhang, Y., and Gladyshev, V. N. (2005). An algorithm for identification of bacterial selenocysteine insertion sequence elements and selenoprotein genes. *Bioinformatics* **21**, 2580–2589.

82. Zavacki, A. M., Mansell, J. B., Chung, M., Klimovitsky, B., Harney, J. W., and Berry, M. J. (2003). Coupled tRNA(Sec)-dependent assembly of the selenocysteine decoding apparatus. *Mol. Cell* **11,** 773–781.
83. Kinzy, S. A., Caban, K., and Copeland, P. R. (2005). Characterization of the SECIS binding protein 2 complex required for the co-translational insertion of selenocysteine in mammals. *Nucleic Acids Res.* **33,** 5172–5180.
84. Chavatte, L., Brown, B. A., and Driscoll, D. M. (2005). Ribosomal protein L30 is a component of the UGA-selenocysteine recoding machinery in eukaryotes. *Nat. Struct. Mol. Biol.* **12,** 408–416.
85. Berry, M. J. (2005). Knowing when to stop. *Nat. Struct. Mol. Biol.* **12,** 389–390.
86. Howard, M. T., Aggarwal, G., Anderson, C. B., Khatri, S., Flanigan, K. M., and Atkins, J. F. (2005). Recoding elements located adjacent to a subset of eukaryal selenocysteine-specifying UGA codons. *EMBO J.* **24,** 1596–1607.
87. Ma, S., Hill, K. E., Caprioli, R. M., and Burk, R. F. (2002). Mass spectrometric characterization of full-length rat selenoprotein P and three isoforms shortened at the C terminus. Evidence that three UGA codons in the mRNA open reading frame have alternative functions of specifying selenocysteine insertion or translation termination. *J. Biol. Chem.* **277,** 12749–12754.
88. McCaughan, K. K., Brown, C. M., Dalphin, M. E., Berry, M. J., and Tate, W. P. (1995). Translational termination efficiency in mammals is influenced by the base following the stop codon. *Proc. Natl. Acad. Sci. USA* **92,** 5431–5435.
89. Nasim, M. T., Jaenecke, S., Belduz, A., Kollmus, H., Flohe, L., and McCarthy, J. E. G. (2000). Eukaryotic selenocysteine incorporation follows a nonprocessive mechanism that competes with translational termination. *J. Biol. Chem.* **275,** 14846–14852.
90. Grundner-Culemann, E., Martin, G. W., III, Tujebajeva, R., Harney, J. W., and Berry, M. J. (2001). Interplay between termination and translation machinery in eukaryotic selenoprotein synthesis. *J. Mol. Biol.* **310,** 699–707.
91. Copeland, P. R., Stepanik, V. A., and Driscoll, D. M. (2001). Insight into mammalian selenocysteine insertion: Domain structure and ribosome binding properties of Sec insertion sequence binding protein 2. *Mol. Cell. Biol.* **21,** 1491–1498.
92. Fletcher, J. E., Copeland, P. R., Driscoll, D. M., and Krol, A. (2001). The selenocysteine incorporation machinery: Interactions between the SECIS RNA and the SECIS-binding protein SBP2. *RNA* **7,** 1442–1453.
93. Low, S. C., Grundner-Culemann, E., Harney, J. W., and Berry, M. J. (2000). SECIS-SBP2 interactions dictate selenocysteine incorporation efficiency and selenoprotein hierarchy. *EMBO J.* **19,** 6882–6890.
94. Allan, C.B, Lacourciere, G. M., and Stadtman, T. C. (1999). Responsiveness of selenoproteins to dietary selenium. *Annu. Rev. Nutr.* **19,** 1–16.
95. Brigelius-Flohe, R. (1999). Tissue-specific functions of individual glutathione peroxidases. *Free Radic. Biol. Med.* **27,** 951–965.
96. Behne, D., Hilmet, H., Scheid, S., Gessner, H., and Elger, W. (1988). Evidence for specific selenium target tissues and new biologically important selenoproteins. *Biochim. Biophys. Acta.* **996,** 12–21.
97. Hill, K. E., Lyons, P. R., and Burk, R. F. (1992). Differential regulation of rat liver selenoprotein mRNAs in selenium deficiency. *Biochem. Biophys. Res. Commun.* **185,** 260–263.
98. Mitchell, J. H., Nicol, F., Beckett, G. J., and Arthur, J. R. (1997). Selenium and iodine deficiencies: Effects on brain and brown adipose tissue selenoenzyme activity and expression. *J. Endocrinol.* **155,** 255–263.
99. Christensen, M. J., and Burgener, K. W. (1992). Dietary selenium stabilized glutathione peroxidase mRNA in rat liver. *J. Nutr.* **122,** 1620–1626.

100. Lei, X. G., Evenson, J. K., Thompson, K. M., and Sunde, R. A. (1995). Glutathione peroxidase and phospholipid hydroperoxide glutathione peroxidase are differentially regulated in rats by dietary selenium. *J. Nutr.* **125,** 1438–1446.
101. Saedi, M. S., Smith, C. G., Frampton, J., Chambers, I., Harrison, P. R., and Sunde, R. A. (1988). Effect of selenium status on mRNA levels for glutathione peroxidase in rat liver. *Biochem. Biophys. Res. Commun.* **153,** 855–861.
102. Moriarty, P. M., Reddy, C. C., and Maquat, L. E. (1998). Selenium deficiency reduces the abundance of mRNA for Se-dependent glutathione peroxidase 1 by a GUA-dependent mechanism likely to be nonsense codon-mediated decay of cytoplasmic mRNA. *Mol. Cell. Biol.* **18,** 2932–2939.
103. Weiss, S. L., and Sunde, R. A. (1998). Cis-acting elements are required for selenium regulation of glutathione peroxidase-1 mRNA levels. *RNA* **4,** 816–827.
104. Sun, X., Li, X., Moriarty, P. M., Henics, T., LaDuca, J. P., and Maquat, L. E. (2001). Nonsense-mediated decay of mRNA for the selenoprotein phospholipid hydroperoxide glutathione peroxidase is detectable in cultured cells but masked or inhibited in rat tissues. *Mol. Biol. Cell* **12,** 1009–1017.
105. Sun, X., Moriarty, P. M., and Maquat, L. E. (2000). Nonsense-mediated decay of glutathione peroxidase 1 mRNA in the cytoplasm depends on intron position. *EMBO J.* **19,** 4734–4744.
106. Castellano, S., Novoselov, S. V., Kryukov, G. V., Lescure, A., Blanco, E., Krol, A., Gladyshev, V. N., and Guigo, R. (2004). Reconsidering the evolution of eukaryotic selenoproteins: A novel nonmammalian family with scattered phylogenetic distribution. *EMBO Rep.* **5,** 71–77.
107. Kryukov, G. V., Kryukov, V. M., and Gladyshev, V. N. (1999). New mammalian selenocysteine-containing proteins identified with an algorithm that searches for selenocysteine insertion sequence elements. *J. Biol. Chem.* **274,** 33888–33897.
108. Lescure, A., Gautheret, D., Carbon, P., and Krol, A. (1999). Novel selenoproteins identified *in silico* and *in vivo* by using a conserved RNA structural motif. *J. Biol. Chem.* **274,** 38147–38154.
109. Castellano, S., Morozova, N., Morey, M., Berry, M. J., Serras, F., Corominas, M., and Guigo, R. (2001). *In silico* identification of novel selenoproteins in the Drosophila melanogaster genome. *EMBO Rep.* **2,** 697–702.
110. Martin-Romero, F. J., Kryukov, G. K., Lobanov, A. V., Carlson, B. A., Lee, B. J., Gladyshev, V. N., and Hatfield, D. L. (2001). Selenium metabolism in Drosophila: Selenoproteins, selenoprotein mRNA expression, fertility and mortality. *J. Biol. Chem.* **276,** 29798–29804.
111. Taskov, K., Chapple, C., Kryukov, G. V., Castellano, S., Lobanov, A. V., Korotkov, K. V., Guigó, R., and Gladyshev, V. N. (2005). Nematode selenoproteome: The use of the selenocysteine insertion system to decode one codon in an animal genome? *Nucleic Acids Res.* **33,** 2227–2238.
112. Kryukov, G. V., and Gladyshev, V. N. (2004). The prokaryotic selenoproteome. *EMBO Rep.* **5,** 538–543.
113. Zhang, Y., Fomenko, D. E., and Gladyshev, V. N. (2005). The microbial selenoproteome of the Sargasso Sea. *Genome Biol.* **6,** 37.
114. Gladyshev, V. N., Kryukov, G. V., Fomenko, D. E., and Hatfield, D. L. (2004). Identification of trace element-containing proteins in genomic databases. *Annu. Rev. Nutr.* **24,** 579–596.
115. Kryukov, G. V., and Gladyshev, V. N. (2002). Mammalian selenoprotein gene signature: Identification and functional analysis of selenoprotein genes using bioinformatics methods. *Methods Enzymol.* **347,** 84–100.
116. Wilting, R., Schorling, S., Persson, B. C., and Bock, A. (1997). Selenoprotein synthesis in archaea: Identification of an mRNA element of *M. jannaschii* probably directing selenocysteine insertion. *J. Mol. Biol.* **266,** 637–641.

117. Zhang, Y., Baranov, P. V., Atkins, J. F., and Gladyshev, V. N. (2005). Pyrrolysine and selenocysteine use dissimilar decoding strategies. *J. Biol. Chem.* **280**, 20740–20751.
118. Srinivasan, G., James, C. M., and Krzycki, J. A. (2002). Pyrrolysine encoded by UAG in Archaea: Charging of a UAG-decoding specialized tRNA. *Science* **296**, 1459–1462.
119. Castellano, S., Lobanov, A. V., Chapple, C., Novoselov, S. V., Albrecht, M., Hua, D., Lescure, A., Lengauer, T., Krol, A., Gladyshev, V. N., and Guigó, R. (2005). Diversity and functional plasticity of eukaryotic selenoproteins: Identification and characterization of the SelJ family. *Proc. Natl. Acad. Sci. USA* **102**, 16188–16193.
120. Kryukov, G. V., and Gladyshev, V. N. (2000). Selenium metabolism in zebrafish: Multiplicity of selenoprotein genes and expression of a protein containing seventeen selenocysteine residues. *Genes Cells* **5**, 1049–1060.
121. Yant, L. J., Ran, Q., Rao, L., Van Remmen, H., Shibatani, T., Belter, J. G., Motta, L., Richardson, A., and Prolla, T. A. (2003). The selenoprotein GPX4 is essential for mouse development and protects from radiation and oxidative damage insults. *Free Radic. Biol. Med.* **34**, 496–502.
122. Jakupoglu, C., Przemeck, G. K., Schneider, M., Moreno, S. G., Mayr, N., Hatzopoulos, A. K., de Angelis, M. H., Wurst, W., Bornkamm, G. W., Brielmeier, M., and Conrad, M. (2005). Cytoplasmic thioredoxin reductase is essential for embryogenesis but dispensable for cardiac development. *Mol. Cell. Biol.* **25**, 1980–1988.
123. Conrad, M., Jakupoglu, C., Moreno, S. G., Lippl, S., Banjac, A., Schneider, M., Beck, H., Hatzopoulos, A. K., Just, U., Sinowatz, F., Schmahl, W., Chien, K. R. *et al.* (2004). Essential role for mitochondrial thioredoxin reductase in hematopoiesis, heart development, and heart function. *Mol. Cell. Biol.* **24**, 9414–9423.
124. Ho, Y. S., Magnenat, J. L., Bronson, R. T., Cao, J., Gargano, M., Sugawara, M., and Funk, C. D. (1997). Mice deficient in cellular glutathione peroxidase develop normally and show no increased sensitivity to hyperoxia. *J. Biol. Chem.* **272**, 16644–16651.
125. Esworthy, R. S., Aranda, R., Martin, M. G., Doroshow, J. H., Binder, S. W., and Chu, F. F. (2001). Mice with combined disruption of Gpx1 and Gpx2 genes have colitis. *Am. J. Physiol. Gastrointest. Liver Physiol.* **281**, G848–G855.
126. Hill, K. E., Zhou, J., McMahan, W. J., Motley, A. K., Atkins, J. F., Gesteland, R. F., and Burk, R. F. (2003). Deletion of selenoprotein P alters distribution of selenium in the mouse. *J. Biol. Chem.* **278**, 13640–13646.
127. Schomburg, L., Schweizer, U., Holtmann, B., Flohe, L., Sendtner, M., and Kohrle, J. (2003). Gene disruption discloses role of selenoprotein P in selenium delivery to target tissues. *Biochem J.* **370**, 397–402.
128. Christoffolete, M. A., Linardi, C. C., de Jesus, L., Ebina, K. N., Carvalho, S. D., Ribeiro, M. O., Rabelo, R., Curcio, C., Martins, L., Kimura, E. T., and Bianco, A. C. (2004). Mice with targeted disruption of the Dio2 gene have cold-induced overexpression of the uncoupling protein 1 gene but fail to increase brown adipose tissue lipogenesis and adaptive thermogenesis. *Diabetes* **53**, 577–584.
129. Ng, L., Goodyear, R. J., Woods, C. A., Schneider, M. J., Diamond, E., Richardson, G. P., Kelley, M. W., Germain, D. L., Galton, V. A., and Forrest, D. (2004). Hearing loss and retarded cochlear development in mice lacking type 2 iodothyronine deiodinase. *Proc. Natl. Acad. Sci. USA* **101**, 3474–3479.
130. de Jesus, L. A., Carvalho, S. D., Ribeiro, M. O., Schneider, M., Kim, S. W., Harney, J. W., Larsen, P. R., and Bianco, A. C. (2001). The type 2 iodothyronine deiodinase is essential for adaptive thermogenesis in brown adipose tissue. *J. Clin. Invest.* **108**, 1379–1385.
131. Petit, N., Lescure, A., Rederstorff, M., Krol, A., Moghadaszadeh, B., Wewer, U. M., and Guicheney, P. (2003). Selenoprotein N: An endoplasmic reticulum glycoprotein with an early developmental expression pattern. *Hum. Mol. Genet.* **12**, 1045–1053.

132. Tajsharghi, H., Darin, N., Tulinius, M., and Oldfors, A. (2005). Early onset myopathy with a novel mutation in the Selenoprotein N gene (SEPN1). *Neuromuscul. Disord.* **15**, 299–302.
133. Dumitrescu, A. M., Liao, X.-H., Abdullah, M. S. Y., Lado-Abeal, J., Majed, F. A., Moeller, L. C., Boran, G., Schomburg, L., Weiss, R. E., and Refetoff, S. (2005). Mutations in the SBP2 gene produce abnormal thyroid hormone metabolism in man. *Nat. Med.* **37**, 1247–1252.
134. Conrad, M., Moreno, S. G., Sinowatz, F., Ursini, F., Kolle, S., Roveri, A., Brielmeier, M., Wurst, W., Maiorino, M., and Bornkamm, G. W. (2005). The nuclear form of phospholipid hydroperoxide glutathione peroxidase is a protein thiol peroxidase contributing to sperm chromatin stability. *Mol. Cell. Biol.* **25**, 7637–7644.
135. Schweizer, U., Streckfuss, F., Pelt, P., Carlson, B. A., Hatfield, D. L., Kohrle, J., and Schomburg, L. (2005). Hepatically derived selenoprotein P is a key factor for kidney but not for brain selenium supply. *Biochem. J.* **386**, 221–226.
136. Gladyshev, V. N., and Kryukov, G. V. (2001). Evolution of selenocysteine-containing proteins: Significance of identification and functional characterization of selenoproteins. *Biofactors* **14**, 87–92.
137. Nirenberg, M., Caskey, T., Marshall, R., Brimacombe, R., Kellog, D., Doctor, B., Hatfield, D., Levin, J., Rothman, F., Pestka, S., Wilcox, M., and Anderson, F. (1966). The RNA code in protein synthesis. *Cold Spring Harb. Symp. Quant. Biol.* **31**, 11–24.
138. Meyer, F., Schmidt, H. J., Plumper, E., Hasilik, A., Mersmann, G., Meyer, H. E., Engstrom, A., and Heckmann, K. (1991). UGA is translated as cysteine in pheromone 3 of *Euplotes octocarinatus*. *Proc. Natl. Acad. Sci. USA* **88**, 3758–3761.
139. Osawa, S., Jukes, T. H., Watanabe, K., and Muto, A. (1992). Recent evidence for evolution of the genetic code. *Microbiol. Rev.* **56**, 229–264.
140. Watanabe, K., and Osawa, S. (1995). tRNA Sequences and Variations in the Genetic Code. In "tRNA: Structure, Biosynthesis and Function" (D. Söll and U. RajBhandary, Eds.), Chap 13, pp. 225–250. American Society for Microbiology, Washington, DC.
141. Lovett, P. S., Ambulos, N.P, Jr., Mulbry, W., Noguchi, N., and Rogers, E. J. (1991). UGA can be decoded as tryptophan at low efficiency in *Bacillus subtilis*. *J. Bacteriol.* **173**, 1810–1812.
142. Weiner, A. M., and Weber, K. (1973). A single UGA codon functions as a natural termination signal in the coliphage beta coat protein cistron. *J. Mol. Biol.* **80**, 837–855.
143. Chittum, H. S., Lane, W. S., Carlson, B. A., Roller, P. P., Lung, F. T., Lee, B. J., and Hatfield, D. L. (1998). Rabbit β-globin is extended beyond its UGA stop codon by multiple suppressions and translation reading gaps. *Biochemistry* **37**, 10866–10870.
144. Heider, J., Baron, C., and Bock, A. (1992). Coding from a distance: Dissection of the mRNA determinants required for the incorporation of selenocysteine into protein. *EMBO J.* **11**, 3759–3766.
145. Berry, M. J., Harney, J. W., Ohama, T., and Hatfield, D. L. (1994). Selenocysteine insertion or termination factors affecting UGA codon fate and complementary anticodon:codon mutations. *Nucleic Acids Res.* **22**, 3753–3759.
146. Hornberger, T. A., McLoughlin, T. J., Leszczynske, J. K., Armstrong, D. D., Jameson, R. R., Bowen, P. E., Hwang, E. S., Hou, H., Moustafa, M. E., Carlson, B. A., Hatfield, D. L., Diamond, A. M., and Esser, K. A. (2003). Selenoprotein-deficient transgenic mice exhibit enhanced exercise-induced muscle growth. *J. Nutrition* **133**, 3091–3097.

Indirect Readout of DNA Sequence by Proteins: The Roles of DNA Sequence-Dependent Intrinsic and Extrinsic Forces

GERALD B. KOUDELKA,
STEVEN A. MAURO,[1] AND
MIHAI CIUBOTARU[2]

*Department of Biological Sciences,
University at Buffalo, Cooke Hall,
North Campus, Buffalo, New York 14260*

I. Introduction	144
II. Indirect Readout is a Common DNA Recognition Mechanism	145
III. Indirect Readout Strategies	146
A. Altering the Average Structure of the DNA Helix	146
B. Altering the Flexibility of the DNA	147
C. Modulating the Conformation of Protein–DNA Complex	147
IV. Structural and Physicochemical Basis for DNA Sequence-Dependent Structural Polymorphisms	148
A. DNA Conformation Displays Sequence-Dependent Polymorphism	148
B. Sequence-Dependent Polymorphisms and Indirect Readout	150
C. Intrinsic Forces and DNA Structure—Steric Repulsion and DNA Conformation and Flexibility	150
D. Instrinsic Forces and DNA Structure—Base Stacking and DNA Conformation and Flexibility	151
E. Extrinsic Forces and DNA Structure—Sequence-Dependent Interactions of Solvent with DNA and Its Effects—DNA Conformation and Flexibility	152
V. Indirect Readout by Bacteriophage Repressor Proteins	154
A. Bacteriophage Developmental Decisions	154
B. Lambdoid Phage Repressor Structure	156
C. Evidence for Indirect Readout of DNA Sequence by Lambdoid Bacteriophage Repressors	157
D. DNA Structure and Indirect Readout of DNA Sequence by 434 Repressor	159

---

[1]Present Address: Department of Biology, Mercyhurst College, 501 East 38th Street, Erie, Pennsylvania 16546.

[2]Present Address: Department of Immunobiology, Yale University, New Haven, Connecticut 06520.

    E. Role of Extrinsic and Intrinsic Forces in Indirect Readout
        by 434 Repressor 162
    F. Insights from Structural Studies... 165
VI. *In Vivo* Studies of Cation-Dependent 434 Repressor Gene
    Regulatory Activity.. 169
VII. Summary and Prospects 170
    References 171

I. Introduction

The binding of proteins to specific DNA sequences plays a central role in the regulation of gene expression in all organisms. These proteins regulate gene expression by binding DNA at specific sites and activating or repressing transcription. Structural and biochemical studies have provided a detailed insight into how the intimate contacts between proteins and DNA enable proteins to bind specifically and with high affinity only to their cognate DNA-binding sites. One conclusion of these studies is that sequence specific DNA recognition involves both direct and indirect readout of the binding site sequence. The direct reading of a DNA sequence by a protein involves the close approach of amino acid side chains to functional groups on the DNA bases and backbone. The ability of the groups on the protein and DNA to form a functionally relevant interaction is governed by their chemical complementarity, for example, between hydrogen bonds, donor and acceptor pairs, or ionic interactions between groups of opposite charge. Direct, nonpolar interactions between thymine—CH_3 groups and hydrophobic side chains or portions of side chains also play an important role in direct readout of DNA sequence by proteins. The formation and strength of these interactions is affected by geometrical constraints, that is, the distance between the interacting groups and/or the angle of approach. Hence, the physicochemical nature of direct protein–DNA contacts is well understood.

In indirect readout, the stability and specificity of a protein–DNA complex is regulated by the sequence of bases not in contact with the protein. These noncontacted bases can inhibit or prevent the contacted DNA from being properly juxtaposed with protein groups. DNA sequence-dependent differences in the structure and flexibility of noncontacted bases lead to alterations in the strength and/or ease of forming protein–DNA contacts. The DNA sequence-limited geometry changes thereby indirectly alter the affinity and/or specificity of a protein for its cognate binding site. Hence, indirect effects of DNA sequence on protein–DNA complex formation occur by a modulation of the structural complementarity between the interacting molecules.

Failure to properly align the functional groups of the protein with those in DNA compromises the strength of the protein–DNA complex, the specificity of the protein for a particular DNA sequence, or both of these. In order to ensure functional alignment of the relevant groups in a protein–DNA complex, structural adjustments of the protein, DNA, or both must occur. Indirect effects of DNA sequence on the formation of a protein–DNA complex could, therefore, arise from either an alteration of the flexibility of DNA, which may modulate the ease with which the DNA can be distorted into the proper configuration for complex formation, and/or differences in the structure of the bound DNA. Moreover, if the DNA site must be distorted in order to form a complex, since the energy of DNA deformation is proportional to the degree of deformation, noncontacted bases can influence the affinity of a site for its cognate-binding protein by changing the structure of the unbound DNA as well. While it is clear that indirect readout of DNA sequence is an important component of the DNA sequence recognition mechanisms of many proteins, until recently, it was unclear how DNA sequence directs changes in DNA structure and/or flexibility. The focus of this chapter is to detail recent insights into how DNA sequence effects DNA structure and outline how solvent-mediated alterations in DNA structure may play a role in gene regulation.

II. Indirect Readout is a Common DNA Recognition Mechanism

The term indirect readout stems from the ability of a protein to distinguish certain sequences over others, based solely on bases not in direct contact with the protein. In indirect readout, sequence-dependent differences in the structure and flexibility of noncontacted bases in DNA regulate the stability and sequence-specificity of a protein–DNA complex. TATA Binding Protein (TBP), 434 repressor, and trp repressor were among the first proteins shown to use an indirect readout mechanism to recognize their DNA-binding sites (*1*, *2*). The number of proteins shown to rely on indirect readout, either solely or partially, to recognize DNA-binding sites has greatly increased over the past decade. Some examples include met repressor, integration host factor (IHF), P22 repressor, c-myb, mar A, metJ, papillomavirus E2 proteins, and estrogen receptors (*3–11*). In all these cases, the protein binds to single or multiple DNA sequences with high affinity, sensing DNA structural differences to distinguish the target-binding site(s) over the preponderance of potential sites in the cell.

Proteins that utilize indirect readout are ubiquitous among organisms in all domains of life. However, these proteins show very little structural homology between one another. For example, the papillomavirus E2 proteins utilize a $\beta1$-$\alpha1$-$\beta2$-$\beta3$-$\alpha2$-$\beta4$ DNA-binding domain, the first α-helix inserting into the major groove and participating in direct contact to the DNA bases (12). The DNA-binding domain of 434 repressor consists entirely of α-helices, using a helix-turn-helix motif to contact base functional groups (13). In *Escherichia coli*, IHF makes contacts with bases in the major groove with antiparallel β-turns (5). The wealth of disparate proteins in a variety of organisms that recognize their respective sequences by indirect readout affirm that this binding strategy is a universal scheme used by many proteins to achieve and/or modulate sequence-specific recognition of DNA-binding sites.

III. Indirect Readout Strategies

The ability of a protein to recognize, and/or coerce, DNA into a variety of noncanonical structures is an absolute prerequisite for many of the important DNA transactions that take place in the genome, including replication, transcription, repair, and recombination. Detailed analysis of the sequence dependence of the structures of DNA both alone and in complex with proteins shows that DNA has design elements within it that allow it to assume a wide variety of structural variants. Although this structural polymorphism is crucial to the biological functions of DNA and protein–DNA complexes, precise knowledge of how base sequence modulates DNA structure, and thereby indirect readout, is lacking. Nonetheless as illustrated later, sequence-dependent differences in DNA structure does influence gene regulation by affecting the formation, stability, and/or conformation of protein–DNA complexes.

A. Altering the Average Structure of the DNA Helix

Given that a protein may prefer to bind to a specific conformation of DNA, the affinity of a protein for its site is affected by the average conformation of the free DNA. Hence, for DNAs with identical flexibilities, an unbound DNA whose conformation resembles that found in complex with protein will bind with a higher affinity than one whose unbound structure is dissimilar. In one of the best-studied examples of this phenomenon, Widlund *et al*. (14–16) showed that a nucleosome preparation preferentially binds DNA sequences that contain A-tract–induced static bends interspersed with regions of DNA sequence that are overwound relative to native DNA. Conversely relatively underwound DNAs and/or DNAs containing G-tracts or G/C-rich DNA sequences bind poorly to nucleosomes. Sequence-induced alterations in the average structure of DNA also influence the affinity of many other proteins for their particular binding sites. Among these are the cyclic AMP receptor protein of *E. coli* (CRP) (17), IHF (5, 18–20), nuclear factor I (NF-1) (21), NF-κB (22), and bacteriophage 434 repressor.

B. Altering the Flexibility of the DNA

The strength of the contacts between a protein and DNA critically depends on the alignment between the interacting groups in the two molecules. Since DNA is structurally polymorphic, most proteins alter the conformation of their DNA-binding sites to facilitate the formation of these contacts. Consequently, the affinity of many proteins for their DNA-binding sites depends on the ease with which the DNA can be distorted into the proper conformation for complex formation. Among these are the 434 repressor, the nuclear receptors (23), the λ-integrase protein (24, 25), E. coli RNA polymerase (26, 27), and the lac repressor (28, 29). In these and other cases, noncontacted bases within or adjacent to the contacted region of the binding site modulate the structural complementarity between protein and DNA. For example, changing the composition of the four noncontacted bases at the center of the binding site for the cancer-associated human papilloma virus E2 protein from an A/T-rich sequence to one that is G/C rich decreases the affinity protein for DNA by ≥ 100-fold (12, 30). These sequence changes do not alter the final conformation of the protein–DNA complex. Instead G/C-rich central sequences inhibit E2 DNA binding by decreasing the DNA bending flexibility of the E2 binding sites (31).

C. Modulating the Conformation of Protein–DNA Complex

DNA sequence-dependent alterations in the conformation of a protein–DNA complex can alter its stability by changing the geometry of the contacts between protein groups and DNA bases. An example of this effect is found in the ETS family of proteins (32). In each of these proteins, two conserved Arg residues within the "recognition helix" of their "winged-helix" motif make direct hydrogen bonds with the bases of a conserved G G A sequence, but the pattern of hydrogen bonds from these conserved arginines to DNA, and their relative importance in stabilizing the complex, differs between the different complexes (33–36). Sequence-dependent differences in the degree of protein-induced DNA bending, correlated with the A/T vs G/C content of the DNA in noncontacted regions of the different binding sites, are responsible for the variation in the geometry of specific protein–DNA contacts.

DNA sequence-dependent differences in the structure of protein–DNA complexes directly affect their function. For example, inserting G/C-rich sequences into the binding site of the TBP changes the structure of the complex (37). These structural alterations decrease the efficiency of assembling the preinitiation complex of eukaryotic RNA polymerase II. Similar DNA sequence-dependent differences in protein–DNA complex structure affects the affinity and transcriptional regulatory mechanisms of fos/jun heterodimers (38), bacteriophage P22 repressor, *E. coli* RNA polymerase (27, 39), the glucocorticoid hormone receptor (40–43), and certain cAMP-responsive enhancers (44) among others.

IV. Structural and Physicochemical Basis for DNA Sequence-Dependent Structural Polymorphisms

A. DNA Conformation Displays Sequence-Dependent Polymorphism

For more than 20 years after Watson and Crick elaborated in 1953 their double helical model of DNA (45), the only available data regarding DNA structure came from X-ray diffraction studies of fiber DNA (46). These types of structures only gave an overall low-resolution picture that reinforced the idea that the structure of DNA was uniform along its length. The only structural variation envisioned from these studies was that a B→A transition occurs at lowered relative humidity (46, 47).

The first data that challenged the picture of the uniform DNA helical ladder proposed by Watson and Crick (45) came from X-ray diffraction on crystals that were obtained with two free nitrogen bases (48). These studies showed major differences in stacking geometries adopted by bases depending on their chemical nature, suggesting that DNA conformation may be sequence dependent, adopting locally different structures in order to accommodate such different stacking geometries.

The first double helical B-DNA type structure to be solved from X-ray diffraction pattern of single crystals was the dodecamer of the sequence d(CGCGAATTCGCG)$_2$ (49, 50). This structure confirmed the main features of the Watson–Crick model: double-stranded antiparallel helix where complementary bases A/T, G/C pair with a pseudodyad symmetry. However, the dodecamer structure also showed that as predicted (48), the individual base stacking interactions are subject to major variations in their local parameters.

To better understand the detailed structure of DNA, I will now introduce the reader to the basic structural parameters that characterize local geometry of DNA base–base interactions (Fig. 1). First if one considers the outer surface of DNA, delimited by sugar-phosphate backbones, continuous as the surface of a cylinder the main axis of DNA is considered to be the axis attached to this cylinder. The following parameters' definitions delimit the minimum DNA characteristic units that are between or affected by two consecutive base pairs.

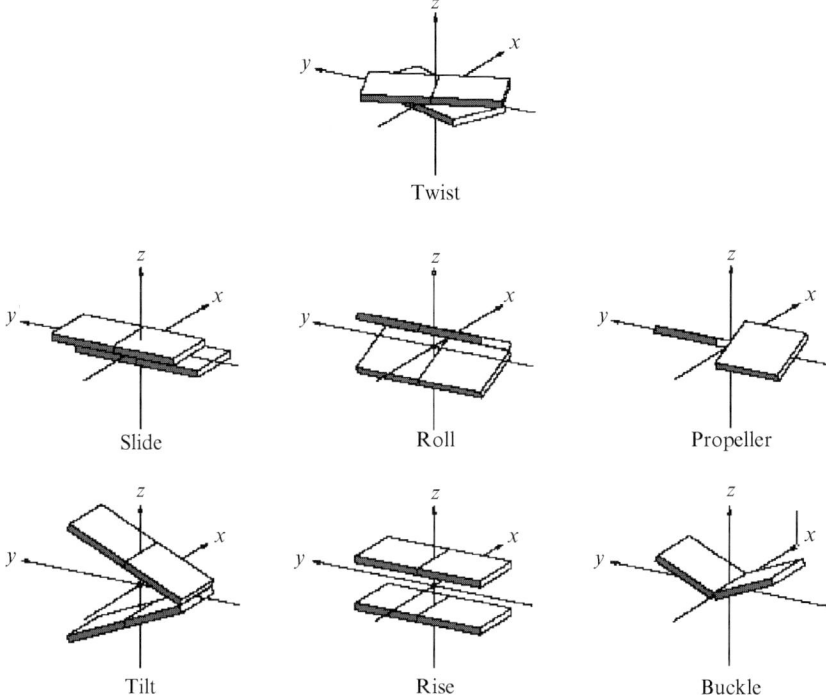

FIG. 1. Illustration of parameters that describe interrelation of two adjacent base pairs in double helical DNA. Diagrams were drawn using 3DNA (128).

- *Helical twist* is the angle formed between two intrastrand adjacent base glycosidic bonds 5′ C_1' $N_{(9)(1)n}$ C_1' $N_{(9)(1)n\ +1}$ 3′ seen in projection on a plane perpendicular to DNA main axis.
- *Roll angle*: If one considers for each individual base pair an average plane between each individual base plane, the roll angle is considered to be the dihedral angle formed between two adjacent average planes corresponding to two consecutive base pairs. By convention, a roll angle is considered positive if it opens toward the minor groove of DNA.
- *Tilt* is the angle made between the average plane of a base pair and a plane considered perpendicular to the DNA main axis.
- *Lateral displacement (slide)* is considered to be the distance measured along the average base pair plane C_8C_6-long axis, from the actual center of the base pair to the conventional main axis of DNA.
- *Helical rise* represents the distance along main axis between two average consecutive base pairs planes.

- *Propeller twist* represents the dihedral angle between individual base planes within a base pair along the long (y-axis) of the base pair.
- *Buckle* represents the dihedral angle between individual base planes within a base pair along the short (x-axis or pseudo-symmetry axis) of the base pair.

B. Sequence-Dependent Polymorphisms and Indirect Readout

It is now generally agreed that sequence-dependent intrinsic and extrinsic forces both contribute to the structural and dynamic polymorphism of DNA. Intrinsic forces derive from the electronic properties of the bases and mechanical properties of helices. These forces include short-range nonelectrostatic interactions between bases, including base-stacking interactions (51) and propeller-twisting effects resulting from steric repulsion among exocyclic groups on the bases (52). Extrinsic forces derive from the electrostatic interactions of DNA with its environment. The different electrostatic potentials of the various DNA base and backbone functional groups direct the positions of cations and solvent. The positioned cations subsequently influence DNA conformation through electrostatic interactions, causing sequence-dependent variation in groove width, DNA twist, and axial bending (53). Despite the demonstrated biological relevance of indirect readout of DNA sequence, and therefore, sequence-dependent differences in DNA structure, little is known about how a given DNA sequence generates a structure with a particular conformation or conformational flexibility.

C. Intrinsic Forces and DNA Structure—Steric Repulsion and DNA Conformation and Flexibility

Several investigators have examined whether the pattern of sequence-dependent clashes that are proposed to occur in particular DNA sequences can form the basis for a series of predictive rules that could anticipate the distribution of structures of DNA of known sequence. The first attempts at this type of DNA structural prediction were Calladine's rules (54). The Calladine rules are derived from observation that N2 amines of two G residues that are located on opposite strands in 5'-GC 3' steps clash in the minor groove of DNA. Similarly, either N6 amines of adenine residues or O6 carbonyl substituents of guanine residues located on opposite strands could clash in the major groove of 5' purine-pyrimidine 3' base steps. In order to avoid such steric hindrances, Calladine proposed that strain can be ameliorated by several strategies:

1. Reduce the propeller twist in one or both pairs by rotating bases closer to their mean base pair plane.
2. Open the roll angle between base pairs at the side of steric hindrances. In the case of 5′-pyrimidine, purine 3′ makes a more positive roll by opening it toward the minor groove, and vice versa for 5′ purine-pyrimidine 3′ major groove clash.
3. Move by translation of one or both base pairs along their long C6C8 axis in a direction that moves pyrimidines out of stack (slide).
4. Decrease the helical twist by rotation of one or both base pairs thus reducing the interstrand purine clash.

Calladine's rules are of great use in understanding the structural basis for some peculiar features of the DNA structure formed by DNA sequences containing alternating purine–pyrimidine steps. However, outside of this limited sequence context, the proposed "escape" strategies could lead to a wide range of structural changes that would be limited by surrounding sequence context. Calladine's rules raised another unexpected issue in DNA structure, that is, the importance of the flexibility of a dinucleotide base pair step. Parameter changes like those proposed by Calladine's rules in a base pair step may be influenced by how flexible the adjacent flanking DNA sequences are in order to accommodate them (55–59). Hence, predictions of DNA structure and flexibility that are based solely on analysis of clashes that occur within individual dinucleotide steps may have limited value in understanding how DNA sequence contributes to DNA structure and function.

D. Instrinsic Forces and DNA Structure—Base Stacking and DNA Conformation and Flexibility

In aqueous solution, the individual nucleic acid bases interact and pile up like coins in a roll. Both in solution and in double helices, the bases are stacked parallel to one another such that the base planes are separated by their *van der Waals* distance, 3.4 Å. Formation of base-stacking interactions is a diffusion controlled, additive process, in which base stacks are stabilized by weak forces by two separate forces: the sequence-independent hydrophobic effect and DNA sequence-dependent London dispersion forces (60).

The sequence dependence of base stacking interactions is driven by interactions between the partial electronic dipoles present on the DNA bases. The direction and strength of an individual base's fixed partial dipole depends on the electron withdrawing or donating potential of the base substituents. Hence, the primary feature that regulates base-stacking strength and the relative orientations of the bases within a stack is the electronic structure of the bases. Since all bases have different substituents, the stacking potential of the bases and their preferred relative stacking geometries are all different. Consequently, stacking strength depends both on base composition and base sequence. In general, the stacking interactions of base paired nucleotide dimers containing G+C base pairs are more stable than those containing A+T base pairs. Another generalization that can be made is that

$$3'\text{pyrimidine} \cdot \text{purine} 5'$$
$$5'\text{purine} \cdot \text{pyrimidine} 3'$$

is more stable than

$$3'\text{purine} \cdot \text{pyrimidine} 5'$$
$$5'\text{pyrimidine} \cdot \text{purine} 3'$$

The variation of stacking energy with sequence occurs because in alternating purine, pyrimidine sequences, overlap between adjacent base pairs in a stack is much greater in B-DNA than the pyrimidine, purine alternation. In particular, the polar groups are more precisely centered within the π-electron "cloud" of the adjacent bases in purine, pyrimidine, rather than in the pyrimidine–purine stack.

Measured base-stacking energies have been used accurately to calculate the stability of both RNA and DNA double helices. However, due to their low energy and geometrical constraints imposed by the sugar phosphate backbone and the consequent stereochemistry limitations on the nucleotide unit, these values have been of little use in calculations aimed at predicting the precise conformation and flexibility of particular DNA sequences. Hence, while sequence-dependent stacking interactions may play a role in regulating indirect readout by proteins, it precise role and importance remains unresolved.

E. Extrinsic Forces and DNA Structure—Sequence-Dependent Interactions of Solvent with DNA and Its Effects—DNA Conformation and Flexibility

For many years models put forth to explain the origins of sequence-specific variations in DNA structure focused exclusively on the intrinsic properties of the DNA double helix in isolation, that is, base-stacking interactions, groove-hydration patterns, and the preferred geometries of A•T vs G•C base pairs (e.g., propeller twist) (50, 54, 61–65). However the evidence currently in hand shows that sequence-specific interactions of the solvent and, more particularly, cation localization within the grooves is a significant factor in driving the sequence-dependent polymorphism of DNA structure (53, 66–71).

That metal cations are an intricate part of nucleic acid structure and can occupy particular sites, reflecting both sequence and topological specificity, is not a recent observation. Na^+ and Mg^{2+} ions as well as the polyamine spermine were identified in some of the first crystallographic structures of RNA (72, 73). In 1973, Rosenberg et al. identified sodium ion near the floor of an abbreviated minor groove of an A/T-tract in DNA (72). However, the more recent insights into the sequence determinants of cation–DNA interactions has led a revision of thought on the role of these interactions in determining DNA conformation existing B-from DNA structures (74). One idea that is derived from these observations is that DNA conformation is not an intrinsic property of the double helix in isolation, but is the response of DNA to sequence-dependent asymmetries in the distribution of counterions. These findings force us to reconsider the role of cations in mediating sequence-specific DNA structure and the consequences this may have on protein binding. Before considering a specific case of cation-mediated effects on DNA structure and consequently protein–DNA interactions, we will first consider how cations bind to, and help shape the structure of, DNA double helices.

1. MONOVALENT CATIONS AND THE MINOR GROOVE

A number of studies demonstrate that monovalent cations have the ability to localize to specific positions along a DNA helix depending on base pair combinations. The positioning of these charged molecules alter the structure of the DNA molecule at the area of localization. Studies with K^+, Na^+, Cs^+, Li^+, NH_4^+, Rb^+, and Tl^+ solutions demonstrate that monovalent cations reside mostly in the minor groove of DNA (68, 69). In all cases, the monovalent cation is found near the floor of the minor groove, specifically interacting with the functional groups on the bases. Monovalent cations bind preferentially at TA/AT or AT/TA dinucleotide steps, demonstrating a sequence-dependent localization of monovalent cations along a DNA molecule. For example, using a 12-base pair (bp) oligonucleotide, Minasov et al. observed that Rb^+ or K^+ are found to preferentially occupy a position in the minor groove of AT/TA base pair steps (75). Regardless of the type of monovalent cation, there is a high preference in all cases for localization of monovalent cations at T•A over G•C base pairs.

In the minor groove, bound monovalent cations participate in cross-strand phosphate and O4 thymine charge neutralization (67) leading to narrowing of the minor groove in T•A-rich sequences. Monovalent cation-dependent electrostatic collapse and consequent narrowing of the minor groove at positions where T•A bases are present has been proposed to explain structural alterations exhibited by A-tract sequences (76, 77).

The finding that monovalent cations narrow a minor groove has immediate implications in protein recognition and binding. One of the underlying ideas of indirect readout is that the protein displays its highest affinity for sites that most closely resembles the final structure in the complex. This observation provokes consideration of monovalent cation localization along a DNA molecule as a factor contributing to the binding affinity of proteins that utilize indirect readout to locate its target sequence.

2. DIVALENT CATIONS AND THE MAJOR GROOVE

Due to their larger atomic radii, locating divalent cations at specific locations along a helix has been easier than locating monovalent cations. Experiments aimed at elucidating divalent cation residency along B-form DNA demonstrate a sequence-specific localization within the major groove of the DNA helix (78). Similar to monovalent-cation binding, the location of divalent cations depends both on DNA sequence and identity of the divalent cation. For example, a Mg^{2+} ion binds preferentially to GT/CA base pair steps, while both Ca^{2+} and Mg^{2+} localize to AG/TC dinucleotide sequences. Regardless of the identity of the divalent cation, the sequence specific localization in the major groove causes bending of the DNA molecule toward the major groove by base roll compression. Similarly, the presence of divalent cations at sequence specific locations in the minor groove may also predispose a DNA molecule to bend in a manner that allows easier recognition by protein. This feature appears to be important in mediating nucleosome interactions with G/C-rich sequences. This DNA sequence binds poorly to nucleosomes. Consequently, it appears that divalent cation(s) chelated by minor groove functional groups assist bending into this groove and facilitate DNA binding by this protein (79, 80).

V. Indirect Readout by Bacteriophage Repressor Proteins

The work in our laboratory focuses on the factors that regulate the developmental decisions of lambdoid bacteriophages. One aspect of this work concerns the DNA recognition mechanism by the repressor protein. This protein regulates the central developmental decisions of this class of prokaryotic viruses.

A. Bacteriophage Developmental Decisions

Members of the lambdoid class of bacteriophages infect a wide variety of bacterial host strains. Despite this broad host distribution, all these phages share a common developmental program. On infection of a bacterial cell, the lambdoid phages choose between two developmental fates. The phage can grow lytically, thereby killing the host. Alternatively in lysogenic growth, the phage chromosome is inserted into the host chromosome and replicated along with it, until a signal to revert to the lytic growth mode is perceived by the lysogenized phage. Central to the phage's decision between lytic and lysogenic growth is the activity of the bacteriophage-encoded repressor protein.

Every lambdoid bacteriophage codes for this DNA-binding transcriptional regulatory protein [(81) and R. Yocum, unpublished results].

In all lambdoid phages, the repressor directs the establishment and maintenance of the lysogenic state by binding DNA and simultaneously repressing transcription of the genes needed for lytic phage growth and activating transcription of a gene needed for lysogen formation (82). Each phage genome contains two operator regions O_L and O_R. Both regions contain two divergently transcribed promoters whose expression is controlled by repressor binding to three closely spaced binding sites (Fig. 2). The sequences of these sites display incomplete rotational symmetry and are partially conserved between them. The sequence of 434 O_R is typical of these regions (Fig. 3). In each phage, the repressor binds as a dimer of identical subunits to each of these sites. Efficient functioning of the genetic switch between lysis and lysogeny depends on ability of repressor to bind with different affinity to each of these sites.

The importance of repressor's differential affinities for its binding sites to bacteriophage development is illustrated by considering 434 repressor's DNA binding and gene regulatory activities at 434 O_R. At O_R, the relative transcriptional activities of two promoters, P_R and P_{RM}, depend on which of the three operators, O_R1, O_R2, or O_R3, is bound by repressor. When the three sites are present on separate DNA fragments, 434 repressor binds with lowest affinity to O_R2, twofold more tightly to O_R3, and with yet a sixfold higher affinity to O_R1. In contrast, in intact O_R repressor binds to O_R1 and O_R2 with almost equal affinity and subsequently to O_R3 with eightfold lower affinity. Thus in intact O_R repressor, dimers do not bind independently to the three sites. Instead, they bind cooperatively to two adjacent sites, while binding the third independently. Hence, in a lysogenic cell, the integrated prophage bears one repressor dimer bound at each O_R1 and O_R2 (Fig. 3). In this configuration, the P_R promoter, which directs transcription of the genes required for lytic growth,

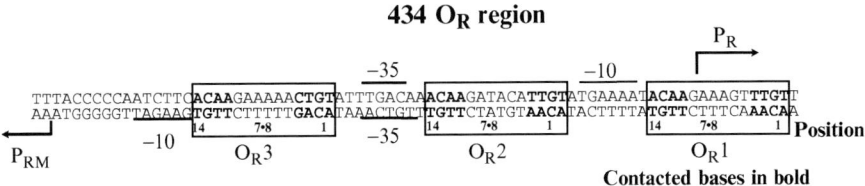

FIG. 2. The sequence of the O_R region of bacteriophage 434. The O_R1, O_R2, and O_R3 sites are enclosed in boxes. The bases that are directly contacted by 434 repressor in 434 repressor–DNA complexes are highlighted in bold. The transcription start sites of P_{RM} and P_R are indicated by bent arrows. The −35 and −10 regions of P_R and P_{RM} are underlined.

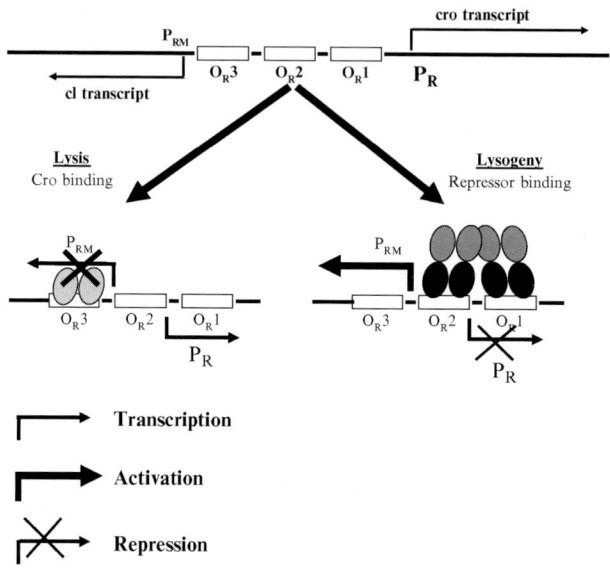

Fig. 3. The lysis-lysogeny switch of temperate lambdoid bacteriophages. Upon infection of a susceptible host, the relative amounts of phage repressor and cro and their relative binding affinities for sites in O_R and O_L (not shown) determine whether the bacteriophage develops lytically or enters the lysogenic pathway.

is repressed, while the P_{RM} promoter directing cI (repressor) transcription is activated (83).

The O_R-bound repressors regulate the relative amounts of transcripts initiating at P_R and P_{RM}. The O_R2-bound repressor stimulates P_{RM} transcription by directly contacting the σ^{70} subunit of the P_{RM}-bound RNA polymerase (84–86). These DNA-bound repressors also activate P_{RM} transcription indirectly, by removing an interfering RNA polymerase bound at P_R (87, 88). In line with classical models (89), early studies suggested that repressor inhibits P_R transcription by blocking access of RNA polymerase to the promoter sequence (90). Our recent studies show that the phage repressors act subsequent to formation of the initial RNA polymerase–promoter complex (83, 91).

B. Lambdoid Phage Repressor Structure

The bacteriophage repressors can specifically bind to their DNA sites only as dimers of identical subunits [(92, 93) and G. Koudelka, unpublished results]. Each repressor monomer contains two structurally and functionally distinct domains (93–95). The amino (N-) terminal domain comprises approximately the first third of the protein. Structural analysis (96–98) and sequence homologies (81, 99) show that the N-terminal domains of all the bacteriophage repressors contain a helix-turn-helix structural motif. This structure is responsible for mediating all specific and nonspecific contacts between protein and DNA. The twofold related "recognition" α-helices in this element, one from each protein monomer, lie in successive major grooves on one face of the DNA. Side chains extending from each recognition helix are positioned in the major groove to make base-specific contacts with the outermost base pairs of the partially twofold rotationally symmetric binding site sequence [(100), Figs. 2 and 4]. The other helix of the helix-turn-helix unit makes nonspecific contacts with the DNA-phosphate backbone. The carboxyl (C-) terminal domain encompasses approximately the last half of the repressor polypeptide. This domain stabilizes the formation of repressor dimers and tetramers (93, 95, 101). A tetramer forms readily between two repressor dimers bound at adjacent sites on DNA and is required for regulating DNA binding by repressor (Fig. 3). The N- and C-terminal domains are joined by a "linker" of ∼40 amino acids. Recent evidence suggests that this linker region packs tightly against the core of the C-terminal domain (102–104).

C. Evidence for Indirect Readout of DNA Sequence by Lambdoid Bacteriophage Repressors

The picture of the bacteriophage repressor–DNA complex is described earlier and highlights the role of base-specific contacts in DNA recognition by proteins. This is, however, an incomplete portrait of the determinants of sequence specificity of these proteins. The suggestion that base pair contacts made by amino acids, contained within the recognition helix, are solely responsible for mediating sequence-specific binding by these proteins does not account for the effects that mutations in bases in the center of the binding site have on the stability of several protein–DNA complexes (3, 17, 105).

A closer inspection of the sequences of the binding sites of several phage repressor proteins reveals a provocative pattern; the symmetrically arrayed outer base pairs in these sites are highly conserved within each phage, but the innermost bases are highly divergent. Despite the fact that these highly conserved bases are the only sites of specific protein–DNA contacts, these proteins must and do display a range of affinities for these sites.

To gain a more complete understanding of the mechanisms by which indirectly read bases control affinity of a binding site for its cognate-binding protein and consequently the function(s) of the protein–DNA complexes, we are examining the role of DNA sequence-dependent structure in regulating the conformation, stability, and function of lambdoid bacteriophage

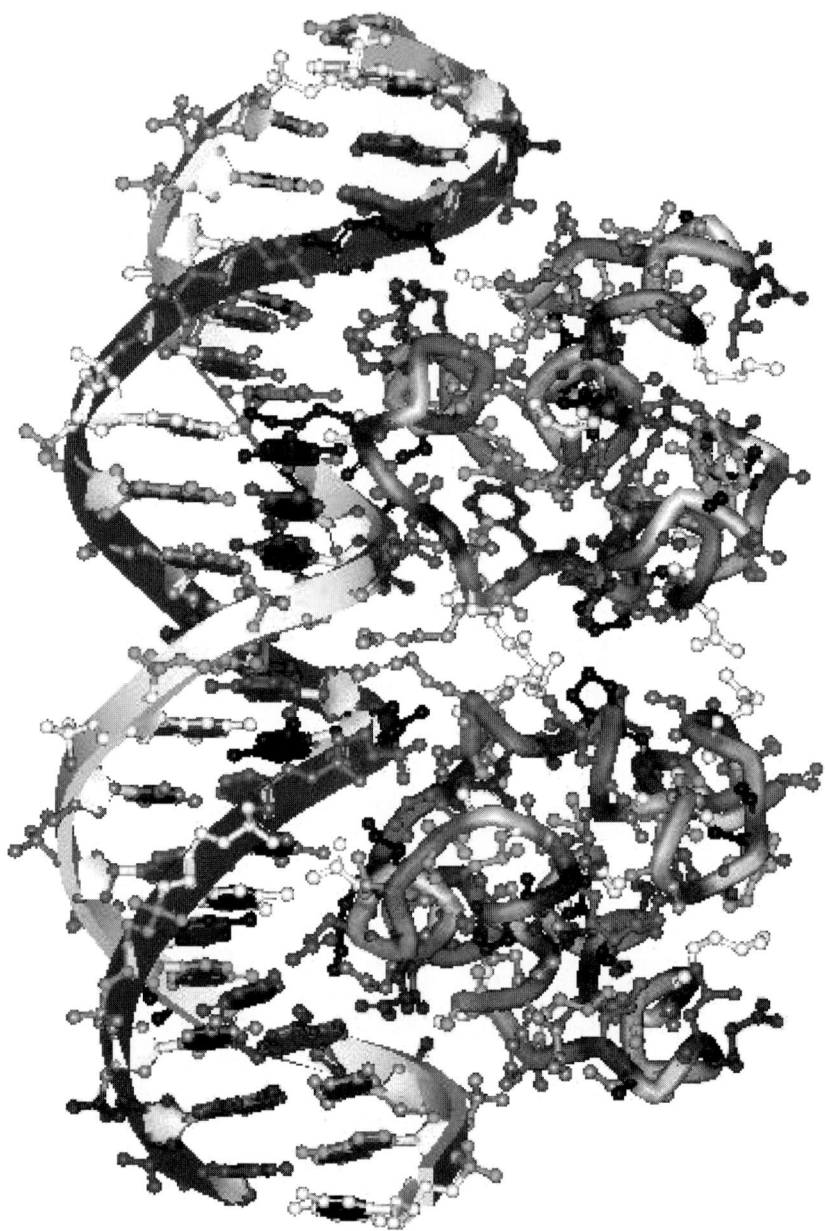

FIG. 4. Structure of the N-terminal DNA-binding domain of 434 repressor in complex with the O_R2 binding site (108).

repressor–DNA complexes, in particular here, the repressor of bacteriophage 434. The availability of the three dimensional structures of wild-type 434 repressor and mutant variants in complex with DNAs bearing different non-contacted base sequences provides an unequaled opportunity to explore the structural bases of indirect readout.

The sequence of bases at the center of the binding sites for 434 repressor has a significant role in determining the affinity of operator for 434 repressor, despite X-ray crystallographic (26, 27) and biochemical (22) evidence showing that these bases are not contacted by the protein. For example, binding site bearing A•T or T•A base pairs at these central positions bind 434 repressor 60-fold more strongly than those bearing C•G or G•C. X-ray crystallographic analyses reveal that, in complex with 434 repressor, the DNA in noncontacted region the 434 repressor-binding site is deformed relative to canonical B-DNA (96, 106). These DNA deformations are required to align the two halves of the binding site with the DNA-contacting surfaces present on each monomer of the bound repressor dimer. Hence indirect readout of DNA sequence at the center of the 434 binding site is an essential feature of 434 repressor's DNA recognition mechanism. Since the naturally occurring repressor-binding sites are distinguished only by differences in the noncontacted bases, indirect readout plays a central role in the lysis–lysogeny decision of this organism (13, 96, 97, 105, 107, 108).

D. DNA Structure and Indirect Readout of DNA Sequence by 434 Repressor

In the 434 repressor–DNA complex, the DNA at the center of the binding site is overtwisted (106, 107) and the minor groove is narrowed (13, 96, 108) (Figs. 5 and 7). Relative to canonical B-DNA, the net overtwisting of the central four bases is about 30° (107). Additionally, there is a slight (25°) bending of the binding site. Thus, sequence changes at the center of the 434 operator could affect operator strength by influencing DNA bending, DNA twisting, or both of these deformations. Moreover, since the energy of DNA deformation increases with the square of the extent of deformation (109), sequence-dependent alterations of DNA structure could influence operator strength by affecting the properties of the unbound and/or bound DNA.

As presented in (110), several lines of evidence indicate that the DNA bending observed in 434 repressor–DNA complexes is unrelated to and unaffected by central base sequence changes. First, three operators bearing different central sequences, whose affinity for repressor vary over 200-fold do not display intrinsic bends. Hence, central base composition does not affect operator affinity for protein by influencing the extent of bending of the unbound DNA. Second, when 434 repressor binds to these same operators, each protein

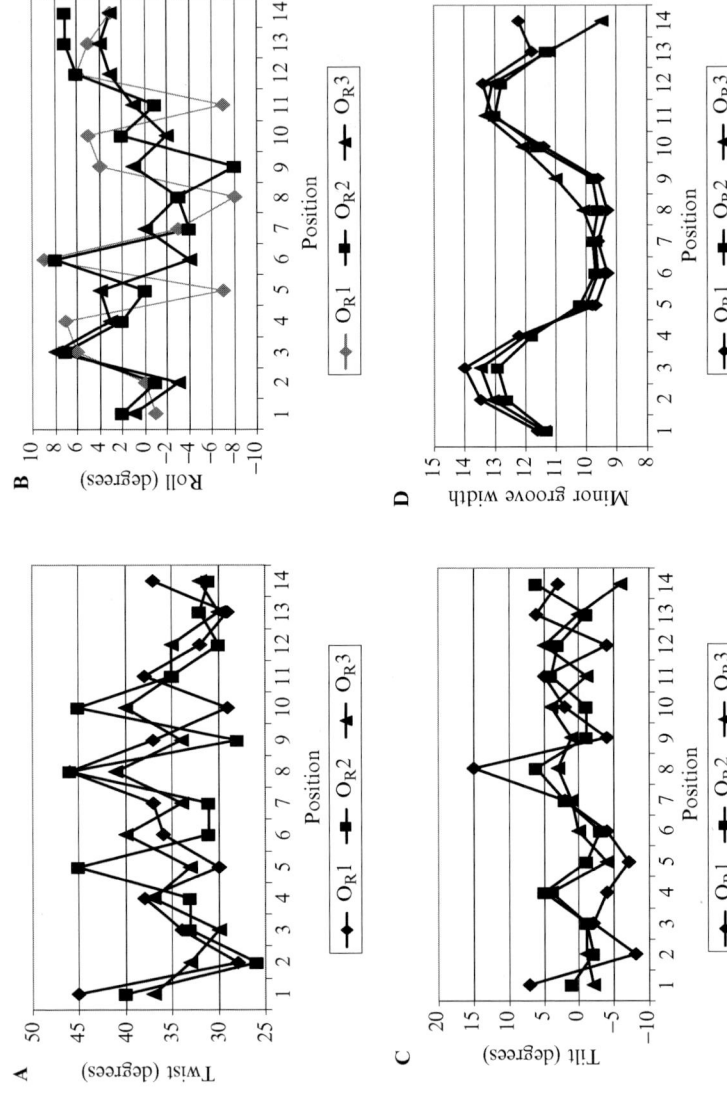

FIG. 5. Polymorphism of DNA structures in 434 repressor–DNA complexes with the values of twist (A), roll (B), tilt (C), and minor groove widths (D) of O_R1 (♦), O_R2 (■), and O_R3 (▲) are plotted as a function of position in the DNA sequence (see Fig. 2 for numbering). Values were calculated as described in Ref. (128).

induces the same degree of bending in all the operators. Thus, variations in the base composition at the operator's center do not influence affinity for protein by differentially altering the degree of operator bending in the protein–DNA complexes. Third, two different 434 repressor proteins, each bearing mutations in their amino terminal dimer interfaces that render them 8–15-fold less sensitive than wild-type repressor to central base changes, bend DNA to precisely the same extent as does the wild-type protein (*111*). Taken together, these observations rule out models in which the central base pairs are imagined to exert their effect on the affinity of operator for protein by modulating the extent of DNA bending or bending flexibility (*107, 112*).

Therefore, central base composition affects the strength of 434 repressor–DNA interactions by influencing DNA twist and/or DNA twisting flexibility. Support for this view is derived from the observed direct effect DNA sequence and 434 repressor binding on the twist of DNA. The twist of several unbound synthetic 434 binding sites varies in a noncontacted base sequence-dependent manner, which parallels their affinities for 434 repressor. On binding DNA, 434 repressor increases the twist of all synthetic and naturally occurring 434 binding sites examined thus far (*107, 113*). Also, biochemical studies show that in complex with 434 repressor, the twist of all binding-site DNA is identical. This finding confirms observation from X-ray crystallographic studies showing that the sugar-phosphate backbone structures of three different naturally occurring binding sites and one synthetic-binding site, when complexed with 434 repressor, are essentially indistinguishable (*13, 96, 106, 108*).

The free energy (ΔG) twisting (torsion) of DNA is given by

$$\Delta G = Cx\phi^2/2l$$

where C is the torsional spring constant, which represents the torsional rigidity of DNA, ϕ is the angle of twist (torsion), and l is the length of DNA over which the twist is changed. Hence, noncontacted base sequence-dependent effects on DNA twist could influence binding-site affinity for 434 repressor by (a) altering the twisting or torsional flexibility of DNA thereby modulating the ease with which the DNA can be distorted into the properly overtwisted configuration for complex formation, (b) influencing the equilibrium twist of the unbound DNA, or (c) limiting the degree that the DNA can be twisted on the protein–DNA complex. The third hypothesis is inconsistent with the observation that the twist of the DNA in all tested 434 repressor–DNA complexes is the same. Consistent with the second hypothesis, the weaker 434 binding sites appear to be undertwisted relative to the stronger sites and that all repressor-liganded sites have identical twists (*107, 111*). Repressor must, therefore, induce larger changes in the twist of the weaker binding sites than it does in the stronger binding sites (*107*). This observation indicates that the difference in affinity between various binding sites must depend, at least in part, on the central sequence dependence of the equilibrium twist of the uncomplexed binding site. Calculations of the free energy for twisting each of the central base sequence binding site variants show that changes in twist cannot completely account for variation in affinity of these 434 binding sites for repressor. Thus, consistent with hypothesis 1 earlier, the measured torsional spring constants of 434 binding sites varies with the sequence of noncontacted bases. Calculations of the free energy of twisting of the binding-site DNAs, using the experimentally determined torsional spring constants together with the difference in twist between unbound and bound DNAs, allow accurate prediction of the differences in binding site strength (*107*). Thus the sequence of noncontacted bases at the center of the 434 binding site influences its affinity for 434 repressor by affecting both the equilibrium twist of the unbound site and its twisting flexibility.

E. Role of Extrinsic and Intrinsic Forces in Indirect Readout by 434 Repressor

The observation that varying the sequence of noncontacted bases influences the affinity of the 434 binding site for 434 repressor by affecting DNA twist leads to the question of how sequence determines DNA twist and/or twisting flexibility. Several observations indicate that the $N2-NH_2$ group of the purine bases present at the center of the 434 repressor-binding site plays a significant role. First, introducing this group at the center of the binding site by changing the noncontacted central base sequences from one that is A/T-rich to one that is G/C-rich decreases the affinity of the DNA for repressor by 60-fold. Second, binding sites bearing I•C base pairs at the central positions bind 434 repressor as well as do binding sites bearing A/T-rich sequences at their centers. Inosine is identical to guanine, except that it lacks the $N2-NH_2$ group on the minor groove surface of the base and thereby resembles A•T base pairs in the minor groove (Fig. 6). Hence, the presence of the $N2-NH_2$ group appears to decrease the affinity of a binding site for 434 repressor.

The substitution of G/C-rich for A/T-rich sequences profoundly affects the structure and/or stability of virtually all protein–DNA complexes whose DNA recognition mechanisms involve indirect readout. In many systems, the effect of A/T→G/C substitution on DNA structure is related to the presence or absence of the $N2-NH_2$ group on guanine. The mechanism(s) by which these sequence changes exert their effects on DNA structure and protein binding are largely unknown. How might the $N2-NH_2$ group exert this deleterious effect on DNA structure and thereby affinity of a binding site for protein? It is possible that (i) the third base pair hydrogen bond mediated by the $N2-NH_2$ group on guanine renders binding sites bearing G/C base pairs at

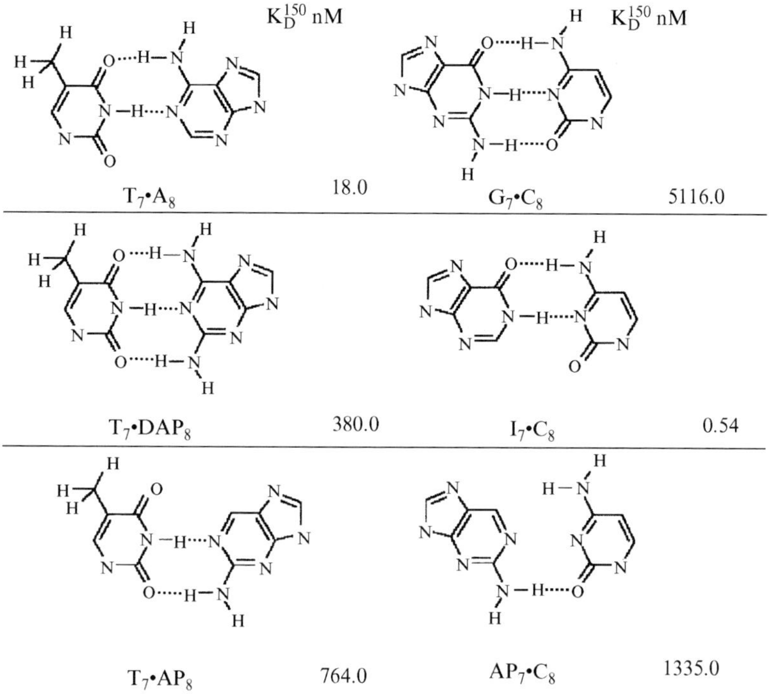

FIG. 6. The role of the N2-NH$_2$ group and number of base pair hydrogen bonds at positions 6 and 7 of the 434 repressor-binding site on the affinity of repressor for DNA. The indicated bases were substituted that the central two positions of a rotationally symmetric binding site sequence and their affinity for 434 repressor was determined in a filter-binding experiment [see (114) for details].

their center less readily deformable than those containing A/T (or I/C) pairs at these positions or (ii) the steric or electrostatic properties of the N2-NH$_2$ group of guanine opposes the repressor-induced DNA overwinding of this region of the DNA. In the 434 repressor case, the repressor's DNA affinity is independent of the number of H-bonds between the noncontacted bases at the center of the binding site (Fig. 6). By contrast, repressor binds with higher affinity to DNAs that do not contain an N2-NH$_2$ group at the central positions than to those that do, regardless of the number of base pair H-bonds. Thus in this case, the steric or electrostatic properties of the N2-NH$_2$ group are responsible for the deleterious effect of G/C base substitution. The affinity of repressor for DNA sites bearing the N2-NH$_2$ group in the central positions and the conformation of these complexes is more sensitive to changes in the salt concentration than it is for sites lacking this group at these positions (114).

Thus it seems that electrostatic properties of the $N2$-NH_2 group on the noncontacted base are responsible for its deleterious effect on repressor–DNA complex formation.

The observation that salt concentration dependence of 434 repressor's affinity for DNA varies with the sequence of noncontacted bases at the center of the 434 binding site, combined with the fact that indirect readout by this protein is driven by sequence-dependent differences in the intrinsic mechanical "stiffness" of the DNA (107, 114, 115), suggests that salt influences the stability of the repressor–DNA complex by interacting directly or indirectly with the DNA bases in the unbound and/or bound DNA. The finding that the functional groups that regulate and salt-dependent indirect readout are located in the minor groove and the knowledge that monovalent cations bind to functional groups in this groove suggests a model in which, in certain DNA molecules, cations bound in a sequence-specific fashion are *required* to facilitate the establishment of the DNA conformation that is compatible with repressor–DNA complex formation.

Several lines of evidence support this model. First, the finding that the stability of the 434 repressor–DNA complex is independent of both the concentration and type of anion in the binding reaction, regardless of DNA sequence is compatible with this model, and focuses our attention on the roles of cations in indirect readout. Second, our earlier studies demonstrated that repressor's affinity for O_R1 depends strongly on the concentration of K^+, whereas its affinity for O_R3 is essentially unaffected by salt concentration (116, 117), although the sequences of these two binding sites differ at only three positions (Fig. 2). Third, changing cation type affects the affinity of 434 repressor for O_R1, but not O_R3. Fourth, while the overall affinities of repressor for both O_R1 and O_R3 are generally lower in the presence of only divalent cations, adding K^+ to a solution that contains only 10 mM of a divalent cation significantly increases the affinity of repressor for O_R1. Under the same conditions, the affinity of repressor for O_R3 is unaffected or slightly decreases.

The sequences of O_R1 and O_R3 differ at only three positions (Fig. 2). This feature allows us to explore the characteristics of noncontacted base sequence that confer cation type sensitivity to a repressor–DNA complex. The noncontacted base sequence at the centers of both O_R1 and O_R3 is A/T rich. However, the noncontacted central region of the O_R3 site is composed of a 5-bp poly dA•dT-tract, whereas the central poly dA•dT-tract in O_R1 site is only 3-bp long (Fig. 2).

Swapping the sequence of the noncontacted position 6 from O_R3 into O_R1 (a C•G→T•A mutations) increases the length of the poly dA•dT-tract and decreases the cation dependence of repressor's affinity for O_R1. The converse T•A→C•G swap at position 6 in O_R3 decreases the length of the poly dA•dT-tract in this site and increases the cation type dependence of repressor's affinity for O_R3. These sequence changes alter both the length of the polydA tract in the noncontacted regions of O_R1 and O_R3 and the composition of these bases. Thus, we cannot be certain whether the noncontacted base sequence or composition is important in mediating the cation type dependence of 434 repressor's affinity for DNA. However, recent evidence indicates that monovalent cations specifically bind poly dA•dT-tract containing high occupancy and facilitate the overwinding and minor groove collapse that is characteristic of this DNA sequence. As discussed earlier, these characteristics are essential for the formation of stable specific repressor–DNA complexes. Regardless of whether the length of an A-tract or the A/T-base composition of the noncontacted region determines the monovalent cation type and/or concentration dependence of repressor's affinity for DNA, our data demonstrate that the identity of the noncontacted bases directs the cation sensitivity repressor's DNA affinity. In addition, these observations are, to our knowledge, the first to document a role for cation-dependent differences in DNA structure in regulating the binding of a protein and may provide clues as to how indirect readout of DNA sequence is regulated in other systems.

F. Insights from Structural Studies

Examination of the structures of the 434 repressor bound to DNAs containing various noncontacted base sequences, several possible explanations into how changes in these sequences site lead to alterations in repressor affinity, and the role of cations in regulating indirect readout of noncontacted bases.

1. ROLE OF ARGININE 43

All 434 repressor–DNA structures indicate that two arginine residues, located at position 43 in each repressor subunit, approach the minor groove of the noncontacted bases at the center of the binding site (see Figs. 4 and 7). These arginine residues are positioned at the mouth of the minor groove, near the phosphate backbone of these bases. It has been proposed (106, 118) that these residues facilitate compression of this minor groove, a structural alteration that is necessary for high affinity–repressor binding. If this idea is correct, the protein-facilitated DNA structural change would then explain indirect readout by this protein. However, careful examination of these structures reveals that the precise positioning of these arginine residues is remarkably influenced by the sequence of the noncontacted bases (Fig. 7); yet, the width of the central minor groove in complexes of 434 repressor with DNA of varying central sequence is identical (Fig. 5D). Thus, these residues (i) are not responsible for regulating central minor groove width and; (ii) may not have an identical role in formation of all repressor–DNA complexes. Moreover, removal of this arginine by mutation to alanine or lysine does not affect repressor's ability to discriminate between binding sites bearing changes in

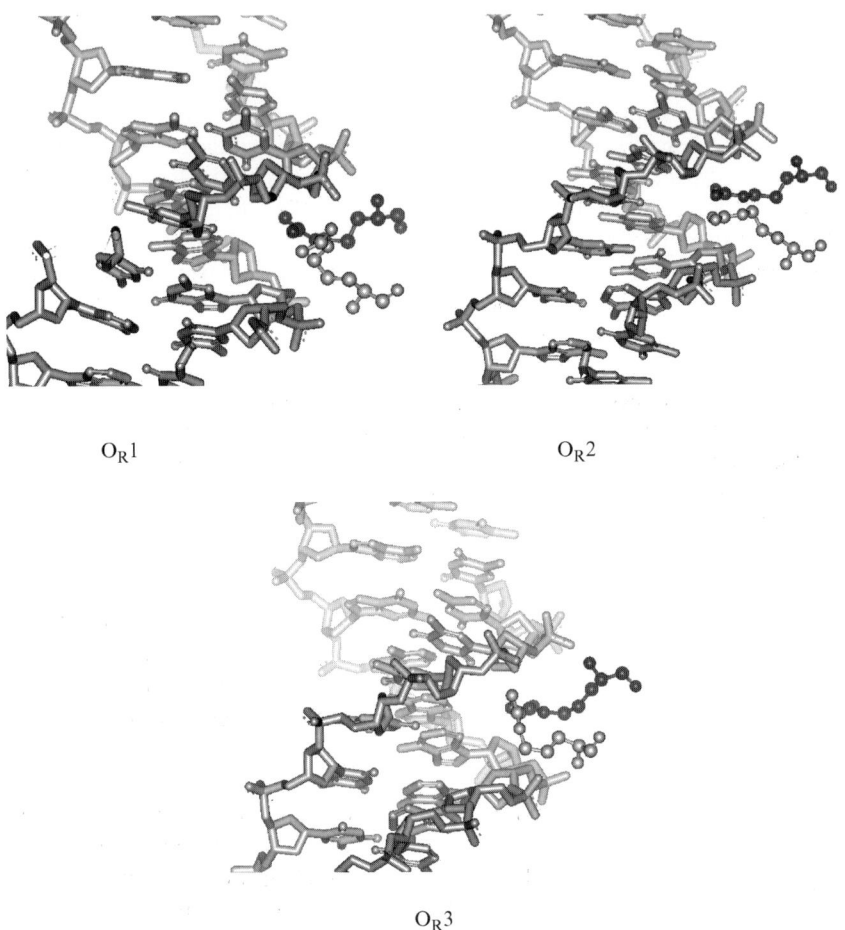

FIG. 7. Position of Arg-43 in complexes between 434 repressor and O_R1, O_R2, and O_R3. Structural data were obtained from Refs. (*13, 96,* and *108*) respectively.

the noncontacted base sequence (*105, 115*). Taken together, these observations show that indirect readout by 434 repressor is unaffected by the presence or orientation of arginine 43.

2. Role of Solvent-Mediated Protein–DNA Contacts

Despite the moderate resolution of the extant 434 repressor-DNA structures, examination of these structures reveals a series of significant sequence dependent differences in the presence of solvent- and protein-mediated contacts with the DNA phosphate backbone that may help explain the existence and salt dependence of indirect readout by 434 repressor. As shown in Fig. 2, the sequences of O_R1 and O_R3 differ in only three positions in one half-site of the DNA, meaning the sequence of the opposite half-sites are identical in the two binding sites. A third naturally occurring site, O_R2, shares this conserved half-site sequence. In the conserved half-site of all these binding sites, the Nε of Gln-33 donates a hydrogen bond to the thymine O4 at position 4, while the Oε makes a solvent-mediated contact with the phosphate located between bases 5 and 6 on the bottom strand. This solvent also interacts with the Oγ of Ser-30 (Fig. 8A).

In the nonconserved half-site of O_R1 and O_R2, the arrangement of DNA contacting residues and solvent are identical to that found in the conserved half-site. However, in the nonconserved half-site of O_R3, which bears a different noncontacted base sequence and a different base sequence at position 4, the repressor contact to the phosphate located between bases 5 and 6 on the bottom strand is made by Thr-39, while the residue at position 33 makes no contact to the DNA bases or backbone (Fig. 8B). In addition, Ser-30 makes a direct contact to the DNA phosphate between positions 4 and 5 in this

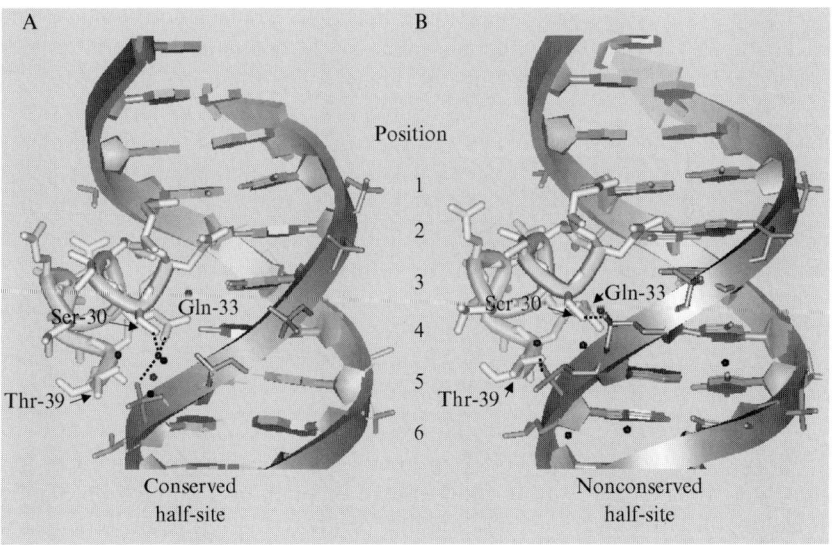

FIG. 8. Noncontacted base sequence effects on repressor–DNA contacts. Shown is a closeup of the interactions between 434 repressor residues 28–39 and the conserved (A) and nonconserved (B) half-sites of O_R3. Dotted lines indicate the positions of proposed hydrogen bonds protein and/or solvent molecules and the DNA phosphate backbone. Solvent molecules are represented as black spheres.

half-site. From this analysis it is unclear whether the differences in protein- and solvent-mediated DNA contacts observed in the nonconserved half-site of O_R1 and O_R3 are consequence of changes in the noncontacted base sequence or the contacted base at position 4. These alternatives were distinguished by examining of the structure of repressor in complex with an O_R1 site bearing a G•C base pair at position 4 and the complex between O_R1 and a Q33→Ala mutant repressor, both of which alter the status of repressor contacts to position 4 but not noncontacted base sequence (Harrison, S. C., Rodgers, D., personal communication). This analysis reveals that, except for the status of the contact at position 4, the DNA contacts are identical to that seen in the nonconserved half-site of wild-type repressor in complex with wild-type O_R1. This observation indicates that the pattern of solvent and protein-mediated contacts to the DNA phosphate backbone is determined by the sequence of the noncontacted bases in that half-site.

As a result of the limited resolution of the current 434 repressor-DNA complex structures, we are certain that we have not identified all important noncontacted base-dependent differences between O_R1 and O_R3. However, the current structures definitively show that changing the noncontacted bases sequence from one containing an oligo-dA tract (O_R3) to one that does not (O_R1 and O_R2) causes a change in the conformation of DNA, leading to an alteration in the identity of the residues that make contacts to the phosphate backbone. In O_R3, contact with a phosphate at the center of the binding site is made by Thr-39. In sites lacking the longer oligo-dA tract, this contact is replaced by a solvent-mediated interaction, with the solvent molecule liganded to Ser-30 and Gln-33. It is not yet possible to definitively identify the bridging solvent molecule in the O_R1 structures as a monovalent cation. However, the existence of a cation mediated at this position in the nonconserved half-sites would help explain why the affinity of repressor for O_R1, but not for O_R3, depends on the type and concentration of monovalent cation.

Thus the analysis of the repressor–DNA complex structures shows that the juxtaposition of protein residues and solvent molecules with the DNA phosphate backbone in the noncontacted region of the binding site varies markedly with the sequence of these bases (Fig. 8). We assert that these sequence dependent differences underlie indirect readout of noncontacted bases by 434 repressor. We suggest that the presence of certain central sequences will facilitate strong and/or solvent-independent interactions leading to formation of stable salt type- and concentration-independent 434 repressor–DNA complexes, whereas other will not. That is, indirect readout by 434 repressor is determined by the intrinsic ability of the noncontacted base sequence to readily assume a particular conformation, with the precise conformation structure of the binding site in the protein–DNA complex potentially being modulated by bound cations. In the case of repressor–DNA complexes that are incapable of assuming a favorable conformation for repressor–DNA complex formation, solvent- and likely cation-mediated interactions facilitate binding of repressor to DNA.

VI. *In Vivo* Studies of Cation-Dependent 434 Repressor Gene Regulatory Activity

All free living organisms are subject to a wide variety of rapidly changing conditions, many of which directly or indirectly alter the ionic composition of their environment. For example, increasing external salt concentration causes accumulation of monovalent cations within the cell and an immediate plasmolysis (cell shrinkage). These responses alter the concentrations of all cytoplasmic solutes. In addition, this treatment causes a rapid increase in negative DNA supercoiling. The increased negative supercoiling alters the activity of several osmoregulated genes (*119–122*). These acute responses are followed by a distinct sequence of cellular events that includes active potassium influx and glutamate synthesis to restore intracellular water balance, and finally, replacement of potassium glutamate with osmoprotectants more compatible with cell growth [see Refs. (*123, 124*) for review].

Changing the type and/or concentration monovalent cations differentially alters the affinities of the 434 repressor for its various binding sites (*114, 125*). Thus for 434 repressor, and potentially other regulatory proteins, the genes they control may also be differentially regulated by changes in salt type and/or concentration. As discussed earlier, the ability of 434 repressor to maintain the lysogenic state depends on its relative affinities for O_R1 and O_R3, which in turn are influenced by changes in cation type and concentration. Together with the known effects of salt concentration intracellular ionic composition, these observations provide beginnings for a new general model for coupling changes in cellular physiology to alterations in the DNA-binding affinities of proteins, and ultimately gene expression patterns.

We tested this idea by determining whether changing internal salt type and concentration affected the gene regulatory activities bacteriophage 434 repressor *in vivo*, independent of osmotic effects. Since inactivation of repressor DNA binding or alteration of the relative affinities for its binding sites affect repressor's gene regulatory activities lead to derepression of lytic genes and lytic growth (*93, 126, 127*), the effect of salt on repressor-DNA binding can be detected by an increase in spontaneous induction frequency of bacteriophage 434 lysogens when these cells are grown in high salt. When these cells are grown to stationary phase and subsequently incubated in media of increasing salt concentration, we observe a salt-type and concentration-dependent increase in the frequency of bacteriophage spontaneous induction. The salt-dependent increase in phage production is unaffected by RecA mutation indicating that the salt-dependent increase in phage production is not caused by activation of the SOS pathway. Furthermore, the effect of changing cation type on spontaneous induction frequency mirror those seen on 434 repressor's affinity for O_R1, that is, 434 lysogens grown in media containing 200 mM LiCl produce ~50-fold more phage/viable cell than lysogens grown in 200 mM KCl. Also, the presence of the osmolyte, betaine, which should block the osmotic effect of high salt but not the change in intracellular salt concentration, does *not* inhibit the salt-induced increase in frequency of spontaneous induction. Together, these observations are consistent with the idea that salt stress induces lysogenic 434 bacteriophage by interfering with repressor–DNA interactions. Thus it appears that salt-mediated indirect readout of DNA structure can play a crucial role in modulating the growth of an organism.

VII. Summary and Prospects

The stability and sequence specificity of many protein–DNA complexes is remarkably dependent on the sequences of bases that are not in contact with protein. In all cases, these noncontacted bases inhibit or prevent the contacted regions of the DNA-binding site from being properly juxtaposed with protein groups. Hence, noncontacted bases affect protein–DNA complex formation via sequence-dependent differences in their structure and/or flexibility. Initial explanations of how noncontacted bases influenced the affinity of a protein for DNA focused on sequence-dependent differences in electrostatic interactions (stacking) between the base and the intrinsic mechanical properties of different DNA sequence. While it is clear that these features play a role in limiting the overall types of structural deformations tolerated by a particular sequence of DNA, within the limits of deformation induced by most proteins, intrinsic forces are too weak to play a role in determining the affinity of a protein for its DNA-binding site. We argue here that at least in the case of 434 repressor, the affinity, base specificity, and conformation of 434 repressor–DNA complexes are influenced by monovalent cation-type in a process that depends on the noncontacted base sequence. Our findings suggest that indirect readout is mediated by extrinsic forces, at least in part *via* sequence-dependent solvent–DNA interactions.

The sequence dependence of the effect of salt on a protein's affinity for DNA implies that for a given regulatory protein, the genes it controls can be differentially regulated by changes in salt type and/or concentration. An increasing number of examples exist of gene regulatory proteins whose affinity for DNA-binding sites are dependent on cation composition. Thus, cations may have a larger role in differentially regulating gene expression than has thus far been recognized.

REFERENCES

1. Kim, Y., Geiger, J. H., Hahn, S., and Sigler, P. B. (1993). Crystal structure of a yeast TBP/TATA-box complex. *Nature* **365,** 512–520.
2. Scheviz, R. W., Otwinowski, Z., Joachimiak, A., Lawson, C. L., and Sigler, P. B. (1985). The three dimensional structure of trp repressor. *Nature* **317,** 782–786.
3. Wu, L., Vertino, A., and Koudelka, G. B. (1992). Non-contacted bases affect the affinity of synthetic P22 operators for P22 repressor. *J. Biol. Chem.* **267,** 9134–9135.
4. Hegde, R. S., Grossman, S. R., Laimins, L. A., and Sigler, P. B. (1992). Crystal structure at 1.7 Å of the bovine papillomavirus-1 E2 DNA-binding domain bound to its DNA target. *Nature* **359,** 505–512.
5. Yang, C. C., and Nash, H. A. (1989). The interaction of *E. coli* IHF protein with its specific binding sites. *Cell* **57,** 869–880.
6. Somers, W. S., and Phillips, S. E. V. (1992). Crystal structure of the *met* repressor-operator complex at 2.8 Å resolution reveals DNA recognition by β-strands. *Nature* **359,** 387–393.
7. Gabrielsen, O. S., Sentenac, A., and Fromageot, P. (1991). Specific DNA binding by c-Myb: Evidence for a double helix- turn-helix-related motif. *Science* **253,** 1140–1143.
8. Dangi, B., Pelupessey, P., Martin, R. G., Rosner, J. L., Louis, J. M., and Gronenborn, A. M. (2001). Structure and dynamics of MarA-DNA complexes: An NMR investigation. *J. Mol. Biol.* **314,** 113–127.
9. Schwabe, J. W. R., Chapman, L., Finch, J. T., and Rhodes, D. (1993). The crystal structure of the estrogen receptor DNA-binding domain bound to DNA: How receptors discriminate between their response elements. *Cell* **75,** 567–578.
10. Jeltsch, A., Alves, J., Wolfes, H., Maass, G., and Pingoud, A. (1994). Pausing of the restriction endonuclease EcoRI during linear diffusion on DNA. *Biochemistry* **33,** 10215–10219.
11. Jeltsch, A., Wenz, C., Stahl, F., and Pingoud, A. (1996). Linear diffusion of the restriction endonuclease EcoRV on DNA is essential for the *in vivo* function of the enzyme. *EMBO J.* **15,** 5104–5111.
12. Hegde, R. S., and Androphy, E. J. (1998). Crystal structure of the E2 DNA-binding domain from human papillomavirus type 16: Implications for its DNA binding-site selection mechanism. *J. Mol. Biol.* **284,** 1479–1489.
13. Rodgers, D. W., and Harrison, S. C. (1993). The complex between phage 434 repressor DNA-binding domain and operator site OR3: Structural differences between consensus and non-consensus half-sites. *Structure* **1,** 227–240.
14. Widlund, H. R., Cao, H., Simonsson, S., Magnusson, E., Simonsson, T., Nielsen, P. E., Kahn, J. D., Crothers, D. M., and Kubista, M. (1997). Identification and characterization of genomic nucleosome-positioning sequences. *J. Mol. Biol.* **267,** 807–817.
15. Thastrom, A., Lowary, P. T., Widlund, H. R., Cao, H., Kubista, M., and Widom, J. (1999). Sequence motifs and free energies of selected natural and non-natural nucleosome positioning DNA sequences. *J. Mol. Biol.* **288,** 213–229.

16. Widlund, H. R., Kuduvalli, P. N., Bengtsson, M., Cao, H., Tullius, T. D., and Kubista, M. (1999). Nucleosome structural features and intrinsic properties of the TATAAACGCC repeat sequence. *J. Biol. Chem.* **274,** 31847–31852.
17. Gartenberg, M. R., and Crothers, D. M. (1986). DNA sequence determinants of CAP-induced bending and protein binding affinity. *Nature* **333,** 824–829.
18. Craig, N. L., and Nash, H. A. (1984). E. coli integration host factor binds to specific sites in DNA. *Cell* **39,** 707–716.
19. Goodman, S. D., Nicholson, S. C., and Nash, H. A. (1992). Deformation of DNA during site-specific recombination of bacteriophage lambda: Replacement of IHF protein by HU protein or sequence-directed bends. *Proc. Natl. Acad. Sci. USA* **89,** 11910–11914.
20. Segall, A. M., Goodman, S. D., and Nash, H. A. (1994). Architectural elements in nucleo-protein complexes: Interchangeability of specific and non-specific DNA binding proteins. *EMBO J.* **13,** 4536–4548.
21. Blomquist, P., Belikov, S., and Wrange, O. (1999). Increased nuclear factor 1 binding to its nucleosomal site mediated by sequence-dependent DNA structure. *Nucleic Acids Res.* **27,** 517–525.
22. Tisne, C., Delepierre, M., and Hartmann, B. (1999). How NF-kappaB can be attracted by its cognate DNA. *J. Mol. Biol.* **293,** 139–150.
23. Sierk, M. L., Zhao, Q., and Rastinejad, F. (2001). DNA deformability as a recognition feature in the reverb response element. *Biochemistry* **40,** 12833–12843.
24. Moitoso de Vargas, L., Pargellis, C. A., Hasan, N. M., Bushman, E. W., and Landy, A. (1988). Autonomous DNA binding domains of lambda integrase recognize two different sequence families. *Cell* **54,** 923–929.
25. Kwon, H. J., Tirumalai, R., Landy, A., and Ellenberger, T. (1997). Flexibility in DNA recombination: Structure of the lambda integrase catalytic core. *Science* **276,** 126–131.
26. Werel, W., Schickor, P., and Heumann, H. (1991). Flexibility of the DNA enhances promoter affinity of *Escherichia coli* RNA polymerase. *EMBO J.* **10,** 2589–2594.
27. Auble, D. T., Allen, T. L., and DeHaseth, P. L. (1986). Promoter recognition by *Escherichia coli* RNA polymerase. *J. Biol. Chem.* **261,** 11202–11206.
28. Sasmor, H. M., and Betz, J. L. (1990). Specific binding of *lac* repressor to linear versus circular polyoperator molecules. *Biochemistry* **29,** 9023–9028.
29. Sasmor, H. M., and Betz, J. L. (1990). Symmetric *lac* operator derivatives: Effects of half-operator sequence and spacing on repressor affinity. *Gene* **89,** 1–6.
30. Hines, C. S., Meghoo, C., Shetty, S., Biburger, M., Brenowitz, M., and Hegde, R. S. (1998). DNA structure and flexibility in the sequence-specific binding of papillomavirus E2 proteins. *J. Mol. Biol.* **276,** 809–818.
31. Rohs, R., Sklenar, H., and Shakked, Z. (2005). Structural and energetic origins of sequence-specific DNA bending: Monte carlo simulations of papillomavirus E2-DNA binding sites. *Structure (Camb.)* **13,** 1499–1509.
32. Szymczyna, B. R., and Arrowsmith, C. H. (2000). DNA binding specificity studies of four ETS proteins support an indirect read-out mechanism of protein-DNA recognition. *J. Biol. Chem.* **275,** 28363–28370.
33. Kodandapani, R., Pio, F., Ni, C. Z., Piccialli, G., Klemsz, M., McKercher, S., Maki, R. A., and Ely, K. R. (1996). A new pattern for helix-turn-helix recognition revealed by the PU.1 ETS-domain-DNA complex. *Nature* **380,** 456–460.
34. Batchelor, A. H., Piper, D. E., de la Brousse, F. C., McKnight, S. L., and Wolberger, C. (1998). The structure of GABPalpha/beta: An ETS domain-ankyrin repeat heterodimer bound to DNA. *Science* **279,** 1037–1041.

35. Mo, Y., Vaessen, B., Johnston, K., and Marmorstein, R. (1998). Structures of SAP-1 bound to DNA targets from the E74 and c-fos promoters: Insights into DNA sequence discrimination by Ets proteins. *Mol. Cell* **2**, 201–212.
36. Mo, Y., Vaessen, B., Johnston, K., and Marmorstein, R. (2000). Structure of the elk-1-DNA complex reveals how DNA-distal residues affect ETS domain recognition of DNA. *Nat. Struct. Biol.* **7**, 292–297.
37. Bareket-Samish, A., Cohen, I., and Haran, T. E. (2000). Signals for TBP/TATA box recognition. *J. Mol. Biol.* **299**, 965–977.
38. Patel, L. R., Curran, T., and Kerppola, T. K. (1994). Energy transfer analysis of Fos-Jun dimerization and DNA binding. *Proc. Natl. Acad. Sci. USA* **91**, 7360–7364.
39. Hershberger, P. A., and DeHaseth, P. L. (1993). Interference by P_R-bound RNA polymerase with P_{RM} function *in vitro*. Modulation by the bacteriophage λ *c*I protein. *J. Biol. Chem.* **268**, 8943–8948.
40. La Baer, J., and Yamamoto, K. R. (1994). Analysis of the DNA-binding affinity, sequence specificity and context dependence of the glucocorticoid receptor zinc finger region. *J. Mol. Biol.* **239**, 664–688.
41. Lefstin, J. A., Thomas, J. R., and Yamamoto, K. R. (1994). Influence of a steroid receptor DNA-binding domain on transcriptional regulatory functions. *Genes Dev.* **8**, 2842–2856.
42. Lefstin, J. A., and Yamamoto, K. R. (1998). Allosteric effects of DNA on transcriptional regulators. *Nature* **392**, 885–888.
43. van Tilborg, M. A., Lefstin, J. A., Kruiskamp, M., Teuben, J., Boelens, R., Yamamoto, K. R., and Kaptein, R. (2000). Mutations in the glucocorticoid receptor DNA-binding domain mimic an allosteric effect of DNA. *J. Mol. Biol.* **301**, 947–958.
44. Spiro, C., Bazett-Jones, D. P., Wu, X., and McMurray, C. T. (1995). DNA structure determines protein binding and transcriptional efficiency of the proenkephalin cAMP-responsive enhancer. *J. Biol. Chem.* **270**, 27702–27710.
45. Watson, J. D., and Crick, F. H. (1953). Molecular structure of nucleic acids; a structure for deoxyribose nucleic acid. *Nature* **171**, 737–738.
46. Arnott, S., and Hukins, D. W. (1973). Refinement of the structure of B-DNA and implications for the analysis of x-ray diffraction data from fibers of biopolymers. *J. Mol. Biol.* **81**, 93–105.
47. Fuller, W., Wilkins, M. H., Wilson, H. R., and Hamilton, L. D. (1965). The molecular configuration of deoxyribonucleic acid. IV. X-ray diffraction study of the A form. *J. Mol. Biol.* **12**, 60–76.
48. Bugg, C. E., Thomas, J. M., Sundaralingam, M., and Rao, S. T. (1971). Stereochemistry of nucleic acids and their constituents. X. Solid-state base-stacking patterns in nucleic acid constituents and polynucleotides. *Biopolymers* **10**, 175–219.
49. Wing, R., Drew, H., Takano, T., Broka, C., Tanaka, S., Itakura, K., and Dickerson, R. E. (1980). Crystal structure analysis of a complete turn of B-DNA. *Nature* **287**, 755–758.
50. Drew, H. R., Wing, R. M., Takano, T., Broka, C., Tanaka, S., Itakura, K., and Dickerson, R. E. (1981). Structure of a B-DNA dodecamer: Conformation and dynamics. *Proc. Natl. Acad. Sci. USA* **78**, 2179–2183.
51. Grzeskowiak, K., Yanagi, K., Prive, G. G., and Dickerson, R. E. (1991). The structure of B-helical C-G-A-T-C-G-A-T-C-G and comparison with C-C-A-A-C-G-T-T-G-G. The effect of base pair reversals. *J. Biol. Chem.* **266**, 8861–8883.
52. Chiu, T. K., Kaczor-Grzeskowiak, M., and Dickerson, R. E. (1999). Absence of minor groove monovalent cations in the crosslinked dodecamer C-G-C-G-A-A-T-T-C-G-C-G. *J. Mol. Biol.* **292**, 589–608.
53. Shui, X., McFail-Isom, L., Hu, G. G., and Williams, L. D. (1998). The B-DNA dodecamer at high resolution reveals a spine of water on sodium. *Biochemistry* **37**, 8341–8355.

54. Calladine, C. R. (1982). Mechanics of sequence-dependent stacking of bases in B-DNA. *J. Mol. Biol.* **161,** 343–352.
55. Packer, M. J., Dauncey, M. P., and Hunter, C. A. (2000). Sequence-dependent DNA structure: Tetranucleotide conformational maps. *J. Mol. Biol.* **295,** 85–103.
56. Packer, M. J., Dauncey, M. P., and Hunter, C. A. (2000). Sequence-dependent DNA structure: Dinucleotide conformational maps. *J. Mol. Biol.* **295,** 71–83.
57. Dlakic, M., and Harrington, R. E. (1995). Bending and torsional flexibility of G/C-rich sequences as determined by cyclization assays. *J. Biol. Chem.* **270,** 29945–29952.
58. Dlakic, M., and Harrington, R. E. (1996). The effects of sequence context on DNA curvature. *Proc. Natl. Acad. Sci. USA* **93,** 3847–3852.
59. Dlakic, M., and Harrington, R. E. (1998). Unconventional helical phasing of repetitive DNA motifs reveals their relative bending contributions. *Nucleic Acids Res.* **26,** 4274–4279.
60. Saenger, W. (1984). "Principles of Nucleic Acid Structure." Springer-Verlag, Inc., New York.
61. Nelson, H. C. M., Finch, J. T., Luisi, B. F., and Klug, A. (1987). The structure of an oligo(dA). oligo(dT) tract and its biological implications. *Nature* **330,** 221–226.
62. el Hassan, M. A., and Calladine, C. R. (1995). The assessment of the geometry of dinucleotide steps in double-helical DNA; a new local calculation scheme. *J. Mol. Biol.* **251,** 648–664.
63. el Hassan, M. A., and Calladine, C. R. (1996). Propeller-twisting of base-pairs and the conformational mobility of dinucleotide steps in DNA. *J. Mol. Biol.* **259,** 95–103.
64. el Hassan, M. A., and Calladine, C. R. (1996). Structural mechanics of bent DNA. *Endeavour* **20,** 61–67.
65. Suzuki, M., Amano, N., Kakinuma, J., and Tateno, M. (1997). Use of a 3D structure data base for understanding sequence-dependent conformational aspects of DNA. *J. Mol. Biol.* **274,** 421–435.
66. Hud, N. V., and Feigon, J. (1997). Localization of divalent metal ions in the minor groove of DNA A-tracts. *J. Am. Chem. Soc.* **119,** 5756–5757.
67. Shui, X., Sines, C. C., McFail-Isom, L., VanDerveer, D., and Williams, L. D. (1998). Structure of the potassium form of CGCGAATTCGCG: DNA deformation by electrostatic collapse around inorganic cations. *Biochemistry* **37,** 16877–16887.
68. Hud, N. V., Sklenar, V., and Feigon, J. (1999). Localization of ammonium ions in the minor groove of DNA duplexes in solution and the origin of DNA A-tract bending. *J. Mol. Biol.* **286,** 651–660.
69. McFail-Isom, L., Sines, C. C., and Williams, L. D. (1999). DNA structure: Cations in charge? *Curr. Opin. Struct. Biol.* **9,** 298–304.
70. Rouzina, I., and Bloomfield, V. A. (1998). DNA bending by small, mobile multivalent cations. *Biophys. J.* **74,** 3152–3164.
71. Rouzina, I., and Bloomfield, V. A. (2001). Force-induced melting of the DNA double helix. 2. Effect of solution conditions. *Biophys. J.* **80,** 894–900.
72. Rosenberg, J. M., Seeman, N. C., Kim, J. J., Suddath, F. L., Nicholas, H. B., and Rich, A. (1973). Double helix at atomic resolution. *Nature* **243,** 150–154.
73. Quigley, G. J., Teeter, M. M., and Rich, A. (1978). Structural analysis of spermine and magnesium ion binding to yeast phenylalanine transfer RNA. *Proc. Natl. Acad. Sci. USA* **75,** 64–68.
74. Howerton, S. B., Sines, C. C., VanDerveer, D., and Williams, L. D. (2001). Locating monovalent cations in the grooves of B-DNA. *Biochemistry* **40,** 10023–10031.
75. Minasov, G., Tereshko, V., and Egli, M. (1999). Atomic-resolution crystal structures of B-DNA reveal specific influences of divalent metal ions on conformation and packing. *J. Mol. Biol.* **291,** 83–99.

76. Woods, K., McFail-Isom, L., Sines, C. C., Howerton, S. B., Stephens, R. K., and Williams, L. D. (2001). Monovalent cations sequester within the A-Tract minor groove of [d(CGCGAATTCGCG)]2. *J. Am. Chem. Soc.* **122**, 1546–1547.

77. Hud, N. V., and Polak, M. (2001). DNA-cation interactions: The major and minor grooves are flexible ionophores. *Curr. Opin. Struct. Biol.* **11**, 293–301.

78. Chiu, T. K., and Dickerson, R. E. (2000). 1 angstrom crystal structures of B-DNA reveal sequence-specific binding and groove-specific bending of DNA by magnesium and calcium. *J. Mol. Biol.* **301**, 915–945.

79. Davey, C. A., and Richmond, T. J. (2002). DNA-dependent divalent cation binding in the nucleosome core particle. *Proc. Natl. Acad. Sci. USA* **99**, 11169–11174.

80. Davey, C. A., Sargent, D. F., Luger, K., Maeder, A. W., and Richmond, T. J. (2002). Solvent mediated interactions in the structure of the nucleosome core particle at 1.9 a resolution. *J. Mol. Biol.* **319**, 1097–1113.

81. Sauer, R. T., Yocum, R. R., Doolittle, R. F., Lewis, M., and Pabo, C. O. (1982). Homology among DNA-binding proteins suggests use of a conserved supersecondary structure. *Nature* **298**, 447–451.

82. Ptashne, M. (1986). "A Genetic Switch." Blackwell Press, Palo Alto.

83. Xu, J., and Koudelka, G. B. (2001). Repression of transcription initiation at 434 P(R) by 434 repressor: Effects on transition of a closed to an open promoter complex. *J. Mol. Biol.* **309**, 573–587.

84. Shang, Z., Isaac, V. E., Li, H., Patel, L., Catron, K. M., Curran, T., Montelione, G. T., and Abate, C. (1994). Design of a "minimal" homeodomain: The N-terminal arm modulates DNA binding affinity and stabilizes homeodomain structure. *Proc. Natl. Acad. Sci. USA* **91**, 8373–8377.

85. Hochschild, A., Irwin, N., and Ptashne, M. (1983). Repressor structure and the mechanism of positive control. *Cell* **32**, 319–325.

86. Xu, J., and Koudelka, G. B. (2000). DNA sequence requirements for the activation of 434 P_{RM} transcription by 434 repressor. *DNA Cell Biol.* **19**, 621–630.

87. Hershberger, P. A., and DeHaseth, P. L. (1991). RNA polymerase bound to the P_R promoter of bacteriophage lambda inhibits open complex formation at the divergently transcribed P_{RM} promoter. Implications for an indirect mechanism of transcriptional activation by lambda repressor. *J. Mol. Biol.* **222**, 479–494.

88. Xu, J., and Koudelka, G. B. (2000). Mutually exclusive utilization of P(R) and P(RM) promoters in bacteriophage 434 O(R). *J. Bacteriol.* **182**, 3165–3174.

89. Jacob, F., and Monod, J. (1961). Genetic regulatory mechanisms in the synthesis of proteins. *J. Mol. Biol.* **3**, 318–356.

90. Bushman, F. D. (1993). The bacteriophage 434 right operator. Roles of O(R)1, O(R)2 and O(R)3. *J. Mol. Biol.* **230**, 28–40.

91. Choy, H. E., and Adhya, S. (1993). RNA polymerase idling and clearance in gal promoters: Use of supercoiled minicircle DNA template made *in vivo*. *Proc. Natl. Acad. Sci. USA* **90**, 472–476.

92. Johnson, A. D., Pabo, C. O., and Sauer, R. T. (1982). Bacteriophage lambda repressor and cro protein: Interactions with operator DNA. *Methods Enzymol.* **65**, 839–856.

93. DeAnda, J., Poteete, A. R., and Sauer, R. T. (1983). P22 c2 repressor-domain structure and function. *J. Biol. Chem.* **258**, 10536–10542.

94. Pabo, C. O., Sauer, R. T., Sturtevant, J. M., and Ptashne, M. (1979). The lambda repressor contains two domains. *Proc. Natl. Acad. Sci. USA* **76**, 1608–1612.

95. Anderson, J. E. (1984). Ph.D. thesis. Harvard University, Cambridge.

96. Aggarwal, A., Rodgers, D. W., Drottar, M., Ptashne, M., and Harrison, S. C. (1988). Recognition of a DNA operator by the repressor of phage 434: A view at high resolution. *Science* **242,** 899–907.
97. Anderson, J. E., Harrison, S. C., and Ptashne, M. (1987). A phage repressor-operator complex at 7# resolution. *Nature* **326,** 888–891.
98. Mondragon, A., Subbiah, S., Almo, S. C., Drottar, M., and Harrison, S. C. (1989). Structure of the amino-terminal domain of phage 434 repressor at 2.0 A resolution. *J. Mol. Biol.* **205,** 189–200.
99. Fattah, K. R., Mizutani, S., Fattah, F. J., Matsushiro, A., and Sugino, Y. (2000). A comparative study of the immunity region of lambdoid phages including shiga-toxin-converting phages: Molecular basis for cross immunity. *Genes Genet. Syst.* **75,** 223–232.
100. Harrison, S. C. (1991). A structural taxonomy of DNA-binding domains. *Nature* **353,** 715–719.
101. Johnson, A. D., Meyer, B. J., and Ptashne, M. (1979). Interaction between DNA-bound repressors govern regulation by the lambda repressor. *Proc. Natl. Acad. Sci. USA* **76,** 5061–5065.
102. Luo, Y., Pfuetzner, R. A., Mosimann, S., Paetzel, M., Frey, E. A., Cherney, M., Kim, B., Little, J. W., and Strynadka, N. C. (2001). Crystal structure of LexA: A conformational switch for regulation of self-cleavage. *Cell* **106,** 585–594.
103. Donner, A. L., and Koudelka, G. B. (1998). Carboxyl-teminal domain dimer interface mutant 434 repressors have altered dimerization and DNA binding specificities. *J. Mol. Biol.* **283,** 931–946.
104. Ciubotaru, M., Bright, F. V., Ingersoll, C. M., and Koudelka, G. B. (1999). DNA-induced conformational changes in bacteriophage 434 repressor. *J. Mol. Biol.* **294,** 859–873.
105. Koudelka, G. B., Harrison, S. C., and Ptashne, M. (1987). Effect of non-contacted bases on the affinity of 434 operator for 434 repressor and Cro. *Nature* **326,** 886–888.
106. Anderson, J. E., Ptashne, M., and Harrison, S. C. (1987). Structure of the repressor-operator complex of bacteriophage 434. *Nature* **326,** 846–852.
107. Koudelka, G. B., and Carlson, P. (1992). DNA twisting and the effects of non-contacted bases on affinity of 434 operator for 434 repressor. *Nature* **355,** 89–91.
108. Shimon, L. J. W., and Harrison, S. C. (1993). The phage 434 O_R2/R1-69 complex at 2•5 Å resolution. *J. Mol. Biol.* **232,** 826–838.
109. Barkley, M. D., and Zimm, B. H. (1979). Theory of twisting and bending of chain macromolecules; analysis of the fluorescence depolarization. *J. Chem. Phys.* **70,** 2991–3007.
110. Koudelka, G. B. (1991). Bending of synthetic bacteriophage 434 operators by bacteriophage 434 proteins. *Nucleic Acids Res.* **19,** 4115–4119.
111. Koudelka, G. B. (1998). Recognition of DNA structure by 434 repressor. *Nucleic Acids Res.* **26,** 669–675.
112. Fujimoto, B. S., and Schurr, J. M. (1990). Dependence of the torsional rigidity of DNA on base composition. *Nature* **344,** 175–178.
113. Donner, A. L., Carlson, P. A., and Koudelka, G. B. (1997). Dimerization specificity of P22 and 434 repressors is determined by multiple polypeptide segments. *J. Bacteriol.* **179,** 1253–1261.
114. Mauro, S. A., Pawlowski, D., and Koudelka, G. B. (2004). The role of the minor groove substituents in indirect readout of DNA sequence by 434 repressor. *J. Biol. Chem.* **278,** 12955–12960.
115. Koudelka, G. B., Harbury, P. H., Harrison, S. C., and Ptashne, M. (1988). DNA twisting and the affinity of bacteriophage 434 operator for bacteriophage 434 repressor. *Proc. Natl. Acad. Sci. USA* **85,** 4633–4637.
116. Bell, A. C., and Koudelka, G. B. (1995). How 434 repressor discriminates between O_R1 and O_R3. The influence of contacted and noncontacted base pairs. *J. Biol. Chem.* **270,** 1205–1212.

117. Bell, A. C., and Koudelka, G. B. (1993). Operator sequence context influences amino acid-base-pair interactions in 434 repressor-operator complexes. *J. Mol. Biol.* **234,** 542–553.
118. Duong, T. H., and Zakrzewska, K. (1998). Sequence specificity of bacteriophage 434 repressor-operator complexation. *J. Mol. Biol.* **280,** 31–39.
119. Higgins, C. F., Dorman, C. J., Stirling, D. A., Waddell, L., Booth, I. R., May, G., and Bremer, E. (1988). A physiological role for DNA supercoiling in the osmotic regulation of gene expression in S. typhimurium and E. coli. *Cell* **52,** 569–584.
120. Jordi, B. J., and Higgins, C. F. (2000). The downstream regulatory element of the proU operon of Salmonella typhimurium inhibits open complex formation by RNA polymerase at a distance. *J. Biol. Chem.* **275,** 12123–12128.
121. Conter, A., Menchon, C., and Gutierrez, C. (1997). Role of DNA supercoiling and rpoS sigma factor in the osmotic and growth phase-dependent induction of the gene osmE of *Escherichia coli* K12. *J. Mol. Biol.* **273,** 75–83.
122. Graeme-Cook, K. A., May, G., Bremer, E., and Higgins, C. F. (1989). Osmotic regulation of porin expression: A role for DNA supercoiling. *Mol. Microbiol.* **3,** 1287–1294.
123. Record, M. T., Courtenay, E. S., Cayley, D. S., and Guttman, H. J. (1998). Responses of *E-coli* to osmotic stress: Large changes in amounts of cytoplasmic solutes and water. *Trends Biochem. Sci.* **23,** 143–148.
124. Record, M. T., Courtenay, E. S., Cayley, S., and Guttman, H. J. (1998). Biophysical compensation mechanisms buffering *E-coli* protein-nucleic acid interactions against changing environments. *Trends Biochem. Sci.* **23,** 190–194.
125. Mauro, S. A., and Koudelka, G. B. (2004). Monovalent cations regulate DNA sequence recognition by 434 repressor. *J. Mol. Biol.* **340,** 445–457.
126. Daniels, D. I., Schroeder, J. L., Szybalski, W., Sanger, F., Coulson, A. R., Hong, G. F., Hill, D. F., Petersen, G. F., and Blattner, F. R. (1983). Lambda II, pp. 519–676. Cold Spring Harbor Laboratory, Cold Spring Harbor.
127. Sauer, R. T., Nelson, H. C., Hehir, K., Hecht, M. H., Gimble, F. S., DeAnda, J., and Poteete, A. R. (1983). The lambda and P22 phage repressors. *J. Biomol. Struct. Dyn.* **1,** 1011–1022.
128. Lu, X. J., and Olson, W. K. (2003). 3DNA: A software package for the analysis, rebuilding and visualization of three-dimensional nucleic acid structures. *Nucleic Acids Res.* **31,** 5108–5121.

Repair of Topoisomerase I-Mediated DNA Damage

YVES POMMIER,*
JUANA M. BARCELO,*
V. ASHUTOSH RAO,*
OLIVIER SORDET,*
ANDREW G. JOBSON,*
LAURENT THIBAUT,*
ZE-HONG MIAO,*
JENNIFER A. SEILER,*
HONGLIANG ZHANG,*
CHRISTOPHE MARCHAND,*
KELI AGAMA,* JOHN L. NITISS,[†]
AND CHRISTOPHE REDON*

*Laboratory of Molecular Pharmacology, Center for Cancer Research, National Cancer Institute, National Institutes of Health, DHHS, Bethesda, Maryland 20892

[†]Molecular Pharmacology Department, St. Jude Children Research Hospital, Memphis, TN 38105

I. Introduction: Mammalian Topoisomerase Families, Top1 Functions, and Catalytic Mechanisms.	180
II. Induction and Stabilization of Top1 Cleavage Complexes by Camptothecin and Anticancer Drugs and by Carcinogens and Endogenous DNA Lesions.	184
III. Conversion of Top1 Cleavage Complexes into DNA Damage.	188
IV. Repair of Top1-Associated DNA Damage	189
A. Reversal of Top1–DNA Covalent Complexes by 5′-end Religation (Fig. 5A).	190
B. Top1 Excision by Tyrosyl-DNA-Phosphodiesterase (Fig. 5B).	190
C. Top1 Excision by Endonucleases (Fig. 5C).	206
V. Checkpoint Response to Top1-Associated DNA Damage.	210
A. Chk1 Activation by Camptothecin.	211
B. Chk2 Activation by Camptothecin.	212
C. Chk1 vs Chk2 Activation by Camptothecin and Top1-Mediated DNA Damage.	213
VI. Conclusion and Perspective.	213
References.	214

Topoisomerase I (Top1) is an abundant and essential enzyme. Top1 is the selective target of camptothecins, which are effective anticancer agents. Top1–DNA cleavage complexes can also be trapped by various endogenous and exogenous DNA lesions including mismatches, abasic sites, and carcinogenic adducts. Tyrosyl-DNA phosphodiesterase (Tdp1) is one of the repair enzymes for Top1–DNA covalent complexes. Tdp1 forms a multiprotein complex that includes poly(ADP–ribose) polymerase (PARP). PARP-deficient cells are hypersensitive to camptothecins and functionally deficient for Tdp1. We will review developments in several pathways involved in the repair of Top1 cleavage complexes and the role of Chk1 and Chk2 checkpoint kinases in the cellular responses to Top1 inhibitors. The genes conferring camptothecin hypersensitivity are compiled for humans, budding yeast, and fission yeast.

I. Introduction: Mammalian Topoisomerase Families, Top1 Functions, and Catalytic Mechanisms

Seven topoisomerase genes are encoded in the human nuclear genome (1). The enzymes (abbreviated Topo or Top) have been numbered in the order of their discovery except for the enzyme mitochondrial topoisomerase I (Top1mt) (2, 3). Vertebrate cells contain two Top1 (Top1 for the nuclear genome and Top1mt for the mitochondrial genome), two Top2 (Top2α and β), and two Top3 (Top3α and β). The seventh topoisomerase is Spo11 whose expression is restricted to germ cells. Top3α forms heterodimers with *BLM* (the gene product deficient in Bloom syndrome) and is functionally related to the resolution of postreplicative hemicatenanes and recombination intermediates (4, 5). Top1 proteins belong to the family of the tyrosine recombinases (which includes λ-integrase, Flip and Cre recombinases), and Top2 is related to bacterial gyrase and Topo IV, which are the targets of quinolone antibiotics.

Topoisomerases and tyrosine recombinases nick and religate DNA by forming a covalent enzyme-DNA intermediate between an enzyme catalytic tyrosine residue and the end of the broken DNA (Fig. 1). These covalent intermediates are generally referred to as "cleavage (or cleavable) complexes" (Fig. 2). Topoisomerases have also been classified in two groups depending on whether they cleave and religate one strand (type I) or both strands (type II) of the DNA duplex. Type I enzymes include Top1 (nuclear), Top1mt, Top3α and β, and type II enzymes include Top2α and β and Spo11.

Top1 is essential in vertebrates and flies but not in yeast. Knocking out the *TOP1* gene results in early embryonic lethality in mouse (6) and fly (7). By contrast, yeast survives in the absence of *TOP1* (8). Top1 is expressed constitutively throughout the cell cycle (9) and is concentrated in the nucleolus (10, 11). Its main function is to relieve both positive and negative DNA supercoiling

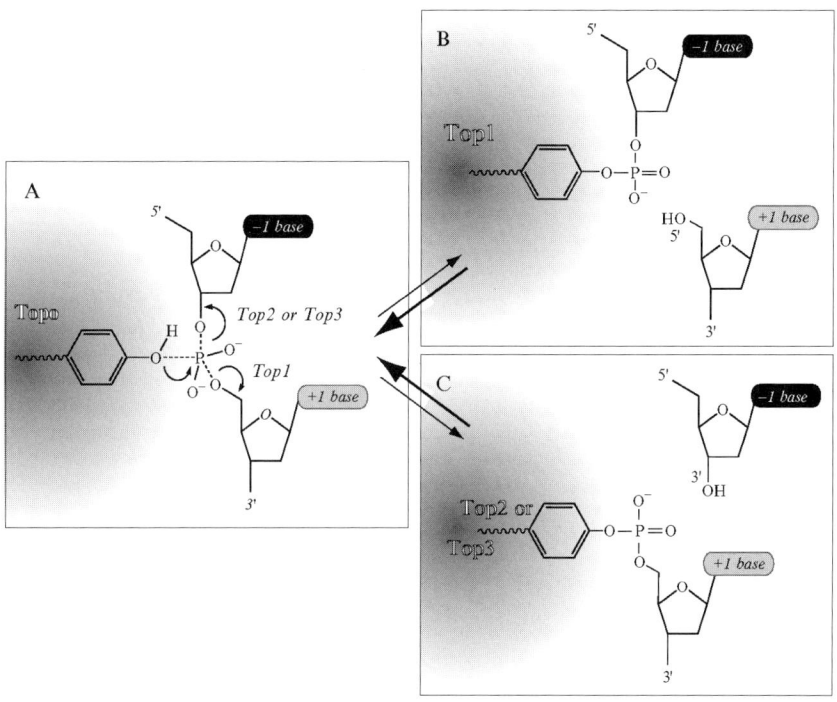

FIG. 1. Topoisomerase cleavage complexes. Topoisomerases (abbreviated Topo in panel A) utilize a catalytic tyrosine residue for nucleophilic attack and breakage of a DNA phosphoester bond. (A) The polarity depends on the Topo (only human enzymes are considered here). (B) Topoisomerases I (nuclear and mitochondrial Top1) form a covalent bond with the 3′-DNA end and generate a 5′-hydroxyl end. This cleavage intermediate allows controlled rotation of the 5′ end around the intact DNA strand (see Fig. 3B). Under normal conditions, the reaction is reversible. Religation (back arrow from B to A) is favored over cleavage and requires the alignment of the 5′-hydroxyl end with the phosphoester tyrosyl DNA bond for nucleophilic attack. (C) All other human Topo enzymes (Top2 and Top3) have an opposite polarity compared to Top1 (see Fig. 2). They form covalent bonds with the 5′ end of the break and generate 3′-hydroxyl ends.

generated by transcription and replication and possibly DNA repair and chromatin remodeling (1, 12–14). The mechanistic similarities between Top1 and other tyrosine recombinases suggest that Top1 may also play a role in DNA recombinations (15, 16). The Top1 recombinase activity has been proposed for the replication of *vaccinia* (17) and *hepadnaviruses* (16). Top1 may also contribute to RNA splicing by phosphorylating SR proteins (18, 19).

Top1 relaxes DNA supercoiling in the absence of energy cofactor by nicking the DNA and allowing rotation of the broken strand around the Top1-bound DNA strand (Fig. 3B—curved arrow). Crystal structures of Top1 (20–22) show the enzyme encircling the DNA tightly like a clamp

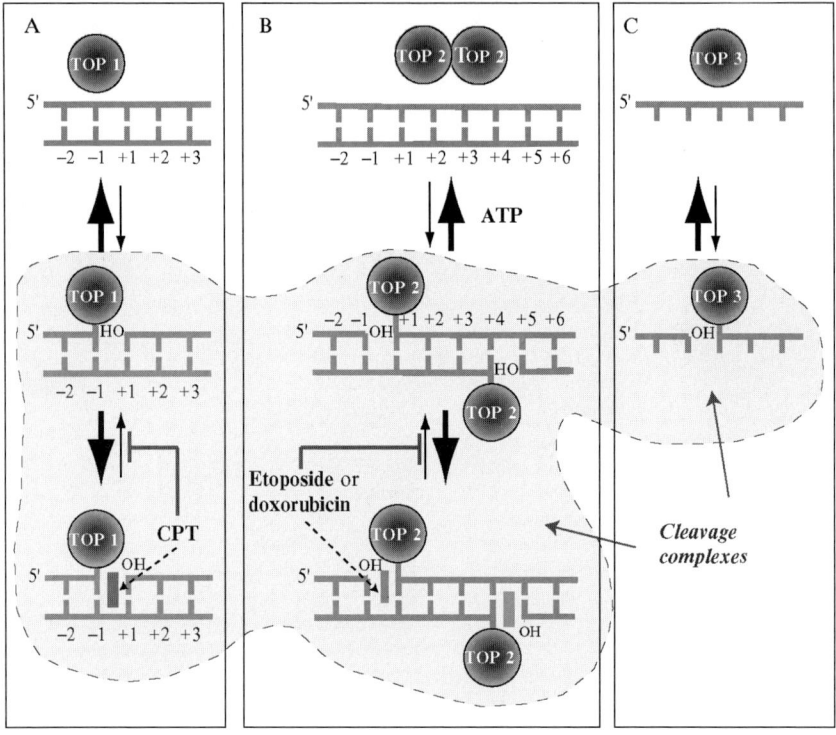

FIG. 2. Schematic architecture of the topoisomerase cleavage complexes. (A) Topoisomerases I (Top1 nuclear and Top1mt) bind to double-stranded DNA and form covalent complexes at the 3' end of the breaks. All other topoisomerases form covalent complexes at the 5' end of the breaks. Top1 cleavage complexes are selectively stabilized by the natural alkaloid camptothecin (CPT). (B) Topoisomerase II homodimers (Top2α and Top2β) bind to double-stranded DNA and form cleavage complexes with a canonical four-base pair overhang. Top2 binds and hydrolyze ATP during catalysis. Top2 inhibitors stabilize the Top2 cleavage complexes and are potent anticancer drugs. (C) Topoisomerases III (Top3α and Top3β) bind as monomers to noncanonical DNA structures (single-stranded DNA) (194) in association with a RecQ helicase (BLM in humans, Sgs1 in budding yeast, Rhq1 in fission yeast). Top3 has been proposed to resolve double-holiday junctions arising from stalled replication forks (see Fig. 5A and the corresponding text). Top3 inhibitors have not been reported.

(Fig. 3D), which accounts for the fact that Top1 controls the processive relaxation of supercoiled DNA (20, 23–25). Once the DNA is relaxed, Top1 religates the breaks by reversing its covalent binding. Religation requires the DNA end 5'-hydroxyl group to be aligned with the tyrosine-DNA phosphodiester bond. Under normal conditions, the cleavage intermediates (Figs. 2A and 3B) are transient and religation is favored over cleavage.

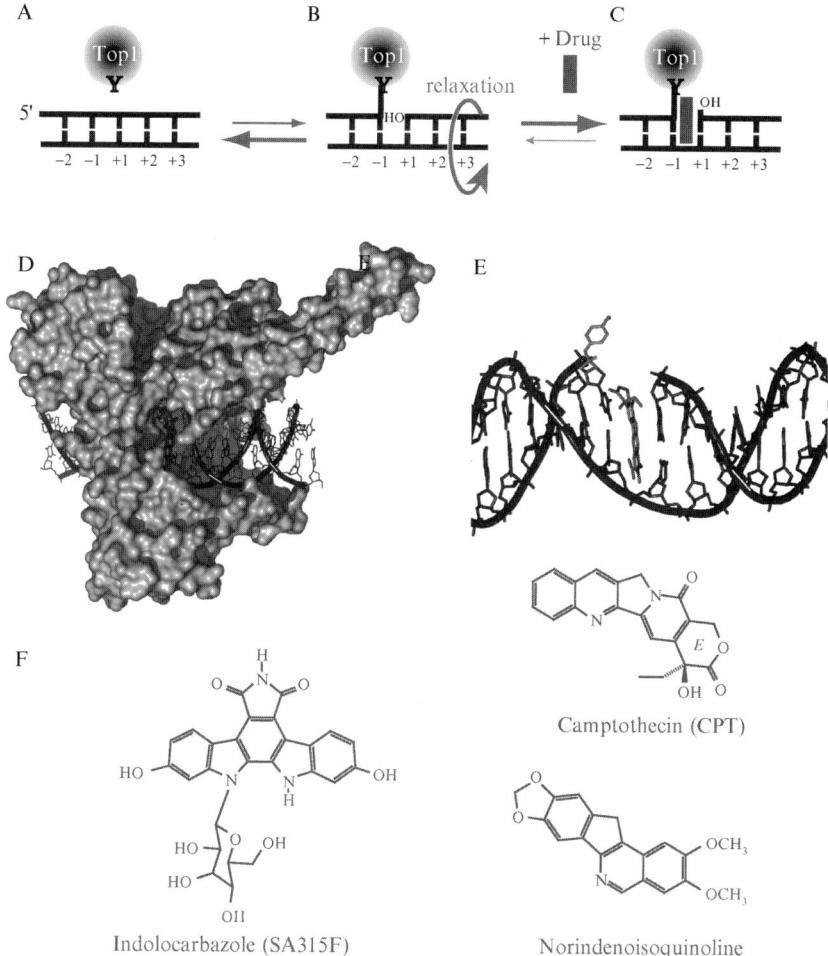

Fig. 3. Trapping of Top1 cleavage complexes by camptothecin and noncamptothecin inhibitors. (A) Under physiological conditions, Top1 is associated with chromatin in noncovalent complexes. (B) A small fraction of Top1 forms cleavage complexes that relax DNA supercoiling by controlled rotation of the cleaved strand around the intact strand (green curved arrow). (C) Anticancer drugs, such as those shown in panel F, reversibly trap the Top1 cleavage complex by inhibiting religation. (D) Crystal structure of camptothecin bound to the Top1–DNA cleavage complex [from (29)] showing "interfacial inhibition" (26, 27) of the Top1 cleavage complex by camptothecin. Interfacial inhibition also applies to noncamptothecin Top1 inhibitors shown in panel F (29, 30). (E) Same structure as in panel D. The Top1 has been removed except for the catalytic tyrosine (in orange). Camptothecin is shown intercalated between the base pairs flanking the Top1 cleavage site. (F) Structures of three Top1 inhibitors. (See Color Insert.)

II. Induction and Stabilization of Top1 Cleavage Complexes by Camptothecin and Anticancer Drugs and by Carcinogens and Endogenous DNA Lesions

The normally transient Top1 cleavage complexes can be converted into potential DNA lesions. Stabilization of the cleavage complexes generally results from misalignment of the 5′-hydroxyl-DNA end. Misalignments can be generated by drugs bound at the interface of the enzyme and broken DNA (26, 27) and by alterations of the DNA substrate (Table I and Fig. 4A).

Camptothecins and noncamptothecin Top1 inhibitors trap Top1 cleavage complexes by binding at the enzyme–DNA interface (Fig. 3D and E) (22, 28–30). Hence, Top1 inhibitors represent a paradigm for "interfacial inhibitors" (26, 27).

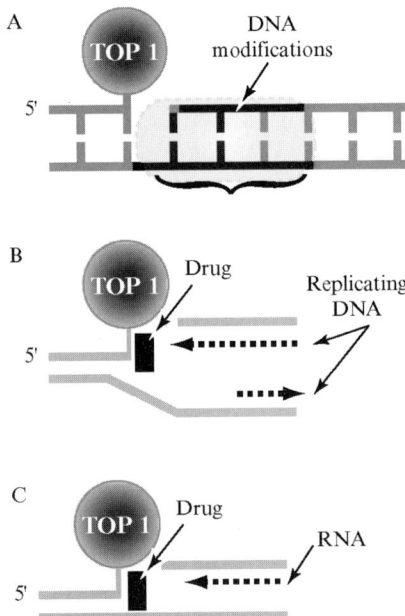

FIG. 4. Conversion of reversible Top1 cleavage complexes into DNA damage. (A) Irreversible (suicide) Top1 cleavage complexes are produced when Top1 cleaves previously damaged DNA (DNA modifications that trap Top1 are detailed in Table I). (B) Top1 cleavage complexes can be converted to irreversible complexes upon replication fork collision when the Top1 cleavage complex is on the leading strand for DNA synthesis. The drug is shown as the initiating event for the collision. However, once the replication-mediated DNA double strand break (Rep-DSB) is formed, dissociation of the drug has no impact on the irreversible covalent complex. (C) Conversion of Top1 cleavage complexes into irreversible covalent complexes by transcription complexes.

TABLE I
EXOGENOUS AND ENDOGENOUS FACTORS PRODUCING Top1 CLEAVAGE COMPLEXES

Drugs[a]	Mechanism[b]	Rev.[c]	Notes	References
Camptothecins	T	r	Derived from the natural alkaloid	(196)
Indenoisoquinolines	T	r	Synthetic; in preclinical development	(32)
Indolocarbazoles (NB-506)	T	r	Semisynthetic; in clinical development	(32)
Actinomycin D	T	r	Other effects: DNA, RNA polymerase	(196)
Hoechst minor groove	T	r	Other effects: DNA	(32)
Ecteinascidin 743 (Yondelis®)	T	r	Other effects: traps TC-NER complex	(32)
Cytosine Arabinoside	T	r	Other effects: blocks DNA synthesis	(197, 198)
Gemcitabine	T	r	Other effects: blocks DNA synthesis	(199)
Endogenous DNA lesions				
Single base mismatches	T	r	Polymerase and mismatch defects	(37, 44)
Mismatched loops	T	ir	Mismatch deficiencies	(37, 40)
Abasic sites	T	ir	AP sites; base excision repair	(40)
8-Oxoguanosine	B	r	Free radicals	(21)
5-Hydroxycytosine	ND	r	Free radicals	(21)
Single-strand breaks	T	ir	Free radicals; base excision repair	(41)
Cytosine methylation	F + T	r	Physiological	(200)
Triple helix formation	F + T	r	ND	(201)
Apoptotic chromatin fragmentation	B + T	ir	Appears ubiquitous during apoptosis	(42–45)

(*Continues*)

TABLE I (Continued)

Drugs[a]	Mechanism[b]	Rev.[c]	Notes	References
Exogenous DNA lesions				
UV lesions	ND	ND	Dimers and 6,4-photoproducts	(37, 252)
IR-induced DNA breaks	T	ir	Both single- and double-strand breaks	(202, 203)
O^6-Methylguanine	T	r	Produced by alkylating drugs (MNNG)	(41, 237)
O^6-dA-benzo[a]pyrene adducts	T	r	Intercalated carcinogenic adducts	(204)
N^2-dG-benzo[a]pyrene adducts	F	ir	Minor groove carcinogenic adducts	(205)
N^6-dA-benzo[c]phenanthrene adducts	T	r	Intercalated carcinogenic adducts	(206, 207)
N^6-Ethenoadenine	T	r	Carcinogenic vinyl adduct	(207)
N^2-dG-ethyl adducts	T	r	Produced by acetaldehyde (alcohol)	(208)
				(209)

[a] For detailed review on noncamptothecin inhibitors see (32).
[b] Mechanism for Top1 cleavage complex production: T, trapping of the Top1 cleavage complexes (i.e., inhibition of religation) (see Fig. 3C); B, enhancement of binding; F, enhancement of the forward (cleavage) reaction; ND, not determined.
[c] Reversibility of the Top1 cleavage complexes: r, reversible; ir, irreversible.

Interfacial inhibition accounts for the molecular mechanism of inhibition of many natural products that block specific conformational states of macromolecular complexes. Aphidicolin and Top2 inhibitors have also been proposed to follow the interfacial inhibition paradigm (26, 27).

Two camptothecin derivatives are used in cancer therapy: hycamtin (Topotecan®) and CPT-11 (Irinotecan; Camptosar®) (31). CPT-11 is an inactive prodrug. It needs to be converted to its active metabolite SN-38. Hence, it is preferable to use SN-38, topotecan, or camptothecin for pharmacological studies. Additional camptothecin derivatives are in preclinical and clinical developments (32).

Two key pharmacological properties of camptothecins need to be stressed. First, camptothecins bind reversibly to the Top1 cleavage complexes. Under pharmacological conditions, a rapid equilibrium is established between the ternary drug–enzyme–DNA complex and the dissociated complex. Hence, once camptothecins are diluted out and removed from cell culture, the cleavage complexes reverse rapidly (33). The equilibrium can be shifted toward religation by increasing the temperature to 65°C or the salt concentration ($\geq$0.35 M NaCl) in biochemical reactions. Salt reversal is commonly used to study the "on" and "off" rates of Top1 cleavage complexes under various conditions [e.g., see (34)]. A second key pharmacological feature of camptothecins is the trapping of only a subset of the existing Top1 cleavage complexes, that is, those with a guanine at the 5′ end of the break (+1 position–see Fig. 3) (35, 36). Indenoisoquinolines (Fig. 3F) on the other hand tend to stabilize those cleavage complexes with a cytosine at the 3′ end of the breaks (–1 position) (34). This sequence selectivity explains why camptothecins are relatively poor Top1 catalytic inhibitors as a fraction of the Top1 cleavage complexes (those not bearing a guanine +1) are immune to the drugs. It also explains why camptothecins only reveal a subset of the Top1 sites and should not be used alone to map all the Top1 cleavage complexes in a given DNA or chromatin segment.

Top1 cleavage complexes can be trapped by endogenous and frequent DNA lesions, including abasic sites, mismatches, oxidized bases, nicks, and carcinogenic DNA adducts (37) (Table I and references therein). For instance, DNA modifications, such as those associated with oxidative damage [thousands per cell per day (38)], can produce Top1 cleavage complexes (39). By contrast to camptothecins and other Top1 inhibitory drugs, these DNA modifications can produce irreversible cleavage complexes when the 5′ end of the DNA is irreversibly misaligned as in the case of abasic sites (40) or DNA breaks (41) (Table I, Fig. 4A). Finally, we discovered the formation of Top1 cleavage complexes during apoptosis (42–45), which we explained by the trapping of Top1 by chromatin modifications (primarily due to reactive oxygen species).

III. Conversion of Top1 Cleavage Complexes into DNA Damage

Cleavage complexes induced by DNA modifications that induce pronounced DNA structural alterations (abasic sites, mismatches, breaks) can be irreversible (Fig. 4A, Table I). Such irreversible cleavage complexes have been referred to as "suicide complexes" (46, 47). They constitute composite DNA lesions, consisting of disruption of the DNA backbone (break on one or both strands) in association with a large protein covalently bound to the 3' end of the broken DNA (Top1 is a 100-kDa protein). Reversible cleavage complexes can also produce irreversible cleavage complexes after processing by DNA and RNA polymerases (Fig. 4B and C). Thus, both DNA and RNA synthesis convert reversible cleavage complexes into DNA lesions. The relative contribution of DNA replication and transcription depends on the camptothecin concentration and the cell type. In highly proliferative cancer cells, replication-induced DNA damage contributes to most of the cytotoxicity at low camptothecin doses, whereas transcription-induced DNA damage contributes to the cytotoxicity of high doses of camptothecin (48, 49). In nondividing cells (neurons and lymphocytes) transcription-induced damage can kill cells at pharmacological concentrations (50, 51).

Camptothecin-induced Top1 cleavage complexes can readily be converted into replication double-strand breaks (Rep-DSB) (Fig. 4B) as demonstrated by: (i) analyses of the broken ends by ligation-mediated PCR (52) showing the extension of the leading strand up to last nucleotide [leading to blunt-ended DSB by "replication run-off" (52)], and (ii) rapid phosphorylation of histone H2AX (referred to as γ-H2AX) (53), which is a hallmark for double-strand breaks (54, 55). Inhibition of DNA synthesis occurs within minutes following camptothecin treatment. It is intense ($\geq 80\%$) and persists for several hours following drug removal (56). At least two mechanisms lead to DNA synthesis inhibition: (i) direct block of replication forks that have collided with the Top1 cleavage complexes (Fig. 4B) and (ii) indirect replication arrest by S-phase checkpoint activation. The checkpoint implication is consistent with the fact that the checkpoint abrogator 7-hydroxystaurosporine (UCN-01) prevents inhibition of DNA synthesis by camptothecins (56–58). The lethality of the Rep-DSB stems from the fact that when DNA synthesis is inhibited by aphidicolin, a specific inhibitor of replicative DNA polymerases (59), cells become immune to camptothecin in spite of their ability to form reversible cleavage complexes (48, 49). Similarly, aphidicolin prevents the formation of γ-H2AX foci in camptothecin-treated cells (53).

Camptothecin is a potent inhibitor of both nucleoplasmic (mRNA) and nucleolar (rRNA) transcription (60–62). Although the overall level of transcripts decreases rapidly following Top1 inhibition, specific genes are differentially affected. For example, camptothecin causes a strong holdback of the endogenous *c-myc* gene at the P2 promoter, whereas it produces minimal effect on an episomal *c-myc* gene or on the basal transcription of the *Hsp70* and *Gadph* genes (63). Camptothecin also enhances the expression of a large number of genes, including *c-fos* (63–66). Transcription inhibition is primarily due to transcription elongation blocks by trapped Top1 cleavage complexes (Fig. 4C), which is a high probability event considering that Top1 is associated with transcription complexes (12). Camptothecin has little effect on transcription initiation (11, 67). It has been proposed that the elongating RNA polymerase collides with trapped Top1 cleavage complexes on the transcribed strand, resulting in the conversion of reversible Top1 cleavage complexes into irreversible strand breaks (Fig. 4C) (68, 69). Inhibition of Top1 catalytic activity by camptothecins might also inhibit transcription by producing an accumulation of positive supercoils upstream from the elongating RNA polymerase (63, 70) and by compacting chromatin domains (70–72). The transcriptional effects of camptothecins could also be related to functions of Top1 besides its DNA nicking-closing activity. Top1 regulates transcription initiation by interacting with TATA binding proteins (73, 74), and phosphorylates/activates RNA splicing factors from the SR family (19).

By contrast to replication (56), transcription inhibition recovers rapidly following camptothecin treatment (62, 67, 75). Recovery of RNA synthesis depends both on degradation of Top1 and functional transcription-coupled nucleotide excision repair (TC-NER) (75). Tumor cells that are deficient in Top1 degradation following camptothecin treatment and Cockayne syndrome cells that are deficient in TC-NER are hypersensitive to camptothecin (76, 77), suggesting the importance of transcription-coupled DNA repair for RNA-synthesis recovery and cell survival in response to Top1-mediated DNA damage.

IV. Repair of Top1-Associated DNA Damage

The molecular mechanisms/pathways involved in the repair of Top1-associated DNA damage are better understood than for Top2. Because camptothecin can be readily used in yeast, multiple pathways have been uncovered. We will consider three main repair pathways: (i) reversal of the covalent Top1–DNA complexes by 5′-end religation, (ii) Top1 excision by Tdp1, and (iii) Top1 excision by endonucleases. In spite of an apparent redundancy, it remains to be determined which pathways are preferred or selective for the Top1-associated DNA damages represented in Fig. 4.

A. Reversal of Top1–DNA Covalent Complexes by 5′-end Religation (Fig. 5A)

Top1-mediated DNA religation requires that the intact 5′-hydroxyl end be aligned with the 3′ end bonded to Top1 for nucleophilic attack against the tyrosyl-phosphoester bond (see Fig. 1B). Thus, this pathway/mechanism excludes the Top1 suicide complexes generated by DNA lesions affecting the 5′ end of the broken DNA (Fig. 4A and Table I) unless they are repaired first. Top1 religation can be envisaged for the replication and transcription breaks (Fig. 4B and C) following regression (pull-back) of the replication or transcription complexes (Fig. 5A).

Replication fork regression could generate a "chicken foot" [see Fig. 6 in (78)], which is topologically equivalent to a holiday junction. RecQ helicases (the Bloom syndrome helicase BLM in mammals and Sgs1 in yeast) and Top3 (Top3α in mammals) form helicase–Top3 complexes, which have been implicated in the regression of replication forks and their restart (4, 79). A plausible intermediate is the conversion of the chicken foot into a double-holiday junctions catalyzed by Rad51, which can then be resolved by Top3 (80, 81). The role of the RecQ helicases in processing Top1-mediated DNA damage is demonstrated by the hypersensitivity of yeast *Sgs1* and *Rhq1* mutants (see Tables III and IV) and the hypersensitivity of Bloom syndrome cells to camptothecin (79, 82) (see Table II).

In the case of transcription, blocked RNA complexes might be displaced (pull-back mechanism) from the Top1 cleavage complexes without removal from the transcribed DNA. Transcription elongation restart can finish the incomplete mRNA. Rad26 (the yeast homologue of CSB) and TFIIS have been implicated in the backtracking of RNA polymerase II (83). RNA polymerase II can also be degraded following camptothecin treatment, but this process is limited to some cell types and delayed as compared to RNA-synthesis recovery (75).

Because Top1 is very effective in joining a 5′-hydroxyl end from a nonhomologous substrate to the Top1 covalent complex, Top1 cleavage complexes might reverse by religation of a nonhomologous end, which leads to DNA recombinations (84). Camptothecin is a potent inducer of sister chromatid exchanges and chromosomal abnormalities (85–87). Top1-mediated 5′-end ligation with *vaccinia* Top1 is commonly used for cloning recombinant genomes (TOPO cloning kit; Invitrogen, Carlsbad, CA).

B. Top1 Excision by Tyrosyl-DNA-Phosphodiesterase (Fig. 5B)

Tdp1 was discovered by Nash and coworkers (88) as the enzyme capable of hydrolyzing the covalent bond between a Top1 catalytic tyrosine and the 3′ end of the DNA (89). Tdp1 generates a 3′-phosphate, which is further processed by a 3′-phosphatase, such as PNKP (hPNK) (Fig. 5B).

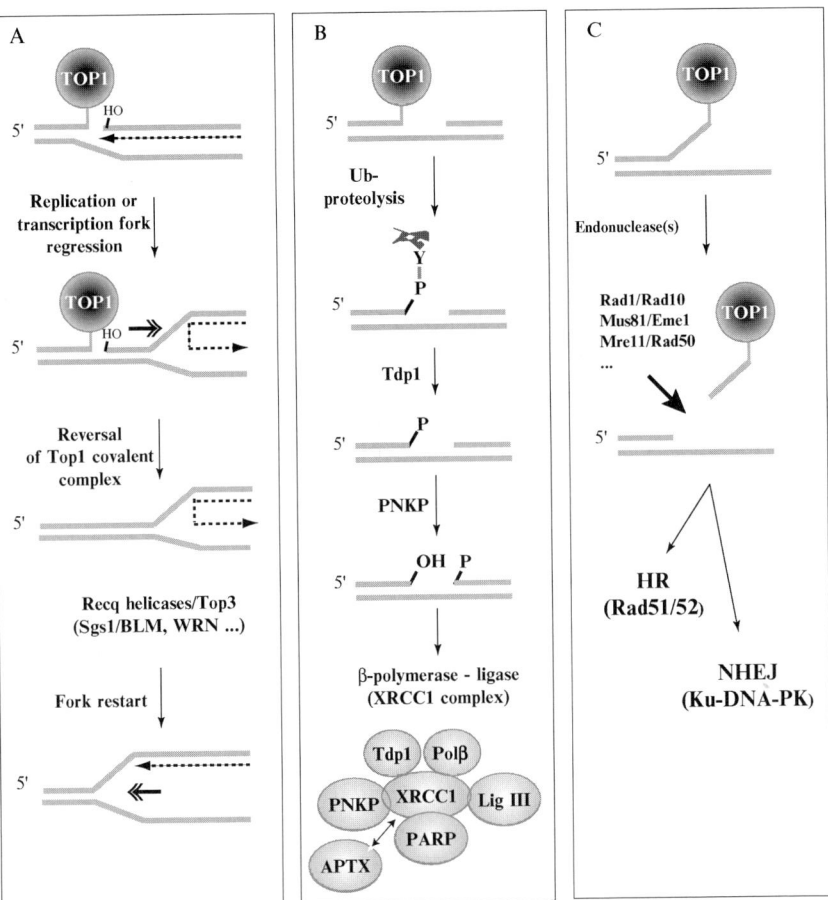

FIG. 5. Schematic representation of three main pathways for the repair of Top1 covalent complexes. (A) 5′-end religation requires realignment of the 5′-hydroxyl with the end of the broken DNA bonded to Top1. This would require a "pull-back" (regression) of the replication or transcription complexes (see Fig. 4B and C). Fork regression and restart require helicase activities (in particular RecQ helicases such as Sgs1 in budding yeast and BLM or WRN in humans) in association with Top3. (B) Top1 excision by Tdp1 requires prior proteolysis of Top1 (106) or denaturation of Top1 (105) to expose the phosphotyrosyl bond to be attacked (see Fig. 6A–C). Tdp1 generates a 3′-phosphate DNA end, which needs to be hydrolyzed by polynucleotide kinase phosphatase (PNKP = hPNK). PNKP also catalyzes the phosphorylation of the 5′ end of the DNA. Tdp1 and PNKP are part of the XRCC1 complex (shown at the bottom). (C) Excision of the Top1–DNA covalent complex by 3′-endonucleases. Studies in budding yeast have implicated at least three endonuclease families. The human orthologs are listed: Rad1/Rad10, Mus81/Eme1, and Mre11/Rad50. The resulting DNA lesion is probably processed by homologous recombination initiated by the Rad51, Rad52 complexes or by nonhomologous end joining (Ku-DNA-PK pathway).

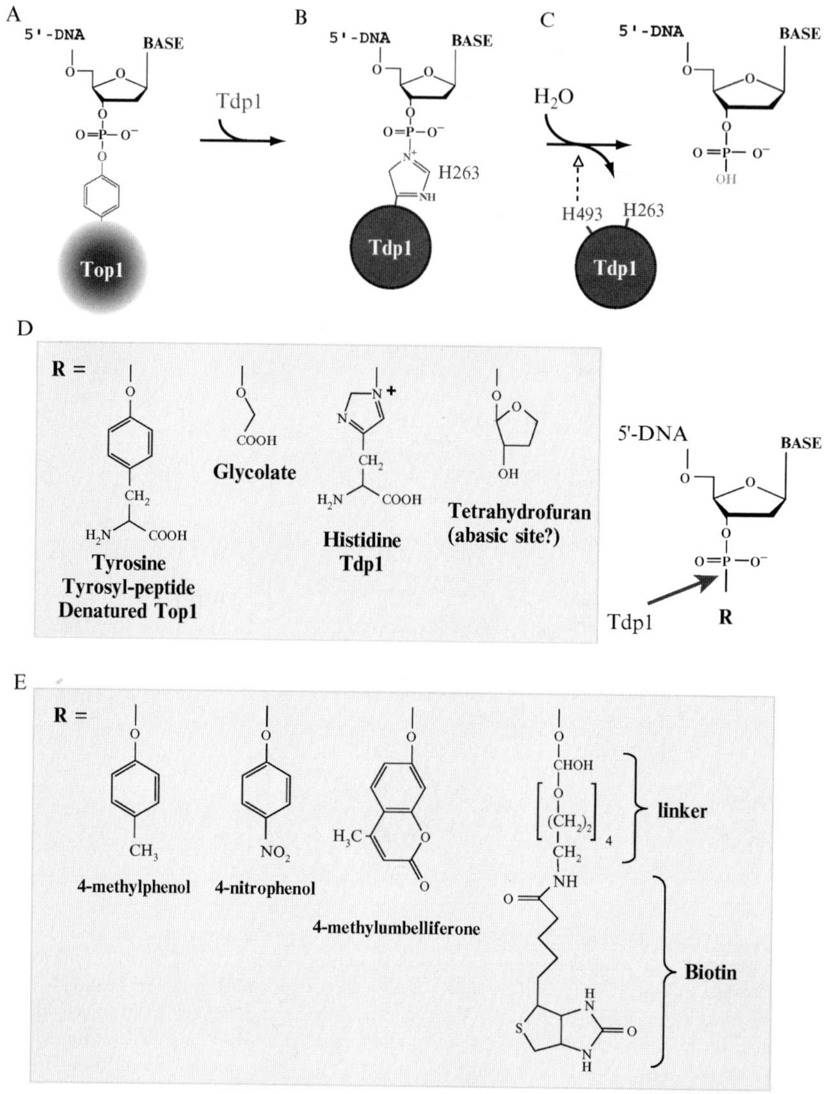

FIG. 6. Tdp1-mediated reactions and substrates. (A) Structure of the Top1–DNA covalent intermediate (see Fig. 1B). (B) Tdp1 releases the Top1 by forming a covalent bond between its active histidine 263 and the DNA end. (C) Histidine 493 from the second HKD motif of Tdp1 (and which is mutated H493R in SCAN1) promotes the hydrolysis of the Tdp1–DNA intermediate and frees the 3′-phosphate DNA end. Tdp1 is regenerated for another catalytic cycle. (D) Physiological substrates for Tdp1. (E) Tdp1 substrates used for biochemical assays.

TABLE II
Genetic Alterations Sensitizing Mammalian Cells to Top1 Poisons[a]

Genes[b]	Functions	References
APTX	Mutated in AOA1; encodes aprataxin, which associate XRCC1 (Fig. 5B)	(210)
ATM[c]	Mutated in AT; protein kinase from the PI3K family implicated in DSB response	(211–214)
ATR	Protein kinase from the PI3K family; implicated in replication stress	(163)
BCL-2	Mutated in B-cell lymphoma; apoptosis	(215)
BLM	Mutated in BS; helicase from the RecQ family involved in genomic stability	(79, 82)
BRCA1	Mutated in familial breast cancers; DNA damage response; TC-NER	(216)
BRCA2	Mutated in familial breast cancers; Rad51 loading; homologous recombination	(217)
CSA/CSB	Mutated in CS; TCR/BER	(76)
Chk1	Checkpoint kinase phosphorylated/activated by PI3K (primarily ATR)	(162, 165)
Chk2	Checkpoint kinase phosphorylated/activated by PI3K (primarily ATM and DNA-PK)	(162, 218)
DNA-PKcs	Protein kinase from the PI3K family; implicated in DSB response	(56, 219, 220)
FEN-1	Flap and gap endonuclease; processing of stalled replication forks	(133)
H2AX	Core histon; phosphorylated in response to DSB (γ-H2AX foci)	(53)
NBS1	Mutated in NBS; scaffolding protein forming a complex with Mre11 and Rad50 (MRN complex); DSB repair and recombination pathways	(212, 221)
PARP	BER (see Fig. 7)	(118, 222)
PNKP/hPNK	Processing of DNA ends: 3′-DNA-phosphatase + 5′-DNA-kinase	(115)

(Continues)

TABLE II (*Continued*)

Genes[b]	Functions	References
Rad51C	One of the five Rad51 paralogs; implicated in DNA strand exchange/homologous recombination	(223)
TDP1	Mutated in SCAN1; hydrolysis of 3′-phosphotyrosyl and phosphoglycolate) and phosphamides (Tdp1 cleavage complex)	(93, 94, 224)
TP53	Mutated in Li-Fraumeni syndrome; encodes p53; checkpoints; apoptosis	(225, 226)
WRN	Mutated in Werner syndrome; RecQ helicase involved in genomic stability	(227–229)
XRCC1	BER; binds to Tdp1, PARP, β-polymerase, ligase III, and aprataxin	(116, 117, 119, 230–233)
XRCC2	One of the five Rad51 paralogs: Rad51B, Rad51C, Rad51D, XRCC2, and XRCC3; implicated in DNA strand exchange/homologous recombination	(219, 230, 234)
XRCC3	One of the five Rad51 paralogs; Implicated in DNA strand exchange/homologous recombination	(167, 219)

[a]Note: AOA1, ataxia-oculomotor apraxia 1; ATM, ataxia telangiectasia mutant; ATR, ataxia telangiectasia and Rad3-related; BER, base excision repair; BLM, Bloom syndrome (BS); CSA/CSB, Cockayne Syndrome (CS) complementation groups A and B; DNA-PKcs, DNA-dependent protein kinase catalytic subunit; DSB, DNA double-strand breaks; NBS, Nijmegen breakage syndrome; NER, nucleotide excision repair; PARP, poly(ADP-ribose) polymerase; PI3K, phosphatidyl inositol 3 kinase; PNKP, polynucleotide kinase phosphatase; SCAN, spino cerebellar ataxia axonal neuropathy; TCR, transcription-coupled repair; WRN, Werner syndrome.

[b]Genes are in alphabetic order.

[c]The contribution of ATM has not been found consistently: ATM-siRNA cells are not hypersensitive to camptothecin (162) and we found that AT-complemented cells are not more resistant to camptothecin (our unpublished results).

Tdp1 belongs to the phospholipase D superfamily (90) of phospholipid hydrolyzing enzymes. It is ubiquitous and highly conserved in eukaryotes (from yeast to humans—see Tables II–IV). Tdp1 is physiologically important since the homozygote mutation *H493R* causes spinocerebellar ataxia with axonal neuropathy (SCAN1) (91). SCAN1 cells are hypersensitive to camptothecin (92–94) (Table II) and ionizing radiation (95) but not to etoposide or bleomycin (94). The budding yeast *TDP1* knockout is viable (88) and hypersensitive to high levels of Top1 cleavage complexes generated by overexpression of a toxic Top1 (96). It is hypersensitive to camptothecin only when the checkpoint gene *Rad9* is simultaneously inactivated (96) or when some endonuclease pathways (Rad1/Rad10 and Slx1/Slx4) are inactive (97–99) (Table III). However, Tdp1 function is not limited to the repair of Top1 cleavage complexes. Deletion of the *Tdp1* gene in yeast confers hypersensitivity to Top2 targeting agents and yeast Tdp1 is able to remove a peptide derived from yeast Top2 that is covalently bound to DNA by a 5′ phosphotyrosyl linkage (263). Tdp1 can remove 3′-phosphoglycolate (Fig. 6D) generated by oxidative DNA damage, which suggests a broader role for Tdp1 in the maintenance of genomic stability (100).

Human Tdp1 acts as a monomer and crystal structures demonstrate the presence of two catalytic domains related by a pseudo-twofold axis of symmetry (101–103). Each domain contains three conserved HKN residues (H263, K266, and N283 and H493, K495, and N516; equivalent to the "HKD motifs" of phospholipases) (101, 102, 104), forming a catalytic network with two water molecules critical for Tdp1 activity (90, 91). Tdp1 hydrolyzes the DNA–phosphotyrosine bond in two consecutive SN2 reactions (Fig. 6B and C). In the first reaction, H263 of the first HKD motif releases the Top1 tyrosine from the phosphodiester by forming a transient covalent phosphoamide bond between the Nε2 atom of the nucleophilic H263 and the 3′ end of the DNA (Fig. 6B). Mutating H263 totally abolishes Tdp1 catalytic activity (90). In the second reaction, H493 from the second HKD motif catalyzes the nucleophilic attack of the phosphoamide bond by a water molecule. This regenerates Tdp1 and produces a 3′-phosphate DNA end (90) (Figs. 6C and 5B). The SCAN1 mutation H493R affects preferentially the second step of the reaction, which leads to an accumulation of Tdp1–DNA covalent intermediate and reduces Tdp1 catalytic activity ≈25-fold (94, 105). Interthal, Champoux, and coworkers have shown that wild-type Tdp1 can hydrolyze the phosphoamide Tdp1–DNA covalent intermediate (Fig. 6B) (105), and proposed that lack of symptoms in SCAN1 heterozygote carriers might be due to the release of these covalent intermediates by the coexisting wild-type Tdp1 (Fig. 6D). They also proposed that the covalent Tdp1–DNA intermediates rather than deficient Tdp1 catalytic activity might be responsible for DNA damage leading to SCAN1 (105).

TABLE III
GENETIC ALTERATIONS CONFERRING HYPERSENSITIVITY TO TOP1 POISONING IN BUDDING YEAST

Yeast Saccharomyces cerevisiae				Humans		
Gene	Effect	References	Function	Gene	Effect	References
AAT2	S	(*)	Aspartate aminotransferase			
AKR1	S	(131)	Protein palmitoylation			
ASM4	S	(254)	Nuclear pore complex subunit			
APN1	CS[b,c]	(98, 236)	AP endonuclease (endo IV family)			
APN2	CS[b,c]	(97, 98)	AP endonuclease (exo III family)	APE1	ND	
ASF1	SS	(98, 131, 245)	Chromatin assembly	ASF1B	ND	
BEM1	S	(131)	Cell polarity, SH3 domain protein			
BCK1	S	(131)	MapKKK			
BIK1	S	(131)	Microtubule associated protein			
BMH1	SS	(254)	14-3-3 protein/Signaling			
BUR2	S	(131)	Cyclin partner for Bur1	Cyclin H	ND	
CCR4	S	(131)	Transcription	KIAA1194	ND	
CDC45	S	(134)	Initiation of DNA replication	CDC45L	ND	
CHL1	S	(131)	Chromosome segregation	Similarity to BACH1	ND	
CLB5	S	(99)	B type cyclin in G1/S transition			
CLC1	S	(131)	Intracellular protein transport, endocytosis			
CNM67	S	(131)	Cytoskeleton, nuclear migration			
CSM3	S	(99, 242)	Meiotic chromosome segregation			

Gene		Ref	Function	Homolog	
CTF4	SS	(98, 131)	Sister chromatid cohesion and segregation	AND-1	ND
CTF8	SS	(131)	Sister chromatid cohesion		
CTF18	S	(°)	Sister chromatid cohesion		
DBF2	S	(259)	Protein Ser/Thr kinase	NDR1/NDR2	ND
DCC1	S	(131, 242)	Sister chromatid cohesion		
DDC1	MS	(98)	Replication/Repair Clamp, "9-1-1"	RAD9	ND
DEF1/ VID31	S	(131)	DNA repair		
DHH1	S	(131)	mRNA decapping		
DOA4	SS	(135, 246)	De-Ubiquitinating enzyme		
DOC1	S	(131)	Ubiquitination of APC		
DPB11	S	(134)	Replication initiation/checkpoint	TOPBP1	ND
ESA1	S	(243)	Histone H4 acetyltransferase	MYST1/HAT	ND
ESC4/ RTT107	SSS	(260)	4 BRCT domains protein, target of Mec1, interact with Mms22		
FUN12	S	(131)	Translation initiation		
FUN30	S	(99)	Overexpression affects chromosome stability		
GCN5	S	(244)	Histone H3 acetyltransferase	PCAF	ND
GRR1	S	(131)	Transcription		
HFI1/ADA1	S	(131)	Transcription, SAGA complex		
HHF1/2	S	(243)	Histone H4	H4	ND
HHT1/2	SSS	(261)	Histone H3	H3	ND
HMO1	S	(99)	Involved in rDNA transcription		

(Continues)

TABLE III (Continued)

Gene	Yeast Saccharomyces cerevisiae			Humans		
	Effect	References	Function	Gene	Effect	References
HOF1	S	(131)	SH3 domain protein			
HPR1	S	(131)	Transcription and recombination	THOC1	ND	
HTA1/2	S	(237)	Histone H2A	H2AX	S	(53)
HTZ1	S	(99)	Transcription, Histone H2AZ			
ILM1	S	(99)	Unknown			
LOC1	S	(131)	Constituent of 60S pre-ribosomal particles			
LSM1	S	(99)	Controls mRNA decay			
MCD1	S	(240)	Chromatin cohesion	RAD21	ND	
MEC1	HS	(236, 237)	PI3LK checkpoint sensor kinase	ATR	HS	(163)
MEC3	MS	(131)	Replication/Repair Clamp, "9-1-1"	HUS1	ND	
MMS1	SS	(99, 238)	Replication repair (epistatic to Rad52)			
MMS4	HS	(98, 99, 126)	Partner for Mus81 endonuclease	EME1	ND	
MMS22	SS	(131)	DNA repair (may be epistatic to Rad52)			
MPH1	MS	(131, 241)	RNA helicase	MPH1	ND	
MRC1	CS	(242)	DNA replication checkpoint	CLASPIN		
MRE11	HS[e]	(97, 98, 99, 237)	MRXN complex, endonuclease	MRE11	ND	
MUS81	HS	(98, 99, 126, 131)	3′-flap endonuclease with Mms4	MUS81	NS	(137)

Gene	Ref	Function	Human homolog		Ref
NPL6	S	(°)	Transcription		
NUP60	S	(99)	Nucleocytoplasmic transport		
NUP84	S	(131)	Subunit of the nuclear pore complex		
NUP120	S	(131)	Subunit of the nuclear pore complex		
NUP133	S	(131)	Subunit of the nuclear pore complex		
PAT1	S	(99)	Controls mRNA decay		
PPH3	S	(99)	Protein Ser/Thr phosphatase		
POL32	MS	(131)	Small subunit for Polδ	TEX14	ND
PSY2	S	(99)	Unknown		
RAD1	CS[b]	(98, 99, 236)	3'-flap endonuclease with Rad10	XPF	NS (our observations)
RAD6	MS	(131, 236, 241)	PRR,[d] Ubiquitin conjugating enzyme	RAD6A, B	ND
RAD9	MS	(236, 237)	Adaptor for checkpoint kinases	MDC1	ND
			BRCA1	HS	(216)
RAD10	CS[b]	(97, 98)	Partner for Rad1	ERCC1	NS
RAD17	MS	(131, 236, 239)	Replication/Repair Clamp, "9-1-1"	RAD1	ND
RAD18	S	(131, 236)	PRR,[d] loads Rad6	RAD18	ND
RAD24	MS	(237)	Clamp loader for 9-1-1	RAD17	ND
RAD27	MS	(98, 99, 133)	5'-flap endonuclease	FEN1	HS	(133)
RAD50	HS	(99, 131, 236)	MRX/N complex, scaffold	RAD50	ND
RAD51	HS	(98, 99, 131, 236)	RecA homolog, strand invasion	RAD51C	HS	(223)
RAD52[a]	HS	(8, 98, 99, 131, 235, 236, 237)	Strand annealing	RAD52	ND
RAD53	MS	(236)	Checkpoint effector kinase	CHK2	S	(162, 218)

(*Continues*)

TABLE III (Continued)

Yeast Saccharomyces cerevisiae				Humans		
Gene	Effect	References	Function	Gene	Effect	References
RAD54	HS	(98, 99)	ATPase			
RAD55	HS	(98, 99, 131)	Strand annealing, exchange	XRCC2	HS	(219, 234)
RAD57	HS	(98, 99, 131)	Strand annealing, exchange	XRCC3	HS	(219)
RAD59	SS	(98, 99)	Rad52-related recombination			
REF2	SS	(131)	RNA binding protein			
RPB9	S	(131)	RNA polymerase subunit	POLR21	ND	
RRM3/ RTT104	MS	(265)	DNA Helicase			
RTT101	SS	(99)	DNA repair (may be epistatic to Rad52)			
RTT109	S	(131, 259)	Regulation of Ty1 transposition			
RVS161	S	(131)	Cytoskeletal protein	Amphiphysin	ND	
RVS167	S	(131)	Cytoskeletal protein	Amphiphysin	ND	
SAE2	HS	(99, 254)	Activates Mre11 endonuclease, meiotic and mitotic recombination			
SCP160	S	(131)	Mating response, pathway similar to vigilins			
SFP1	S	(131)	Transcription	REQ	ND	
SGS1	MS	(126, 131)	Top3-associated helicase	WRN, BLM	HS	(82, 227)
SLA1	SS	(246, 135)	Cytoskeletal protein			
SLA2	SS	(246, 135)	Cytoskeletal protein			
SLX1	MS	(99)	Form a complex with Slx4			

SLX4	MS	(99)	5′ flap endonuclease		
SLX8	S	(254)	Replication/Repair		
SOD1	S	(259)	Superoxide dismutase	SOD1	ND
SPT10	S	(131)	Transcription		
SPT20	SS	(254)	Transcription		
SPT21	S	(99)	Transcription		
SRS2	HS	(98, 99, 131)	Rad51-associated helicase		ND
SWI6	S	(°)	Transcription		
TAF14	S	(°)	Transcription		
TAF47	SS	(246)	Transcription		
TAH11	SS	(246)	DNA replication		
TAH18	SS	(246)	DNA replication		
TDP1	CS[b]	(97, 98, 99)	Tyrosyl-DNA phosphodiesterase	TDP1	ND (94, 224)
TEL1	S	(237)	PI3LK checkpoint sensor kinase	ATM	HS (211, 214)
			PI3LK checkpoint sensor kinase	DNA-PK	HS (56, 219)
TOF1	S	(99, 242)	Chromatid cohesion	TIMELESS	ND
TOP3	S	(98, 126)	Replication/recombination, type 1A topoisomerase	TOP3α	ND
				TOP3β	ND
TPP1	CS[b,c]	(98, 110)	Polynucleotide 3′-phosphatase	PNKP[b]	HS (115)
TRF4	S	(240)	DNA polymerase	POLS	ND
TUP1	S	(254)	Transcription		
UBC4	S	(99)	Ubiquitin conjugating enzyme		
UBC9	SS	(136, 246)	SUMO conjugating enzyme	UBE2I	ND
ULA1	S	(99)	RUB1-protein conjugation		
UME6	S	(131)	Transcription		

(*Continues*)

TABLE III (Continued)

Yeast Saccharomyces cerevisiae				Humans		
Gene	Effect	References	Function	Gene	Effect	References
VAC14	S	(99)	Vacuole inheritance			
VID21/ EAF1	S	(131, 259)	Component of NuA4 acetyltransferase complex			
XRS2	HS	(131)	MRX/N complex, signaling	NBS1	HS	(212, 221)
YDJ1	S	(254)	Protein chaperone			
YNG2	S	(244)	Component of NuA4 acetyltransferase complex	ING1-5	ND	
YBR094W	S	(259)	Hypothetical protein			
YJL184W	S	(131)	Cell wall			
YLR435W	S	(131)	Potential role in pre-rRNA processing			

Abbreviations for effects: HS, hypersensitivity; S, sensitivity; MS, moderate sensitivity; CS, conditional sensitivity to camptothecin; NS, no hypersensitivity; ND, not determined.

[a]The Rad52 epistasis group includes the *RAD 50, 51, 52, 54, 55, 57, 59, MRE11* and *XRS2* genes.
[b]Tdp1 deficiency results in HS only in the presence of Rad1/Rad10 deficiency (97, 98); conversely Rad1 deficiency does not confer hypersensitivity to CPT (126) unless the Tdp1-Apn1 pathway is defective (97). Tpp1, Apn1+Apn2+Tpp1 need to be inactivated to confer full camptothecin hypersensitivity (111); see Fig. 3A.
[c]PNKP possesses both 3'-phosphatase and 5'-kinase activities, whereas the yeast ortholog, Tpp1 only possesses 3'-phosphatase activity. Neither Apn1, Apn2 or Tpp1 possess AP endonuclease activity (111).
[d]PRR: post-replication repair. Deficiency of Rhp6 or Rhp18 (YSP orthologs of Rad6 and Rad18) does not confer CPT hypersensitivity (247).
[e]Mre1 deficiency results in sensitivity to CPT in the presence of Rad9, Tof1 or Csm3 deficiencies (242).
(*): Nitiss, J. L., unpublished data.

Both the structure of the DNA segment bound to Top1 (88, 106) and the length of the Top1 polypeptide chain determine Tdp1's activity (106). Optimum Tdp1 activity requires: (i) a DNA segment consisting of at least a few nucleotides (106); (ii) an exposed phosphotyrosyl bond at the Top1–DNA junction [a tyrosyl group linked to the 3′ end of a nick is a poor substrate (96), indicating that Tdp1 acts after the 5′ end of the broken DNA has been either digested or displaced to provide access to the 3′-phosphotyrosyl bond]; and (iii) a short Top1 polypeptide segment, as the effectiveness of Tdp1 decreases with the length of the Top1 polypeptide (106). In fact, Top1 needs to be proteolyzed or denatured for efficient Tdp1 activity (89, 105, 106). Top1 ubiquitination and degradation have been observed following camptothecin treatment (107, 108). The Top1 degradation pathway appears selectively deficient in transformed cells, although not all transformed cells appear equally able to proteolyze Top1 following camptothecin treatment (77). Such differences have been proposed to contribute to camptothecin resistance (77). A structure of Tdp1 bound to a tyrosine-containing peptide demonstrates that both the DNA and the Top1 polypeptide need to adapt their structure to bind Tdp1 in the crystal structure (103). The DNA binds in a narrow groove that fits a single-stranded substrate, and the short Top1 polypeptide is folded differently from the native Top1 (103). An alternative model has been presented for duplex DNA, which can also be processed effectively by Tdp1 (88, 106, 109).

The 3′-phosphate ends generated by Tdp1 need to be hydrolyzed to 3′-hydroxyl for further processing by DNA polymerases and/or ligases (Fig. 5B). In budding yeast, this 3′-phosphatase activity is carried out by Tpp1 (110) and by the two functionally overlapping multifunctional apurinic (AP) endonucleases, Apn1 and Apn2 (97) (see Fig. 8A). Apn1 is the ortholog of *Escherichia coli* endonuclease IV and represents the major yeast AP endonuclease. Apn2 (also called Eth1) belongs to the second family of AP endonuclease (the *E. coli* exonuclease III family) and is the ortholog of Ape1 in humans. Simultaneous inactivation of Tpp1, Apn1, and Apn2 is required to sensitize yeast to camptothecin (110), indicating the functional redundancy of the 3′-phosphatase pathways. Noticeably, the hypersensitivity of the *tpp1 apn1 apn2* triple mutant is rescued by inactivation of Tdp1 (111), which indicates that in the absence of Tdp1 budding yeast uses an alternative endonuclease pathway for removal of the Top1 covalent complexes (see Figs. 5C and 8A, and next section). The 3′-phosphatase orthologs of Tpp1 are Pnk1 in fission yeast (112) and PNKP (also referred to as hPNK) in humans (110, 113, 114) (see Tables II–IV). In addition to their 3′-phosphatase activity, Pnk1 (112) and PNKP/hPNK (113, 114) possess 5′-kinase activity, which is missing for Tpp1. Stable downregulation of human PNKP (hPNK) sensitizes to camptothecin (Table II), ionizing- and UV-radiation, H_2O_2 and UV, and increases spontaneous mutation frequency (115).

TABLE IV
GENETIC ALTERATIONS CONFERRING HYPERSENSITIVITY TO TOP1 POISONING IN FISSION YEAST

Yeast *Saccharomyces pombe*				Humans		
Gene	Effect	References	Function	Gene	Effect	References
cdt2	HS	(258, 262)	Cell division cycle			
chk1	S	(155, 249)	Checkpoint effector kinase	CHK1	S	(162, 165)
crb2	S	(249)	Adaptor for checkpoint kinases and checkpoint proteins	53BP1	ND	
				BRCA1	HS	(216)
eme1	S	(129, 257)	Partner for mus81 nuclease	MUS81	ND	
hus1	S	(°)	DNA damage checkpoint	HUS1	ND	
mus81	HS	(129, 257)	3'-flap endonuclease with Eme1; meiotic recombination	MUS81	NS	(137)
pnk1	S[b]	(112)	Polynucleotide kinase phosphatase	PNKP	HS	(115)
rad1	HS	(255)	DNA damage checkpoint	RAD1	ND	
rad2	MS	(255)	Flap endonuclease			
rad8	HS	(255)	Unknown			
rad9	HS	(255)	DNA damage checkpoint	RAD9	ND	
rad13	MS	(255)	Excision repair, SS annealing			

Gene	Effect	Ref	Function	Mammalian homolog	Effect	Ref
rad22	HS	(248)	Rec A homolog; functions with Mus81			
rad32	HS	(255)	Homologous recombination	MRE11	ND	
rad50	HS	(129)	MRX/N complex; scaffold	RAD50	ND	
rfc1	S	(253)	Replication	RFC1	ND	
rhp22A	MS	(129)	Homologous recombination	XRCC3	HS	(219)
rhp51	MS	(248)	Homologous recombination	RAD51C	HS	(223)
rhp54	HS	(129, 255)	Homologous recombination	RAD52	ND	
rhp55	MS	(129)	Homologous recombination	XRCC2	HS	(219, 234)
rlp1	S	(256)				
rqh1	MS	(129, 247)	Top3-associated helicase	WRN	HS	(227, 229)
				BLM	HS	(82)
rusA	RS[a]	(129)	HJ resolvase			
set9	HS	(266)	Histone H4 methyl transferase			
srs2	HS	(247)	Helicase			
swi1	HS	(251)	Mating-type switching	TIMELESS	ND	

Abbreviations for effect: HS, hypersensitivity of *Mus81/Eme1*; S, sensitivity; MS, moderate sensitivity; CS, conditional sensitivity to camptothecin; NS, no hypersensitivity; ND, not determined.

[a]*rusA* suppresses hypersensitivity of *Mus81/Eme1*− but does not reverse sensitivity of *rqh1*−; rusA also suppresses the lethality of double mutants for *Mus81/Eme1* + *rqh1* (129). RusA expressed in budding yeast partially suppresses hypersensitivity to CPT in Mms4-deficient cells (126).

[b]*Pnk1*− cells are hypersensitive to CPT in the absence of additional defects, indicating difference from budding (see [a]) and importance of this pathway in fission yeast, which like mammals possesses a gene that has both 3'-phosphatase and 5'-kinase activity (112).

(*): Nitiss, J. L., unpublished data.

In humans, Tdp1 and PNKP form a multiprotein complex with XRCC1, poly(ADP-ribose) polymerase (PARP), β-polymerase, and ligase III (*116*, *117*) (Fig. 5B, bottom). This complex contains the critical elements for base excision repair. Mammalian cells deficient for XRCC1 or PARP are also hypersensitive to camptothecin (Fig. 7) (Table I) (*78*, *116*, *118*). The hypersensitivity of PARP–/– cells to camptothecin (Fig. 7A) can be related to a functional defect in Tdp1 (Fig. 7B and C). It is not explained by abnormal levels of Tdp1 or Top1 proteins (Barcelo and Pommier, unpublished). A novel protein, aprataxin, has been found to associate with the XRCC1 complex (Fig. 5B) (*119*). Aprataxin is a 342 amino acid protein encoded by the *APTX* gene whose homozygote mutation produces ataxia-oculomotor apraxia (AOA1; the most common autosomal recessive ataxia in Japan) (*120*, *121*). AOA1 cells are hypersensitive to camptothecin (*122*) (Table I).

There is no pharmacological inhibitor of Tdp1 reported to date. Vanadate and tungstate act as phosphate mimetic in cocrystal structures and block Tdp1 activity at millimolar concentrations (*102*). It would, however, be rational to develop Tdp1 inhibitors for cancer treatment in combination with camptothecins. The anticancer activity of Tdp1 inhibitors may prove dependent on the presence of cancer-related genetic abnormalities, since camptothecin hypersensitivity in Tdp1-defective yeast is conditional for deficiencies in the checkpoint (Rad9) (*88*, *97*, *98*). A Rad9 defect in a Tdp1-deficient background confers marked sensitization to camptothecin (*88*), and it is tempting to speculate that Tdp1 is primarily required when the checkpoints are deficient as in the case of the yeast *RAD9* mutant. A second group of conditional genes (with respect to Tdp1 deficiencies) includes three sets of genes from the 3′-flap endonuclease pathway: *Rad1/Rad10*, *Mre11/Rad50*, and *Mus81/Eme1* (*88*, *97*, *98*). Mutations in each of these genes renders Tdp1-deficient cells highly sensitive to camptothecin (see Table III). Hence, colon cancers that are commonly mutated for Mre11 might be selectively sensitive to the combination of camptothecin with a Tdp1 inhibitor. The potent activity of Tdp1 against a number of artificial substrates (Fig. 6E) (*105*) can be used to design high throughput screens (our unpublished observations) (*123*, *124*), and Tdp1 inhibitors will be reported in the near future.

C. Top1 Excision by Endonucleases (Fig. 5C)

Studies in yeast demonstrate the existence of alternative pathways besides Tdp1 for removing the Top1 covalent complexes. Figure 8 summarizes the multiple endonuclease genetic pathways implicated in the repair of Top1 cleavage complexes (*97*–*99*).

Rad1/Rad10 and Slx1/Slx4 appear to function in parallel and redundant pathways with Tdp1 (*97*–*99*) (Fig. 8A). Rad1/Rad10 is the ortholog of the

Gene	Effect	Ref	Function	Mammalian homolog	Effect	Ref
rad22	HS	(248)	Rec A homolog; functions with Mus81			
rad32	HS	(255)	Homologous recombination	MRE11	ND	
rad50	HS	(129)	MRX/N complex; scaffold	RAD50	ND	
rfc1	S	(253)	Replication	RFC1	ND	
rhp22A	MS	(129)	Homologous recombination	XRCC3	HS	(219)
rhp51	MS	(248)	Homologous recombination	RAD51C	HS	(223)
rhp54	HS	(129, 255)	Homologous recombination	RAD52	ND	
rhp55	MS	(129)	Homologous recombination	XRCC2	HS	(219, 234)
rlp1	S	(256)				
rqh1	MS	(129, 247)	Top3-associated helicase	WRN	HS	(227, 229)
				BLM	HS	(82)
rusA	RS[a]	(129)	HJ resolvase			
set9	HS	(266)	Histone H4 methyl transferase			
srs2	HS	(247)	Helicase			
swi1	HS	(251)	Mating-type switching	TIMELESS	ND	

Abbreviations for effect: HS, hypersensitivity; S, sensitivity; MS, moderate sensitivity; CS, conditional sensitivity to camptothecin; NS, no hypersensitivity; ND, not determined.

[a] rusA suppresses hypersensitivity of Mus81/Eme1− but does not reverse sensitivity of rqh1−; rusA also suppresses the lethality of double mutants for Mus81/Eme1 + rqh1 (129). RusA expressed in budding yeast partially suppresses hypersensitivity to CPT in Mms4-deficient cells (126).

[b] Pnk1− cells are hypersensitive to CPT in the absence of additional defects, indicating difference from budding (see [a]) and importance of this pathway in fission yeast, which like mammals possesses a gene that has both 3′-phosphatase and 5′-kinase activity (112).

(*): Nitiss, J. L., unpublished data.

In humans, Tdp1 and PNKP form a multiprotein complex with XRCC1, poly(ADP-ribose) polymerase (PARP), β-polymerase, and ligase III (*116, 117*) (Fig. 5B, bottom). This complex contains the critical elements for base excision repair. Mammalian cells deficient for XRCC1 or PARP are also hypersensitive to camptothecin (Fig. 7) (Table I) (*78, 116, 118*). The hypersensitivity of PARP–/– cells to camptothecin (Fig. 7A) can be related to a functional defect in Tdp1 (Fig. 7B and C). It is not explained by abnormal levels of Tdp1 or Top1 proteins (Barcelo and Pommier, unpublished). A novel protein, aprataxin, has been found to associate with the XRCC1 complex (Fig. 5B) (*119*). Aprataxin is a 342 amino acid protein encoded by the *APTX* gene whose homozygote mutation produces ataxia-oculomotor apraxia (AOA1; the most common autosomal recessive ataxia in Japan) (*120, 121*). AOA1 cells are hypersensitive to camptothecin (*122*) (Table I).

There is no pharmacological inhibitor of Tdp1 reported to date. Vanadate and tungstate act as phosphate mimetic in cocrystal structures and block Tdp1 activity at millimolar concentrations (*102*). It would, however, be rational to develop Tdp1 inhibitors for cancer treatment in combination with camptothecins. The anticancer activity of Tdp1 inhibitors may prove dependent on the presence of cancer-related genetic abnormalities, since camptothecin hypersensitivity in Tdp1-defective yeast is conditional for deficiencies in the checkpoint (Rad9) (*88, 97, 98*). A Rad9 defect in a Tdp1-deficient background confers marked sensitization to camptothecin (*88*), and it is tempting to speculate that Tdp1 is primarily required when the checkpoints are deficient as in the case of the yeast *RAD9* mutant. A second group of conditional genes (with respect to Tdp1 deficiencies) includes three sets of genes from the 3′-flap endonuclease pathway: *Rad1/Rad10*, *Mre11/Rad50*, and *Mus81/Eme1* (*88, 97, 98*). Mutations in each of these genes renders Tdp1-deficient cells highly sensitive to camptothecin (see Table III). Hence, colon cancers that are commonly mutated for Mre11 might be selectively sensitive to the combination of camptothecin with a Tdp1 inhibitor. The potent activity of Tdp1 against a number of artificial substrates (Fig. 6E) (*105*) can be used to design high throughput screens (our unpublished observations) (*123, 124*), and Tdp1 inhibitors will be reported in the near future.

C. Top1 Excision by Endonucleases (Fig. 5C)

Studies in yeast demonstrate the existence of alternative pathways besides Tdp1 for removing the Top1 covalent complexes. Figure 8 summarizes the multiple endonuclease genetic pathways implicated in the repair of Top1 cleavage complexes (*97–99*).

Rad1/Rad10 and Slx1/Slx4 appear to function in parallel and redundant pathways with Tdp1 (*97–99*) (Fig. 8A). Rad1/Rad10 is the ortholog of the

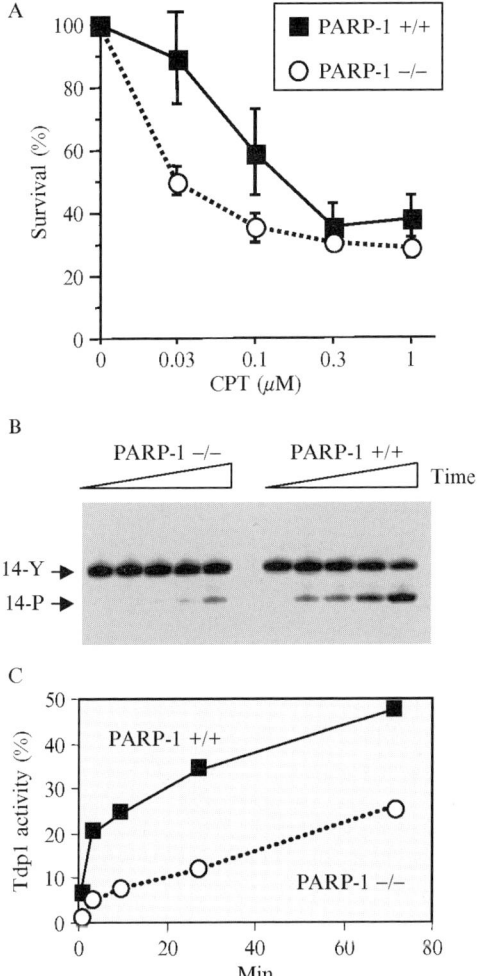

FIG. 7. Hypersensitivity of PARP-1–/– cells to camptothecin and functional Tdp1 deficiency in PARP-1–/– cells. (A) Mouse fibroblasts were exposed to camptothecin (concentrations indicated on X axis) for 1 hr, washed, kept in culture for 4 days, and counted. The curves represent the averages and standard deviations from five independent experiments. (B) A 14-mer single-stranded oligonucleotide with a phosphotyrosine at the 3′ end (see Fig. 6) (14-Y) (106) was incubated with nuclear extracts obtained from PARP-1–/– and PARP-1+/+ cells (195) in the presence of 50 mM EDTA and absence of $MgCl_2$ to eliminate PNKP activity, which requires $MgCl_2$. A representative experiment is shown. TDP1 activity was determined as a shift in band position (from 14-Y to 14-P). (C) Quantitation of the results shown in panel B using ImageQuant (Molecular Dynamics, Sunnyvale, CA).

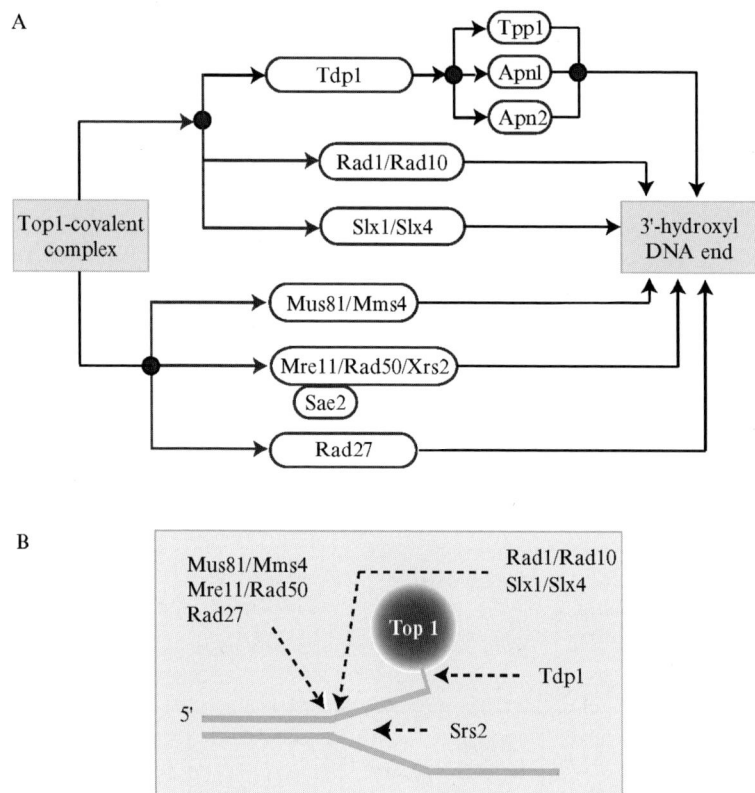

FIG. 8. Pathways involved in the repair of Top1 covalent complexes in budding yeast. (A) Schematic representation of the genetic pathways implicated in the removal of the Top1–DNA covalent complexes. Tdp1 appear to function in alternative pathways with Rad1/Rad10 and Slx1/Slx4. The other endonucleases (Mus81/Mms4, Mre11/Rad50/Xrs2, and Rad27) appear to function in parallel. Sae2 is a cofactor for the endonuclease activity of Mre11. Tpp1, Apn1, and Apn2 are 3′-phosphatases that remove the 3′-phosphate left after Tdp1 hydrolyzes the tyrosyl-DNA adduct (see Figs. 5B and 6C). (B) Schematic representation of the sites of attack for Tdp1 and the endonucleases described in panel A. Srs2 helicase is also shown.

human endonuclease XPF/ERCC1, which cleaves DNA 3′ from bulky adduct during nucleotide excision repair (*125*). Like Tdp1, Rad1/Rad10 requires a single-stranded gap between the 3′ end to be processed and the 5′ end of the DNA (Fig. 8B) (*126*), suggesting that Tdp1 and Rad1/Rad10 share common substrates. *SLX1* and *SLX4* were originally identified as genes synthetically lethal with mutations in *SGS1* and *TOP3* (*127*). The dimeric complex has strong endonuclease activity with a wide range of substrates (*128*), and

Slx4/Slx1 appears to function as an alternative pathway in the absence of Tdp1 (99) (Fig. 8).

Mus81/Mms4 (the ortholog of budding yeast Mms4 is Eme1 in humans and fission yeast—see Tables III and IV) preferentially cleaves broken replication forks and requires the presence of duplex DNA near the 3′ end to be processed [see Figs. 8B and 4B in (78)] (126, 129, 130). Mus81- and Rad50-deficient yeasts are highly sensitive to camptothecin (Tables III and IV) (98, 99, 126, 129, 131).

The Mre11/Rad50/Xrs2 complex (MRX) [the human orthologs are Mre11/Rad50/Nbs1 (MRN)] preferentially cleaves gapped substrates (Fig. 8B) and hairpin structures (132). MRX appears to function independently from the Tdp1 pathway, which is also the case for Mus81/Mms4 and Rad27 (97–99) (Fig. 8A). Sae2 is required for the endonuclease activity of MRX, and *sae2* deletion strains are among the strains most sensitive to camptothecin (99). The MRN complex also possesses checkpoint functions, which probably contribute to cell survival in response to camptothecin (97, 98).

Surprisingly, the 5′-flap endonuclease, Rad27 (the human ortholog is FEN-1) also contributes to the repair of Top1 covalent complexes (Fig. 8). Deletion of *RAD27* causes mild sensitivity to camptothecin (Table III) (98, 99, 133). The apparent discrepancy between the established 5′-flap endonuclease activity of Rad27 (FEN-1) and its role in Top1 repair can be explained by a study demonstrating that Rad27 in coordination with the Werner syndrome protein and replication protein A (RPA) possesses gap 5′-endonuclease (GEN) activity and can process substrates that mimic stalled replication forks. Human FEN-1 is able to rescue the defect in resistance to camptothecin and UV in a yeast *FEN-1* null mutant (133).

Recent yeast genome-wide screen to detect novel genes that are important for protection against growth inhibition and killing produced by camptothecin identified a large number of genes for further exploration. In interpreting the results of genomic screens (such as much of the data shown in Table III), several caveats need to be kept in mind. First, most of the genome wide screens have used the yeast deletion set, and therefore failed to identify potential roles of genes that are essential for viability. Second, many of the screens that identified genes important for surviving damage induced by camptothecin were originally designed to identify genes that are important for surviving other types of DNA damage, such as damage from ionizing radiation (131) or from simple alkylating agents (253). Camptothecin hypersensitivity was assessed as a secondary screen following the primary screen. Finally, camptothecin sensitivity in yeast has been assessed in a variety of strain backgrounds using different assay conditions. For example, some experiments have been conducting using strains with mutations that have enhanced drug accumulation or in strains with ectopic overexpression of Top1. Differences in strain background or in screening approaches understandably have led to differing assessments of the importance of individual genes in camptothecin sensitivity. Therefore, the conclusions regarding camptothecin sensitivity listed in Table III represent our consensus judgment in cases where authors reach differing conclusions.

Nonetheless, the recent yeast genetic screens for hypersensitivity to camptothecin have identified a large number of genes that merit further exploration (99). For example, *SRS2* (also termed HPR5) was among the most critical genes required for growth in the presence of camptothecin (98, 99). SRS2 encodes a DNA helicase that might contribute to the local unwinding (Fig. 8B) that would provide access to the endonucleases described earlier. A number of other genes involved in transcription, replication, ubiquitination, and protein degradation await further studies (Table III) (99, 134–136). However, extrapolating the results obtained in budding yeast to human cells is not straightforward. Cells from Mus81 knockout mice have been tested and failed to show significant hypersensitivity to camptothecin in spite of their hypersensitivity to DNA cross-linking agents (cisplatin and nitrogen mustards) (137). Similarly, XPF cells do not demonstrate hypersensitivity to camptothecin (Pourquier and Pommier, unpublished). These apparent discrepancies may be related to the presence of additional repair pathways and different checkpoint controls in humans.

V. Checkpoint Response to Top1-Associated DNA Damage

Cellular responses to Top1 poisons determine both tumor response and host toxicity. Efficient repair is probably coupled with checkpoint activation. Cell cycle arrest following checkpoint activation would have two beneficial consequences: (i) it would give time for the repair of DNA damage, and (ii) it would prevent further replication-dependent DNA damage. Both the S-phase and the G2 checkpoints, as well as the p53/p21 pathways are activated by Top1-mediated DNA damage (57, 138). Because cell cycle checkpoints are also connected to the apoptosis machinery, it is likely that extensive DNA damage activates apoptosis by involving the same DNA damage sensors and checkpoints (42). Thus, an exciting challenge is to elucidate the relationships between sensor proteins, checkpoints, DNA repair, and apoptosis. Integration of these pathways should explain the cellular determinants of response to Top1 poisons. We will focus on the Chk1 and Chk2 pathways/responses elicited by Top1 poisons, and how defects in these pathways can sensitize tumors to Top1-mediated DNA damage. We will not review the roles of p53, c-Abl, and the stress kinase (JNK/SAPK) pathways, which have been detailed elsewhere (42, 78, 139).

A. Chk1 Activation by Camptothecin

Chk1 is an evolutionarily conserved kinase and essential member of the DNA damage checkpoint (140–142). Deletion of Chk1 is embryonic lethal with massive apoptosis in stem cells (143, 144). Chk1 deletion is not lethal in budding and fission yeast or chicken somatic cells (DT40). However, these cells display an inability to recover from replication blockade and are hypersensitive to ionizing radiation (145–147). In budding yeast, the upstream kinase Tel1 activates Chk1 in concert with the "9-1-1 PCNA–like" clamp proteins: Rad17, Mec3, and Rad24. Strains defective for the Tel1 or the 9-1-1 clamp proteins are sensitive to camptothecin (Table III). In xenopus and humans, ATR (the Tel1 ortholog) with the Rad9-Rad1-Hus1 (9-1-1) complex regulate Chk1 via phosphorylation. The DNA-binding protein claspin also associates with Chk1 and is required for Chk1 activation (143, 148–150). Other kinases, including ATM (151, 152), might also activate Chk1 by phosphorylation (152–154).

The first report of Chk1 activation by camptothecin was obtained in the fission yeast (155). Blocking DNA synthesis with a hydroxyurea pretreatment blocked camptothecin-induced Chk1 phosphorylation/activation, and cells lacking functional Chk1 were hypersensitive to camptothecin (155–157). Phosphorylation of Chk1 is rapid following exposure to camptothecin and reaches a plateau between 60 min and 3 hr (155). Chk1 activation requires phosphorylation by Rad3 and is required for cell cycle arrest and interaction with Rad24 and Rad25 (14–3-3 orthologs), which sequester (and functionally inactivate) Cdc25 (158–164). In mammalian cells, camptothecin also activates Chk1, and Chk1 antisense oligonucleotides abolish the checkpoint response to camptothecin (165). Chk1 activation requires phosphorylation by ATR on S345. G2/M arrest is mediated by phosphorylation of Cdc25C by Chk1 and subsequent inactivation of Cdc25C (161–164), and S-phase arrest is mediated by Chk1-mediated phosphorylation of Cdc25A and degradation of Cdc25A (166). Chk1 can also phosphorylate Rad51 on T309. Rad51 is involved in homologous recombination and provides a link between Chk1 activation, DNA repair, and survival after replicative stress (167).

Camptothecin also induces Chk1 degradation by the ubiquitin-proteosome pathway (168) via the E4 ligase complex containing cullins (Cul1 and Cul4A). Phosphorylation at S345 marks Chk1 for proteolytic degradation and may be indicative of a negative feedback mechanism that may promote the checkpoint termination. Chk1 degradation had been previously suggested to occur via a proteosomal pathway after exposure to geldanamycin, an Hsp90 binding agent that stimulate proteasome-mediated degradation (169). Downregulation of Chk1 by siRNA potentiates the cytotoxicity of camptothecin in cancer cell lines (162, 170). Consistently, Chk1 inhibition by UCN-01 (7-hydroxystaurosporine) markedly potentiates the cytoxicity of camptothecin via abrogation of the S-phase checkpoint (57). Preincubation with UCN-01 prevents camptothecin-induced degradation of Cdc25A and cyclin E in a p53-dependent manner, which is consistent with an abrogation of S-phase arrest (171). In another study, concurrent treatment with camptothecins and UCN-01 resulted in S-phase checkpoint abrogation, increased phosphorylation of γ-H2AX (a marker of DNA double-strand breaks), and cell killing independently of p53 (172).

B. Chk2 Activation by Camptothecin

Chk2 shares no sequence homology with Chk1 [for review see (173, 254)]. Chk2 knockout mice are viable and are defective in p53 stabilization following DNA double-strand breaks (174, 175). Chk2 functions as a checkpoint and apoptotic kinase after being activated via phosphorylation primarily by ATM (141, 176, 177), and also by ATR (178), DNA-PK (via interactions with Ku70/80) (179), Polo-like kinases, Plk-1 and Plk-3 (180–184), and TTK/hMPS1 kinases (185). Following initial phosphorylation on T68, Chk2 undergoes dimerization via the FHA domain of a second Chk2 molecule (186). This interaction is followed by a cascade of autophosphorylation steps on T387/T383 and S516, which are required for full activation of Chk2 [reviewed in (173, 187, 264) and (http://discover.nci.nih.gov/mim/) (chk2mim)].

Human kidney embryonic cells expressing antisense Chk2 display defective S-phase delay and enhanced cell killing in response to replication-mediated DNA damage induced camptothecin (188). Chk2 siRNA experiments also demonstrate a role for Chk2 for cell survival after camptothecin (162). However, the phenotype of Chk2-deficient cells may depend on the cell type. In cortical neurons, ATM deficiency, but not Chk2 deficiency, attenuates cell death and significantly inhibits the induction of p53 phosphorylation on S15 and p53 levels induced by camptothecin (189). Further analyses in cells with fully functional replication machinery are warranted to determine the potential role of Chk2 in Top1-induced DNA replication stress. More specifically, the relative roles of ATR and ATM remain to be elucidated. Chk2 is rapidly phosphorylated by ATM in camptothecin-treated cells (190), and phosphorylated Chk2 forms nuclear foci that are associated with sites of DNA damage marked by 53BP1, γ-H2AX, and NBS1 (191). Camptothecin also reduces the amount of chromatin-associated Chk2, especially the active T68-phosphorylated forms of Chk2 (179). Soluble Chk2 might be implicated in the transmission of signals to remote chromatin sites from the DNA damage.

Chk2 has a dual role on cell cycle checkpoints and apoptosis via phosphorylation of its downstream substrates including Cdc25A, Cdc25C, p53, BRCA1, E2F1, and PML [reviewed in (173, 264)] (http://discover.nci.nih.gov/mim/). Chk2 provides an unexplored therapeutic target in cancer cells with inherent defects in G1 checkpoint function. By virtue of Chk2's role in both cell cycle checkpoint regulation and apoptosis, selective inhibition of Chk2 could improve the therapeutic index of DNA-damaging agents such as camptothecin [reviewed in (173)]. This may be especially true in p53-deficient tumors where the p53-dependent apoptotic response is deficient. In normal tissues Chk2 may act as a proapoptotic effector, thus Chk2 inhibitors may protect normal tissues. However, no clinical agent presents selective Chk2 inhibition.

C. Chk1 vs Chk2 Activation by Camptothecin and Top1-Mediated DNA Damage

As discussed earlier, Top1-induced DNA damage activates both the Chk1 and Chk2 kinases. However, the relative timing of such activations has not been described in detail. Understanding the kinetics of activation of each has the potential for presenting a better target for inhibition in combination with camptothecins. A study (192) reported that both Chk1 and Chk2 are rapidly activated following low dose exposure to camptothecin, suggesting concomitant activation of both the ATR-Chk1 and ATM-Chk2 pathways. Additionally, a deficiency of ATM kinase prevented the activation of Chk2 with no effects on Chk1 activity or the degradation of Cdc25A (192). The authors proposed that the ATR-Chk1 pathway is sufficient, though not the only pathway, to induce checkpoint-mediated degradation of Cdc25A. One explanation for the ATM-independent analysis might be the cautionary finding of a highly activated ATR-Chk1 pathway in AT cells treated with ionizing radiation (193). Reports identifying multiple phosphorylation sites on Chk1 and Chk2 and their unraveling functions also provide a rationale for further studies that are needed to detail the kinetics of Chk1 and Chk2 activation by camptothecin. Overall, the redundancy in activation of both Chk1 and Chk2 by camptothecin may be indicative of distinct functions for each kinase in collecting, sustaining, and deploying the DNA damage signal to its various substrates for favorable use by the cell. Abrogation of these kinases, and consequently the cell cycle arrest checkpoint, thus presents novel opportunities for enhancing drug efficacy.

VI. Conclusion and Perspective

Studies performed in yeast have provided major insights in the DNA damage and repair of Top1 cleavage complexes. These insights include the demonstration that Top1 is the selective target of camptothecin, the discovery of Tdp1 and the identification of genetic defects (in recombination pathways and endonucleases) that sensitize yeast to Top1-mediated DNA damage. It is clear that more lessons remain to be learned from yeast, including the roles of chromatin structure, as well as ubiquitin and other protein modification pathways. However, not all yeast pathways can be translated to human cells even when the orthologs are present in both. For instance, the *Tdp1* mutant in budding yeast is only sensitive to camptothecin if a checkpoint pathway (Rad9) or an endonuclease pathway (Rad1/Rad10) is also defective. By contrast, human cells deficient for Tdp1 (cells from SCAN1 patients) appear to be sensitive with this single alteration. Mus81-deficient yeast is highly sensitive to camptothecin whereas murine Mus81-knockout cells are not. These differences must be related to the "genetic background" of the cells.

The divergent phenotypes (and genotypes) of mammalian cells in culture provide therapeutic opportunities for cancer treatment. Programmed genetic and pharmacological deficiencies should have different consequences with respect to cellular response to Top1-mediated DNA damage depending on the genomic context of the cell. This challenge brings the opportunity to find out which genetic contexts provide selective sensitivity or resistance to Top1 inhibitors, and the rationale for developing therapies tailored to the genetic deficiencies selective to particular tumors. It is in this context that one could foresee the use of inhibitors of Tdp1 and Chk1/2 in combination with Top1 inhibitors once it is established which tumor-specific deficiencies provide the greatest sensitization to Top1 inhibitors.

Acknowledgments

This research is supported by the Intramural Research Program of the NIH, National Cancer Institute, Center for Cancer Research.

We wish to thank Dr. Mark Smulson for providing us with the PARP−/− and PARP+/+ cells. We wish to thank the past and current member of the Laboratory of Molecular Pharmacology (Drs. Smitha Antony, Laurent Debethune, Takahisa Furuta, Linghua Meng, Philippe Pourquier, Haruyuki Takemura, and Qiang Yu), and more particularly Dr. Kurt W. Kohn for outstanding contribution to the topoisomerase inhibitor studies and for sustained input and encouragements.

References

1. Wang, J. C. (2002). Cellular roles of DNA topoisomerases: A molecular perspective. *Nat. Rev. Mol. Cell. Biol.* **3**, 430–440.
2. Zhang, H., Barcelo, J. M., Lee, B., Kohlhagen, G., Zimonjic, D. B., Popescu, N. C., and Pommier, Y. (2001). Human mitochondrial topoisomerase I. *Proc. Natl. Acad. Sci. USA* **98**, 10608–10613.
3. Zhang, H., Meng, L. H., Zimonjic, D. B., Popescu, N. C., and Pommier, Y. (2004). Thirteen-exon-motif signature for vertebrate nuclear and mitochondrial type IB topoisomerases. *Nucleic Acids Res.* **32**, 2087–2092.
4. Hickson, I. D. (2003). RecQ helicases: Caretakers of the genome. *Nat. Rev. Cancer* **3**, 169–178.

5. Harmon, F. G., Brockman, J. P., and Kowalczykowski, S. C. (2003). RecQ helicase stimulates both DNA catenation and changes in DNA topology by topoisomerase III. *J. Biol. Chem.* **278,** 42668–42678.
6. Morham, S., Kluckman, K. D., Voulomanos, N., and Smithies, O. (1996). Targeted disruption of the mouse topoisomerase I gene by camptothecin selection. *Mol. Cell. Biol.* **16,** 6804–6809.
7. Zhang, C. X., Chen, A. D., Gettel, N. J., and Hsieh, T. S. (2000). Essential functions of DNA topoisomerase I in *Drosophila melanogaster. Dev. Biol.* **222,** 27–40.
8. Eng, W. K., Faucette, L., Johnson, R. K., and Sternglanz, R. (1988). Evidence that DNA topoisomerase I is necessary for the cytotoxic effects of camptothecin. *Mol. Pharmacol.* **34,** 755–760.
9. Baker, S. D., Wadkins, R. M., Stewart, C. F., Beck, W. T., and Danks, M. K. (1995). Cell cycle analysis of amount and distribution of nuclear DNA topoisomerase I as determined by fluorescence digital imaging microscopy. *Cytometry* **19,** 134–145.
10. Muller, M. T., Pfund, W. P., Mehta, V. B., and Trask, D. K. (1985). Eukaryotic type I topoisomerase is enriched in the nucleolus and catalytically active on ribosomal DNA. *EMBO. J.* **4,** 1237–1243.
11. Zhang, H., Wang, J. C., and Liu, L. F. (1988). Involvement of DNA topoisomerase I in transcription of human ribosomal RNA genes. *Proc. Natl. Acad. Sci. USA* **85,** 1060–1064.
12. Pommier, Y., Kohlhagen, G., Wu, C., and Simmons, D. T. (1998). Mammalian DNA topoisomerase I activity and poisoning by camptothecin are inhibited by simian virus 40 large T antigen. *Biochemistry* **37,** 3818–3823.
13. Champoux, J. J. (2001). DNA topoisomerases: Structure, function, and mechanism. *Annu. Rev. Biochem.* **70,** 369–413.
14. Leppard, J. B., and Champoux, J. J. (2005). Human DNA topoisomerase I: Relaxation, roles, and damage control. *Chromosoma* **114,** 75–85.
15. Xu, C.-J., Grainge, I., Lee, J., Harshey, R. M., and Jayaram, M. (1998). Unveiling two distinct ribonuclease activities and a topoisomerase activity in a site-specific DNA recombinase. *Mol. Cell* **1,** 729–739.
16. Pourquier, P., Jensen, A. D., Gong, S. S., Pommier, Y., and Rogler, C. E. (1999). Human DNA topoisomerase I-mediated cleavage and recombination of duck hepatitis B virus DNA *in vitro. Nucleic Acids Res.* **27,** 1919–1923.
17. Cheng, C., and Shuman, S. (2000). Recombinogenic flap ligation pathway for intrinsic repair of topoisomerase IB-induced double-strand breaks. *Mol. Cell. Biol.* **20,** 8059–8068.
18. Soret, J., Gabut, M., Dupon, C., Kohlhagen, G., Stevenin, J., Pommier, Y., and Tazi, J. (2003). Altered serine/arginine-rich protein phosphorylation and exonic enhancer-dependent splicing in Mammalian cells lacking topoisomerase I. *Cancer Res.* **63,** 8203–8211.
19. Rossi, F., Labourier, E., Forne, T., Divita, G., Derancourt, J., Riou, J. F., Antoine, E., Cathala, G., Brunel, C., and Tazi, J. (1996). Specific phosphorylation of SR proteins by mammalian DNA topoisomerase I. *Nature* **381,** 80–82.
20. Stewart, L., Redinbo, M. R., Qiu, X., Hol, W. G. J., and Champoux, J. J. (1998). A model for the mechanism of human topoisomerase I. *Science* **279,** 1534–1541.
21. Lesher, D. T., Pommier, Y., Stewart, L., and Redinbo, M. R. (2002). 8-Oxoguanine rearranges the active site of human topoisomerase I. *Proc. Natl. Acad. Sci. USA* **99,** 12102–12107.
22. Staker, B. L., Hjerrild, K., Feese, M. D., Behnke, C. A., Burgin, A. B., Jr., and Stewart, L. (2002). The mechanism of topoisomerase I poisoning by a camptothecin analog. *Proc. Natl. Acad. Sci. USA* **99,** 15387–15392.
23. Carey, J. F., Schultz, S. J., Sisson, L., Fazzio, T. G., and Champoux, J. J. (2003). DNA relaxation by human topoisomerase I occurs in the closed clamp conformation of the protein. *Proc. Natl. Acad. Sci. USA* **100,** 5640–5645.

24. Woo, M. H., Losasso, C., Guo, H., Pattarello, L., Benedetti, P., and Bjornsti, M. A. (2003). Locking the DNA topoisomerase I protein clamp inhibits DNA rotation and induces cell lethality. *Proc. Natl. Acad. Sci. USA* **100**, 13767–13772.
25. Koster, D. A., Croquette, V., Dekker, C., Shuman, S., and Dekker, N. H. (2005). Friction and torque govern the relaxation of DNA supercoils by eukaryotic topoisomerase IB. *Nature* **434**, 671–674.
26. Pommier, Y., and Cherfils, J. (2005). Interfacial protein inhibition: A nature's paradigm for drug discovery. *Trends Pharmacol. Sci.* **28**, 136–145.
27. Pommier, Y., and Marchand, C. (2005). Interfacial inhibitors of protein-nucleic acid interactions. *Curr. Med. Chem. Anti-Canc. Agents* **5**, 421–429.
28. Chrencik, J. E., Staker, B. L., Burgin, A. B., Pourquier, P., Pommier, Y., Stewart, L., and Redinbo, M. R. (2004). Mechanisms of camptothecin resistance by human topoisomerase I mutations. *J. Mol. Biol.* **339**, 773–784.
29. Staker, B. L., Feese, M. D., Cushman, M., Pommier, Y., Zembower, D., Stewart, L., and Burgin, A. B. (2005). Structures of three classes of anticancer agents bound to the human topoisomerase I-DNA covalent complex. *J. Med. Chem.* **48**, 2336–2345.
30. Ioanoviciu, A., Antony, S., Pommier, Y., Staker, B. L., Stewart, L., and Cushman, M. (2005). Synthesis and mechanism of action studies of a series of norindenoisoquinoline topoisomerase I poisons reveal an inhibitor with a flipped orientation in the ternary DNA-enzyme-inhibitor complex as determined by x-ray crystallographic analysis. *J. Med. Chem.* **48**, 4803–4814.
31. Garcia-Carbonero, R., and Supko, J. G. (2002). Current perspectives on the clinical experience, pharmacology, and continued development of the camptothecins. *Clin. Cancer Res.* **8**, 641–661.
32. L.-H., Meng, Z.-Y., Liao, and Pommier, Y. (2003). Non-camptothecin DNA topoisomerase I inhibitors in cancer chemotherapy. *Curr. Topics Med. Chem.* **3**, 305–320.
33. Covey, J. M., Jaxel, C., Kohn, K. W., and Pommier, Y. (1989). Protein-linked DNA strand breaks induced in mammalian cells by camptothecin, an inhibitor of topoisomerase I. *Cancer Res.* **49**, 5016–5022.
34. Antony, S., Jayaraman, M., Laco, G., Kohlhagen, G., Kohn, K. W., Cushman, M., and Pommier, Y. (2003). Differential induction of topoisomerase I-DNA cleavage complexes by the indenoisoquinoline MJ-III-65 (NSC 706744) and camptothecin: Base sequence analysis and activity against camptothecin-resistant topoisomerases I. *Cancer Res.* **63**, 7428–7435.
35. Jaxel, C., Capranico, G., Kerrigan, D., Kohn, K. W., and Pommier, Y. (1991). Effect of local DNA sequence on topoisomerase I cleavage in the presence or absence of camptothecin. *J. Biol. Chem.* **266**, 20418–20423.
36. Tanizawa, A., Kohn, K. W., Kohlhagen, G., Leteurtre, F., and Pommier, Y. (1995). Differential stabilization of eukaryotic DNA topoisomerase I cleavable complexes by camptothecin derivatives. *Biochemistry* **34**, 7200–7206.
37. Pourquier, P., and Pommier, Y. (2001). Topoisomerase I-mediated DNA damage. *Adv. Cancer Res.* **80**, 189–216.
38. Sokhansanj, B. A., and Wilson, D. M., III (2004). Oxidative DNA damage background estimated by a system model of base excision repair. *Free Radic. Biol. Med.* **37**, 422–427.
39. Pourquier, P., Ueng, L.-M., Fertala, J., Wang, D., Park, H.-J., Essigman, J. M., Bjornsti, M.-A., and Pommier, Y. (1999). Induction of reversible complexes between eukaryotic DNA topoisomerase I and DNA-containing oxidative base damages. *J. Biol. Chem.* **274**, 8516–8523.
40. Pourquier, P., Ueng, L.-M., Kohlhagen, G., Mazumder, A., Gupta, M., Kohn, K. W., and Pommier, Y. (1997). Effects of uracil incorporation, DNA mismatches, and abasic sites on cleavage and religation activities of mammalian topoisomerase I. *J. Biol. Chem.* **272**, 7792–7796.

41. Pourquier, P., Pilon, A., Kohlhagen, G., Mazumder, A., Sharma, A., and Pommier, Y. (1997). Trapping of mammalian topoisomerase I and recombinations induced by damaged DNA containing nicks or gaps: Importance of DNA end phosphorylation and camptothecin effects. *J. Biol. Chem.* **272,** 26441–26447.
42. Sordet, O., Khan, Q., Kohn, K. W., and Pommier, Y. (2003). Apoptosis induced by topoisomerase inhibitors. *Curr. Med. Chem. Anti-canc. Agents* **3,** 271–290.
43. Sordet, O., Khan, Q. A., Plo, I., Pourquier, P., Urasaki, Y., Yoshida, A., Antony, S., Kohlhagen, G., Solary, E., Saparbaev, M., Laval, J., and Pommier, Y. (2004). Apoptotic topoisomerase I-DNA complexes induced by staurosporine-mediated oxygen radicals. *J. Biol. Chem.* **279,** 50499–50504.
44. Sordet, O., Khan, Q. A., and Pommier, Y. (2004). Apoptotic topoisomerase I-DNA complexes induced by oxygen radicals and mitochondrial dysfunction. *Cell Cycle* **3,** 1095–1097.
45. Sordet, O., Liao, Z., Liu, H., Antony, S., Stevens, E. V., Kohlhagen, G., Fu, H., and Pommier, Y. (2004). Topoisomerase I-DNA complexes contribute to arsenic trioxide-induced apoptosis. *J. Biol. Chem.* **279,** 33968–33975.
46. Burgin, A. B., Huizenga, B. N., and Nash, H. A. (1995). A novel suicide substrate for DNA topoisomerases and site-specific recombinases. *Nucleic Acids Res.* **23,** 2973–2979.
47. Shuman, S. (1989). Vaccinia DNA topoisomerase I promotes illegitimate recombination in *Escherichia coli*. *Proc. Natl. Acad. Sci. USA* **86,** 3489–3493.
48. Holm, C., Covey, J. M., Kerrigan, D., and Pommier, Y. (1989). Differential requirement of DNA replication for the cytotoxicity of DNA topoisomerase I and II inhibitors in Chinese hamster DC3F cells. *Cancer Res.* **49,** 6365–6368.
49. Hsiang, Y.-H., Lihou, M. G., and Liu, L. F. (1989). Arrest of DNA replication by drug-stabilized topoisomerase I-DNA cleavable complexes as a mechanism of cell killing by camptothecin. *Cancer Res.* **49,** 5077–5082.
50. Stefanis, L., Park, D. S., Friedman, W. J., and Greene, L. A. (1999). Caspase-dependent and -independent death of camptothecin-treated embryonic cortical neurons. *J. Neurosci.* **19,** 6235–6247.
51. Morris, E. J., and Geller, H. M. (1996). Induction of neuronal apoptosis by camptothecin, an inhibitor of DNA topoisomerase-I: Evidence for cell cycle-independent toxicity. *J. Cell Biol.* **134,** 757–770.
52. Strumberg, D., Pilon, A. A., Smith, M., Hickey, R., Malkas, L., and Pommier, Y. (2000). Conversion of topoisomerase I cleavage complexes on the leading strand of ribosomal DNA into 5′-phosphorylated DNA double-strand breaks by replication runoff. *Mol. Cell. Biol.* **20,** 3977–3987.
53. Furuta, T., Takemura, H., Liao, Z. Y., Aune, G. J., Redon, C., Sedelnikova, O. A., Pilch, D. R., Rogakou, E. P., Celeste, A., Chen, H. T., Nussenzweig, A., Aladjem, M. I. *et al.* (2003). Phosphorylation of histone H2AX and activation of Mre11, Rad50, and Nbs1 in response to replication-dependent DNA-double-strand breaks induced by mammalian DNA topoisomerase I cleavage complexes. *J. Biol. Chem.* **278,** 20303–20312.
54. Rogakou, E. P., Pilch, D. R., Orr, A. H., Ivanova, V. S., and Bonner, W. M. (1998). DNA double-stranded breaks induce histone H2AX phosphorylation on serine 139. *J. Biol. Chem.* **273,** 5858–5868.
55. Redon, C., Pilch, D., Rogakou, E., Sedelnikova, O., Newrock, K., and Bonner, W. (2002). Histone H2A variants H2AX and H2AZ. *Curr. Opin. Genet. Dev.* **12,** 162–169.
56. Shao, R.-G., Cao, C.-X., Zhang, H., Kohn, K. W., Wold, M. S., and Pommier, Y. (1999). Replication-mediated DNA damage by camptothecin induces phosphorylation of RPA by DNA-dependent protein kinase and dissociates RPA:DNA-PK complexes. *EMBO J.* **18,** 1397–1406.

57. Shao, R.-G., Cao, C.-X., Shimizu, T., O'Connor, P., Kohn, K. W., and Pommier, Y. (1997). Abrogation of an S-phase checkpoint and potentiation of camptothecin cytotoxicity by 7-hydroxystaurosporine (UCN-01) in human cancer cell lines, possibly influenced by p53. *Cancer Res.* **57,** 4029–4035.
58. Kohn, E. A., Ruth, N. D., Brown, M. K., Livingstone, M., and Eastman, A. (2002). Abrogation of the S phase DNA damage checkpoint results in S phase progression or premature mitosis depending on the concentration of 7-hydroxystaurosporine and the kinetics of Cdc25C activation. *J. Biol. Chem.* **277,** 26553–26564.
59. Huberman, J. A. (1981). New views of the biochemistry of eucaryotic DNA replication revealed by aphidicolin, an unusual inhibitor of DNA polymerase alpha. *Cell* **23,** 647–648.
60. Horwitz, S. B., Chang, C. K., and Grollman, A. P. (1971). Studies on camptothecin. I. Effects of nucleic acid and protein synthesis. *Mol. Pharmacol.* **7,** 632–644.
61. Kessel, D. (1971). Effects of camptothecin on RNA synthesis in leukemia cells. *Biochim. Biophys. Acta* **246,** 225–232.
62. Kann, H. E., Jr., and Kohn, K. W. (1972). Effects of deoxyribonucleic acid-reactive drugs on ribonucleic acid synthesis in leukemia L1210 cells. *Mol. Pharmacol.* **8,** 551–560.
63. Collins, I., Weber, A., and Levens, D. (2001). Transcriptional consequences of topoisomerase inhibition. *Mol. Cell. Biol.* **21,** 8437–8451.
64. Stewart, A. F., Herrera, R. E., and Nordheim, A. (1990). Rapid induction of c-fos transcription reveals quantitive linkage of RNA polymerase II and DNA topoisomerase I enzyme activities. *Cell* **60,** 141–146.
65. Zhou, Y., Gwadry, F. G., Reinhold, W. C., Miller, L. D., Smith, L. H., Scherf, U., Liu, E. T., Kohn, K. W., Pommier, Y., and Weinstein, J. N. (2002). Transcriptional regulation of mitotic genes by camptothecin-induced DNA damage: Microarray analysis of dose- and time-dependent effects. *Cancer Res.* **62,** 1688–1695.
66. Daoud, S. S., Munson, P. J., Reinhold, W., Young, L., Prabhu, V., Yu, Q., LaRose, J., Kohn, K. W., Weinstein, J. N., and Pommier, Y. (2003). Impact of p53 knockout and topotecan treatment on gene expression profiles in Human colon carcinoma cells: A pharmacogenomic study. *Cancer Res.* **63,** 2782–2793.
67. Ljungman, M., and Hanawalt, P. C. (1996). The anti-cancer drug camptothecin inhibits elongation but stimulates initiation of RNA polymerase II transcription. *Carcinogenesis* **17,** 31–35.
68. Bendixen, C., Thomsen, B., Alsner, J., and Westergaard, O. (1990). Camptothecin-stabilized topoisomerase I-DNA adducts cause premature termination of transcription. *Biochemistry* **29,** 5613–5619.
69. Wu, J., and Liu, L. F. (1997). Processing of topoisomerase I cleavable complexes into DNA damage by transcription. *Nucleic Acids Res.* **25,** 4181–4186.
70. Mondal, N., and Parvin, J. D. (2001). DNA topoisomerase IIalpha is required for RNA polymerase II transcription on chromatin templates. *Nature* **413,** 435–438.
71. Duann, P., Sun, M., Lin, C. T., Zhang, H., and Liu, L. F. (1999). Plasmid linking number change induced by topoisomerase I-mediated DNA damage. *Nucleic Acids Res.* **27,** 2905–2911.
72. Sun, M., Duann, P., Lin, C. T., Zhang, H., and Liu, L. F. (2000). Rapid chromatin reorganization induced by topoisomerase I-mediated DNA damage. *Ann. N. Y. Acad. Sci.* **922,** 340–342.
73. Merino, A., Madden, K. R., Lane, W. S., Champoux, J. J., and Reinberg, D. (1993). DNA topoisomerase I is involved in both repression and activation of transcription. *Nature* **365,** 227–232.

74. Shykind, B. M., Kim, J., Stewart, L., Champoux, J. J., and Sharp, P. A. (1997). Topoisomerase I enhances TFIID-TFIIA complex assembly during activation of transcription. *Genes Dev.* **11**, 397–407.
75. Desai, S. D., Zhang, H., Rodriguez-Bauman, A., Yang, J. M., Wu, X., Gounder, M. K., Rubin, E. H., and Liu, L. F. (2003). Transcription-dependent degradation of topoisomerase I-DNA covalent complexes. *Mol. Cell. Biol.* **23**, 2341–2350.
76. Squires, S., Ryan, A. J., Strutt, H. L., and Johnson, R. T. (1993). Hypersensitivity of Cockayne's syndrome cells to camptothecins is associated with the generation of abnormally high levels of double strand breaks in nascent DNA. *Cancer Res.* **53**, 2012–2019.
77. Desai, S. D., Li, T. K., Rodriguez-Bauman, A., Rubin, E. H., and Liu, L. F. (2001). Ubiquitin/26S proteasome-mediated degradation of topoisomerase I as a resistance mechanism to camptothecin in tumor cells. *Cancer Res.* **61**, 5926–5932.
78. Pommier, Y., Redon, C., Rao, A., Seiler, J. A., Sordet, O., Takemura, H., Antony, S., Meng, L.-H., Liao, Z. Y., Kohlhagen, G., Zhang, H., and Kohn, K. W. (2003). Repair of and checkpoint response to topoisomerase I-mediated DNA damage. *Mutat. Res.* **532**, 173–203.
79. Rao, A., Fan, A. M., Meng, L., Doe, C. L., North, P. S., Hickson, I., and Pommier, Y. (2005). Phosphorylation of BLM, dissociation for topoisomerase IIIα and colocalization with γ-H2AX after topoisomerase I-induced replication damage. *Mol. Cell. Biol.* **25**, 8925–8937.
80. Wu, L., and Hickson, I. D. (2003). The Bloom's syndrome helicase suppresses crossing over during homologous recombination. *Nature* **426**, 870–874.
81. Bjergbaek, L., Cobb, J. A., Tsai-Pflugfelder, M., and Gasser, S. M. (2005). Mechanistically distinct roles for Sgs1p in checkpoint activation and replication fork maintenance. *EMBO J.* **24**, 405–417.
82. Imamura, O., Fujita, K., Itoh, C., Takeda, S., Furuichi, Y., and Matsumoto, T. (2002). Werner and Bloom helicases are involved in DNA repair in a complementary fashion. *Oncogene* **21**, 954–963.
83. Boom, V., Jaspers, N. G., and Vermeulen, W. (2002). When machines get stuck-obstructed RNA polymerase II: Displacement, degradation or suicide. *Bioessays* **24**, 780–784.
84. Pommier, Y., Jenkins, J., Kohlhagen, G., and Leteurtre, F. (1995). DNA recombinase activity of eukaryotic DNA topoisomerase I; effects of camptothecin and other inhibitors. *Mutat. Res.* **337**, 135–145.
85. Huang, C. C., Han, C. S., Yue, X. F., Shen, C. M., Wang, S. W., Wu, F. G., and Xu, B. (1983). Cytotoxicity and sister chromatid exchanges induced *in vitro* by six anticancer drugs developed in the People's Republic of China. *J. Natl. Cancer Inst.* **71**, 841–847.
86. Pinero, J., Baena, M., Ortiz, T., and Cortes, F. (1996). Sister chromatid exchange induced by DNA topoisomerases poisons in late replicating heterochromatin: Influence of inhibition of replication and transcription. *Mutat. Res.* **354**, 195–201.
87. Fasullo, M., Zeng, L., and Giallanza, P. (2004). Enhanced stimulation of chromosomal translocations by radiomimetic DNA damaging agents and camptothecin in *Saccharomyces cerevisiae rad9* checkpoint mutants. *Mutat. Res.* **547**, 123–132.
88. Pouliot, J. J., Yao, K. C., Robertson, C. A., and Nash, H. A. (1999). Yeast gene for a Tyr-DNA phosphodiesterase that repairs topo I covalent complexes. *Science* **286**, 552–555.
89. Yang, S.-W., Burgin, A. B., Huizenga, B. N., Robertson, C. A., Yao, K. C., and Nash, H. A. (1996). A eukaryotic enzyme that can disjoin dead-end covalent complexes between DNA and type I topoisomerases. *Proc. Natl. Acad. Sci. USA* **93**, 11534–11539.
90. Interthal, H., Pouliot, J. J., and Champoux, J. J. (2001). The tyrosyl-DNA phosphodiesterase Tdp1 is a member of the phospholipase D superfamily. *Proc. Natl. Acad. Sci. USA* **98**, 12009–12014.

91. Takashima, H., Boerkoel, C. F., John, J., Saifi, G. M., Salih, M. A., Armstrong, D., Mao, Y., Quiocho, F. A., Roa, B. B., Nakagawa, M., Stockton, D. W., and Lupski, J. R. (2002). Mutation of TDP1, encoding a topoisomerase I-dependent DNA damage repair enzyme, in spinocerebellar ataxia with axonal neuropathy. *Nat. Genet.* **32,** 267–272.
92. Barthelmes, H. U., Grue, P., Feineis, S., Straub, T., and Boege, F. (2000). Active DNA topoisomerase IIalpha is a component of the salt-stable centrosome core. *J. Biol. Chem.* **275,** 38823–38830.
93. Nivens, M. C., Felder, T., Galloway, A. H., Pena, M. M., Pouliot, J. J., and Spencer, H. T. (2004). Engineered resistance to camptothecin and antifolates by retroviral coexpression of tyrosyl DNA phosphodiesterase-I and thymidylate synthase. *Cancer Chemother. Pharmacol.* **53,** 107–115.
94. Interthal, H., Chen, H. J., Khel-Fie, T. E., Zotzmann, J., Leppard, J. B., and Champoux, J. J. (2005). SCAN1 mutant Tdp1 accumulates the enzyme–DNA intermediate and causes camptothecin hypersensitivity. *EMBO J.* **24,** 2224–2233.
95. Zhou, T., Lee, J. W., Tatavarthi, H., Lupski, J. R., Valerie, K., and Povirk, L. F. (2005). Deficiency in 3′-phosphoglycolate processing in human cells with a hereditary mutation in tyrosyl-DNA phosphodiesterase (TDP1). *Nucleic Acids Res.* **33,** 289–297.
96. Pouliot, J. J., Robertson, C. A., and Nash, H. A. (2001). Pathways for repair of topoisomerase I covalent complexes in *Saccharomyces cerevisiae*. *Genes Cells* **6,** 677–687.
97. Liu, C., Pouliot, J. J., and Nash, H. A. (2002). Repair of topoisomerase I covalent complexes in the absence of the tyrosyl-DNA phosphodiesterase Tdp1. *Proc. Natl. Acad. Sci. USA* **99,** 14970–14975.
98. Vance, J. R., and Wilson, T. E. (2002). Yeast Tdp1 and Rad1-Rad10 function as redundant pathways for repairing Top1 replicative damage. *Proc. Natl. Acad. Sci. USA* **99,** 13669–13674.
99. Deng, C., Brown, J. A., You, D., and Brown, J. M. (2005). Multiple endonucleases function to repair covalent topoisomerase I complexes in *Saccharomyces cerevisiae*. *Genetics* **170,** 591–600.
100. Inamdar, K. V., Pouliot, J. J., Zhou, T., Lees-Miller, S. P., Rasouli-Nia, A., and Povirk, L. F. (2002). Conversion of phosphoglycolate to phosphate termini on 3′ overhangs of DNA double-strand breaks by the human tyrosyl-DNA phosphodiesterase hTdp1. *J. Biol. Chem.* **277,** 27162–27168.
101. Davies, D. R., Interthal, H., Champoux, J. J., and Hol, W. G. (2002). The crystal structure of human tyrosyl-DNA phosphodiesterase, Tdp1. *Structure (Camb.)* **10,** 237–248.
102. Davies, D. R., Interthal, H., Champoux, J. J., and Hol, W. G. (2002). Insights into substrate binding and catalytic mechanism of human tyrosyl-DNA phosphodiesterase (Tdp1) from vanadate and tungstate-inhibited structures. *J. Mol. Biol.* **324,** 917–932.
103. Davies, D. R., Interthal, H., Champoux, J. J., and Hol, W. G. (2003). Crystal structure of a transition state mimic for tdp1 assembled from vanadate, DNA, and a topoisomerase I-derived peptide. *Chem. Biol.* **10,** 139–147.
104. Stuckey, J. A., and Dixon, J. E. (1999). Crystal structure of a phospholipase D family member. *Nat. Struct. Biol.* **6,** 278–284.
105. Interthal, H., Chen, H. J., and Champoux, J. J. (2005). Human Tdp1 cleaves a broad spectrum of substrates including phosphoamide linkages. *J. Biol. Chem.* **280,** 36518–36528.
106. Debethune, L., Kohlhagen, G., Grandas, A., and Pommier, Y. (2002). Processing of nucleopeptides mimicking the topoisomerase I-DNA covalent complex by tyrosyl-DNA phosphodiesterase. *Nucleic Acids Res.* **30,** 1198–1204.
107. Beidler, D. R., Chang, J.-Y., Zhou, B.-S., and Cheng, Y.-C. (1995). Camptothecin resistance involving steps subsequent to the formation of protein-linked DNA breaks in human camptothecin-resistant KB cell lines. *Cancer Res.* **56,** 345.

108. Desai, S. D., Liu, L. F., Vazquez-Abad, D., and D'Arpa, P. (1997). Ubiquitin-dependent destruction of topoisomerase I is stimulated by the antitumor drug camptothecin. *J. Biol. Chem.* **272,** 24159–24164.
109. Raymond, A. C., Staker, B. L., and Burgin, A. B., Jr. (2005). Substrate specificity of tyrosyl-DNA phosphodiesterase I (Tdp1). *J. Biol. Chem.* **280,** 22029–22035.
110. Vance, J. R., and Wilson, T. E. (2001). Uncoupling of 3'-phosphatase and 5'-kinase functions in budding yeast. Characterization of *Saccharomyces cerevisiae* DNA 3'-phosphatase (TPP1). *J. Biol. Chem.* **276,** 15073–15081.
111. Vance, J. R., and Wilson, T. E. (2001). Repair of DNA strand breaks by the overlapping functions of lesion-specific and non-lesion-specific DNA 3' phosphatases. *Mol. Cell. Biol.* **21,** 7191–7198.
112. Meijer, M., Karimi-Busheri, F., Huang, T. Y., Weinfeld, M., and Young, D. (2002). Pnk1, a DNA kinase/phosphatase required for normal response to DNA damage by gamma-radiation or camptothecin in *Schizosaccharomyces pombe*. *J. Biol. Chem.* **277,** 4050–4055.
113. Karimi-Busheri, F., Daly, G., Robins, P., Canas, B., Pappin, D. J. C., Sgouros, J., Miller, G. G., Fakhrai, H., Davis, E. M., Beau, M. M., and Weinfeld, M. (1999). Molecular Characterization of a Human DNA Kinase. *J. Biol. Chem.* **274,** 24187–24194.
114. Jilani, A., Ramotar, D., Slack, C., Ong, C., Yang, X. M., Scherer, S. W., and Lasko, D. D. (1999). Molecular cloning of the human gene, PNKP, encoding a polynucleotide kinase 3'-phosphatase and evidence for its role in repair of DNA strand breaks caused by oxidative damage. *J. Biol. Chem.* **274,** 24176–24186.
115. Rasouli-Nia, A., Karimi-Busheri, F., and Weinfeld, M. (2004). Stable down-regulation of human polynucleotide kinase enhances spontaneous mutation frequency and sensitizes cells to genotoxic agents. *Proc. Natl. Acad. Sci. USA* **101,** 6905–6910.
116. Plo, I., Liao, Z. Y., Barcelo, J. M., Kohlhagen, G., Caldecott, K. W., Weinfeld, M., and Pommier, Y. (2003). Association of XRCC1 and tyrosyl DNA phosphodiesterase (Tdp1) for the repair of topoisomerase I-mediated DNA lesions. *DNA Repair (Amst.)* **2,** 1087–1100.
117. El-Khamisy, S. F., Saifi, G. M., Weinfeld, M., Johansson, F., Helleday, T., Lupski, J. R., and Caldecott, K. W. (2005). Defective DNA single-strand break repair in spinocerebellar ataxia with axonal neuropathy-1. *Nature* **434,** 108–113.
118. Chatterjee, S., Cheng, M.-F., Trivedi, D., Petzold, S. J., and Berger, N. A. (1989). Camptothecin hypersensitivity in poly (adenosine diphosphate-ribose) polymerase-deficient cell lines. *Cancer Commun.* **1,** 389–394.
119. Clements, P. M., Breslin, C., Deeks, E. D., Byrd, P. J., Ju, L., Bieganowski, P., Brenner, C., Moreira, M C., Taylor, A. M., and Caldecott, K. W. (2004). The ataxia-oculomotor apraxia 1 gene product has a role distinct from ATM and interacts with the DNA strand break repair proteins XRCC1 and XRCC4. *DNA Repair (Amst.)* **3,** 1493–1502.
120. Moreira, M. C., Barbot, C., Tachi, N., Kozuka, N., Uchida, E., Gibson, T., Mendonca, P., Costa, M., Barros, J., Yanagisawa, T., Watanabe, M., Ikeda, Y. *et al.* (2001). The gene mutated in ataxia-ocular apraxia 1 encodes the new HIT/Zn-finger protein aprataxin. *Nat. Genet.* **29,** 189–193.
121. Paulson, H. L., and Miller, V. M. (2005). Breaks in coordination: DNA repair in inherited ataxia. *Neuron* **46,** 845–848.
122. Mosesso, P., Piane, M., Palitti, F., Pepe, G., Penna, S., and Chessa, L. (2005). The novel human gene aprataxin is directly involved in DNA single-strand-break repair. *Cell. Mol. Life Sci.* **62,** 485–491.
123. Cheng, T. J., Rey, P. G., Poon, T., and Kan, C. C. (2002). Kinetic studies of human tyrosyl-DNA phosphodiesterase, an enzyme in the topoisomerase I DNA repair pathway. *Eur. J. Biochem.* **269,** 3697–3704.

124. Rideout, M. C., Raymond, A. C., and Burgin, A. B., Jr. (2004). Design and synthesis of fluorescent substrates for human tyrosyl-DNA phosphodiesterase I. *Nucleic Acids Res.* **32**, 4657–4664.
125. Hoeijmakers, J. H. (2001). Genome maintenance mechanisms for preventing cancer. *Nature* **411**, 366–374.
126. Bastine-Shanower, S. A., Fricke, W. M., Mullen, J. R., and Brill, S. J. (2003). The mechanism of mus81-mms4 cleavage site selection distinguishes it from the homologous endonuclease rad1-rad10. *Mol. Cell. Biol.* **23**, 3487–3496.
127. Mullen, J. R., Kaliraman, V., Ibrahim, S. S., and Brill, S. J. (2001). Requirement for three novel protein complexes in the absence of the Sgs1 DNA helicase in *Saccharomyces cerevisiae*. *Genetics* **157**, 103–118.
128. Fricke, W. M., and Brill, S. J. (2003). Slx1-Slx4 is a second structure-specific endonuclease functionally redundant with Sgs1-Top3. *Genes Dev.* **17**, 1768–1778.
129. Doe, C. L., Ahn, J. S., Dixon, J., and Whitby, M. C. (2002). Mus81-Eme1 and Rqh1 involvement in processing stalled and collapsed replication forks. *J. Biol. Chem.* **277**, 32753–32759.
130. Ciccia, A., Constantinou, A., and West, S. C. (2003). Identification and characterization of the human Mus81/Eme1 endonuclease. *J. Biol. Chem.* **278**, 25172–25178.
131. Bennett, C. B., Lewis, L. K., Karthikeyan, G., Lobachev, K. S., Jin, Y. H., Sterling, J. F., Snipe, J. R., and Resnick, M. A. (2001). Genes required for ionizing radiation resistance in yeast. *Nat. Genet.* **29**, 426–434.
132. D'Amours, D., and Jackson, S. P. (2002). The Mre11 complex: At the crossroads of DNA repair and checkpoint signalling. *Nat. Rev. Mol. Cell. Biol.* **3**, 317–327.
133. Zheng, L., Zhou, M., Chai, Q., Parrish, J., Xue, D., Patrick, S. M., Turchi, J. J., Yannone, S. M., Chen, D., and Shen, B. (2005). Novel function of the flap endonuclease 1 complex in processing stalled DNA replication forks. *EMBO Rep.* **6**, 83–89.
134. Reid, R. J., Fiorani, P., Sugawara, M., and Bjornsti, M. A. (1999). CDC45 and DPB11 are required for processive DNA replication and resistance to DNA topoisomerase I-mediated DNA damage. *Proc. Natl. Acad. Sci. USA* **96**, 11440–11445.
135. Fiorani, P., Reid, R. J., Schepis, A., Jacquiau, H. R., Guo, H., Thimmaiah, P., Benedetti, P., and Bjornsti, M. A. (2004). The deubiquitinating enzyme Doa4p protects cells from DNA topoisomerase I poisons. *J. Biol. Chem.* **279**, 21271–21281.
136. Jacquiau, H. R., Waardenburg, R. C., Reid, R. J., Woo, M. H., Guo, H., Johnson, E. S., and Bjornsti, M. A. (2005). Defects in SUMO (small ubiquitin-related modifier) conjugation and deconjugation alter cell sensitivity to DNA topoisomerase I-induced DNA damage. *J. Biol. Chem.* **280**, 23566–23575.
137. Dendouga, N., Gao, H., Moechars, D., Janicot, M., Vialard, J., and McGowan, C. H. (2005). Disruption of murine Mus81 increases genomic instability and DNA damage sensitivity but does not promote tumorigenesis. *Mol. Cell. Biol.* **25**, 7569–7579.
138. Nelson, W. G., and Kastan, M. B. (1994). DNA strand breaks: The DNA template alterations that trigger p53-dependent DNA damage response pathways. *Mol. Cell. Biol.* **14**, 1815–1823.
139. Kohn, K. W., and Pommier, Y. (2005). Molecular interaction map of the p53 and Mdm2 logic elements, which control the Off-On switch of p53 in response to DNA damage. *Biochem. Biophys. Res. Commun.* **331**, 816–827.
140. Chen, Y., and Sanchez, Y. (2004). Chk1 in the DNA damage response: Conserved roles from yeasts to mammals. *DNA Repair (Amst.)* **3**, 1025–1032.
141. Matsuoka, S., Huang, M., and Elledge, S. J. (1998). Linkage of ATM to cell cycle regulation by the Chk2 protein kinase. *Science* **282**, 893–897.
142. Ljungman, M. (2005). Activation of DNA damage signaling. *Mutat. Res.* **577**, 203–216.

143. Liu, Q., Guntuku, S., Cui, X. S., Matsuoka, S., Cortez, D., Tamai, K., Luo, G., Carttini-Rivera, S., DeMayo, F., Bradley, A., Donehower, L. A., and Elledge, S. J. (2000). Chk1 is an essential kinase that is regulated by Atr and required for the G(2)/M DNA damage checkpoint. *Genes Dev.* **14,** 1448–1459.
144. Takai, H., Tominaga, K., Motoyama, N., Minamishima, Y. A., Nagahama, H., Tsukiyama, T., Ikeda, K., Nakayama, K., and Nakanishi, M. (2000). Aberrant cell cycle checkpoint function and early embryonic death in Chk1(−/−) mice. *Genes Dev.* **14,** 1439–1447.
145. Zachos, G., Rainey, M. D., and Gillespie, D. A. (2003). Chk1-deficient tumour cells are viable but exhibit multiple checkpoint and survival defects. *EMBO J.* **22,** 713–723.
146. Garvik, B., Carson, M., and Hartwell, L. (1995). Single-stranded DNA arising at telomeres in cdc13 mutants may constitute a specific signal for the RAD9 checkpoint. *Mol. Cell. Biol.* **15,** 6128–6138.
147. Sanchez, Y., Bachant, J., Wang, H., Hu, F., Liu, D., Tetzlaff, M., and Elledge, S. J. (1999). Control of the DNA damage checkpoint by chk1 and rad53 protein kinases through distinct mechanisms. *Science* **286,** 1166–1171.
148. Chini, C. C., and Chen, J. (2004). Claspin, a regulator of Chk1 in DNA replication stress pathway. *DNA Repair (Amst.)* **3,** 1033–1037.
149. Sorensen, C. S., Syljuasen, R. G., Lukas, J., and Bartek, J. (2004). ATR, Claspin and the Rad9-Rad1-Hus1 complex regulate Chk1 and Cdc25A in the absence of DNA damage. *Cell Cycle* **3,** 941–945.
150. Gottifredi, V., and Prives, C. (2005). The S phase checkpoint: When the crowd meets at the fork. *Semin. Cell. Dev. Biol.* **16,** 355–368.
151. Gatei, M., Sloper, K., Sorensen, C., Syljuasen, R., Falck, J., Hobson, K., Savage, K., Lukas, J., Zhou, B. B., Bartek, J., and Khanna, K. K. (2003). ATM and NBS1 dependent phosphorylation of CHK1 on S317 in response to IR. *J. Biol. Chem.* **278,** 14806–14811.
152. Bartek, J., and Lukas, J. (2003). Chk1 and Chk2 kinases in checkpoint control and cancer. *Cancer Cell* **3,** 421–429.
153. Guerra, B., Issinger, O. G., and Wang, J. Y. (2003). Modulation of human checkpoint kinase Chk1 by the regulatory beta-subunit of protein kinase CK2. *Oncogene* **22,** 4933–4942.
154. Shtivelman, E., Sussman, J., and Stokoe, D. (2002). A role for PI 3-kinase and PKB activity in the G2/M phase of the cell cycle. *Curr. Biol.* **12,** 919–924.
155. Wan, S., Capasso, H., and Walworth, N. C. (1999). The topoisomerase I poison camptothecin generates a Chk1-dependent DNA damage checkpoint signal in fission yeast. *Yeast* **15,** 821–828.
156. Tsao, Y.-P., D'Arpa, P., and Liu, L. F. (1992). The involvement of active DNA synthesis in camptothecin-induced G2 arrest: Altered regulation of p34cdc2/cyclin B. *Cancer Res.* **52,** 1823–1829.
157. Wan, S., and Walworth, N. C. (2001). A novel genetic screen identifies checkpoint-defective alleles of *Schizosaccharomyces pombe* chk1. *Curr. Genet.* **38,** 299–306.
158. Dunaway, S., Liu, H. Y., and Walworth, N. C. (2005). Interaction of 14-3-3 protein with Chk1 affects localization and checkpoint function. *J. Cell. Sci.* **118,** 39–50.
159. Chen, L., Liu, T.-H., and Walworth, N. C. (1999). Association of Chk1 with 14.3.3 proteins is stimulated by DNA damage. *Genes Dev.* **13,** 675–685.
160. Capasso, H., Palermo, C., Wan, S., Rao, H., John, U. P., O'Connell, M. J., and Walworth, N. C. (2002). Phosphorylation activates Chk1 and is required for checkpoint-mediated cell cycle arrest. *J. Cell. Sci.* **115,** 4555–4564.
161. Yin, M. B., Hapke, G., Wu, J., Azrak, R. G., Frank, C., Wrzosek, C., and Rustum, Y. M. (2002). Chk1 signaling pathways that mediated G(2)M checkpoint in relation to the cellular resistance to the novel topoisomerase I poison BNP1350. *Biochem. Biophys. Res. Commun.* **295,** 435–444.

162. Flatten, K., Dai, N. T., Vroman, B. T., Loegering, D., Erlichman, C., Karnitz, L. M., and Kaufmann, S. H. (2005). The role of checkpoint kinase 1 in sensitivity to topoisomerase I poisons. *J. Biol. Chem.* **280**, 14349–14355.
163. Cliby, W. A., Lewis, K. A., Lilly, K. K., and Kaufmann, S. H. (2002). S phase and G2 arrests induced by topoisomerase I poisons are dependent on ATR kinase function. *J. Biol. Chem.* **277**, 1599–1606.
164. Hapke, G., Yin, M. B., Wu, J., Frank, C., and Rustum, Y. M. (2002). Phosphorylation of chk1 at serine-345 affected by topoisomerase I poison SN-38. *Int. J. Oncol.* **21**, 1059–1066.
165. Wang, J. L., Wang, X., Wang, H., Iliakis, G., and Wang, Y. (2002). CHK1-regulated S-phase checkpoint response reduces camptothecin cytotoxicity. *Cell Cycle* **1**, 267–272.
166. Xiao, Z., Chen, Z., Gunasekera, A. H., Sowin, T. J., Rosenberg, S. H., Fesik, S., and Zhang, H. (2003). Chk1 mediates S and G2 arrests through Cdc25A degradation in response to DNA-damaging agents. *J. Biol. Chem.* **278**, 21767–21773.
167. Sorensen, C. S., Hansen, L. T., Dziegielewski, J., Syljuasen, R. G., Lundin, C., Bartek, J., and Helleday, T. (2005). The cell-cycle checkpoint kinase Chk1 is required for mammalian homologous recombination repair. *Nat. Cell. Biol.* **7**, 195–201.
168. Zhang, Y. W., Otterness, D. M., Chiang, G. G., Xie, W., Liu, Y. C., Mercurio, F., and Abraham, R. T. (2005). Genotoxic stress targets human Chk1 for degradation by the ubiquitin-proteasome pathway. *Mol. Cell.* **19**, 607–618.
169. Nomura, M., Nomura, N., and Yamashita, J. (2005). Geldanamycin-induced degradation of Chk1 is mediated by proteasome. *Biochem. Biophys. Res. Commun.* **335**, 900–905.
170. Xiao, Z., Xue, J., Sowin, T. J., Rosenberg, S. H., and Zhang, H. (2005). A novel mechanism of checkpoint abrogation conferred by Chk1 downregulation. *Oncogene* **24**, 1403–1411.
171. Levesque, A. A., Kohn, E. A., Bresnick, E., and Eastman, A. (2005). Distinct roles for p53 transactivation and repression in preventing UCN-01-mediated abrogation of DNA damage-induced arrest at S and G2 cell cycle checkpoints. *Oncogene* **24**, 3786–3796.
172. Tse, A. N., and Schwartz, G. K. (2004). Potentiation of cytotoxicity of topoisomerase i poison by concurrent and sequential treatment with the checkpoint inhibitor UCN-01 involves disparate mechanisms resulting in either p53-independent clonogenic suppression or p53-dependent mitotic catastrophe. *Cancer Res.* **64**, 6635–6644.
173. Pommier, Y., Sordet, O., Rao, A., Zhang, H., and Kohn, K. W. (2005). Targeting Chk2 kinase: Molecular interaction maps and therapeutic rationale. *Curr. Pharm. Des.* **11**, 2855–2872.
174. Hirao, A., Kong, Y. Y., Matsuoka, S., Wakeham, A., Ruland, J., Yoshida, H., Liu, D., Elledge, S. J., and Mak, T. W. (2000). DNA damage-induced activation of p53 by the checkpoint kinase Chk2. *Science* **287**, 1824–1827.
175. Takai, H., Naka, K., Okada, Y., Watanabe, M., Harada, N., Saito, S., Anderson, C. W., Appella, E., Nakanishi, M., Suzuki, H., Nagashima, K., Sawa, H. *et al.* (2002). Chk2-deficient mice exhibit radioresistance and defective p53-mediated transcription. *EMBO J.* **21**, 5195–5205.
176. Ahn, J., Urist, M., and Prives, C. (2004). The Chk2 protein kinase. *DNA Repair (Amst.)* **3**, 1039–1047.
177. Ahn, J.-Y., Schwartz, J. K., Piwnica-Worms, H., and Canman, C. E. (2000). Threonine 68 phosphorylation by ataxia telangiectasia mutated is required for efficient activation of Chk2 in response to ionizing radiation. *Cancer Res.* **60**, 5934–5936.
178. Helt, C. E., Cliby, W. A., Keng, P. C., Bambara, R. A., and O'Reilly, M. A. (2005). Ataxia telangiectasia mutated (ATM) and ATM and Rad3-related protein exhibit selective target specificities in response to different forms of DNA damage. *J. Biol. Chem.* **280**, 1186–1192.
179. Li, J., and Stern, D. F. (2005). Regulation of CHK2 by DNA-dependent protein kinase. *J. Biol. Chem.* **280**, 12041–12050.

180. Bahassi, El M., Conn, C. W., Myer, D. L., Hennigan, R. F., McGowan, C. H., Sanchez, Y., and Stambrook, P. J. (2002). Mammalian Polo-like kinase 3 (Plk3) is a multifunctional protein involved in stress response pathways. *Oncogene* **21**, 6633–6640.
181. Myer, D. L., Bahassi el, M., and Stambrook, P. J. (2005). The Plk3-Cdc25 circuit. *Oncogene* **24**, 299–305.
182. Tsvetkov, L., and Stern, D. F. (2005). Phosphorylation of Plk1 at S137 and T210 is inhibited in response to DNA damage. *Cell Cycle* **4**, 166–171.
183. Tsvetkov, L., Xu, X., Li, J., and Stern, D. F. (2003). Polo-like kinase 1 and Chk2 interact and co-localize to centrosomes and the midbody. *J. Biol. Chem.* **278**, 8468–8475.
184. Tsvetkov, L. M., Tsekova, R. T., Xu, X., and Stern, D. F. (2005). The Plk1 Polo box domain mediates a cell cycle and DNA damage regulated interaction with Chk2. *Cell Cycle* **4**, 609–617.
185. Wei, J. H., Chou, Y. F., Ou, Y. H., Yeh, Y. H., Tyan, S. W., Sun, T. P., Shen, C. Y., and Shieh, S. Y. (2005). TTK/hMps1 participates in the regulation of DNA damage checkpoint response by phosphorylating CHK2 on threonine 68. *J. Biol. Chem.* **280**, 7748–7757.
186. Ahn, J. Y., Li, X., Davis, H. L., and Canman, C. E. (2002). Phosphorylation of threonine 68 promotes oligomerization and autophosphorylation of the Chk2 protein kinase via the forkhead-associated domain. *J. Biol. Chem.* **277**, 19389–19395.
187. Schwarz, J. K., Lovly, C. M., and Piwnica-Worms, H. (2003). Regulation of the Chk2 protein kinase by oligomerization-mediated cis- and trans-phosphorylation. *Mol. Cancer Res.* **1**, 598–609.
188. Yu, Q., Rose, J. H., Zhang, H., and Pommier, Y. (2001). Inhibition of Chk2 activity and radiation-induced p53 elevation by the cell cycle checkpoint abrogator 7-hydroxystaurosporine (UCN-01). *Proc. Am. Assoc. Cancer Res.* **42**, 800.
189. Keramaris, E., Hirao, A., Slack, R. S., Mak, T. W., and Park, D. S. (2003). Ataxia telangiectasia-mutated protein can regulate p53 and neuronal death independent of Chk2 in response to DNA damage. *J. Biol. Chem.* **278**, 37782–37789.
190. Chaturvedi, P., Eng, W. K., Zhu, Y., Mattern, M. R., Mishra, R., Hurle, M. R., Zhang, X., Annan, R. S., Lu, Q., Faucette, L. F., Scott, G. F., Li, X. *et al*. (1999). Mammalian Chk2 is a downstream effector of the ATM-dependent DNA damage checkpoint pathway. *Oncogene* **18**, 4047–4054.
191. Ward, I. M., and Chen, J. (2001). Histone H2AX is phosphorylated in an ATR-dependent manner in response to replicational stress. *J. Biol. Chem.* **276**, 47759–47762.
192. Agner, J., Falck, J., Lukas, J., and Bartek, J. (2005). Differential impact of diverse anticancer chemotherapeutics on the Cdc25A-degradation checkpoint pathway. *Exp. Cell. Res.* **302**, 162–169.
193. Wang, X., Khadpe, J., Hu, B., Iliakis, G., and Wang, Y. (2003). An overactivated ATR/CHK1 pathway is responsible for the prolonged G2 accumulation in irradiated AT cells. *J. Biol. Chem.* **278**, 30869–30874.
194. Goulaouic, H., Roulon, T., Flamand, O., Grondard, L., Lavelle, F., and J.-F., Riou (1999). Purification and characterization of human DNA topoisomerase IIIα. *Nucleic Acids Res.* **27**, 2443–2450.
195. Rosenthal, D. S., Simbulan-Rosenthal, C. M., Liu, W. F., Velena, A., Anderson, D., Benton, B., Wang, Z. Q., Smith, W., Ray, R., and Smulson, M. E. (2001). PARP determines the mode of cell death in skin fibroblasts, but not keratinocytes, exposed to sulfur mustard. *J. Invest. Dermatol.* **117**, 1566–1573.
196. Pommier, Y., Pourquier, P., Fan, Y., and Strumberg, D. (1998). Mechanism of action of eukaryotic DNA topoisomerase I and drugs targeted to the enzyme. *Biochim. Biophys. Acta* **1400**, 83–105.

197. Pourquier, P., Takebayashi, Y., Urasaki, Y., Gioffre, C., Kohlhagen, G., and Pommier, Y. (2000). Induction of topoisomerase I cleavage complexes by 1-β-D-arabinofuranosylcytosine (Ara-C) in vitro and in ara-C-treated cells. *Proc. Natl. Acad. Sci. USA* **97**, 1885–1890.
198. Chrencik, J. E., Burgin, A. B., Pommier, Y., Stewart, L., and Redinbo, M. R. (2003). Structural impact of the leukemia drug Ara-C on the covalent human topoisomerase I DNA complex. *J. Biol. Chem.* **278**, 12461–12466.
199. Pourquier, P., Gioffre, C., Kohlhagen, G., Urasaki, Y., Goldwasser, F., Hertel, L. W., Yu, S., Pon, R. T., Gmeiner, W. H., and Pommier, Y. (2002). Gemcitabine (2′,2′-difluoro-2′-deoxycytidine), an antimetabolite that poisons topoisomerase I. *Clin. Cancer Res.* **8**, 2499–2504.
200. Leteurtre, F., Kohlhagen, G., Fesen, M. R., Tanizawa, A., Kohn, K. W., and Pommier, Y. (1994). Effects of DNA methylation on topoisomerase I and II cleavage activities. *J. Biol. Chem.* **269**, 7893–7900.
201. Antony, S., Arimondo, P. B., Sun, J. S., and Pommier, Y. (2004). Position- and orientation-specific enhancement of topoisomerase I cleavage complexes by triplex DNA structures. *Nucleic Acids Res.* **32**, 5163–5173.
202. Lanza, A., Tornatelli, S., Rodolfo, C., Scanavini, M. C., and Pedrini, A. M. (1996). Human DNA topoisomerase I-mediated cleavages stimulated by ultraviolet light-induced DNA damage. *J. Biol. Chem.* **271**, 6978–6986.
203. Subramanian, D., Rosenstein, B. S., and Muller, M. T. (1998). Ultraviolet-induced DNA damage stimulates topoisomerase I-DNA complex formation in vivo: Possible relationship with DNA repair. *Cancer Res.* **58**, 976–984.
204. Pourquier, P., Waltman, J. L., Urasaki, Y., Loktionova, N. A., Pegg, A. E., Nitiss, J. L., and Pommier, Y. (2001). Topoisomerase I-mediated cytotoxicity of N-methyl-N′-nitro-N-nitrosoguanidine: Trapping of topoisomerase I by the O^6-methylguanine. *Cancer Res.* **61**, 53–58.
205. Pommier, Y., Laco, G. S., Kohlhagen, G., Sayer, J. M., Kroth, H., and Jerina, D. M. (2000). Position-specific trapping of topoisomerase I-DNA cleavage complexes by intercalated benzo[a]-pyrene diol epoxide adducts at the 6-amino group of adenine. *Proc. Natl. Acad. Sci. USA* **97**, 10739–10744.
206. Pommier, Y., Kohlhagen, G., Pourquier, P., Sayer, J. M., Kroth, H., and Jerina, D. M. (2000). Benzo[a]pyrene epoxide adducts in DNA are potent inhibitors of a normal topoisomerase I cleavage site and powerful inducers of other topoisomerase I cleavages. *Proc. Natl. Acad. Sci. USA* **97**, 2040–2045.
207. Pommier, Y., Kohlhagen, G., Laco, G. S., Kroth, H., Sayer, J. M., and Jerina, D. M. (2002). Different effects on human topoisomerase I by minor groove and intercalated deoxyguanosine adducts derived from two polycyclic aromatic hydrocarbon diol epoxides at or near a normal cleavage site. *J. Biol. Chem.* **277**, 13666–13672.
208. Pourquier, P., Bjornsti, M.-A., and Pommier, Y. (1998). Induction of topoisomerase I cleavage complexes by the vinyl chloride adduct, 1, N6-ethenoadenine. *J. Biol. Chem.* **273**, 27245–27249.
209. Antony, S., Theruvathu, J. A., Brooks, P. J., Lesher, D. T., Redinbo, M., and Pommier, Y. (2004). Enhancement of camptothecin-induced topoisomerase I cleavage complexes by the acetaldehyde adduct N2-ethyl-2′-deoxyguanosine. *Nucleic Acids Res.* **32**, 5685–5692.
210. Moser, B. A., Brondello, J. M., Baber-Furnari, B., and Russell, P. (2000). Mechanism of caffeine-induced checkpoint override in fission yeast. *Mol. Cell. Biol.* **20**, 4288–4294.
211. Smith, P. J., Makinson, T. A., and Watson, J. V. (1989). Enhanced sensitivity to camptothecin in ataxia telangiectasia cells and its relationship with the expression of DNA topoisomerase I. *Int. J. Radiat. Biol.* **55**, 217–231.
212. Johnson, M. A., and Jones, N. J. (1999). The isolation and genetic analysis of V79-derived etoposide sensitive Chinese hamster cell mutants: Two new complementation groups of etoposide sensitive mutants. *Mutation Res.* **435**, 271–282.

213. Jones, N. J., Ellard, S., Waters, R., and Parry, E. M. (1993). Cellular and chromosomal hypersensitivity to DNA crosslinking agents and topoisomerase inhibitors in the radiosensitive Chinese hamster irs mutants: Phenotypic similarities to ataxia telangiectasia and Fanconi's anaemia cells. *Carcinogenesis* **14**, 2487–2494.
214. Johnson, M. A., Bryant, P. E., and Jones, N. J. (2000). Isolation of camptothecin-sensitive Chinese hamster cell mutants: Phenotypic heterogeneity within the ataxia telangiectasia-like XRCC8 (irs2) complementation group. *Mutagenesis* **15**, 367–374.
215. Walton, M. I., Whysong, D., O'Connor, P. M., Hockenbery, D., Korsmeyer, S. J., and Kohn, K. W. (1993). Constitutive expression of human Bcl-2 modulates nitrogen mustard and camptothecin induced apoptosis. *Cancer Res.* **53**, 1853–1861.
216. Fedier, A., Steiner, R. A., Schwarz, V. A., Lenherr, L., Haller, U., and Fink, D. (2003). The effect of loss of Brca1 on the sensitivity to anticancer agents in p53-deficient cells. *Int. J. Oncol.* **22**, 1169–1173.
217. Rahden-Staron, I. I., Szumilo, M., Grosicka, E., Zwet, M., and Zdzienicka, M. Z. (2003). Defective Brca2 influences topoisomerase I activity in mammalian cells. *Acta Biochim. Pol.* **50**, 139–144.
218. Yu, Q., Rose, J. H., Zhang, H., and Pommier, Y. (2001). Antisense inhibition of Chk2/hCds1 expression attenuates DNA damage-induced S and G2 checkpoints and enhances apoptotic activity in HEK-293 cells. *FEBS Lett.* **505**, 7–12.
219. Hinz, J. M., Helleday, T., and Meuth, M. (2003). Reduced apoptotic response to camptothecin in CHO cells deficient in XRCC3. *Carcinogenesis* **24**, 249–253.
220. Culmsee, C., Bondada, S., and Mattson, M. P. (2001). Hippocampal neurons of mice deficient in DNA-dependent protein kinase exhibit increased vulnerability to DNA damage, oxidative stress and excitotoxicity. *Brain Res. Mol. Brain Res.* **87**, 257–262.
221. Zwet, M., Overkamp, W. J., Friedl, A. A., Klein, B., Verhaegh, G. W., Jaspers, N. G., Midro, A. T., Eckardt-Schupp, F., Lohman, P. H., and Zdzienicka, M. Z. (1999). Immortalization and characterization of Nijmegen Breakage syndrome fibroblasts. *Mutat. Res.* **434**, 17–27.
222. Chatterjee, S., Cheng, M. F., and Berger, N. A. (1990). Hypersensitivity to clinically useful alkylating agents and radiations in poly(ADP-ribose) polymerase-deficient cell lines. *Cancer Commun.* **2**, 401–407.
223. Godthelp, B. C., Wiegant, W. W., van Duijn-Goedhart, A., Scharer, O. D., Buul, P. P., Kanaar, R., and Zdzienicka, M. Z. (2002). Mammalian Rad51C contributes to DNA cross-link resistance, sister chromatid cohesion and genomic stability. *Nucleic Acids Res.* **30**, 2172–2182.
224. Barthelmes, H. U., Habermeyer, M., Christensen, M. O., Mielke, C., Interthal, H., Pouliot, J. J., Boege, F., and Marko, D. (2004). TDP1 overexpression in human cells counteracts DNA damage mediated by topoisomerases I and II. *J. Biol. Chem.* **279**, 55618–55625.
225. Gupta, M., Fan, S., Zhan, Q., Kohn, K. W., O'Connor, P. M., and Pommier, Y. (1997). Inactivation of p53 increases the cytotoxicity of camptothecin in human colon HCT116 and breast MCF-7 cancer cells. *Clin. Cancer Res.* **3**, 1653–1660.
226. Han, Z., Wei, W., Dunaway, S., Darnowski, J. W., Calabresi, P., Sedivy, J., Hendrickson, E. A., Balan, K. V., Pantazis, P., and Wyche, J. H. (2002). Role of p21 in apoptosis and senescence of human colon cancer cells treated with camptothecin. *J. Biol. Chem.* **277**, 17154–17160.
227. Lebel, M., and Leder, P. (1998). A deletion within the murine Werner syndrome helicase induces sensitivity to inhibitors of topoisomerase and loss of proliferative capacity. *Proc. Natl. Acad. Sci. USA* **95**, 13097–13102.
228. Poot, M., Gollahon, K. A., and Rabinovitch, P. S. (1999). Werner syndrome lymphoblastoid cells are sensitive to camptothecin-induced apoptosis in S-phase. *Hum. Genet.* **104**, 10–14.
229. Pichierri, P., Franchitto, A., Mosesso, P., and Palitti, F. (2000). Werner's syndrome cell lines are hypersensitive to camptothecin-induced chromosomal damage. *Mutat. Res.* **456**, 45–57.

230. Caldecott, K., and Jeggo, P. (1991). Cross-sensitivity of gamma-ray-sensitive hamster mutants to cross-linking agents. *Mutat Res.* **255**, 111–121.
231. Barrows, L. R., Holden, J. A., Anderson, M., and D'Arpa, P. (1998). The CHO XRCC1 mutant, EM9, deficient in DNA ligase III activity, exhibits hypersensitivity to camptothecin independent of DNA replication. *Mutat. Res.* **408**, 103–110.
232. Park, S. Y., Lam, W., and Cheng, Y. C. (2002). X-ray repair cross-complementing gene I protein plays an important role in camptothecin resistance. *Cancer Res.* **62**, 459–465.
233. Gueven, N., Becherel, O. J., Kijas, A. W., Chen, P., Howe, O., Rudolph, J. H., Gatti, R., Date, H., Onodera, O., Taucher-Scholz, G., and Lavin, M. F. (2004). Apraxatin, a novel protein that protects against genotoxic stress. *Hum. Mol. Genet.* **13**, 1081–1093.
234. Thacker, J., and Ganesh, A. N. (1990). DNA-break repair, radioresistance of DNA synthesis, and camptothecin sensitivity in the radiation-sensitive irs mutants: Comparisons to ataxia-telangiectasia cells. *Mutat. Res.* **235**, 49–58.
235. Nitiss, J., and Wang, J. C. (1988). DNA topoisomerase-targeting antitumor drugs can be studied in yeast. *Proc. Natl. Acad. Sci. USA* **85**, 7501–7505.
236. Simon, J. A., Szankasi, P., Nguyen, D. K., Ludlow, C., Dunstan, H. M., Roberts, C. J., Jensen, E. L., Hartwell, L. H., and Friend, S. H. (2000). Differential toxicities of anticancer agents among DNA repair and checkpoint mutants of *Saccharomyces cerevisiae*. *Cancer Res.* **60**, 328–333.
237. Redon, C., Pilch, D., Rogakou, E., Orr, A. H., Lowndes, N. F., and Bonner, W. M. (2003). Yeast histone 2A serine 129 is essential for the efficient repair of checkpoint-blind DNA damage. *EMBO Rep.* **4**, 1–7.
238. Hryciw, T., Tang, M., Fontanie, T., and Xiao, W. (2002). MMS1 protects against replication-dependent DNA damage in *Saccharomyces cerevisiae*. *Mol. Genet. Genomics* **266**, 848–857.
239. Zhang, H., and Siede, W. (2003). Validation of a novel assay for checkpoint responses: Characterization of camptothecin derivatives in *Saccharomyces cerevisiae*. *Mutat. Res.* **527**, 37–48.
240. Walowsky, C., Fitzhugh, D. J., Castaño, I. B., Ju, J. Y., Levin, N. A., and Christman, M. F. (1999). The topoisomerase-related function gene TRF4 affects cellular sensitivity to the antitumor agent camptothecin. *J. Biol. Chem.* **274**, 7302–7308.
241. Scheller, J., Schurer, A., Rudolph, C., Hettwer, S., and Kramer, W. (2000). MPH1, a yeast gene encoding a DEAH protein, plays a role in protection of the genome from spontaneous and chemically induced damage. *Genetics* **155**, 1069–1081.
242. Redon, C., Pilch, D. R., and Bonner, W. M. (2006). Genetic analysis of *Saccharomyces cerevisiae* H2A serine 129 mutant suggests a functional relationship between H2A and the sister chromatid cohesion partners Csm3-Tof1 for the repair of topoisomerase I-induced DNA damage. *Genetics* **172**, 67–76.
243. Bird, A. W., Yu, D. Y., Pray-Grant, M. G., Qiu, Q., Harmon, K. E., Megee, P. C., Grant, P. A., Smith, M. M., and Christman, M. F. (2002). Acetylation of histone H4 by Esa1 is required for DNA double-strand break repair. *Nature* **419**, 411–415.
244. Choy, J. S., and Kron, S. J. (2002). NuA4 subunit Yng2 function in intra-S-phase DNA damage response. *Mol. Cell. Biol.* **22**, 8215–8225.
245. Lewis, L. K., Karthikeyan, G., Cassiano, J., and Resnick, M. A. (2005). Reduction of nucleosome assembly during new DNA synthesis impairs both major pathways of double-strand break repair. *Nucleic Acids Res.* **33**, 4928–4939.
246. Fiorani, P., and Bjornsti, M. A. (2000). Mechanisms of DNA topoisomerase I-induced cell killing in the yeast *Saccharomyces cerevisiae*. *Ann. N. Y. Acad. Sci.* **922**, 65–75.
247. Doe, C. L., and Whitby, M. C. (2004). The involvement of Srs2 in post-replication repair and homologous recombination in fission yeast. *Nucleic Acids Res.* **32**, 1480–1491.
248. Doe, C. L., Osman, F., Dixon, J., and Whitby, M. C. (2004). DNA repair by a Rad22-Mus81-dependent pathway that is independent of Rhp51. *Nucleic Acids Res.* **32**, 5570–5581.

249. Collura, A., Blaisonneau, J., Baldacci, G., and Francesconi, S. (2005). The fission yeast Crb2/Chk1 pathway coordinates the DNA damage and spindle checkpoint in response to replication stress induced by topoisomerase I inhibitor. *Mol. Cell. Biol.* **25,** 7889–7899.
250. Francesconi, S., Smeets, M., Grenon, M., Tillit, J., Blaisonneau, J., and Baldacci, G. (2002). Fission yeast chk1 mutants show distinct responses to different types of DNA damaging treatments. *Genes Cells* **7,** 663–673.
251. Dalgaard, J. Z., and Klar, A. J. (2000). swi1 and swi3 perform imprinting, pausing, and termination of DNA replication in S. pombe. *Cell* **102,** 745–751.
252. Nitiss, J. L., Nitiss, K. C., Rose, A., and Waltman, J. L. (2001). Overexpression of type I topoisomerases sensitizes yeast cells to DNA damage. *J. Biol. Chem.* **276,** 26708–26714.
253. Kim, J., Robertson, K., Mylonas, K. J. L., Gray, F. C., Charapitsa, I., and MacNeill, S. A. (2005). Contrasting effects of Elg1-RFC and Ctf18-RFC inactivation in the absence of fully functional RFC in fission yeast. *Nucleic Acids Res.* **33,** 4078–4089.
254. Westmoreland, T. J., Marks, J. R., Olson, J. A., Jr., Thompson, E. M., Resnick, M. A., and Bennet, C. B. (2004). Cell cycle progression in G1 and S phases is CCR4 dependent following ionizing radiation or replication stress in Saccharomyces cerevisiae. *Eukaryot. Cell* **3,** 430–446.
255. Malik, M., and Nitiss, J. L. (2004). DNA repair functions that control sensitivity to topoisomerase-targeting drugs. *Eukaryot. Cell* **3,** 82–90.
256. Khasanov, F. K., Salakhova, A. F., Chepurnaja, O. V., Korolev, V. G., and Bashkirov, V. I. (2004). Identification and characterization of the rlp1+, the novel Rad51 paralog in the fission yeast Schizosaccharomyces pombe. *DNA Repair (Amst)* **3,** 1363–1374.
257. Lambert, S., Mason, S. J., Barber, L. J., Hartley, J. A., Pearce, J. A., Carr, A. M., and McHugh, P. J. (2003). Schizosaccharomyces pombe checkpoint response to DNA interstrand cross-links. *Mol. Cell. Biol.* **23,** 4728–4737.
258. Yoshida, S. H., Al-Amodi, H., Nakamura, T., McInerny, C. J., and Shimoda, C. (2003). The Schizosaccharomyces pombe cdt2(+) gene, a target of G1-S phase-specific transcription factor complex DSC1, is required for mitotic and premeiotic DNA replication. *Genetics* **164,** 881–893.
259. Parsons, A. B., Brost, R. L., Ding, H., Li, Z., Zhang, C., Sheikh, B., Brown, G. W., Kane, P. M., Hughes, T. R., and Boone, C. (2004). Integration of chemical-genetic and genetic interaction data links bioactive compounds to cellular target pathways. *Nat. Biotechnol.* **22,** 62–69.
260. Rouse, J. (2004). Esc4p, a new target of Mec1p (ATR), promotes resumption of DNA synthesis after DNA damage. *EMBO J.* **23,** 1188–1197.
261. Masumoto, H., Hawke, D., Kobayashi, R., and Verreault, A. (2005). A role for cell-cycle-regulated histone H3 lysine 56 acetylation in the DNA damage response. *Nature* **436,** 294–298.
262. Liu, C., Poitelea, M., Watson, A., Yoshida, S. H., Shimoda, C., Holmberg, C., Nielsen, O., and Carr, A. M. (2005). Transactivation of Schizosaccharomyces pombe cdt2+ stimulates a Pcu4-Ddb1-CSN ubiquitin ligase. *EMBO J.* **24,** 3940–3951.
263. Nitiss, K. C., Malik, M., He, X., White, S. W., and Nitiss, J. L. (2006). Tyrosyl-DNA phosphodiesterase (Tdp1) participates in the repair of Top2-mediated DNA damage. *Proc. Natl. Acad. Sci. USA* **103,** 8953–8958.
264. Pommier, Y., Weinstein, J. N., Aladjem, M. I., and Kohn, M. (2006). Chk2 molecular interaction map and rationale for Chk2 inhibitors. *Clin. Cancer Res.* **12,** 2657–2661.
265. Torres, J. Z., Schnakenberg, S. L., and Zakian, V. A. (2004). Saccharomyces cerevisiae Rrm3p DNA helicase promotes genome integrity by preventing replication fork stalling: Viability of rrm3 cells requires the intra-S-phase checkpoint and fork restart activities. *Mol. Cell. Biol.* **24,** 3198–3212.
266. Sanders, S. L., Portoso, M., Mata, J., Bahler, J., Allshire, R. C., and Kouzarides, T. (2004). Methylation of histone H4 lysine 20 controls recruitment of Crb2 to sites of DNA damage. *Cell* **119,** 603–614.

Regulation of L-Histidine Decarboxylase and Its Role in Carcinogenesis

WANDONG AI,[*] SHIGEO TAKAISHI,[*] TIMOTHY C. WANG,[*] AND JOHN V. FLEMING[†]

[*]Division of Digestive and Liver Diseases, Department of Medicine, Columbia University, Irving Cancer Research Center, New York, New York 10032

[†]Department of Biochemistry and School of Pharmacy, University College Cork, Lee Maltings, Prospect Row, Cork, Ireland

I. Introduction 232
II. Transcriptional Regulation of HDC 233
   A. Epigenetic Regulation of *HDC* Gene Expression 234
   B. Signal Transduction Pathways Involved in *HDC* Gene Expression 235
   C. Transcription Factors and *cis*-Elements Involved in *HDC* Gene Expression 238
III. Posttranscriptional Regulation of HDC 239
   A. HDC Is Translated as a $\sim$74-kDa Protein That Undergoes Posttranslational Processing 240
   B. HDC Processing 242
   C. Regulation of HDC Processing 244
   D. Computer Modeling of the HDC Active Site and Insights into the Catalytic Mechanism 245
   E. Flexible Loop Domain and Structural Compaction of the Dimer 246
   F. The Next 15 Years 249
IV. Potential Role of HDC in Cancer Development 251
   A. HDC Expression in Different Cancers 251
   B. Link of Angiogenesis in HDC/Histamine Mediated Cancer Development 254
   C. Link of Inflammation Induced by Histamine/HDC for Cancer Development 255
   D. Role of Histamine/HDC in Development of Gastric Cancer 256
V. Concluding Remarks 258
   References 260

Histamine is a bioamine whose roles in allergy, inflammation, neurotransmission, and gastric acid secretion have been well described. Animal studies using mice deficient in histamine production have confirmed these important functions and identified a number of new ones. This includes regulation of angiogenesis, bone morphogenesis, and neutrophil recruitment. A single enzyme, L-histidine decarboxylase (EC 4.1.1.22, HDC), is responsible for histamine biosynthesis in mammals. It is expressed in the liver of the developing fetus and in the stomach, brain, thymus, spleen, and bones of adults. Successive studies have identified regulation of gene transcription and posttranslational processing as the two principal levels at which control of HDC expression and function can be exercised. At the transcription level, *HDC* gene expression is regulated by several stimuli, including gastrin (a stomach peptide hormone), LPS, phorbol 12-myristate-13-acetate (PMA), oxidative stress, and *Helicobacter pylori* infection. At the posttranslational level on the other hand, processing into multiple truncated isoforms provides a cellular mechanism for controlling activity of the actual enzyme. While these aspects of regulation are important for normal physiological function, histamine is increasingly being recognized as a contributory factor in the development of some cancer types. Given these advances, and our improved understanding of histamine biosynthesis and function, in this chapter we will review important findings relating to the regulation of HDC at the transcriptional and posttranslational levels, and discuss in detail the role of HDC/histamine in carcinogenesis.

I. Introduction

Histamine is one of the most important bioamines being traditionally recognized as a regulator of smooth muscle contraction, immune response, vascular permeability, neurotransmission, and the stimulation of gastric acid secretion. Studies have also described a number of novel functions, including regulatory roles in proliferation and angiogenesis. It is produced by a variety of specialized cell types for either immediate use or for storage in granules for delayed secretion. Four G-protein–coupled histamine receptors have thus far been described to mediate its actions, and allow for the regulation of many different functions within receptor expressing target cells (1, 2). For example, well-established H1 and H2 receptors mediate allergic response and gastric acid secretion, respectively, and H3 receptors are responsible for the inhibition of neuronal histamine release. The functional role of the H4 receptor is a little less clear, although expression has been reported in human bone marrow and colon and there is some evidence for involvement in immune response.

Histamine is synthesized by a single enzyme called L-histidine decarboxylase (HDC), which decarboxylates the amino acid L-histidine in a catalytic reaction that additionally requires pridoxal phosphate (PLP) as an essential cofactor. HDC is present in mast cells, basophils, macrophages (3, 4), neutrophils (5), and lymphocytes (6), as well as exhibiting a particularly high level of expression in the gastric enterochromaffinlike (ECL) cells of adult mammals (7). At the transcriptional level, expression of HDC is regulated by many different stimuli, including gastrin (8), phorbol ester phorbol 12-myristate-13-acetate (PMA) (8–11), oxidative stress (12), thrombopoietin (13), and *Helicobacter pylori* infection (14, 15). The molecular basis for this regulation has not been well characterized, however, with only a small number of *cis*- and *trans*-regulatory factors identified as mediators. Certainly at a higher chromatin level we are only now beginning to explore the manner in which *HDC* gene expression can be regulated epigenetically.

Following transcription, mammalian HDC mRNAs are translated to produce proteins with a molecular weight of about 74 kDa (16–18). While this finding is inconsistent with early biochemical studies, which most frequently described the mammalian HDC subunit as 53–55 kDa in size, it is now well established that the primary translation product undergoes significant post-translational processing (19–21). This allows for the production of multiple HDC isoforms that exhibit markedly different patterns of regulation being differentially localized and differentially degraded within the cell (22). Enzymatic activity also appears to depend on processing, although findings with a mutant enzyme deficient in production of the ∼54-kDa isoform demonstrated that other isoforms contribute toward histamine production (23).

Histamine has for sometime been implicated in the regulation of proliferation in both normal and neoplastic tissues. For example, the rate of proliferation in wound repair, embrogenesis, hematopoiesis, and malignant growth has been found to correlate closely with levels of histamine production (24, 25). Furthermore, the expressions of HDC and histamine biosynthesis have been detected in a number of different tumors, including human small cell lung carcinomas (26), pancreatic tumors (27), colonic cancer cells (28), and human melanoma cells (29). There may however be some species and cell-type specificity of this effect. For example, in a mastomys model, the rapid induction of ECLoma was induced by using the H2 receptor antagonist loxtidine to block acid secretion (30), suggesting a negative role for histamine in tumor development. In contrast, our studies where tumor development was achieved by long-term *Helicobacter felis* infection or gastrinemia in a mouse model (INS-GAS mice), application of loxtidine ameliorated the progression of the gastric hyperplasia/dysplasia (31).

II. Transcriptional Regulation of HDC

The *HDC* gene, with a size of around 24 kb, contains 12 exons that are located on chromosome 15 in humans and chromosome 2 in mouse (8, 32, 33). The first human exon contains 5′-untranslated region (UTR) sequences, the translation start codon, and the coding region for the first 10 amino acid residues of HDC. In contrast almost half of the mRNA (encoding the 248 carboxyl-terminus residues) is derived from the final exon. Analysis of the human HDC promoter shows no obvious canonical TATA or CCAAT boxes. Instead, a TATA-like sequence (TAAATAAA), a GC box, and four CACC boxes can be found upstream from a transcriptional start site that was identified by primer extension analysis and S1 nuclease mapping (8, 33). Not all of these upstream sites may be of importance, however, with experimental studies focusing mainly on the GC box.

Transcriptional regulation of HDC appears to occur via a core promoter region located around this upstream GC box. While it is known that this element binds Sp1 to induce promoter activation (34, 35), there are now studies showing that binding of other factors, such as Kruppel-like factor 4 (KLF4) (34) and Yin Yang 1 (YY1) (36), can act to repress transcription. Additional enhancer or repressor elements located outside of the core promoter may also be of importance. This includes downstream noncoding sequences that will be discussed in detail later (35, 37, 38).

A. Epigenetic Regulation of *HDC* Gene Expression

Mammalian genomic DNA is wrapped with histone proteins to form nucleosomes, the basic repeating unit of chromatin. Nucleosomal DNA is compacted to a higher order chromatin structure by looping and folding of the chromatin fiber. The organization of chromatin was originally thought to be essential for protecting the underlying DNA from potentially harmful nuclear factors. It is now clear that chromatin structure is dynamic and the modifications of histone proteins can alter chromatin configuration and play a direct regulatory role in gene expression. This regulation is called epigenetic regulation since it is not mediated at the DNA sequence level (39). Molecular mechanisms that mediate epigenetic regulation include DNA methylation and chromatin/histone modifications. Covalent modifications to the amino terminal tails of histones, including acetylation, methylation, and phosphorylation, regulate the packing of chromatin into transcriptionally active or transcriptionally inactive states (40). Histone acetyltransferases (HATs) catalyze the transfer of acetyl group to the N-terminal conserved amino acid and are associated with transcriptional activation, whereas histone deacetylases (HDACs) catalyze the opposite reaction and are correlated with transcriptionally inaccessible chromatin. Given the importance of histamine as a physiological regulator it is hardly surprising that there is now evidence for the epigenetic regulation of HDC, as demonstrated by the involvement of DNA methylation and HATs/HDACs.

In hematopoietic cell lineages, *HDC* gene expression has been reported in mast cells and basophils. In an attempt to discover how *HDC* gene expression is regulated in these cells, Kuramasu *et al.* found out that the human HDC-promoter region in HDC-expressing cell lines is selectively unmethylated (*34*). A correlation between HDC expression and hypomethylation was also found in primary mast cells, and methylation of a HDC promoter-luciferase reporter construct resulted in decreased luciferase activity in transient expression analysis. Later on, the same group confirmed the importance of methylation in regulating the activity of the mouse promoter (*32*). *HDC* gene expression is strongly induced in the mouse immature mast cell line P815 after incubation in the peritoneal cavity. The induction of gene expression is correlated with the demethylation of the promoter. Consistently, forced demethylation by 5-azacytidine treatment induced high expression of HDC mRNA in P815. These data suggest that DNA methylation might be one of the mechanisms that regulate *HDC* gene expression, and might be of functional importance in the development of mast cells.

Besides DNA methylation, there is also evidence that HATs and HDACs might regulate *HDC* gene expression. Having demonstrated that KLF4 binds and represses the *HDC* gene promoter, (*35*), we have identified Tip60, a protein that possesses HAT activity, as a KLF4-interacting protein. Transfection experiments suggest that Tip60 inhibits HDC-promoter activity. Since Tip60 has been proposed as a corepressor for STAT3 with HDAC7 (*41*), the function of HDAC7 in regulation of HDC expression has also been tested. Overexpression of HDAC7 inhibits HDC-promoter activity by transient cotransfection assay. Our data suggest that Tip60 functions as a corepressor of Tip60 with HDAC7 to regulate gene expression of *HDC* (DDW 2005, Abstract #395). How Tip60, a protein with HAT activity, inhibits HDC-promoter activity, and the functional importance of a putative Tip60/HDAC7 corepressor complex in regulation of HDC transcription, remains to be fully determined.

Despite these evidences for the epigenetic control of *HDC* gene expression additional experimentation is required to determine the methylase(s) responsible for methylation and to establish whether there is any coordinated function between methylation and HAT/HDAC activity at the promoter. Only after this information is available, it will be possible to fully understand the physiological importance of this epigenetic regulation.

B. Signal Transduction Pathways Involved in *HDC* Gene Expression

The regulation of *HDC* gene expression by several different stimuli has been reported, including growth factors like gastrin and PMA (*8–10, 42*), oxidative stress (*12*), and *H. pylori* infection (*14, 15*). These different stimulations share some common signal transduction pathways. In the following discussion, the signaling pathway by gastrin treatment, in the hematopoietic lineages, and by *H. pylori* infection will be reviewed in more detail.

Gastrin is a stomach peptide hormone that signals through a cell-surface G-protein–coupled CCK-B/gastrin receptor that has high affinity for both gastrin and the cholecystokinin octapeptide (*43, 44*). While gastrin receptors have been detected in both the brain and the pancreas of most mammalian species, the major gastrointestinal target for gastrin is the ECL cell of the gastric corpus. Gastrin stimulation of ECL cells leads to elevated HDC activity and increased secretion of histamine, the major gastric acid secretogogue. The gastrin-stimulated increase in HDC activity has been shown to result in part from an increase in *HDC* gene transcription in the gastric corpus and isolated ECL cells (*45, 46*). The involvement of PKC in this regulation has been implied since early studies demonstrated that application of a PKC activator, called TPA to mouse skin, resulted in a rapid increase in HDC activity (*47*). Subsequent studies have confirmed this pattern of regulation in mastocytoma cells (*11*), gastric cancer cells (*8, 9*), and basophilic cells (*42*). In gastric cancer cells (AGS-B cells where the gastrin receptor is stably transfected), gastrin-mediated HDC-promoter activation was also significantly inhibited by a down-regulation of PKC pathways that resulted from either pretreatment with PMA, or addition of H7, a specific PKC inhibitor. This provided supplementary evidence that PKC pathways transduce the gastrin signal from cell-surface CCK B receptors to the nucleus, resulting in activation of the *HDC* gene promoter (*8*).

Cell stimulation with either gastrin or PMA is known to increase the phosphorylation of extracellular signal-regulated kinases (ERKs) and increase ERK activity. Not surprisingly then overexpression of Erk-1 and Erk-2, or activation of endogenous ERKs using activated MEK-1 (ERK kinase-1), stimulated HDC-promoter activity. This demonstrated that ERK Map kinase pathway signaling is involved in regulation of *HDC* gene expression by both factors. Moreover, while activation of the HDC promoter by gastrin was not influenced by expression of dominant negative Ras (N15 or N17) proteins, gastrin-stimulated Raf-1 kinase activity, and activation of the HDC promoter was blocked by coexpression of a dominant negative Raf-1 construct. Collectively, these data demonstrate that gastrin regulates HDC transcription in a Raf-dependent, Ras-independent fashion through activation of the ERK-related pathway (*9, 10, 48*). Oxidative stress (H_2O_2 treatment in AGS-B cells)-mediated activation of the HDC promoter shares similar signal transduction pathways from Raf-1 to Erk Map kinase. However, it appears that the upstream signaling molecules are different from that of gastrin treatment. The oxidative stress-mediated HDC-promoter activation was blocked by dominant negative Ras (N15 and N17) proteins, and partially by a dominant negative *EGFP* mutant, but was not influenced by PKC pathway blockade (*12*).

In hematopoietic lineage, where histamine plays an important role in allergy and inflammation, HDC expression is observed in mast cells (49, 50), T-lymphocytes (6, 51), basophils (42), and monocytes/macrophages (52, 53). Mast cells, in particular, exhibit high levels of expression, and their prominent localization to surfaces where pathogens are frequently encountered (e.g., skin, airways, and gastrointestinal tract) makes them one of the first cells to interact with environmental allergens and other antigens (54). Mice deficient in HDC exhibit reduced differentiation of mast cells from bone marrow-derived stem cells, and have mast cells with an abnormal phenotype (49, 55).

In other histamine-producing cells it has been reported that ConA treatment potentiated the production and release of histamine from CD4+ or CD8+ T cells that had been purified from the spleens of mast cell-deficient mice. In separate studies it has also been demonstrated that granulocyte/macrophage colony-stimulating factor (GM-CSF) or interleukin (IL)-3 strongly enhanced the ConA-induced histamine formation that resulted from the activation of HDC (51). In a monocyte cell line U937, GM-CSF similarly activated *HDC* gene expression and histamine release (53). It appears that basal but not GM-CSF–inducible transcription of *HDC* gene is regulated by AP-1, since *dKR*, a *trans*-dominant negative mutant of Fos and Jun (56), showed a suppressive effect compared with c-Jun and c-Fos coexpression, but it showed only a slight inhibition of GM-CSF–induced HDC-promoter activity. Another group has demonstrated a mast cell-independent release of histamine. It has been shown that lipopolysaccharide (LPS), a bacterial product, and the proinflammatory cytokine IL-1 induce a marked elevation of HDC mRNA and activity in various tissues and organs. In nonstimulated mice, the histamine in the lung and spleen was contained largely within mast cells. However, the LPS-stimulated increase in HDC activity in a given organ was similar between wild-type and mast cell-deficient mice, and between IL-1$\alpha\beta$/TNFαKO and wild-type mice. Given that LPS stimulates gene expression of IL-1, these data suggest that in innate immunity (liver, lung, spleen, and bone marrow), either the major cells that supply histamine via HDC induction in response to LPS or IL-1 are not mast cells or a contribution by mast cells is not a prerequisite for HDC induction. In addition, neither IL-1 nor TNF-α is a prerequisite for the induction of HDC by LPS. Rather, LPS might directly target IL-1 receptors, since Toll4 receptors, the targets of LPS, and IL-1 receptors and their signaling pathways share many similarities (57, 58).

H. pylori has been identified as a major pathogen associated with the development of chronic gastritis, gastroduodenal ulcer diseases, gastric adenocarcinoma, and mucosa-associated lymphoid tissue lymphoma (59–61). Histamine is known to play a role in inflammation, and consistently it has been shown that *H. pylori* infection activated HDC-promoter activity (14). Although activation of the HDC promoter required the activation of ERK Map kinase pathway for both gastrin treatment and *H. pylori* infection, the upstream signaling pathways are different. While Raf-1 signaling is involved in gastrin-induced HDC-promoter activation, it was not involved in *H. pylori* infection-induced promoter activation. Instead, B-Raf and Rap1 were required in this process. Furthermore, it appears that *H. pylori* released molecule(s) communicate the signal to Gαs, a component of the heterotrimeric G-protein complex. This then activates adenylate cyclase and results in the accumulation of cAMP. The increased level of cAMP accordingly activates Rap1 through Epac (15). Similar to gastrin treatment, the *H. pylori*-dependent signal transduction leading to HDC-promoter activity may contribute to gastric disorders.

C. Transcription Factors and *cis*-Elements Involved in *HDC* Gene Expression

As discussed previously, *HDC* gene expression has been shown to be regulated by several different stimuli, including gastrin, PMA, LPS, IL-1, GM-CSF, oxidative stress, and *H. pylori* infection. However, the transcriptional factors that mediate these effects, and the *cis* DNA elements in the promoter to which they bind, have not been well characterized. Analysis of the structure of the HDC promoter revealed no obvious TATA box. Instead, an upstream GC box plays an important role in both basal and regulated promoter activity. Sp1 has been shown to activate promoter activity through this DNA element by both transient transfection and DNA–protein interaction assays (34, 35). We wished to identify additional nuclear factors that might regulate *HDC* gene expression through this DNA element and have demonstrated that a multifunctional transcriptional factor YY1 was recruited and inhibited the promoter through this site. Furthermore, we found that a YY1 interacting protein, SREBP-1a, appears to inhibit the HDC-promoter activity by interacting with YY1 or by a similar mechanism as YY1 (36).

Gastrin has been shown to activate *HDC* gene expression through three *cis*-regulatory elements (37, 38). These gastrin responsive elements are GC rich but fail to exhibit obvious homology with the DNA-binding motifs of previously described transcription factors. Similar GC-rich sequences have also been reported to mediate the effect of gastrin on the chromogranin A promoter by Sp1/Egr1 and on the vesicular monoamine transporter-2 promoter by AP-2/Sp1 (62, 63). Although most of these gastrin responsive elements are GC-rich, they do not share significant homology with each other. Using yeast-one-hybrid screening, one GC-rich sequence-binding transcriptional factor, KLF4, was found to bind gastrin responsive elements in the HDC promoter (35). Electrophoretic mobility shift assay confirmed the binding of KLF4 with all three gastrin responsive elements. In addition, transient transfection assay showed that overexpression of KLF4 dose-dependently and specifically inhibited HDC-promoter activity. Regulation of HDC transcription by KLF4 was confirmed by changes in endogenous HDC mRNA levels by KLF4 small interference RNA and KLF4 overexpression. Furthermore, KLF4 was shown to inhibit HDC-promoter activity by competing with Sp1 at the upstream GC box and also independently by binding the three downstream gastrin responsive elements. Although KLF4 binds the three downstream gastrin responsive elements, it does not appear to be the main transcriptional factor that regulates gastrin-induced *HDC* gene expression, since it inhibits HDC-promoter activity mainly through the upstream GC box. It is possible that gastrin activates *HDC* gene expression by downregulating potential transcriptional inhibitors, including KLF4, through these elements. In any case, the key transcriptional factors that mediate the transcriptional activation by gastrin remain to be identified.

The three downstream gastrin responsive elements not only mediate gastrin-induced *HDC* gene expression, but also other stimuli, including PMA, oxidative stress, and *H. pylori* infection. This observation on the one hand confirmed the importance of these DNA *cis*-elements in regulation of *HDC* gene expression, but on the other hand raised the question: why different upstream stimuli activate the gene expression through the same DNA elements? One possible reason is that these stimuli share a similar ERK Map kinase signal transduction pathway leading to activation of *HDC* gene expression, as discussed earlier. If this is the case, GM-CSF–induced (and other stimuli like IL-1) *HDC* gene expression may also involve ERK Map kinase signaling transduction pathway, since the downstream gastrin responsive elements are essential for both basal HDC-promoter activity and the response to GM-CSF (53). It should be noted that given the complexity of transcriptional regulation, there should be other DNA elements and corresponding transcriptional factors involved in regulation of *HDC* gene expression which needs further investigation. A model of gene regulation of HDC by different transcription factors has been proposed in Fig. 1.

III. Posttranscriptional Regulation of HDC

While *HDC* gene transcription is the critical first step in the regulation of histamine biosynthesis, other studies performed in the last 15 years have identified additional posttranscriptional steps at which control might be exercised. Most reports have focused on posttranslational events, although there is evidence also for regulation at the mRNA splicing and protein translational levels (22, 64). Posttranslational studies emphasize the importance of protein processing or cleavage and the effect that this has on parameters such as specific activity (16, 18, 65), intracellular localization (22, 66, 67), and

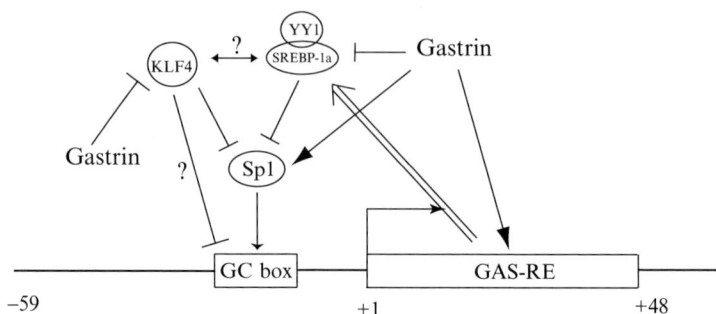

FIG. 1. The upstream GC box in the promoter plays an important role in regulation of *HDC* gene expression. Sp1 has been shown to activate the promoter through this element. Gastrin activates *HDC* gene expression through both downstream gastrin responsive element and Sp1 stabilization. Our data also suggest that YY1 and SREBP-1a form a complex to compete Sp1 in binding the GC box, resulting in the transcriptional inhibition. The double line with arrow indicates that the downstream GAS-RE is required for this inhibition. Furthermore, our published and unpublished results suggest that KLF4 may form a repressive complex with Tip60 and HDAC7 to regulation HDC transcription by competing Sp1 or by binding directly with the GC box.

proteasomal degradation (22, 23, 68–71). Although we still lack a deep understanding of these events, with studies by and large failing to delineate the molecular basis for regulation, they all influence the production of histamine and have significant bearing on its involvement in cancer progression. In the next section, we will review what is known about posttranslational processing to generate HDC isoforms with markedly different cellular and catalytic properties. We will emphasize new data relating to the catalytic mechanism and 3D structure of the enzyme, and describe how these advances might be applied to more accurately examine the regulation of histamine biosynthesis in both normal and diseased cell types.

A. HDC Is Translated as a ~74-kDa Protein That Undergoes Posttranslational Processing

Mammalian *HDC* genes produce primary translation protein products that are about ~74 kDa in size. While this is now universally accepted, it was not the conclusion reached from early studies to purify the enzyme. Hammar and Hjerten, for example, who purified enzymatic activity from the soluble fraction of mouse mastocytoma cells (72), concluded that the active isoform consisted of two ~55-kDa monomers that dimerize to form a ~110-kDa catalytically active unit, and in a subsequent study dismissed additional ~75-kDa and

∼35-kDa proteins as impurities (73). Taguchi et al. (74), who purified the protein from the soluble fraction of fetal rat livers also claimed that the active isoform consisted of a ∼110-kDa dimer with two ∼55-kDa monomers, and described other studies that reported 145-kDa and 66-kDa fragments (75) as flawed. Successive studies to purify the enzyme from the soluble fractions of mouse kidney lysates (76) and mouse stomach lysates (21), all seemed to confirm this interpretation, and while there was evidence that the ∼55-kDa isoform exists in multiple charged states [that possibly reflected different phosphorylation or dephosphorylation states (73, 76–79)], the consensus was that only a single enzyme form existed.

It was not until the rat and mouse (16, 18) cDNA's sequences were cloned in the early 1990s that it became apparent that HDC is in fact translated as a ∼74-kDa protein and that the ∼54-kDa monomer is only a processed isoform. Initially this led to the interpretation that a single posttranslational processing event results in formation of only the 54-kDa isoform (16, 18, 80). The development of newer anti-HDC antisera that recognize additional HDC isoforms has demonstrated that this was still an oversimplification and that posttranslational processing is a lot more complex than originally believed with multiple major and minor bands having been detected (22, 23, 42, 81, 82).

While we might wonder in hindsight how HDC isoforms, such as the ∼74-kDa primary translation product, could have been missed for so long, there are a number of plausible reasons emerging to help explain the discrepancy. These additionally provide important insights into the expression and function of the protein. Three factors in particular are noteworthy. First, early cell fractionation protocols purified the fraction containing the greatest enzymatic activity, and it is now clear that the ∼54-kDa isoform has a much higher specific activity than the apparently inactive ∼74-kDa one (65, 80). Second, the highly active ∼54-kDa isoform is soluble, whereas it is generally recognized that the ∼74-kDa isoform is not (due to ER membrane localization) (22, 66, 80, 83). Early purification protocols that used the activity containing soluble fraction (21, 72, 74–76) may have unintentionally removed the primary translation product from the very first step. Finally, it has become clear that regulatory elements present in the ∼74-kDa isoform target it for proteasomal degradation (22, 23, 64, 68–70, 81, 82). These sequences are removed during processing to the ∼54-kDa isoform. It is hardly surprising then that a stable, soluble, and active form of the enzyme would be isolated in preference to an unstable, insoluble, and inactive form. These observations go someway toward explaining why the primary translation product might initially have been missed, and point to the importance of posttranslational processing on enzyme function and the differential properties of the various HDC isoforms.

B. HDC Processing

1. HDC Processing and Enzyme Activity

One of the first tasks after cloning of mammalian HDC cDNAs was to assess the nature of processing events, and while sequence comparisons with other L-amino acid decarboxylases suggested that the amino terminus would encode the active site, it remained necessary to prove this experimentally (18). Antibodies raised against a range of peptide sequences have now been used in western blots to confirm that production of the major processed isoforms involves removal of the carboxyl terminus (22, 80–82). A series of studies in a variety of cell models have also attempted to assess the effect of processing on enzyme activity. These demonstrate that histidine decarboxylation function does map to the amino terminus, and, in general, they conclude that processing is required for activity (23, 65, 80, 83).

The most comprehensive study to examine the comparative activity of carboxyl-truncated rat HDC isoforms has been performed using *in vitro* coupled transcription-translation reactions, an experimental model where expressed HDC proteins show little or no additional processing. These experiments indicated that the HDC primary translation product, and isoforms carboxyl-truncated to residue 633, is inactive. Inactivity appears to be a function of specific sequences that inhibit substrate binding. Truncations beyond residue 617 result in a gradual increase in enzyme activity and removal of the inhibitory block (65). Other studies have demonstrated that a truncated protein containing residues 1 through 477 (HDC1/477) is the minimal carboxyl-truncated isoform that retains enzymatic activity. Truncations beyond residue 477 inactivate the enzyme (23). Together, these studies demonstrated that processed isoforms between $\sim$54 kDa and $\sim$70 kDa in size (including the $\sim$63-kDa isoform that is generated in many tissue types) have intrinsic activity that is not dependent on further processing, and that isoforms smaller than $\sim$53 kDa in size are inactive. Experimental evidence for low levels of amino-terminal processing has also been presented (22, 23, 68) but its functional importance has thus far not been explored.

2. HDC Processing and Degradation

Computer analysis by Viguera and coworkers identified two PEST domains within the HDC primary protein sequence (71). Similar domains are located in a number of short-lived proteins and while there remains debate as to the molecular basis by which they mediate degradation, there seems to be a strong correlation between their presence in a protein and rate of protein turnover/proteolysis (84–86). Examples of this include ornithine decarboxylase, PTEN, and IκBα (87–92). In the case of HDC, a short PEST domain can be detected between residues 20 and 70 in the amino terminus (PEST1) and a second can be detected between amino acids 500 and 570 (PEST2). It is not surprising that the primary translation product that contains both these elements is extremely unstable and undergoes proteasomal degradation both *in vitro* and *in vivo* (22, 23, 64, 68–70, 81, 82).

This situation becomes more interesting when it is considered that posttranslational processing to remove the carboxyl-terminus PEST2 region would facilitate the production of isoforms with increased cellular stabilities. A number of studies have now confirmed this experimentally with the ~74-kDa isoform but not the 55-kDa isoform identified as a substrate for cellular ubiquitination and proteasomal degradation (22, 23). The ubiquitin-proteasome pathway for the degradation of cellular proteins generally involves the cascade transfer of ubiquitin between activating, conjugating, and ligase enzymes—E1 to E2 to E3. The final transfer of ubiquitin to a lysine residue in the substrate protein and subsequent polyubiquitination is widely regarded as the most specific step in the cascade and targets the protein to the proteasome (93). Very little remains known about the molecular mechanisms that mediate degradation of HDC—the E3 ligase enzyme that promotes ubiquitination has yet to be identified, and lysine residues that get ubiquitinated have not yet been mapped. This aspect of HDC expression will undoubtedly become the focus of future studies, however, as it has been shown that high-molecular-weight isoforms can be differentially stabilized following activation of protein kinase C pathways (22, 23). This property of regulation was initially identified as a feature of HDC expression following gastrin stimulation of the CCK-B/gastrin receptor (22). However, it could be a feature of other protein kinase C regulated events, such as antigen independent induction of histamine biosynthesis, as described in murine bone marrow-derived mast cells (94), and cancer-causing mutations that increase PKC signaling would also likely lead to increased HDC isoform stability and increased histamine biosynthesis.

While carboxyl-terminus PEST2 sequences clearly contribute toward HDC proteasomal degradation, there is some evidence that the amino-terminal PEST1 domain might also be involved. It has been shown that these sequences (rather than PEST2 sequences) mediate proteasomal degradation *in vitro* (68, 69, 95). Other *in vitro* studies have shown that HDC can be degraded by calpains (100). Future studies will be needed, however, to fully address the cellular relevance of these *in vitro* observations.

3. HDC Processing and Intracellular Localization

In addition to encoding sequences that inhibit enzymatic activity and promote proteasomal degradation, the processed carboxyl terminal has now also been shown to influence the cellular localization and ER targeting of the primary translation product. This has long been one of the most contentious and contradictory areas of HDC expression and function with reports of

~74-kDa and ~55-kDa localization to the insoluble and soluble fractions, respectively, and vice versa. The emerging picture appears to be that the ~74-kDa isoform eventually gets targeted to the ER membrane (22, 66, 67). This can happen posttranslationally *in vitro* and it is possible that cytosolic translation followed by ER localization would explain why it has occasionally been detected in the soluble fraction (66, 67). Although the majority of reports now indicate that the processed ~55-kDa isoform is in the soluble fraction, a study suggests it is in the insoluble fraction facing the lumen of the ER (67), while another states that it faces the cytosol of secretory granules (97). There is some evidence then that this feature of expression might be influenced by cell type and cell function, and whether synthesized histamine is to be used for extracellular or intracellular signaling.

C. Regulation of HDC Processing

While there can be general agreement that processing of ~74-kDa HDC is primarily carboxyl-terminus in nature, and that processing of HDC is of critical importance for the differential regulation of multiple enzyme functions as described earlier, relatively little is known about processing events. Only two proteases have been described with HDC processing activity—porcine elastase can mediate production of the ~54-kDa isoform (98), while bovine trypsin promotes production of the ~36-kDa isoform (99). The drawback of these studies is that they have been performed *in vitro* using exogenously added enzymes, and while a HDC converting activity has been detected in crude cell lysates (100), the specific cellular enzymes that mediate cleavage *in vivo* have yet to be identified.

Other studies have addressed the issue of HDC processing by attempting to identify sites that mediate cleavage. One such study has demonstrated that mutation of residues 502/503/504 disrupts production of the ~55-kDa isoform in transfected Cos-7 cells (23).

Given the number of cleavage events that are now known to be involved in HDC processing it is likely that this aspect of expression will cause considerable confusion for some time to come, and the emerging picture of tissue-specific patterns of processing further complicates the situation. For example, while we have previously demonstrated that HDC processing in transfected Cos-7 cells is similar to that observed in fetal rat liver extracts (23), unique patterns of tissue-specific isoform expression are observed when stomach, testes, or kidney lysates from adult rats are fractionated alongside (Fleming and Wang, unpublished observations). Unique patterns of HDC isoform expression have also been described for different regions of the mouse testes (101).

Differences in isoform expression fall into one of the two categories. In the first category the molecular weights of expressed isoforms are different, while in the second only the ratios of the expressed isoforms differ. In the first category where differences in isoform size are small (e.g., Cos-7 vs adult rat stomach), it may reflect tissue-specific patterns of isoform modification such as phosphorylation. Larger differences in isoform sizes on the other hand (e.g., adult rat stomach vs adult rat testes) will almost certainly reflect differences in processing sites and involve tissue-specific proteases.

In the second category where isoform sizes are the same but the ratio of expressed isoforms differs, these differences can arise as a result of differential isoform stability. As previously mentioned, this is a parameter of expression that can be influenced by the physiological conditions that act on a tissue at any given time (22). A differential pattern of isoform ratios could also arise as a result of tissue-specific regulation of cleaving proteases. Instances where this latter type of expression might occur have not yet been reported but as increased numbers of HDC-expressing cells become described and cleaving proteases become identified, this is clearly an element of HDC regulation that will need to be revisited.

D. Computer Modeling of the HDC Active Site and Insights into the Catalytic Mechanism

It is easy to get lost in the intricacies and choreography of intracellular isoform processing and forget that it all ultimately comes back to regulation of histamine production. It has become increasingly apparent in the last few years, therefore, that a deeper understanding of the catalytic mechanism that mediates this, and how it relates to the different isoforms, can provide valuable insights into expression and regulation of the enzyme *in vivo*. Mammalian HDC enzymes belong to the evolutionarily conserved family of PLP-dependent enzymes known as the group II decarboxylases (102–105). Sequence homology with family members L-DOPA decarboxylase (DDC) and the two glutamic acid decarboxylase (GAD) isozymes is greatest in the first 480 amino acids (51% identify with DDC, around 30% identify with GAD isozymes). Based on these similarities it has been possible for sometime to make predictions about the catalytic mechanism of histamine biosynthesis (105–107). These predictions have by and large been confirmed in biophysical and biochemical analysis of amino terminal HDC1/512 and HDC1/516 fragments as described elsewhere (23, 108). One important feature of catalysis identified in these studies that was not previously apparent is that the cofactor-substrate bond experiences a greater torsion than in comparable PLP-dependent enzymes such as DDC. This could in part explain why the enzyme is relatively inefficient compared to the other group II decarboxylases (108).

Despite this increased understanding of the catalytic reaction mammalian HDC enzymes have doggedly proven difficult to purify and hence to crystallize.

Understandably, there was very little known to date about the specific enzyme sequences or structures that mediate catalysis, and only two residues had been identified experimentally to be of importance for catalysis—K308 (which binds the PLP cofactor) (23) and H274 (which is a highly conserved residue in mammalian decarboxylases) (68, 109). The breakthrough came in 2003 when the homologous enzyme DDC was crystallized (110) and for the first time molecular modeling of HDC became feasible. Initially, this involved a simple substitution of residues in the DDC active site for the equivalent residues in HDC, and the level of similarity became apparent. Subsequently, a more comprehensive energy optimized model of an amino-terminal 5/480 fragment in water was generated. Together, these have provided important insights into the 3D structure of the enzyme and the arrangement of the active site relative to the enzyme as a whole (99, 111, 112).

In essence, these modeling studies infer that the dimeric enzyme, which exhibits twofold axial symmetry, has two distinct active sites that consist of amino acid residues from both monomers. Residues N305, D276, H197, K308, and S307 define the cofactor binding pocket, while residues A86, Y83, and L105 (the latter contributed by the second monomer) mediate substrate interactions. The central role of many of these residues in catalysis has now been confirmed by mutation analysis (99, 111).

While some early reports suggested that catalytic activity could be decreased by phosphorylation (79), there is now also evidence that direct phosphorylation of active HDC1/512 by protein kinase A actually increases activity (113). Once sites of phosphorylation have been mapped and the proximity of these residues to the active site has been established in the 3D model, it will undoubtedly provide important new insights into the catalytic mechanism.

E. Flexible Loop Domain and Structural Compaction of the Dimer

In DDC a flexible loop structure located between amino acid residues 326 and 345 has been described. By exploiting the advantageous existence of a trypsin proteolytic site within this loop it has been possible to demonstrate that DDC undergoes a structural change during catalysis (110, 114). The equivalent domain in HDC exhibits 47% identity and it has been proposed that tyrosine residue Y337 located within this region is sufficiently close to tyrosine residue Y83 of the HDC active site as to allow events in the catalytic site to be communicated to the loop (99, 111).

The importance of Y337 and proximal residues has now been confirmed by mutation analysis and a combination of experimental approaches has been used to demonstrate not only that movement of the loop occurs during catalysis but also that this movement is consistent with a significant structural compaction of the whole dimer (65, 99, 111). These approaches have all employed L-histidine analogs that are capable of occupying the HDC active site and have been well described as inhibitors of catalysis. Histidine methylester (HME) enters the active site and forms an external aldimine with PLP but is not decarboxylated, whereas α-fluoromethylhistidine (α-FMH) is decarboxylated before binding irreversibly to the enzyme backbone (105, 115–125). Interactions between the enzyme and HME can be used as an indicator of substrate binding. Interactions with α-FMH can be used as an indicator of binding and additionally decarboxylation.

Based on analogous studies on the flexible loop of DDC (114), it was hypothesized that incubation of recombinant HDC1/516His with substrate analogs HME or α-FMH would lead to occupation of the active site by carboxylated and decarboxylated catalytic intermediates, respectively. This would result in the enzyme becoming "locked" into the compacted dimer configuration that is adopted during catalysis and protect the dimer from proteolytic digestion at a trypsin digestion site located within the loop. This has now been demonstrated experimentally with decreased production of a 36-kDa digestion fragment observed when recombinant enzyme was preincubated with either HME or α-FMH [Fig. 2A and (99)]. It is noteworthy that a 36-kDa isoform is detected in all HDC expression tissues and cell types that we have studied thus far, suggesting that a similar proteolytic event at this exposed loop can occur in a regulated manner *in vivo*.

Two additional experimental approaches have been developed to monitor the molecular compaction that accompanies incubation with substrate analogs. For example, compaction leads to a decreased Stokes' radius of the dimer that can be detected as an increased rate of migration when samples are fractionated on native polyacrylamide gels (Fig. 2B, top panel), or by FPLC (111). A third independent property of the compacted dimer is that it is more resistant to denaturation. When recombinant HDC is preincubated with HME or α-FMH and subsequently fractionated on native gels containing sodium dodecyl sulphate (SDS) (referred to as semidenaturing or nonreducing gels), not all of the compacted dimer becomes dissociated into its monomeric components [(65, 111) and Fig. 2B, middle panel].

Protection against tryptic proteolysis, decreased Stokes' radius, and resistance to SDS denaturation are three independent properties of the compacted dimer that can be demonstrated experimentally. We have exploited this, and the fact that compaction reflects the accumulation of catalytic intermediates in the active site, to assist in the characterization of active and inactive HDC isoforms. Specifically, we have demonstrated that the inactive *in vitro* expressed ∼74-kDa primary translation product is unable to undergo molecular compaction in the presence of either of the substrate analogs pointing to a deficiency in substrate binding (65). The differential interactions of the

A SDS-PAGE

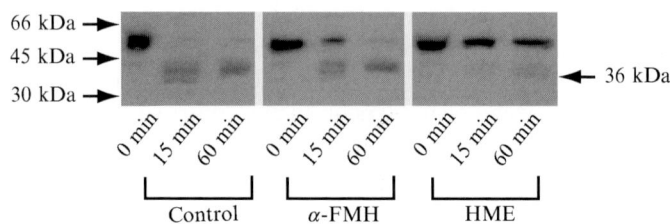

B

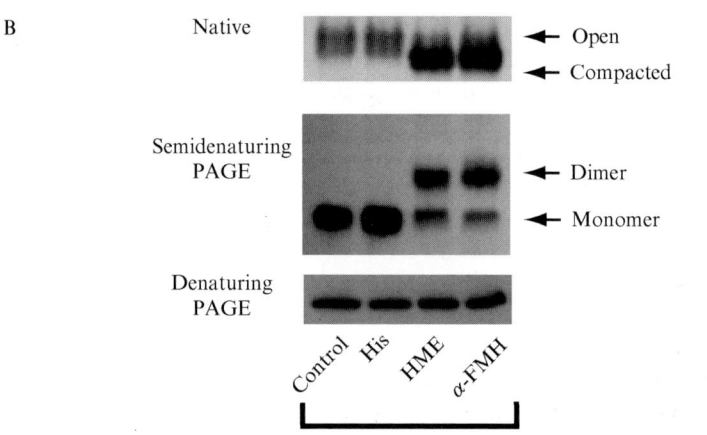

FIG. 2. Recombinant HDC1/516His was incubated with 20 mM histidine, histidine methyl ester (HME), or α-fluoromethyl histidine (α-FMH) as indicated. After 1 hr the preincubated proteins were (A) digested for 1 hr with 2.5 ng/ml bovine trypsin and fractionated on denaturing SDS polyacrylamide gels, or (B) fractionated on native (top panel), semidenaturing (middle panel), or denaturing (lower panel) polyacylamide gels. In all cases (A and B), fractionated proteins were blotted to PVDF membrane and probed with an anti-HDC antibody as described (65, 99, 111). It was frequently noted that the HDC antibody used in these studies bound with greater affinity to the compacted configuration compared to the open configuration on native polyacrylamide gels.

enzyme with HME and α-FMH may also assist in probing the HDC active site to identify the residues that mediate specific steps in the catalytic mechanism. We have described a range of active site mutants and shown, for example, that mutation of residues Y83, H197, and K308 each individually results in decreased catalytic activity (99). Assuming that all inactivating mutations are not equal, the challenge remains as to whether we can exploit differential interactions of the enzyme with the substrate analogs in order to identify mutants that are specifically compromised in substrate binding, substrate decarboxylation, or product release.

F. The Next 15 Years …

Considerable progress has been made in characterizing the expression, properties, and catalytic mechanism of different HDC enzyme isoforms. Despite this a number of very significant problems remain in understanding how this relates to histamine biosynthesis in both normal and diseased cells. Delineating its role in cancer is especially difficult as histamine biosynthesis is often inferred after detection of HDC mRNA or protein in tissue sections. Yet the fact that both active and inactive protein isoforms are produced from HDC mRNA means that it is not always possible to make this inference. In particular, the interpretation of immunohistochemical data obtained with antibodies that recognize multiple active and inactive isoforms is fraught with difficulties. Even in rare cases where western blots are performed we still lack the ability to look at a western blot that shows the expression of multiple HDC isoforms and identify categorically the isoforms that are contributing to histamine biosynthesis at any given time.

Arguably it may never be completely possible to achieve this as it is increasingly apparent that we need to get away from the idea that histamine output is a function only of classically described active HDC isoform homodimers. Fifteen years after cloning of HDC cDNAs, it is surely more realistic to believe that cellular histamine biosynthesis will occur as a result of different active and inactive isoforms interacting with one another. Even for the $\sim$74-kDa primary translation product that is inactive *in vitro*, we cannot yet rule out the possibility that it might be catalytically active if dimerized with a carboxyl-truncated active isoform. The most pertinent question then becomes whether it is possible, in a heterogenous mixture of HDC isoforms, to identify the isoforms that are contributing toward histamine biosynthesis under a certain set of physiological or pathological conditions?

We are investigating the use of α-FMH (and the detection of compacted dimers that demonstrate an ability to bind and decarboxylate substrate) as a means of identifying catalytically relevant isoforms *in vivo*. To this end we have transfected Cos-7 cells to express an active HDC1/516HA isoform and treated the transfected cells with α-FMH. In this model, with cells predominantly expressing the active carboxyl-truncated HDC1/516HA isoform, we can report the detection in anti-HA western blots of compacted dimers that are resistant to SDS denaturation (see arrow for HDC1/516HA: HDC1/516HA homodimer in Fig. 3A, top panel, third lane). To confirm that this represents a HDC dimer, as opposed to a specific interaction with another cellular protein, cells were cotransfected to express HDC1/516HA and additionally HDC1/516GFP in

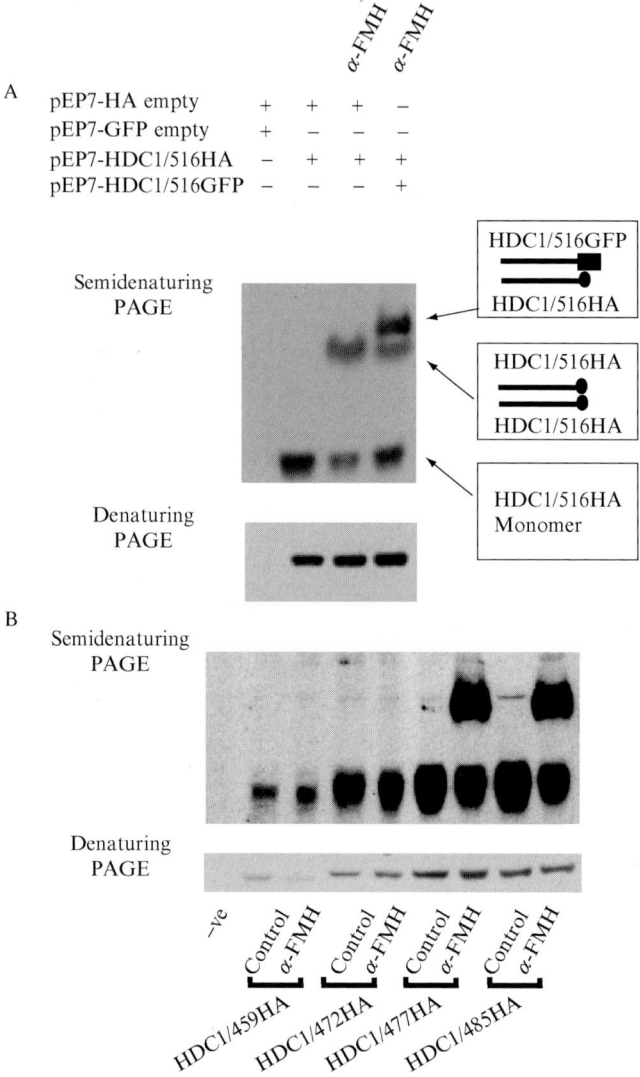

FIG. 3. (A) Cos-7 cells were transfected to transiently express HDC1/516HA and HDC1/516GFP in the presence or absence of 1 mM α-FMH as indicated. Whole cell lysates were fractionated on semidenaturing (top panel) or denaturing (lower panel) polyacrylamide gels. Fractionated proteins were blotted to PVDF and membranes probed with an anti-HA antibody. (B) Cos-7 cells were transfected to express active and inactive HDC isoforms (as described in 27) in the presence or absence of 1 mM α-FMH. Whole cell lysates were fractionated on semidenaturing (top panel) or denaturing (lower panel) polyacrylamide gels. Fractionated proteins were blotted to PVDF and membranes probed with an anti-HA antibody.

the presence of α-FMH. The GFP tag (∼27 kDa) is larger than the HA tag (<1 kDa) so cotransfection with HDC1/516GFP leads to a shift in the dimer band. This can still be detected with an anti-HA antibody as it represents the formation of an HDC-HA and HDC-GFP heterodimer (Fig. 3A, top panel, fourth lane).

Cos-7 cells transfected to express the inactive carboxyl-truncated HDC1/459 and HDC1/472 isoforms, and incubated with α-FMH, failed to generate SDS-resistant dimers (Fig. 3B, top panel). This demonstrated that compaction *in vivo* is a function of only catalytically active HDC isoforms. These early results are encouraging and suggestive that we can use this approach to identify compacted enzyme in cellular lysates. Future studies will address whether it remains informative when applied to a truly heterogenous mixture of HDC isoforms.

IV. Potential Role of HDC in Cancer Development

The involvement of histamine and HDC in carcinogenesis and tumor proliferation was proposed several decades ago (*24, 126*), but its role has remained poorly defined until recent years. Accumulating experimental data now indicate a strong correlation between upregulation of HDC activity with increased incidence of a number of tumor types such as colon cancers (*127, 128*), melanoma (*129, 130*), pancreatic cancers (*27*), breast cancers (*131*), and gastric cancers (*132*). However, the mechanism of how HDC/histamine is involved in tumor development is still not totally clear.

A. HDC Expression in Different Cancers

Several different groups have investigated the role of histamine/HDC in the development of colon cancers. Masini *et al.*, for example, showed that HDC and histamine contents were significantly higher in colon cancer specimens than in the corresponding normal mucosa. These parameters were significantly higher in tumors with lymph node and/or hematogenous metastases (stage III–IV) than in those without any metastases (stage I–II), suggesting a role for histamine/HDC in the acquisition of an invasive and metastatic phenotype for colorectal cancer cells (*128*). Consistent with the elevated histamine/HDC levels in colon cancer cells and adenomatous polyps (*28*), the activity of diamine oxidase, the enzyme that catalyzes the oxidation of histamine, was significantly decreased in tumors compared to normal mucosa in the colon (*133*).

Mechanistically, H2 receptors appear to be the principal mediators of histamine-induced cell growth in colon cancers. *In vitro*, two H2 receptor antagonists—cimetidine and ranitidine—inhibited the cell growth and induced apoptosis of a human colorectal cancer cell line, Caco-2 (*134*). *In vivo* studies appear to confirm this involvement and additionally suggest that H2 receptor antagonists suppress the growth of colon cancer implants in mice by inhibiting angiogenesis via reducing vascular endothelial growth factor (VEGF) expression (*135, 136*). Syngeneic colon cancer (Colon 38) was implanted in C57BL/6 mice, and H2 receptor antagonisits—roxatidine and cimetidine—were orally delivered to the mice for 26–29 days beginning before or after the implantation. Both antagonists significantly suppressed the growth of Colon 38 tumor implants with histological analysis, revealing increased necrotic areas and decreased density of microvessels in tumor tissue. Both H2 receptor antagonists suppressed VEGF levels in tumor tissue and significantly decreased serum VEGF levels in Colon 38–bearing mice (*136*). Similar results were also obtained by implantation of syngeneic colon cancer cells, CMT93 cells, into the subcutaneous space of C57BL/6 mice (*135*).

H1 receptor may also be involved in colon cancer development. Diks *et al.* reported that histamine stimulation causes transactivation of a T-cell factor/β-catenin-responsive construct in HeLa cells and in the SW-480 colon cell line, and pharmacological inhibitors of the histamine H1 receptor counteracted histamine-induced T-cell factor/β-catenin-responsive construct transactivation and the dephosphorylation of β-catenin (*137*). Melanoma represents another well-studied tumor model where HDC expression and activity are correlated with tumor growth. Studies using both western blot and immunohistochemical approaches, for example, showed much higher HDC expression in human melanoma than in control cultured human melanocytes, and suggested that histamine was contributing to cancer development (*138*). This was confirmed in subsequent studies where there was a detectable decrease in the proliferation of melanoma cells by HDC antisense oligonucleotides. Together, these data point toward a significant role for histamine in melanoma growth (*129*).

Melanoma cells have also been shown to release a detectable amount of histamine into the medium without external stimuli. Histamine produced and released by the melanoma cells, and acting in an autocrine or paracrine manner, may therefore combine with exogenous histamine to influence cell proliferation and modulate the *in situ* immune response of the host (*29*). This may involve an increase in expression of Ets-1 (v-ets erythroblastosis virus E26 oncogene homolog 1), a protooncogene that can be regulated by histamine H2 receptor signaling in human melanoma cell lines (*139*).

The most convincing data to establish a correlation between HDC expression and melanoma progression come from animal studies. B16-F10 mouse melanoma cells were used in mouse engraftment experiments where three different transgenic variants constitutively expressing the full-length sense mouse HDC mRNA, a mock control, or an antisense HDC RNA segment, respectively, were generated in triplicate. After establishment of the primary skin tumors and lung metastases in C57BL/6 mice, the nine variants with different histamine-releasing capacities were subjected to comprehensive progression profiling *in vivo*. Markedly accelerated tumor growth and moderately increased metastatic colony-forming potential, along with rising levels of local histamine production, were observed for the group expressing full-length HDC. In addition, intensive tumor progression profiling revealed positive correlations between histamine production, tumor histamine H2 receptor, and rho-C expression (*130*).

Upregulation of HDC expression in pancreatic tumors has also been reported. While expression of HDC in nontumorous parts of the surgically resected pancreas was not apparent, 17 out of 22 pancreatic endocrine neoplasms (77%) were found to be HDC positive by immunohistochemistry (*27*). Although HDC might potentially be used as a marker for diagnosis of pancreatic tumors, it remains to be determined whether elevated HDC expression and the production of histamine actually contribute to human pancreatic tumor progression. Increased HDC activity and histamine content has also been detected in breast cancers. For example, HDC activity is increased in mammary tumor tissue compared to healthy mammary gland, skin, and muscle tissues in breast cancer patients (40 patients) (*140*). In a large group of patients (95 patients) with primary ductal breast cancers, both HDC activity and concentration of histamine in cancerous tissues were significantly higher than adjacent healthy tissues (*131*). In experimental mammary carcinomas, endogenous histamine was shown to be critical for cell proliferation through H2 receptors (*141*). Histamine was also shown to be a major secretory product of human small cell lung carcinomas (SCLCs). The detection of HDC by immunohistochemistry in paraffin sections of SCLC tumors was confirmed by immunoblotting and reverse transcription-polymerase chain reaction experiments using established SCLC cell lines, frozen and paraffin-embedded SCLC tumors (*142*). In a similar study by a different group, HDC antibody staining was observed in the majority of SCLC tumors tested (18 of 23, sensitivity 0.78), and was rarely observed with nonneuroendocrine lung tumors (2 of 44; specificity, 0.95) (*26*).

Besides the tumors discussed earlier, HDC expression in other tumors, like gastric cancers, which will be reviewed in more detail, has also been reported. In addition, the proliferative effect of HDC/histamine has been observed in a number of other cell types. For example, histamine and HDC were detected by immunohistochemistry in osteoarthritic cartilage tissues, and histamine stimulated the proliferation of human articular chondrocytes *in vitro*. This stimulation was blocked by the addition of mepyramine, an antagonist specific to H1 histamine receptor and inhibited by α-FMH, an inhibitor of HDC (*143*). In another case, HDC is induced in a human T-cell line Jurkat cells by anti-CD3. The H1 receptor antagonist triprolidine dose-dependently inhibits proliferation of Jurkat cells, while the H2 antagonist ranitidine was ineffective (*144*).

B. Link of Angiogenesis in HDC/Histamine-Mediated Cancer Development

Angiogenesis is the development of new blood vessels from a preexisting vascular structure, and its role in tumor progression was first suggested several decades ago (*145*). During the progression to malignancy, local tissues switched from an avascular state where cell proliferation is balanced by cell death (apoptosis) to an angiogenic phase marked by the recruitment of new blood vessels. This switch is also thought to be required for tumor growth and metastasis. However, the molecular basis of blood vessel growth is still incompletely understood in spite of the physiological and pathological importance and therapeutic promises. Over the past decade, extensive work has been devoted to the biology of VEGF. It is generally accepted that VEGF signaling often represents a crucial rate-limiting step in physiological angiogenesis, although new vessel growth and maturation are highly complex and coordinated processes that require the sequential activation of a series of receptors by numerous ligands (*146, 147*). Besides the growth factors, like VEGF, the microenvironment of tumor cells is also important for the angiogenesis. Tumor cells are surrounded by an infiltrate of inflammatory cells, including lymphocytes, neutrophils, macrophages, and mast cells, which communicate via a complex network of intercellular signaling pathways. Particularly, the contribution of histamine-storing mast cells to angiogenesis and tumor progression has been studied for sometime and has been reviewed by several investigators (*148–150*).

The involvement of histamine secreted from mast cells in angiogenesis was first studied in 1994, using the rat mesenteric window assay and histamine H1 receptor antagonist brompheniramine maleate (BPA) and H2 receptor antagonist metiamide (*151*). Angiogenesis was effected by i. p. injections of the mast cell secretagogue compound 48/80 for five consecutive days and analyzed 14 days after i. p. injections and antagonist treatment. It was found that BPA significantly suppressed the number of vessel profiles per unit tissue length, one measurement of angiogenesis, and metiamide significantly reduced the number of vessel profiles per unit tissue length and the vascular density, another measurement of angiogenesis. Together these data suggested that endogenous mast cell histamine is angiogenic through both H1 and H2 receptors. The involvement of histamine from nonmast cells in angiogenesis has also been demonstrated using a HDC knockout model (*152*). In this model, angiogenesis was induced by subcutaneous implantation of a cotton thread in the dorsum with formation of granulated tissues. The formation of granulated tissue and angiogenesis in wild-type mice was strongly suppressed by a topical injection of anti-VEGF IgG. This result was in agreement with the expected role of VEGF in angiogenesis. While there was no significant difference in the formation of granulated tissue and angiogenesis between the mast cell-deficient mice and the control mice, there was notably less angiogenesis and granulated tissue formation in HDC knockout mice compared to wild type. Consistently, the levels of VEGF in the granulated tissue were lower in HDC knockout mice. In addition, a topical injection of histamine or the H2 agonist dimaprit rescued the observed deficiencies in angiogenesis and granulated tissue formation in HDC knockout mice, suggesting the involvement of H2 receptor signaling pathway. As discussed earlier, the inhibition of angiogenesis by H2 receptor antagonists, cimetidine and roxatidine, suppresses growth of colon cancer implants in syngeneic mice, which might be the most relevant data, suggesting the existence of a histamine-angiogenesis-tumor growth axis (*135*, *136*). In a similar study, cimetidine was shown to suppress the tumor growth of CMT93-(another type of syngeneic colon cancer cells) bearing mice with reduced neovascularization in the tumor. Although VEGF production in the tumor was not affected by cimetidine, vascularlike tube formation by endothelial cells *in vitro* was significantly impaired by its presence (*135*). These data suggest that the reported antitumor activity of H2 receptor antagonists, including cimetidine, may execute their functions in part by inhibiting tumor-associated angiogenesis.

C. Link of Inflammation Induced by Histamine/HDC for Cancer Development

Inflammation is a crucial function of the innate immune system that protects against pathogens and initiates specific immunity. While acute inflammation is trigged to destroy infectious agents, repair damaged tissue, and initiate a long-term response to specific pathogens, it does not always resolve. In many cases chronic inflammation appears to contribute also toward the progression of physiological and pathological changes, including aging and cancer development (*153*). Accumulating data have been provided to show the links between cancer and inflammation: (i) chronic inflammation increases risk of cancer, and many cancers arise at sites of chronic inflammation; (ii) the immune cells that mediate chronic inflammation are found in cancers and promote tumor growth in cell transfer experiments; (iii) the chemical mediators that regulate inflammation are produced by cancers; (iv) deletion or inhibition of inflammatory mediators inhibits development of experimental cancers; (v) genetic variations in inflammation genes alter susceptibility to and severity of cancer; and (vi) long-term use of nonsteroidal anti-inflammatory agents reduces risk of some cancers (*154*). These links have been confirmed in a number of murine models, including gastric cancers triggered by chronic *H. pylori* infection (155), liver cancers by cholangitis (156), and colon cancers by chronic colitis (157). Although the link between inflammation and cancer development has been well established, the mechanisms involved are not fully understood; it is believed to involve both initiated cells that give rise to the cancer and inflammatory cells in the surrounding stroma.

Histamine is known to be a mediator of inflammation (158). Therefore, it is reasonable to propose that histamine production is associated with some cases of cancer progression through modulation of the inflammatory responses. Histamine was reported in many cell lines to increase the production of IL-6, a multifunctional cytokine with central importance in the induction of systemic inflammation. In Epstein-Barr virus (EBV)-infected human B lymphoma line and the glioblastoma line SK-MG4, the biosynthesis and expression of IL-6 were enhanced by histamine treatment (159). In human endothelial cells, histamine increases IL-6 production in a dose-dependent fashion, an increase that could be inhibited by H1 or H2 receptor antagonists (160). Similar results were also observed in human peripheral blood mononuclear cells (161) and in a stromal cell line (MC3T3-G2/PA6) (162). The role of HDC/histamine in regulation of IL-6 was also demonstrated using a HDC knockout mouse model. Without *in vivo* induction, no IL-6 was measured in plasma. However, after turpentine or LPS treatment, the induction of serum IL-6 in wild-type mice strongly exceeded that observed in HDC knockout mice (163). More convincing data to suggest that histamine/HDC-induced inflammation may play a role in gastric cancer development came from the animal studies using HDC knockout mice that had been infected with *H. pylori*. *H. pylori* infection in humans causes gastritis and gastric cancer. The infection elicits a complex immune response where the activation of mast cells and histamine release is of particular importance. In this animal study, *H. pylori* were administered intragastrically to HDC knockout and wild-type mice for 1 week with three times of infection. Eight weeks after the first intervention, the mice were sacrificed for further analysis. It was found that the local TNF-α and IL-6 cytokine levels in gastric mucosal specimens were significantly higher in the wild-type mice than in the HDC knockout mice. Histological analysis revealed that the grades of inflammation were less severe in the infected HDC knockout animals. These data, therefore, provide further support for the role of HDC/histamine in the pathological mechanism of *H. pylori*-induced gastritis and even gastric cancer (164).

D. Role of Histamine/HDC in Development of Gastric Cancer

In the gastric mucosa, ECL-cell derived histamine is the predominant regulator of gastric acid secretion and a significant contributory factor in the maintenance of mucosal morphology (165, 166). A minor role for gastric mast cells has also been proposed, although its importance in the regulation of gastric mucosal morphology remains uncertain. For ECL cells, histamine synthesis is tightly regulated by gastrin through the CCK-B/gastrin receptor, with receptor stimulation leading to increased *HDC* gene transcription and increased HDC protein stability. The resulting histamine is stored within secretory vesicles until calcium depolarization results in its release from the ECL cells. Binding of secreted histamine to membrane H2 receptors located on parietal cells triggers the release of gastric acid. It appears then that function of histamine/HDC in the stomach is tightly linked to gastrin. In gastrin-deficient mice, the induction of acid secretion was undetectable after stimulation with any of the secretagogues including histamine, suggesting a fundamental requirement for gastrin in acid secretion (*167*). Consistently, in CCK-B/gastrin receptor knockout mice, HDC activity in the oxyntic mucosa was undetectable yet was readily apparent in the mucosa from wild-type animals. Accordingly the mucosal histamine content in the knockout mice is only a minute fraction of that present in wild-type mice (*168*). On the other hand, HDC knockout mice showed hypergastrinemia and did not undergo acid secretion upon treatment with exogenous gastrin (*128*).

The role of histamine/HDC in the development of malignancy in stomach has been suggested from several hypergastrinemic (elevated serum gastrin) animal studies. It has been shown that hypergastrinemic rats, mice, and mastomys all develop gastric malignancy, although the mechanisms involved have not been fully understood. Male cotton rats dosed with the H2-blocker loxtidine for 6 months developed hypergastrinemia and resulted in cancer in the oxyntic mucosa. In the malignant tissue, there were more HDC-positive cells by immunohistochemistry (*169*). In histamine H2 receptor-deficient mice, hypergastrinemia was observed. Although the mice showed normal basal pH, they exhibited a marked hypertrophy with enlarged folds in gastric mucosa with a significant increase of histamine content. Further, immunohistochemistry showed that both the parietal cells and ECL cells were increased (*170*). In 9-month-old HDC knockout mice, where hypergastrinemia was also developed, hyperplasia was observed in the oxyntic glandular base region, as well as an increase in the number of parietal cells and ECL cells (*171*). A similar increase in parietal cell number has been demonstrated independently by a separate study (*172*). Superficially, these data seemed to suggest an inhibitory role of HDC/histamine in the development of the gastric malignancy, but actually these results originated from hypergastrinemia induced by HDC/histamine deficiency.

A different line of animal studies would argue a promotive effect of HDC/histamine in gastric malignancy progression, especially when combined with *H. pylori* infection. It is known that hypergastrinemia in an amidated-gastrin transgenic mouse (INS-GAS mice) model leads to the development of gastric metaplasia, dysplasia, cancinoma *in situ*, and gastric cancer with vascular invasion at 20 months of age. *H. felis* infection led to accelerated (<7 months) development of intramucosal carcinoma with submucosal invasion and intravascular invasion (173). In this model, ECL cell number did not increase at 20 months of age. The parietal cell number increased initially but later decreased. Using this model, the effect of gastrin receptor antagonist YF476 and/or the histamine H2 receptor antagonist loxtidine was evaluated in the development of gastric atrophy and cancer (31). It was found that INS-GAS mice infected with *H. felis* treated with YF476 or loxtidine alone partially suppressed both gastric acid secretion and progression to neoplasia. The combination of these two drugs resulted in nearly complete inhibition of both parameters. The ECL-cell density was significantly increased by either single or combined use of the drugs, whereas the parietal cell loss was prevented only by a combination. Both animal models and human studies suggest involvement of a host immune response, and particularly a strong Th1 response, in the pathogenesis of *Helicobater* infection and the development of gastric atrophy and other metaplasia, neoplastic conditions that are strongly associated with progression to cancer (60). Not surprisingly, in the study discussed earlier, the expression level of Th1-oriented cytokine interferon (IFN)-γ and TNF-α were significantly reduced, and Th2-oriented cytokine IL-4 was significantly increased after the treatment of H2 receptor loxtidine. This cytokine profile was changed by loxtidine, but not by YF476, and suggested that the immune response was mediated by histamine/HDC. In agreement with this hypothesis, similar results were observed in the HDC knockout mouse model infected with *H. pylori*. It was found that the local TNF-α and IL-6 cytokine levels in gastric mucosal specimens were much lower compared to the wild-type mice after 8 weeks' *H. pylori* infection. Histological analysis revealed that the grades of inflammation were less severe in the infected HDC knockout mice, which further confirms the importance of HDC/histamine in inflammation, and potentially the progression to gastric malignancy (164). The role of gastrin and histamine/HDC in the development of gastric neoplasia has been proposed in Fig. 4.

V. Concluding Remarks

In the last decades, the importance of histamine has been confirmed experimentally with a range of data demonstrating important functional roles in a variety of species. This includes gastric acid secretion, vascular permeability, circadian rhythms, immune and allergy response, neurotransmission, angiogenesis, neutrophil recruitment, anaphylaxis, and bone morphogenesis.

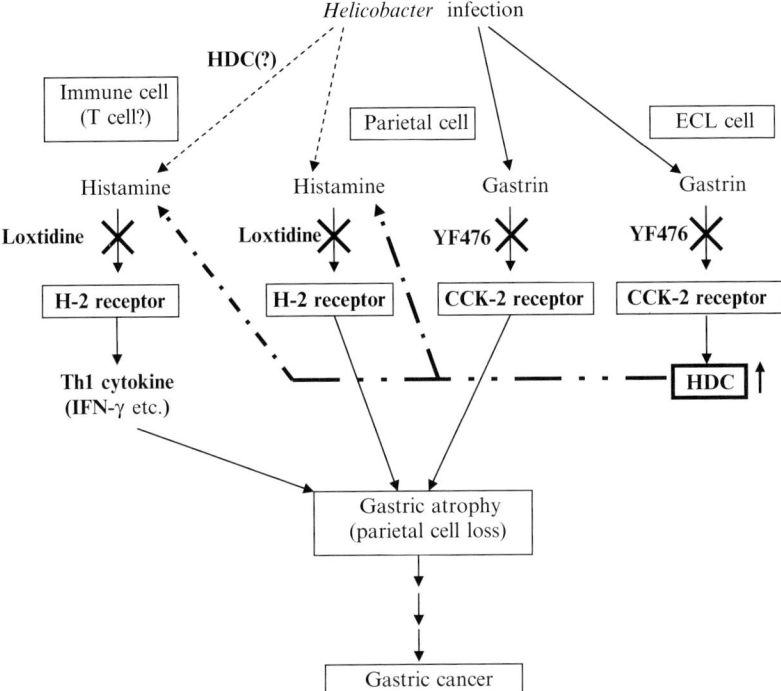

FIG. 4. The role of gastrin-histamine axis in the development of gastric cancer. *Helicobacter* infection causes hypergastrinemia and possibly elevates histamine level by increased HDC activity. In ECL cells, gastrin increases histamine secretion by activating *HDC* gene expression. Long-term presence of histamine (hyperhistaminemia) and/or gastrin (hypergastrinemia) in parietal cells results in gastric atrophy characterized by loss of parietal cells, and eventually gastric cancer. Histamine also activates a secretion of Th1 cytokines from immune cells (possibly T cells) through H2 receptor, and those cytokines may also contribute to gastric atrophy. As shown in this figure, inhibition of CCK-2 receptor by antagonist YF476 and H2 receptor by antagonist loxitidine almost completely blocked the development of gastric atrophy induced by *H. felis* infection to hypergastrinemic (INS-GAS) mice (31).

While the molecular mechanisms that underlie many of these processes remain to be fully delineated, generation of a HDC knockout mouse model and molecular modeling of the 3D structure of the active HDC dimer will undoubtedly continue to provide valuable insights into the production and actions of histamine. Future studies on the regulation of HDC at the transcriptional and posttranscriptional levels will help in our understanding of how histamine is involved in so many important physiological and pathological processes, and highlight new opportunities for the treatment of human diseases where HDC and histamine play a role.

References

1. Hill, S. J., Ganellin, C. R., Timmerman, H., Schwartz, J. C., Shankley, N. P., Young, J. M., Schunack, W., Levi, R., and Haas, H. L. (1997). International Union of Pharmacology. XIII. Classification of histamine receptors. *Pharmacol. Rev.* **49**, 253–278.
2. Coruzzi, G., Morini, G., Adami, M., and Grandi, D. (2001). Role of histamine H3 receptors in the regulation of gastric functions. *J. Physiol. Pharmacol.* **52**, 539–553.
3. Shiraishi, M., Hirasawa, N., Kobayashi, Y., Oikawa, S., Murakami, A., and Ohuchi, K. (2000). Participation of mitogen-activated protein kinase in thapsigargin- and TPA-induced histamine production in murine macrophage RAW 264.7 cells. *Br. J. Pharmacol.* **129**, 515–524.
4. Takamatsu, S., Nakashima, I., and Nakano, K. (1996). Modulation of endotoxin-induced histamine synthesis by cytokines in mouse bone marrow-derived macrophages. *J. Immunol.* **156**, 778–785.
5. Shiraishi, M., Hirasawa, N., Oikawa, S., Kobayashi, Y., and Ohuchi, K. (2000). Analysis of histamine-producing cells at the late phase of allergic inflammation in rats. *Immunology* **99**, 600–606.
6. Aoi, R., Nakashima, I., Kitamura, Y., Asai, H., and Nakano, K. (1989). Histamine synthesis by mouse T lymphocytes through induced histidine decarboxylase. *Immunology* **66**, 219–223.
7. Rangachari, P. K. (1992). Histamine: Mercurial messenger in the gut. *Am. J. Physiol.* **262**, G1–G13.
8. Zhang, Z., Hocker, M., Koh, T. J., and Wang, T. C. (1996). The human histidine decarboxylase promoter is regulated by gastrin and phorbol 12-myristate 13-acetate through a downstream *cis*-acting element. *J. Biol. Chem.* **271**, 14188–14197.
9. Hocker, M., Henihan, R. J., Rosewicz, S., Riecken, E. O., Zhang, Z., Koh, T. J., and Wang, T. C. (1997). Gastrin and phorbol 12-myristate 13-acetate regulate the human histidine decarboxylase promoter through Raf-dependent activation of extracellular signal-regulated kinase-related signaling pathways in gastric cancer cells. *J. Biol. Chem.* **272**, 27015–27024.
10. Hocker, M., Zhang, Z., Fenstermacher, D. A., Tagerud, S., Chulak, M., Joseph, D., and Wang, T. C. (1996). Rat histidine decarboxylase promoter is regulated by gastrin through a protein kinase C pathway. *Am. J. Physiol.* **270**, G619–G633.
11. Ohgoh, M., Yamamoto, J., Kawata, M., Yamamura, I., Fukui, T., and Ichikawa, A. (1993). Enhanced expression of the mouse L-histidine decarboxylase gene with a combination of dexamethasone and 12-O-tetradecanoylphorbol-13-acetate. *Biochem. Biophys. Res. Commun.* **196**, 1113–1119.
12. Hocker, M., Rosenberg, I., Xavier, R., Henihan, R. J., Wiedenmann, B., Rosewicz, S., Podolsky, D. K., and Wang, T. C. (1998). Oxidative stress activates the human histidine decarboxylase promoter in AGS gastric cancer cells. *J. Biol. Chem.* **273**, 23046–23054.
13. Pacilio, M., Debili, N., Arnould, A., Machavoine, F., Rolli-Derkinderen, M., Bodger, M., Arock, M., Dumenil, D., Dy, M., and Schneider, E. (2001). Thrombopoietin induces histidine decarboxylase gene expression in c-mpl transfected UT7 cells. *Biochem. Biophys. Res. Commun.* **285**, 1095–1101.
14. Wessler, S., Hocker, M., Fischer, W., Wang, T. C., Rosewicz, S., Haas, R., Wiedenmann, B., Meyer, T. F., and Naumann, M. (2000). *Helicobacter pylori* activates the histidine decarboxylase

promoter through a mitogen-activated protein kinase pathway independent of pathogenicity island-encoded virulence factors. *J. Biol. Chem.* **275,** 3629–3636.
15. Wessler, S., Rapp, U. R., Wiedenmann, B., Meyer, T. F., Schoneberg, T., Hocker, M., and Naumann, M. (2002). B-Raf/Rap1 signaling, but not c-Raf-1/Ras, induces the histidine decarboxylase promoter in *Helicobacter pylori* infection. *FASEB J.* **16,** 417–419.
16. Yamamoto, J., Yatsunami, K., Ohmori, E., Sugimoto, Y., Fukui, T., Katayama, T., and Ichikawa, A. (1990). cDNA-derived amino acid sequence of L-histidine decarboxylase from mouse mastocytoma P-815 cells. *FEBS Lett.* **276,** 214–218.
17. Yamauchi, K., Sato, R., Tanno, Y., Ohkawara, Y., Maeyama, K., Watanabe, T., Satoh, K., Yoshizawa, M., Shibahara, S., and Takishima, T. (1990). Nucleotide sequence of the cDNA encoding L-histidine decarboxylase derived from human basophilic leukemia cell line, KU-812-F. *Nucleic Acids Res.* **18,** 5891.
18. Joseph, D. R., Sullivan, P. M., Wang, Y. M., Kozak, C., Fenstermacher, D. A., Behrendsen, M. E., and Zahnow, C. A. (1990). Characterization and expression of the complementary DNA encoding rat histidine decarboxylase. *Proc. Natl. Acad. Sci. USA* **87,** 733–737.
19. Taguchi, Y., Watanabe, T., Kubota, H., Hayashi, H., and Wada, H. (1984). Purification of histidine decarboxylase from the liver of fetal rats and its immunochemical and immunohistochemical characterization. *J. Biol. Chem.* **259,** 5214–5221.
20. Ohmori, E., Fukui, T., Imanishi, N., Yatsunami, K., and Ichikawa, A. (1990). Purification and characterization of L-histidine decarboxylase from mouse mastocytoma P-815 cells. *J. Biochem. (Tokyo)* **107,** 834–839.
21. Watabe, A., Fukui, T., Ohmori, E., and Ichikawa, A. (1992). Purification and properties of L-histidine decarboxylase from mouse stomach. *Biochem. Pharmacol.* **43,** 587–593.
22. Fleming, J. V., and Wang, T. C. (2000). Amino- and carboxy-terminal PEST domains mediate gastrin stabilization of rat L-histidine decarboxylase isoforms. *Mol. Cell. Biol.* **20,** 4932–4947.
23. Fleming, J. V., and Wang, T. C. (2003). The production of 53-55-kDa isoforms is not required for rat L-histidine decarboxylase activity. *J. Biol. Chem.* **278,** 686–694.
24. Kahlson, G., and Rosengren, E. (1968). New approaches to the physiology of histamine. *Physiol. Rev.* **48,** 155–196.
25. Tang, Z. C., and Xu, Y. H. (1987). Comparison of the effects of 4-methylhistamine on the cell-cycle response between CFU-s and its subpopulation. *Int. J. Cell. Cloning* **5,** 511–517.
26. Matsuki, Y., Tanimoto, A., Hamada, T., and Sasaguri, Y. (2003). Histidine decarboxylase expression as a new sensitive and specific marker for small cell lung carcinoma. *Mod. Pathol.* **16,** 72–78.
27. Tanimoto, A., Matsuki, Y., Tomita, T., Sasaguri, T., Shimajiri, S., and Sasaguri, Y. (2004). Histidine decarboxylase expression in pancreatic endocrine cells and related tumors. *Pathol. Int.* **54,** 408–412.
28. Boer, K., Darvas, Z., Baki, M., Kaszas, I., Pal, Z., and Falus, A. (2003). Expression of histidine decarboxylase in human colonic cancer cells and adenomatous polyps. *Inflamm. Res.* **52** (Suppl. 1), S76–S77.
29. Darvas, Z., Sakurai, E., Schwelberger, H. G., Hegyesi, H., Rivera, E., Othsu, H., Watanabe, T., Pallinger, E., and Falus, A. (2003). Autonomous histamine metabolism in human melanoma cells. *Melanoma. Res.* **13,** 239–246.
30. Modlin, I. M., Kumar, R. R., Soroka, C. J., Ahlman, H., Nilsson, O., and Goldenring, J. R. (1994). Histamine as an intermediate growth factor in genesis of gastric ECLomas associated with hypergastrinemia in mastomys. *Dig. Dis. Sci.* **39,** 1446–1453.
31. Takaishi, S., Cui, G., Frederick, D. M., Carlson, J. E., Houghton, J., Varro, A., Dockray, G. J., Ge, Z., Whary, M. T., and Rogers, A. B. (2005). Synergistic inhibitory effects of gastrin and

histamine receptor antagonists on *Helicobacter*-induced gastric cancer. *Gastroenterology* **128,** 1965–1983.
32. Suzuki-Ishigaki, S., Numayama-Tsuruta, K., Kuramasu, A., Sakurai, E., Makabe, Y., Shimura, S., Shirato, K., Igarashi, K., Watanabe, T., and Ohtsu, H. (2000). The mouse L-histidine decarboxylase gene: Structure and transcriptional regulation by CpG methylation in the promoter region. *Nucleic Acids Res.* **28,** 2627–2633.
33. Yatsunami, K., Ohtsu, H., Tsuchikawa, M., Higuchi, T., Ishibashi, K., Shida, A., Shima, Y., Nakagawa, S., Yamauchi, K., and Yamamoto, M. (1994). Structure of the L-histidine decarboxylase gene. *J. Biol. Chem.* **269,** 1554–1559.
34. Kuramasu, A., Saito, H., Suzuki, S., Watanabe, T., and Ohtsu, H. (1998). Mast cell-/basophil-specific transcriptional regulation of human L-histidine decarboxylase gene by CpG methylation in the promoter region. *J. Biol. Chem.* **273,** 31607–31614.
35. Ai, W., Liu, Y., Langlois, M., and Wang, T. C. (2004). Kruppel-like factor 4 (KLF4) represses histidine decarboxylase gene expression through an upstream Sp1 site and downstream gastrin responsive elements. *J. Biol. Chem.* **279,** 8684–8693.
36. Ai, W., Liu, Y., and Wang, T. C. (2005). Yin Yang 1 (YY1) represses histidine decarboxylase (HDC) gene expression with SREBP-1a in part through an upstream Sp1 site. *Am. J. Physiol. Gastrointest. Liver Physiol.*
37. Raychowdhury, R., Fleming, J. V., McLaughlin, J. T., Bulitta, C. J., and Wang, T. C. (2002). Identification and characterization of a third gastrin response element (GAS-RE3) in the human histidine decarboxylase gene promoter. *Biochem. Biophys. Res. Commun.* **297,** 1089–1095.
38. Raychowdhury, R., Zhang, Z., Hocker, M., and Wang, T. C. (1999). Activation of human histidine decarboxylase gene promoter activity by gastrin is mediated by two distinct nuclear factors. *J. Biol. Chem.* **274,** 20961–20969.
39. Holliday, R. (1987). The inheritance of epigenetic defects. *Science* **238,** 163–170.
40. Jenuwein, T., and Allis, C. D. (2001). Translating the histone code. *Science* **293,** 1074–1080.
41. Xiao, H., Chung, J., Kao, H. Y., and Yang, Y. C. (2003). Tip60 is a co-repressor for STAT3. *J. Biol. Chem.* **278,** 11197–11204.
42. Fajardo, I., Urdiales, J. L., Medina, M. A., and Sanchez-Jimenez, F. (2001). Effects of phorbol ester and dexamethasone treatment on histidine decarboxylase and ornithine decarboxylase in basophilic cells. *Biochem. Pharmacol.* **61,** 1101–1106.
43. Kopin, A. S., Beinborn, M., Lee, Y. M., McBride, E. W., and Quinn, S. M. (1994). The CCK-B/gastrin receptor. Identification of amino acids that determine nonpeptide antagonist affinity. *Ann. N. Y. Acad. Sci.* **713,** 67–78.
44. Nilsson, O., Kolby, L., Wangberg, B., Wank, S. A., and Ahlman, H. (1994). Expression of CCK-A and CCK-B/gastrin receptors in enterochromaffin-like cell carcinoids of Mastomys natalensis. *Ann. N. Y. Acad. Sci.* **713,** 435–438.
45. Chen, D., Monstein, H. J., Nylander, A. G., Zhao, C. M., Sundler, F., and Hakanson, R. (1994). Acute responses of rat stomach enterochromaffinlike cells to gastrin secretory activation and adaptation. *Gastroenterology* **107,** 18–27.
46. Sandvik, A. K., Dimaline, R., Marvik, R., Brenna, E., and Waldum, H. L. (1994). Gastrin regulates histidine decarboxylase activity and mRNA abundance in rat oxyntic mucosa. *Am. J. Physiol.* **267,** G254–G258.
47. Watanabe, T., Taguchi, Y., Sasaki, K., Tsuyama, K., and Kitamura, Y. (1981). Increase in histidine decarboxylase activity in mouse skin after application of the tumor promoter tetradecanoylphorbol acetate. *Biochem. Biophys. Res. Commun.* **100,** 427–432.
48. Colucci, R., Fleming, J. V., Xavier, R., and Wang, T. C. (2001). L-histidine decarboxylase decreases its own transcription through downregulation of ERK activity. *Am. J. Physiol. Gastrointest. Liver Physiol.* **281,** G1081–G1091.

49. Ohtsu, H., Tanaka, S., Terui, T., Hori, Y., Makabe-Kobayashi, Y., Pejler, G., Tchougounova, E., Hellman, L., Gertsenstein, M., and Hirasawa, N. (2001). Mice lacking histidine decarboxylase exhibit abnormal mast cells. *FEBS Lett.* **502,** 53–56.
50. Endo, Y., Kikuchi, T., Nakamura, M., and Shinoda, H. (1992). Determination of histamine and polyamines in calcified tissues of mice: Contribution of mast cells and histidine decarboxylase to the amount of histamine in the bone. *Calcif. Tissue Int.* **51,** 67–71.
51. Kubo, Y., and Nakano, K. (1999). Regulation of histamine synthesis in mouse CD4+ and CD8+ T lymphocytes. *Inflamm. Res.* **48,** 149–153.
52. Hirasawa, N., Murakami, A., and Ohuchi, K. (2001). Expression of 74-kDa histidine decarboxylase protein in a macrophage-like cell line RAW 264.7 and inhibition by dexamethasone. *Eur. J. Pharmacol.* **418,** 23–28.
53. Murata, Y., Tanimoto, A., Wang, K. Y., Tsutsui, M., Sasaguri, Y., De Corte, F., and Matsushita, H. (2005). Granulocyte macrophage-colony stimulating factor increases the expression of histamine and histamine receptors in monocytes/macrophages in relation to arteriosclerosis. *Arterioscler. Thromb. Vasc. Biol.* **25,** 430–435.
54. Galli, S. J., Nakae, S., and Tsai, M. (2005). Mast cells in the development of adaptive immune responses. *Nat. Immunol.* **6,** 135–142.
55. Wiener, Z., Andrasfalvy, M., Pallinger, E., Kovacs, P., Szalai, C., Erdei, A., Toth, S., Nagy, A., and Falus, A. (2002). Bone marrow-derived mast cell differentiation is strongly reduced in histidine decarboxylase knockout, histamine-free mice. *Int. Immunol.* **14,** 381–387.
56. Ransone, L. J., Visvader, J., Wamsley, P., and Verma, I. M. (1990). Trans-dominant negative mutants of Fos and Jun. *Proc. Natl. Acad. Sci. USA* **87,** 3806–3810.
57. O'Neill, L. A., Dunne, A., Edjeback, M., Gray, P., Jefferies, C., and Wietek, C. (2003). Mal and MyD88: Adapter proteins involved in signal transduction by Toll-like receptors. *J. Endotoxin. Res.* **9,** 55–59.
58. Takeuchi, O., and Akira, S. (2001). Toll-like receptors; their physiological role and signal transduction system. *Int. Immunopharmacol.* **1,** 625–635.
59. Crowe, S. E. (2005). Helicobacter infection, chronic inflammation, and the development of malignancy. *Curr. Opin. Gastroenterol.* **21,** 32–38.
60. Houghton, J., and Wang, T. C. (2005). Helicobacter pylori and gastric cancer: A new paradigm for inflammation-associated epithelial cancers. *Gastroenterology* **128,** 1567–1578.
61. Farinha, P., and Gascoyne, R. D. (2005). Helicobacter pylori and MALT lymphoma. *Gastroenterology* **128,** 1579–1605.
62. Hocker, M., Raychowdhury, R., Plath, T., Wu, H., O'Connor, D. T., Wiedenmann, B., Rosewicz, S., and Wang, T. C. (1998). Sp1 and CREB mediate gastrindependent regulation of chromogranin A promoter activity in gastric carcinoma cells. *J. Biol. Chem.* **273,** 34000–34007.
63. Gerhard, M., Neumayer, N., Presecan-Siedel, E., Zanner, R., Lengyel, E., Cramer, T., Hocker, M., and Prinz, C. (2001). Gastrin induces expression and promoter activity of the vesicular monoamine transporter subtype 2. *Endocrinology* **142,** 3663–3672.
64. Zhao, C. M., Chen, D., Yamada, H., Dornonville de la Cour, C., Lindstrom, E., Persson, L., and Hakanson, R. (2003). Rat stomach ECL cells: Mode of activation of histidine decarboxylase. *Regul. Pept.* **114,** 21–27.
65. Fleming, J. V., Fajardo, I., Langlois, M. R., Sanchez-Jimenez, F., and Wang, T. C. (2004). The C-terminus of rat L-histidine decarboxylase specifically inhibits enzymic activity and disrupts pyridoxal phosphate-dependent interactions with L-histidine substrate analogues. *Biochem. J.* **381,** 769–778.
66. Suzuki, S., Tanaka, S., Nemoto, K., and Ichikawa, A. (1998). Membrane targeting and binding of the 74-kDa form of mouse L-histidine decarboxylase via its carboxyl-terminal sequence. *FEBS Lett.* **437,** 44–48.

67. Tanaka, S., Nemoto, K., Yamamura, E., and Ichikawa, A. (1998). Intracellular localization of the 74- and 53-kDa forms of L-histidine decarboxylase in a rat basophilic/mast cell line, RBL-2H3. *J. Biol. Chem.* **273,** 8177–8182.
68. Engel, N., Olmo, M. T., Coleman, C. S., Medina, M. A., Pegg, A. E., and Sanchez-Jimenez, F. (1996). Experimental evidence for structure-activity features in common between mammalian histidine decarboxylase and ornithine decarboxylase. *Biochem. J.* **320,** 365–368.
69. Olmo, M. T., Urdiales, J. L., Pegg, A. E., Medina, M. A., and Sanchez-Jimenez, F. (2000). In vitro study of proteolytic degradation of rat histidine decarboxylase. *Eur. J. Biochem.* **267,** 1527–1531.
70. Tanaka, S., Nemoto, K., Yamamura, E., Ohmura, S., and Ichikawa, A. (1997). Degradation of the 74 kDa form of L-histidine decarboxylase via the ubiquitin-proteasome pathway in a rat basophilic/mast cell line (RBL-2H3). *FEBS Lett.* **417,** 203–207.
71. Viguera, E., Trelles, O., Urdiales, J. L., Mates, J. M., and Sanchez-Jimenez, F. (1994). Mammalian L-amino acid decarboxylases producing 1,4-diamines: Analogies among differences. *Trends. Biochem. Sci.* **19,** 318–319.
72. Hammar, L., and Hjerten, S. (1980). Purification and immunochemical analysis of histidine decarboxylase from murine mastocytoma. *Agents Actions* **10,** 92–93.
73. Hammar, L. (1983). Mammalian histidine decarboxylase: Effect by protein kinase on mouse mastocytoma histidine decarboxylase. *Agents Actions* **13,** 246–249.
74. Taguchi, Y., Watanabe, T., Kubota, H., Hayashi, H., and Wada, H. (1984). Purification of histidine decarboxylase from the liver of fetal rats and its immunochemical and immunohistochemical characterization. *J. Biol. Chem.* **259,** 5214–5221.
75. Tran, V., and Snyder, S. (1981). Histidine decarboxylase. Purification from fetal rat liver, immunologic properties, and histochemical localization in brain and stomach. *J. Biol. Chem.* **256,** 680–686.
76. Martin, S. A., and Bishop, J. O. (1986). Purification and characterization of histidine decarboxylase from mouse kidney. *Biochem. J.* **234,** 349–354.
77. Savany, A., and Cronenberger, L. (1989). Inactivation of rat gastric mucosal histidine decarboxylase by phosphatase. *Biochem. Int.* **19,** 429–438.
78. Savany, A., and Cronenberger, L. (1982). Isolation and properties of multiple forms of histidine decarboxylase from rat gastric mucosa. *Biochem. J.* **205,** 405–412.
79. Huszti, Z., Magyar, K., and Keleti, J. (1991). Possible regulation of hypothalamus and lung histidine decarboxylase activity by cAMP-dependent protein kinase. *Eur. J. Biochem.* **197,** 191–196.
80. Yamamoto, J., Fukui, T., Suzuki, K., Tanaka, S., Yatsunami, K., and Ichikawa, A. (1993). Expression and characterization of recombinant mouse mastocytoma histidine decarboxylase. *Biochim. Biophys. Acta.* **1216,** 431–440.
81. Dartsch, C., Chen, D., and Persson, L. (1998). Multiple forms of rat stomach histidine decarboxylase may reflect posttranslational activation of the enzyme. *Regul. Pept.* **77,** 33–41.
82. Dartsch, C., Chen, D., Hakanson, R., and Persson, L. (1999). Histidine decarboxylase in rat stomach ECL cells: Relationship between enzyme activity and different molecular forms. *Regul. Pept.* **81,** 41–48.
83. Yatsunami, Y., Tsuchikawa, M., Kamada, M., Hori, K., and Higuchi, T. (1995). Comparative studies of human recombinant 74- and 54-kDa L-histidine decarboxylases. *J. Biol. Chem.* **270,** 30813–30817.
84. Rechsteiner, M. (1990). PEST sequences are signals for rapid intracellular proteolysis. *Semin. Cell Biol.* **1,** 433–440.
85. Rogers, S., Wells, R., and Rechsteiner, M. (1986). Amino acid sequences common to rapidly degraded proteins: The PEST hypothesis. *Science* **234,** 364–368.

86. Rechsteiner, M., and Rogers, S. W. (1996). PEST sequences and regulation by proteolysis. *Trends Biochem. Sci.* **21,** 267–271.
87. Van Antwerp, D. J., and Verma, I. M. (1996). Signal-induced degradation of I(kappa)B (alpha): Association with NF-kappaB and the PEST sequence in I(kappa)B(alpha) are not required. *Mol. Cell Biol.* **16,** 6037–6045.
88. Loetscher, P., Pratt, G., and Rechsteiner, M. (1991). The C terminus of mouse ornithine decarboxylase confers rapid degradation on dihydrofolate reductase. Support for the pest hypothesis. *J. Biol. Chem.* **266,** 11213–11220.
89. Georgescu, M. M., Kirsch, K. H., Akagi, T., Shishido, T., and Hanafusa, H. (1999). The tumor-suppressor activity of PTEN is regulated by its carboxyl-terminal region. *Proc. Natl. Acad. Sci. USA* **96,** 10182–10187.
90. Georgescu, M. M., Kirsch, K. H., Kaloudis, P., Yang, H., Pavletich, N. P., and Hanafusa, H. (2000). Stabilization and productive positioning roles of the C2 domain of PTEN tumor suppressor. *Cancer Res.* **60,** 7033–7038.
91. Schoonbroodt, S., Ferreira, V., Best-Belpomme, M., Boelaert, J. R., Legrand-Poels, S., Korner, M., and Piette, J. (2000). Crucial role of the amino-terminal tyrosine residue 42 and the carboxyl-terminal PEST domain of I kappa B alpha in NF-kappa B activation by an oxidative stress. *J. Immunol.* **164,** 4292–4300.
92. Alvarez-Castelao, B., and Castano, J. G. (2005). Mechanism of direct degradation of Ikappa-Balpha by 20S proteasome. *FEBS Lett.* **579,** 4797–4802.
93. Ciechanover, A. (1994). The ubiquitin-proteasome proteolytic pathway. *Cell.* **79,** 13–21.
94. Tanaka, S., Takasu, Y., Mikura, S., Satoh, N., and Ichikawa, A. (2002). Antigen-independent induction of histamine synthesis by immunoglobulin E in mouse bone marrow-derived mast cells. *J. Exp. Med.* **196,** 229–235.
95. Olmo, M. T., Rodriguez-Agudo, D., Medina, M. A., and Sanchez-Jimenez, F. (1999). The pest regions containing C-termini of mammalian ornithine decarboxylase and histidine decarboxylase play different roles in protein degradation. *Biochem. Biophys. Res. Commun.* **257,** 269–272.
96. Rodriguez-Agudo, D., Olmo, M. T., Sanchez-Jimenez, F., and Medina, M. A. (2000). Rat histidine decarboxylase is a substrate for m-calpain *in vitro*. *Biochem. Biophys. Res. Commun.* **271,** 777–781.
97. Tanaka, S., Deai, K., Konomi, A., Takahashi, K., Yamane, H., Sugimoto, Y., and Ichikawa, A. (2004). Expression of L-histidine decarboxylase in granules of elicited mouse polymorphonuclear leukocytes. *Eur. J. Immunol.* **34,** 1472–1482.
98. Tanaka, S., Fukui, T., Yamamoto, J., Shima, Y., Kume, T., Ohgo, M., and Ichikawa, A. (1995). Processing and activation of recombinant mouse mastocytoma histidine decarboxylase in the particulate fraction of Sf9 cells by porcine pancreatic elastase. *Biochim. Biophys. Acta* **1253,** 9–12.
99. Fleming, J. V., Sanchez-Jimenez, F., Moya-Garcia, A. A., Langlois, M. R., and Wang, T. C. (2004). Mapping of catalytically important residues in the rat L-histidine decarboxylase enzyme using bioinformatic and site-directed mutagenesis approaches. *Biochem. J.* **379,** 253–261.
100. Ichikawa, A., Fukui, T., Yamamoto, J., Ohgoh, M., Tanaka, S., and Funakoshi, S. (1995). De Novo synthesis and posttranslational processing of L-histidine decarboxylase in mice. *Methods Find Exp. Clin. Pharmacol.* **17**(Suppl. C), 5–9.
101. Safina, F., Tanaka, S., Inagaki, M., Tsuboi, K., Sugimoto, Y., and Ichikawa, A. (2002). Expression of L-histidine decarboxylase in mouse male germ cells. *J. Biol. Chem.* **277,** 14211–14215.
102. Christen, P., and Mehta, P. K. (2001). From cofactor to enzymes. The molecular evolution of pyridoxal-5′-phosphate-dependent enzymes. *Chem. Rec.* **1,** 436–447.

103. Jansonius, J. N. (1998). Structure, evolution and action of vitamin B6-dependent enzymes. *Curr. Opin. Struct. Biol.* **8**, 759–769.
104. Sandmeier, E., Hale, T. I., and Christen, P. (1994). Multiple evolutionary origin of pyridoxal-5′-phosphate-dependent amino acid decarboxylases. *Eur. J. Biochem.* **221**, 997–1002.
105. Hayashi, H. (1995). Pyridoxal enzymes: Mechanistic diversity and uniformity. *J. Biochem. (Tokyo)* **118**, 463–473.
106. Momany, C., Ghosh, R., and Hackert, M. L. (1995). Structural motifs for pyridoxal-5′-phosphate binding in decarboxylases: An analysis based on the crystal structure of the Lactobacillus 30a ornithine decarboxylase. *Protein Sci.* **5**, 849–854.
107. Qu, K., Martin, D. L., and Lawrence, C. E. (1998). Motifs and structural fold of the cofactor binding site of human glutamate decarboxylase. *Protein Sci.* **7**, 1092–1105.
108. Olmo, M. T., Sanchez-Jimenez, F., Medina, M. A., and Hayashi, H. (2002). Spectroscopic analysis of recombinant rat histidine decarboxylase. *J. Biochem. (Tokyo)* **132**, 433–439.
109. Capitani, G., Tramonti, A., Bossa, F., Grutter, M. G., and De Biase, D. (2003). The critical structural role of a highly conserved histidine residue in group II amino acid decarboxylases. *FEBS Lett.* **554**, 41–44.
110. Burkhard, P., Dominici, P., Borri-Voltattorni, C., Jansonius, J. N., and Malashkevich, V. N. (2001). Structural insight into Parkinson's disease treatment from drug-inhibited DOPA decarboxylase. *Nat. Struct. Biol.* **8**, 963–967.
111. Rodriguez-Caso, C., Rodriguez-Agudo, D., Moya-Garcia, A. A., Fajardo, I., Medina, M. A., Subramaniam, V., and Sanchez-Jimenez, F. (2003). Local changes in the catalytic site of mammalian histidine decarboxylase can affect its global conformation and stability. *Eur. J. Biochem.* **270**, 4376–4387.
112. Moya-Garcia, A. A., Medina, M. A., and Sanchez-Jimenez, F. (2005). Mammalian histidine decarboxylase: From structure to function. *Bioessays* **27**, 57–63.
113. Torrent, A., Moreno-Delgado, D., Gomez-Ramirez, J., Rodriguez-Agudo, D., Rodriguez-Caso, C., Sanchez-Jimenez, F., Blanco, I., and Ortiz, J. (2005). H3 autoreceptors modulate histamine synthesis through calcium/calmodulin- and cAMP-dependent protein kinase pathways. *Mol. Pharmacol.* **67**, 195–203.
114. Ishii, S., Hayashi, H., Okamoto, A., and Kagamiyama, H. (1998). Aromatic L-amino acid decarboxylase: Conformational change in the flexible region around Arg334 is required during the transaldimination process. *Protein Sci.* **7**, 1802–1810.
115. Kollonitsch, J., Perkins, L. M., Patchett, A. A., Doldouras, G. A., Marburg, S., Duggan, D. E., Maycock, A. L., and Aster, S. D. (1978). Selective inhibitors of biosynthesis of aminergic neurotransmitters. *Nature* **274**, 906–908.
116. Kubota, H., Hayashi, H., Watanabe, T., Taguchi, Y., and Wada, H. (1984). Mechanism of inactivation of mammalian L-histidine decarboxylase by (S)-alpha-fluoromethylhistidine. *Biochem. Pharmacol.* **33**, 983–990.
117. Hammar, L., Henningsson, H., Henningsson, A. C., Appelgren, L. E., Tjalve, H., and Kollonitsch, J. (1990). Autoradiography of 3H-alpha-fluoromethyl histidine in mice: Correlation with the kidney histidine decarboxylase activity. *Pharmacol. Toxicol.* **67**, 61–68.
118. Onodera, K., Yamatodani, A., and Watanabe, T. (1993). Effect of alpha-fluoromethylhistidine on brain histamine and noradrenaline in muricidal rats. *Methods Find Exp. Clin. Pharmacol.* **15**, 423–427.
119. Poulin, R., Lu, L., Ackermann, B., Bey, P., and Pegg, A. E. (1992). Mechanism of the irreversible inactivation of mouse ornithine decarboxylase by alpha-difluoromethylornithine. Characterization of sequences at the inhibitor and coenzyme binding sites. *J. Biol. Chem.* **267**, 150–158.
120. Gallagher, T., Snell, E. E., and Hackert, M. L. (1989). Pyruvoyl-dependent histidine decarboxylase. Active site structure and mechanistic analysis. *J. Biol. Chem.* **264**, 12737–12743.

121. Lindstrom, E., Andersson, K., Chen, D., Monstein, H. J., Boketoft, A., and Hakanson, R. (1997). alpha-Fluoromethylhistidine elevates histidine decarboxylase mRNA and chromogranin A mRNA levels in rat oxyntic mucosa. *Inflamm. Res.* **46**(Suppl. 1), S107–S108.
122. Maycock, A. L., Aster, S. D., and Patchett, A. A. (1980). Inactivation of 3-(3,4-dihydroxyphenyl)alanine decarboxylase by 2-(fluoromethyl)-3-(3,4-dihydroxyphenyl)alanine. *Biochemistry* **19**, 709–718.
123. Watanabe, T., Yamatodani, A., Maeyama, K., and Wada, H. (1990). Pharmacology of alpha-fluoromethylhistidine, a specific inhibitor of histidine decarboxylase. *Trends Pharmacol. Sci.* **11**, 363–367.
124. Hayashi, H., Tanase, S., and Snell, E. E. (1986). Pyridoxal 5′-phosphate-dependent histidine decarboxylase. Inactivation by alpha-fluoromethylhistidine and comparative sequences at the inhibitor- and coenzyme-binding sites. *J. Biol. Chem.* **261**, 11003–11009.
125. Bhattacharjee, M. K., and Snell, E. E. (1990). Pyridoxal 5′-phosphate-dependent histidine decarboxylase. Mechanism of inactivation by alpha-fluoromethylhistidine. *J. Biol. Chem.* **265**, 6664–6668.
126. Bartholeyns, J., and Bouclier, M. (1984). Involvement of histamine in growth of mouse and rat tumors. Antitumoral properties of monofluoromethylhistidine, an enzyme-activated irreversible inhibitor of histidine decarboxylase. *Cancer Res.* **44**, 639–645.
127. Garcia-Caballero, M., Neugebauer, E., Campos, R., Nunez de Castro, I., and Vara-Thorbeck, C. (1988). Increased histidine decarboxylase (HDC) activity in human colorectal cancer: Results of a study on ten patients. *Agents Actions* **23**, 357–360.
128. Masini, E., Fabbroni, V., Giannini, L., Vannacci, A., Messerini, L., Perna, F., Cortesini, C., and Cianchi, F. (2005). Histamine and histidine decarboxylase up-regulation in colorectal cancer: Correlation with tumor stage. *Inflamm. Res.* **54**(Suppl. 1), S80–S81.
129. Hegyesi, H., Somlai, B., Varga, V. L., Toth, G., Kovacs, P., Molnar, E. L., Laszlo, V., Karpati, S., Rivera, E., and Falus, A. (2001). Suppression of melanoma cell proliferation by histidine decarboxylase specific antisense oligonucleotides. *J. Invest. Dermatol.* **117**, 151–153.
130. Pos, Z., Safrany, G., Muller, K., Toth, S., Falus, A., and Hegyesi, H. (2005). Phenotypic profiling of engineered mouse melanomas with manipulated histamine production identifies histamine H2 receptor and rho-C as histamine-regulated melanoma progression markers. *Cancer Res.* **65**, 4458–4466.
131. Sieja, K., Stanosz, S., von Mach-Szczypinski, J., Olewniczak, S., and Stanosz, M. (2005). Concentration of histamine in serum and tissues of the primary ductal breast cancers in women. *Breast* **14**, 236–241.
132. Kolby, L., Wangberg, B., Ahlman, H., Modlin, I. M., Granerus, G., Theodorsson, E., and Nilsson, O. (1996). Histidine decarboxylase expression and histamine metabolism in gastric oxyntic mucosa during hypergastrinemia and carcinoid tumor formation. *Endocrinology* **137**, 4435–4442.
133. Boer, K., Darvas, Z., Bosze, S., Schwelberger, H., Baki, M., Belai, F., Pal, Z., and Falus, A. (2004). Histamine metabolism and CD8(+) T cell infiltration in colon adenomas. *Inflamm. Res.* **53**(Suppl. 1), S83–S84.
134. Rajendra, S., Mulcahy, H., Patchett, S., and Kumar, P. (2004). The effect of H2 antagonists on proliferation and apoptosis in human colorectal cancer cell lines. *Dig. Dis. Sci.* **49**, 1634–1640.
135. Natori, T., Sata, M., Nagai, R., and Makuuchi, M. (2005). Cimetidine inhibits angiogenesis and suppresses tumor growth. *Biomed. Pharmacother.* **59**, 56–60.
136. Tomita, K., Izumi, K., and Okabe, S. (2003). Roxatidine- and cimetidine-induced angiogenesis inhibition suppresses growth of colon cancer implants in syngeneic mice. *J. Pharmacol. Sci.* **93**, 321–330.

137. Diks, S. H., Hardwick, J. C., Diab, R. M., van Santen, M. M., Versteeg, H. H., van Deventer, S. J., Richel, D. J., and Peppelenbosch, M. P. (2003). Activation of the canonical beta-catenin pathway by histamine. *J. Biol. Chem.* **278**, 52491–52496.
138. Haak-Frendscho, M., Darvas, Z., Hegyesi, H., Karpati, S., Hoffman, R. L., Laszlo, V., Bencsath, M., Szalai, C., Furesz, J., and Timar, J. (2000). Histidine decarboxylase expression in human melanoma. *J. Invest. Dermatol.* **115**, 345–352.
139. Hegyesi, H., Horvath, B., Pallinger, E., Pos, Z., Molnar, V., and Falus, A. (2005). Histamine elevates the expression of Ets-1, a protooncogen in human melanoma cell lines through H2 receptor. *FEBS Lett.* **579**, 2475–2479.
140. Garcia-Caballero, M., Neugebauer, E., Rodriguez, F., Nunez de Castro, I., and Vara-Thorbeck, C. (1994). Histamine synthesis and content in benign and malignant breast tumours. Its effects on other host tissues. *Surg. Oncol.* **3**, 167–173.
141. Cricco, G. P., Davio, C. A., Martin, G., Engel, N., Fitzsimons, C. P., Bergoc, R. M., and Rivera, E. S. (1994). Histamine as an autocrine growth factor in experimental mammary carcinomas. *Agents Actions* **43**, 17–20.
142. Graff, L., Frungieri, M., Zanner, R., Pohlinger, A., Prinz, C., and Gratzl, M. (2002). Expression of histidine decarboxylase and synthesis of histamine by human small cell lung carcinoma. *Am. J. Pathol.* **160**, 1561–1565.
143. Tetlow, L. C., and Woolley, D. E. (2003). Histamine stimulates the proliferation of human articular chondrocytes *in vitro* and is expressed by chondrocytes in osteoarthritic cartilage. *Ann. Rheum. Dis.* **62**, 991–994.
144. Radvany, Z., Darvas, Z., Kerekes, K., Prechl, J., Szalai, C., Pallinger, E., Valeria, L., Varga, V. L., Sandor, M., and Erdei, A. (2000). H1 histamine receptor antagonist inhibits constitutive growth of Jurkat T cells and antigen-specific proliferation of ovalbumin-specific murine T cells. *Semin. Cancer Biol.* **10**, 41–45.
145. Folkman, J. (1975). Tumor angiogenesis: A possible control point in tumor growth. *Ann. Intern. Med.* **82**, 96–100.
146. Ferrara, N. (2002). VEGF and the quest for tumour angiogenesis factors. *Nat. Rev. Cancer* **2**, 795–803.
147. McMahon, G. (2000). VEGF receptor signaling in tumor angiogenesis. *Oncologist* **5**(Suppl. 1), 3–10.
148. Ribatti, D., Crivellato, E., Roccaro, A. M., Ria, R., and Vacca, A. (2004). Mast cell contribution to angiogenesis related to tumour progression. *Clin. Exp. Allergy* **34**, 1660–1664.
149. Iamaroon, A., Pongsiriwet, S., Jittidecharaks, S., Pattanaporn, K., Prapayasatok, S., and Wanachantararak, S. (2003). Increase of mast cells and tumour angiogenesis in oral squamous cell carcinoma. *J. Oral Pathol. Med.* **32**, 195–199.
150. Yano, H., Kinuta, M., Tateishi, H., Nakano, Y., Matsui, S., Monden, T., Okamura, J., Sakai, M., and Okamoto, S. (1999). Mast cell infiltration around gastric cancer cells correlates with tumor angiogenesis and metastasis. *Gastric Cancer* **2**, 26–32.
151. Sorbo, J., Jakobsson, A., and Norrby, K. (1994). Mast-cell histamine is angiogenic through receptors for histamine1 and histamine2. *Int. J. Exp. Pathol.* **75**, 43–50.
152. Ghosh, A. K., Hirasawa, N., Ohtsu, H., Watanabe, T., and Ohuchi, K. (2002). Defective angiogenesis in the inflammatory granulation tissue in histidine decarboxylase-deficient mice but not in mast cell-deficient mice. *J. Exp. Med.* **195**, 973–982.
153. Caruso, C., Lio, D., Cavallone, L., and Franceschi, C. (2004). Aging, longevity, inflammation, and cancer. *Ann. N. Y. Acad. Sci.* **1028**, 1–13.
154. Balkwill, F., Charles, K. A., and Mantovani, A. (2005). Smoldering and polarized inflammation in the initiation and promotion of malignant disease. *Cancer Cell* **7**, 211–217.

155. Houghton, J., Stoicov, C., Nomura, S., Rogers, A. B., Carlson, J., Li, H., Cai, X., Fox, J. G., Goldenring, J. R., and Wang, T. C. (2004). Gastric cancer originating from bone marrow-derived cells. *Science.* **306,** 1568–1571.
156. Pikarsky, E., Porat, R. M., Stein, I., Abramovitch, R., Amit, S., Kasem, S., Gutkovich-Pyest, E., Urieli-Shoval, S., Galun, E., and Ben-Neriah, Y. (2004). NF-kappaB functions as a tumour promoter in inflammation-associated cancer. *Nature.* **431,** 461–466.
157. Greten, F. R., Eckmann, L., Greten, T. F., Park, J. M., Li, Z. W., Egan, L. J., Kagnoff, M. F., and Karin, M. (2004). IKKbeta links inflammation and tumorigenesis in a mouse model of colitis-associated cancer. *Cell* **118,** 285–296.
158. MacGlashan, D., Jr. (2003). Histamine: A mediator of inflammation. *J. Allergy Clin. Immunol.* **112,** S53–S59.
159. Falus, A. (1993). Interleukin-6 biosynthesis is increased by histamine in human B-cell and glioblastoma cell lines. *Immunology* **78,** 193–196.
160. Delneste, Y., Lassalle, P., Jeannin, P., Joseph, M., Tonnel, A. B., and Gosset, P. (1994). Histamine induces IL-6 production by human endothelial cells. *Clin. Exp. Immunol.* **98,** 344–349.
161. Mor, S., Nagler, A., Barak, V., Handzel, Z. T., Geller-Bernstein, C., and Fabian, I. (1995). Histamine enhances granulocyte-macrophage colony-stimulating factor and interleukin-6 production by human peripheral blood mononuclear cells. *J. Leukoc. Biol.* **58,** 445–450.
162. Takamatsu, S., and Nakano, K. (1998). Regulation of interleukin-6, and macrophage colony-stimulating factor mRNA levels by histamine in stromal cell line (MC3T3-G2/PA6). *Inflamm. Res.* **47,** 221–226.
163. Horvath, B. V., Falus, A., Toth, S., Szalai, C., Lazar-Molnar, E., Holub, M. C., Buzas, E., Nagy, A., and Fulop, A. K. (2002). Inverse regulation of interleukin-6 (IL-6) and IL-6 receptor in histamine deficient histidine decarboxylase-knock-out mice. *Immunol. Lett.* **80,** 151–154.
164. Klausz, G., Buzas, E., Scharek, P., Tiszlavicz, L., Gyulai, Z., Fulop, A. K., Falus, A., and Mandi, Y. (2004). Effects of *Helicobacter pylori* infection on gastric inflammation and local cytokine production in histamine-deficient (histidine decarboxylase knock-out) mice. *Immunol. Lett.* **94,** 223–228.
165. Chen, D., Zhao, C. M., Andersson, K., Sundler, F., and Hakanson, R. (1996). Ultrastructure of enterochromaffin-like cells in rat stomach: Effects of alpha-fluoromethylhistidine-evoked histamine depletion and hypergastrinemia. *Cell Tissue Res.* **283,** 469–478.
166. Andersson, K., Zhao, C. M., Chen, D., Sundler, F., and Hakanson, R. (1996). Effect of alpha-fluoromethylhistidine-evoked histamine depletion on ultrastructure of endocrine cells in acid-producing mucosa of stomach in mouse, rat and hamster. *Cell Tissue Res.* **286,** 375–384.
167. Friis-Hansen, L., Sundler, F., Li, Y., Gillespie, P. J., Saunders, T. L., Greenson, J. K., Owyang, C., Rehfeld, J. F., and Samuelson, L. C. (1998). Impaired gastric acid secretion in gastrin-deficient mice. *Am. J. Physiol.* **274,** G561–G568.
168. Chen, D., Zhao, C. M., Al-Haider, W., Hakanson, R., Rehfeld, J. F., and Kopin, A. S. (2002). Differentiation of gastric ECL cells is altered in CCK (2) receptor-deficient mice. *Gastroenterology* **123,** 577–585.
169. Fossmark, R., Martinsen, T. C., Bakkelund, K. E., Kawase, S., and Waldum, H. L. (2004). ECL-cell derived gastric cancer in male cotton rats dosed with the H2-blocker loxtidine. *Cancer Res.* **64,** 3687–3693.
170. Kobayashi, T., Tonai, S., Ishihara, Y., Koga, R., Okabe, S., and Watanabe, T. (2000). Abnormal functional and morphological regulation of the gastric mucosa in histamine H2 receptor-deficient mice. *J. Clin. Invest.* **105,** 1741–1749.

171. Nakamura, E., Kataoka, T., Furutani, K., Jimbo, K., Aihara, T., Tanaka, S., Ichikawa, A., Ohtsu, H., and Okabe, S. (2004). Lack of histamine alters gastric mucosal morphology: Comparison of histidine decarboxylase-deficient and mast cell-deficient mice. *Am. J. Physiol. Gastrointest. Liver Physiol.* **287,** G1053–G1061.
172. Hunyady, B., Zolyomi, A., Czimmer, J., Mozsik, G., Kozicz, T., Buzas, E., Tanaka, S., Ichikawa, A., Nagy, A., and Palkovits, M. (2003). Expanded parietal cell pool in transgenic mice unable to synthesize histamine. *Scand. J. Gastroenterol.* **38,** 133–140.
173. Wang, T. C., Dangler, C. A., Chen, D., Goldenring, J. R., Koh, T., Raychowdhury, R., Coffey, R. J., Ito, S., Varro, A., and Dockray, G. J. (2000). Synergistic interaction between hypergastrinemia and *Helicobacter* infection in a mouse model of gastric cancer. *Gastroenterology* **118,** 36–47.

Eukaryotic Initiation Factor 2B and Its Role in Alterations in mRNA Translation That Occur Under a Number of Pathophysiological and Physiological Conditions

NEIL KUBICA,
LEONARD S. JEFFERSON, AND
SCOT R. KIMBALL

Department of Cellular and Molecular Physiology, The Pennsylvania State University College of Medicine, Hershey, Pennsylvania 17033

I. Introduction	272
II. Regulation of eIF2B GEF Activity	272
A. Role of eIF2α Phosphorylation	272
B. Role of eIF2Bε Phosphorylation	274
C. Regulation by Pyridine Dinucleotides	274
III. eIF2B Subunit Composition and Function	275
IV. Role of eIF2B in Contributing to Alterations in mRNA Translation Under Various Pathophysiological and Physiological Conditions	277
A. Diabetes	277
B. Resistance Exercise	280
C. Cancer	282
D. Sepsis	283
E. Vanishing White Matter Disease	285
V. Unanswered Questions	287
References	287

Eukaryotic initiation factor (eIF)2B is a guanine nucleotide exchange factor (GEF) for a second initiation factor, eIF2. During initiation of mRNA translation, eIF2 binds GTP and initiator methionyl-tRNA$_i$ (met-tRNA$_i$) and the resulting ternary complex subsequently binds to the 40S ribosomal subunit. Alterations in ternary complex formation due to changes in eIF2B GEF activity not only produce alterations in global rates of protein synthesis but also cause specific changes in the translation of mRNAs encoding certain proteins such as the transcription factors GCN2 and ATF4. Studies have implicated changes in eIF2B GEF activity in alterations in protein synthesis in a number of pathophysiological and physiological conditions such as diabetes, cancer, sepsis, vanishing white matter disease, and resistance exercise. The goal of the present chapter is to summarize the state of knowledge concerning the function of eIF2B and its role in regulating mRNA translation.

I. Introduction

Eukaryotic initiation factor (eIF)2B is a G-protein that plays a pivotal role in the regulation of mRNA translation. During one of the first steps in the initiation of mRNA translation, the initiation factor eIF2 binds to GTP and initiator methionyl-tRNA$_i$ (met-tRNA$_i$) to form a ternary complex that subsequently binds to a 40S ribosomal subunit to generate a 43S preinitiation complex. Several other initiation factors, including eIF4A, eIF4E, and eIF4G, mediate the binding of mRNA to the 43S preinitiation complex resulting in assembly of a 48S preinitiation complex. Addition of the 60S ribosomal subunit to generate the 80S initiation complex is preceded by the release of translation initiation factors from the 48S preinitiation complex. During the latter step, the GTPase activating protein eIF5 promotes the hydrolysis of GDP bound to eIF2 to GTP and eIF2 is released from the 48S preinitiation complex as an eIF2–GDP binary complex. Before eIF2 can bind to met-tRNA$_i$ and participate in another round of initiation, the GDP bound to the protein must be exchanged for GTP because, compared to the eIF2–GTP complex, the eIF2–GDP complex has only low affinity for met-tRNA$_i$. However, at physiological concentrations of GDP, GTP, and magnesium, the exchange of GDP bound to eIF2 for GTP is slow. Intracellularly, the exchange of eIF2-bound GDP for GTP is catalyzed by eIF2B, a guanine nucleotide exchange factor (GEF). In the absence of eIF2B GEF activity, eIF2 exists primarily in the eIF2–GDP complex, and protein synthesis is inhibited through repression of translation initiation. The goal of the present chapter is to review mechanisms through which eIF2B GEF activity is regulated and the role of the protein in contributing to altered mRNA translation under various physiological and pathophysiological conditions.

II. Regulation of eIF2B GEF Activity

A. Role of eIF2α Phosphorylation

The original identification of eIF2B was a consequence of studies designed to define the mechanism through which protein synthesis is inhibited in rabbit reticulocyte lysate (RRL) deprived of hemin. In contrast to hemin-supplemented reticulocyte lysate, protein synthesis is rapidly and almost completely suppressed in unsupplemented lysate [reviewed in (1)]. The inhibition of protein synthesis associated with hemin deprivation is caused by activation of a protein kinase termed hemin-regulated inhibitor (HRI) that phosphorylates the α-subunit of eIF2 on Ser^{51} [reviewed in (2)]. However, a number of early findings strongly suggested that phosphorylation by HRI did not directly inhibit the function of eIF2 but instead modulated the function of another protein (3). The putative eIF2 regulatory protein was subsequently purified from rabbit reticulocytes and shown to restore protein synthesis in hemin-deprived reticulocyte lysate (4). It was originally referred to as RF (4) and later by various names including anti-inhibitor (5), GEF (6), and, its current name, eIF2B (7). Subsequent studies showed that phosphorylation of the α-subunit of eIF2 on Ser^{51} converts the protein from a substrate of eIF2B into a competitive inhibitor of the guanine nucleotide exchange reaction [reviewed in (8)]. Because the amount of eIF2 in reticulocyte lysate, and in all mammalian cells examined to date, is less than the amount of eIF2B [reviewed in (8)], phosphorylation of only a portion of the total eIF2 is sufficient to repress the exchange activity of all of the eIF2B and, therefore, inhibit protein synthesis.

In addition to HRI, there are three other protein kinases identified in mammalian cells that phosphorylate Ser^{51} on eIF2α, including the double-stranded RNA-activated protein kinase [PKR (9)], the PKR-like endoplasmic reticulum-associated protein kinase [PERK or PEK, (10, 11)], and the general control nonderepressible protein kinase [GCN2 (12–14)]. Each of the eIF2α kinases is activated in response to divergent cellular stresses. For example, PKR is activated in response to infection by viruses that generate double-stranded RNA during replication, PERK is activated by accumulation of unfolded proteins in the lumen of the endoplasmic reticulum, and GCN2 is activated in cells deprived of essential amino acids [reviewed in (14)]. However, regardless of the kinase that phosphorylates eIF2α, inhibition of eIF2B GEF activity is the ultimate response.

Given that all mammalian proteins initiate with met-$tRNA_i$, it is not surprising that the inhibition of eIF2B GEF activity caused by phosphorylation of eIF2α would repress the translation of most mRNAs. However, less intuitive is the observation that phosphorylation of eIF2α enhances the translation of a subset of mRNAs, including those encoding the transcription factors Gcn4p (15) and ATF4 (16). The mRNAs belonging to this subset have multiple upstream open reading frames (uORFs) in the 5′-noncoding region of the mRNA that enhance the translation of the coding region when eIF2B activity is limiting. The mechanism responsible for the preferential increase in translation of such mRNAs is complex, and the reader is encouraged to peruse reviews for more information (e.g., 13). The number of mRNAs in *Saccharomyces cerevisiae* that have been found to contain uORFs is less than 40 (17, 18).

A functional analysis of the number of mammalian mRNAs that contain uORFs has not been performed. However, an analysis of 2195 Refman transcripts reveals that 41% of the transcripts have one or more upstream AUG codons (19). Although the presence of an upstream AUG does not imply the existence of a functional uORF, the finding suggests that in higher eukaryotes regulation of mRNA translation by uORFs may be more prevalent than in yeast.

B. Role of eIF2Bε Phosphorylation

Studies in animals and in cells in culture have demonstrated changes in eIF2B activity independent of alterations in eIF2α phosphorylation (20–24), indicating that other mechanisms must exist for regulating eIF2B GEF activity. In this regard, a variety of studies have shown that the ε-subunit of eIF2B can be phosphorylated by at least four different protein kinases including casein kinase (CK)-I, CK-II, glycogen synthase kinase (GSK)-3, and the dual-specificity tyrosine-phosphorylation kinase (DYRK). The available evidence is inconsistent in support of a role for phosphorylation of eIF2Bε by CK-I or CK-II in modulating eIF2B GEF activity. However, more convincing evidence has been presented in support of changes in eIF2B GEF activity in response to phosphorylation by GSK-3 (25, 26). The results of those studies suggest that phosphorylation of eIF2Bε by GSK-3 is associated with inhibition of GEF activity. Because DYRK phosphorylates the priming site for GSK-3 (27), that is, phosphorylation by DYRK precedes and is required for phosphorylation by GSK-3, DYRK indirectly modulates eIF2B GEF activity. However, reports demonstrating direct changes in eIF2B activity in response to phosphorylation by DYRK are lacking.

C. Regulation by Pyridine Dinucleotides

eIF2B isolated from RRL is associated with NADPH (28). Moreover, in in vitro assays, the guanine nucleotide exchange activity of eIF2B is inhibited by incubation with either NAD^+ or $NADP^+$, an effect that is blocked by equal molar amounts of reduced pyridine dinucleotides (22, 28). However, as noted previously (29), in rat liver the ratio of NAD^+:NADH is approximately 1000:1 (30), suggesting that eIF2B GEF activity would be substantially repressed under basal conditions. Except perhaps during fertilization of sea urchin eggs (31), there is little evidence that eIF2B GEF activity is regulated in vivo by pyridine dinucleotides.

III. eIF2B Subunit Composition and Function

The subunit composition of eIF2B is highly conserved among eukaryotes. In mammalian cells, eIF2B exists as a heteropentameric complex consisting of five subunits termed eIF2Bα, eIF2Bβ, eIF2Bγ, eIF2Bδ, and eIF2Bε that are encoded by the genes *EIF2B1*, *EIF2B2*, *EIF2B3*, *EIF2B4*, and *EIF2B5*, respectively. Each subunit is encoded by a single locus in the human genome (32). In yeast, the corresponding proteins are referred to as Gcn3p, Gcd7p, Gcd2p, and Gcd6p and are encoded by the genes *GCN3*, *GCD7*, *GCD1*, *GCD2*, and *GCD6*, respectively. Note that hereafter, the yeast protein will be referred to as yeIF2Bα, yeIF2Bβ, yeIF2Bγ, yeIF2Bδ, and yeIF2Bε, respectively. The yeast proteins are similar in size to their mammalian counterparts and exhibit 30–40% sequence identity at the amino acid level, suggesting that the function of the individual subunits may be conserved between yeast and mammals.

In yeast, each of the eIF2B subunits except yeIF2Bα is essential for eIF2B function because deletion of the gene encoding any one of the other four proteins is lethal. Deletion of GCN3 has no effect on cell growth and results in a failure to increase translation of the GCN4 mRNA in response to eIF2α phosphorylation, indicating that yeIF2Bα functions to recognize the phosphorylation state of eIF2α. However, other genetic studies performed in yeast suggest that yeIF2Bα does not function alone to recognize eIF2α phosphorylation. A number of point mutations have been identified in both yeIF2Bβ and yeIF2Bδ that suppress the effect of eIF2α phosphorylation on GCN4 expression without causing loss of yeIF2Bα from the yeIF2B holoprotein (33). Most of the mutations that were identified in that study map to regions of strong sequence similarity among the three proteins, a finding that suggests that the homologous regions shared by the proteins may have similar functions. yeIF2Bα, yeIF2Bβ, and yeIF2Bδ form a stable complex independent of the other two yeIF2B subunits (33, 34). A complex of those three subunits has no GEF activity toward yeIF2. However, the three subunit complex is able to bind to phosphorylated yeIF2, although no binding was observed for the individual subunits (34). The latter finding provides further support for the idea that the three subunit complex consisting of yeIF2Bα, yeIF2Bβ, and yeIF2Bδ binds to and recognizes the phosphorylation state of eIF2.

In yeast lacking yeIF2Bα or with mutations in GCN3, GCD2, or GCD7, the failure to induce Gcn4p expression in response to eIF2α phosphorylation could be due either to eIF2B not binding to phosphorylated eIF2 or to eIF2B catalyzing GDP/GTP exchange on phosphorylated eIF2. Evidence supporting the latter possibility is provided by a study showing that wildtype eIF2B exhibits low-GEF activity toward eIF2–[^{3}H]GDP when it is phosphorylated on Ser^{51} of its α-subunit compared to its activity toward the unphosphorylated eIF2–[^{3}H]GDP complex (34). In contrast, eIF2B lacking yeIF2Bα or eIF2B containing mutant yeIF2Bβ exhibits similar GEF activity toward both the phosphorylated and unphosphorylated eIF2–[^{3}H]GDP complexes. Likewise, a two subunit eIF2B complex consisting only of yeIF2Bγ and yeIF2Bε catalyzed GDP/GTP exchange equally well using either phosphorylated or unphosphorylated eIF2–[^{3}H]GDP as substrate.

Similar to yeast eIF2B, the α- and δ-subunits of mammalian eIF2B function to recognize the phosphorylation state of eIF2α. For example, the five subunit rat eIF2B complex is inhibited by incubation with eIF2 phosphorylated on Ser^{51} prior to assay using unphosphorylated eIF2–[^{3}H]GDP as substrate (35). In contrast, the four subunit eIF2B complex lacking eIF2Bα is not inhibited by preincubation with phosphorylated eIF2. Moreover, in contrast to the five subunit eIF2B complex, the four subunit complex lacking eIF2Bα catalyzes GDP/GTP exchange using either phosphorylated or unphosphorylated eIF2 as substrate (36). Overall, the available evidence suggests that, like Gcn3p, one role of eIF2Bα is to sense the phosphorylation state of eIF2α. A caveat to this assumption is the finding that eIF2B purified from rabbit reticulocytes and lacking the α-subunit exhibits little or no catalytic activity toward eIF2 regardless of its phosphorylation state (37). The basis for the different results obtained for the rabbit compared to the rat and yeast proteins is unknown.

The role of the δ-subunit of rat eIF2B in recognizing the phosphorylation state of eIF2α has also been examined using the Sf9 cell-baculovirus expression system (36). In that study, two point mutations corresponding to mutations in the yeast protein, which have been shown to engender insensitivity to eIF2α phosphorylation, were made in the cDNA encoding the δ-subunit of rat eIF2B (33). The mutant eIF2Bδ was coexpressed with the other four subunits of the protein and the eIF2B GEF activity was measured in cell extracts. These studies demonstrated that, similar to the yeast protein, the rat eIF2B complex containing the mutant δ-subunit is relatively insensitive to preincubation with phosphorylated eIF2 prior to GEF assay. However, unlike yeast eIF2B, the rat eIF2B complex containing the mutant δ-subunit catalyzes guanine nucleotide exchange using unphosphorylated, but not phosphorylated, eIF2–[^{3}H]GDP as substrate, suggesting that the affinity of the eIF2B complex containing the mutant δ-subunit for phosphorylated eIF2 is significantly less than for the wild-type protein. Fundamental differences must exist between rat and yeast in the mechanisms through which phosphorylated eIF2 inhibits eIF2B GEF activity.

In yeast, mammals, and *Drosophila* the largest eIF2B subunit (yeIF2Bε, eIF2Bε, and DeIF2Bε, respectively) is the only one to exhibit GEF activity when expressed alone (34, 35, 38). In fact, residues 518–712 of yeIF2Bε are sufficient for catalytic activity (39). Further N-terminal deletion to residue 581 results in loss of GEF activity but not eIF2 binding, suggesting that the minimal catalytic domain is located within a fragment of the protein corresponding to residues 518–580 and that the eIF2-binding domain is located near the C-terminus of the protein. Similarly, the eIF2-binding domain in rat eIF2Bε is also located in a conserved bipartite motif located near the C-terminus of the protein (40). Structural analysis of a fragment consisting of yeIF2Bε residues 544–704 reveals a series of helical structures with eight α-helices arranged in pairs (41). The first pair of helices is contained within residues 518–580 and has been proposed to comprise the catalytic core of the protein whereas the other three pairs constitute the eIF2-binding domain. Residues 518–712 of yeIF2Bε contain two functional domains, a catalytic domain, and an eIF2-binding domain.

Although both yeIF2Bε and eIF2Bε exhibit GEF activity when expressed alone, the specific activity of the isolated subunit is dramatically less than the five subunit holocomplex, suggesting that one or more other subunits are required for maximal activity. This suggestion is supported by the finding that in yeast, coexpression of yeIF2Bγ and yeIF2Bε yields a complex with a specific activity similar to the five subunit holocomplex (34). In contrast, the specific activity of rat eIF2Bε coexpressed with eIF2Bγ is the same as that of eIF2Bε alone and only about 10% of that of the five subunit holocomplex (42). Instead, coexpression of all the rat eIF2B subunits except the α is required for maximal GEF activity. The minimal rat complex with full activity consists of the β-, γ-, δ-, and ε-subunits.

Based on the observation that phosphorylation of eIF2α alters its interaction with eIF2B, it might be expected that the α-subunit of eIF2 would directly interact with one or more subunits of eIF2B. However, *in vitro* studies using purified proteins reveal that neither eIF2α nor eIF2α phosphorylated on Ser^{51} exhibit detectable eIF2B-binding activity (43). Instead, eIF2B binds to the β-subunit of eIF2. Likewise, the β-subunit of yeast eIF2 (encoded by SUI3) binds to yeIF2Bε (44). Examination of the binding of eIF2β to individual subunits of eIF2B reveals that the β-subunit of eIF2 binds to both the γ- and ε-subunits of eIF2B, but not to eIF2Bβ (43). The finding that both the γ- and ε-subunits bind to eIF2 may explain why both yeIF2Bγ and yeIF2Bε are required for maximal GEF activity in yeast, that is, both of those subunits might be required for optimal eIF2 binding. However, the finding does not clarify the requirement for the β- and δ-subunits to generate a mammalian eIF2B complex with full activity. Further studies will be required to elucidate the roles of those subunits in eIF2B function.

IV. Role of eIF2B in Contributing to Alterations in mRNA Translation Under Various Pathophysiological and Physiological Conditions

A. Diabetes

It has long been appreciated that diabetes mellitus is associated with drastic alterations in whole-body protein metabolism resulting in negative nitrogen balance, a condition that is rapidly reversed by exogenous administration of insulin (45). Forker et al. (46, 47) used incorporation of radiolabeled amino acid to establish that the negative nitrogen balance in a depancreatized dog model of diabetes is primarily explained by a decrease in synthesis of new proteins in skeletal muscle. Similar results are observed in perfused liver from normal vs alloxan-treated rats (48) and the diabetic rat heart (49, 50). Increased protein synthesis induced by insulin in isolated rat diaphragm muscle occurs independently of increased glucose (51–53) and/or amino acid uptake (53–55), a result repeated in the heart (56).

These results suggest that a primary cellular process is involved in mediating insulin-induced increases in protein synthesis. Although it is established that insulin induces an increase in rRNA transcription, this process has been excluded from consideration based on the rapid temporal nature of insulin-induced increases in protein synthesis (57) and use of the RNA transcription inhibitor actinomycin (58, 59), which is unable to prevent the observed increase in protein synthesis. Alternatively, ribosomes from diabetic rat muscle are defective in the ability to translate mRNA in vitro (49, 50). Furthermore, an insulin-dependent decrease in large actively translating polysome complexes is observed in diabetic muscle (60, 61) when compared to muscle from normal rats. These data provided the first insight that regulation of mRNA translation initiation may play a crucial role in aberrant protein synthesis observed in diabetes. Such a possibility is supported by studies of Flaim et al. (62) who report that initiation of mRNA translation is impaired in diabetic animals in an insulin-dependant manner. The observed reduction in the initiation of mRNA translation is observed only in muscles composed primarily of mixed fast-twitch muscle fiber (Types IIA and B) composition (i.e., psoas, gastrocnemius), not in muscles with a high percentage of slow-twitch (Type I) fibers (i.e., heart, soleus).

Subsequently, it has been shown that diabetes induces a reduction in eIF2B activity (21, 23) in muscles of diabetic animals, which are primarily composed of Type II fibers, but not in muscle primarily containing Type I fibers, suggesting a molecular mechanism responsible for the previously observed decrement in the initiation of mRNA translation. Analogous to the decline in translation initiation (62), the reduction in eIF2B GEF activity in diabetic animals is also rapidly reversed by insulin treatment (23). The reduction in eIF2B GEF activity is not associated with an alteration in eIF2α phosphorylation (21, 63), suggesting an alternative mode of regulation. The NADPH:NADP$^+$ ratio is unchanged in Type II muscle following induction of diabetes; however, the ratio is increased in the primarily Type I heart (63), suggesting elevated NADPH levels could be protective against reductions in eIF2B activity in vivo. However, the role of NADPH in the regulation of eIF2B GEF activity in vivo has been called into question. Although the tissue-specific differences between fast-twitch and slow-twitch muscles with respect to eIF2B regulation in response to diabetes/insulin availability is not elucidated, a contribution of free fatty acid (FFA) β-oxidation in slow-twitch muscles like the heart is implicated by a number of studies (64, 65).

Early insights into a role for insulin in the regulation of eIF2B GEF activity spurred numerous studies addressing the mechanisms responsible for this control. Welsh and Proud (66) have been able to establish that GSK-3 can phosphorylate eIF2B *in vitro*. Furthermore, insulin treatment of Chinese hamster ovary cells transfected with the human insulin receptor (CHO-T) in culture leads to inactivation of GSK-3 via phosphorylation of serine/threonine residues, as the effect is reversed by protein phosphatase 2A (PP2A). A previously published study (67) has determined that GSK-3 activity is decreased by insulin treatment in adipose tissue. GSK-3α (68) and GSK-3β (69) activities are repressed by Ser-21/Ser-9 phosphorylation, respectively, by various kinases known to be activated by insulin. The pharmacological inhibitor wortmannin (Wm) blocks the effect of insulin on GSK-3 activity (70, 71) implicating phosphatidyl inositol 3 kinase (PI3K) in the process. In contrast rapamycin, an inhibitor of the protein kinase mammalian target of rapamycin (mTOR) has no effect on GSK-3 repression by insulin. These observations have been interpreted to support a role for PI3K in repressing insulin-dependent GSK-3 activity through the mitogen activated protein kinase (MAPK)/ribosomal protein S6 kinase (RSK) signaling pathway.

This explanation began to be revised upon the discovery that Wm (72, 73) and overexpression of a dominant-negative PI3K mutant (72) also cause inhibition of PKB/Akt. Cross *et al.* (74) then established that PKB/Akt is the protein kinase responsible for phosphorylating and inactivating GSK-3 in response to insulin in L6 myoblasts. Furthermore, both rapamycin and PD98059 have no effect on insulin-dependent GSK-3 repression suggesting the exclusion of both the mTOR/S6K1 and MAPK signaling pathways, respectively. Subsequently, similar results were observed in primary adipocytes (75, 76), human myoblasts (77), CHO-T and CHO-IR cells (78), and rat skeletal muscle (75). Finally, Welsh *et al.* established (78) that activation of eIF2B by insulin required signaling through PI3K and identified the GSK-3 phosphorylation site on the rabbit eIF2Bε subunit as Ser540 (Ser535 in the rat sequence) (79). eIF2Bε phosphorylation by GSK-3 has subsequently been shown to reduce eIF2B activity in rabbit reticulate lysates and CHO-T cell extracts and that phosphorylation of eIF2Bε on this site is reduced in response to insulin, an effect blocked by the PI3K inhibitors Wm and LY294002.

In agreement with earlier study on translation initiation and eIF2B GEF activity, Jefferson and Kimball (80) went on to show that incorporation of radioactive phosphate into eIF2Bε is higher in extracts of the fast-twitch psoas muscle of diabetic animals compared to nondiabetic controls; however, no changes are observed in the predominantly slow-twitch heart. Again, similar to diabetes-associated changes in translation initiation and eIF2B GEF activity, kinase activity against eIF2Bε is reduced by treatment of diabetic animals with insulin. Furthermore, based on the phosphorylation patterns of wildtype vs various eIF2Bε mutants, the predominant insulin-dependent eIF2Bε kinase in skeletal muscle is GSK-3. While phosphorylation of the ε-subunit of eIF2B in response to insulin is clearly a promising mechanism for the regulation of protein synthesis in diabetes mellitus, a recent study (81) suggests that dephosphorylation of the aforementioned site on eIF2Bε is insufficient for activation of eIF2B by insulin. As numerous other phosphorylation sites have been described in the catalytic ε-subunit, further research is required to delineate the role of phosphorylation on eIF2B function in the diabetic state.

To this point, the focus of this section has been on the role of eIF2B in protein synthetic regulation in the absence of sufficient levels of insulin, as in Type I diabetes mellitus. However, discoveries in intracellular signaling [reviewed in (82)] suggest the possibility for a role of eIF2B in Type II diabetes as well. Two groups (83, 84) reported that hyperactivation of mTOR and S6K1, the downstream target of mTOR, leads to repression of PI3K signaling to PKB/Akt in response to insulin signaling. The proposed mechanism for this suppression of signaling to PKB involves phosphorylation of insulin receptor substrate (IRS)-1 on Ser-302 (and potentially other residues) by mTOR and/or S6K1, a phosphorylation event that prevents association of IRS-1 with the insulin receptor (84) and promotes IRS-1 degradation (83–86). One of the primary risk factors for Type II diabetes mellitus is obesity resultant from chronic nutrient excess states associated with hyperactivation of mTOR signaling. Wildtype mice fed a high-fat diet exhibit elevated S6K1 phosphorylation, IRS1 hyperphosphorylation, and attenuated PKB phosphorylation in response to insulin treatment, a result not observed in S6K1–/– animals (87). Because of the aforementioned role of PKB in inactivation of GSK-3 and derepression of eIF2B activity, future studies will be required to assess the possibility that this negative feedback inhibition on IRS-1 from mTOR/S6K1 may affect eIF2Bε phosphorylation, eIF2B activity, mRNA translation, and ultimately protein synthesis in a variety of tissues in various models of Type II diabetes mellitus.

B. Resistance Exercise

Skeletal muscle exhibits extraordinary phenotypic plasticity in response to functional demand. Resistance exercise is a complex physiological perturbation that promotes elevated rates of muscle protein synthesis and, when applied over time, skeletal muscle hypertrophy. A model of acute resistance exercise utilized by our group and others results in increased protein synthesis in skeletal muscles of the rat hindlimb (88–94), including muscles composed of primarily Type II fibers (i.e., gastrocnemius) and Type I fibers (i.e., soleus). The increase in muscle protein synthesis appears to require a permissive level of circulating insulin (88, 91). For example, when protein synthesis is measured by bilateral hindlimb perfusion *in situ*, no change in synthesis is observed when insulin is excluded from the perfusion media (88). Insulin availability also appears to be critical *in vivo* in the case of a partial pancreatectomy model of diabetes. Similar to normal control animals (arterial glucose < 180 mg/dl; arterial insulin = 450 pM), moderately diabetic animals (arterial glucose = 450–500 mg/dl; arterial insulin = 200 pM) also exhibit increased protein synthesis (91) following acute resistance exercise. However, severely diabetic animals (arterial glucose > 500 mg/dl; arterial insulin = 40 pM) show no alteration in protein synthetic rates after exercise. The ability of moderately diabetic animals to increase protein synthetic rates appears to require the action of IGF-1 (92).

The alteration in rates of protein synthesis in control, moderately diabetic, and severely diabetic animals correlates to eIF2B GEF activity (90, 93). That is, the gastrocnemius of both normal and moderately diabetic animals exhibit increased eIF2B GEF activity 16 hr following the completion of acute resistance exercise, while severely diabetic animals show no change in eIF2B GEF activity. As described in an earlier section in this chapter, the action of eIF2B promotes recycling of eIF2 and comprises one rate-limiting step in translation initiation. The other rate-limiting step involves delivery of mRNA to the ribosome and this process is regulated by the eIF4F complex, which is composed of eIF4E, eIF4G, and eIF4A. Delivery of mRNA is limited by the availability of eIF4E due to the low level of eIF4E expression and the action of an inhibitory family of eIF4E-binding proteins known as 4E-BPs. Hyperphosphorylation of the 4E-BPs prevent association with eIF4E so the initiation factor can enter the eIF4F complex. 4E-BP1 phosphorylation, 4E-BP1-eIF4E association, and eIF4G-eIF4E association are not altered concomitant with increases in protein synthesis following resistance exercise (95).

There has been an attempt to begin to understand the mechanisms by which eIF2B GEF activity is regulated during the recovery period following exercise. Several models of resistance exercise, muscle contractions, and/or muscle loading suggest that PKB/Akt phosphorylation or activity is increased in the recovery period following these perturbations (96–99). These results suggest a possible role for PKB-mediated GSK-3 repression and ensuing decreased eIF2Bε phosphorylation in modulation of eIF2B activity following exercise. GSK-3α and GSK-3β activity are repressed by endurance exercise (96). GSK-3β Ser-9 phosphorylation, but not GSK-3α Ser-21 phosphorylation, is increased after high-frequency electrical stimulation of rat skeletal muscle and correlated to a decrease in eIF2Bε Ser-535 phosphorylation (100).

However, in a physiological model of resistance exercise (94), no change in GSK-3β Ser-9 or eIF2Bε Ser-535 phosphorylation is observed concomitant with elevated rates of protein synthesis. Alternatively, a significant increase in eIF2Bε protein expression is observed as early as 3 hr postexercise (101) and remains elevated 16 h following resistance exercise (94), coincident with increased rates of protein synthesis. Both the elevation in muscle protein synthesis and eIF2Bε protein expression are prevented by pretreatment with the mTOR-inhibitor rapamycin (94). Furthermore, eIF2Bε mRNA shifts into actively translating polysomes following resistance exercise in an mTOR-dependent manner, suggesting translational control of eIF2Bε protein expression. Increased mRNA translation/protein expression of other eIF2B subunits is not observed. Numerous models of exercise or muscle contraction have established elevated signaling through mTOR (99, 102–104) and genetic activation of mTOR by PKB causes muscle hypertrophy *in vivo* (105–107), presumably by increasing mRNA translation and protein synthesis. In this respect, the focus of the literature is on phosphorylation of 4E-BP1 by mTOR, a subsequent increase in eIF4E availability, and elevated eIF4F complex assembly. However, as stated previously, these parameters are not altered concomitant with exercise-induced protein synthesis. The aforementioned mTOR-dependent increase in eIF2Bε mRNA translation and protein expression suggests a novel mechanism for mTOR control of protein synthesis in skeletal muscle following exercise. Importantly, the ε-subunit of eIF2B contains the catalytic site for guanine nucleotide exchange on eIF2 (108–110). Overexpression of the catalytic fragment (109) or eIF2Bε (111–114) protein alone can increase guanine nucleotide exchange. Overexpression of wildtype or a constitutively active S535A eIF2Bε mutant causes increased protein synthesis and cell size in primary cardiomyocytes (115).

C. Cancer

Disruption of normal mRNA translational control is known to contribute to malignant progression associated with numerous human cancers [reviewed in (116)]. Overexpression of eIF4E induces transformation of NIH3T3 and Rat 2 fibroblasts (117). Furthermore, eIF4E expression is observed in numerous human tumors and is believed to contribute to enhanced cellular proliferation, transformation, and metastasis [reviewed in (118)]. Genetically modified animals have been used to establish the role of eIF4E in oncogenesis *in vivo* (119). In contrast, the role of eIF2B in human cancer is relatively understudied; however, evidence suggests that eIF2B may also be an oncogene. PKR autophosphorylation and phosphorylation of the downstream target of PKR, eIF2α Ser-51, are both elevated in mammary carcinoma cell lines compared to nontransformed mammary epithelial cell lines (120). Phosphorylation of eIF2α would be expected to decrease protein synthetic rates via competitive inhibition of eIF2B; however, no such repression is observed. Increased expression of the α- and β-subunits of eIF2B is also described in this report, leading the authors to conclude that the lack of protein synthetic inhibition by eIF2α phosphorylation is the result of elevated eIF2B expression. Balachandran and Barber (121) demonstrated that transformed mouse embryonic fibroblasts display elevated eIF2B activity, along with a greater than tenfold increase in the relative expression of eIF2Bε compared to genetically matched control cells and an approximately twofold increase in protein synthesis. Similar to the previously mentioned study regarding eIF2Bε expression following resistance exercise (94), the observed increase in ε-subunit expression occurred independent of altered expression of other eIF2B subunits. Finally, eIF2Bε mRNA transcription is upregulated in a variety of human tumor samples (121).

There is also limited evidence to suggest that intracellular signaling to eIF2Bε is altered in various models of cellular transformation and tumorigenesis. Increased GSK-3β phosphorylation and inhibition of the ability of GSK-3β to phosphorylate eIF2Bε are observed early in the malignant phenotype of HT-29 human colon cancer cells (122). Similarly, transformation of hamster fibroblasts with Rous sarcoma virus (RSV) leads to activation of the PI3K/PKB/GSK-3 signaling and increased protein synthesis via activation of eIF2B (123). Mutations in the tumor suppressor phosphatase and tensin homolog (PTEN) (124) results in Cowden disease and loss of PTEN is found in numerous human cancer samples. PTEN acts to antagonize PI3K action by dephosphorylation of phosphatidylinositol-3,4,5-trisphosphate (PIP_3) to phosphatidylinositol-4,5-bisphosphate (PIP_2) (125), repressing phosphorylation and activation of PKB. Theoretically, prevention of PKB activation would lead to an increase in GSK-3 activity and a repression of eIF2B GEF activity. Since Cowden patients are predisposed to a variety of malignant cancers (126–128), the role of eIF2B in such tumor development must be explored. Manning et al. (129) showed that growth of tumors in mice lacking the tumor suppressor TSC2, another upstream suppressor of mTOR, are limited through feedback inhibition of PKB/Akt. These data are reminiscent of the aforementioned feedback inhibition of mTOR/S6K1 on IRS-1 suggested previously with regards to Type II diabetes. This report suggests the possibility that eIF2B action is involved in tumorigenesis and that feedback inhibition of IRS-1/PI3K/PKB/GSK-3/eIF2B could partially explain the benign hamartoma phenotype found in TSC patients.

D. Sepsis

Sepsis is a condition associated with serious infection and the systemic response to infection. The pathogenesis of sepsis involves proliferation of infectious microorganisms at the site of an infection, subsequent release of antigens and endotoxins, and hyperactivation of monocytes, macrophages, neutrophils, and other immune cells. The resulting exaggerated immune response results in hypotension, acute respiratory distress, renal failure, liver dysfunction, and muscle wasting [reviewed in (130)]. Sepsis is the leading cause of death in critically ill patients in the Unites States with over 750,000 cases each year and an ~28% mortality rate [reviewed in (131)].

Chronic sepsis is associated with severe muscle wasting. The observed muscle loss in a 5 day septic rat model is associated with a 50% reduction in protein synthesis compared to healthy control animals (132). Vary and Kimball (133) report that muscle protein synthesis is not reduced in nonseptic (i.e., sterile) inflammation and that the observed decrease in protein synthesis in septic animals is limited to muscles like the gastrocnemius that are primarily composed of Type II fibers. Furthermore, the sepsis-induced inhibition in protein synthesis is a result of decreased translation initiation (133), as a 1.6-fold increase in free 40S and 60S ribosomal subunits are observed. The repression in initiation of mRNA translation is not associated with altered eIF2 content or eIF2α Ser-51 phosphorylation; however, a significant decrease in eIF2B GEF activity is observed (134). The mechanism responsible for altered eIF2B GEF activity is not fully elucidated; however, several observations correlate with this finding. First, the expression of both the β- and ε-subunits of eIF2B are reduced ~50% compared to control levels during the first 5 days of sepsis, an observation associated with a similar decrease in eIF2Bε mRNA levels suggestive of transcriptional control (135). Second, PKB and GSK-3 phosphorylation are decreased in septic rats, while inhibitory phosphorylation of eIF2Bε on Ser-535 is increased (136).

It is well established that both amino acids and insulin can stimulate skeletal muscle protein synthesis in healthy control animals; however, these effects are attenuated in septic rats (137). Insulin fails to stimulate eIF2B GEF activity in gastrocnemius of septic animals despite unimpaired signaling through mTOR to both 4E-BP1 and S6K1 (138). Contrary to this finding, inclusion of IGF-1 in the perfusate in an isolated rat hindlimb preparation in septic rats results in a 2.5-fold increase in gastrocnemius protein synthesis (139) compared to septic animals into which no IGF-1 is perfused. Chronic sepsis leads to decreased circulating and gastrocnemius IGF-1 levels (140), suggesting a role for IGF-1 in sepsis-induced decrements in muscle protein synthesis. This suggestion is supported by the observation that infusion of the IGF-1/IGFBP-3 binary complex into septic rats attenuates inhibition of gastrocnemius protein synthesis (141).

It is widely accepted that elevated cytokine levels are responsible for the pathological effects associated with sepsis. Treatment of septic rats with interleukin (IL)-1 receptor antagonist (IL-1ra) prevents sepsis-induced decreases in plasma and gastrocnemius IGF-1 (140) and inhibition of both gastrocnemius protein synthesis and translation initiation (142). The inability of sepsis to inhibit initiation of mRNA translation in rats treated with IL-1ra is associated with maintenance of eIF2Bε protein content and ε-subunit mRNA (142).

Similarly, administration of tumor necrosis factor (TNF)-binding protein (TNFBP) also prevents decreased gastrocnemius protein synthesis associated with sepsis (143) and reduces eIF2Bε protein content. Taken together the literature suggests that the flood of cytokines, such as IL-1 and TNF, associated with sepsis causes a decrease in local and systemic IGF-1 that reduces protein synthesis in fast-twitch muscles as a result of decreased initiation of mRNA translation. The observed decrease in initiation of mRNA translation is at least partially explained by reduced eIF2B GEF activity that may be a consequence of altered catalytic ε-subunit expression and/or phosphorylation.

E. Vanishing White Matter Disease

Vanishing white matter (VWM) syndrome is an autosomal recessive neurological disease associated with loss of brain myelin and astocytes. Magnetic resonance imaging (144, 145) reveals a loss of white matter in the cerebral hemispheres and replacement with cerebrospinal fluid. While there is a varied clinical presentation, patients typically show normal early development; however, childhood neurological deterioration leads to abnormal gait, spasticity, optic atrophy, moderate cognitive impairment, and eventually fatality. Progression of the disease is chronic progressive with periods of rapid progression associated with head trauma or infection.

In 1999, Leegwater et al. (146) used genome linkage analysis in 19 families with members exhibiting VWM to map a gene associated with the disorder to a 5-cM interval on chromosome 3q27. A subsequent study (147) revealed this gene to be EIF2B5 encoding eIF2Bε and identified an additional VWM mutation at chromosome 14q24 corresponding to EIF2B2, the gene that encodes eIF2Bβ. It is now known that mutations in each of the five subunits of eIF2B can cause VWM (148) with varying degrees of severity. The role of eIF2B in VWM and a list of mutations in eIF2B subunits (currently 77 mutations in 148 patients) has been the reported in several reviews (149–151). The consequence of 90% of the known mutations is a homozygous missense substitution; however, heterozygous frameshifts and terminations are also observed. The most commonly mutated eIF2B subunit is the catalytic ε-subunit, with a total of 21 known mutations. Among these are Y309L (152) and R195H (153) substitutions that lead to a severe phenotype associated with early onset and rapid fatality. The later missense mutation is also observed in Cree leukoencephalopathy, another autosomal recessive leukodystrophy seen in North American Cree and Chippewayan populations.

The heterogeneity of the genetic background in VWM and the resulting differential impairment of eIF2B function is likely explanatory of the varied severity in clinical presentation. Several model systems including mutated yeast strains (154), HEK 293 human embryonic kidney cells (155), and transformed lymphocytes from leukodystrophy patients (156) have been used to assess the functional consequences of eIF2B mutations. Taken together these studies suggest that all the mutations associated with VWM cause partial loss of eIF2B function. However, different mutations alter eIF2B GEF activity by different means including reduction in protein expression of the affected subunit, reduced eIF2B holocomplex stability or inability to form complexes, compromised intrinsic catalytic activity, and inability to bind the substrate eIF2. These findings suggest that the variable phenotype observed in different patients may be reflective of these distinct mechanisms for disruption of eIF2B GEF activity and the extent of such aberrations.

One final mechanism that can be altered by eIF2B mutations associated with VWM is altered response to eIF2α Ser-51 phosphorylation. As discussed in an earlier section, eIF2α is phosphorylated in response to a variety of cellular stresses. This phosphorylation event causes eIF2 to be converted from a substrate to a competitive inhibitor of eIF2B activity reducing global rates of protein synthesis. Despite the global decrease in protein synthesis, translation of a subset of mRNAs containing elements, such as uORFs and internal ribosomal entry sites (IRESs) in the 5′-noncoding region, is increased under these conditions. mRNAs including GCN4 in yeast (157) and ATF4 (158) in mammalian cells have been shown to be regulated in this manner. Richardson et al. (154) showed that the eIF2Bβ^{V341D} mutant yeast strain displayed significantly elevated GCN4 expression, reduced rates of global translation, and severe growth impairment. Overexpression of two eIF2Bβ point mutants in HEK 293 cells (155) was also shown to increase expression of a ATF4 5′UTR-EGFP fusion construct. Primary fibroblasts from VWM patients (159) exhibit normal eIF2α phosphorylation and reduced rates of protein synthesis following endoplasmic reticulum (ER) stress (mediated by the ER transmembrane protein PERK). However, fibroblasts from VWM patients exhibit significantly greater induction of ATF4 compared to normal controls, suggesting that eIF2B is more sensitive to inhibition by phosphorylated eIF2 in such cells compared to control cells. Immunohistochemical analysis of glial cells from patients with VWM (160) exhibit increased phosphorylation of PERK and eIF2α accompanied by increased expression of ATF4 and its downstream transcriptional target C/EBP homologous protein (CHOP). Translational control of brain-specific messages in response to eIF2α phosphorylation and/or transcriptional control of genes downstream of transcription factors, such as ATF4, could explain the confined phenotype in VWM patients despite ubiquitous expression of mutated eIF2B subunits. An alternative explanation is that eIF2B expression levels may be low in the cells affected in VWM and the eIF2B mutations may make these cells predisposed to apoptosis when exposed to stress conditions. In support of this hypothesis, Inamura et al. (161) demonstrated that protein expression of the β-, γ-, δ-, and ε-subunits of eIF2B in the rat hippocampus peak at embryonic day 18 and decline throughout development.

Not only are the consequences of eIF2B mutations primarily confined to the brain but they also appear to be confined to particular cell-types within the brain. Patients with VWM actually display an increased number of oligodendrocytes (the CNS myelinating cells). Astrocyte abnormalities are observed in severe early-onset VWM patients (153) and primary cell cultures derived from a VWM patient with two heterozygous mutations in eIF2Bε (162) yielded normal oligodendrocytes but relatively few astrocytes. Additionally, knockdown of eIF2Bε using small interference RNA (siRNA) technology inhibited induction of astrocytes from human glial progenitor cells. Decrements in glial cell differentiation as a consequence of mutation in eIF2B could at least partially explain the distinct regional loss of white matter in VWM syndrome.

V. Unanswered Questions

The studies described previously suggest a critical role for eIF2B in a number of pathophysiological as well as physiological conditions. However, a number of important questions are still unanswered including:

- What are the functional consequences of altered eIF2Bε mRNA translation?
- What is the mechanism through which eIF2Bε mRNA translation is preferentially upregulated under conditions of enhanced signaling through mTOR?
- Is enhanced signaling through mTOR sufficient to increase eIF2Bε expression and, if so, what is the identity of the upstream activators and downstream targets of mTOR involved in the effect?
- Why are the cells comprising the cerebral white matter particularly sensitive to mutations in eIF2B subunits?

The solution to these and other questions will hopefully be answered in the next few years.

References

1. Safer, B., Jagus, R., Konieczny, A., and Crouch, D. (1982). The mechanism of translational inhibition in hemin-deficient lysates. *In* "Interactions of Translational and Transcriptional Controls in the Regulation of Gene Expression" (M. Grunberg-Manago and B. Safer, Eds.), pp. 311–325. Elsevier Science Publishing Co., Inc., NY, NY.
2. Chen, J.-J. (2000). Heme-regulated eIF2α kinase. *In* "Translational Control of Gene Expression" (N. Sonenberg, J. W. B. Hershey, and M. B. Mathews, Eds.), pp. 529–546. Cold Spring Harbor Laboratory Press, Cold Spring Harbor, NY.

3. Ochoa, S. (1983). Regulation of protein synthesis initiation in eucaryotes. *Arch. Biochem. Biophys.* **223**, 325–349.
4. Ralston, R. O., Das, A., Dasgupta, A., Roy, R., Palmieri, S., and Gupta, N. K. (1978). Protein synthesis in rabbit reticulocytes: Characteristics of a ribosomal factor that reverses inhibition of protein synthesis in heme-deficient lysates. *Proc. Natl. Acad. Sci. USA* **75**, 4858–4862.
5. Amesz, H., Goumans, H., Haubrich-Morree, T., Voorma, H. O., and Benne, R. (1979). Purification and characterization of a protein factor that reverses the inhibition of protein synthesis by the heme-regulated translational inhibitor in rabbit reticulocyte lysates. *Eur. J. Biochem.* **98**, 513–520.
6. Panniers, R., and Henshaw, E. C. (1983). A GDP/GTP exchange factor essential for eukaryotic initiation factor 2 cycling in ehrlich ascites tumor cells and its regulation by eukaryotic initiation factor 2 phosphorylation. *J. Biol. Chem.* **258**, 7928–7934.
7. Konieczny, A., and Safer, B. (1983). Purification of the eukaryotic initiation factor 2-eukaryotic initiation factor 2B complex and characterization of its guanine nucleotide exchange activity during protein synthesis initiation. *J. Biol. Chem.* **258**, 3402–3408.
8. Hinnebusch, A. G. (2000). Mechanism and regulation of initiator methionyl-tRNA binding to ribosomes. *In* "Translational Control of Gene Expression" (N. Sonenberg, J. W. B. Hershey, and M. B. Mathews, Eds.), pp. 185–243. Cold Spring Harbor Laboratory Press, Cold Spring Harbor, NY.
9. Farrell, P. J., Balkow, K., Hunt, T., Jackson, R. J., and Trachsel, H. (1977). Phosphorylation of initiation factor eIF-2 and the control of reticulocyte protein synthesis. *Cell* **11**, 187–200.
10. Harding, H. P., Zhang, Y., and Ron, D. (1999). Protein translation and folding are coupled by an endoplasmic-reticulum-resident kinase. *Nature* **397**, 271–274.
11. Shi, Y., Vattem, K. M., Sood, R., An, J., Liang, J., Stramm, L., and Wek, R. C. (1998). Identification and characterization of pancreatic eukaryotic initiation factor 2 α-subunit kinase, PEK, involved in translational control. *Mol. Cell. Biol.* **18**, 7499–7509.
12. Berlanga, J. J., Santoyo, J., and de Haro, C. (1999). Characterization of a mammalian homolog of GCN2 eukaryotic initiation factor 2α kinase. *Eur. J. Biochem.* **265**, 754–762.
13. Hinnebusch, A. G., and Natarajan, K. (2002). Gcn4p, a master regulator of gene expression, is controlled at multiple levels by diverse signals of starvation and stress. *Eukaryot. Cell* **1**, 22–32.
14. Proud, C. G. (2005). eIF2 and the control of cell physiology. *Semin. Cell Develop. Biol.* **16**, 3.
15. Hinnebusch, A. G. (1994). Translational control of *GCN4*: An *in vivo* barometer of initiation factor activity. *Trends Biochem. Sci.* **19**, 409–414.
16. Vattem, K. M., and Wek, R. C. (2004). Reinitiation involving upstream ORFs regulates ATF4 mRNA translation in mammalian cells. *Proc. Natl. Acad. Sci. USA* **101**, 11269–11274.
17. Vilela, C., and McCarthy, J. E. G. (2003). Regulation of fungal gene expression via short open reading frames in the mRNA 5′untranslated region. *Mol. Microbiol.* **49**, 859–867.
18. Zhang, Z., and Dietrich, F. S. (2005). Identification and characterization of upstream open reading frames (uORF) in the 5′-untranslated regions (UTR) of genes in *Saccharomyces cerevisiae*. *Curr. Genet.* **48**, 77–87.
19. Peri, S., and Pandey, A. (2001). A reassessment of the translation initiation codon in vertebrates. *Trends Genet.* **17**, 685–687.
20. Cox, S., Redpath, N. T., and Proud, C. G. (1988). Regulation of polypeptide-chain initiation in rat skeletal muscle starvation does not alter the activity or phosphorylation state of initiation factor eIF-2. *FEBS Lett.* **239**, 333–338.
21. Jeffrey, I. W., Kelly, F. J., Duncan, R., Hershey, J. W. B., and Pain, V. M. (1990). Effect of starvation and diabetes on the activity of the eukaryotic initiation factor eIF-2 in rat skeletal muscle. *Biochimie* **72**, 751–757.

22. Karinch, A. M., Kimball, S. R., Vary, T. C., and Jefferson, L. S. (1993). Regulation of eukaryotic initiation factor 2B activity in muscle of diabetic rats. *Am. J. Physiol.* **264,** E101–E108.
23. Kimball, S. R., and Jefferson, L. S. (1988). Effect of diabetes on guanine nucleotide exchange factor activity in skeletal muscle and heart. *Biochem. Biophys. Res. Commun.* **156,** 706–711.
24. Welsh, G. I., and Proud, C. G. (1992). Regulation of protein synthesis in Swiss 3T3 fibroblasts. Rapid activation of the guanine nucleotide exchange factor by insulin and growth factors. *Biochem. J.* **284,** 19–23.
25. Wang, X., Janmaat, M., Beugnet, A., Paulin, F. E., and Proud, C. G. (2002). Evidence that the dephosphorylation of Ser(535) in the epsilon-subunit of eukaryotic initiation factor (eIF) 2B is insufficient for the activation of eIF2B by insulin. *Biochem. J.* **367,** 475–481.
26. Welsh, G. I., Miller, C. M., Loughlin, A. J., Price, N. T., and Proud, C. G. (1998). Regulation of eukaryotic initiation factor eIF2B: Glycogen synthase kinase-3 phosphorylates a conserved serine which undergoes dephosphorylation in response to insulin. *FEBS Lett.* **421,** 125–130.
27. Woods, Y. L., Cohen, P., Becker, W., Jakes, R., Goedert, M., Wang, X., and Proud, C. G. (2001). The kinase DYRK phosphorylates protein-synthesis initiation factor eIF2Bepsilon at Ser539 and the microtubule-associated protein tau at Thr212: Potential role for DYRK as a glycogen synthase kinase 3-priming kinase. *Biochem. J.* **355,** 609–615.
28. Dholakia, J. N., Mueser, T. C., Woodley, C. L., Parkhurst, L. J., and Wahba, A. J. (1986). The association of NADPH with the guanine nucleotide exchange factor from rabbit reticulocytes: A role of pyridine dinucleotides in eukaryotic polypeptide chain initiation. *Proc. Natl. Acad. Sci. USA* **83,** 6746–6750.
29. Kimball, S. R., Mellor, H., Flowers, K. M., and Jefferson, L. S. (1996). Role of translation initiation factor eIF-2B in the regulation of protein synthesis in mammalian cells. *Prog. Nucleic Acid Res. Mol. Biol.* **54,** 165–196.
30. Greenbaum, A. L., Gumaa, K. A., and McLean, P. (1971). The distribution of hepatic metabolites and the control of the pathways of carbohydrate metabolism in animals of different dietary and hormonal status. *Arch. Biochem. Biophys.* **143,** 617–663.
31. Akkaraju, G. R., Hansen, L. J., and Jagus, R. (1991). Increase in eukaryotic initiation factor 2B activity following fertilization reflects changes in redox potential. *J. Biol. Chem.* **266,** 24451–24459.
32. Abbott, C. M., and Proud, C. G. (2004). Translation factors: In sickness and in health. *Trends Biochem. Sci.* **29,** 25–31.
33. Pavitt, G. D., Yang, W., and Hinnebusch, A. G. (1997). Homologous segments in three subunits of the guanine nucleotide exchange factor eIF2B mediate translational regulation by phosphorylation of eIF2. *Mol. Cell. Biol.* **17,** 1298–1313.
34. Pavitt, G. D., Ramaiah, K. V. A., Kimball, S. R., and Hinnebusch, A. G. (1998). eIF2 independently binds two distinct eIF2B subcomplexes that catalyze and regulate guanine-nucleotide exchange. *Genes Dev.* **12,** 514–526.
35. Fabian, J. R., Kimball, S. R., Heinzinger, N. D., and Jefferson, L. S. (1997). Subunit assembly and guanine nucleotide exchange activity of eukaryotic initiation factor-2B (eIF-2B) expressed in Sf9 cells. *J. Biol. Chem.* **272,** 12359–12365.
36. Kimball, S. R., Fabian, J. R., Pavitt, G. D., Hinnebusch, A. G., and Jefferson, L. S. (1998). Regulation of guanine nucleotide exchange through phosphorylation of eukaryotic initiation factor eIF2α. Role of the α- and δ-subunits of eIF2B. *J. Biol. Chem.* **273,** 12841–12845.
37. Craddock, B. L., and Proud, C. G. (1996). The α-subunit of the mammalian guanine nucleotide-exchange factor eIF-2B is essential for catalytic activity *in vitro. Biochem. Biophys. Res. Commun.* **220,** 843–847.

38. Williams, D. D., Pavitt, G. D., and Proud, C. G. (2001). Characterization of the initiation factor eIF2B and its regulation in *Drosophila melanogaster*. *J. Biol. Chem.* **276**, 3733–3742.
39. Gomez, E., Mohammad, S. S., and Pavitt, G. D. (2002). Characterization of the minimal catalytic domain within eIF2B: The guanine-nucleotide exchange factor for translation initiation. *EMBO J.* **21**, 5292–5301.
40. Anthony, T. G., Fabian, J. R., Kimball, S. R., and Jefferson, L. S. (2000). Identification of domains within the ε-subunit of the translation initiation factor eIF2B that are necessary for guanine nucleotide exchange activity and eIF2B holoprotein formation. *Biochim. Biophys. Acta* **1492**, 56–62.
41. Boesen, T., Mohammad, S. S., Pavitt, G. D., and Andersen, G. R. (2004). Structure of the catalytic fragment of translation initiation factor 2B and identification of a critically important catalytic residue. *J. Biol. Chem.* **279**, 10584–10592.
42. Fabian, J. R., Kimball, S. R., and Jefferson, L. S. (1998). Reconstitution and purification of eukaryotic initiation factor 2B (eIF2B) expressed in Sf21 insect cells. *Prot. Express. Purif.* **13**, 16–22.
43. Kimball, S. R., Heinzinger, N. K., Horetsky, R. L., and Jefferson, L. S. (1998). Identification of interprotein interactions between the subunits of eukaryotic initiation factors eIF2 and eIF2B. *J. Biol. Chem.* **273**, 3039–3044.
44. Asano, K., Krishnamoorthy, T., Phan, L., Pavitt, G. D., and Hinnebusch, A. G. (1999). Conserved bipartite motifs in yeast eIF5 and eIF2Bε, GTPase-activating and GDP-GTP exchange factors in translation initiation, mediate binding to their common substrate eIF2. *EMBO J.* **18**, 1673–1688.
45. Chaikoff, I. L., and Forker, L. L. (1950). The antidiabetic action of insulin on nitrogen metabolism. *Endocrinology* **46**, 319–326.
46. Forker, L. L., Chiakoff, I. L., Entenman, C., and Tarver, H. (1951). Oxidation of methionine S35 to sulfate by the eviscerated dog. *J. Biol. Chem.* **188**, 31–35.
47. Forker, L. L., and Chaikoff, I. L. (1952). Turnover of serum proteins in diabetes as studied with S35 labeled proteins. *J. Biol. Chem.* **196**, 829–840.
48. Green, M., and Miller, L. L. (1960). Protein catabolism and protein synthesis in perfused livers of normal and alloxan-diabetic rats. *J. Biol. Chem.* **235**, 3202–3208.
49. Rampersad, O. R., and Wool, I. G. (1965). Protein synthesis by ribosomes from heart muscle: Effect of insulin and diabetes. *Science* **149**, 1102–1103.
50. Wool, I. G., Rampersad, O. R., and Moyer, A. N. (1966). Effect of insulin and diabetes on protein synthesis by ribosomes from heart muscle. Significance for theories of the hormone's mechanism of action. *Am. J. Med.* **40**, 716–723.
51. Manchester, K. L., and Young, F. G. (1958). The effect of insulin on incorporation of amino acids into protein of normal rat diaphragm *in vitro*. *Biochem. J.* **70**, 353–358.
52. Wool, I. G., and Krahl, M. E. (1959). Incorporation of C14-amino acids into protein of isolated diaphragms: An effect of insulin independent of glucose entry. *Am. J. Physiol.* **196**, 961–964.
53. Wool, I. G., and Krahl, M. E. (1959). An effect of insulin on peptide synthesis independent of glucose or amino-acid transport. *Nature* **183**, 1399–1400.
54. Manchester, K. L., and Krahl, M. E. (1959). Effect of insulin on the incorporation of C14 from C14-labeled carboxylic acids and bicarbonate into the protein of isolated rat diaphragm. *J. Biol. Chem.* **234**, 2938–2942.
55. Wool, I. G. (1965). Relation of effects of insulin on amino acid transport and on protein synthesis. *Fed. Proc.* **24**, 1060–1070.
56. Stirewalt, W. S., and Wool, I. G. (1966). Protein synthesis by heart muscle ribosomes: An effect of insulin independent of substrate transport. *Science* **154**, 284–285.
57. Wool, I. G. (1961). Effect of insulin on distribution of radioactivity in protein of cell fractions from isolated rat diaphragm. *Biochim. Biophys. Acta* **52**, 574–576.

58. Eboue Bonis, D., Chambaut, A. M., Volfin, P., and Clauser, H. (1963). Action of insulin on the isolated rat diaphragm in the presence of actinomycin D and puromycin. *Nature* **199**, 1183–1184.
59. Wool, I. G., and Moyer, A. N. (1964). Effect of actinomycin and insulin on the metabolism of isolated rat diaphragm. *Biochim. Biophys. Acta* **91**, 248–256.
60. Fahmy, L. H., and Leader, D. P. (1980). The effect of diabetes and insulin on the polyadenylic acid-containing RNA of rat skeletal muscle. *Biochim. Biophys. Acta* **608**, 344–357.
61. Stirewalt, W. S., Wool, I. G., and Cavicchi, P. (1967). The relation of RNA and protein synthesis to the sedimentation of muscle ribosomes: Effect of diabetes and insulin. *Proc. Natl. Acad. Sci. USA* **57**, 1885–1892.
62. Flaim, K. E., Copenhaver, M. E., and Jefferson, L. S. (1980). Effects of diabetes on protein synthesis in fast- and slow-twitch rat skeletal muscle. *Am. J. Physiol.* **239**, E88–E95.
63. Karinch, A. M., Kimball, S. R., Vary, T. C., and Jefferson, L. S. (1993). Regulation of eukaryotic initiation factor-2B activity in muscle of diabetic rats. *Am. J. Physiol.* **264**, E101–E108.
64. Crozier, S. J., Bolster, D. R., Reiter, A. K., Kimball, S. R., and Jefferson, L. S. (2002). β-Oxidation of free fatty acids is required to maintain translational control of protein synthesis in heart. *Am. J. Physiol.* **283**, E1144–E1150.
65. Crozier, S. J., Anthony, J. C., Schworer, C. M., Reiter, A. K., Anthony, T. G., Kimball, S. R., and Jefferson, L. S. (2003). Tissue-specific regulation of protein synthesis by insulin and free fatty acids. *Am. J. Physiol. Endocrinol. Metab.* **285**, E754–E762.
66. Welsh, G. I., and Proud, C. G. (1993). Glycogen synthase kinase-3 is rapidly inactivated in response to insulin and phosphorylates eukaryotic initiation factor eIF-2B. *Biochem. J.* **294**, 625–629.
67. Ramakrishna, S., and Benjamin, W. (1988). Insulin action rapidly decreases multifunctional protein kinase activity in rat adipose tissue. *J. Biol. Chem.* **263**, 12677–12681.
68. Sutherland, C., and Cohen, P. (1994). The alpha-isoform of glycogen synthase kinase-3 from rabbit skeletal muscle is inactivated by p70 S6 kinase or MAP kinase-activated protein kinase-1 in vitro. *FEBS Lett.* **338**, 37–42.
69. Sutherland, C., Leighton, I., and Cohen, P. (1993). Inactivation of glycogen synthase kinase-3 beta by phosphorylation: New kinase connections in insulin and growth-factor signalling. *Biochem. J.* **296**, 15–19.
70. Cross, D. A., Alessi, D. R., Vandenheede, J. R., McDowell, H. E., Hundal, H. S., and Cohen, P. (1994). The inhibition of glycogen synthase kinase-3 by insulin or insulin-like growth factor 1 in the rat skeletal muscle cell line L6 is blocked by wortmannin, but not by rapamycin: Evidence that wortmannin blocks activation of the mitogen-activated protein kinase pathway in L6 cells between Ras and Raf. *Biochem. J.* **303**, 21–26.
71. Welsh, G. I., Foulstone, E. J., Young, S. W., Tavare, J. M., and Proud, C. G. (1994). Wortmannin inhibits the effects of insulin and serum on the activities of glycogen synthase kinase-3 and mitogen-activated protein kinase. *Biochem. J.* **303**, 15–20.
72. Burgering, B. M., and Coffer, P. J. (1995). Protein kinase B (c-Akt) in phosphatidylinositol-3-OH kinase signal transduction. *Nature* **376**, 599–602.
73. Franke, T. F., Yang, S. I., Chan, T. O., Datta, K., Kazlauskas, A., Morrison, D. K., Kaplan, D. R., and Tsichlis, P. N. (1995). The protein kinase encoded by the Akt proto-oncogene is a target of the PDGF-activated phosphatidylinositol 3-kinase. *Cell* **81**, 727–736.
74. Cross, D. A. E., Alessi, D. R., Cohen, P., Andjelkovich, M., and Hemmings, B. A. (1995). Inhibition of glycogen synthase kinase-3 by insulin mediated by protein kinase B. *Nature* **378**, 785–789.
75. Cross, D. A. E., Watt, P. W., Shaw, M., van der Kaay, J., Downes, C. P., Holder, J. C., and Cohen, P. (1997). Insulin activates protein kinase B, inhibits glycogen synthase kinase-3 and

activates glycogen synthase by rapamycin-insensitive pathways in skeletal muscle and adipose tissue. *FEBS Lett.* **406,** 211–215.

76. Moule, S. K., Edgell, N. J., Welsh, G. I., Diggle, T. A., Proud, C. G., and Denton, R. M. (1995). Studies with wortmannin indicate that insulin stimulation of lipogenesis involves multiple signalling pathways. *Biochem. Soc. Trans.* **23,** 206S.

77. Hurel, S. J., Rochford, J. J., Borthwick, A. C., Wells, A. M., Vandenheede, J. R., Turnbull, D. M., and Yeaman, S. J. (1996). Insulin action in cultured human myoblasts: Contribution of different signalling pathways to regulation of glycogen synthesis. *Biochem. J.* **320,** 871–877.

78. Welsh, G. I., Stokes, C. M., Wang, X., Sakaue, H., Ogawa, W., Kasuga, M., and Proud, C. G. (1997). Activation of translation initiation factor eIF2B by insulin requires phosphatidyl inositol 3-kinase. *FEBS Lett.* **410,** 418–422.

79. Welsh, G. I., Miller, C. M., Loughlin, A. J., Price, N. T., and Proud, C. G. (1998). Regulation of eukaryotic initiation factor eIF2B: Glycogen synthase kinase-3 phosphorylates a conserved serine which undergoes dephosphorylation in response to insulin. *FEBS Lett.* **421,** 125–130.

80. Jefferson, L. S., Fabian, J. R., and Kimball, S. R. (1999). Glycogen synthase kinase-3 is the predominant insulin-regulated eukaryotic initiation factor 2B kinase in skeletal muscle. *Int. J. Biochem. Cell. Biol.* **31,** 191–200.

81. Wang, X., Janmaat, M., Beugent, A., Paulin, F. E., and Proud, C. G. (2002). Evidence that the dephosphorylation of Ser(535) in the epsilon-subunit of eukaryotic initiation factor (eIF) 2B is insufficient for the activation of eIF2B by insulin. *Biochem. J.* **367,** 475–481.

82. Fisher, T. L., and White, M. F. (2004). Signaling pathways: The benefits of good communication. *Curr. Biol.* **14,** R1005–R1007.

83. Shah, O. J., Wang, Z., and Hunter, T. (2004). Inappropriate activation of the TSC/Rheb/ mTOR/S6K cassette induces IRS1/2 depletion, insulin resistance, and cell survival deficiencies. *Curr. Biol.* **14,** 1650–1656.

84. Harrington, L. S., Findlay, G. M., Gray, A., Tolkacheva, T., Wigfield, S., Rebholz, H., Barnett, J., Leslie, N. R., Cheng, S., Shepherd, S., Gout, I., Downes, C. P. *et al.* (2004). The TSC1–2 tumor suppressor controls insulin-PI3K signaling via regulation of IRS proteins. *J. Cell Biol.* **166,** 213–223.

85. Mothe, I., and Van Obberghen, E. (1996). Phosphorylation of insulin receptor substrate-1 on multiple serine residues, 612, 632, 662, and 731, modulates insulin action. *J. Biol. Chem.* **271,** 11222–11227.

86. Rui, L., Fisher, T. L., Thomas, J., and White, M. F. (2001). Regulation of insulin/insulin-like growth factor-1 signaling by proteasome-mediated degradation of insulin receptor substrate-2. *J. Biol. Chem.* **276,** 40362–40367.

87. Um, S. H., Frigerio, F., Watanabe, M., Picard, F., Joaquin, M., Sticker, M., Fumagalli, S., Allegrini, P. R., Kozma, S. C., Auwerx, S. C., and Thomas, G. (2004). Absence of S6K1 protects against age- and diet-induced obesity while enhancing insulin sensitivity. *Nature* **431,** 200–205.

88. Fluckey, J. D., Vary, T. C., Jefferson, L. S., and Farrell, P. A. (1996). Augmented insulin action on rates of protein synthesis after resistance exercise in rats. *Am. J. Physiol.* **270,** E313–E319.

89. Farrell, P. A., Fedele, M. J., Vary, T. C., Kimball, S. R., and Jefferson, L. S. (1998). Effects of intensity of acute-resistance exercise on rates of protein synthesis in moderately diabetic rats. *J. Appl. Physiol.* **85,** 2291–2297.

90. Farrell, P. A., Fedele, M. J., Vary, T. C., Kimball, S. R., Lang, C. H., and Jefferson, L. S. (1999). Regulation of protein synthesis after acute resistance exercise in diabetic rats. *Am. J. Physiol.* **276,** E721–E727.

91. Fedele, M. J., Hernandez, J. M., Lang, C. H., Vary, T. C., Kimball, S. R., Jefferson, L. S., and Farrell, P. A. (2000). Severe diabetes prohibits elevations in muscle protein synthesis after acute resistance exercise in rats. *J. Appl. Physiol.* **88,** 102–108.

92. Fedele, M. J., Lang, C. H., and Farrell, P. A. (2001). Immunization against IGF-I prevents increases in protein synthesis in diabetic rats after resistance exercise. *Am. J. Physiol.* **280,** E877–E885.
93. Kostyak, J. C., Kimball, S. R., Jefferson, L. S., and Farrell, P. A. (2001). Severe diabetes inhibits resistance exercise-induced increase in eukaryotic initiation factor 2B activity. *J. Appl. Physiol.* **91,** 79–84.
94. Kubica, N., Bolster, D. R., Farrell, P. A., Kimball, S. R., and Jefferson, L. S. (2005). Resistance exercise increases muscle protein synthesis and translation of eukaryotic initiation factor 2Bε mRNA in a mammalian target of rapamycin-dependent manner. *J. Biol. Chem.* **280,** 7570–7580.
95. Farrell, P. A., Hernandez, J. M., Fedele, M. J., Vary, T. C., Kimball, S. R., and Jefferson, L. S. (2000). Eukaryotic initiation factors and protein synthesis after resistance exercise in rats. *J. Appl. Physiol.* **88,** 1036–1042.
96. Markuns, J. F., Wojtaszewski, J. F. P., and Goodyear, L. J. (1999). Insulin and exercise decrease glycogen synthase kinase-3 activity by different mechanisms in rat skeletal muscle. *J. Biol. Chem.* **274,** 24896–24900.
97. Turinsky, J., and Damrau-Abney, A. (1999). Akt kinases and 2-deoxyglucose uptake in rat skeletal muscles *in vivo*: Study with insulin and exercise. *Am. J. Physiol.* **276,** R277–R282.
98. Sakamoto, K., Aschenbach, W. G., Hirshman, M. F., and Goodyear, L. J. (2003). Akt signaling in skeletal muscle: Regulation by exercise and passive stretch. *Am. J. Physiol.* **285,** E1081–E1088.
99. Bolster, D. R., Kubica, N., Crozier, S. J., Williamson, D. L., Farrell, P. A., Kimball, S. R., and Jefferson, L. S. (2003). Immediate response of mammalian target of rapamycin (mTOR)-mediated signalling following acute resistance exercise in rat skeletal muscle. *J. Physiol.* **553,** 213–220.
100. Atherton, P. J., Babraj, J., Smith, K., Singh, J., Rennie, M. J., and Wackerhage, H. (2005). Selective activation of AMPK-PGC-1α or PKB-TSC2-mTOR signaling can explain specific adaptive responses to endurance or resistance training-like electrical muscle stimulation. *FASEB J.* **19,** 786–788.
101. Kubica, N., Kimball, S. R., Jefferson, L. S., and Farrell, P. A. (2004). Alterations in the expression of mRNAs and proteins that code for species relevant to eIF2B activity after an acute bout of resistance exercise. *J. Appl. Physiol.* **96,** 679–687.
102. Hernandez, J. M., Fedele, M. J., and Farrell, P. A. (2000). Time course evaluation of protein synthesis and glucose uptake after acute resistance exercise in rats. *J. Appl. Physiol.* **88,** 1142–1149.
103. Nader, G. A., and Esser, K. A. (2001). Intracellular signaling specificity in skeletal muscle in response to different modes of exercise. *J. Appl. Physiol.* **90,** 1936–1942.
104. Parkington, J. D., Siebert, A. P., LeBrasseur, N. K., and Fielding, R. A. (2003). Differential activation of mTOR signaling by contractile activity in skeletal muscle. *Am. J. Physiol.* **285,** R1086–R1090.
105. Bodine, S. C., Stitt, T. N., Gonzalez, M., Kline, W. O., Stover, G. L., Bauerlein, R., Zlotchenko, E., Scrimgeous, A., Lawrence, J. C., Glass, D. J., and Yancopoulos, G. D. (2001). Akt/mTOR pathway is a crucial regulator of skeletal muscle hypertrophy and can prevent muscle atrophy *in vivo*. *Nat. Cell Biol.* **3,** 1014–1019.
106. Pallafacchina, G., Calabria, E., Serrano, A. L., Kalhovde, J. M., and Schiaffino, S. (2002). A protein kinase B-dependent and rapamycin-sensitive pathway controls skeletal muscle growth but not fiber type specification. *Proc. Natl. Acad. Sci. USA* **99,** 9213–9218.
107. Lai, K.-M. V., Gonzalez, M., Poueymirou, W. T., Kline, W. O., Na, E., Zlotchenko, E., Stitt, T. N., Economides, A. N., Yancopoulos, G. D., and Glass, D. J. (2004). Conditional activation of Akt in adult skeletal muscle induces rapid hypertrophy. *Mol. Cell. Biol.* **24,** 9295–9304.

108. Gomez, E., and Pavitt, G. D. (2000). Identification of domains and residues within the ε subunit of eukaryotic translation initiation factor 2B (eIF2Bε) required for guanine nucleotide exchange reveals a novel activation function promoted by eIF2B complex formation. *Mol. Cell. Biol.* **20,** 3965–3976.
109. Gomez, E., Mohammad, S. S., and Pavitt, G. D. (2002). Characterization of the minimal catalytic domain within eIF2B: The guanine-nucleotide exchange factor for translation initiation. *EMBO J.* **21,** 5292–5301.
110. Boesen, T., Mohammad, S. S., Pavitt, G. D., and Andersen, G. R. (2004). Structure of the catalytic fragment of translation initiation factor 2B and identification of a critically important catalytic residue. *J. Biol. Chem.* **279,** 10584–10592.
111. Fabian, J. R., Kimball, S. R., Heinzinger, N. K., and Jefferson, L. S. (1997). Subunit assembly and guanine nucleotide exchange activity of eukaryotic initiation factor-2B expressed in Sf9 Cells. *J. Biol. Chem.* **272,** 12359–12365.
112. Fabian, J. R., Kimball, S. R., and Jefferson, L. S. (1998). Reconstitution and purification of eukaryotic initiation factor 2B (eIF2B) expressed in Sf21 insect cells. *Prot. Express. Purif.* **13,** 16–22.
113. Pavitt, G. D., Ramaiah, K. V. A., Kimball, S. R., and Hinnebusch, A. G. (1998). eIF2 independently binds two distinct eIF2B subcomplexes that catalyze and regulate guanine-nucleotide exchange. *Genes Dev.* **12,** 514–526.
114. Williams, D. D., Pavitt, G. D., and Proud, C. G. (2001). Characterization of the initiation factor eIF2B and its regulation in *Drosophila melanogaster*. *J. Biol. Chem.* **276,** 3733–3742.
115. Hardt, S. E., Tomita, H., Katus, H. A., and Sadoshima, J. (2004). Phosphorylation of eukaryotic translation initiation factor 2Bε by glycogen synthase kinase-3β regulates β-adrenergic cardiac myocyte hypertrophy. *Circ. Res.* **94,** 926–935.
116. Ruggero, D., and Pandolfi, P. P. (2003). Does the ribosome translate cancer? *Nat. Rev. Cancer* **3,** 179–192.
117. Lazaris-Karatzas, A., Montine, K. S., and Sonenberg, N. (1990). Malignant transformation by a eukaryotic initiation factor subunit that binds to mRNA 5′cap. *Nature* **345,** 544–547.
118. De Benedetti, A., and Graff, J. R. (2004). eIF-4E expression and its role in malignancies and metastases. *Oncogene* **23,** 3189–3199.
119. Ruggero, D., Montanaro, L., Ma, L., Xu, W., Londei, P., Cordon-Cardo, C., and Pandolfi, P. P. (2004). The translation factor eIF-4E promotes tumor formation and cooperates with c-myc in lymphomagenesis. *Nat. Med.* **10,** 484–486.
120. Kim, S. H., Forman, A. P., Mathews, M. B., and Gunnery, S. (2000). Human breast cancer cells contain elevated levels and activity of the protein kinase, PKR. *Oncogene* **19,** 3086–3094.
121. Balachandran, S., and Barber, G. N. (2004). Defective translational control facilitates vesicular stomatitis virus oncolysis. *Cancer Cell* **5,** 51–65.
122. Tuhackova, Z., Sloncova, E., Hlavacek, J., Sovova, V., and Velek, J. (1999). Activity of glycogen synthase kinase-3beta is down-regulated during transient differentiation of human colon cancer HT-29 cells. *Oncol. Rep.* **6,** 827–832.
123. Vojtechova, M., Sloncova, E., Kucerova, D., Jiricka, J., Sovova, V., and Tuhackova, Z. (2003). Initiation factor eIF2B not p70 S6 kinase is involved in the activation of the PI-3K signalling pathway induced by the v-src oncogene. *FEBS Lett.* **543,** 81–86.
124. Myers, M. P., Pass, I., Batty, I. H., Van der Kaay, J., Stolarov, J. P., Hemmings, B. A., Wigler, M. H., Downes, C. P., and Tonks, N. K. (1998). The lipid phosphatase activity of PTEN is critical for its tumor supressor function. *Proc. Natl. Acad. Sci. USA* **95,** 13513–13518.
125. Maehama, T., and Dixon, J. E. (1998). The tumor suppressor, PTEN/MMAC1, dephosphorylates the lipid second messenger, phosphatidylinositol 3,4,5-trisphosphate. *J. Biol. Chem.* **273,** 13375–13378.

126. Cantley, L. C., and Neel, B. G. (1999). New insights into tumor suppression: PTEN suppresses tumor formation by restraining the phosphoinositide 3-kinase/AKT pathway. *Proc. Natl. Acad. Sci. USA* **96**, 4240–4245.
127. Li, J., Yen, C., Liaw, D., Podsypanina, K., Bose, S., Wang, S. I., Puc, J., Miliaresis, C., Rodgers, L., McCombie, R., Bigner, S. H., Giovanella, B. C. *et al.* (1997). PTEN, a putative protein tyrosine phosphatase gene mutated in human brain, breast, and prostate cancer. *Science* **275**, 1943–1947.
128. Liaw, D., Marsh, D. J., Li, J., Dahia, P. L., Wang, S. I., Zheng, Z., Bose, S., Call, K. M., Tsou, H. C., Peacocke, M., Eng, C., and Parsons, R. (1997). Germline mutations of the PTEN gene in Cowden disease, an inherited breast and thyroid cancer syndrome. *Nat. Genet.* **16**, 64–67.
129. Manning, B. D., Logsdon, M. N., Lipovsky, A. I., Abbott, D., Kwiatkowski, D. J., and Cantley, L. C. (2005). Feedback inhibition of Akt signaling limits the growth of tumors lacking Tsc2. *Genes Dev.* **19**, 1773–1778.
130. Parrillo, J. E. (1993). Pathogenetic mechanisms of septic shock. *N. Engl. J. Med.* **328**, 1471–1478.
131. Hotchkiss, R. S., and Karl, I. E. (2003). The pathophysiology and treatment of sepsis. *N. Engl. J. Med.* **348**, 138–150.
132. Vary, T. C., Siegel, J. H., Tall, B. D., Morris, J. G., and Smith, J. A. (1988). Inhibition of skeletal muscle protein synthesis in septic intra-abdominal abscess. *J. Trauma-Injury Infect. Crit. Care* **28**, 981–988.
133. Vary, T. C., and Kimball, S. R. (1992). Sepsis-induced changes in protein synthesis: Differential effects on fast- and slow-twitch muscles. *Am. J. Physiol.* **262**, C1513–C1519.
134. Vary, T. C., Jurasinski, C. V., Karinch, A. M., and Kimball, S. R. (1994). Regulation of eukaryotic initiation factor-2 expression during sepsis. *Am. J. Physiol.* **266**, E193–E201.
135. Voisin, L., Gray, K., Flowers, K. M., Kimball, S. R., Jefferson, L. S., and Vary, T. C. (1996). Altered expression of eukaryotic initiation factor 2B in skeletal muscle during sepsis. *Am. J. Physiol.* **270**, E43–E50.
136. Vary, T. C., Deiter, G., and Kimball, S. R. (2002). Phosphorylation of eukaryotic initiation factor eIF2Bepsilon in skeletal muscle during sepsis. *Am. J. Physiol.* **283**, E1032–E1039.
137. Jurasinski, C. V., and Vary, T. C. (1995). Insulin-like growth factor I accelerates protein synthesis in skeletal muscle during sepsis. *Am. J. Physiol.* **269**, E977–E981.
138. Vary, T. C., Jefferson, L. S., and Kimball, S. R. (2001). Insulin fails to stimulate muscle protein synthesis in sepsis despite unimpaired signaling to 4E-BP1 and S6K1. *Am. J. Physiol.* **281**, E1045–E1053.
139. Jurasinski, C. V., and Vary, T. C. (1995). Insulin-like growth factor I accelerates protein synthesis in skeletal muscle during sepsis. *Am. J. Physiol.* **269**, E977–E981.
140. Lang, C. H., Fan, J., Cooney, R., and Vary, T. C. (1996). IL-1 receptor antagonist attenuates sepsis-induced alterations in the IGF system and protein synthesis. *Am. J. Physiol.* **270**, E430–E437.
141. Svanberg, E., Frost, R. A., Lang, C. H., Isgaard, J., Jefferson, L. S., Kimball, S. R., and Vary, T. C. (2000). IGF-I/IGFBP-3 binary complex modulates sepsis-induced inhibition of protein synthesis in skeletal muscle. *Am. J. Physiol.* **279**, E1145–E1158.
142. Vary, T. C., Voisin, L., and Cooney, R. N. (1996). Regulation of peptide-chain initiation in muscle during sepsis by interleukin-1 receptor antagonist. *Am. J. Physiol.* **271**, E513–E520.
143. Cooney, R., Kimball, S. R., Eckman, R., Maish, G., III, Shumate, M., and Vary, T. C. (1999). TNF-binding protein ameliorates inhibition of skeletal muscle protein synthesis during sepsis. *Am. J. Physiol.* **276**, E611–E619.
144. van der Knaap, M. S., Smit, L. M., Barth, P. G., Catsman-Berrevoets, C. E., Brouwer, O. F., Begeer, J. H., de Coo, I. F., and Valk, J. (1997). Magnetic resonance imaging in classification of congenital muscular dystrophies with brain abnormalities. *Ann. Neurol.* **42**, 50–59.

145. van der Knaap, M. S., Kamphorst, W., Barth, P. G., Kraaijeveld, C. L., Gut, E., and Valk, J. (1998). Phenotypic variation in leukoencephalopathy with vanishing white matter. *Neurology* **51,** 540–547.
146. Leegwater, P. A., Konst, A. A., Kuyt, B., Sandkuijl, L. A., Naidu, S., Oudejans, C. B., Schutgens, R. B., Pronk, J. C., and van der Knaap, M. S. (1999). The gene for leukoencephalopathy with vanishing white matter is located on chromosome 3q27. *Am. J. Hum. Genet.* **65,** 728–734.
147. Leegwater, P. A., Vermeulen, G., Konst, A. A., Naidu, S., Mulders, J., Visser, A., Kersbergen, P., Mobach, D., Fonds, D., van Berkel, C. G., Lemmers, R. J., Frants, R. R. *et al.* (2001). Subunits of the translation initiation factor eIF2B are mutant in leukoencephalopathy with vanishing white matter. *Nat. Genet.* **29,** 383–388.
148. van der Knaap, M. S. (2001). Magnetic resonance in childhood white-matter disorders. *Dev. Med. Child Neurol.* **43,** 705–712.
149. Abbott, C. M., and Proud, C. G. (2004). Translation factors: In sickness and in health. *Trends Biochem. Sci.* **29,** 25–31.
150. Fogli, A., and Boespflug-Tanguy, O. (2006). The large spectrum of eIF2B-related diseases. *Biochem. Soc. Trans.* **34,** 22–29.
151. Pavitt, G. D. (2005). eIF2B, a mediator of general and gene-specific translational control. *Nucl. Gene Express.* **33,** 1487–1492.
152. Fogli, A., Dionisi-Vici, C., Deodato, F., Bartuli, A., Boespflug-Tanguy, O., and Bertini, E. (2002). A severe variant of childhood ataxia with central hypomyelination/vanishing white matter leukoencephalopathy related to EIF21B5 mutation. *Neurology* **59,** 1966–1968.
153. Fogli, A., Wong, K., Eymard-Pierre, E., Wenger, J., Bouffard, J. P., Goldin, E., Black, D. N., Boespflug-Tanguy, O., and Schiffmann, R. (2002). Cree leukoencephalopathy and CACH/VWM disease are allelic at the EIF2B5 locus. *Ann. Neurol.* **52,** 506–510.
154. Richardson, J. P., Mohammad, S. S., and Pavitt, G. D. (2004). Mutations causing childhood ataxia with central nervous system hypomyelination reduce eukaryotic initiation factor 2B complex formation and activity. *Mol. Cell. Biol.* **24,** 2352–2363.
155. Li, W., Wang, X., van der Knaap, M. S., and Proud, C. G. (2004). Mutations linked to leukoencephalopathy with vanishing white matter impair the function of the eukaryotic initiation factor 2B complex in diverse ways. *Mol. Cell. Biol.* **24,** 3295–3306.
156. Fogli, A., Schiffmann, R., Hugendubler, L., Combes, P., Bertini, E., Rodriguez, D., Kimball, S. R., and Boespflug-Tanguy, O. (2004). Decreased guanine nucleotide exchange factor activity in eIF2B-mutated patients. *Eur. J. Hum. Genet.* **12,** 561–566.
157. Hinnebusch, A. G. (2000). Mechanism and regulation of initiator methionyl-tRNA binding to ribosomes. *In* "Translational Control" (N. Sonenberg, J. W. B. Hershey, and M. B. Mathews, Eds.), pp. 185–243. Cold Spring Harbor Laboratory Press, Cold Spring Harbor, NY.
158. Harding, H. P., Novoa, I., Zhang, Y., Zeng, H., Wek, R., Schapira, M., and Ron, D. (2000). Regulated translation initiation controls stress-induced gene expression in mammalian cells. *Mol. Cell* **6,** 1099–1108.
159. Kantor, L., Harding, H. P., Ron, D., Schiffmann, R., Kaneski, C. R., Kimball, S. R., and Elroy-Stein, O. (2005). Heightened stress response in primary fibroblasts expressing mutant eIF2B genes from CACH/VWM leukodystrophy patients. *Hum. Genet.* **118,** 99–106.
160. van der Voorn, J. P., van Kollenburg, B., Bertrand, G., Van Haren, K., Scheper, G. C., Powers, J. M., and van der Knaap, M. S. (2005). The unfolded protein response in vanishing white matter disease. *J. Neuropathol. Exp. Neurol.* **64,** 770–775.
161. Inamura, N., Nawa, H., and Takei, N. (2003). Developmental changes of eukaryotic initiation factor 2B subunits in rat hippocampus. *Neurosci. Lett.* **346,** 117–119.
162. Dietrich, J., Lacagnina, M., Gass, D., Richfield, E., Mayer-Proschel, M., Noble, M., Torres, C., and Proschel, C. (2005). EIF2B5 mutations compromise GFAP+ astrocyte generation in vanishing white matter leukodystrophy. *Nat. Med.* **11,** 277–283.

Role of Protein Tyrosine Phosphatases in Cancer

Tasneem Motiwala[*] and
Samson T. Jacob[*,†]

[*]Department of Molecular and Cellular
Biochemistry, The Ohio State University,
College of Medicine, Columbus,
Ohio 43210

[†]Comprehensive Cancer Center,
The Ohio State University,
Columbus, Ohio 43210

I. Introduction	298
II. Genetic and Epigenetic Alterations of *PTP* Genes	299
A. PTPs Localized to Regions of Loss/Gain in Human Cancers	300
B. PTPs Mutated in Cancers	300
C. Epigenetic Regulation of PTPs	303
III. Transformation-Related Phenotypes Attributable to PTPs	304
IV. PTPs as Drug Targets	311
A. Gene Therapy	312
B. Small Molecule Inhibitors	314
C. Epigenetic Therapy	315
V. Concluding Remarks	316
References	320

Protein phosphorylation and dephosphorylation are complex enzymatic reactions that are performed by the concerted action of protein kinases and phosphatases, respectively. Deregulation of such coordination due to loss or gain of a single component of the process can result in disease conditions that include, but are not limited to, neoplastic transformation, developmental, autoimmune, and metabolic disorders. Unlike many protein tyrosine kinases that function as oncoproteins, protein tyrosine phosphatases (PTPs) could impart positive or negative effect on cell proliferation. Although past studies have suggested a potential role for PTPs in cancer, the molecular mechanisms of the altered activity/level of these enzymes and the pathological manifestations of these modifications in diseases, particularly in cancer, have not been critically analyzed. This chapter is a comprehensive survey of the alterations of PTPs and the implications of the growth, proliferation, and apoptosis phenotypes attributable to the altered function of this family of phosphatases in cancer. Further, the potential applications of different therapeutic approaches to rectify the adverse effects of alterations in expression of the phosphatase genes and of the phosphatase activity in cancer are discussed.

I. Introduction

Protein phosphorylation plays an important role in several cellular processes, including differentiation, cell growth, adhesion, motility, and apoptosis. Cascading events involving phosphorylation and dephosphorylation of proteins are responsible for transfer of signals from a cell's exterior to its ultimate target in the cytoplasm or nucleus. The membrane proximal signaling generally involves tyrosine phosphorylation, which is regulated by the concerted actions of protein tyrosine kinases (PTKs) and protein tyrosine phosphatases (PTPs). Aberrations in this fine-tuned regulation of protein phosphorylation can result in altered cellular processes like uncontrolled cell growth, a dedifferentiated phenotype, defective apoptosis (all characteristics of neoplastic disease), and in some cases also increased cell migration (characteristic of metastatic disease). PTKs comprise a majority of the dominant known oncogenes. Further, somatic mutations in tyrosine kinases account for a large number of cancers (1). PTKs have thus been implicated in oncogenic transformation. Since PTPs catalyze the reverse reaction, it was logical to assume that some PTPs would act as tumor suppressors. However, unlike the PTKs, PTPs can act as positive or negative regulators of signal transduction pathways. They can either activate tyrosine kinases or counteract their activity by dephosphorylating the kinase itself or its downstream target. PTPs are, therefore, a complex group of enzymes whose function is dependent on the availability of their functional partners.

The sequencing of the human genome has helped identify 107 PTP-coding genes of which only 81 are active protein phosphatases. Similarly, of the 90 genes coding for PTKs only 85 are catalytically active. Further, both PTPs and PTKs are distributed almost equally in tissues. It is, therefore, conceivable that both group of enzymes share some substrate specificities and that both are equally important in maintaining optimal protein phosphorylation levels. Although the function of a few PTKs as oncogenes has been accepted, there is still no defined role for PTPs in cancer. The PTP superfamily can be subdivided into three major families based on their structure, function, and sequence: (i) tyrosine-specific or "classical" phosphatases, (ii) dual-specificity phosphatases (DUSP), and (iii) low-molecular-weight phosphatases (LMW-PTP). In addition to their phosphatase activity on tyrosine and serine/threonine residues of the same protein, some phosphatases with structural similarity to DUSPs also dephosphorylate lipids (2). In fact the phosphatase PTEN (phosphatase and tensin homologue), often classified as a DUSP, derives its well-established tumor suppressor property from its phospholipid phosphatase activity (3). On the contrary, the cdc25 family of proteins also classified as DUSPs is frequently overexpressed in several different cancers and is thought to circumvent the cell cycle checkpoints facilitating cell proliferation [See (4) for a review]. Additionally, several other members of the DUSPs function as either tumor suppressors or oncogenes [See (5) for a review]. There is only one known *LMW-PTP* gene that gives rise to four different isoforms as a result of alternate splicing. Its upregulation during contact inhibition and the antagonistic role in PDGF stimulated cell growth suggest that it is a protein capable of inducing growth arrest [reviewed in (6)]. The classical PTPs can be further divided into two groups, receptor-type protein tyrosine phosphatases (RPTPs) and nonreceptor-type protein tyrosine phosphatases (NRPTPs), depending on whether they are transmembrane or cytosolic proteins. These are also segregated into several subtypes on the basis of sequences or functional domains outside their catalytic domain. The DUSPs and LMW-PTPs have been reviewed in detail for their involvement in neoplastic disease and potential for pharmacological intervention (6, 7). Although the classical PTPs are receiving increased attention for their role in cancer, there is a lack of critical analysis of each PTP subtype and each individual PTP therein, comparison of the similarities and differences between the action of closely related PTPs, and how such information can be applied to the development of novel therapeutics. This chapter will discuss the implications of genetic and epigenetic alterations of the classical PTPs in cancer and their potential as novel molecular targets in cancer therapy.

II. Genetic and Epigenetic Alterations of *PTP* Genes

Tumor suppressor genes are characterized by loss of gene function, which can occur as a result of deletions, inactivating mutations, and epigenetic alterations. Knudson's two-hit hypothesis coined in the early 70s called for two genetic hits that would inactivate both alleles of a gene. However, it is found that inactivation of a single allele (haploinsufficiency) can result in cellular phenotype leading to tumorigenesis. On the contrary, oncogenes are characterized by gain of gene function as a result of gene amplification, activating mutations, and translocations leading to aberrant expression. The next sections discuss the alterations in expression of *PTP* genes and their functional involvement in growth and apoptosis, the processes that are deregulated in cancer.

A. PTPs Localized to Regions of Loss/Gain in Human Cancers

A review on the genetic variations of PTPs in human diseases has implicated locations of 19 genes (of the 38 classical PTP-encoding genes) as regions frequently deleted in different human cancers (8). This assignment is, however, based essentially on reports of loss of heterozygosity (LOH) within the particular cytogenetic band. Such a large region could harbor many genes, which could potentially include the *PTP*. Demonstration of the loss of a single or both copies of a specific PTP associated with a particular or several different types of cancers will prove its involvement in carcinogenesis. To date, there are reports of specific loss of *RPTPs PTPRG, PTPRJ, PTPRD,* and *PTPRK* in several solid tumors (refer to Table I for additional information). In most cases, loss of these genes is associated with reduced expression (9–11). Further, *PTPRG* and *PTPRJ* also exhibit growth suppressive potential in cancer cell lines (11, 12) and are the commonly accepted tumor suppressive genes amongst the classical tyrosine phosphatase family. Additionally, using specific sequence-tagged site (STS) markers, deletions observed in some human cancers have been narrowed to specific smaller areas of minimal losses, thus, facilitating the identification of candidate tumor suppressors associated with cancer. Such analysis has implicated *PTPRF* in neuroectodermal cancer (13), *PTPN6* in acute lymphoblastic leukemia (14), *PTPN12* and *PTPN23* in many different cancers (15, 16). These genes can, therefore, be considered candidate tumor suppressors. Besides gene loss there is report of amplification of the gene encoding an intracellular tyrosine phosphatase PTPN1 in several different solid tumors, including gastric cancer, Barrett's adenocarcinoma, ovarian cancer, hepatocellular carcinoma (HCC), and breast cancer (17–23). Consistent with gene amplification, the PTPN1 transcript was upregulated in ovarian cancer (17, 22). Amplification of the *PTPN1* gene in solid tumors of different origins suggests that it has oncogenic potential.

B. PTPs Mutated in Cancers

Activating mutations are the most common mechanism resulting in deregulation of *PTKs* (24) that function as oncogenes. If a majority of *PTPs* are considered potential tumor suppressors, it is anticipated that at least some of these genes would carry inactivating mutations. It is, however, surprising that a detailed analysis for mutations within the coding regions of the tyrosine phosphatase superfamily was only recently performed in colorectal cancers (25). This analysis identified a total of 83 nonsynonymous, somatic mutations in 3 RPTPs (PTPRF, PTPRG, and PTPRT) and 3 nonreceptor-type tyrosine phosphatases (PTPN3, PTPN13, and PTPN14). At least 15 of these mutations were predicted to result in loss of gene function as a result of nonsense, frame

TABLE I
PTPs Localized to Regions of Loss/Gain in Human Cancers

Gene	PTP	Chr location	Aberration	Type of cancer	Gene/Region[a]	References
PTPRF	hLAR	1p34.2	LOH	Neuroectodermal cancer	Region	(13)
PTPRG	hPTPγ	3p14.2	Homozygous deletion	Benign proliferative breast disease	Gene	(136)
			LOH	Primary lung carcinomas, clear cell renal carcinoma, and renal cell carcinoma cell lines	Gene	(137–139)
PTPN23	hHDPTP	3p21.31	LOH	Lung adenocarcinoma and other solid tumors	Region	(16, 140)
PTPRK	hPTPκ	6q22.33	LOH	Primary central nervous system lymphomas (PCNSL)	Gene	(9)
			Deletion	Hematological neoplasms, melanomas, ovary carcinomas, and other solid tumors	Region	(65)
PTPN12	hPEST	7q11.23	Chromosomal rearrangement	Malignant melanoma	Region	(15)
			Chromosomal break	Hematopoietic disorders	Region	(15)
PTPRD	hPTPδ	9p24.1	Homozygous deletion	Small cell and non-small cell lung cancer	Gene	(141)
PTPRJ	hDEP1	11p11.2	LOH	Colon, lung, breast, and thyroid cancer	Gene	(31, 93)
PTPN6	hSHP-1	12p13.31	LOH	Acute lymphoblastic leukemia	Gene	(14)
PTPN1	hPTP1B	20q13.13	Amplification	Barrett's adenocarcinoma, HCC, gastric, ovarian, and breast cancer	Gene	(17–23)

[a]Gene/region indicates whether the gene itself or the chromosomal location of the gene is implicated in the cancer.

shift or splice-site alterations. Based on the evidence discussed in the study it appears that these mutations are functional, exerting growth advantage to the tumors (25). In this respect, it is possible that other missense mutations resulting in amino acid substitution could alter protein structure and thereby its function. Further analysis of these six PTPs in other solid tumors identified some mutations in lung, gastric, and breast cancers but not in medulloblastomas, glioblastomas, pancreatic, and ovarian cancers (refer to Table II for additional information on mutation of PTPs in cancer). Among these PTPs, PTPRG exhibits growth suppressive potential and has been implicated as a tumor suppressor in breast and ovarian cancer (12). Overexpression of PTPRT also resulted in growth suppression of colon cancer cell lines suggesting a role in tumorigenesis. On the contrary, increased expression of PTPRF correlates with metastatic potential in breast and prostate cancer (26, 27), suggesting a positive role in transformation. However, a study investigating the relationship between PTPRF/LAR and E-cadherin demonstrated cell density dependent parallel increase of both LAR and E-cadherin and requirement of E-cadherin mediated cell—cell contact for upregulation of LAR (28). This suggests that the increased expression of PTPRF observed in metastatic breast cancer and prostate cancer cell lines could be a consequence rather than the cause of uncontrolled cell proliferation. While the roles of other PTPs mutated in colon cancer have not been fully explored, preliminary studies have shown alteration in growth phenotype upon overexpression of PTPN3 and PTPN14 in NIH/3T3 and HeLa cells, respectively (29, 30). It is noteworthy that additional PTPs could be mutated in other cancers. In fact PTPRJ, a tyrosine phosphatase with

TABLE II
PTPs Mutated in Cancer

Gene	PTP	Type of cancer	References
PTPRF	hLAR	Colon, breast, and lung cancer	(25)
PTPN14	hPTPD2	Colon cancer	(25)
PTPRG	hPTPγ	Colon	(25)
PTPN23	hHDPTP	Small cell lung cancer cell line	(16)
PTPN13	hPTPBAS	Colon cancer	(25)
PTPN12	hPEST	Colon cancer cell line	(15)
PTPN3	hPTPH1	Colon cancer	(25)
PTPRJ	hDEP1	Colon, lung, and breast cancer	(31)
PTPN11	hSHP-2	Leukemia	(32–34)
PTPRT	hPTPρ	Colon, lung, and gastric cancer	(25)

tumor suppressor properties, was mutated in colon, lung, and breast cancer (*31*). Additionally, PTPN23 and PTPN12 are also mutated in small cell lung cancer cell line and colon cancer cell line, respectively (*15, 16*). To date, there is report of the contribution of an activating mutation in only one NRPTP, PTPN11 to leukemogenesis (*32–34*). Functional analysis to investigate the role of this leukemia-associated mutation demonstrated that its overexpression induced aberrant growth in multiple hematopoietic compartments. This study thus confirmed the involvement of hyperactive PTPN11, a positive regulator of Ras, in the pathogenesis of juvenile myelomonocytic leukemia (*35*).

C. Epigenetic Regulation of PTPs

Epigenetic alterations are inheritable changes affecting gene expression that are mediated by effects on chromatin structure and not on DNA sequence. These alterations include DNA methylation, histone modifications (acetylation, methylation, and phosphorylation), and chromatin remodeling and are emerging as factors contributing to loss of gene expression specifically affecting the wild-type allele. Among these processes, DNA methylation is the most common modification in cancer occurring on the 5' position of cytosines of CpG dinucleotides. PTPN6 (SHP-1) was the first classical PTP demonstrated to be methylated in cutaneous T-cell lymphoma (*36*). Subsequently, methylation of the gene encoding this hematopoietic PTP was found in a small subset of leukemia and lymphoma cases, including anaplastic large cell lymphoma (ALCL), multiple myeloma, and acute myeloid leukemia (AML) (*37–41*) (refer to Table III for effect of methylation on PTPs). Our laboratory was the first to demonstrate methylation-mediated suppression of an RPTP, protein tyrosine phosphatase receptor-type O (PTPRO), in rat HCCs (*42*). The *PTPRO* gene encodes six different transcript variants that produce two major protein isoforms: (i) full-length (PTPRO-FL, GLEPP1, and PTP-U2) and

TABLE III
PTPs METHYLATED IN CANCER

Gene	PTP	Type of cancer	Gene expression	References
PTPRG	hPTPγ	Cutaneous T-cell lymphoma	Suppressed	(*47*)
PTPRZ1	hPTPζ	CLL	Not tested	(*46*)
PTPRN2	hPTPIA2β	CLL	Not tested	(*46*)
PTPN6	hSHP-1	ALCL, multiple myeloma, acute myeloid leukemia, cutaneous T-cell lymphoma	Suppressed	(*36–41*)
PTPRO	hGLEPP1	Rat hepatocellular carcinoma, lung and colon cancer, CLL	Suppressed	(*42, 44, 45*)

(ii) truncated (PTPROt) (42). The full-length form is expressed in epithelial cells of tissues like the brain, kidney, lung, and breast while the truncated variant is of hematopoietic origin with high-level expression in B-lymphoid cells (43). Our studies demonstrated that methylation of the CpG island located in the promoter region affects expression of the full-length isoform in solid tumors (lung, HCC, breast) [(42, 44) and unpublished data] as well as the truncated isoform in leukemia, specifically chronic lymphocytic leukemia (CLL) (T. Motiwala, S. Majumder, J. Byrd, M. Grever, D. Lucas, and S. Jacob, unpublished data). These results are consistent with an observation that found increased methylation of PTPRO in right-sided MSI+ primary colon tumors (45). Global epigenetic profiling of CLL using restriction landmark genomic scanning (RLGS) identified PTPRN2 and PTPRZ2 as candidate methylation targets in 30% CLL cases tested relative to neutrophils from same individuals and $CD19^+$ selected B lymphocytes from normal individuals (46). Similarly, epigenetic screening of cutaneous T-cell lymphomas using differential hybridization on CpG island microarray identified increased methylation of PTPRG relative to benign T cells, which correlated with its transcriptional suppression (47). It is noteworthy that except for PTPN6/SHP-1, methylation of other PTPs was identified during genome-wide search for epigenetic alterations in different types of cancers. It is conceivable that differential methylation of other PTP-coding genes was not identified in cancer due to technical limitations of screening techniques and screening of only a single specific type of cancer. In this context, it is of interest that genomic sequence analysis showed presence of CpG island in the proximal promoters of all PTPs except PTPN22, PTPRC, PTPN7, PTPRR, PTPRQ, and PTPRA (48). Methylation status of the CpG island could, thus, be involved in regulating the expression of these PTPs in normal cells, and its alteration may lead to disease conditions particularly cancer.

Gene expression can also be altered by epigenetic mechanisms that involve changes in posttranslational modification of histones associated with the gene (49–53) as well as differential association of chromatin remodeling complexes (54, 55). These factors structurally modify the promoter in chromatin context, making it either accessible or inaccessible to the normal transcriptional machinery. Such epigenetic alterations could also be responsible for the alteration in the expression of *PTP* genes in cancer. Exploration of this possibility could reveal a novel mechanism for the regulation of this class of enzymes in disease states and its application to promising epigenetic therapy (discussed later).

III. Transformation-Related Phenotypes Attributable to PTPs

The first PTP was identified and cloned in the late 1980s (56, 57). The identification of any new proteins or class of proteins calls for the study of their function in a specific cell or tissue type. Because PTPs appear to play a role in several diseases, the most common being cancer and diabetes, their involvement in cellular processes related to these and other diseases have been explored either by overexpression or knockdown studies. Although knockout animal models have facilitated identification of tumor suppressor or oncogenic function of a gene, most studies have been performed using cells in culture. There are several drawbacks in relying completely on animal models. First, in some cases knockout animals may be embryonic lethal in which case an inducible knockout may be appropriate. Second, whole animal knockouts may not be as beneficial as tissue-specific knockout to study the role of a gene product. Third, several *PTP* genes express multiple transcripts, which could functionally compensate for the loss of another isoform. Alternatively, the function of one PTP may be compensated by another closely related PTP. Fourth, cancer is a complex disease with multiple players involved in its etiology. These factors may be intrinsic (genetic alterations, which in turn could be hereditary or acquired) or extrinsic (environmental). It is, therefore, highly unlikely that knocking out a particular gene would result in spontaneous effect on tumor formation in a complex whole animal system. In this respect, although 22 of the 38 class I cysteine-based classical PTPs have exhibited either positive or negative effects and in a few cases both manifestations on growth and apoptosis, these alterations were not always demonstrated using animal models (refer to Table IV for details on the growth and apoptosis related phenotypes attributable to PTPs). Loss of gene function associated with particular types of cancer prompted investigators to study specific contribution of these genes to cancer. This was accomplished by ectopically expressing them in cancer cell lines of different origins and studying the alterations in phenotypic properties (e.g., proliferation, cell cycle, anchorage independence, contact inhibition, motility, and apoptosis), some or all of which are deregulated in a cancer cell. In this context, overexpression of PTPN6 and PTPRT inhibited growth of hematopoietic and colorectal cells, respectively (*25, 41*). Similarly, overexpression of PTPRJ suppressed the malignant phenotype of transformed rat thyroid cells (*11*) and inhibited growth of breast cancer cell lines (*58*). Overexpression of PTPRH altered the morphology, reduced growth rate and migratory activity of a HCC cell line (*59*), and induced apoptosis in NIH3T3 fibroblasts (*60*). PTPRH mediated impediment of the migratory activity of the cell suggests its role in influencing the metastatic/invasive potential of a cancer cell. In addition to PTPRH, overexpression of PTPN12 inhibited cell migration in Rat1 fibroblast-derived stable cell lines (*61*) and rat aortic smooth muscle cells (*62*). More direct evidence of the role of PTPs in determining the metastatic potential of a cell stems from the observation that injection of PTPRA overexpressing mammary tumor cells transgenic for HER2/neu into mice causes reduced tumor growth and delayed lung metastasis (*63*).

TABLE IV
PHENOTYPIC CHARACTERISTICS ATTRIBUTABLE TO THE FUNCTION OF PTPs

Gene	PTP	Technique	Observation	TS/Oncogene	References
PTPN7	hHePTP	Knockdown	Increased apoptosis	Oncogene	(64)
		Overexpression	Altered cell morphology, anchorage-independent, and disorganized growth	Oncogene	(77, 133)
PTPN14	hPTPD2	Overexpression	Diminished cell adherence and slow growth	TS[a]	(29)
PTPRG	hPTPγ	Overexpression	Reduced proliferation and anchorage-independent growth	TS	(12)
		Knockdown	Increased proliferation and anchorage-independent growth		
PTPN13	hPTPBAS	Endogenous upregulation	Resistance to Fas-mediated apoptosis	Oncogene	(80–83, 142)
		Knockdown	Loss of 4-hydroxy-tamoxifen induced apoptosis (Fas independent)	TS	(84)
PTPRK	hPTPκ	Knockdown	Increased apoptosis	Oncogene	(64)
PTPN12	hPEST	Overexpression	Inhibition of cell migration	TS	(61, 62)
PTPRZ1	hPTPζ	Knockdown	Increased apoptosis	Oncogene	(64)
		Knockdown	Reduced cell motility	Oncogene	(79)
PTPRD	hPTPδ	Knockdown	Resistance to apoptosis-inducing agents	TS	(64)
PTPN3	hPTPH1	Overexpression	Inhibition of cell growth by altering reentry into cell cycle and not induction of apoptosis	TS	(30)
PTPRE	hPTPε	Knockout	Larger and flatter cells with slow proliferation	Oncogene	(89, 143)
		Transgenic mice	Hyperplastic mammary tissue resulting in increased incidence of mammary tumors	Oncogene	(144)

Gene	Protein	Manipulation	Phenotype	TS/Oncogene	Ref.
PTPRJ	hDEP1	Overexpression	Cells are more adherent, arrested in G1 phase, unable to grow in soft agar, and form tumors when injected into nude mice	TS	(11)
		Overexpression	Growth inhibition	TS	(58)
		Knockdown	Resistance to apoptosis-inducing agents	TS	(64)
PTPN5	hSTEP	Knockdown	Increased apoptosis	Oncogene	(64)
PTPN6	hSHP-1	Overexpression	Growth inhibition	TS	(41, 145)
PTPRO	hGLEPP1	Overexpression	Reduced proliferation and growth in soft agar, delayed reentry into cell cycle, increased sensitivity to apoptosis-inducing agent	TS	(43, 44)
		Overexpression	Increased adherence, differentiated phenotype, and subsequent apoptosis in the presence of TPA	TS	(67)
PTPRQ/PTPGMC1	hPTPS31	Overexpression	PIPase activity (and not PTPase activity) resulted in reduced proliferation and increased apoptosis in low serum environment	TS	(94)
PTPN11	hSHP-2	Overexpression (gain-of-function mutant)	Aberrant growth in multiple hematopoietic compartments	Oncogene	(35)
PTPN9	hMEG2	Overexpression (CS and DA mutants)	Suppressed growth in semisolid media	Oncogene	(78)
		Knockdown	Increased apoptosis	Oncogene	(64)

(*Continues*)

TABLE IV (Continued)

Gene	PTP	Technique	Observation	TS/Oncogene	References
PTPN2	hTCPTP	Overexpression	Increased rate of cell division, lower serum requirement, larger colonies in soft agar, loss of contact inhibition, and altered morphology	Oncogene	(71, 72)
		Knockout	Lower proliferation rate	Oncogene	(73)
		Overexpression	Increased p53-dependent apoptosis	TS	(75)
		Overexpression	Restoration of imatinib mesylate (Gleevec) sensitivity in resistant chronic myeloid leukemia (CML) cells	TS	(76)
PTPRS	hPTPσ	Knockdown	Resistance to apoptosis-inducing agents	TS	(64)
PTPRH	hSAP1	Overexpression	Altered morphology, contact inhibition, reduced growth rate and migratory activity, increased apoptosis	TS	(59, 60, 146)
PTPRA	hPTPα	Overexpression	Reduced growth as a result of G0/G1 arrest and not apoptosis, reduced tumor growth and delayed lung metastasis in nude mice	TS	(63)
PTPRT	hPTPρ	Overexpression	Growth suppression	TS	(25)

[a]TS, tumor suppressor.

In recent years RNA interference (RNAi) and small interfering RNA (siRNA) technologies to knockdown expression of genes have emerged as alternate and perhaps the most reliable tools to study their functions. Knocking down the expression of a gene that is normally expressed in a particular cell is more physiologically relevant over the artifacts that may be induced particularly on expression of a gene at a much higher level than its normal expression. Studies to demonstrate tumor suppressive properties of *PTPRG*, a gene whose function is lost in cancers as a result of LOH, mutation, or methylation, have used both overexpression and antisense-mediated knockdown approaches. Both techniques showed that PTPRG can inhibit anchorage dependent and independent growth of MCF-7 cells (*12*). Many studies are now recognizing phosphatases as an important family of enzymes involved in regulation of phosphorylation in practically every cellular process. The complex nature of this regulation by protein phosphatases (i.e., both positive and negative effects on signal transduction) is overriding the importance of kinases, which participate as partners of phosphatases in the same process. A study has, thus, employed RNAi screening for protein phosphatases to determine their role in cancer. Since loss of balance between cell death and survival can result in development of cancer, this study used apoptosis and chemoresistance as a means to measure the effect of systematic knockdown of each individual PTP in HeLa cells (*64*). Phosphatases whose knockdown resulted in resistance to the drugs cisplatin, taxol, and etoposide were classified as "cell death phosphatases" that include three classical PTPs: PTPRD, PTPRJ, and PTPRS. On the contrary, knockdown of certain phosphatases increased spontaneous apoptosis in HeLa cells. Such phosphatases designated as "survival phosphatases" consist of several classical PTPs. Considering a twofold increase in apoptosis as significant, a few PTPs with tumor suppressor characteristics (e.g., PTPRK, PTPRO, PTPRG, and PTPRA) were categorized as survival phosphatases. Although this may be a true phenotype in HeLa cells, these PTPs tend to lean towards the tumor suppressor group for the following reasons. First, loss of PTPRK function is implicated in several liquid and solid tumors (*65, 66*). Further, suppression of PTPRK in primary central nervous system lymphomas (PCNSL) as a result of LOH appears to be relevant to its pathogenesis and unfavorable prognosis (*9*). Second, since our initial observation of extensive *PTPRO* methylation in a large number of primary lung tumors (relative to their normal adjacent tissue) (*44*), we have demonstrated reduced proliferation and anchorage-independent growth, delayed reentry into cell cycle, and increased susceptibility to apoptosis inducing agent of a nonexpressing human lung cancer cell line ectopically expressing *PTPRO* (compared to the vector transfected cells) (*44*). We subsequently extended this observation to PTPROt that is predominantly expressed in hematopoietic cells and demonstrated methylation and suppression of PTPROt in primary CLL cells as well as leukemia cell lines of different origins. The gene was, however, essentially unmethylated and functional in control lymphocytes (T. Motiwala, S. Majumder, J. Byrd, M. Grever, D. Lucas, and S. Jacob, unpublished data). Further, expression of *PTPROt* in the nonexpressing cells increased their susceptibility to fludarabine, a drug commonly used in the treatment of CLL. The results are in agreement with earlier findings of the involvement of PTPROt in cell cycle arrest (*43*) and of PTPU2L/PTPRO in apoptosis subsequent to TPA induced differentiation (*67*). Third, while PTPRA can transform NIH/3T3 cells by dephosphorylating and activating oncogenic c-Src (*68*), its overexpression in breast cancer cells suppressed proliferation in both *in vitro* and *in vivo* assays (*63*). This disparity in the function of PTPRA can be reconciled in several different ways, which could also be extrapolated to some other PTPs showing conflicting phenotypes. First, although PTPRA can activate c-Src and transform fibroblasts, its action in breast cancer may be mediated by some as yet unknown substrate (*63, 69*). Second, the association of increased PTPRA expression with low tumor grade (*63*) may suggest an involvement in tumor initiation with eventual loss in aggressive disease. Alternatively, increased cell density as a result of uncontrolled proliferation could induce PTPRA consistent with density-dependent upregulation of other PTPs (*70*). Additionally, while some of the "survival" protein phosphatases may facilitate cell survival by reducing spontaneous apoptosis [as observed by MacKeigan (*64*)], their effect on growth inhibition may predominate over the cell survival function. PTPN2/TCPTP, a nonreceptor PTP, is an example of the differential effects of a PTP on growth and apoptosis based on a series of detailed studies using overexpression and knockdown approaches. To investigate the role of this PTP, which is upregulated upon mitogenic stimulation of several cell lines, it was overexpressed in HeLa cells. Proliferation and cell cycle assays demonstrated lower serum requirement, formation of larger colonies in soft agar, growth in multiple layers, altered morphology, rapid progress through G1 and S phases, and increased rate of cell division for PTPN2 overexpressing cells relative to the control cells (*71, 72*). An independent study demonstrated reduced proliferation rate for TCPTP(-/-) lymphocytes compared to TCPTP(+/+) lymphocytes (*73*). Although cell cycle and apoptosis are discrete processes, these could be mediated by common regulators. A typical example being *c-myc*, a mitogenic oncogene that can also induce apoptosis (*74*). Similarly, overexpression of PTPN2 induces apoptosis in the p53+ A549 and MCF-7 cells but not in p53- HeLa cells (*75*). Unfortunately, this study did not assess the effect of PTPN2 overexpression on proliferation of A549 and MCF-7 cells to rule out the possibility of cell-type specific effect on proliferation and apoptosis. Another study showed that restoration of TCPTP, which is downregulated in imatinib mesylate resistant human chronic myeloid leukemia (CML) cell line, could revive sensitivity by increasing apoptosis (*76*). A few classical "survival

PTPs", like PTPN5, PTPN9, PTPN7, and PTPRZ1, demonstrate greater than threefold increase in apoptosis upon knockdown (64). Other independent studies have revealed oncogenic potential of these PTPs for the following reasons. First, overexpression of PTPN7 in cell culture resulted in altered cell morphology, anchorage independent and disorganized growth (77). Second, overexpression of the catalytically inactive $C \to S$ and $D \to A$ mutants of PTPN9 but not wild-type PTPN9 in erythroid cells suppressed growth in semisolid media (78). Third, siRNA mediated knockdown of PTPRZ1 in glioblastoma cells decreased the migration capacity of the cells indicating that PTPRZ1 is involved in the invasiveness of glioblastomas (79). Similar to the effect of PTPRA on cell growth and proliferation (63, 68), the function of certain PTPs in apoptosis also depends on cellular context as well as the pathway involved in mediating this effect. One such effect is observed with PTPN13/FAP-1, which is referred as "antiapoptotic" tyrosine phosphatase. Its increased expression in ovarian cancer, HCC, and pancreatic cancer has been associated with resistance to Fas-mediated apoptosis (80–83). On the contrary, its expression is upregulated by the antiestrogen 4-hydroxy-tamoxifen in breast cancer cells and this expression was necessary for the Fas-independent apoptotic action of tamoxifen (84). PTPN13 is thus considered "proapoptotic" in breast cancer. Apart from the observed phenotypic effects of PTPs on growth and apoptosis, a few PTPs are involved in signaling via adherens junction proteins and are thus implicated in cell—cell contacts, cell shape, and motility (85, 86). The extracellular domains of RPTPs, although diverse, are characteristic of cell adhesion molecules. This coupled with the lack of known ligands for the majority of RPTPs reinforces their role in contact inhibition (87). Any disruption in the normal functioning of such PTPs could thus result in uncontrolled growth and metastatic invasion of cancer cells.

IV. PTPs as Drug Targets

PTPs control protein tyrosine phosphorylation status in concert with tyrosine kinases, which comprise 80% of oncoproteins. Oncogenic tyrosine kinases are either overexpressed in wild-type form or expressed as mutated constitutively active enzymes in cancer. It is, therefore, tempting to postulate a tumor suppressive role for the PTPs. Many PTPs are underexpressed in cancers as a result of LOH, inactivating mutations or epigenetic alterations. It is, however, also becoming increasingly clear that PTPs are not simply negative regulators reversing the action of tyrosine kinases. Instead they may also act in synergy with tyrosine kinases to enhance protein phosphorylation. In fact, certain PTPs are upregulated in cancers and function by activating the oncogenic src kinase (88, 89). Consistent with this complex regulation of protein phosphorylation by

PTPs, these can act as either negative or positive regulators of cell signaling. Additionally, the same PTPs can act in both ways depending on the cellular context (90). Thus, while it is simple to use kinases as targets for anticancer drugs, the application of PTPs as drug targets is a rather complex strategy. In the following paragraphs we will review options for targeting PTPs in anticancer therapy.

A. Gene Therapy

Gene therapy is a means of restoring the function of an inactivated gene either by gene repair (in case of mutated genes) or by introduction of a functional copy of the gene (gene augmentation in case of mutated as well as lost genes). Although gene repair has several advantages over gene augmentation (e.g., it is applicable to both dominant and recessive mutations and the gene expression is regulated by its natural elements), gene therapy has conventionally focused on gene augmentation (91). Such gene transfers into human cells are accomplished by means of a viral vector, which is capable of infecting human cells and introducing the viral DNA (manipulated to carry the gene of interest) into these cells. Other approaches include direct introduction of the DNA into human cells, liposome-mediated transfer, and receptor-mediated internalization of ligand-linked DNA. Nonviral approaches, however, are less efficient and their lack of integration does not allow for stable long-term expression of the gene of interest. Viral vectors are, therefore, preferred as the gene delivery system. Among the commonly used viral vectors are retrovirus, adenovirus, adeno-associated virus, and herpes simplex virus. Because all viruses except adeno-associated virus can cause human diseases, these viral vectors are genetically engineered to remove any disease-causing genes and introduce the "therapeutic gene". Despite such measures to manipulate the viral genome it is possible that they regain their disease potential. Further, it can generate fatal immune response and such a response if not fatal prevents repetitive gene transfers, which may be necessary for long-lived effect of gene therapy. Additionally, in most cases of gene therapy the introduced genes cannot be directed for integration at specific sites. Their aberrant insertion at random sites can cause secondary damage such as initiation of cancer upon insertion at an oncogenic locus. Since diseases such as cancer are caused by multiple genetic and epigenetic variations, gene therapy can pose serious problems. Despite such challenges more than half of the over 400 gene therapy clinical trials worldwide are related to cancer (92).

Although loss of PTP function may not be the sole cause of a particular cancer, numerous studies have demonstrated that their overexpression can revert the transformed phenotype of cells and in some cases also limit their metastatic potential. Use of gene therapy to restore the function of a PTP that is lost in cancer, therefore, appears to be promising. Several *in vitro* and *in vivo* analyses have demonstrated tumor suppressor functions for DEP-1 or its rat homologue rPTPη (11, 31) in thyroid, colorectal, breast, and lung carcinomas. A subsequent study has proposed the use of gene therapy for thyroid cancers specifically in undifferentiated anaplastic thyroid carcinomas that are refractory to radiotherapy and chemotherapy. This study has successfully used a replication deficient adenoviral vector (ONYX-015) to transfer the rPTPη gene into thyroid cancer cell lines. Adenovirus-mediated expression of rPTPη inhibited proliferation of the cells and their growth when engrafted into nude mice (93). A few studies have, however, reported difficulty in obtaining cells stably expressing PTPs (94, 95) probably because of its enzymatic involvement in a cascading pathway. To overcome this problem a transposon based "Sleeping Beauty" system was adapted to overexpress an osteoclastic PTP (PTP-oc/PTPROt) in osteoclast cells. This technology enabled to retain stable expression of the protein for over a year without having to maintain selection pressure (95). Transposable elements are natural, relatively safe, nonviral alternatives as gene delivery vehicles and result in stable gene expression. These approaches had been used widely for invertebrate systems (96) but were not available for vertebrate systems until recently. The reconstruction of the vertebrate transposon element Sleeping Beauty from the fossils of the fish genome (97) thus revived interest in using such elements as efficient gene transfer technologies (98). The concerted efforts of different laboratories has made it possible to use this technology in gene therapy as demonstrated by its ability to treat human diseases modeled in mice (99–102) and its applicability to transfer genes into human tissues (103). Unlike viruses that exhibit the tendency to integrate into genes rather than nongenomic sites transposon-mediated integration tends to occur at intergenic sites and repeat elements (104, 105) rather than transcribed genes or their regulatory elements (106). They are, therefore, potentially devoid of secondary deleterious effects and considered safer than viral vectors (107, 108). A transposable system, which has been proven to be efficient in long-term expression of a PTP gene (95), thus, appears to be a promising tool for PTP gene therapy.

Apart from PTPs that are suppressed in cancer, a few PTPs are upregulated or carry activating mutations and demonstrate oncogenic potential. Such PTPs can be targeted using inhibitors as described in the following section. Alternatively, gene repair approach (for PTP genes carrying activating mutations) or antisense oligonucleotide (ASO) therapy can be used for targets that are resistant to small molecule inhibitors. ASOs primarily function by forming a complex with the target mRNA, thus, leading to RNaseH-mediated degradation. Alternatively, they may alter mRNA transport, splicing, or even inhibit transcription by forming a triple helix complex with double-stranded DNA. There are several clinical trials testing this strategy to fight cancer. Antiapoptotic proteins appear to be the most popular targets for such therapy [reviewed in (109)]. Since PTPs are involved in regulating apoptosis and some are implicated in protecting cells from apoptosis, these could emerge as potential targets for ASO therapy.

B. Small Molecule Inhibitors

The large family of tyrosine phosphatases with a common CX_5R catalytic motif is involved in a variety of physiological processes by positively or negatively regulating signal transduction pathways. While this family of enzymes may include promising drug targets, they may also include "antitargets" (5). Inhibition of these PTPs may be detrimental to normal functioning of cells. It, therefore, becomes very critical to test the specificity of any inhibitor designed for a particular PTP. These protein phosphatases are inhibited by several mechanisms, including oxidative inactivation of catalytic site cysteine (110, 111), covalent modification of the conserved active site arginine side chain (112), or by noncompetitive inhibitors, which act outside the catalytic site but impair phosphatase activity (113). Competitive inhibitors for PTPs rely on the pTyr-binding site. In some PTPs with two pTyr-binding sites, the use of two pTyr mimetics joined by a linker imparts better specificity (5, 114). It is noteworthy that several PTPs have been implicated in diabetes because of their role as negative regulators of insulin receptor signaling. Among these PTPs, PTP1B/PTPN1, TCPTP/PTPN2, and LAR/PTPRF appear to be the most promising drug targets for Type II diabetes and obesity, and PTP1B has been the most widely studied protein tyrosine phosphatase for this purpose. To avoid any unwarranted effects on other PTPs, screens for testing the specificity of inhibitors developed for PTP1B include several other PTPs. Inhibitors of several other PTPs were discovered in such screens. *PTP1B* gene is known to be amplified and upregulated in several different cancers (17–23), which also makes it a promising target for treatment with small molecule inhibitors developed for diabetes and obesity. On the contrary, inhibition of TCPTP and LAR could result in increased proliferation (73, 115) making them less attractive as therapeutic targets. Further, several PTPs namely CD45/PTPRC, SHP-1/PTPN6, and SHP-2/PTPN11 are targets for immune-related diseases, for example, transplant rejection and infectious diseases (5). It is important to explore these interesting targets for inhibitor development. Activating mutations of PTPN11 have been implicated in hematopoietic malignancies (32–34). This observation provides an impetus to use SHP-2 inhibitors for treatment of leukemia associated with SHP-2 mutation. Thus far, efforts have been expended to develop inhibitors for these few PTPs. The involvement of certain other PTPs in oncogenic transformation and metastatic potential offers the rationale for further development of specific inhibitors that could be used as anticancer drugs.

C. Epigenetic Therapy

Reactivation of genes silenced by epigenetic modifications, particularly DNA methylation, is usually referred to as epigenetic therapy. A classical agent used for this purpose is 5-azacytidine (5-azaC) or its congener 5-aza-deoxycytidine (5-aza-CdR or decitabine). Both are potent DNA hypomethylating agents used in cancer clinical trials (116–121). Unlike most anticancer drugs available to date, epigenetic therapy is reversible and can target reactions that are unique to cancer. Despite the importance and distinct advantage of this therapeutic concept, this mode of therapy has not been explored extensively.

Unlike 5-azaC that is incorporated into RNA and DNA, decitabine is incorporated only into DNA and is, therefore, much more selective in its action and less toxic than 5-azaC. These drugs can also be deaminated into the respective uridines and their triphosphates, which interfere with *de novo* thymidylate synthesis. A noteworthy mechanism is based on studies that showed the inability of DNA methyltransferases (DNMTs) to methylate DNA following the incorporation of decitabine into DNA. Animal cells contain three functional DNMTs. Among these enzymes, DNMT3a and DNMT3b exhibit predominantly *de novo* methyltransferase activity whereas DNMT1 (the predominant enzyme) is exclusively involved in the methylation of hemimethylated DNA. Previous study in our laboratory (122) suggested that 5-azaC or decitabine may have differential effects on DNMTs. A recent study substantiated this notion by demonstrating that DNMT1 is rapidly and selectively degraded. We elucidated the molecular mechanism of this selective action of decitabine by demonstrating that DNMT1 is degraded by a proteasomal pathway that requires the KEN box, a signature motif missing in DNMT3a and 3b. Further, the degradation occurred rapidly and independent of its catalytic function, which did not require incorporation of decitabine into DNA (123). These DNA hypomethylating agents were used to reactivate PTPRO in human lung cancer cells (44). Our laboratory has also shown reexpression of (PTPROt) that is predominantly expressed in hematopoietic cells (T. Motiwala, S. Majumder, J. Byrd, M. Grever, D. Lucas, and S. Jacob, unpublished data) from an independent downstream promoter (124) but appears to be regulated by methylation at the CpG island located upstream. The expression of PTPN6/SHP-1 and PTPRG could also be revived upon treatment with the hypomethylating agent 5-aza-CdR (36, 47).

Unmethylated genes can also be suppressed as a result of the association of specific posttranslationally modified histones such as hypoacetylated histones H3 or H4 with their promoters. In this case, the gene can be reexpressed by treatment of cells with inhibitors of histone deacetylases (HDACs) (125). Several such inhibitors, including suberoyl anilide hydroxamic acid (SAHA), LAQ-824, PXD-101, depsipeptide, valproic acid, phenylbutyrate, MS-275, and

CI-994, have been or are being tested for HDAC inhibition in clinical trials (126). Since HDACs can alter cell growth, death, and differentiation, they can play important roles in cancer. HDACs belonging to Class I appear to be involved in regulating the proliferation of cancer cells (127). Class I HDACs could, thus, be promising targets for development of specific inhibitors. Many studies have also shown that combined treatment with inhibitors of DNMTs and HDACs could impede tumor growth (122, 128, 129). This mode of therapy can cause synergistic activation of some genes (122, 130), specifically those that are suppressed by DNA methylation. This combination therapy with inhibitors against DNMTs and HDACs can result in increased potency at lower doses and is being tested in anticancer trials [See (131, 132) for reviews].

V. Concluding Remarks

Unlike PTKs that have been the subject of intense investigations for many years because of their oncogenic potential, PTPs have not received much attention until recently. Several observations strongly suggest an important role for PTPs in cancer. First, half of the classical *PTP* genes are located in regions frequently deleted in different human cancers. Second, at least 10 *PTP* genes are mutated in some cancers including activating mutation in one of them. It is conceivable that many other PTPs could also be mutated in cancers and some mutations could be cancer type specific. Third, genes encoding some PTPs are silenced by promoter methylation, a hallmark of many established tumor suppressor genes. To this date, only five *PTP* genes are shown to be methylated of which some are also suppressed upon methylation. In addition to the loss/gain of function by genetic or epigenetic mechanisms, some PTPs also exhibit growth suppressor or oncogenic characteristics (refer to Table V for concise information on all the classical PTPs). Only a few of them have, however, been extensively studied with respect to their role in cancer (11, 12, 63, 71, 72, 77, 133). One of these PTPs, PTPRO, has been extensively studied by our laboratory with respect to its tumor suppressor potential and ability to induce apoptosis (44). Identification and characterization of additional PTPs with tumor suppressor characteristics will be an important challenge, particularly, if a relationship exists between reduced expression of PTPs and specific cancer types. The role of other factors in the epigenetic machinery, namely posttranslational modifications of histones (histone code) and chromatin remodeling, in the regulation of PTP expression should also be studied in order to elucidate the exact molecular mechanism underlying the altered expression of PTPs in cancer.

Unlike most known tumor suppressors, PTPs are emerging as a unique class of tumor suppressors with inherent enzymatic activities that utilize

TABLE V
SUMMARY OF THE TUMOR SUPPRESSOR AND ONCOGENIC PROPERTIES OF THE CLASSICAL PTPs GROUPED BY SUBTYPES

Subtype	Gene	PTP	Genetic/Epigenetic evidence	Expression evidence[a]	Phenotypic evidence
R1/R6	PTPRC	hCD45	ND[b]		ND
R2A	PTPRF	hLAR	TS[b]	Oncogene (147)	Oncogene
	PTPRD	hPTPδ	TS	TS (148)	TS
	PTPRS	hPTPσ	ND		TS
R2B	PTPRU	hPTPλ	ND		ND
	PTPRK	hPTPκ	TS	TS (9, 66)	Oncogene
	PTPRM	hPTPμ	ND		ND
	PTPRT	hPTPρ	TS		TS
R3	PTPRJ	hDEP1	TS	TS (11)	TS
	PTPRO	hGLEPP1	TS		TS
	PTPRB	hPTPβ	ND		ND
	PTPRQ/PTPGMC1	hPTPS31	ND		TS
	PTPRH	hSAP1	ND	TS (59)	TS
R4	PTPRE	hPTPε	ND	Oncogene (144)	Oncogene
	PTPRA	hPTPα	ND	Oncogene (149, 150)	TS
R5	PTPRG	hPTPγ	TS	TS (10, 151)	TS
	PTPRZ1	hPTPζ	TS	Oncogene (152, 153)	Oncogene
R7	PTPN7	hHePTP	ND	Oncogene (77, 133)	Oncogene
	PTPN5	hSTEP	ND		Oncogene
	PTPRR	hPCPTP1	ND		

(Continues)

TABLE V (Continued)

Subtype	Gene	PTP	Genetic/Epigenetic evidence	Expression evidence[a]	Phenotypic evidence
R8	PTPRN	hPTPIA2	ND		ND
	PTPRN2	hPTPIA2β	TS		ND
NT1	PTPN2	hTCPTP	ND	Oncogene (154)	TS/oncogene
	PTPN1	hPTP1B	Oncogene	Oncogene (17, 22, 155) TS (156)	Oncogene (src activation)
NT2	PTPN6	hSHP-1	TS	TS (41) oncogene (157, 158)	TS
	PTPN11	hSHP-2	Oncogene	Oncogene (34)	Oncogene
NT3	PTPN9	hMEG2	ND	Oncogene (78)	Oncogene
NT4	PTPN22	hLyPTP	ND		ND
	PTPN18	hBDP1	ND		ND
	PTPN12	hPEST	TS		TS
NT5	PTPN4	hMEG1	ND		ND
	PTPN3	hPTPH1	TS	Oncogene (156)	TS
NT6	PTPN14	hPTPD2	TS		TS
	PTPN21	hPTPD1	ND		ND
NT7	PTPN13	hPTPBAS	TS	Oncogene (80, 81, 83, 159, 160) TS (161)	TS/oncogene
NT8	PTPN20	hPTPTyp	ND		ND
NT9	PTPN23	hHDPTP	TS		ND

[a]The gene is denoted as TS or oncogene based on observed, reduced, or increased expression, respectively, in cancer. It is, however, not known whether altered expression is a cause or consequence of transformation. References for expression of the PTP are provided in parentheses.
[b]ND, not determined; TS, tumor suppressor.

proteins as substrates. This characteristic provides another level of regulation by these proteins namely modification of the phosphorylation status of their specific substrates. PTPs could themselves promote signaling (positive effect) by dephosphorylation and activation of PTKs, thus, coordinating rather than antagonizing the functions of PTKs (*134*). The physiological functions of PTPs could depend on the phosphorylated state of a specific PTP substrate(s) that could be critical to the tumor initiation and/or progression. It is, therefore, essential that the substrates of PTPs with suspected tumor suppressor function are identified, and the potential role of the phosphorylated status of these substrates in tumorigenesis is explored.

Most studies addressing the involvement of PTPs in cancer focused on a specific PTP rather than the PTP family or at least PTPs belonging to the same subtype. It is possible that the loss or gain of the function of one PTP is counterbalanced by the gain or loss of another PTP, particularly, if the two PTPs share critical substrates. To address this issue, it is important to develop highly sensitive techniques, like gene arrays, that will help to determine the expression levels of all PTPs simultaneously in specific cancer types. An exhaustive analysis of the levels of these enzymes in different cell types could reveal a specific profile of expressing and nonexpressing PTPs in specific cancers. Further, such comparisons will also assist in explaining the differential phenotypes observed between two different cell types that either express or do not express a particular PTP. Finally, reexpression of the wild-type PTPs in cells where a specific PTP is lost or mutated and reactivation of the PTPs silenced by promoter methylation pose exciting challenges and offer novel molecular targets in cancer therapy. Although not extensively studied, small molecule inhibitors can also be used to compensate for the loss of a gene. In fact, one study has identified inhibitors capable of reversing the consequences of the loss of PTEN by targeting its downstream effector (*135*). Development of such inhibitors, however, requires a detailed understanding of the cellular functions of the protein. Loss of function of several PTPs in cancer coupled with the ease of developing, screening, and applying small molecule inhibitors to therapy further reinforces the necessity for identifying the critical substrates and ensuing pathways mediating the function of PTPs.

Acknowledgments

We thank Drs. Kalpana Ghoshal and Sarmila Majumder for their useful comments. Although every effort was made to search several databases (OMIM, PubMed, Gene, Source, and so on) using different search criteria, we regret any inadvertent omission of relevant papers. The work in the authors' laboratory was supported by grants (CA 81024 and CA 86978) from the National Cancer Institute and ES 10874 from the National Institute of Environmental Health Science.

References

1. Blume-Jensen, P., and Hunter, T. (2001). Oncogenic kinase signalling. *Nature* **411,** 355–365.
2. Maehama, T., Taylor, G. S., and Dixon, J. E. (2001). PTEN and myotubularin: Novel phosphoinositide phosphatases. *Annu. Rev. Biochem.* **70,** 247–279.
3. Furnari, F. B., Huang, H. J., and Cavenee, W. K. (1998). The phosphoinositol phosphatases activity of PTEN mediates a serum-sensitive G1 growth arrest in glioma cells. *Cancer Res.* **58,** 5002–5008.
4. Kristjansdottir, K., and Rudolph, J. (2004). Cdc25 phosphatases and cancer. *Chem. Biol.* **11,** 1043–1051.
5. Hoffman, B. T., Nelson, M. R., Burdick, K., and Baxter, S. M. (2004). Protein tyrosine phosphatases: Strategies for distinguishing proteins in a family containing multiple drug targets and anti-targets. *Curr. Pharm. Des.* **10,** 1161–1181.
6. Raugei, G., Ramponi, G., and Chiarugi, P. (2002). Low molecular weight protein tyrosine phosphatases: Small, but smart. *Cell. Mol. Life Sci.* **59,** 941–949.
7. Ducruet, A. P., Vogt, A., Wipf, P., and Lazo, J. S. (2005). Dual specificity protein phosphatases: Therapeutic targets for cancer and Alzheimer's disease. *Annu. Rev. Pharmacol. Toxicol.* **45,** 725–750.
8. Andersen, J. N., Jansen, P. G., Echwald, S. M., Mortensen, O. H., Fukada, T., Del Vecchio, R., Tonks, N. K., and Moller, N. P. (2004). A genomic perspective on protein tyrosine phosphatases: Gene structure, pseudogenes, and genetic disease linkage. *FASEB J.* **18,** 8–30.
9. Nakamura, M., Kishi, M., Sakaki, T., Hashimoto, H., Nakase, H., Shimada, K., Ishida, E., and Konishi, N. (2003). Novel tumor suppressor loci on 6q22–23 in primary central nervous system lymphomas. *Cancer Res.* **63,** 737–741.
10. van Niekerk, C. C., and Poels, L. G. (1999). Reduced expression of protein tyrosine phosphatase gamma in lung and ovarian tumors. *Cancer Lett.* **137,** 61–73.
11. Trapasso, F., Iuliano, R., Boccia, A., Stella, A., Visconti, R., Bruni, P., Baldassarre, G., Santoro, M., Viglietto, G., and Fusco, A. (2000). Rat protein tyrosine phosphatase eta suppresses the neoplastic phenotype of retrovirally transformed thyroid cells through the stabilization of p27(Kip1). *Mol. Cell. Biol.* **20,** 9236–9246.
12. Liu, S., Sugimoto, Y., Sorio, C., Tecchio, C., and Lin, Y. C. (2004). Function analysis of estrogenically regulated protein tyrosine phosphatase gamma (PTPgamma) in human breast cancer cell line MCF-7. *Oncogene* **23,** 1256–1262.
13. Longo, F. M., Martignetti, J. A., Le Beau, J. M., Zhang, J. S., Barnes, J. P., and Brosius, J. (1993). Leukocyte common antigen-related receptor-linked tyrosine phosphatase. Regulation of mRNA expression. *J. Biol. Chem.* **268,** 26503–26511.
14. Oka, T., Ouchida, M., Koyama, M., Ogama, Y., Takada, S., Nakatani, Y., Tanaka, T., Yoshino, T., Hayashi, K., Ohara, N., Kondo, E., Takahashi, K. et al. (2002). Gene silencing of the tyrosine phosphatase SHP1 gene by aberrant methylation in leukemias/lymphomas. *Cancer Res.* **62,** 6390–6394.
15. Takekawa, M., Itoh, F., Hinoda, Y., Adachi, M., Ariyama, T., Inazawa, J., Imai, K., and Yachi, A. (1994). Chromosomal localization of the protein tyrosine phosphatase G1 gene and characterization of the aberrant transcripts in human colon cancer cells. *FEBS Lett.* **339,** 222–228.
16. Toyooka, S., Ouchida, M., Jitsumori, Y., Tsukuda, K., Sakai, A., Nakamura, A., Shimizu, N., and Shimizu, K. (2000). HD-PTP: A novel protein tyrosine phosphatase gene on human chromosome 3p21.3. *Biochem. Biophys. Res. Commun.* **278,** 671–678.
17. Tanner, M. M., Grenman, S., Koul, A., Johannsson, O., Meltzer, P., Pejovic, T., Borg, A., and Isola, J. J. (2000). Frequent amplification of chromosomal region 20q12–q13 in ovarian cancer. *Clin. Cancer Res.* **6,** 1833–1839.

18. Tanner, M. M., Tirkkonen, M., Kallioniemi, A., Isola, J., Kuukasjarvi, T., Collins, C., Kowbel, D., Guan, X. Y., Trent, J., Gray, J. W., Meltzer, P., and Kallioniemi, O. P. (1996). Independent amplification and frequent co-amplification of three nonsyntenic regions on the long arm of chromosome 20 in human breast cancer. *Cancer Res.* **56**, 3441–3445.
19. Zondervan, P. E., Wink, J., Alers, J. C., IJzermans, J. N., Schalm, S. W., de Man, R. A., and van Dekken, H. (2000). Molecular cytogenetic evaluation of virus-associated and non-viral hepatocellular carcinoma: Analysis of 26 carcinomas and 12 concurrent dysplasias. *J. Pathol.* **192**, 207–215.
20. van Dekken, H., Geelen, E., Dinjens, W. N., Wijnhoven, B. P., Tilanus, H. W., Tanke, H. J., and Rosenberg, C. (1999). Comparative genomic hybridization of cancer of the gastroesophageal junction: Deletion of 14Q31–32.1 discriminates between esophageal (Barrett's) and gastric cardia adenocarcinomas. *Cancer Res.* **59**, 748–752.
21. Yang, S. H., Seo, M. Y., Jeong, H. J., Jeung, H. C., Shin, J., Kim, S. C., Noh, S. H., Chung, H. C., and Rha, S. Y. (2005). Gene copy number change events at chromosome 20 and their association with recurrence in gastric cancer patients. *Clin. Cancer Res.* **11**, 612–620.
22. Watanabe, T., Imoto, I., Katahira, T., Hirasawa, A., Ishiwata, I., Emi, M., Takayama, M., Sato, A., and Inazawa, J. (2002). Differentially regulated genes as putative targets of amplifications at 20q in ovarian cancers. *Jpn. J. Cancer Res.* **93**, 1114–1122.
23. Albrecht, B., Hausmann, M., Zitzelsberger, H., Stein, H., Siewert, J. R., Hopt, U., Langer, R., Hofler, H., Werner, M., and Walch, A. (2004). Array-based comparative genomic hybridization for the detection of DNA sequence copy number changes in Barrett's adenocarcinoma. *J. Pathol.* **203**, 780–788.
24. Paul, M. K., and Mukhopadhyay, A. K. (2004). Tyrosine kinase-role and significance in cancer. *Int. J. Med. Sci.* **1**, 101–115.
25. Wang, Z., Shen, D., Parsons, D. W., Bardelli, A., Sager, J., Szabo, S., Ptak, J., Silliman, N., Peters, B. A., van der Heijden, M. S., Parmigiani, G., Yan, H. *et al.* (2004). Mutational analysis of the tyrosine phosphatome in colorectal cancers. *Science* **304**, 1164–1166.
26. Trojan, L., Schaaf, A., Steidler, A., Haak, M., Thalmann, G., Knoll, T., Gretz, N., Alken, P., and Michel, M. S. (2005). Identification of metastasis-associated genes in prostate cancer by genetic profiling of human prostate cancer cell lines. *Anticancer Res.* **25**, 183–191.
27. Levea, C. M., McGary, C. T., Symons, J. R., and Mooney, R. A. (2000). PTP LAR expression compared to prognostic indices in metastatic and non-metastatic breast cancer. *Breast Cancer Res. Treat.* **64**, 221–228.
28. Symons, J. R., LeVea, C. M., and Mooney, R. A. (2002). Expression of the leucocyte common antigen-related (LAR) tyrosine phosphatase is regulated by cell density through functional E-cadherin complexes. *Biochem. J.* **365**, 513–519
29. Ogata, M., Takada, T., Mori, Y., Oh-hora, M., Uchida, Y., Kosugi, A., Miyake, K., and Hamaoka, T. (1999). Effects of overexpression of PTP36, a putative protein tyrosine phosphatase, on cell adhesion, cell growth, and cytoskeletons in HeLa cells. *J. Biol. Chem.* **274**, 12905–12909.
30. Zhang, S. H., Liu, J., Kobayashi, R., and Tonks, N. K. (1999). Identification of the cell cycle regulator VCP (p97/CDC48) as a substrate of the band 4.1-related protein-tyrosine phosphatase PTPH1. *J. Biol. Chem.* **274**, 17806–17812.
31. Ruivenkamp, C. A., van Wezel, T., Zanon, C., Stassen, A. P., Vlcek, C., Csikos, T., Klous, A. M., Tripodis, N., Perrakis, A., Boerrigter, L., Groot, P. C., Lindeman, J. *et al.* (2002). Ptprj is a candidate for the mouse colon-cancer susceptibility locus Scc1 and is frequently deleted in human cancers. *Nat. Genet.* **31**, 295–300.
32. Chantrain, C. F., Jijon, P., De Raedt, T., Vermylen, C., Poirel, H. A., Legius, E., and Brichard, B. (2005). Therapy-related acute myeloid leukemia in a child with Noonan syndrome and clonal duplication of the germline PTPN11 mutation. *Pediatr. Blood Cancer*.

33. Tartaglia, M., and Gelb, B. D. (2005). Germ-line and somatic PTPN11 mutations in human disease. *Eur. J. Med. Genet.* **48,** 81–96.
34. Xu, R., Yu, Y., Zheng, S., Zhao, X., Dong, Q., He, Z., Liang, Y., Lu, Q., Fang, Y., Gan, X., Xu, X., Zhang, S. *et al.* (2005). Overexpression of Shp2 tyrosine phosphatase is implicated in leukemogenesis in adult human leukemia. *Blood* **106,** 3142–3149.
35. Schubbert, S., Lieuw, K., Rowe, S. L., Lee, C. M., Li, X., Loh, M. L., Clapp, D. W., and Shannon, K. M. (2005). Funtional analysis of leukemia-associated PTPN11 mutations in primary hematopoietic cells. *Blood* **106,** 311–317.
36. Zhang, Q., Raghunath, P. N., Vonderheid, E., Odum, N., and Wasik, M. A. (2000). Lack of phosphotyrosine phosphatase SHP-1 expression in malignant T-cell lymphoma cells results from methylation of the SHP-1 promoter. *Am. J. Pathol.* **157,** 1137–1146.
37. Khoury, J. D., Rassidakis, G. Z., Medeiros, L. J., Amin, H. M., and Lai, R. (2004). Methylation of SHP1 gene and loss of SHP1 protein expression are frequent in systemic anaplastic large cell lymphoma. *Blood* **104,** 1580–1581.
38. Chim, C. S., Fung, T. K., Cheung, W. C., Liang, R., and Kwong, Y. L. (2004). SOCS1 and SHP1 hypermethylation in multiple myeloma: Implication for epigenetic activation of the Jak/STAT pathway. *Blood* **103,** 4630–4635.
39. Johan, M. F., Bowen, D. T., Frew, M. E., Goodeve, A. C., and Reilly, J. T. (2005). Aberrant methylation of the negative regulators RASSFIA, SHP-1 and SOCS-1 in myelodysplastic syndromes and acute myeloid leukaemia. *Br. J. Haematol.* **129,** 60–65.
40. Oka, T., Yoshino, T., Hayashi, K., Ohara, N., Nakanishi, T., Yamaai, Y., Hiraki, A., Sogawa, C. A., Kondo, E., Teramoto, N., Takahashi, K., Tsuchiyama, J. *et al.* (2001). Reduction of hematopoietic cell-specific tyrosine phosphatase SHP-1 gene expression in natural killer cell lymphoma and various types of lymphomas/leukemias: Combination analysis with cDNA expression array and tissue microarray. *Am. J. Pathol.* **159,** 1495–1505.
41. Koyama, M., Oka, T., Ouchida, M., Nakatani, Y., Nishiuchi, R., Yoshino, T., Hayashi, K., Akagi, T., and Seino, Y. (2003). Activated proliferation of B-cell lymphomas/leukemias with SHP1 gene silencing by aberrant CpG methylation. *Lab. Invest.* **83,** 1849–1858.
42. Motiwala, T., Ghoshal, K., Das, A., Majumder, S., Weichenhan, D., Wu, Y. Z., Holman, K., James, S. J., Jacob, S. T., and Plass, C. (2003). Suppression of the protein tyrosine phosphatase receptor type 0 gene (PTPRO) by methylation in hepatocellular carcinomas. *Oncogene* **22,** 6319–6331.
43. Aguiar, R. C., Yakushijin, Y., Kharbanda, S., Tiwari, S., Freeman, G. J., and Shipp, M. A. (1999). PTPROt: An alternatively spliced and developmentally regulated B-lymphoid phosphatase that promotes GO/G 1. *Blood* **94,** 2403–2413.
44. Motiwala, T., Kutay, H., Ghoshal, K., Bai, S., Seimiya, H., Tsuruo, T., Suster, S., Morrison, C., and Jacob, S. T. (2004). Protein tyrosine phosphatase receptor-type 0 (PTPRO) exhibits characteristics of a candidate tumor suppressor in human lung cancer. *Proc. Natl. Acad. Sci. USA* **101,** 13844–13849.
45. Mori, Y., Yin, J., Sato, F., Sterian, A., Simms, L. A., Selaru, F. M., Schulmann, K., Xu, Y., Olaru, A., Wang, S., Deacu, E., Abraham, J. M. *et al.* (2004). Identification of genes uniquely involved in frequent microsatellite instability colon carcinogenesis by expression profiling combined with epigenetic scanning. *Cancer Res.* **64,** 2434–2438.
46. Rush, L. J., Raval, A., Funchain, P., Johnson, A. J., Smith, L., Lucas, D. M., Bembea, M., Liu, T. H., Heerema, N. A., Rassenti, L., Liyanarachchi, S., Davuluri, R. *et al.* (2004). Epigenetic profiling in chronic lymphocytic leukemia reveals novel methylation targets. *Cancer Res.* **64,** 2424–2433.
47. van Doorn, R., Zoutman, W. H., Dijkman, R., de Menezes, R. X., Commandeur, S., Mulder, A. A., van der Velden, P. A., Vermeer, M. H., Willemze, R., Yan, P. S., Huang, T. H., and Tensen, C. P. (2005). Epigenetic profiling of cutaneous T-cell lymphoma: Promoter hypermethylation

of multiple tumor suppressor genes including BCL 7a, PTPRG, and *p73*. *J. Clin. Oncol.* **23**, 3886–3896.
48. Jacob, S. T., and Motiwala, T. (2005). Epigenetic regulation of protein tyrosine phosphatases: Potential molecular targets for cancer therapy. *Cancer Gene Ther.* **12**, 665–672.
49. Cheung, P., Allis, C. D., and Sassone-Corsi, P. (2000). Signaling to chromatin through histone modifications. *Cell* **103**, 263–271.
50. Hake, S. B., Xiao, A., and Allis, C. D. (2004). Linking the epigenetic 'language' of covalent histone modifications to cancer. *Br. J. Cancer* **90**, 761–769.
51. Jenuwein, T., and Allis, C. D. (2001). Translating the histone code. *Science* **293**, 1074–1080.
52. Peterson, C. L., and Laniel, M. A. (2004). Histones and histone modifications. *Curr. Biol.* **14**, R546–R551.
53. Wang, Y., Wysocka, J., Perlin, J. R., Leonelli, L., Allis, C. D., and Coonrod, S. A. (2004). Linking covalent histone modifications to epigenetics: The rigidity and plasticity of the marks. *Cold Spring Harb. Symp. Quant. Biol.* **69**, 161–169.
54. Peterson, C. L. (2002). Chromatin remodeling: Nucleosomes bulging at the seams. *Curr. Biol.* **12**, R245–R247.
55. Peterson, C. L. (2002). Chromatin remodeling enzymes: Taming the machines. Third in review series on chromatin dynamics. *EMBO Rep.* **3**, 319–322.
56. Guan, K. L., Haun, R. S., Watson, S. J., Geahlen, R. L., and Dixon, J. E. (1990). Cloning and expression of a protein-tyrosine-phosphatase. *Proc. Natl. Acad. Sci. USA* **87**, 1501–1505.
57. Charbonneau, H., Tonks, N. K., Kumar, S., Diltz, C. D., Harrylock, M., Cool, D. E., Krebs, E. G., Fischer, E. H., and Walsh, K. A. (1989). Human placenta protein-tyrosine-phosphatase: Amino acid sequence and relationship to a family of receptor-like proteins. *Proc. Natl. Acad. Sci. USA* **86**, 5252–5256.
58. Keane, M. M., Lowrey, G. A., Ettenberg, S. A., Dayton, M. A., and Lipkowitz, S. (1996). The protein tyrosine phosphatase DEP-1 is induced during differentiation and inhibits growth of breast cancer cells. *Cancer Res.* **56**, 4236–4243.
59. Nagano, H., Noguchi, T., Inagaki, K., Yoon, S., Matozaki, T., Itoh, H., Kasuga, M., and Hayashi, Y. (2003). Downregulation of stomach cancer-associated protein tyrosine phosphatase-1 (SAP-1) in advanced human hepatocellular carcinoma. *Oncogene* **22**, 4656–4663.
60. Takada, T., Noguchi, T., Inagaki, K., Hosooka, T., Fukunaga, K., Yamao, T., Ogawa, W., Matozaki, T., and Kasuga, M. (2002). Induction of apoptosis by stomach cancer-associated protein-tyrosine phosphatase-1. *J. Biol. Chem.* **277**, 34359–34366.
61. Garton, A. J., and Tonks, N. K. (1999). Regulation of fibroblast motility by the protein tyrosine phosphatase PTP-PEST. *J. Biol. Chem.* **274**, 3811–3818.
62. Lin, Y., Ceacareanu, A. C., and Hassid, A. (2003). Nitric oxide-induced inhibition of aortic smooth muscle cell motility: Role of PTP-PEST and adaptor proteins p130cas and Crk. *Am. J. Physiol. Heart Circ. Physiol.* **285**, H710–H721.
63. Ardini, E., Agresti, R., Tagliabue, E., Greco, M., Aiello, P., Yang, L. T., Menard, S., and Sap, J. (2000). Expression of protein tyrosine phosphatase alpha (RPTPalpha) in human breast cancer correlates with low tumor grade, and inhibits tumor cell growth *in vitro* and *in vivo*. *Oncogene* **19**, 4979–4987.
64. MacKeigan, J. P., Murphy, L. O., and Blenis, J. (2005). Sensitized RNAi screen of human kinases and phosphatases identifies new regulators of apoptosis and chemoresistance. *Nat. Cell Biol.* **7**, 591–600.
65. Zhang, Y., Siebert, R., Matthiesen, P., Yang, Y., Ha, H., and Schlegelberger, B. (1998). Cytogenetical assignment and physical mapping of the human R-PTP-kappa gene (PTPRK) to the putative tumor suppressor gene region 6q22.2–q22.3. *Genomics* **51**, 309–311.

66. McArdle, L., Rafferty, M., Maelandsmo, G. M., Bergin, O., Farr, C. J., Dervan, P. A., O'Loughlin, S., Herlyn, M., and Easty, D. J. (2001). Protein tyrosine phosphatase genes downregulated in melanoma. *J. Invest. Dermatol.* **117,** 1255–1260.
67. Seimiya, H., and Tsuruo, T. (1998). Functional involvement of PTP-U2L in apoptosis subsequent to terminal differentiation of monoblastoid leukemia cells. *J. Biol. Chem.* **273,** 21187–21193.
68. Zheng, X. M., Resnick, R. J., and Shalloway, D. (2000). A phosphotyrosine displacement mechanism for activation of Src by PTPalpha. *EMBO J.* **19,** 964–978.
69. Egan, C., Pang, A., Durda, D., Cheng, H. C., Wang, J. H., and Fujita, D. J. (1999). Activation of Src in human breast tumor cell lines: Elevated levels of phosphotyrosine phosphatase activity that preferentially recognizes the Src carboxy terminal negative regulatory tyrosine 530. *Oncogene* **18,** 1227–1237.
70. Ostman, A., Yang, Q., and Tonks, N. K. (1994). Expression of DEP-l, a receptor-like protein-tyrosine-phosphatase, is enhanced with increasing cell density. *Proc. Natl. Acad. Sci. USA* **91,** 9680–9684.
71. Ganapati, U., Gupta, S., Radha, V., Sudhakar, C., Manogaran, P. S., and Swarup, G. (2001). A nuclear protein tyrosine phosphatase induces shortening of G 1 phase and increase in c-Myc protein level. *Exp. Cell Res.* **265,** 1–10.
72. Radha, V., Nambirajan, S., and Swarup, G. (1997). Overexpression of a nuclear protein tyrosine phosphatase increases cell proliferation. *FEBS Lett.* **409,** 33–36.
73. Dupuis, M., De Jesus Ibarra-Sanchez, M., Tremblay, M. L., and Duplay, P. (2003). Gr-l+ myeloid cells lacking T cell protein tyrosine phosphatase inhibit lymphocyte proliferation by an IFN-gamma- and nitric oxide-dependent mechanism. *J. Immunol.* **171,** 726–732.
74. Evan, G. I., Wyllie, A. H., Gilbert, C. S., Littlewood, T. D., Land, H., Brooks, M., Waters, C. M., Penn, L. Z., and Hancock, D. C. (1992). Induction of apoptosis in fibroblasts by c-myc protein. *Cell* **69,** 119–128.
75. Radha, V., Sudhakar, C., and Swarup, G. (1999). Induction of p53 dependent apoptosis upon overexpression of a nuclear protein tyrosine phosphatase. *FEBS Lett.* **453,** 308–312.
76. Shimizu, T., Miyakawa, Y., Iwata, S., Kuribara, A., Tiganis, T., Morimoto, C., Ikeda, Y., and Kizaki, M. (2004). A novel mechanism for imatinib mesylate (STI571) resistance in CML cell line KT-1: Role of TC-PTP in modulating signals downstream from the BCR-ABL fusion protein. *Exp. Hematol.* **32,** 1057–1063.
77. Zanke, B., Squire, J., Griesser, H., Henry, M., Suzuki, H., Patterson, B., Minden, M., and Mak, T. W. (1994). A hematopoietic protein tyrosine phosphatase (HePTP) gene that is amplified and overexpressed in myeloid malignancies maps to chromosome 1q32.1. *Leukemia* **8,** 236–244.
78. Xu, M. J., Sui, X., Zhao, R., Dai, C., Krantz, S. B., and Zhao, Z. J. (2003). PTP-MEG2 is activated in polycythemia vera erythroid progenitor cells and is required for growth and expansion of erythroid cells. *Blood* **102,** 4354–4360.
79. Muller, S., Kunkel, P., Lamszus, K., Ulbricht, U., Lorente, G. A., Nelson, A. M., von Schack, D., Chin, D. J., Lohr, S. C., Westphal, M., and Melcher, T. (2003). A role for receptor tyrosine phosphatase zeta in glioma cell migration. *Oncogene* **22,** 6661–6668.
80. Elnemr, A., Ohta, T., Yachie, A., Kayahara, M., Kitagawa, H., Fujimura, T., Ninomiya, I., Fushida, S., Nishimura, G. I., Shimizu, K., and Miwa, K. (2001). Human pancreatic cancer cells disable function of Fas receptors at severallevels in Fas signal transduction pathway. *Int. J. Oncol.* **18,** 311–316.
81. Ungefroren, H., Kruse, M. L., Trauzold, A., Roeschmann, S., Roeder, C., Arlt, A., Henne-Bruns, D., and Kalthoff, H. (2001). F AP-1 in pancreatic cancer cells: Functional and mechanistic studies on its inhibitory role in CD95-mediated apoptosis. *J. Cell Sci.* **114,** 2735–2746.

82. Lee, S. H., Shin, M. S., Lee, H. S., Bae, J. H., Lee, H. K., Kim, H. S., Kim, S. Y., Jang, J. J., Joo, M., Kang, Y. K., Park, W. S., Park, J. Y. et al. (2001). Expression of Fas and Fas-related molecules in human hepatocellular carcinoma. *Hum. Pathol.* **32,** 250–256.
83. Meinhold-Heerlein, I., Stenner-Liewen, F., Liewen, H., Kitada, S., Krajewska, M., Krajewski, S., Zapata, J. M., Monks, A., Scudiero, D. A., Bauknecht, T., and Reed, J. C. (2001). Expression and potential role of Fas-associated phosphatase-1 in ovarian cancer. *Am. J. Pathol.* **158,** 1335–1344.
84. Bompard, G., Puech, C., Prebois, C., Vignon, F., and Freiss, G. (2002). Protein-tyrosine phosphatase PTPL1/FAP-1 triggers apoptosis in human breast cancer cells. *J. Biol. Chem.* **277,** 47861–47869.
85. Larsen, M., Tremblay, M. L., and Yamada, K. M. (2003). Phosphatases in cell-matrix adhesion and migration. *Nat. Rev. Mol. Cell Biol.* **4,** 700–711.
86. Stoker, A. W. (2005). Protein tyrosine phosphatases and signaling. *J. Endocrinol.* **185,** 19–33.
87. Fischer, E. H. (1999). Cell signaling by protein tyrosine phosphorylation. *Adv. Enzyme Regul.* **39,** 359–369.
88. Ponniah, S., Wang, D. Z., Lim, K. L., and Pallen, C. J. (1999). Targeted disruption of the tyrosine phosphatase PTPalpha leads to constitutive downregulation of the kinases Src and Fyn. *Curr. Biol.* **9,** 535–538.
89. Granot-Attas, S., and Elson, A. (2004). Protein tyrosine phosphatase epsilon activates Yes and Fyn in Neu-induced mammary tumor cells. *Exp. Cell Res.* **294,** 236–243.
90. Pallen, C. J. (2003). Protein tyrosine phosphatase alpha (PTPalpha): A Src family kinase activator and mediator of multiple biological effects. *Curr. Top. Med. Chem.* **3,** 821–835.
91. Richardson, P. D., Augustin, L. B., Kren, B. T., and Steer, C. J. (2002). Gene repair and transposon- mediated gene therapy. *Stem Cells* **20,** 105–118.
92. Li, C., Bowles, D. E., van Dyke, T., and Samulski, R. J. (2005). Adeno-associated virus vectors: Potential applications for cancer gene therapy. *Cancer Gene Ther.* **12,** 913–925.
93. Iuliano, R., Trapasso, F., Le Pera, I., Schepis, F., Sama, I., Clodomiro, A., Dumon, K. R., Santoro, M., Chiariotti, L., Viglietto, G., and Fusco, A. (2003). An adenovirus carrying the rat protein tyrosine phosphatase eta suppresses the growth of human thyroid carcinoma cell lines in vitro and in vivo. *Cancer Res.* **63,** 882–886.
94. Oganesian, A., Poot, M., Daum, G., Coats, S. A., Wright, M. B., Seifert, R. A., and Bowen-Pope, D. F. (2003). Protein tyrosine phosphatase RQ is a phosphatidylinositol phosphatase that can regulate cell survival and proliferation. *Proc. Natl. Acad. Sci. USA* **100,** 7563–7568.
95. Amoui, M., Suhr, S. M., Baylink, D. J., and Lau, K. H. (2004). An osteoclastic protein-tyrosine phosphatase may play a role in differentiation and activity of human monocytic U-937 cell-derived, osteoclast-like cells. *Am. J. Physiol. Cell Physiol.* **287,** C874–C884.
96. Cooley, L., Kelley, R., and Spradling, A. (1988). Insertional mutagenesis of the Drosophila genome with single P elements. *Science* **239,** 1121–1128.
97. Ivics, Z., Hackett, P. B., Plasterk, R. H., and Izsvak, Z. (1997). Molecular reconstruction of Sleeping Beauty, a Tc1-like transposon from fish, and its transposition in human cells. *Cell* **91,** 501–510.
98. Izsvak, Z., and Ivics, Z. (2004). Sleeping beauty transposition: Biology and applications for molecular therapy. *Mol. Ther.* **9,** 147–156.
99. He, C. X., Shi, D., Wu, W. J., Ding, Y. F., Feng, D. M., Lu, B., Chen, H. M., Yao, J. H., Shen, Q., Lu, D. R., and Xue, J. L. (2004). Insulin expression in livers of diabetic mice mediated by hydrodynamics-based administration. *World J. Gastroenterol.* **10,** 567–572.
100. Ohlfest, J. R., Frandsen, J. L., Fritz, S., Lobitz, P. D., Perkinson, S. G., Clark, K. J., Nelsestuen, G., Key, N. S., McIvor, R. S., Hackett, P. B., and Largaespada, D. A. (2005). Phenotypic correction and long-term expression of factor VIII in hemophilic mice by immunotolerization and nonviral gene transfer using the Sleeping Beauty transposon system. *Blood* **105,** 2691–2698.

101. Montini, E., Held, P. K., Noll, M., Morcinek, N., Al-Dhalimy, M., Finegold, M., Yant, S. R., Kay, M. A., and Grompe, M. (2002). In vivo correction of murine tyrosinemia type I by DNA-mediated transposition. *Mol. Ther.* **6**, 759–769.
102. Yant, S. R., Meuse, L., Chiu, W., Ivics, Z., Izsvak, Z., and Kay, M. A. (2000). Somatic integration and long-term transgene expression in normal and haemophilic mice using a DNA transposon system. *Nat. Genet.* **25**, 35–41.
103. Ortiz-Urda, S., Lin, Q., Yant, S. R., Keene, D., Kay, M. A., and Khavari, P. A. (2003). Sustainable correction of junctional epidermolysis bullosa via transposon-mediated nonviral gene transfer. *Gene Ther.* **10**, 1099–1104.
104. Liu, G., Geurts, A. M., Yae, K., Srinivasan, A. R., Fahrenkrug, S. C., Largaespada, D. A., Takeda, J., Horie, K., Olson, W. K., and Hackett, P. B. (2005). Target-site preferences of Sleeping Beauty transposons. *J. Mol. Biol.* **346**, 161–173.
105. Vigdal, T. J., Kaufman, C. D., Izsvak, Z., Voytas, D. F., and Ivics, Z. (2002). Common physical properties of DNA affecting target site selection of sleeping beauty and other Tc1/mariner transposable elements. *J. Mol. Biol.* **323**, 441–452.
106. Yant, S. R., Wu, X., Huang, Y., Garrison, B., Burgess, S. M., and Kay, M. A. (2005). High-resolution genome-wide mapping of transposon integration in mammals. *Mol. Cell. Biol.* **25**, 2085–2094.
107. Hackett, P. B., Ekker, S. C., Largaespada, D. A., and McIvor, R. S. (2005). Sleeping beauty transposon-mediated gene therapy for prolonged expression. *Adv. Genet.* **54**, 189–232.
108. Essner, J. J., McIvor, R. S., and Hackett, P. B. (2005). Awakening gene therapy with Sleeping Beauty transposons. *Curr. Opin. Pharmacol.* **5**, 513–519.
109. Gleave, M. E., and Monia, B. P. (2005). Antisense therapy for cancer. *Nat. Rev. Cancer* **5**, 468–479.
110. Huyer, G., Liu, S., Kelly, J., Moffat, J., Payette, P., Kennedy, B., Tsaprailis, G., Gresser, M. J., and Ramachandran, C. (1997). Mechanism of inhibition of protein-tyrosine phosphatases by vanadate and pervanadate. *J. Biol. Chem.* **272**, 843–851.
111. Denu, J. M., and Tanner, K. G. (1998). Specific and reversible inactivation of protein tyrosine phosphatases by hydrogen peroxide: Evidence for a sulfenic acid intermediate and implications for redox regulation. *Biochemistry* **37**, 5633–5642.
112. Fu, H., Park, J., and Pei, D. (2002). Peptidyl aldehydes as reversible covalent inhibitors of protein tyrosine phosphatases. *Biochemistry* **41**, 10700–10709.
113. Liljebris, C., Martinsson, J., Tedenborg, L., Williams, M., Barker, E., Duffy, J. E., Nygren, A., and James, S. (2002). Synthesis and biological activity of a novel class of pyridazine analogues as non-competitive reversible inhibitors of protein tyrosine phosphatase 1B (PTP1B). *Bioorg. Med. Chem.* **10**, 3197–3212.
114. Puius, Y. A., Zhao, Y., Sullivan, M., Lawrence, D. S., Almo, S. C., and Zhang, Z. Y. (1997). Identification of a second aryl phosphate-binding site in protein-tyrosine phosphatase 1B: A paradigm for inhibitor design. *Proc. Natl. Acad. Sci. USA* **94**, 13420–13425.
115. Mooney, R. A., and LeVea, C. M. (2003). The leukocyte common antigen-related protein LAR: Candidate PTP for inhibitory targeting. *Curr. Top. Med. Chem.* **3**, 809–819.
116. Saunthararajah, Y., Hillery, C. A., Lavelle, D., Molokie, R., Dorn, L., Bressler, L., Gavazova, S., Chen, Y. H., Hoffman, R., and DeSimone, J. (2003). Effects of 5-aza-2'-deoxycytidine on fetal hemoglobin levels, red cell adhesion, and hematopoietic differentiation in patients with sickle cell disease. *Blood* **102**, 3865–3870.
117. Leone, G., Teofili, L., Voso, M. T., and Lubbert, M. (2002). DNA methylation and demethylating drugs in myelodysplastic syndromes and secondary leukemias. *Haematologica* **87**, 1324–1341.
118. Issa, J. P. (2003). Decitabine. *Curr. Opin. Oncol.* **15**, 446–451.

119. Claus, R., and Lubbert, M. (2003). Epigenetic targets in hematopoietic malignancies. *Oncogene* **22**, 6489–6496.
120. Christman, J. K. (2002). 5-Azacytidine and 5-aza-2'-deoxycytidine as inhibitors of DNA methylation: Mechanistic studies and their implications for cancer therapy. *Oncogene* **21**, 5483–5495.
121. Baylin, S. B. (2004). Reversal of gene silencing as a therapeutic target for cancer—roles for DNA methylation and its interdigitation with chromatin. *Novartis Found. Symp.* **259**, 226–233; discussion 234–227, 285–228.
122. Ghoshal, K., Datta, J., Majumder, S., Bai, S., Dong, X., Parthun, M., and Jacob, S. T. (2002). Inhibitors of histone deacetylase and DNA methyltransferase synergistically activate the methylated metallothionein I promoter by activating the transcription factor MTF-1 and forming an open chromatin structure. *Mol. Cell. Biol.* **22**, 8302–8319.
123. Ghoshal, K., Datta, J., Majumder, S., Bai, S., Kutay, H., Motiwala, T., and Jacob, S. T. (2005). 5-Aza-deoxycytidine induces selective degradation of DNA methyltransferase 1 by a proteasomal pathway that requires the KEN box, bromo-adjacent homology domain, and nuclear localization signal. *Mol. Cell. Biol.* **25**, 4727–4741.
124. Amoui, M., Baylink, D. J., Tillman, J. B., and Lau, K. H. (2003). Expression of a structurally unique osteoclastic protein-tyrosine phosphatase is driven by an alternative intronic, cell type-specific promoter. *J. Biol. Chem.* **278**, 44273–44280.
125. Peterson, C. L. (2002). HDAC's at work: Everyone doing their part. *Mol. Cell* **9**, 921–922.
126. Kelly, W. K., and Marks, P. A. (2005). Drug insight: Histone deacetylase inhibitors—development of the new targeted anticancer agent suberoylanilide hydroxamic acid. *Nat. Clin. Pract. Oncol.* **2**, 150–157.
127. Glaser, K. B., Li, J., Staver, M. J., Wei, R. Q., Albert, D. H., and Davidsen, S. K. (2003). Role of class I and class II histone deacetylases in carcinoma cells using siRNA. *Biochem. Biophys. Res. Commun.* **310**, 529–536.
128. Shaker, S., Bernstein, M., and Momparler, R. L. (2004). Antineoplastic action of 5-aza-2'-deoxycytidine (Dacogen) and depsipeptide on Raji lymphoma cells. *Oncol. Rep.* **11**, 1253–1256.
129. Belinsky, S. A., Klinge, D. M., Stidley, C. A., Issa, J. P., Herman, J. G., March, T. H., and Baylin, S. B. (2003). Inhibition of DNA methylation and histone deacetylation prevents murine lung cancer. *Cancer Res.* **63**, 7089–7093.
130. Cameron, E. E., Bachman, K. E., Myohanen, S., Herman, J. G., and Baylin, S. B. (1999). Synergy of demethylation and histone deacetylase inhibition in the re-expression of genes silenced in cancer. *Nat. Genet.* **21**, 103–107.
131. Rosenfeld, C. S. (2005). Clinical development of decitabine as a prototype for an epigenetic drug program. *Semin. Oncol.* **32**, 465–472.
132. Murgo, A. J. (2005). Innovative approaches to the clinical development of DNA methylation inhibitors as epigenetic remodeling drugs. *Semin. Oncol.* **32**, 458–464.
133. Pettiford, S. M., and Herbst, R. (2000). The MAP-kinase ERK2 is a specific substrate of the protein tyrosine phosphatase HePTP. *Oncogene* **19**, 858–869.
134. Hermiston, M. L., Xu, Z., Majeti, R., and Weiss, A. (2002). Reciprocal regulation of lymphocyte activation by tyrosine kinases and phosphatases. *J. Clin. Invest.* **109**, 9–14.
135. Kau, T. R., Schroeder, F., Ramaswamy, S., Wojciechowski, C. L., Zhao, J. J., Roberts, T. M., Clardy, J., Sellers, W. R., and Silver, P. A. (2003). A chemical genetic screen identifies inhibitors of regulated nuclear export of a Forkhead transcription factor in PTEN-deficient tumor cells. *Cancer Cell* **4**, 463–476.
136. Panagopoulos, I., Pandis, N., Thelin, S., Petersson, C., Mertens, F., Borg, A., Kristoffersson, U., Mitelman, F., and Aman, P. (1996). The FHIT and PTPRG genes are deleted in benign

proliferative breast disease associated with familial breast cancer and cytogenetic rearrangements of chromosome band 3p14. *Cancer Res.* **56**, 4871–4875.
137. Druck, T., Kastury, K., Hadaczek, P., Podolski, J., Toloczko, A., Sikorski, A., Ohta, M., LaForgia, S., Lasota, J., McCue, P., Lubinski, J., and Huebner, K. (1995). Loss of heterozygosity at the familial RCC t(3;8) locus in most clear cell renal carcinomas. *Cancer Res.* **55**, 5348–5353.
138. LaForgia, S., Morse, B., Levy, J., Barnea, G., Cannizzaro, L. A., Li, F., Nowell, P. C., Boghosian-Sell, L., Glick, J., Weston, A., Harris, C. C., Drabkin, H. et al. (1991). Receptor protein-tyrosine phosphatase gamma is a candidate tumor suppressor gene at human chromosome region 3p21. *Proc. Natl. Acad. Sci. USA* **88**, 5036–5040.
139. Pitterle, D. M., Jolicoeur, E. M., and Bepler, G. (1998). Hot spots for molecular genetic alterations in lung cancer. *In Vivo* **12**, 643–658.
140. Kok, K., Naylor, S. L., and Buys, C. H. (1997). Deletions of the short arm of chromosome 3 in solid tumors and the search for suppressor genes. *Adv. Cancer Res.* **71**, 27–92.
141. Zhao, X., Weir, B. A., LaFramboise, T., Lin, M., Beroukhim, R., Garraway, L., Beheshti, J., Lee, J. C., Naoki, K., Richards, W. G., Sugarbaker, D., Chen, F. et al. (2005). Homozygous deletions and chromosome amplifications in human lung carcinomas revealed by single nucleotide polymorphism array analysis. *Cancer Res.* **65**, 5561–5570.
142. Lee, S. H., Shin, M. S., Lee, J. Y., Park, W. S., Kim, S. Y., Jang, J. J., Dong, S. M., Na, E. Y., Kim, C. S., Kim, S. H., and Yoo, N. J. (1999). In vivo expression of soluble Fas and FAP-1: Possible mechanisms of Fas resistance in human hepatoblastomas. *J. Pathol.* **188**, 207–212.
143. Gil-Henn, H., and Elson, A. (2003). Tyrosine phosphatase-epsilon activates Src and supports the transformed phenotype of Neu-induced mammary tumor cells. *J. Biol. Chem.* **278**, 15579–15586.
144. Elson, A. (1999). Protein tyrosine phosphatase epsilon increases the risk of mammary hyperplasia and mammary tumors in transgenic mice. *Oncogene* **18**, 7535–7542.
145. Bruecher-Encke, B., Griffin, J. D., Neel, B. G., and Lorenz, U. (2001). Role of the tyrosine phosphatase SHP-1 in K562 cell differentiation. *Leukemia* **15**, 1424–1432.
146. Noguchi, T., Tsuda, M., Takeda, H., Takada, T., Inagaki, K., Yamao, T., Fukunaga, K., Matozaki, T., and Kasuga, M. (2001). Inhibition of cell growth and spreading by stomach cancer-associated protein-tyrosine phosphatase-1 (SAP-1) through dephosphorylation of p130cas. *J. Biol. Chem.* **276**, 15216–15224.
147. Yang, T., Zhang, J. S., Massa, S. M., Han, X., and Longo, F. M. (1999). Leukocyte common antigen-related tyrosine phosphatase receptor: Increased expression and neuronal-type splicing in breast cancer cells and tissue. *Mol. Carcinog.* **25**, 139–149.
148. Urushibara, N., Karasaki, H., Nakamura, K., Mizuno, Y., Ogawa, K., and Kikuchi, K. (1998). The selective reduction in PTPdeita expression in hepatomas. *Int. J. Oncol.* **12**, 603–607.
149. Tabiti, K., Smith, D. R., Goh, H. S., and Pallen, C. J. (1995). Increased mRNA expression of the receptor-like protein tyrosine phosphatase alpha in late stage colon carcinomas. *Cancer Lett.* **93**, 239–248.
150. Berndt, A., Luo, X., Bohmer, F. D., and Kosmehl, H. (1999). Expression of the transmembrane protein tyrosine phosphatase RPTPalpha in human oral squamous cell carcinoma. *Histochem. Cell Biol.* **111**, 399–403.
151. Liu, S., Kulp, S. K., Sugimoto, Y., Jiang, J., Chang, H. L., and Lin, Y. C. (2002). Involvement of breast epithelial-stromal interactions in the regulation of protein tyrosine phosphatase-gamma (PTPgamma) mRNA expression by estrogenically active agents. *Breast Cancer Res. Treat.* **71**, 21–35.
152. Lu, K. V., Jong, K. A., Kim, G. Y., Singh, J., Dia, E. Q., Yoshimoto, K., Wang, M. Y., Cloughesy, T. F., Nelson, S. F., and Mischel, P. S. (2005). Differential induction of glioblastoma migration and growth by two forms of pleiotrophin. *J. Biol. Chem.* **280**, 26953–26964.

153. Ulbricht, U., Brockmann, M. A., Aigner, A., Eckerich, C., Muller, S., Fillbrandt, R., Westphal, M., and Lamszus, K. (2003). Expression and function of the receptor protein tyrosine phosphatase zeta and its ligand pleiotrophin in human astrocytomas. *J. Neuropathol. Exp. Neurol.* **62,** 1265–1275.
154. Lu, X., Nechushtan, H., Ding, F., Rosado, M. F., Singal, R., Alizadeh, A. A., and Lossos, I. S. (2005). Distinct IL-4-induced gene expression, proliferation, and intracellular signaling in germinal center B-cell-like and activated B-cell-like diffuse large-cell lymphomas. *Blood* **105,** 2924–2932.
155. LaMontagnes, K. R., Jr., Flint, A. J., Franza, B. R., Jr., Pandergast, A. M., and Tonks, N. K. (1998). Protein tyrosine phosphatase IB antagonizes signalling by oncoprotein tyrosine kinase p210 bcr-abl *in vivo*. *Mol. Cell. Biol.* **18,** 2965–2975.
156. Warabi, M., Nemoto, T., Ohashi, K., Kitagawa, M., and Hirokawa, K. (2000). Expression of protein tyrosine phosphatases and its significance in esophageal cancer. *Exp. Mol. Pathol.* **68,** 187–195.
157. Mok, S. C., Kwok, T. T., Berkowitz, R. S., Barrett, A. J., and Tsui, F. W. (1995). Overexpression of the protein tyrosine phosphatase, nonreceptor type 6 (PTPN6), in human epithelial ovarian cancer. *Gynecol. Oncol.* **57,** 299–303.
158. Wu, C., Sun, M., Liu, L., and Zhou, G. W. (2003). The function of the protein tyrosine phosphatase SHP-l in cancer. *Gene* **306,** 1–12.
159. Abaan, O. D., Levenson, A., Khan, O., Furth, P. A., Uren, A., and Toretsky, J. A. (2005). PTPLl is a direct transcriptional target of EWS-FLII and modulates Ewing's Sarcoma tumorigenesis. *Oncogene* **24,** 2715–2722.
160. Itakura, M., Mori, S., Park, N. H., and Bonavida, B. (2000). Both HPV and carcinogen contribute to the development of resistance to apoptosis during oral carcinogenesis. *Int. J. Oncol.* **16,** 591–597.
161. Mundle, S. D., Mativi, B. Y., Bagai, K., Feldman, G., Cheema, P., Gautam, U., Reza, S., Cartlidge, J. D., Venugopal, P., Shetty, V., Gregory, S. A., Robin, E. *et al.* (1999). Spontaneous down-regulation of Fas-associated phosphatase-I may contribute to excessive apoptosis in myelodysplastic marrows. *Int. J. Hematol.* **70,** 83–90.

The Case for mRNA 5′ and 3′ End Cross Talk During Translation in a Eukaryotic Cell

> Anastassia V. Komarova,
> Michèle Brocard, and
> Katherine M. Kean
>
> Unité Postulante Régulation de la
> Traduction Eucaryote et Virale,
> Institut Pasteur, CNRS URA 1966,
> 75724 Paris cedex 15, France

I. Introduction 331
II. Classical Eukaryotic Cellular mRNAs 337
III. 5′-3′ RNA–RNA Interactions 339
IV. Examples of RNA–Protein Interactions... 344
   A. The Case of Histone Protein Synthesis 344
   B. The Case of Picornavirus Translation 350
V. Protein–Protein Interaction Stories.. 354
   A. The Example of DAZ-Family Proteins... 354
   B. The Example of NSP3 of Rotaviruses... 356
VI. Concluding Remarks.. 358
   References.. 358

I. Introduction

Rather than considering mRNA as a simple linear molecule, there is now a wealth of evidence to support the concept that mRNA is in fact a pseudocircular

Abbreviations: BYDV, barley yellow dwarf virus; CHO cells, Chinese hamster ovary cells; CITE, cap-independent translational enhancer; CrPV, cricket paralysis virus; CSFV, classical swine fever virus; CS, 5′-3′ cyclization sequence; Ct, C-terminal part of a protein; DAZ, deleted in AZoospermia; Dcp1, dipeptidyl carboxypeptidase; eIF, eukaryotic [translation] initiation factor; EMCV, encephalomyocarditis virus; eRF, eukaryotic release factor; gRNA, genomic RNA; HBP, hairpin-binding protein, the same as SLBP; IRES, internal ribosome entry segment/site; ITAFs, IRES-transactivating factors; La, Lupus autoantigen; NSP3, nonstructural protein 3; Nt, N-terminal part of a protein; PABP, poly(A)-binding protein; Paip, PABP-interacting protein; PTB, polypyrimidine-tract-binding protein; PV, poliovirus IRES; RBD, an RNA-binding domain; RNP complex, ribonucleoprotein complex; RRL, rabbit reticulocyte lysate; RoXaN, for Rotavirus X protein–associated with NSP3; RRM, RNA-recognition motif; SD, Shine-Dalgarno domain; SL, stem-loop structure; SLBP, stem-loop-binding protein; sgRNA, subgenomic RNAs; SLIP1, an SLBP-interacting protein; TBSV, tomato bushy stunt virus; TE, translational enhancer; TNV, tobacco necrosis virus; UTR, untranslated region.

nucleoprotein complex in which the ends of the RNA are held together by a combination of RNA–RNA, RNA–protein, and protein–protein interactions. Accepted models of gene expression from prokaryotes to eukaryotes emphasize the importance of cross talk between the 5' and 3' ends of the mRNA. Data show that pseudocircularization of RNA molecules could extend to include viral RNA genomes. For example, for Uukuniemi virus (a tick-borne arbovirus) the circularization of ribonucleoproteins (RNP) has been revealed by electron microscopy, and the probability of maintaining base pairing between inverted complementary sequences at the 3' and 5' ends of linear RNA molecules has been examined (*1*). Another example comes from the analysis of genomic RNAs (gRNA) of influenza virus, which are held in a circular conformation both in virions and in infected cells by a terminal panhandle (*2*). Thus, depending on the model studied, end-to-end RNA communication has been proposed to play a role in the different steps of gene expression involving RNA, that is, each of mRNA transcription, replication, translation, and degradation (Fig. 1).

In prokaryotes, transcription and translation are not separated by a nuclear membrane, and the translation of an mRNA begins long before its transcription is finished (Fig. 2A). So in many cases, transcription and translation are tightly coupled (*3*, *4*), with RNA polymerase cited on the nascent 3' end of an mRNA being pursued by ribosomes from its 5' end. It has been shown that in the case of inefficient/slowed translation, which results in an extended spacing between the RNA polymerase and the elongating ribosome, this information concerning translation efficacy is somehow transmitted from the mRNA 5' end to the RNA polymerase at the 3' end of the same mRNA leading to an increase in the extent of premature termination of transcription [(*4*), for a review see (*5*)]. This is the so-called polarity effect. Thus, the communication between the two ends of the mRNA ensures a quantitative link between the two global processes of transcription and translation in a prokaryotic cell.

In eukaryotes, it is rather that mRNA posttranscriptional modification or translation are linked to degradation, and it is particularly for the mature

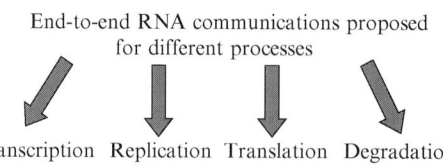

FIG. 1. End-to-end communication in important different cellular processes.

THE CASE FOR mRNA 5′ AND 3′ END CROSS TALK

A Prokaryotes

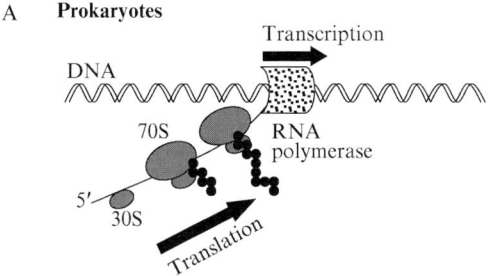

B (−)-Strand RNA viruses
unimolecular

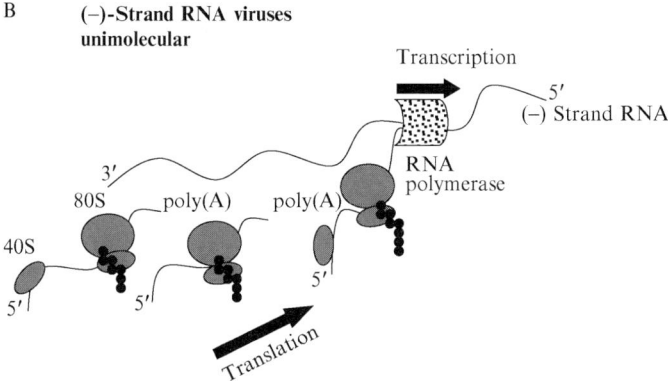

C (+)-Strand RNA viruses
flavi- and picornaviruses

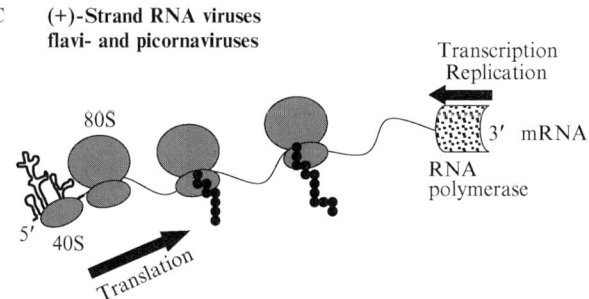

FIG. 2. Interplays between transcription and translation on prokaryotic cellular mRNAs (A) and viral mRNAs in eukaryotic cells (B and C). RNA is shown as a single line with the 5′ and 3′ ends indicated. DNA is shown as a helical double line and protein as a string of black beads. RNA polymerases are represented as stippled bullet forms and ribosomal subunits by gray ovals. The relative directions of movement of these latter elements during transcription and translation are indicated by arrows.

mRNA that has been exported from the cell nucleus where circularization has been reported, that is mRNA that will be translated or degraded. Thus, the RNA in question carries an m^{7}GpppN cap at its 5' end and a poly(A) tail at its 3' end, and these are the key features that ensure circularization of the RNA by a multiprotein complex composed of different translation initiation factors (Fig. 3). The binding capacities of these different factors, such as eIF4E for the RNA cap structure, PABP for the poly(A) tail and eRF3, and both eIF4E and PABP for eIF4G, form a strong argument for the existence of circularized mRNA (6, 7). On mixing purified components together, circular forms of RNA have been demonstrated by atomic force microscopy (8). However, the biological consequences of such circularization still remain somewhat controversial, and its potential role in increasing the efficiency of translation initiation is discussed in more detail later. Suffice it to say here that interactions within the yeast mRNA 5'-3' complex involving dipeptidyl carboxypeptidase (Dcp1), a protein that plays a key role in mRNA decay by cleaving off the 5' cap structure to leave a 5' end that is susceptible to exonucleolytic degradation, have been proposed to modulate the interface between translation and mRNA decay (9). Effectively, it has been shown that Dcp1 binds to eIF4G and PABP, either when these proteins are alone or in the 5'-3' translation complex involving eIF4F and PABP. Thus, Dcp1 fixation to complexed eIF4G/PABP would site Dcp1 to the correct place for it to be active. However, the presence of eIF4E bound to eIF4G prevented access of Dcp1 to the m^{7}GpppN cap, and decapping did not occur unless eIF4E was released from the eIF4F complex (9). It has been postulated that activation of Dcp1 would correlate with abrogation of mRNA 5'-3' end communication on total disruption of the RNP complex and release of PABP as well as eIF4E (9). In this scenario, mRNA 5'-3' end communication should serve to retain intact, competent mRNA within the actively translated pool.

For RNA viruses, it is of interest that their RNAs often differ from classical cellular mRNAs at one or the other of their ends due to the dictates of signals to ensure specific efficient RNA replication, transcription, or packaging into the virion. Consequent to this, viruses have often subverted the host cell translation machinery somewhat, and developed alternative translation strategies (10, 11). Despite their different end sequences, all mRNAs studied to date seem to encompass alternative circularization strategies that would conform to the idea of efficient translation on a circular RNA [see, e.g. (12, 13)]. Furthermore, these viral RNAs resemble prokaryotic RNAs, rather than eukaryotic RNAs, by initial positioning of the ribosome at the 5' end and the presence of RNA polymerase at the 3' end (Fig. 2). Therefore, it seems intuitive that 5'-3' end communication should also have consequences on a link between transcription and translation. Effectively, for a negative-strand RNA virus the similitude with prokaryotes even extends as far as the direction of travel of

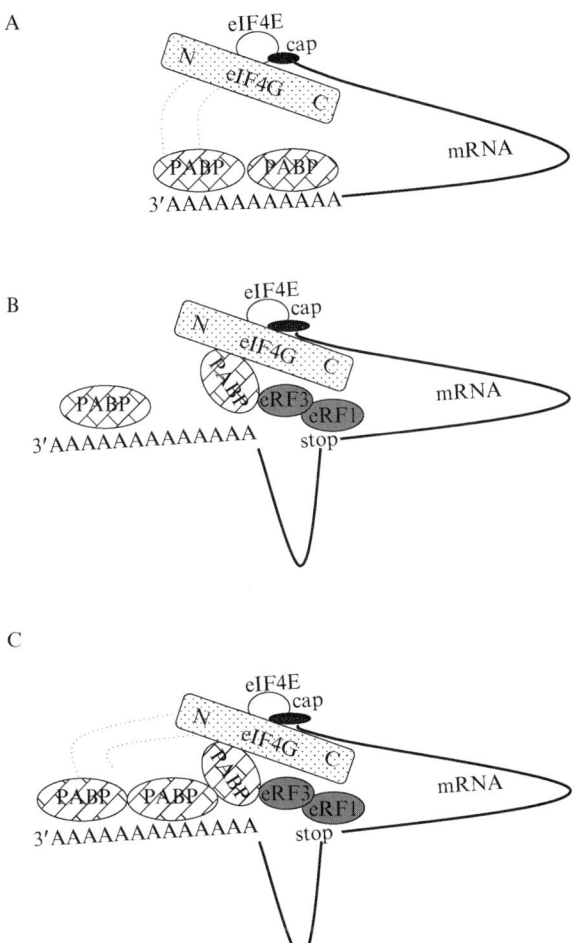

FIG. 3. Schematic representations of possible ways for 5'-3'-UTRs cross talk to occur in the eukaryotic cell on classical (Kozak-like) mRNAs. mRNA is shown as a thick line, with the 5' m⁷GpppN cap structure as a black oval. eIF4E is shown as an open circle, and eIF4G as a stippled box. The poly(A)-binding protein is depicted by brick-filled ovals, with its interaction with eIF4G represented by dotted lines. The translation termination factors eRF1 and eRF3 are indicated as gray ovals. The three forms (A, B, C) of the model differ in the finer details of the 3'-interacting partners, as discussed in the text.

the ribosome and polymerase, relative to the RNA (Fig. 2A and B). Hence, it can be postulated that as for prokaryotes, communication between the two ends of the RNA in the case of negative-strand RNA viruses would ensure a direct link between transcription and translation efficiencies. In contrast, the interplay between transcription (or replication at the step of negative strand synthesis) and translation of positive-strand virus RNA should be antagonistic to ensure that translation is correctly superceded by replication during the virus life cycle (Fig. 2C). For the moment, rather than attempting to study such coordinated processes, the available data concern the simple requirement for RNA circularization in viral RNA transcription. Depending on the virus, different mechanisms have been proposed to account for circularization. For transcription and replication, tombusvirus RNAs [eukaryotic (+)-strand RNA viruses] utilize long-distance RNA–RNA interactions that loop the RNA without truly circularizing it, since the interactions implicated rely on internal sequences rather than end-to-end communication, and would leave free tails at either end of the RNA (14).

In the case of flaviviruses, several lines of evidence support the notion of RNA–RNA interactions between the 5' and 3' ends of the genome. The flavivirus genome is a 5'm7G-capped positive-sense RNA with a nonpolyadenylated 3'-UTR. Contiguous base pairings have been proposed for West Nile, dengue, Japanese encephalitis, Murray Valley, and yellow fever virus RNAs (15, 16). For Kunjin virus the thermodynamic feasibility of a 5'-3' cyclization sequence (CS) base pairing has been indicated by computer analysis (17). Moreover, the visualization of isolated dengue virus RNA molecules in a circular conformation has been achieved by atomic force microscopy (18). Furthermore, the biological importance of the presence of the 5' and 3' CS for efficient RNA synthesis by the dengue virus RNA-dependent RNA polymerase has been demonstrated (19). Similarly, mutations in the context of Kunjin virus replicons that abolish 5'-3' CS complementarity also abrogate RNA amplification (20). Finally, UV-induced cross-linking assays showed that some cellular proteins are likely to be involved in this 3'-5' interaction. For example, Lupus autoantigen (La) binds to the 5'- and 3'-UTRs of dengue virus (21).

At the other extreme, poliovirus (PV)–a picornavirus–RNA genome circularization in the configuration compatible with RNA replication requires not only protein–RNA interactions, but also a protein–protein scaffold bridge. The genome of PV is an uncapped positive-sense RNA with a poly(A) tail. The 5'-UTR consists of a long element involved in cap-independent initiation of translation [internal ribosome entry site, IRES (22–24)] and a shorter 5'-terminal structure (the RNA cloverleaf) involved in viral RNA replication (25). It is known that removal or shortening of the 3' poly(A) tail results in a defect in RNA replication (26, 27). The 5' cloverleaflike structure interacts with both the viral polymerase precursor (polypeptide 3CD) (25, 28, 29) and a cellular poly(rC)-binding protein 2 (29, 30). This 3CD-cloverleaf interaction leads to a switch in the use of the gRNA from translation to replication (31). The 5' RNP complex, in its turn, interacts indirectly with the poly(A) tail of the genome. Thus, the 5′ and 3′ end of the PV RNA remain in a noncovalent juxtaposition via an RNA–protein–protein–RNA bridge that leads to the circularization of the gRNA (32).

This chapter will concentrate on evidence and hypotheses concerning the place of RNA circularization during mRNA translation. Obviously, we cannot cover all the examples found in the literature, but we will provide a short list that is representative of the conundrums facing the researcher in the field. Our particular aim is to present different RNAs according to the chemical nature of the interactions involved in mRNA 5′-3′ end-to-end communication: RNA–RNA, RNA–protein, and protein–protein interactions. However, we will begin with a section on the widely known classical eukaryotic mRNA model, where we will focus on the potential biological consequences of mRNA circularization.

II. Classical Eukaryotic Cellular mRNAs

A unified closed-loop model for eukaryotic cellular mRNA during translation initiation was initially proposed over 20 years ago (33, 34). This concept was extremely speculative at the time since the interacting component(s) that would serve to close the circle on all eukaryotic mRNAs were largely undefined, and this remained true even as late as 1996 (35). However, the model gained widespread interest throughout the translation field, and in 1998 it was shown by biophysical techniques that for capped polyadenylated mRNAs, circular RNA molecules could be evidenced in the presence of eIF4E, PABP, and a fragment from the N-terminal domain of eIF4G (8). Furthermore, the meeting point of the RNA ends was found to encompass all three of these proteins. From what was then known concerning the binding capacities of these proteins, it was presumed that the RNP complexes had the form RNA 5′ cap–eIF4E–eIF4G–PABP–RNA 3′ poly(A) (Fig. 3A). However, it has been shown that PABP is not only capable of fixing eIF4G and poly(A), but also the cellular translation termination factor eRF3 (6). eRF3 indirectly associates with the stop codon of an RNA through its binding to eRF1 (Fig. 3B). Thus, two alternative, nonexclusive, models of default eukaryotic mRNA circularization now exist. Since PABP also multimerizes, it is possible that both the stop codon and the poly(A) are used as 3′ interaction sites on a single RNA molecule to give a triangular tether network in terms of the key RNA–protein interaction points (Fig. 3C). Despite the highly suggestive data cornered from known binding capacities of eIF4E and PABP, Kozak published a review that stated "there is no evidence that mRNA actually circularizes in the course of translation" (36). This point of view that mRNA somehow remains linear during translation seems to us more than unlikely. It would require that isolated eIF4E rather than eIF4E associated in the form of the eIF4F complex fixes the 5′ cap structure, or that PABP linked to the poly(A) tail or eRF3 does not concomitantly fix eIF4G. All available evidences indicate that the contrary is true, since studies concerning binding affinities would favor the participation of the complexed form of eIF4F in RNA binding. For example, the interaction between PABP and eIF4G increases not only the affinity of PABP for the poly(A) tail (37) but also that of eIF4E for the cap by 20–40-fold (38). Finally, any argument pleading in favor of eukaryotic translation uniquely on linear mRNA molecules fails to adequately explain the visualization of circular polysomes *in vivo* by electron microscopy (39).

To our mind, the available biochemical evidence is sufficient to warrant the acceptance of the existence of mRNA circularization during translation initiation. The value of reviews (36, 40) lie rather in their speculation and questioning of whether such circularization has any effect on translation initiation *per se*, and if so, how and at exactly what step translation initiation could be stimulated by juxtaposing the 5′ and 3′ ends of the mRNA. It is possible that it could actually be that mRNA stability is increased by the formation of an RNP complex protecting the two ends of the mRNA, as discussed in some detail (36), and that apparent increased translation initiation is merely a secondary phenotypic effect of this phenomenon. Nevertheless, several studies have aimed to address more or less directly the question of the role of the 3′ poly(A) tail at different steps of translation initiation with a certain degree of success. It was concluded that in yeast, PABP promotes 60S ribosomal subunit joining (41, 42) but can also increase 40S ribosomal subunit recruitment (43). Similarly, studies in mammalian models have implicated the 3′ poly(A) tail as functioning at these two steps of translation initiation. Different approaches suggested that a principal effect was to increase the affinity of eIF4E for the RNA cap, which would increase 40S ribosomal subunit recruitment. We measured the inhibition of translation efficiency in rabbit reticulocyte lysates (RRL) as a function of cap analog concentration and showed that abrogating the PABP–eIF4G interaction severely reduced the functional affinity of eIF4E for the capped 5′ end of polyadenylated mRNA (44). We thus proposed that the PABP–eIF4G interaction is important to promote 40S ribosomal subunit recruitment. Chemical cross-linking experiments confirmed that PABP depletion of translation extracts reduced the efficiency of binding of eIF4E, and also eIF4A, to a polyadenylated mRNA (45). Taken together, these experiments suggest that in mammals concomitant binding of PABP to eIF4G and to the poly(A) tail increases the affinity of the eIF4F complex as a whole for the mRNA. The expected significant reduction in the amount of mRNA engaged in 40S ribosomal subunit preinitiation complexes upon PABP depletion has now been evidenced directly by sucrose gradient centrifugation of the appropriate Krebs-2 cell extracts [at least 30% reduction, (45)]. This work also demonstrated a twofold more pronounced effect of PABP depletion on 80S ribosome initiation complex formation than on 40S complex formation. The authors concluded that this provides evidence that 60S subunit joining is also a target for the 3' poly(A) tail effect. However, it should be appreciated that the method used to draw this conclusion is extremely indirect.

Taken together, the body of available data would suggest that more than one mechanism intervenes in the maintenance of efficient translation initiation attributed to the 3' poly(A) tail and the different protein–RNA and protein–protein interactions affected by this sequence on the mRNA. However, to our knowledge, no convincing data have yet been published to support the original appealing hypothesis that bringing the 5' and 3' mRNA ends together serves to retain ribosomes on the same mRNA for successive rounds of translation (46). Effectively, this hypothesis proposes that circularization allows ribosome recycling by their posttranslational transfer back to the 5' end of the same mRNA molecule. Thus, the next experiments concerning the closed-loop model of translation should be destined to rule out categorically differences in mRNA stability according to the nature of the mRNA in terms of the presence or absence of a 3' poly(A), and to prove or refute the participation of direct ribosome recycling in eukaryotic translation initiation. None of these mechanistic studies address the wider issues of the functional consequences in terms of cell biology of translation initiation preferentially on a pseudocircular molecule (even if only indirectly due to increased stability of such an mRNA) rather than a linear one. The different possibilities have been outlined (40), and this aspect of the subject is not reiterated here, although obviously it is from an understanding of the biological roles of pseudocircularization that the major information about eukaryotic cell translation will come out of the "classical" circular model in the future.

III. 5'-3' RNA–RNA Interactions

In the scenario of classical eukaryotic cellular mRNA summarized above, circularization of the mRNA relies on RNA sequence-specific RNA–protein interactions and protein–protein binding through a bridging molecule (eIF4G). Simpler mechanisms can be envisaged, which do not depend on a bridging protein, or require merely RNA–RNA interactions. The role of RNA–RNA interactions via canonical base pairings in prokaryotic translation initiation is well known and ubiquitously accepted (47–49). Furthermore, their possible roles in eukaryotic translation are under investigation: starting from looking for Shine-Dalgarno (SD)-like motifs in mRNA sequences (50, 51) up to assessing the potential for RNA–RNA interactions via stem-loops (SLs) at the 5' and 3' extremities.

The positive-strand RNA genomes of the members of the *Luteovirus*, *Necrovirus*, and *Dianthovirus* genera (*Luteoviridae* and *Tombusviridae* families) are translated in a cap and poly(A)-independent manner. This is promoted by sequences in the RNA 3'-UTR. An mRNA end-to-end interaction can be formed by a kissing-loop structure. This involves canonical base pairings between the loop/sequence of an RNA hairpin in the 5'-UTR of the viral gRNA or subgenomic RNAs (sgRNA) and that of another hairpin in the 3'-terminus of the same RNA (52–55). The potential kissing SL interaction between the mRNA 3' and 5' ends is conserved in all mRNAs known or suspected to harbor a so-called 3' translational enhancer (TE). The most studied cases are the sgRNA or gRNA of barley yellow dwarf virus (BYDV) and tobacco necrosis virus (TNV), which have some common structural features (52–54, 56, 57). A less examined example is that of tomato bushy stunt virus (TBSV), which has a structurally different RNA from those of BYDV and TNV (55).

A 3'-terminal RNA sequence was initially defined, which acts as a cap-independent translational enhancer (3'CITE or TE). This element shares some common structural features for all RNAs of members of the *Luteoviridae* and *Tombusviridae* families. The length of the effective potential TE varies depending whether experiments are carried out *in vitro* or *in vivo*, and usually a longer sequence is required for translational enhancement *in vivo* than *in vitro*. For example, the length of the BYDV 3'TE is 109 nts *in vitro* and 869 nts *in vivo* (56, 57). Phylogenetic and secondary structure analyses predict the presence of similar structures within the different TEs, which would all comprise several SL structures and a conserved 18-nt tract. This 18-nt sequence lies near the beginning of the 3'-UTRs of all subgroup I luteovirus genomes, RNA1 of bipartite dianthoviruses, and TNV RNA (56). The 18-nt sequence includes complementarity to the 3' end of the 18S rRNA and is located in the vicinity of a loop that has the capacity to base pair with a sequence within the 5'-UTR (Fig. 4) (52–54). The potential secondary structures within 5'- and 3'-UTRs are usually predicted using the Mfold program based on mRNA folding according to thermodynamic criteria (http://www.bioinfo.rpi.edu/applications/mfold/old/rna/), and the predictions are often supported by experimental structure probing assays in solution (54, 55, 58). Thus a complex structure is derived. For example, for TNV-A the TE is composed of four independent SLs. The second of these (II) harbors the sequence that is complementary to the 3' end of 18S rRNA and has secondary short SLs, including SL-Z that ends in a loop with the capacity to base pair with a sequence within the 5'-UTR, the potential kissing interaction loop (Fig. 4). Both the complex secondary structure and the existence of a loop with the capacity to form a kissing interaction with a sequence in the 5'-UTR are phylogenetically conserved between different viruses.

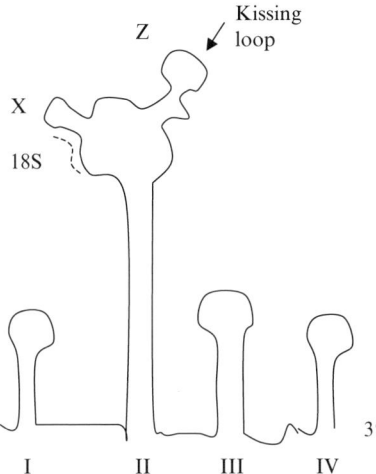

FIG. 4. Schematic illustration of the predicted folding for the 3′-UTR of TNV-A [after (53)]. The four domains of predicted secondary structure are shown in a cartoon form, and the internal SL X and Z, which ends in the potential kissing interaction loop are indicated. The position of the 18-nt sequence that includes complementarity to the 18S rRNA is shown as a dotted line.

The TE SL-Z loops of BYDV or TNV base pair to a loop in the 5′-UTR with 5-bp complementarity (52–54). In contrast, in TBSV this interaction is 9-bp long (interaction between SL3 in the 5′-UTR and SL-B in the so-called R3.5 structure of the 3′CITE). However, the 3′TE of TBSV lacks the conserved 18-nt sequence that is present in the TE of BYDV, necroviruses, and diantho-viruses (55). Despite their relatively short length, it seems probable that these kissing loops are sufficient to ensure viral mRNA end-to-end contact. Effectively, kissing stem loops are thermodynamically and kinetically favored compared to the equivalent base pairing of linear RNAs. Thus, the base pairing between the 3′TE and the 5′-UTR should be significantly more stable than that predicted by Watson-Crick interactions alone (59).

Despite the kissing-loop interaction, a mechanism of translation initiation by 40S ribosomal subunit scanning from the 5′ end of the mRNA seems likely. First of all, the 5′-UTRs vary from several nucleotides in length (for sgRNA) up to 169 nts (TBSV). Second, translation initiates at the 5′-proximal AUG in BYDV gRNA, TBSV gRNA or sgRNA, TNV-A sgRNA2 (53, 55, 56), and so on. Moreover, a stable 5′-proximal SL that blocks ribosome entry on a capped mRNA has the same negative effect on *in vitro* translation of a reporter construct carrying the BYDV UTRs (54). Thus, the 40S ribosomal subunit is predicted to traverse the whole BYDV 5′-UTR from the very 5′ end in a manner analogous to scanning on capped, polyadenylated mRNAs. Certainly, these data provide no support for a model of IRES-driven translation (see chapter later). However, it could not be completely excluded that inserting a 5'-proximal stable hairpin merely induced a secondary structure change in the 5'-UTR, disrupting the kissing-loop interaction, and reducing the efficacy of translation.

Deletion analyses showed that it is in conjunction with the 5'-UTR and that the 3'TE eliminates the need for a 5' cap for efficient translation *in vivo* and *in vitro* (52–55). The ability of these 5'- and 3'-UTRs to drive efficient translation is independent of viral proteins, since reporter constructs that harbor the same terminal sequences behave similarly to viral gRNA and sgRNA (52–54, 56, 57). It is remarkable that in infected tobacco protoplasts the translation efficiency of TNV sgRNA2 (the major template of the coat protein) is as high as that of an efficiently translated mammalian mRNA. In contrast, a cellular type mRNA that is uncapped and nonpolyadenylated is translated some 425-fold less efficiently than the capped, polyadenylated equivalent (60). Thus, the TNV sgRNA2 UTRs contain elements that enable an efficient cap- and poly(A)-independent translation (53).

Because both the 3'TE and the 5'-UTR are necessary for efficient cap-independent translation on viral RNA or a heterologous gene, it has been proposed that the recruitment of the host translation machinery is facilitated through 3'-5' interactions, delivering either ribosomes or initiation factors to the 5'-UTR or to the start codon on viral RNAs (54). In most cases, the implication of the 3'-5' kissing-loop interaction has been examined using mutation analysis, disrupting and restoring Watson-Crick base pairings in the loop sequences. Typically, a disruption in base pairing between the UTRs by introducing single mutations in the loops severely decreases the translation efficacy on either gRNAs or sgRNAs or reporter constructs *in vivo* and *in vitro* (52, 54, 55). Nevertheless, if it is a loop-to-loop interaction that is the driving force for efficient cap and poly(A)-independent translation, restoring the mutated base pairings could be expected to restore the level of translation to the wild type. This prediction has been validated (54, 55). Unfortunately, this is not always the case (52). Thus, it has to be invoked that in certain cases the primary sequence of the kissing loops could be a critical functional determinant. This may reflect sequence-specific binding of host proteins to strengthen a long-distance interaction in certain of the models studied. The examination of possible RNA–protein interactions is an important future orientation in this field and such studies are under way. However, the 5'-3' interaction model may actually have to be called into question. Effectively, moving the 3'TE to the 5'-UTR allows translation *in vitro* at a similar efficiency to that of the natural combination of virus 5' and 3' termini, at least for BYDV and TNV-D TEs (52, 58). Finally, it has not been ruled out completely that the 3'-UTR secondary structures act to prevent degradation of the RNA, since often adequate *in vitro* or *in vivo* RNA stability assays are not carried out. The fact that the length of the 3′TE element necessary to drive efficient translation *in vivo* is frequently longer than that which can function *in vitro* would support this hypothesis (52, 56, 57). Alternatively, it could be that *in vivo* a more elaborate secondary/ternary structure is required to correctly present or maintain the kissing-loop structure.

Chimeric constructs have been made in which the 5′-UTR and the 3′TE come from distinct viruses, one from a member of the *Luteoviridae* family and the other from a virus of the *Tombusviridae* family. It was found that the TNV-D 3′TE could assure cap-independent translation initiation from the BYDV 5′-UTR and that translation from the TNV sgRNA2 5′-element is stimulated synergistically by the BYDV 3′TE (52, 53). This strongly suggests that these viruses use related mechanisms of translation initiation. Furthermore, it indicates plasticity of the RNA–RNA interaction, since the authentic kissing loops do not have the same sequence.

In the BYDV virus genome, the 3′TE element lies 5-kb downstream from the 5′ terminus of the gRNA, in the 3′-UTR of sgRNA1 but in the 5′-UTR of sgRNA2. It has been shown that sgRNA2 added in *trans* efficiently inhibits the translation of gRNA (less than ninefold excess of sgRNA results in 50% inhibition) and less efficiently that of sgRNA1 (30-fold excess is needed for 50% inhibition). This inhibition is independent of the translation of sgRNA2 and of the product of this translation (ORF6) (57), suggesting that inhibition is due to competition of the sgRNA2 sequence for some component of the translational apparatus. The ratios of sgRNA2 to gRNA which lead to inhibition of translation are quite physiological, since sgRNA2 accumulates up to a 20–40-fold molar excess over gRNA in infected cells (61–63).

Another intriguing information, discovered in the early phylogenetic comparisons of subgroup 1 luteovirus, necrovirus, and dianothovirus genome sequences, is the presence of the 18-nt conserved sequence CGGAUCCUGGGAAACACC (Fig. 4) that includes complementarity to 18S rRNA (underlined) (56). This sequence is near the kissing loop in the 3′-UTR and could somehow resemble the bacterial mRNA SD domain. The SD domain base-pairs with its complementary sequence (aSD) near the 3′ end of 16S rRNA during the formation of the small ribosomal subunit-mRNA-initiator tRNA initiation complex (64–67). The number of potential base pairings in the 3′TE (6 bp) is very close to that within the prokaryotic ribosome binding site (47). A duplication of the potential SD-like sequence within the 3′TE to a GAUCGAUCCU sequence completely abolishes the stimulatory activity of the 3′TE *in vivo* and *in vitro* (52, 54, 56, 57). Moreover, this duplication in the BYDV sgRNA2 SD-like sequence abolishes the transinhibition effect on the translation of gRNA and sgRNA1 (57). Finally, as in the case of prokaryotic 16S aSD sequence, the possible aSD-like element in the 18S rRNA is not involved in any obvious secondary structures.

The presence of an SD-like sequence in a highly conserved element at a similar location in the different viral genomes, combined with the fact that duplication of this sequence destroys 3′TE function, suggests that the SD-like sequence could be directly involved in the mechanism of cap-independent translation. A possible effect of this duplication on mRNA stability has been excluded (56). One could imagine that the 3′TE SD-like sequence could serve to keep the small ribosomal subunit attached to the 3′-UTR after translation termination and the kissing loop would help direct it to the 5′-UTR for a subsequent round of translation. This model, therefore, approaches mechanistically ideas concerning translation of capped, polyadenylated mRNAs.

IV. Examples of RNA–Protein Interactions

A. The Case of Histone Protein Synthesis

Replication of the eukaryotic chromosome involves DNA condensation by histone protein binding. Histone protein synthesis itself correlates with DNA replication, and histone mRNA translation is very tightly cell cycle regulated. These RNAs form the so-called replication-dependent class of histone mRNAs. In somatic cells, cell cycle-dependent variations in the level of histone mRNAs translation correlate with variations in the amount of this mRNA. This increases about 35-fold when a cell enters the S-phase (68), whereas histone mRNAs are rapidly degraded in the absence of DNA synthesis during the G2-phase of the cell cycle (69). In contrast to somatic cells, the production of histone protein from highly abundant mRNAs in oocytes is not coupled to DNA replication during development (70–72). A further interest of histone mRNAs is provided by their nature. For the moment, they are the only known metazoan transcripts that are capped but after maturation terminate in an SL structure that is highly conserved from *Caenorhabditis elegans* to man (73, 74). Several reviews have covered in depth this example of translation regulation, covering overall the question of potential circularization of the RNA (36, 68, 75, 76). However, recent publications and data from Gray and Marzluff's laboratories highlight new aspects of histone mRNA translation so we have chosen to discuss this subject again in the present chapter.

To facilitate several steps of histone gene expression in somatic cells, a number of factors associate with the 3′-UTR of replication-depended histone mRNA, notably the stem-loop binding protein (SLBP; or hairpin-binding protein, HBP) (77, 78). SLBP is an essential protein (79–82) and in proliferating cells it is expressed only during the S-phase of the cell cycle (83). It is believed that in somatic cells a stable association between a newly transcribed histone mRNA and SLBP occurs, which is maintained during mRNA processing in the nucleus, transport to the cytoplasm and translation. Effectively, binding of SLBP, a zinc-finger protein (hZPF100), and U7 snRNP on a histone pre-mRNA is required for its 3′-end processing (84, 85). This results in the mRNA ending with the SL structure that is the recognition motif of SLBP. Whereas hZPF100 and U7 snRNP are released upon cleavage of the pre-mRNA, SLBP remains bound to the mature histone mRNA and accompanies it to the cytoplasm (86). SLBP even continues to be associated with histone mRNA on polyribosomes as a component of the histone mRNP (87–89). The somatic histone mRNA 3′SL structure SLBP RNP complex provides a positive effect on histone mRNA translation. Thus, it was proposed that SLBP binding to SL recapitulates the observed effects of PABP binding to the poly(A) tail to ensure efficient translation (68, 90–93). Finally, the RNP complex on the 3′SL sequence influences histone mRNA stability (69, 91). The rapid histone mRNA degradation during the G2-phase of the cell cycle mentioned earlier could be correlated with a lack of expression of SLBP, since this protein could prevent the degradation of the RNA 3′ end by a 3′ exonuclease (69).

In contrast to somatic cells, and as for histones themselves, the production of SLBP during development is not coupled to DNA replication (71, 72). In mammalian oocytes SLBP is present at the G2-phase, where it is enriched in the nucleus, and at the M-phase of the cell cycle. SLBP accumulation during meiotic maturation is necessary for normal histone synthesis in the oocyte, but it has no effect on histone mRNA stability. Moreover, the stable association of a histone mRNA with SLBP suggested in somatic cells is likely to be replaced by a dynamic one during development (70). This latter conclusion could be drawn based on studies of developmental regulation of histone translation in the *Xenopus laevis* model, since in *Xenopus* oocytes there are two proteins that fix the histone mRNA SL sequence: xSLBP1 and xSLBP2 (72). xSLBP1 is the homologue of mammalian SLBP. It is present throughout oogenesis, is active in the processing of histone mRNA, and its quantity increases approximately twofold at oocyte maturation. In contrast, xSLBP2 is an exclusively cytoplasmic protein that is present in immature oocytes and is degraded upon oocyte maturation (72). The binding of xSLBP2 to the SL of histone mRNA does not have an effect on translation (92). After histone pre-mRNA processing and export to the cytoplasm, xSLBP1 is replaced by xSLBP2 to sequester histone mRNA during oogenesis and maintain the histone mRNA in a translationally silent state. Only upon oocyte maturation, when xSLBP2 is degraded, does xSLBP1 bind the mature cytoplasmic histone mRNA and activate its translation (72, 92). The details of these cellular localization and temporal switches in the protein bound to the histone mRNA SL are unknown, but they raise the possibility of an equilibrium between free and SLBP-associated histone mRNA in the oocyte.

If SLBP binds the translation initiation machinery on the 5′ end of histone mRNA, either directly or indirectly, then the 3′SL structure and SLBP could represent a potential RNA–protein interaction model, which provides a key component of mRNA end-to-end communication. To consider this possibility, first of all, the 3′SL motif structural and functional characteristics will be discussed, and then the same aspects concerning SLBP will be highlighted.

The 3′SL structure is the terminal element of a rather short 3′-UTR of 30–75 bases after the stop codon on a histone mRNA (94, 95). SL is formed by a 6-bp stem, which presents a four-base loop. The stem consists of two GC base pairs at the base, a set of the three pyrimidine/purine bases forming the central portion of the stem, and a UA base pair at the top (74). Internalization of the SL results in 60% decrease in the expression efficiency of a reporter mRNA carrying it compared with the wild-type configuration, when assessed in cultured cells, namely Chinese hamster ovary (CHO) cells (91). This agrees with the results obtained concerning the effects of a poly(A) tail on general translation (96). Moreover, the specific rapid degradation of histone mRNAs in the absence of DNA synthesis as observed in HeLa cells requires the SL structure to be positioned the correct distance relative to the translation terminaton codon (95).

Mutations of conserved residues within either the stem or loop regions of SL reduced the ability of this structure to enhance protein expression from reporter mRNAs transfected into CHO cells, suggesting that the SL-mediated increase in histone expression is SL-sequence dependent (91). The mutations studied influence ribosome loading onto mRNAs, and therefore, their ability to be recruited promptly for translation. It has been shown that SLBP affinity for SL structures that bear the translation-defective mutations is reduced, compared to that for the wild-type SL structure (97).

Histone mRNAs are naturally capped, and studies in CHO cells and in yeast showed a codependence on a 5′ cap in order to establish efficient levels of translation on mRNA bearing the SL (90, 91). However, *in vitro* in RRL a capped mRNA did not show increased translation efficiency if it carried the histone 3′SL structure (91). In contrast, if the same reporter mRNA was uncapped, a three- to sixfold increase in translation was observed in an RRL reaction in the presence of xSLBP if the mRNA ended in the histone SL (92). These results are perhaps not surprising, since it is well known that several translational characteristics seen *in vivo* cannot be depicted *in vitro* in RRL. This failure of the RRL includes the effects of other mRNA 3′ regulatory elements such as poly(A) tail [see, e.g. (98)]. It would be interesting to assess the effects of the histone mRNA SL structure on translation efficiency *in vitro* in translation systems that do recapitulate the translational stimulation by the poly(A) tail observed *in vivo*. Such systems include certain HeLa cell translation extracts (99), Krebs-2 cell extract (100), and an RRL that has been partially depleted of ribosomes and associated factors (44). These translation systems have in common that they address a major difference between translation in intact cells and that of a single RNA in a classical RRL, namely that the latter is inordinately efficient because of an unusual richness in the translational machinery and a lack of competition from heterologous mRNAs.

The addition of the histone SL structure to a nonpolyadenylated mRNA 3′ end increases the expression of a reporter gene some 13-fold in CHO cells, which is comparable to, if slightly less efficient than, adding a poly(A) tail to the same mRNA (91). To drive translation efficiently both *in vivo* and *in vitro* the histone SL structure requires SLBP to be present (90, 92, 93). Since SLBP interacts with the 3′SL structure, the codependency between the RNA terminal regulatory elements observed in experiments in cells could be a result of an interaction via trans-acting factors, for example, the cap-binding translation initiation factors and the histone SLBP.

Placing the 3′ histone SL structure on a reporter mRNA increases its stability (91, 101–103). Effectively, the presence of the histone SL structure increases a reporter mRNA half-life twofold in CHO cells compared to that of a control reporter mRNA with a 3′-UTR of the same length but lacking either a poly(A) tail or the SL structure. The functional half-life of an mRNA carrying the SL structure is similar to one with a poly(A) tail, measured in the same experiment (91). However, as has already been mentioned earlier, a drop in histone mRNP level correlates with, and is thought to trigger, histone mRNA degradation in cultured cells such as HeLa cells. Alternatively, it could be hypothesized that rather than a drop in the level, it is a change in the composition of the protein complex that assures a balance between mRNA translation and degradation. Either way, the most recent model is that in somatic cells, on inhibition of DNA replication the terminating ribosome receives a signal from the 3′ end of the histone mRNP. However, degradation of histone mRNA due to an excess of soluble histone proteins following an inhibition of DNA synthesis has not been completely excluded to date (95).

Stem-loop binding protein is an RNA-binding protein that can be divided into three parts: an N-terminus (Nt); a central RNA-binding domain (RBD), through which it interacts with the SL structure on histone mRNA (78); and the C-terminus (Ct) (76). By expression of *xSLBP* mutants, Sanchez and Marzuluff showed that 10 residues within the N-terminal domain of xSLBP are required to activate the translation of SL structure containing reporter RNAs in *Xenopus* oocytes (stage VI) and *in vitro* (92), even though this region of the protein does not correspond to the SL-binding domain. In agreement with this, a report based on a tethered-assay approach (see below) demonstrated that the truncated N-terminal domain of SLBP was sufficient to achieve a higher level of translation activity than that of the intact full-length SLBP (93). This is in contradiction with the results of another group, who showed that the Ct of human SLBP is required for a positive effect of SLBP on translation efficiency in a heterologous system, that of yeast (90).

Early assays in the heterologous yeast system showed that translation in the presence of hSLBP of reporter mRNA carrying the SL 3'-terminal structure was dependent on eIF4G, eIF4E, and the eIF3 complex. Moreover, SLBP copurified with the eIF4F complex on an m^7GTP-Sepharose column and was recovered on a poly(A)-agarose column in association with PABP. An interaction between eIF4G and SLBP was confirmed by coimmunoprecipitation, using endogenous SLBP in mammalian cells extracts (90). On the other hand, when the yeast-two-hybrid system approach was used, a direct interaction between either xSLBP or hSLBP and eIF4G could not be shown (93). However, this approach allowed direct interactions of SLBP with other translation initiation factors to be demonstrated. Thus, the eIF3 h subunit interacts with the RNA-binding domain of SLBP, and Paip1 (PABP-interacting protein) interacts with the C-terminal domain of both xSLBP and hSLBP. These results were confirmed using the GST-pulldown assay with purified protein and in the mammalian cell extracts. Furthermore, eIF3h–SLBP and Paip1–SLBP interactions are not RNA dependent, as RNaseA did not abrogate either of the contacts (93).

None of the factors fished out using the two-hybrid approach were found to interact with the SLBP N-terminal domain, as would perhaps have been expected, to explain the role of this domain in translation stimulation of SL structure containing RNAs. Thus, there could be another factor/factors implicated in the histone mRNP. At an EMBO Conference on Protein Synthesis and Translational Control held in Heidelberg in September 2005, the results of a second yeast-two-hybrid screen were presented, which identified a 222 amino acid heat-domain protein present in all metazoans (termed SLIP1) as interacting with SLBP (N. Gulseren-Cakmakci and W. Marzluff, unpublished results). SLIP1 was shown to bind to 15 amino acids in the N-terminal domain of SLBP, which covers the 10 residues within the Nt portion of xSLBP that had previously been shown to be required to activate the translation of SL structure containing reporter RNAs (92). Furthermore, SLIP1 binds to a site in the Nt of either eIF4G1 or eIF4G2, and data were presented to show that SLIP1 cooperates with SLBP to stimulate the translation of SL structure containing mRNAs (N. Gulseren-Cakmakci and W. Marzluff, unpublished results). Therefore, a new bridge of protein–protein interactions can now be suggested where SLIP1 binds both to SLBP and eIF4F complex.

Recent work has compared the translational activity of both hSLBP and xSLBP1, as well as different domain truncations of these proteins, in *Xenopus* oocytes (stage VI) using the tethered function assay (93). This assay has the potential advantage that it allows the segregation of the role of proteins/domains in translation from their RNA-binding activity. The principle of the assay is simple. A protein of interest is tethered to the bacteriophage MS2 coat protein by the construction of a fusion protein. This is then brought to the desired region of an RNA by the presentation of the MS2 recognition site at the appropriate position on an injected reporter mRNA. Thus, the interaction of the MS2 coat protein with its RNA-binding site targets the protein of interest to the reporter mRNA. Negative controls are extremely important in this assay, including another RNA-binding protein (neutral for translation) fused to MS2, the MS2 protein alone, a reporter mRNA lacking an MS2 RNA-binding site, and a reporter mRNA with a poly(A) tail (104).

Taking into account the proposed model whereby the interaction of SLBP with both the 3′SL structure and the 5′-UTR via a SLIP1/eIF4F interaction would lead to the formation of an end-to-end RNP complex, the efficacy of the translation of a reporter mRNA could then be estimated more readily and malleably in *Xenopus* oocytes. This relies on an SLBP–MS2–MS2 binding site interaction to assess the role of SLBP in potential mRNA circularization. Moreover, specific focused changes of the 5′-UTR sequence enabled studies to examine which of the eIFs are required to drive efficient translation in the presence of SLBP at the 3′ end of the mRNA. By the tethered function assay, it was shown that both xSLBP and hSLBP stimulate the translation of an m7G-capped luciferase reporter mRNA carrying the MS2-binding site (93). Furthermore, the translational enhancement activity of SLBPs increased up to threefold for hSLBP and sevenfold for xSLBP on oocyte maturation (stage VI vs mature *Xenopus* oocytes). The stimulation of translation efficiency by SLBP is likely to occur at the level of translation initiation, since introducing the cricket paralysis virus (CrPV) IRES into the 5′-UTR of the reporter mRNA abrogated the effect of xSLBP on translation. Effectively, this IRES is renowned for ensuring an alternative mechanism of translation initiation in which the IRES interacts with the P-site of the ribosome, and protein synthesis starts directly from the codon in the A-site without a requirement for any of the canonical translation initiation factors (105). The ability of SLBP to drive translation from the classical swine fever virus (CSFV) IRES has also been tested using the same approach. Translation initiation on this IRES does not require eIF4E, eIF4G, eIF4A, eIF4B, eIF1, or eIF1A (106). Use of the CSFV IRES resulted in a lack of translation stimulation by MS2-xSLBP1, so for its role in translation SLBP requires the presence of translation initiation factors and notably one/or more of the factors dispensable for CSFV IRES activity. This suggests that SLBP acts at the level of 40S ribosomal subunit recruitment or scanning. Finally, the same approach was used to verify how the PV IRES or a nonfunctional AppG cap were able to drive the translation of the reporter RNAs, which showed a requirement for eIF4E. This is consistent with earlier results (90, 91).

In conclusion, a combination of modern approaches, such as the MS2 tethered assay and the yeast-two-hybrid system, strengthens the models previously derived pertaining to the role of SLBP in translation initiation. Above all, with the demonstration of the existence of SLIP1, a full model for a closed loop on histone mRNA now exists, which would be of the type, for example: 5′ RNA cap–eIF4E–eIF4G–SLIP1–SLBP–3′ RNA SL structure. This could be stabilized by protein–protein interactions involving other initiation factors such as eIF3 h subunit.

B. The Case of Picornavirus Translation

Picornaviruses are positive-strand RNA viruses, including many well-known examples of important pathogens of man and animals, such as PV, foot-and-mouth disease virus, hepatitis A virus, and so on. The genome is polyadenylated but uncapped, and instead has a viral VPg protein linked to the 5′ end, although it is accepted that this is cleaved off the RNA before the genome serves as the template for translation. We have chosen to discuss the example of picornaviruses because they were the first RNAs for which it was proposed nearly 20 years ago that translation was by internal entry of the ribosome on an IRES. This conclusion was drawn following experiments using dicistronic RNA assays (22, 24). The intellectual arguments against 5′ end dependent translation initiation on picornavirus RNAs were covered some 10 years ago in an excellent review (107). Nevertheless, the concept of the IRES was not universally accepted by the scientific community until 5′ end independent translation was unarguably shown by constructing an artificial circular RNA that carried the EMCV IRES (108).

The state of the art concerning picornavirus IRES function *per se* was comprehensively documented in 2000 (11). At that time, three distinct groups were distinguished on the basis of structural and functional criteria: IRESs from entero/rhino- (including polio-), cardio/aphtho-, or hepatitis A virus. The structures are summarized in Ref. (109), and properties of the three types of IRES are compared directly in Refs. (110, 111). In the partially ribosome-depleted RRL translation system mentioned earlier, the cardio/aphtho- group can be subdivided, with the efficiency of a cardiovirus IRES being 8–20-fold lower than that of an aphthovirus IRES. This contrasts with only a twofold difference between the two in a classical *in vitro* translation system (112). A major discovery since 2000 is that of a fourth type of picornavirus IRES, which shows significant structural and functional similarity to that of the flavivirus, hepatitis C virus (113).

The hunt for IRES-trans activating factors (ITAFs) was founded on observations pertaining to the effects of ribosomal salt wash on PV (an enterovirus) RNA translation *in vitro* in RRL (*114*). Effectively, it was shown that under optimal conditions, aberrant initiation site usage was corrected and translation efficacy could be stimulated significantly. With an approximate length of 450 nt, a picornavirus IRES is large enough to bind multiple factors, and this has turned out to be the case. Initially, it was shown that cellular proteins bind multiple sites within the PV IRES (*115*). One of the earliest examples of a protein shown to bind the PV IRES was p52 (*116*), which later turned out to be La protein (*117*). Many proteins not previously known to function in translation initiation have now been shown to bind IRESs [reviewed in (*118*)]. The details of these studies transcend the scope of this chapter; suffice it to say here that different IRESs bind different proteins, and where a protein binds to several IRESs, it may act as an ITAF in one case, but have no detectable effect on translation in another case. The reader should also be warned that functional studies are still rather incomplete. Translational enhancement, or no, remains to be formally demonstrated in many cases, and the precise function of these noncanonical translation factors remains to be determined. Many of them are proposed to act as chaperones to modulate IRES structure such that a ribosome-friendly environment is created. They could play an important role in cell or tissue tropism. For example, ITAF-45, which is only expressed by proliferating cells, was identified as a protein required for aphthovirus IRES function (*119*). Furthermore, of the other ITAFs that have been identified to date, central nervous system cells are deficient in polypyrimidine-tract binding (PTB) protein and contain a neural-specific homologue nPTB (*120, 121*). Elegant studies on a neurovirulent strain of Theiler's murine encephalomyelitis virus that requires PTB led to the conclusion that differences in function and tissue-specific distribution of PTB and nPTB are important determinants of cell-specific translational control and viral pathogenesis (*122*).

Not surprisingly, canonical translation initiation factors also play a key role in picornavirus IRES activity, but these often function in a somewhat altered fashion with respect to their mode of action during classical cap- and 5'-end-dependent translation initiation. Thus, in terms of the eIF4F complex, eIF4E interacts with the 5' end of a capped mRNA, whereas eIF4G itself interacts specifically with the different picornavirus IRESs. Footprinting assays revealed that it is the central domain of eIF4G that binds to the cardiovirus IRES, only approximately 50 nucleotides upstream of the translation initiation codon (*123, 124*). This IRES also seems to bind eIF4E (*124*), although eIF4E is not a prerequisite for IRES activity (*110*). Similarly, type I (*125*) and type III (*126*) IRESs can also bind eIF4G directly, and in the case of the PV IRES the binding site has been mapped to within domain V (*125*). Future work should be directed toward a more precise identification of the binding site for eIF4G on the type I and type III IRESs, and the verification that it is the same region of eIF4G that binds the different IRESs.

Nevertheless, enough is already known to allow a model of mRNA circularization that is analogous to that of cellular mRNAs to be proposed. This would rely on concomitant fixation of the middle domain of eIF4G to the IRES (instead of eIF4E to the 5′ cap structure) and of eIF4G-linked PABP to the 3′ poly(A) tail or indirectly to the translation stop codon, exactly as for a capped, polyadenylated mRNA (see Fig. 5A). It will be interesting to determine whether, at least for IRESs of the entero/rhino-type, ITAFs participate in the formation of this RNP complex. Effectively, it is tempting to look for similitudes with the suggested function of SLIP1 in histone protein translation initiation (see earlier). Furthermore, it will be particularly interesting to study the recently described teschovirus IRES, since this can probably not be accommodated within the confines of such a model. Effectively, from the functional analysis and by comparison with hepatitis C virus, it is most likely that these elements do not bind or require eIF4G [(*113*), see also (*127*)].

However, at least in the case of entero-, rhino-, and aphthoviruses, this particular model of circular RNA could not be maintained throughout the virus life cycle. Effectively, these viruses all encode proteases that ensure cleavage of eIF4G in the same hinge region (*128, 129*) in such a way that the presumed IRES-binding site is scinded from the PABP-binding site (see Fig. 5B). Certainly, these IRESs are capable of functioning efficiently with only the C-terminal cleavage product of eIF4G (*111, 130*), that is, in the absence of the PABP interaction domain. However, we have suggested that this mode of translation is only operational after the shut off of host cell translation, and that the circular model of RNA is important in the highly competitive cellular environment encountered during the first rounds of translation [Fig. 5 (*13*)].

Certainly, in favor of a model of 5′-3′ RNA interaction, it has been demonstrated in several *in vitro* translation systems that RNA polyadenylation stimulates significantly (between approximately three- and tenfold) the efficiency of translation from each of the three established types of picornavirus IRES (*13, 99, 100, 131*). We and others have shown that the poly(A) stimulation is lost on cleavage of eIF4G by the viral proteases, in the sense that the efficiency of nonpolyadenylated mRNA translation is then increased to reach that of polyadenylated mRNA (*13, 100*). In contrast, the poly(A) effect is abrogated by lowering the efficiency of translation of polyadenylated mRNA upon the displacement of PAPB from eIF4G by the rotavirus nonstructural protein 3 (NSP3) protein (*13*), or by sequestration of PABP by Paip2 (*100*). The significance of these differences is not yet known, although in both cases the 2A viral protease effect is dominant to the latter effects. It has been proposed that the viral 2A protease also cleaves a translation inhibitor (*130*).

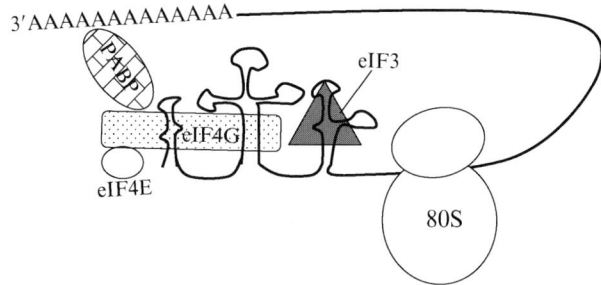

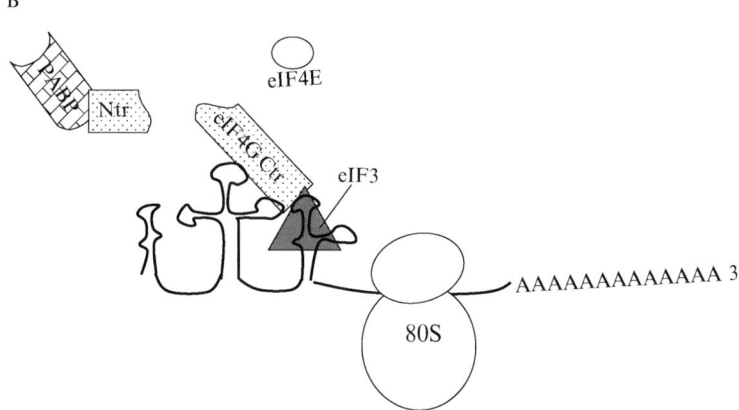

FIG. 5. Models depicting the circularization and translatability of viral mRNA during the infection cycle of certain picornaviruses. Proteins are represented as for Fig. 3, and additionally, the eIF3 complex is shown as a gray triangle, and the ribosome as an open cartoon. (A) Immediately after infection, the RNA of all picornaviruses are circularized via the eIF4G-PABP interaction, relying on eIF4G binding to the 5′ IRES and PABP binding to the 3′ poly(A) tail. This series of interactions is probably required to allow picornavirus RNAs to compete with the actively translating circularized capped-polyadenylated cellular mRNA. (B) After a primary round of viral protein synthesis, the entero-, rhino-, and aphthoviruses induce a dramatic inhibition of host cell protein synthesis, primarily via viral protease-mediated cleavage, first of eIF4G and later of PABP. These cleavage events will both block eukaryotic mRNA circularization and will also have profound effects on the viral RNP complex, as shown.

Alternatively, it could be postulated that there is some redundancy in circularization mechanisms on these RNAs, the large size of the IRES allowing the binding of many factors could be in favor of this idea. Further work should be directed toward determining the true mode(s) of IRES-driven translation initiation in the infected cell.

V. Protein–Protein Interaction Stories

A. The Example of DAZ-Family Proteins

An example of a protein–protein bridge that is a key feature of formation of 5′-3′ circularized mRNA is provided by the DAZ family of proteins (*Deleted in AZoospermia*). The interaction in question is that between DAZ and PABP, and the 5′-interacting partners to complete the mRNA pseudocircle are PABP-eIF4G-eIF4E as described earlier. The molecular mechanisms of the DAZ-PABP model are very intriguing, although studies are perhaps less advanced than for the other examples considered in this chapter. A publication raises the possibility that DAZ-family proteins could stimulate the translation of certain mRNAs by recruiting PABP to their 3′-UTR, irrespective of the length of the poly(A) tail (*132*). Thus, a DAZ-family protein recruits PABP to the 3′-UTR of mRNA via a protein–protein interaction if the latter protein cannot itself efficiently bind to the mRNA.

Genes coding DAZ-family proteins are widely distributed from worms to human. They play important roles in gametogenesis (*133–135*). They are RNA-binding proteins containing an RNA-recognition motif (RRM) and a DAZ motif (Fig. 6). The RRM is believed to determine the specificity of DAZ-family proteins with respect to the target RNA sequence (*136*), whereas, the DAZ motif is thought to mediate protein–protein interactions (*137*). Initially, it was found that the *DAZ* gene is located on the Y chromosome, and DAZ was reported to be one of the candidates responsible for human azoospermia (*134*). *DAZL* (DAZ-like) and *BOULE*, two autosomal homologues of *DAZ*, also encode members of the DAZ family of proteins. Genes homologous to *DAZ* are highly conserved and are expressed exclusively in germ cells (*138*).

DAZ-family proteins all contain an N-terminal RRM, then DAZL and BOULE proteins possess a single copy of a DAZ motif, whereas DAZ proteins

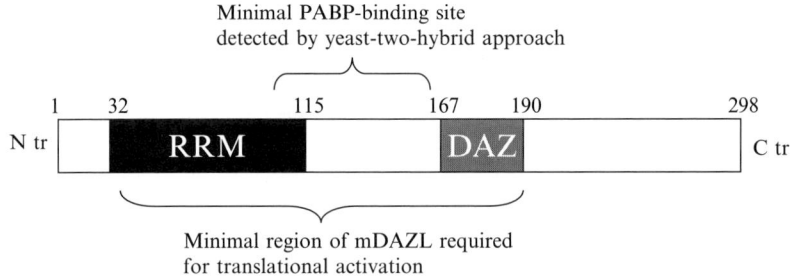

Fig. 6. Schematic diagram depicting the different structural domains of mouse DAZL protein [after (*132*)].

contain 8–24 copies of this motif (*134, 139*). The DAZ-family proteins are conserved, with most of the similarity in the RRM region, and homology ranges from 37% up to 82% within the same organism (*132*). Depending on the stage of the cell cycle, the proteins could be located in the nucleus or in the cytoplasm (*140–142*).

There is a list of potential mRNA targets that would engage in protein–RNA interactions with the DAZ-family proteins. In most cases the DAZ-recognition motifs are sequences within the 3′-UTR (*136, 143, 144*), although potential targets in which DAZ-family proteins would bind the 5′-UTR have also been proposed (*145*). Moreover, an experiment with a reporter mRNA carrying the 3′-UTR of *twine* mRNA (one of the candidates bound by *Drosophila* BOULE) showed that reporter mRNA translation was dependent on BOULE (*143*). Analogous results were obtained using zebra fish DAZL (zDAZL) protein and a reporter mRNA with a "GUUC" sequence in its 3′-UTR (*136*).

Potential protein partners that could bind to the DAZ-family proteins are also known. Thus, the yeast-two-hybrid approach directed specifically toward a panel of eukaryotic translation initiation factors revealed an interaction between the C-terminal part of PABP (PABP1-Ct) or ePABP and *Xenopus*, mouse or human DAZL (mDazl or hDAZL), or human DAZ (hDAZ) (*132*). Coimmunoprecipitation analysis using *Xenopus* DAZL (Xdazl) supported these data. The interactions are RNA independent, since they are not sensitive to RNase treatment, and furthermore, PABP1-Ct cannot bind RNA (*146*). Mapping of the mDazl protein, using a deletion mutation analysis, showed that the DAZ motif was neither sufficient nor required for binding to PABP. Rather, the interaction between mDazl and PABP correlated with the presence of a region of Dazl immediately C-terminal to, and encroaching several amino acids into, the RRM (*132*) (see Fig. 6). These data do not agree with the previous proposition that it is the DAZ motif that assures protein–protein interactions.

Cytoplasmic mDazl and zDAZL are associated with polysomes (*136, 147*). This observation, along with the presence of an RRM, and the interaction of the DAZ-family proteins with PABP (*132*) suggest that one of their roles could be the regulation of expression of specific genes at a posttranscriptional level. This could be regulation of translation or degradation of the mRNA targets. Since the presence of Xdazl does not affect mRNA stability (*132*), the first concept is more probable.

Earlier, we have explained the principle of the MS2 tethered assay in an *X. laevis* oocyte model system. This approach was used to show that the DAZ-family proteins stimulate translation of reporter mRNA transcripts (*132*). The positive effect on reporter mRNA translation compared to negative controls, such as the MS2 protein alone, is up to eightfold in the case of Xdazl, and equivalent stimulation can also be seen of eight- to tenfold for the heterologous mDazl. Significant stimulation was also evidenced for hDAZL (three- to fourfold), for hDAZ (approximately sixfold), and hBOULE (slightly less than fivefold). Importantly, the relative positive effect of MS2-mDazl on the translation of nonpolyadenylated mRNA is greater than on its polyadenylated counterpart. The tethered assay also revealed an analogy to the effects of PABP on polyadenylated mRNAs: the potential to bind multiple molecules of mDazl results in further enhancement of translation of a target mRNA (132). Amino acids 33 and 190 of mDazl were sufficient to activate translation, and the deletion of this region abrogates the positive effect of mDazl on translation in the tethered assay (132). Finally, sucrose gradient analysis shows that DAZL proteins stimulate translation by enhancing the rate of recruitment of the ribosome/translation initiation on a reporter mRNA (132).

It is known that the *Drosophila twine* gene plays an important role in meiosis. Lack of the *twine* gene causes a meiotic arrest during spermatogenesis (148, 149). This phenotype resembles that of a *boule* mutant (135). Phenotypic rescue shows that the proteins may contribute to gametogenesis by related molecular mechanisms. Thus, Xdazl and hBOULE partially rescue the *Drosophila boule* phenotype (150, 151); and both hDAZL and hDAZ partially rescue a DAZL knockout mouse (152, 153).

There is a discrepancy concerning the proposed structural and functional motifs of the DAZL proteins. More precisely, yeast-two-hybrid analysis showed that the DAZ motif is not required for PABP binding, whereas the tethered function assay demonstrated that a full-length translation activation domain includes the DAZ motif (132) (see Fig. 6). In addition, the zDAZL protein associates with polysomes through its DAZ motif, and this association is required for its translational activation (136). Therefore, future work should be orientated to clarify the link between the binding of DAZ-family proteins to PABP and their role as translational transactivators.

B. The Example of NSP3 of Rotaviruses

A well-studied example of protein–protein interactions in viruses that is thought to provide mRNA end-to-end cross talk involves NSP3 of rotaviruses, members of the *Reoviridae* family. Since this subject is frequently discussed in reviews (40, 154), it should be familiar to the reader. Hence we would not cover it deeply and will emphasize only recently published aspects, providing just enough general information to explain how viral mRNAs are subjected to circularization to assure their own translation.

Rotavirus mRNAs comprise 5' and 3' regions of variable length. They are capped, but nonpolyadenylated (155). They comprise a strictly conserved 3'-consensus sequence; 5'-AUGUGACC-3' for serogroup A rotaviruses and AUGUGGCU for the group C rotaviruses (156).

In rotavirus-infected cells, NSP3 binds specifically to the last four nucleotides of the 3′ consensus sequence (*157, 158*). Early studies suggested that two molecules of NSP3 could bind to one molecule of RNA (*159*), and an X-ray structural analysis showed that the protein binds RNA in the form of an asymmetrical homodimer (*158*) and that this interaction with the mRNA 3′-UTR is very stable. NSP3 contains three structural and functional domains. The N-terminal domain (1–150 aa) is an RNA-binding domain through which the interaction with the 3′ consensus sequence occurs (*158, 159*). The central part of the protein is the so-called dimerization motif (150–241 aa), which leads to formation of the homodimer (*159*). The yeast-two-hybrid system approach showed that the carboxy-terminal domain of NSP3 (C-terminal 110 aa) is a protein-interacting domain that binds to the N-terminal part of eIF4GI (*159, 160*). A yeast-two-hybrid screen revealed that the central domain of NSP3 also interacts with a cellular protein, RoXaNI (for rotavirus X protein associated with NSP3) (*161*). RoXaNI protein is suggested to be involved in the regulation of translation, at least during rotavirus infection. For both eIF4GI and RoXaNI, the interactions with NSP3 were confirmed using coimmunoprecipitation analysis (*160, 161*). Thus, two cellular partners have been revealed to interact with NSP3, and these interactions could lead to the formation of a ternary complex during rotavirus infection: eIF4GI-NSP3-RoXaN. In addition, a yeast sandwich-hybrid assay showed that the interaction of RoXaNI with NSP3 does not impair its binding to eIF4GI. In other words, the NSP3 dimer interacts simultaneously with RoXaNI and eIF4GI, and an indirect association of RoXaNI with eIF4GI occurs in the presence of NSP3 (*161*).

Since NSP3 interacts with the same region of eIF4G as does PABP (*160–163*), but with a higher affinity (*163*), the interaction of NSP3 C-terminal domain with eIF4G displaces the binding of the initiation factor to PABP (*160*). Furthermore, in our laboratory it was shown that the recombinant C-terminal fragment of NSP3 alone is sufficient to abolish cap-poly(A) translation, which would explain the shut off of host cell translation seen during rotavirus infection (*131*). On the other hand, translation of capped mRNAs carrying the rotavirus consensus 3′ end sequence is considerably enhanced in the presence of intact recombinant NSP3 protein both *in vitro* and in tissue cultures, and both the RNA-binding and eIF4G-interaction domains of the protein are required for stimulation of translation [(*12*) and unpublished results from our laboratory].

Thus, the proposed model is that during rotavirus infection NSP3 protein functions as a PABP analog. Due to its high affinity for eIF4G it titers the eIF4F complex to satisfy the viral requirements. The PABP–eIF4G interaction is replaced by a (RoXaNI)–NSP3–eIF4G interaction and rotavirus mRNA is circularized via a cap-eIF4E-eIF4G-(RoXaNI)-NSP3-3′ consensus sequence interaction, providing its efficient translation. It will be interesting to determine whether each molecule of the NSP3 homodimer binds eIF4GI and RoXaNI or whether tasks are shared within the asymmetric homodimer, and one molecule of NSP3 interacts with proteins, and the other with RNA, for example.

VI. Concluding Remarks

From these examples, we hope that the reader is convinced that different mRNAs have conserved the principle of being able to assure 5′-3′ end interactions, using widely different means depending on the case. We have tried in this overview of a few examples to show that this field of research is an exciting, young, dynamic one with a richness of choice as to the problem facing the researcher. We have also tried to underline that even the cases that would seem to the nonspecialist to be resolved are in fact far from being cut-and-dried, and much work remains to be done in all cases considered. Therefore the choice comes down to the specialities of individual laboratories, and it seems highly likely that significant progress can now only be made by collaborative efforts in many areas. We hope that this will increasingly be the case in the near future.

Acknowledgments

We are particularly grateful to Ivan Shatsky and Irina Boni for stimulating discussions and to Sylvie Paulous for helpful suggestions. We also thank Bill Marzluff for sharing unpublished results with us. Work in our laboratory is supported by grants from the INSERM (ATC-Hépatite C) and the Agence Nationale de Recherches sur le SIDA (ANRS). Anastassia V. Komarova is funded by Fondation de Recherche Medicale, and Michèle Brocard by the Agence Nationale de Recherches sur le SIDA (ANRS).

References

1. Hewlett, M. J., Pettersson, R. F., and Baltimore, D. (1977). Circular forms of Uukuniemi virion RNA: An electron microscopic study. *J. Virol.* **21**(3), 1085–1093.
2. Hsu, M. T., Parvin, J. D., Gupta, S., Krystal, M., and Palese, P. (1987). Genomic RNAs of influenza viruses are held in a circular conformation in virions and in infected cells by a terminal panhandle. *Proc. Natl. Acad. Sci. USA* **84**(22), 8140–8144.
3. Yarchuk, O., Jacques, N., Guillerez, J., and Dreyfus, M. (1992). Interdependence of translation, transcription and mRNA degradation in the *lacZ* gene. *J. Mol. Biol.* **226**(3), 581–596.
4. Stanssens, P., Remaut, E., and Fiers, W. (1986). Inefficient translation initiation causes premature transcription termination in the *lacZ* gene. *Cell* **44**(5), 711–718.

5. Deana, A., and Belasco, J. G. (2005). Lost in translation: The influence of ribosomes on bacterial mRNA decay. *Genes Dev.* **19**(21), 2526–2533.
6. Uchida, N., Hoshino, S., Imataka, H., Sonenberg, N., and Katada, T. (2002). A novel role of the mammalian GSPT/eRF3 associating with poly(A)-binding protein in Cap/Poly(A)-dependent translation. *J. Biol. Chem.* **277**(52), 50286–50292.
7. Sachs, A. B. (2000). Physical and functional interactions between the mRNA cap structure and the poly(A) tail. *In* "Translational Control of Gene Expression" (N. Sonenberg, H. J. Hershey, and M. B. Mathews, Eds.), pp. 447–466. Cold Spring Harbor Laboratory, Cold Spring Harbor, New York.
8. Wells, S. E., Hillner, P. E., Vale, R. D., and Sachs, A. B. (1998). Circularization of mRNA by eukaryotic translation initiation factors. *Mol. Cell.* **2**(1), 135–140.
9. Vilela, C., Velasco, C., Ptushkina, M., and McCarthy, J. E. (2000). The eukaryotic mRNA decapping protein Dcp1 interacts physically and functionally with the eIF4F translation initiation complex. *EMBO J.* **19**(16), 4372–4382.
10. Pe'ery, T., and Mathews, M. B. (2000). Viral translational strategies and host defense mechanisms. *In* "Translational Control of Gene Expression" (N. Sonenberg, H. J. Hershey, and M. B. Mathews, Eds.), pp. 371–424. Cold Spring Harbor Laboratory, Cold Spring Harbor, New York.
11. Belsham, G. J., and Jackson, R. J. (2000). Translational initiation on picornavirus RNA. *In* "Translational Control of Gene Expression" (N. Sonenberg, H. J. Hershey, and M. B. Mathews, Eds.), pp. 869–900. Cold Spring Harbor Laboratory, Cold Spring Harbor, New York.
12. Vende, P., Piron, M., Castagne, N., and Poncet, D. (2000). Efficient translation of rotavirus mRNA requires simultaneous interaction of NSP3 with the eukaryotic translation initiation factor eIF4G and the mRNA 3' end. *J. Virol.* **74**(15), 7064–7071.
13. Michel, Y. M., Borman, A. M., Paulous, S., and Kean, K. M. (2001). Eukaryotic initiation factor 4G-poly(A) binding protein interaction is required for poly(A) tail-mediated stimulation of picornavirus internal ribosome entry segment-driven translation but not for X-mediated stimulation of hepatitis C virus translation. *Mol. Cell. Biol.* **21**(13), 4097–4109.
14. Lin, H. X., and White, K. A. (2004). A complex network of RNA–RNA interactions controls subgenomic mRNA transcription in a tombusvirus. *EMBO J.* **23**(16), 3365–3374.
15. Corver, J., Lenches, E., Smith, K., Robison, R. A., Sando, T., Strauss, E. G., and Strauss, J. H. (2003). Fine mapping of a cis-acting sequence element in yellow fever virus RNA that is required for RNA replication and cyclization. *J. Virol.* **77**(3), 2265–2270.
16. Markoff, L. (2003). 5'- and 3'-noncoding regions in flavivirus RNA. *Adv. Virus Res.* **59**, 177–228.
17. Khromykh, A. A., Kondratieva, N., Sgro, J. Y., Palmenberg, A., and Westaway, E. G. (2003). Significance in replication of the terminal nucleotides of the flavivirus genome. *J. Virol.* **77**(19), 10623–10629.
18. Alvarez, D. E., Lodeiro, M. F., Luduena, S. J., Pietrasanta, L. I., and Gamarnik, A. V. (2005). Long-range RNA-RNA interactions circularize the dengue virus genome. *J. Virol.* **79**(11), 6631–6643.
19. You, S., Falgout, B., Markoff, L., and Padmanabhan, R. (2001). *In vitro* RNA synthesis from exogenous dengue viral RNA templates requires long range interactions between 5'- and 3'-terminal regions that influence RNA structure. *J. Biol. Chem.* **276**(19), 15581–15591.
20. Khromykh, A. A., Meka, H., Guyatt, K. J., and Westaway, E. G. (2001). Essential role of cyclization sequences in flavivirus RNA replication. *J. Virol.* **75**(14), 6719–6728.
21. Garcia-Montalvo, B. M., Medina, F., and del Angel, R. M. (2004). La protein binds to NS5 and NS3 and to the 5' and 3' ends of Dengue 4 virus RNA. *Virus Res.* **102**(2), 141–150.

22. Pelletier, J., Kaplan, G., Racaniello, V. R., and Sonenberg, N. (1988). Cap-independent translation of poliovirus mRNA is conferred by sequence elements within the 5' noncoding region. *Mol. Cell. Biol.* **8**(3), 1103–1112.
23. Trono, D., Pelletier, J., Sonenberg, N., and Baltimore, D. (1988). Translation in mammalian cells of a gene linked to the poliovirus 5' noncoding region. *Science* **241**(4864), 445–448.
24. Jang, S. K., Krausslich, H. G., Nicklin, M. J., Duke, G. M., Palmenberg, A. C., and Wimmer, E. (1988). A segment of the 5' nontranslated region of encephalomyocarditis virus RNA directs internal entry of ribosomes during in vitro translation. *J. Virol.* **62**(8), 2636–2643.
25. Andino, R., Rieckhof, G. E., and Baltimore, D. (1990). A functional ribonucleoprotein complex forms around the 5' end of poliovirus RNA. *Cell* **63**(2), 369–380.
26. Spector, D. H., Villa-Komaroff, L., and Baltimore, D. (1975). Studies on the function of polyadenylic acid on poliovirus RNA. *Cell* **6**(1), 41–44.
27. Sarnow, P. (1989). Role of 3'-end sequences in infectivity of poliovirus transcripts made in vitro. *J. Virol.* **63**(1), 467–470.
28. Andino, R., Rieckhof, G. E., Achacoso, P. L., and Baltimore, D. (1993). Poliovirus RNA synthesis utilizes an RNP complex formed around the 5'-end of viral RNA. *EMBO J.* **12**(9), 3587–3598.
29. Silvera, D., Gamarnik, A. V., and Andino, R. (1999). The N-terminal K homology domain of the poly(rC)-binding protein is a major determinant for binding to the poliovirus 5'-untranslated region and acts as an inhibitor of viral translation. *J. Biol. Chem.* **274**(53), 38163–38170.
30. Gamarnik, A. V., and Andino, R. (1997). Two functional complexes formed by KH domain containing proteins with the 5' noncoding region of poliovirus RNA. *RNA* **3**(8), 882–892.
31. Gamarnik, A. V., and Andino, R. (1998). Switch from translation to RNA replication in a positive-stranded RNA virus. *Genes Dev.* **12**(15), 2293–2304.
32. Herold, J., and Andino, R. (2001). Poliovirus RNA replication requires genome circularization through a protein–protein bridge. *Mol. Cell* **7**(3), 581–591.
33. Jacobson, A., and Favreau, M. (1983). Possible involvement of poly(A) in protein synthesis. *Nucleic Acids Res.* **11**(18), 6353–6368.
34. Palatnik, C. M., Wilkins, C., and Jacobson, A. (1984). Translational control during early Dictyostelium development: Possible involvement of poly(A) sequences. *Cell* **36**(4), 1017–1025.
35. Jacobson, A. (1996). Poly(A) metabolism and translation: The closed-loop model. *In* "Translational Control" (H. J. Hershey, M. B. Mathews, and N. Sonenberg, Eds.), pp. 451–480. Cold Spring Harbor Laboratory, Cold Spring Harbor, New York.
36. Kozak, M. (2004). How strong is the case for regulation of the initiation step of translation by elements at the 3' end of eukaryotic mRNAs? *Gene* **343**(1), 41–54.
37. Le, H., Tanguay, R. L., Balasta, M. L., Wei, C. C., Browning, K. S., Metz, A. M., Goss, D. J., and Gallie, D. R. (1997). Translation initiation factors eIF-iso4G and eIF-4B interact with the poly(A)-binding protein and increase its RNA binding activity. *J. Biol. Chem.* **272**(26), 16247–16255.
38. Wei, C. C., Balasta, M. L., Ren, J., and Goss, D. J. (1998). Wheat germ poly(A) binding protein enhances the binding affinity of eukaryotic initiation factor 4F and (iso)4F for cap analogues. *Biochemistry* **37**(7), 1910–1916.
39. Hsu, M. T., and Coca-Prados, M. (1979). Electron microscopic evidence for the circular form of RNA in the cytoplasm of eukaryotic cells. *Nature* **280**(5720), 339–340.
40. Kean, K. M. (2003). The role of mRNA 5'-noncoding and 3'-end sequences on 40S ribosomal subunit recruitment, and how RNA viruses successfully compete with cellular mRNAs to ensure their own protein synthesis. *Biol. Cell* **95**(3–4), 129–139.
41. Searfoss, A., Dever, T. E., and Wickner, R. (2001). Linking the 3' poly(A) tail to the subunit joining step of translation initiation:Relations of Pab1p, eukaryotic translation initiation factor 5b (Fun12p), and Ski2p-Slh1p. *Mol. Cell. Biol.* **21**(15), 4900–4908.

42. Sachs, A. B., and Davis, R. W. (1989). The poly(A) binding protein is required for poly(A) shortening and 60S ribosomal subunit-dependent translation initiation. *Cell* **58**(5), 857–867.
43. Tarun, S. Z., Jr., and Sachs, A. B. (1995). A common function for mRNA 5' and 3' ends in translation initiation in yeast. *Genes Dev.* **9**(23), 2997–3007.
44. Borman, A. M., Michel, Y. M., and Kean, K. M. (2000). Biochemical characterisation of cap-poly(A) synergy in rabbit reticulocyte lysates: The eIF4G–PABP interaction increases the functional affinity of eIF4E for the capped mRNA 5'-end. *Nucl. Acids Res.* **28**(21), 4068–4075.
45. Kahvejian, A., Svitkin, Y. V., Sukarieh, R., M'Boutchou, M. N., and Sonenberg, N. (2005). Mammalian poly(A)-binding protein is a eukaryotic translation initiation factor, which acts via multiple mechanisms. *Genes Dev.* **19**(1), 104–113.
46. Jacobson, A., and Peltz, S. W. (1996). Interrelationships of the pathways of mRNA decay and translation in eukaryotic cells. *Annu. Rev. Biochem.* **65**, 693–739.
47. Ma, J., Campbell, A., and Karlin, S. (2002). Correlations between Shine-Dalgarno sequences and gene features such as predicted expression levels and operon structures. *J. Bacteriol.* **184**(20), 5733–5745.
48. Komarova, A. V., Tchufistova, L. S., Supina, E. V., and Boni, I. V. (2002). Protein S1 counteracts the inhibitory effect of the extended Shine-Dalgarno sequence on translation. *RNA* **8**(9), 1137–1147.
49. Jackson, R. J. (2000). Comparative view of initiation site selection mechanisms. In "Translational Control of Gene Expression" (N. Sonenberg, H. J. Hershey, and M. B. Mathews, Eds.), pp. 127–184. Cold Spring Harbor Laboratory, Cold Spring Harbor, New York.
50. Yang, D., Cheung, P., Sun, Y., Yuan, J., Zhang, H., Carthy, C. M., Anderson, D. R., Bohunek, L., Wilson, J. E., and McManus, B. M. (2003). A Shine-Dalgarno-like sequence mediates in vitro ribosomal internal entry and subsequent scanning for translation initiation of coxsackievirus B3 RNA. *Virology* **305**(1), 31–43.
51. Chappell, S. A., Edelman, G. M., and Mauro, V. P. (2004). Biochemical and functional analysis of a 9-nt RNA sequence that affects translation efficiency in eukaryotic cells. *Proc. Natl. Acad. Sci. USA* **101**(26), 9590–9594.
52. Shen, R., and Miller, W. A. (2004). The 3' untranslated region of tobacco necrosis virus RNA contains a barley yellow dwarf virus-like cap-independent translation element. *J. Virol.* **78**(9), 4655–4664.
53. Meulewaeter, F., van Lipzig, R., Gultyaev, A. P., Pleij, C. W., van Damme, D., Cornelissen, M., and van Eldik, G. (2004). Conservation of RNA structures enables TMV and BYDV 5' and 3' elements to cooperate synergistically in cap-independent translation. *Nucleic Acids Res.* **32**(5), 1721–1730.
54. Guo, L., Allen, E. M., and Miller, W. A. (2001). Base-pairing between untranslated regions facilitates translation of uncapped, nonpolyadenylated viral RNA. *Mol. Cell.* **7**(5), 1103–1109.
55. Fabian, M. R., and White, K. A. (2004). 5'-3' RNA-RNA interaction facilitates cap- and poly(A) tail-independent translation of tomato bushy stunt virus mRNA: A potential common mechanism for tombusviridae. *J. Biol. Chem.* **279**(28), 28862–28872.
56. Wang, S., Browning, K. S., and Miller, W. A. (1997). A viral sequence in the 3'-untranslated region mimics a 5' cap in facilitating translation of uncapped mRNA. *EMBO J.* **16**(13), 4107–4116.
57. Wang, S., Guo, L., Allen, E., and Miller, W. A. (1999). A potential mechanism for selective control of cap-independent translation by a viral RNA sequence in *cis* and in *trans*. *RNA* **5**(6), 728–738.
58. Guo, L., Allen, E., and Miller, W. A. (2000). Structure and function of a cap-independent translation element that functions in either the 3' or the 5' untranslated region. *RNA* **6**(12), 1808–1820.

59. Zeiler, B. N., and Simons, R. W. (1997). Antisense RNA structure and function. In "RNA Structure and Function" (R. W. Simons and M. Grunberg-Manago, Eds.), pp. 437–464. Cold Spring Harbor Laboratory, Cold Spring Harbor, New York.
60. Gallie, D. R. (1991). The cap and poly(A) tail function synergistically to regulate mRNA translational efficiency. *Genes Dev.* **5**(11), 2108–2116.
61. Mohan, B. R., Dinesh-Kumar, S. P., and Miller, W. A. (1995). Genes and *cis*-acting sequences involved in replication of barley yellow dwarf virus-PAV RNA. *Virology* **212**(1), 186–195.
62. Kelly, L., Gerlach, W. L., and Waterhouse, P. M. (1994). Characterisation of the subgenomic RNAs of an Australian isolate of barley yellow dwarf luteovirus. *Virology* **202**(2), 565–573.
63. Koev, G., Mohan, B. R., Dinesh-Kumar, S. P., Torbert, K. A., Somers, D. A., and Miller, W. A. (1998). Extreme reduction of disease in oats transformed with the 5(prime) half of the barley yellow dwarf virus-PAV genome. *Phytopathology* **88**, 1013–1019.
64. Shine, J., and Dalgarno, L. (1974). The 3′-terminal sequence of *Escherichia coli* 16S ribosomal RNA: Complementarity to nonsense triplets and ribosome binding sites. *Proc. Natl. Acad. Sci. USA* **71**(4), 1342–1346.
65. Steitz, J. A., and Jakes, K. (1975). How ribosomes select initiator regions in mRNA: Base pair formation between the 3′ terminus of 16S rRNA and the mRNA during initiation of protein synthesis in *Escherichia coli*. *Proc. Natl. Acad. Sci. USA* **72**(12), 4734–4738.
66. Hui, A., and de Boer, H. A. (1987). Specialized ribosome system: Preferential translation of a single mRNA species by a subpopulation of mutated ribosomes in *Escherichia coli*. *Proc. Natl. Acad. Sci. USA* **84**(14), 4762–4766.
67. Jacob, W. F., Santer, M., and Dahlberg, A. E. (1987). A single base change in the Shine-Dalgarno region of 16S rRNA of *Escherichia coli* affects translation of many proteins. *Proc. Natl. Acad. Sci. USA* **84**(14), 4757–4761.
68. Marzluff, W. F., and Duronio, R. J. (2002). Histone mRNA expression: Multiple levels of cell cycle regulation and important developmental consequences. *Curr. Opin. Cell Biol.* **14**(6), 692–699.
69. Dominski, Z., Yang, X. C., Kaygun, H., Dadlez, M., and Marzluff, W. F. (2003). A 3′ exonuclease that specifically interacts with the 3′ end of histone mRNA. *Mol. Cell.* **12**(2), 295–305.
70. Allard, P., Yang, Q., Marzluff, W. F., and Clarke, H. J. (2005). The stem-loop binding protein regulates translation of histone mRNA during mammalian oogenesis. *Dev. Biol.* **286**(1), 195–206.
71. Allard, P., Champigny, M. J., Skoggard, S., Erkmann, J. A., Whitfield, M. L., Marzluff, W. F., and Clarke, H. J. (2002). Stem-loop binding protein accumulates during oocyte maturation and is not cell-cycle-regulated in the early mouse embryo. *J. Cell Sci.* **115**(Pt. 23), 4577–4586.
72. Wang, Z. F., Ingledue, T. C., Dominski, Z., Sanchez, R., and Marzluff, W. F. (1999). Two *Xenopus* proteins that bind the 3′ end of histone mRNA: Implications for translational control of histone synthesis during oogenesis. *Mol. Cell. Biol.* **19**(1), 835–845.
73. Marzluff, W. F. (1992). Histone 3′ ends: Essential and regulatory functions. *Gene Expr.* **2**(2), 93–97.
74. Dominski, Z., and Marzluff, W. F. (1999). Formation of the 3′ end of histone mRNA. *Gene* **239**(1), 1–14.
75. Colegrove-Otero, L. J., Minshall, N., and Standart, N. (2005). RNA-binding proteins in early development. *Crit. Rev. Biochem. Mol. Biol.* **40**(1), 21–73.
76. Jaeger, S., Barends, S., Giege, R., Eriani, G., and Martin, F. (2005). Expression of metazoan replication-dependent histone genes. *Biochimie* **87**(9–10), 827–834.
77. Martin, F., Schaller, A., Eglite, S., Schumperli, D., and Muller, B. (1997). The gene for histone RNA hairpin binding protein is located on human chromosome 4 and encodes a novel type of RNA binding protein. *EMBO J.* **16**(4), 769–778.

78. Wang, Z. F., Whitfield, M. L., Ingledue, T. C., 3rd, Dominski, Z., and Marzluff, W. F. (1996). The protein that binds the 3' end of histone mRNA: A novel RNA-binding protein required for histone pre-mRNA processing. *Genes Dev.* **10**(23), 3028–3040.
79. Zhao, X., McKillop-Smith, S., and Muller, B. (2004). The human histone gene expression regulator HBP/SLBP is required for histone and DNA synthesis, cell cycle progression and cell proliferation in mitotic cells. *J. Cell Sci.* **117**(Pt. 25), 6043–6051.
80. Sullivan, E., Santiago, C., Parker, E. D., Dominski, Z., Yang, X., Lanzotti, D. J., Ingledue, T. C., Marzluff, W. F., and Duronio, R. J. (2001). *Drosophila* stem loop binding protein coordinates accumulation of mature histone mRNA with cell cycle progression. *Genes Dev.* **15**(2), 173–187.
81. Pettitt, J., Crombie, C., Schumperli, D., and Muller, B. (2002). The *Caenorhabditis elegans* histone hairpin-binding protein is required for core histone gene expression and is essential for embryonic and postembryonic cell division. *J. Cell Sci.* **115**(Pt. 4), 857–866.
82. Kodama, Y., Rothman, J. H., Sugimoto, A., and Yamamoto, M. (2002). The stem-loop binding protein CDL-1 is required for chromosome condensation, progression of cell death and morphogenesis in *Caenorhabditis elegans*. *Development* **129**(1), 187–196.
83. Whitfield, M. L., Zheng, L. X., Baldwin, A., Ohta, T., Hurt, M. M., and Marzluff, W. F. (2000). Stem-loop binding protein, the protein that binds the 3' end of histone mRNA, is cell cycle regulated by both translational and posttranslational mechanisms. *Mol. Cell. Biol.* **20**(12), 4188–4198.
84. Cotten, M., Gick, O., Vasserot, A., Schaffner, G., and Birnstiel, M. L. (1988). Specific contacts between mammalian U7 snRNA and histone precursor RNA are indispensable for the *in vitro* 3' RNA processing reaction. *EMBO J.* **7**(3), 801–808.
85. Dominski, Z., Erkmann, J. A., Yang, X., Sanchez, R., and Marzluff, W. F. (2002). A novel zinc finger protein is associated with U7 snRNP and interacts with the stem-loop binding protein in the histone pre-mRNP to stimulate 3'-end processing. *Genes Dev.* **16**(1), 58–71.
86. Williams, A. S., Ingledue, T. C., 3rd, Kay, B. K., and Marzluff, W. F. (1994). Changes in the stem-loop at the 3' terminus of histone mRNA affects its nucleocytoplasmic transport and cytoplasmic regulation. *Nucleic Acids Res.* **22**(22), 4660–4666.
87. Whitfield, M. L., Kaygun, H., Erkmann, J. A., Townley-Tilson, W. H., Dominski, Z., and Marzluff, W. F. (2004). SLBP is associated with histone mRNA on polyribosomes as a component of the histone mRNP. *Nucleic Acids Res.* **32**(16), 4833–4842.
88. Hanson, R. J., Sun, J., Willis, D. G., and Marzluff, W. F. (1996). Efficient extraction and partial purification of the polyribosome-associated stem-loop binding protein bound to the 3' end of histone mRNA. *Biochemistry* **35**(7), 2146–2156.
89. Sun, J., Pilch, D. R., and Marzluff, W. F. (1992). The histone mRNA 3' end is required for localization of histone mRNA to polyribosomes. *Nucleic Acids Res.* **20**(22), 6057–6066.
90. Ling, J., Morley, S. J., Pain, V. M., Marzluff, W. F., and Gallie, D. R. (2002). The histone 3'-terminal stem-loop-binding protein enhances translation through a functional and physical interaction with eukaryotic initiation factor 4G (eIF4G) and eIF3. *Mol. Cell. Biol.* **22**(22), 7853–7867.
91. Gallie, D. R., Lewis, N. J., and Marzluff, W. F. (1996). The histone 3'-terminal stem-loop is necessary for translation in Chinese hamster ovary cells. *Nucleic Acids Res.* **24**(10), 1954–1962.
92. Sanchez, R., and Marzluff, W. F. (2002). The stem-loop binding protein is required for efficient translation of histone mRNA *in vivo* and *in vitro*. *Mol. Cell. Biol.* **22**(20), 7093–7104.
93. Gorgoni, B., Andrews, S., Schaller, A., Schumperli, D., Gray, N. K., and Muller, B. (2005). The stem-loop binding protein stimulates histone translation at an early step in the initiation pathway. *RNA* **11**(7), 1030–1042.

94. Marzluff, W. F., Gongidi, P., Woods, K. R., Jin, J., and Maltais, L. J. (2002). The human and mouse replication-dependent histone genes. *Genomics* **80**(5), 487–498.
95. Kaygun, H., and Marzluff, W. F. (2005). Translation termination is involved in histone mRNA degradation when DNA replication is inhibited. *Mol. Cell. Biol.* **25**(16), 6879–6888.
96. Gallie, D. R., Lucas, W. J., and Walbot, V. (1989). Visualizing mRNA expression in plant protoplasts: Factors influencing efficient mRNA uptake and translation. *Plant Cell* **1**(3), 301–311.
97. Williams, A. S., and Marzluff, W. F. (1995). The sequence of the stem and flanking sequences at the 3' end of histone mRNA are critical determinants for the binding of the stem-loop binding protein. *Nucleic Acids Res.* **23**(4), 654–662.
98. Gallie, D. R., and Tanguay, R. (1994). Poly(A) binds to initiation factors and increases cap-dependent translation *in vitro*. *J. Biol. Chem.* **269**(25), 17166–17173.
99. Bergamini, G., Preiss, T., and Hentze, M. W. (2000). Picornavirus IRESes and the poly(A) tail jointly promote cap-independent translation in a mammalian cell-free system. *RNA* **6**(12), 1781–1790.
100. Svitkin, Y. V., Imataka, H., Khaleghpour, K., Kahvejian, A., Liebig, H. D., and Sonenberg, N. (2001). Poly(A)-binding protein interaction with eIF4G stimulates picornavirus IRES-dependent translation. *RNA* **7**(12), 1743–1752.
101. Capasso, O., Bleecker, G. C., and Heintz, N. (1987). Sequences controlling histone H4 mRNA abundance. *EMBO J.* **6**(6), 1825–1831.
102. Levine, B. J., Chodchoy, N., Marzluff, W. F., and Skoultchi, A. I. (1987). Coupling of replication type histone mRNA levels to DNA synthesis requires the stem-loop sequence at the 3' end of the mRNA. *Proc. Natl. Acad. Sci. USA* **84**(17), 6189–6193.
103. Pandey, N. B., and Marzluff, W. F. (1987). The stem-loop structure at the 3' end of histone mRNA is necessary and sufficient for regulation of histone mRNA stability. *Mol. Cell. Biol.* **7**(12), 4557–4559.
104. Gray, N. K., Coller, J. M., Dickson, K. S., and Wickens, M. (2000). Multiple portions of poly(A)-binding protein stimulate translation *in vivo*. *EMBO J.* **19**(17), 4723–4733.
105. Wilson, J. E., Pestova, T. V., Hellen, C. U., and Sarnow, P. (2000). Initiation of protein synthesis from the A site of the ribosome. *Cell* **102**(4), 511–520.
106. Pestova, T. V., Shatsky, I. N., Fletcher, S. P., Jackson, R. J., and Hellen, C. U. (1998). A prokaryotic-like mode of cytoplasmic eukaryotic ribosome binding to the initiation codon during internal translation initiation of hepatitis C and classical swine fever virus RNAs. *Genes Dev.* **12**(1), 67–83.
107. Jackson, R. J., Hunt, S. L., Gibbs, C. L., and Kaminski, A. (1994). Internal initiation of translation of picornavirus RNAs. *Mol. Biol. Rep.* **19**(3), 147–159.
108. Chen, C. Y., and Sarnow, P. (1995). Initiation of protein synthesis by the eukaryotic translational apparatus on circular RNAs. *Science* **268**(5209), 415–417.
109. Kean, K. M., Michel, Y. M., and Borman, A. M. (1999). Viral exceptions to 5'-end dependent initiation of translation: Is there really a difference in the mechanism of ribosome recruitment to capped and IRES containing mRNAs? *Curr. Top. Virol.* **1**, 191–201.
110. Borman, A. M., Bailly, J. L., Girard, M., and Kean, K. M. (1995). Picornavirus internal ribosome entry segments: Comparison of translation efficiency and the requirements for optimal internal initiation of translation *in vitro*. *Nucleic Acids Res.* **23**(18), 3656–3663.
111. Borman, A. M., Kirchweger, R., Ziegler, E., Rhoads, R. E., Skern, T., and Kean, K. M. (1997). eIF4G and its proteolytic cleavage products: Effect on initiation of protein synthesis from capped, uncapped, and IRES-containing mRNAs. *RNA* **3**(2), 186–196.
112. Paulous, S., Malnou, C. E., Michel, Y. M., Kean, K. M., and Borman, A. M. (2003). Comparison of the capacity of different viral internal ribosome entry segments to direct

translation initiation in poly(A)-dependent reticulocyte lysates. *Nucleic Acids Res.* **31**(2), 722–733.
113. Pisarev, A. V., Chard, L. S., Kaku, Y., Johns, H. L., Shatsky, I. N., and Belsham, G. J. (2004). Functional and structural similarities between the internal ribosome entry sites of hepatitis C virus and porcine teschovirus, a picornavirus. *J. Virol.* **78**(9), 4487–4497.
114. Brown, B. A., and Ehrenfeld, E. (1979). Translation of poliovirus RNA *in vitro*: Changes in cleavage pattern and initiation sites by ribosomal salt wash. *Virology* **97**(2), 396–405.
115. del Angel, R. M., Papavassiliou, A. G., Fernandez-Tomas, C., Silverstein, S. J., and Racaniello, V. R. (1989). Cell proteins bind to multiple sites within the 5′ untranslated region of poliovirus RNA. *Proc. Natl. Acad. Sci. USA* **86**(21), 8299–8303.
116. Meerovitch, K., Pelletier, J., and Sonenberg, N. (1989). A cellular protein that binds to the 5′-noncoding region of poliovirus RNA: Implications for internal translation initiation. *Genes Dev.* **3**(7), 1026–1034.
117. Meerovitch, K., Svitkin, Y. V., Lee, H. S., Lejbkowicz, F., Kenan, D. J., Chan, E. K., Agol, V. I., Keene, J. D., and Sonenberg, N. (1993). La autoantigen enhances and corrects aberrant translation of poliovirus RNA in reticulocyte lysate. *J. Virol.* **67**(7), 3798–3807.
118. Malnou, C. E., Michel, Y. M., Paulous, S., Borman, A. M., and Kean, K. M. (2002). Internal ribosome recruitment: An alternative way that many animal viruses with positive-strand RNA genomes initiate protein synthesis. *Res. Adv. Virol.* **2**, 55–68.
119. Pilipenko, E. V., Pestova, T. V., Kolupaeva, V. G., Khitrina, E. V., Poperechnaya, A. N., Agol, V. I., and Hellen, C. U. (2000). A cell cycle-dependent protein serves as a template-specific translation initiation factor. *Genes Dev.* **14**(16), 2028–2045.
120. Lillevali, K., Kulla, A., and Ord, T. (2001). Comparative expression analysis of the genes encoding polypyrimidine tract binding protein (PTB) and its neural homologue (brPTB) in prenatal and postnatal mouse brain. *Mech. Dev.* **101**(1–2), 217–220.
121. Kikuchi, T., Ichikawa, M., Arai, J., Tateiwa, H., Fu, L., Higuchi, K., and Yoshimura, N. (2000). Molecular cloning and characterization of a new neuron-specific homologue of rat polypyrimidine tract binding protein. *J. Biochem. (Tokyo)* **128**(5), 811–821.
122. Pilipenko, E. V., Viktorova, E. G., Guest, S. T., Agol, V. I., and Roos, R. P. (2001). Cell-specific proteins regulate viral RNA translation and virus-induced disease. *EMBO J.* **20**(23), 6899–6908.
123. Kolupaeva, V. G., Pestova, T. V., Hellen, C. U., and Shatsky, I. N. (1998). Translation eukaryotic initiation factor 4G recognizes a specific structural element within the internal ribosome entry site of encephalomyocarditis virus RNA. *J. Biol. Chem.* **273**(29), 18599–18604.
124. Pestova, T. V., Shatsky, I. N., and Hellen, C. U. (1996). Functional dissection of eukaryotic initiation factor 4F: The 4A subunit and the central domain of the 4G subunit are sufficient to mediate internal entry of 43S preinitiation complexes. *Mol. Cell. Biol.* **16**(12), 6870–6878.
125. Ochs, K., Zeller, A., Saleh, L., Bassili, G., Song, Y., Sonntag, A., and Niepmann, M. (2003). Impaired binding of standard initiation factors mediates poliovirus translation attenuation. *J. Virol.* **77**(1), 115–122.
126. Borman, A. M., Michel, Y. M., and Kean, K. M. (2001). Detailed analysis of the requirements of hepatitis A virus internal ribosome entry segment for the eukaryotic initiation factor complex eIF4F. *J. Virol.* **75**(17), 7864–7871.
127. Hellen, C. U., and Sarnow, P. (2001). Internal ribosome entry sites in eukaryotic mRNA molecules. *Genes Dev.* **15**(13), 1593–1612.
128. Kirchweger, R., Ziegler, E., Lamphear, B. J., Waters, D., Liebig, H. D., Sommergruber, W., Sobrino, F., Hohenadl, C., Blaas, D., Rhoads, R. E., and Skern, T. (1994). Foot-and-mouth disease virus leader proteinase: Purification of the Lb form and determination of its cleavage site on eIF-4 gamma. *J. Virol.* **68**(9), 5677–5684.

129. Lamphear, B. J., Yan, R., Yang, F., Waters, D., Liebig, H. D., Klump, H., Kuechler, E., Skern, T., and Rhoads, R. E. (1993). Mapping the cleavage site in protein synthesis initiation factor eIF-4 gamma of the 2A proteases from human Coxsackievirus and rhinovirus. *J. Biol. Chem.* **268**(26), 19200–19203.
130. Roberts, L. O., Seamons, R. A., and Belsham, G. J. (1998). Recognition of picornavirus internal ribosome entry sites within cells; influence of cellular and viral proteins. *RNA* **4**(5), 520–529.
131. Michel, Y. M., Poncet, D., Piron, M., Kean, K. M., and Borman, A. M. (2000). Cap-Poly(A) synergy in mammalian cell-free extracts. Investigation of the requirements for poly(A)-mediated stimulation of translation initiation. *J. Biol. Chem.* **275**(41), 32268–32276.
132. Collier, B., Gorgoni, B., Loveridge, C., Cooke, H. J., and Gray, N. K. (2005). The DAZL family proteins are PABP-binding proteins that regulate translation in germ cells. *EMBO J.* **24**(14), 2656–2666.
133. Ruggiu, M., Speed, R., Taggart, M., McKay, S. J., Kilanowski, F., Saunders, P., Dorin, J., and Cooke, H. J. (1997). The mouse *Dazla* gene encodes a cytoplasmic protein essential for gametogenesis. *Nature* **389**(6646), 73–77.
134. Reijo, R., Lee, T. Y., Salo, P., Alagappan, R., Brown, L. G., Rosenberg, M., Rozen, S., Jaffe, T., Straus, D., Hovatta, O., de la Chapelle, A., Silber, S. *et al.* (1995). Diverse spermatogenic defects in humans caused by Y chromosome deletions encompassing a novel RNA-binding protein gene. *Nat. Genet.* **10**(4), 383–393.
135. Eberhart, C. G., Maines, J. Z., and Wasserman, S. A. (1996). Meiotic cell cycle requirement for a fly homologue of human Deleted in Azoospermia. *Nature* **381**(6585), 783–785.
136. Maegawa, S., Yamashita, M., Yasuda, K., and Inoue, K. (2002). Zebrafish DAZ-like protein controls translation via the sequence 'GUUC'. *Genes Cells* **7**(9), 971–984.
137. Tsui, S., Dai, T., Roettger, S., Schempp, W., Salido, E. C., and Yen, P. H. (2000). Identification of two novel proteins that interact with germ-cell-specific RNA-binding proteins DAZ and DAZL1. *Genomics* **65**(3), 266–273.
138. Haag, E. S. (2001). Rolling back to BOULE. *Proc. Natl. Acad. Sci. USA* **98**(13), 6983–6985.
139. Yen, P. H., Chai, N. N., and Salido, E. C. (1997). The human *DAZ* genes, a putative male infertility factor on the Y chromosome, are highly polymorphic in the DAZ repeat regions. *Mamm. Genome* **8**(10), 756–759.
140. Mita, K., and Yamashita, M. (2000). Expression of *Xenopus* Daz-like protein during gametogenesis and embryogenesis. *Mech. Dev.* **94**(1–2), 251–255.
141. Cheng, M. H., Maines, J. Z., and Wasserman, S. A. (1998). Biphasic subcellular localization of the DAZL-related protein boule in *Drosophila* spermatogenesis. *Dev. Biol.* **204**(2), 567–576.
142. Reijo, R. A., Dorfman, D. M., Slee, R., Renshaw, A. A., Loughlin, K. R., Cooke, H., and Page, D. C. (2000). DAZ family proteins exist throughout male germ cell development and transit from nucleus to cytoplasm at meiosis in humans and mice. *Biol. Reprod.* **63**(5), 1490–1496.
143. Maines, J. Z., and Wasserman, S. A. (1999). Post-transcriptional regulation of the meiotic Cdc25 protein Twine by the Dazl orthologue Boule. *Nat. Cell Biol.* **1**(3), 171–174.
144. Jiao, X., Trifillis, P., and Kiledjian, M. (2002). Identification of target messenger RNA substrates for the murine deleted in azoospermia-like RNA-binding protein. *Biol. Reprod.* **66**(2), 475–485.
145. Venables, J. P., Ruggiu, M., and Cooke, H. J. (2001). The RNA-binding specificity of the mouse Dazl protein. *Nucleic Acids Res.* **29**(12), 2479–2483.
146. Kuhn, U., and Pieler, T. (1996). Xenopus poly(A) binding protein: Functional domains in RNA binding and protein–protein interaction. *J. Mol. Biol.* **256**(1), 20–30.
147. Tsui, S., Dai, T., Warren, S. T., Salido, E. C., and Yen, P. H. (2000). Association of the mouse infertility factor DAZL1 with actively translating polyribosomes. *Biol. Reprod.* **62**(6), 1655–1660.

148. Alphey, L., Jimenez, J., White-Cooper, H., Dawson, I., Nurse, P., and Glover, D. M. (1992). Twine, a cdc25 homolog that functions in the male and female germline of Drosophila. *Cell* **69**(6), 977–988.
149. Courtot, C., Fankhauser, C., Simanis, V., and Lehner, C. F. (1992). The *Drosophila* cdc25 homolog *twine* is required for meiosis. *Development* **116**(2), 405–416.
150. Xu, E. Y., Lee, D. F., Klebes, A., Turek, P. J., Kornberg, T. B., and Reijo Pera, R. A. (2003). Human *BOULE* gene rescues meiotic defects in infertile flies. *Hum. Mol. Genet.* **12**(2), 169–175.
151. Houston, D. W., Zhang, J., Maines, J. Z., Wasserman, S. A., and King, M. L. (1998). A *Xenopus DAZ*-like gene encodes an RNA component of germ plasm and is a functional homologue of *Drosophila boule*. *Development* **125**(2), 171–180.
152. Vogel, T., Speed, R. M., Ross, A., and Cooke, H. J. (2002). Partial rescue of the *Dazl* knockout mouse by the human *DAZL* gene. *Mol. Hum. Reprod.* **8**(9), 797–804.
153. Slee, R., Grimes, B., Speed, R. M., Taggart, M., Maguire, S. M., Ross, A., McGill, N. I., Saunders, P. T., and Cooke, H. J. (1999). A human *DAZ* transgene confers partial rescue of the mouse *Dazl* null phenotype. *Proc. Natl. Acad. Sci. USA* **96**(14), 8040–8045.
154. Bushell, M., and Sarnow, P. (2002). Hijacking the translation apparatus by RNA viruses. *J. Cell Biol.* **158**(3), 395–399.
155. Imai, M., Akatani, K., Ikegami, N., and Furuichi, Y. (1983). Capped and conserved terminal structures in human rotavirus genome double-stranded RNA segments. *J. Virol.* **47**(1), 125–136.
156. Qian, Y. A., Jiang, B. M., Saif, L. J., Kang, S. Y., Ojeh, C. K., and Green, K. Y. (1991). Molecular analysis of the gene 6 from a porcine group C rotavirus that encodes the NS34 equivalent of group A rotaviruses. *Virology* **184**(2), 752–757.
157. Poncet, D., Aponte, C., and Cohen, J. (1993). Rotavirus protein NSP3 (NS34) is bound to the 3′ end consensus sequence of viral mRNAs in infected cells. *J. Virol.* **67**(6), 3159–3165.
158. Deo, R. C., Groft, C. M., Rajashankar, K. R., and Burley, S. K. (2002). Recognition of the rotavirus mRNA 3′ consensus by an asymmetric NSP3 homodimer. *Cell* **108**(1), 71–81.
159. Piron, M., Delaunay, T., Grosclaude, J., and Poncet, D. (1999). Identification of the RNA-binding, dimerization, and eIF4GI-binding domains of rotavirus nonstructural protein NSP3. *J. Virol.* **73**(7), 5411–5421.
160. Piron, M., Vende, P., Cohen, J., and Poncet, D. (1998). Rotavirus RNA-binding protein NSP3 interacts with eIF4GI and evicts the poly(A) binding protein from eIF4F. *EMBO J.* **17**(19), 5811–5821.
161. Vitour, D., Lindenbaum, P., Vende, P., Becker, M. M., and Poncet, D. (2004). RoXaN, a novel cellular protein containing TPR, LD, and zinc finger motifs, forms a ternary complex with eukaryotic initiation factor 4G and rotavirus NSP3. *J. Virol.* **78**(8), 3851–3862.
162. Imataka, H., Gradi, A., and Sonenberg, N. (1998). A newly identified N-terminal amino acid sequence of human eIF4G binds poly(A)-binding protein and functions in poly(A)-dependent translation. *EMBO J.* **17**(24), 7480–7489.
163. Groft, C. M., and Burley, S. K. (2002). Recognition of eIF4G by rotavirus NSP3 reveals a basis for mRNA circularization. *Mol. Cell* **9**(6), 1273–1283.

Interferon Action and the Double-Stranded RNA-Dependent Enzymes ADAR1 Adenosine Deaminase and PKR Protein Kinase

Ann M. Toth, Ping Zhang,
Sonali Das, Cyril X. George,
and Charles E. Samuel

*Department of Molecular,
Cellular, and Developmental Biology,
University of California,
Santa Barbara, California 93106*

I. Introduction	369
II. Adenosine Deaminase Acting on RNA	370
A. *ADAR1* Gene and Transcripts	370
B. ADAR1 Proteins	375
C. Biological Activities of ADAR1	385
III. Protein Kinase Regulated by RNA	390
A. *PKR* Gene and Transcripts	390
B. PKR Protein	395
C. Biological Activities of PKR	405
IV. Possible Roles of ADAR1 and PKR in Human Genetic and Infectious Diseases	409
A. ADAR1 and Human Diseases	410
B. PKR and Human Diseases	411
References	411

I. Introduction

Interferons were discovered as antiviral agents (*1, 2*). However, it is now clear that these cytokines possess multiple activities that include the ability to affect cell growth, differentiation, and death, in addition to their hallmark

Abbreviations: ADAR1, adenosine deaminase acting on RNA; bp, base pair; dsRNA, double-stranded RNA; dsRBD, double-stranded RNA-binding domain; EMCV, encephalomyocarditis virus; HBV, hepatitis B virus; HCV, hepatitis C virus; HDV, hepatitis delta virus; HIV, human immunodeficiency virus; IFN, interferon; ISRE, interferon-stimulated response element; KCS, kinase conserved sequence; LPS, lipopolysaccharide; NES, nuclear export signal; NLS, nuclear localization signal; nt, nucleotide; ORF, open reading frame; PKR; protein kinase regulated by RNA; ssRNA, single-stranded RNA; UTR, untranslated region; VSV, vesicular stomatitis virus.

ability to interfere with virus multiplication (3–6). Interferons exert their actions by regulating the expression of genes whose products mediate the varied physiologic changes seen in interferon-treated cells and animals. Two important genes regulated by interferons are *ADAR1* and *PKR*.

ADAR1 and *PKR* encode dsRNA-binding proteins that are responsible, in part, for the biochemical and mechanistic actions of interferons. ADAR1 is an RNA-specific C-6 adenosine deaminase that functions as an RNA-editing enzyme. Two different forms of ADAR1, one inducible and the other constitutively expressed, are encoded by the *ADAR1* gene. Both forms of ADAR1, p150 and p110, function to modify the expression of genetic information by changing cellular and viral RNAs through substitution of an inosine, which is recognized as guanine, for adenine. PKR is an RNA-dependent protein kinase that controls the translational pattern in cells through phosphorylation of the α subunit of protein synthesis initiation factor eIF-2; PKR also modulates signal transduction processes. Translational control by eIF-2α phosphorylation is now recognized as a universal mechanism by which eukaryotic cells respond to environmental stresses, including infection with pathogens. Considerable progress has been made toward understanding the regulation and function of the PKR and ADAR enzymes, and their roles in uninfected, as well as infected cells and animals. The availability of cDNA and genomic DNA clones for ADAR1 and PKR have permitted biochemical, biophysical, and biological studies involving purified recombinant proteins, and cells and animals that either overexpress ADAR1 or PKR or lack the proteins as the result of targeted gene disruptions or RNAi knock-downs. Studies of ADAR1 and PKR have contributed significantly to our understanding of the molecular mechanisms of interferon action, and also to our knowledge of fundamentally important cellular processes that are modulated by translational control and RNA editing.

This chapter focuses on the organization and regulated expression of the *ADAR1* and *PKR* genes, the biochemical and biophysical properties of the ADAR1 and PKR proteins, the mechanisms by which ADAR1 and PKR modulate the physiology of cultured cells and intact animals, and finally the roles that these proteins might play in genetic and infectious human diseases.

II. Adenosine Deaminase Acting on RNA

A. *ADAR1* Gene and Transcripts

1. GENE MULTIPLICITY AND CHROMOSOME ASSIGNMENT

Extensive Southern blot hybridization, direct nucleotide sequencing, and PCR analyses of human and mouse ADAR1 genomic and cDNA clones reveal a single mammalian *ADAR1* gene (7–10). The human *ADAR1* gene was mapped by fluorescence *in situ* hybridization (FISH) of genomic clones to a single locus on human chromosome 1 band q21.1–21.2 (*10, 11*). The mouse *Adar1* gene was mapped by FISH to a single locus on mouse chromosome 3 band F2 (*12*). These results are in agreement with human–mouse homology maps localizing other genes from human chromosome 1q to the region of mouse chromosome 3F.

The knockout of the mouse *Adar1* gene has been reported (*13–15*). Homozygosity for *Adar1* null caused embryonic death between E11.5 and E12.5 (*14*). Heterozygosity for *Adar1* null was initially reported to be lethal (*13*), but survival and normal appearance was subsequently observed when additional targeted disruptions were analyzed (*14, 15*).

2. EXON–INTRON ORGANIZATION

The human *ADAR1* gene spans ~40 kbp of DNA and consists of 17 exons (*9, 10, 16, 17*) as shown in Fig. 1. Exon 1 occurs in at least three alternative forms (1A, B, and C), each flanked by their respective promoter, P_A, P_B, and P_C (*8, 16, 17*). Exon 1A is located ~5.4-kbp upstream of exon 2, whereas exon 1B is located ~14.5-kbp upstream of exon 2. Exon 1C is located ~3.1-kbp upstream of exon 2. Of these alternative exon 1 forms, only the exon 1A form (which is found in IFN-inducible transcripts) contains a translation initiation codon AUG (*17*). Exons 1B and C, which are found in constitutively expressed transcripts, lack a translation initiation codon, and for these transcripts, translation begins at an AUG codon in exon 2 (AUG296) that is the next available methionine codon within the 3678 nt long ORF (*9, 18*). With this coding arrangement, transcripts with exon 1A encode a protein of 1200 amino acids, while transcripts with either exon 1B or 1C encode a protein of 931 amino acids (Fig. 1). Exon 15 is the largest exon (2984 nt) and includes the UAG translation termination codon and 3′-UTR structure. Exon 2 is unusually large (1586 nt) for an exon that is an ORF. Introns vary in size from 0.1 kbp (intron V) to ~8.9 kbp (intron IB). All splice site junctions conform to the GT·AG rule (*9*).

The consensus sequence for the human ADAR1 cDNA is of 6474 nt in size (*18, 19*). The cDNA possesses an extensive 3′-UTR of 2749 nt and an ORF of 3678 nt. The nucleotide present at the −3/+4 positions flanking the AUG1 (GCA<u>AUG</u>A) translation start codon in the human exon 1A sequence are both purines (*8*), which is characteristic of a strong translation initiation site in vertebrates (*20*). With transcripts containing either exon 1B or exon 1C, translation initiates from AUG296 present in exon 2 and the downstream ORF is 2793 nt in length. Here too the –3/+4 nt flanking AUG296 likewise are both purines, providing a strong context for translation initiation. The 3′-UTR includes the AATAAA polyadenylation signal found 17 nt upstream of a poly(A) tail. In addition, the 3′-UTR includes a region of 227 nt of unknown function

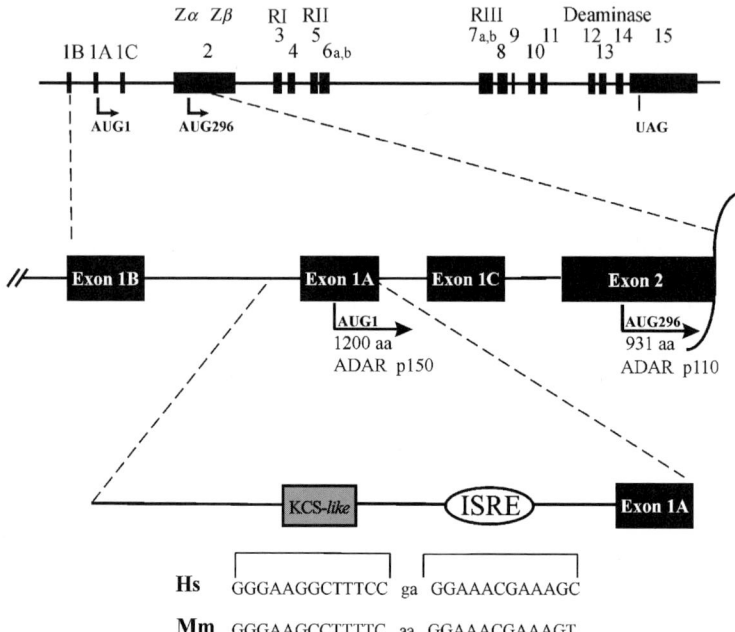

FIG. 1. Organization of the human *ADAR1* gene structure. Organization of introns and exons with regard to the structure of the *ADAR1* gene is represented in the top part of the figure. Exons are indicated by the numbers 1–15 and are represented by the filled boxes; introns and the 5'- and 3'-flanking regions are shown by the solid lines. The entire gene spans about ~40 kbp and includes 15 exons that are present in the mature mRNAs. There are three alternative exon 1 structures—exon 1A, found in transcripts derived from the IFN-inducible promoter, and, exons 1B and C, found in transcripts derived from constitutively active alternative promoters. In addition, there are two alternatively spliced forms of exon 7—7a and b. The translation initiation site of the IFN-inducible p150 protein (1200 amino acids) is present in exon 1A; the constitutively expressed p110 protein (931 amino acids) initiates at AUG296 present in exon 2. The lower diagram depicts an enlargement of the IFN-inducible promoter region of the human and mouse *ADAR1* genes, showing the KCS-like element and the 12-bp ISRE. Human: Hs, *Homo sapiens*; mouse: Mm, *Mus musculus*.

that is repeated with 87% identity, as well as three copies of the ATTTA sequence associated with mRNA instability (*18, 21*).

In addition to the three different forms of exon 1 present in human ADAR1 transcripts that encode the IFN-inducible ~150-kDa protein (exon 1A-containing RNA) and the constitutive ~110-kDa protein (exon 1B and 1C-containing RNAs), at least two other exons have variant forms. Splice variant transcripts possessing alternative forms of exon 6, exon 7, or both exons 6 and 7 were found in cDNA libraries prepared from human kidney and human placenta (9). Exons 6 and 7 each exist in two alternative forms, a and b (9). Exon 6a is 191 nt, whereas exon 6b is 134 nt due to alternative splicing that deletes 57 nt from the 5'-end. Exon 7a is 226 nt, whereas exon 7b is 148 nt, again due to alternative splicing, which in this case deletes 78 nt from the 3'-end. The splice variant transcripts are identified as *form* a (with exons 6a and 7a), *form* b (6a and 7b), or *form* c (6b and 7b). The theoretically possible fourth variant carrying exon 6b with 7a (*form* d) has not yet been detected by reverse transcription-polymerase chain reaction (RT-PCR) analysis of RNA from cultured cells or tissues. A consequence of the use of 6b and 7b alternative exon forms would be a reduction of the linear spacing between the RNA-binding domains themselves or between the RNA-binding domains and the catalytic domain. This structure could potentially affect substrate recognition, editing efficiency, or other properties important for ADAR1 protein function.

The mouse *Adar1* gene structure and organization is similar to that of the human *ADAR1* gene (7, 9). The mouse *Adar1* gene also consists of 17 exons. As with human *ADAR1*, at least three alternate exon 1 forms (exon 1A, B, and C), three alternative promoters as well as two alternative exon 7 structures have been found in mouse cells and tissues (7, 22). Additional cDNAs have been reported for mouse ADAR1 (but have not yet been described for human ADAR1), including those in which both exon 1 and exon 2 are absent and the ORF begins from a translation initiation codon located in the middle of exon 3 (AUG521), and a cDNA splice variant that has exon 1A linked to exon 3, with the entire exon 2 deleted (23, 24). Alternative cDNA forms have been reported in mice induced for acute inflammation, where either exon 3 or exon 7 was excluded (24). This splicing pattern is curious, as the RNA-binding domains RI or RIII present in exons 3 and 7 would hence be lost, and RI and RIII have been found to be the most important for deaminase activity (25, 26).

The rat ADAR1 cDNA shows an ORF of 3582 nt (27), which is equivalent to the mouse and human ORFs for transcripts containing exon 1A. A comparison of the ADAR1 cDNA sequences of human, mouse, and rat shows extensive homology among them. Between mouse and human, and between rat and human, there is >85% identity at the nucleotide level; between mouse and rat, the identity is >90% at the nucleotide level.

Tissue and cell line differences are observed in the occurrence of alternative splice variants of exons 6 and 7, in both the mouse and human systems (7, 9, 22). Whereas a human kidney cDNA library yielded both 7a and b, as well as 6a and b, human placenta showed only the 6a form, with both 7a and b present. Mouse brain shows predominantly 7a, whereas mouse liver shows 7b. The presence of alternative exon 7b has been linked predominantly to IFN-inducible transcripts that contain exon 1A (p150 coding form), whereas exon 7a is found in transcripts that originate from the constitutively active promoter containing exon 1B (p110 coding form) (7). Transcripts possessing 7b, the short form of exon 7, are apparently most abundantly expressed in response to inflammation or interferon treatment (7, 23, 24).

3. Transcriptional Regulation

The ADAR1 cDNA was isolated in a screen for IFN-inducible genes (18). Northern gel blot hybridization analysis established that human ADAR1 cDNA probes derived from exons 2 to 15 hybridized to a single major IFN-inducible mRNA ∼6.7-kb in size (8, 18, 28). However, significant basal levels of a similarly sized RNA were seen in blots with RNA prepared from cells in the absence of IFN treatment. When northern analyses were carried out with exon 1A and exon 1B specific probes, both detected a single major RNA of about 6.7 kb with RNA from IFN-treated cells. The transcript detected with the exon 1A probe was IFN inducible, whereas the transcript detected with the exon 1B probe was abundant and present in comparable amounts in untreated and IFN-treated human U cells (17). Mouse ADAR1 cDNA probes likewise hybridized to a major IFN-inducible mRNA of comparable size, ∼6.5 kb (7, 22). The ∼6.5-kb RNA is abundantly expressed in the liver following *Salmonella* infection and in cultured mouse fibroblasts following IFN treatment. A smaller and less abundant transcript is also increased in liver following infection and in L cells by IFN treatment. Finally, RT-PCR analysis using primers specific for alternative exon 1A or exon 1B, paired with primers in exon 2, revealed that the amount of exon 1A–exon 2 product is low with RNA from untreated relative to IFN-treated mouse L or human U cells (7, 17). After IFN treatment, the exon 1A-containing product is increased several fold, whereas no increase in the abundant exon 1B–exon 2 product is seen following IFN treatment in either L or U cells (7, 17). Transcripts containing exon 1C, when quantitated using an RNase protection assay, also were found not to differ in their steady state level between untreated and IFN-treated cells (16).

The different exon 1-containing ADAR1 transcripts that are not readily resolved by northern gel blot analysis, but which are distinguishable by RT-PCR analysis, initiate from different promoters, one IFN inducible and the others constitutively active (7, 8, 17). The expression of the IFN-inducible exon 1A-containing transcript is driven by an IFN-inducible promoter that possesses an ISRE (7, 8, 17). The IFN-inducible promoters of the human and mouse *ADAR1* genes both possess a 12-bp ISRE (Fig. 1) that is conserved and in concurrence with the consensus nucleotide sequence for ISREs of other type I IFN-regulated genes (29). Also conserved in position within the IFN-inducible promoter is an element designated KCS-like (30, 31) because of homology with an element identified in the human and mouse *PKR* gene promoters designated *k*inase *c*onserved *s*equence (KCS) (32). While the KCS element of the *PKR* gene is required both for basal and IFN-inducible transcription, mutational analyses revealed that the ADAR1 KCS-like and the PKR KCS elements are functionally distinct from each other (30, 31). The exon 1B- and 1C-containing transcripts are regulated by promoters specific to each of them (16, 17). These *ADAR1* gene promoters lack ISRE, and they are not inducible by IFN. Rather, they are constitutively active and their transcripts encode the constitutively expressed p110 protein (7, 18). Except for the ISRE/KCS-like region, there is no significant sequence homology between the 5'-flanking regions of the human and mouse IFN-inducible promoters that drive expression of the exon 1A-containing transcripts.

ADAR1 is ubiquitously expressed. However, tissue-selective expression of the exon 1A and 1B-containing transcripts is observed in mice using RNA isolated from different organs, including brain, liver, spleen, Peyer's patches, kidney, heart, lung, and cecum (7, 22). Employing RT-PCR and using exon 1 specific primers, exon 1A-containing transcripts were detected in mouse organs such as liver, spleen, Peyer's patches, kidney, heart, lung, and cecum. The level of exon 1A transcripts increased two- to threefold at 6 days after infection with *Salmonella typhimurium* (7). Only low levels of the exon 1A-containing transcript were detectable in the brain, and no increase was seen even after infection. All tissues examined also possessed exon 1B-containing transcripts, but they were not found to increase despite infection. High levels of exon 1B-containing transcripts were detected in the mouse brain, but no further increase was seen after infection (7, 22).

B. ADAR1 Proteins

1. SEQUENCE AND FUNCTIONAL DOMAINS

Two different size forms of catalytically active ADAR1 protein are expressed from the *ADAR1* gene [Fig. 2; (18)]: an IFN-inducible p150 form and a constitutively expressed p110 form (Figs. 1 and 2). In addition to the single copy *ADAR1* gene, two more mammalian *ADAR* genes and proteins have been identified (Fig. 2): ADAR2, which is catalytically active and constitutively expressed; and, ADAR3, which is not yet known to possess deaminase activity. Computational analysis of the amino acid sequence deduced from the human ADAR1 cDNA revealed regions of homology with functional domains associated with deaminase catalytic activity, nucleic acid-binding activity, and subcellular localization (Fig. 2). Subsequent mutational and biochemical analyses established the functional significance of these domains. Exon 1A-containing transcripts encode an IFN-inducible protein of 1200 (human) or 1152 (mouse) amino acids, while transcripts containing either exons 1B or 1C encode a constitutively expressed protein of 931 (human) or 903 (mouse) amino acids (7, 17, 18). The deduced molecular weight from the cDNA sequences gives size values for the ADAR1 proteins somewhat smaller than those observed by sodium dodecyl sulfate-polyacrylamide gel electrophoresis (7, 18), for

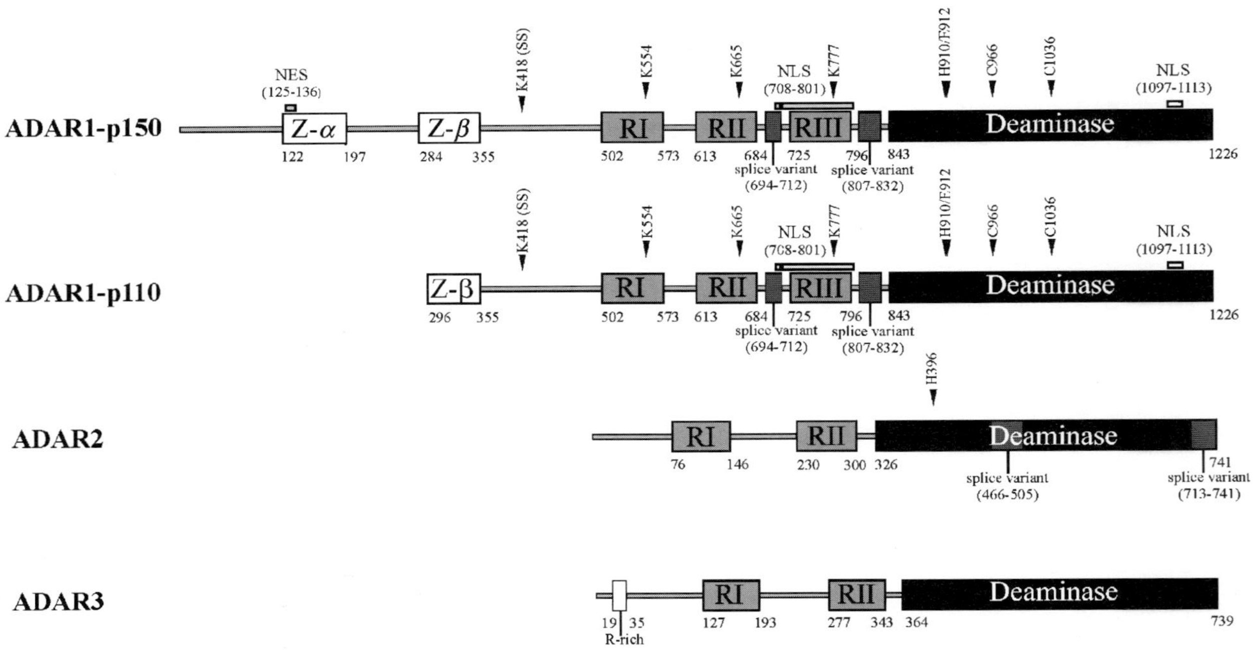

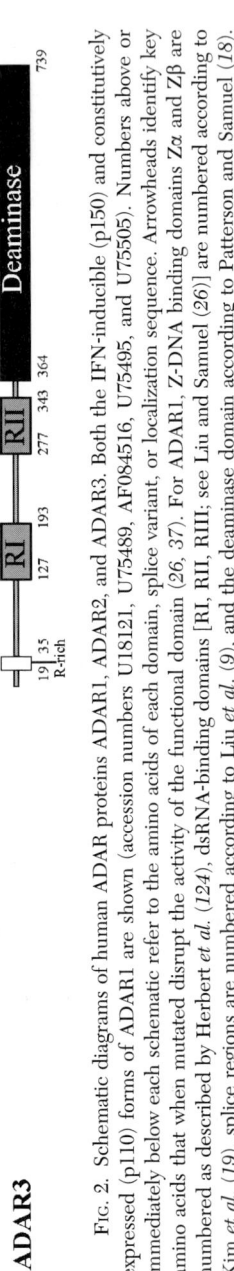

FIG. 2. Schematic diagrams of human ADAR proteins ADAR1, ADAR2, and ADAR3. Both the IFN-inducible (p150) and constitutively expressed (p110) forms of ADAR1 are shown (accession numbers U18121, U75489, AF084516, U75495, and U75505). Numbers above or immediately below each schematic refer to the amino acids of each domain, splice variant, or localization sequence. Arrowheads identify key amino acids that when mutated disrupt the activity of the functional domain (26, 37). For ADAR1, Z-DNA binding domains Zα and Zβ are numbered as described by Herbert et al. (124), dsRNA-binding domains [RI, RII, RIII; see Liu and Samuel (26)] are numbered according to Kim et al. (19), splice regions are numbered according to Liu et al. (9), and the deaminase domain according to Patterson and Samuel (18).

example, ~133 kDa for the human p150 protein and ~103 kDa for the p110 protein (18). Both mammalian and *Drosophila* ADARs appear to be dimers in the catalytically active form, but whether dimerization requires RNA may depend upon the specific enzyme in question (33–35). Some of the domains found within the N-terminally truncated constitutive p110 protein are the same as those of the inducible p150 protein; however, the p150 ADAR1 protein possesses an extended N-terminal region that includes additional functional domains (Zα, NES) that are absent in p110 (Fig. 2).

The catalytic domain of the ADAR1 deaminase proteins is located within the C-terminal region that shows high-sequence homology with known deaminases (Fig. 2). For ADAR1, the catalytic domain corresponding to the C-terminal ~380 amino acids was originally identified to have extensive homology with the *Caenorhabditis elegans* T20 H4.4, murine Tenr and *Saccharomyces cerevisiae HRA400* gene products (18), and most importantly with proteins known at the time to possess cytosine deaminase activity (18, 19, 27, 36–39). His910 and Glu912 (Fig. 2) are key amino acids in the CHAE motif of the catalytic core, which is invariably found in other deaminases. X-ray crystal structures of murine adenosine deaminase (40), *Escherichia coli* cytidine deaminase (41) and the related RNA-specific adenosine deaminase ADAR2 (42) have established that the histidine residue in the CHAE motif is involved in the binding of zinc and the glutamine residue is important in proton transfer. O-phenanthroline, a chelator of zinc ions, has been shown to inhibit the enzymatic activity of purified natural ADAR1 (19). Mutation of the highly conserved CHAE sequence of ADAR1 impairs catalytic activity (26, 37); substitution of His910 with Gln, and Glu912 with Ala, abolishes the A-to-I editing activity of ADAR1 with both synthetic and natural RNA substrates, providing functional evidence that ADAR1 deaminase requires these residues for catalytic activity (26, 37). Two additional amino acids Cys966 and Cys1036, which are believed to play a role in zinc coordination along with His910, were found indispensable for deaminase activity by mutagenesis (37). While inositol hexakisphosphate is bound in the ADAR2 catalytic core and required for RNA editing (42), there is no evidence for a similar cofactor requirement by the ADAR1 enzyme.

An N-terminal nuclear export (NES) signal (62), two nuclear localization (NLS) signals (59, 60), and a sumoylation site (SS) (55) are shown. A nucleolar localization signal (61) is represented by a small black box in the NLS in RIII. For ADAR2 (accession number U76420–22), the arginine-rich domains, RNA-binding domains, and the splice sites are numbered according to Lai *et al.* (384). For ADAR3 (accession number AF034837), which is not known to possess catalytic activity, the R-rich domain and the RNA-binding domains are numbered according to Chen *et al.* (385). Additional splice variants are known for both ADAR1 (9) and ADAR2 (16, 386–388). The boundaries of the catalytic domains of ADAR2 and ADAR3 were determined by alignment with the catalytic domain of ADAR1.

The human ADAR1 proteins, p150 and p110, both possess three copies of the dsRBDs (or Rs) located in the central region of the protein (Fig. 2). Comparison of cDNA and genomic sequences from human and mouse revealed that the coding regions for these three dsRBDs, designated as RI, RII, and RIII, are located in exons 3, 5, and 7, respectively (Fig. 1). dsRNA-binding activity of the human ADAR1 protein initially was demonstrated by poly (rI): poly (rC) Sepharose affinity chromatography and northwestern blot analysis (9, 18). Whereas ADAR1 proteins bound efficiently to poly (rI):poly (rC), they bound poorly to poly(rA) ssRNA. Comparison of the ADAR1 dsRBD sequences with each other and with those of other mammalian dsRNA-binding proteins including PKR showed striking conservation of the core sequence residues required for RNA-binding activity (18). A key lysine residue conserved in all three dsRBDs was shown by mutagenesis to be necessary for RNA-binding activity (26, 43, 44). Of the three dsRBDs in ADAR1, RI and RIII contributed greatly to dsRNA-binding activity, but RII was less essential (9, 26).

The importance of the dsRBDs in deaminase activity more or less correlate with their importance in dsRNA-binding activity. Mutation of the critical lysine residues in RI and especially RIII considerably reduced deaminase activity, whereas mutation in RII did not appreciably affect enzymatic activity (26). Similar conclusions were drawn from results obtained with mutants in which the dsRBD had been deleted (37). The functional difference of each of the three dsRBDs was further investigated by generating mutations in the R motifs of the various splice variants (a, b, and c forms) of human ADAR1. In all the six splice variants studied (three each for p150 and p110), RIII was identified as the most critically important for deaminase activity. RII-binding function was dispensable, and curiously, deaminase activity was increased modestly by rendering RII nonfunctional. Chimeric proteins in which the dsRBDs of ADAR1 are substituted with those of PKR retain significant RNA adenosine deaminase activity measured with synthetic dsRNA, but this substitution dramatically reduces site-specific editing activity in natural substrates (45). These results suggest a difference in RNA-binding selectivity between the dsRBDs of ADAR1 and PKR (45).

Computational analysis revealed that the N-terminal region of the p150 ADAR1 protein possesses a repeated region (amino acid positions 170–195 and 328–353) that displayed significant homology with the N-terminal region of the vaccinia virus E3L protein (31% identity, 58% similarity) (28). This repeated domain within ADAR1 is now known to specify two copies of a Z-DNA-binding domain (46), designated as Zα and Zβ (Fig. 2). Zα is sufficient to bind Z-DNA; Zβ does not bind Z-DNA as a separate entity (47). The helix-turn-helix motif, frequently used to recognize B-DNA, is used by Zα to contact Z-DNA (48). Recognition of Z-DNA by ADAR1 has been hypothesized to allow editing of nascent transcripts immediately after their synthesis, thereby helping to ensure that editing and splicing are performed in the correct temporal order (49). Finally, in the poxvirus E3L interferon resistance protein, the Z-DNA-binding domain plays a role in the pathogenesis of the virus in the mouse model, but the biochemical basis of the effect is not yet elucidated (50).

The presence of Z-DNA domains Zα and Zβ clearly is not an obligate requirement either for dsRNA-binding or for deaminase activity of ADAR1. Whereas the p150 ADAR1 protein has both these domains, the p110 protein retains only the Zβ domain; both proteins, however, possess robust dsRNA-binding activity and A-to-I deaminase activity (9, 18). The ADAR2 protein that has neither of the two Z-DNA-binding domains (Fig. 2) possesses significant deaminase activity. However, mutation of the Z-DNA-binding domains of ADAR1 decreases the efficiency of editing of short (15 bp) dsRNA substrates (51), and the *in vitro* ability of p150 to edit a 50-bp segment of dsRNA is markedly enhanced in presence of a sequence favoring Z-RNA (52). Using ADAR1 from which the Z-DNA-binding domains had been deleted, it was concluded that these sites are not required for chromosomal targeting of the ADAR1 protein (53).

Human ADAR1 is modified at Lys418 by sumoylation (Fig. 2), and this modification alters ADAR1 editing activity (54). An arginine substitution at position Lys418 abolishes SUMO-1 conjugation, and although the K418R mutation does not interfere with localization of ADAR1, it stimulates the ability of the enzyme to edit RNA both *in vivo* and *in vitro* (54). Furthermore, modification of wild-type recombinant ADAR1 by SUMO-1 reduces the editing activity of the enzyme *in vitro*. These results suggest a novel role for sumoylation in regulating RNA-editing activity.

2. SUBCELLULAR LOCALIZATION

Immunostaining and biochemical fractionation studies initially revealed that ADAR1 is primarily a nuclear protein (27, 55–57). Subsequent immunofluorescence microscopy and Western immunoblot analyses demonstrated that the constitutively expressed p110 form of ADAR1 was localized predominantly if not exclusively to the nucleus, whereas the IFN-inducible p150 protein isoform of ADAR1 was localized to both the cytoplasm and the nucleus (18). Bipartite NLSs within the N-terminus of mammalian (19, 27) and *Xenopus* (58) ADAR proteins were predicted from cDNA sequence data, however, more extensive studies revealed that these putative NLSs were biologically inactive (59). ADAR1 is both a nuclear and cytoplasmic protein and as described subsequently, its subcellular localization is determined by multiple nuclear localization and export signal sequences (Fig. 2).

At least two, and possibly three, functional NLSs exist in mammalian ADAR1 proteins. An NLS resides within RIII of human ADAR1 (59, 60), but despite its location in RIII, the nuclear localization activity does not depend on dsRNA-binding (59). In fact, RNA-binding involving RI and RIII might actually mask the localization activity of the NLS within RIII (60). Another NLS was reported in the C-terminal 39 amino acid residues of mouse ADAR1 (61), and an N-terminal NLS might also exist in the first 269 amino acids of ADAR1 (62). Some evidence suggests that ADAR1 specifically localizes to the nucleolus (18, 24, 63). A nucleolar localization signal consisting of a monopartite cluster of basic residues has been identified within the NLS in RIII (61). Nucleolar localization of ADARs may imply that RNA editing occurs cotranscriptionally. LMB, a drug that inhibits the nuclear export protein Crm1 (64, 65), stimulates nuclear accumulation of ADAR1, and ADAR1 has been shown to be a shuttling protein that moves between the cytoplasm and the nucleus (59, 62). Nuclear export of ADAR1 is mediated by a Rev-like NES located in the Zα domain, and this region of ADAR1 specifically interacts with Crm1 and RanGTP *in vitro* (62). The NES is present in p150, but not in p110, ADAR1 (Fig. 2), thus providing an explanation for the differential subcellular localization seen for these two ADAR1 isoforms.

Shuttling of ADAR1 between the cytoplasm and the nucleus may affect which dsRNA substrates are edited. Editing of GluR-B RNA at the R/G site (see Sections II.B.3 and II.C.3) was efficiently catalyzed by N-terminally truncated p150 proteins or p150 proteins with mutations in the NES that accumulated in the nucleus, but the GluR-B substrate was barely edited by wild-type ADAR1 that was localized primarily in the cytoplasm (62). Presumably, in the nucleus one of the key biological functions of ADAR1 may be to edit cellular pre-mRNAs, whereas in the cytoplasm ADAR1 may function to edit other substrates, such as viral RNAs. However, ADAR1 may conceivably play roles within the nucleus that are dependent upon protein–protein interactions rather than editing activity. For example, members of the conserved Vigilin class of proteins have a high affinity for I-containing RNAs. Vigilins localize to heterochromatin and nuclear Vigilin is found in complexes containing not only ADAR1 but also RNA helicase A and Ku86/70 (66). A novel role for ADAR1 in the regulation of gene expression has been reported, whereby ADAR1 interacts with nuclear factor 90 (NF90) proteins, known regulators that bind the antigen response recognition element. ADAR1 regulation of NF90-mediated gene expression occurs independently of RNA editing activity (67).

3. SUBSTRATES

A-to-I editing catalyzed by ADARs is of broad biologic significance. The editing may occur selectively, at one or a few positions, or more frequently at a large number of sites (hyperediting or hypermutation) in a transcript with significant double-stranded character. Among the best-characterized cellular and viral RNAs that undergo A-to-I editing with high selectivity are transcripts encoding glutamate receptor (GluR) channels, the serotonin-2C receptor (5-HT$_{2C}$R), and hepatitis delta virus (HDV) RNA. In addition, A-to-I(G) hypermutation has also been observed for several negative-stranded RNA animal virus genomes during lytic and permissive infection and for transcripts of the polyoma virus, a small circular double-stranded DNA virus.

Measles virus represents the first negative-stranded RNA virus in which A-to-I(G) hypermutation was observed (68, 69), but parainfluenza (70) and vesicular stomatitis (71) viruses likewise are implicated as targets of ADAR1 deaminase activity based on observed sequence changes. These viral RNAs represent hyperedited substrates that are examples of nonspecific editing at a large number of sites. In the case of measles virus, the mutations result in an extensively modified matrix (M) mRNA in the brain with many clustered A-to-I(G) mutations, thereby preventing synthesis of functional M protein. Measles virus M protein is required for virion assembly and release (72) and in the absence of a functional M protein, persistent measles infection may occur in the brain, causing a rare but often fatal necropathic response known as subacute sclerosing panencephalitis (SSPE) (69, 73). When the measles virus from patients was characterized, biased hypermutations were found in the M gene and, albeit less frequently, in other measles virus genes (74; 75). Biased hypermutations were also observed in human parainfluenza virus 3 (HPIV3) RNA isolated from persistently infected LLC-MK2 cells in culture. These A-to-I mutations are seen toward the 3'-end of the HPIV viral RNA along with U-to-C transitions (70). Editing also is reported in transcripts of polyoma virus (76), in which extensive A-to-I modification is seen on viral antisense RNA produced late in viral infection. Adenovirus encodes a Pol III gene product, VA RNA, that antagonizes ADAR activity (77). ADAR is reported to catalyze the A-to-I editing of HCV replicon RNA, facilitating its clearance from infected cells (see Section II.C.1) (78). The editing of HCV RNA was impaired by adenovirus VA RNA.

The editing of HDV RNA is an example of editing of a viral RNA that occurs with high selectivity and at a single specific site. HDV [reviewed in (79, 80)] is a subviral pathogen that depends on a helper virus, HBV, for replication. HDV has three components: a circular RNA genome that is highly structured, the HBV surface antigen that is supplied by the HBV helper virus, and hepatitis delta antigen (HDAg). There are two forms of the HDAg. One, HDAg-S, is a short 195-amino acid protein, which is necessary for viral RNA replication (81), and the other, HDAg-L, is a longer 214-amino acid protein that mediates genome packaging (82). RNA editing of the HDV antigenome at a site called the "amber/W" changes a UAG amber stop codon to a UIG tryptophan (W) codon, allowing for the synthesis of the HDAg-L (83). Without editing, only HDAg-S is made. HDAg-L can act as a potent trans-dominant inhibitor of viral RNA replication; HDAg-S forms a homodimer, and HDAg-L, which has 19 extra N-terminal amino acids, heterodimerizes with HDAg-S and acts in a dominant negative manner, causing a decrease in viral RNA replication (84, 85). The conserved RNA secondary structure around the HDV genotype I amber/W site has been selected not for the highest-editing efficiency but for optimal viral replication and secretion (86). Two groups have independently shown that the constitutively expressed p110 form of ADAR1 is the deaminase that is primarily responsible for editing of the amber/W site (87, 88). One study further indicates that specifically the exon 7a splice variant of p110 is the primary ADAR protein that edits HDV RNA (88).

RNA editing of the neurotransmitter receptor transcript substrates is specified by imperfect duplex RNA structures formed between adjacent exonic and intronic sequences, and the editing leads to selective amino acid substitutions in the encoded receptor proteins that alter either conductance properties in the case of the GluR ion channels or G-protein coupling function in the case of the 5-HT$_{2C}$ receptor (89–91) (see Section II.C.3). Sixteen previously unknown ADAR target genes involved in rapid electrical and chemical neurotransmission in *Drosophila* and one (potassium channel Kv1.1) in humans were identified and experimentally verified (92). Many of the edited sites recode conserved and functionally important amino acids. Active ADAR1 and ADAR2 proteins are found associated with spliceosomal Sm and SR proteins within 200S lnRNP complexes that constitute the natural pre-mRNA processing machinery (93). Such a presence of deaminase activity in lnRNP complexes is intriguing because the A-to-I editing of certain substrates, such as GluR-B and 5-HT$_{2C}$R RNAs, must occur prior to splicing. The critical duplex dsRNA structure that confers editing selectivity at the designated exonic site requires downstream inverted complementary intronic sequence (94–98).

Additional cellular RNAs implicated to undergo A-to-I editing include hepatic α2,6-sialyltransferase transcript (99), oxytocin receptor (100), and endothelin B receptor (101), but the specific ADAR enzyme(s) responsible for the editing events largely remain uncertain. In the case of hepatic α2,6-sialyltransferase, two transcripts were identified that differed in a single codon for a residue within the catalytic domain; the two isoforms of the enzyme displayed different processing and turnover rates (99). Comparison of cDNA sequences for the oxytocin receptor in sheep indicated different receptor populations in ovine endometrium with changes at four amino acid positions as well as the 3'-UTR. Editing in ovine endometrium is proposed to generate the different isoforms of the receptor that account for differences observed in response to an oxytocin challenge and consequent prostaglandin release in the uterus (100). A-to-I editing also is implicated for human endothelin B receptor transcripts, resulting in a change from Q to R at one position in the protein (101).

The endothelin B gene has been shown to be involved in causing Hirschprung disease, a heterogeneous genetic disorder (99).

While most cellular substrates of selective A-to-I editing by ADAR, including the well characterized GluR and 5-HT$_{2C}$R substrates, were identified serendipitously, several thousand candidate A-to-I editing sites were discovered by a computational search of the human transcriptome using large numbers of expressed sequences (102–105). These newly identified sites, together with a more limited number found using a cloning strategy for I-containing RNAs in C. elegans (106), occur primarily in noncoding regions of RNA and often in Alu repeats (102–105). Using a more sensitive bioinformatics approach, a novel multiply edited transcript, BC10, was identified and A-to-I editing verified experimentally (107). BC10 is highly conserved across a range of metazoa and has been implicated in renal and bladder cancer (107). Finally, using comparative genomics and expressed sequence analysis, four human substrates of ADAR-mediated editing were identified and experimentally verified: FLNA, BLCAP, CYFIP2, and IGFBP7 (102). Editing of three of these substrates was experimentally verified in mouse and two in chicken. Although none of these substrates coded for a receptor protein, two of them were strongly expressed in the CNS and appeared necessary for proper nervous system function.

4. ANTAGONISTS

Two viral gene products—the adenovirus VAI RNA and the vaccinia virus E3L protein—have been shown to antagonize ADAR1 (77, 108). Both of these gene products help mediate IFN resistance, and both are antagonists of the IFN-inducible RNA-dependent protein kinase PKR, which like ADAR1, contains more than one copy of the dsRBD (44, 109) (see Section III.B.1).

Adenovirus VAI RNA is a short (~160 bp), highly structured RNA that is needed for efficient translation of cellular and viral mRNAs at late times after infection (110). The structural features of VAI RNA include two long, imperfectly base-paired stem regions, the terminal stem and the apical stem-loop, which are linked by a central domain (111, 112). VAI RNA is found abundantly in the cytoplasm of cells during infection, and it mediates IFN-resistance in adenovirus infected cells (113). When added to extracts from IFN-treated U cells or extracts from monkey COS-1 cells transfected with wild-type ADAR1 cDNAs, VAI RNA inhibited the deamination of a synthetic substrate in a dose-dependent manner, and both the p110 and p150 forms of ADAR1 were antagonized by VAI RNA (77). Analysis of VAI RNA mutants revealed that the central domain of the RNA molecule was important for inhibition of both ADAR and PKR. Although the interactions of VAI with ADAR1 and PKR were similar, they were not identical. VAI RNA antagonizes PKR activity by binding to its dsRBDs and preventing subsequent autophosphorylation and activation of the kinase (*110, 113–116*). The mechanism by which VAI RNA antagonizes ADAR1 is not yet known, but it has been suggested that VAI RNA competes with the binding of substrate RNAs or affects the conformation of ADAR1 in such a way that deamination of bound substrates is impaired (*77*).

Vaccinia virus E3L protein is a 190-amino acid protein that is expressed early during infection. E3L has an established role in viral resistance to IFN and pathogenesis in the mouse model (*117–119*), and it has the ability to antagonize two IFN-inducible enzymes: PKR, the RNA-dependent kinase that modulates the activity of translation factor eIF-2α (*120, 121*) and OAS, the family of RNA-dependent oligoadenylate synthetases that produce 2′5′-oligoadenylate activators of RNase L (*122, 123*). Like ADAR1, the E3L protein possesses Z-DNA-binding (*18, 34*) and dsRNA binding (*18, 121, 125, 126*) domains and is localized to both the cytoplasm and the nucleus. Nuclear localization is dependent on the N-terminal region of the protein (*126, 127*). In vitro studies using synthetic dsRNA substrates have demonstrated that E3L inhibits the deaminase activity of both forms of ADAR1, p150 and p110 (*108*). Inhibition of ADAR1 required the presence of the C-terminus and a functional dsRBD of E3L, suggesting that one potential mode of ADAR1 inhibition by E3L is to bind and sequester dsRNA substrates. However, reovirus σ3 protein and E3L mutants that retained dsRNA-binding activity did not inhibit ADAR1 activity, indicating that the RNA-binding activity alone of these proteins is not sufficient for antagonism. Mutant E3L proteins with amino acid substitutions in the Z-DNA-binding domain or with deletion of only seven C-terminal amino acids do not antagonize ADAR1, although an N-terminally truncated E3L protein that lacked the Z-DNA-binding domain did impair ADAR1 activity. A specific protein conformation, in addition to dsRNA-binding activity, presumably is required for E3L antagonism of ADAR1. However, no physical interaction between E3L and ADAR1 has yet been demonstrated.

Given their ADAR1 antagonist activities, VAI RNA and E3L protein could have broad effects on replication and pathogenesis of adenoviruses and pox viruses. These gene products could inhibit A-to-I conversion in viral RNA transcripts performed by IFN-inducible ADAR1, thereby preventing degradation of I-containing species by a ribonuclease specific to such RNAs (*128*). By antagonizing ADAR1, these gene products could also alter editing of cellular mRNA transcripts in a manner that changes the virus–host interaction. It is known that VAI RNA antagonizes the antiviral action of IFN, and that mutant adenovirus that does not express VAI RNA shows some IFN sensitivity (*113*). It is also known that vaccinia virus mutants in which E3L is mutated, either in the Z-DNA or dsRNA-binding domain, display altered pathogenicity in the mouse model (*50, 129*). While the existence of viral gene products that antagonize ADAR1 activity provides support for the notion that ADAR1 is an antiviral enzyme, it is still unclear whether the VA and E3L viral gene products directly inhibit ADAR1 during the course of viral infection. Possibly, experiments comparing the replication of wild-type viruses and mutant viruses lacking genes encoding the ADAR1 antagonists in cells that overexpress ADAR1 or are deleted of ADAR1 will help to answer this question.

Finally, in contrast to the VAI and E3L antagonists of ADAR1, there are no known activators or cofactor requirements for ADAR1 deaminase activity. By contrast, inositol hexakisphosphate is bound by ADAR2 and is required for ADAR2 A-to-I RNA editing activity (42).

C. Biological Activities of ADAR1

1. Antiviral Activity

Although A-to-I editing characteristic of ADAR proteins has been found in several viral RNAs (see Section B.3), the precise role of ADAR proteins in virus infection, with few exceptions, is unclear. Because of the IFN-inducible nature of ADAR1, it seems likely that a function of this particular deaminase might be to defend against virus infection; however, evidence of an antiviral activity for ADAR1 is scarce (130). Nevertheless, ADAR1 antiviral activity has been demonstrated for two hepatitis viruses (Table I): HDV and HCV.

ADAR1 promotes HDV virus replication by allowing synthesis of HDAg-L, which stops viral RNA replication and enables viral packaging (see Section II.B.3) (79, 80). Without editing, packaging of HDV would not occur; ADAR1, therefore, is necessary for virus replication. However, high levels of ADAR1 protein have been shown to inhibit HDV replication: overexpression of ADAR1 in Huh-7 cells lead to increased levels of HDV RNA editing and a decrease in HDV production (131). The decrease in virus production was primarily due to an early increase in HDAg-L, but also to the production

TABLE I
Antiviral Activities of PKR and ADAR

Gene product	Virus	Reference
PKR	Encephalomyocarditis virus (EMCV)	(159)
	Hepatitis B virus (HBV)	(389)
	Herpes simplex virus (HSV)	(284, 390)
	Human immunodeficiency virus (HIV)	(277)
	Vaccinia virus (VV)	(279)
	Vesicular stomatitis virus (VSV)	(280)
ADAR	Hepatitis C virus (HCV)	(78)
	Hepatitis delta virus (HDV)	(131–133)

of other HDAg forms that arose from hypermutation at nonamber/W sites and, like HDAg-L, acted as *trans*-dominant inhibitors of HDAg-S. Increased ADAR editing activity has also been shown to decrease HDV replication. A replication-competent mutant HDV virus with enhanced editing at the amber/W site was shown to express elevated levels of HDAg-L early in infection, and replication of this mutant virus terminated prematurely (*132*). These studies lend evidence for an antiviral activity of ADAR1 during HDV replication (*131, 132*).

Anti-HDV activity of ADAR1 could occur during a natural infection if levels of the protein were increased as a result of IFN production or IFN therapy. Treatment of Huh-7 cells and human hepatocytes with IFN-α lead to increased expression of the p150 form of ADAR1, and in the case of Huh-7 cells, increased HDV amber/W editing at 14 days posttransfection with HDV RNA (*133*).

ADAR1 has also been reported to play a role in limiting HCV replication (*78*). IFN-α treatment of Huh-7 cells stably expressing the HCV replicon lead to a decrease in HCV RNA and to an increase in editing of replicon RNA. Expression of VA RNA from adenovirus, which is an inhibitor of ADAR1 (*77*, see Section II.B.4), lead to a decrease in inosine containing RNA in replicon-expressing cells. Small interfering RNA (siRNAs) directed against ADAR1 partially alleviated the IFN-induced decrease in HCV RNA replication, indicating that ADAR1, in addition to PKR, play a role in IFN-mediated antiviral activity in HCV infection (*78*).

2. Apoptosis and Embryogenesis

Gene disruption studies have revealed an important role for ADAR1 in cell survival, especially during embryonic development in mammals. These knockout studies have revealed that disruption of the mouse *Adar1* gene is embryonic lethal and that *Adar1* is necessary for survival of hematopoietic and liver tissue during embryogenesis (*13–15*).

One *Adar1* knockout attempt involved the replacement of exons 12–13 of the *Adar1* gene with PGK-neo (*13*). Chimeric mice with a high contribution of embryonic stem cells that were *Adar1* null died before E14.5, and only a small number of embryos even survived to E12.5 and E13.5. Analysis of chimeric embryos revealed defects in the hematopoietic system (*13*). A second targeted knockout involving disruption of exons 12–15 was subsequently reported, in which heterozygous null mice were generated that were normal (*15*). Crossing of the heterozygotes, however, did not yield any viable $Adar1^{-/-}$ mice. Analysis of $Adar1^{-/-}$ dead embryos revealed high amounts of apoptosis in many tissues, especially the liver. In addition, MEF cells derived from the $Adar1^{-/-}$ mouse were prone to serum-deprivation–induced apoptosis (*15*).

Independent efforts to generate an *Adar1* knockout mouse involved two targeting strategies (*14*), one in which the *Adar1* gene was first truncated by deletion of exons 7–9 and the other that involved replacement of almost the entire gene (exons 2–13) with PGK-neo. Chimeric mice survived to adulthood and were bred to make $Adar1^{+/-}$ mice. These heterozygous mice were normal. However, no viable homozygous $Adar1^{-/-}$ mice could be made by crossing the $Adar1$ heterozygous mice. Analysis of embryos revealed an embryonic lethal phenotype at around E11.5-E12.5 (14). There is general agreement that the ADAR1 null embryos have defects in hematopoiesis in the liver (14, 15).

RT-PCR and Western immunoblot analyses have revealed that the constitutively expressed exon 1B-containing transcript and the corresponding p110 ADAR1 protein are both detectable at E10, whereas the IFN-inducible exon 1A-containing transcript and corresponding p150 protein are not detectable at this stage in embryonic development (7). These findings suggest that p110, but not p150, is important in embryogenesis. Possibly selective disruption of either the IFN-inducible P_A promoter and the corresponding exon 1A region, or the constitutive P_B and P_C promoters and corresponding exons 1B and 1C regions, might provide important information on the form of ADAR1 required during embryogenesis. Conceivably mice homozygous null for one form of ADAR1, but not the other might be viable through adulthood, thus enabling further study of the role of ADAR1 in the mouse model.

The phenotype of $Jak1^{-/-}$ mice (134) has some similarity to $Adar1^{-/-}$ mice. Although $Jak1^{-/-}$ mice survive to birth, they are runted, do not nurse, and die perinatally. $Jak1^{-/-}$ mice are also unresponsive to cytokines that bind class II cytokine receptors, including IFN-α, IFN-γ, and IL-7. As a result of their unresponsiveness to IL-7, these mice have defects in hematopoiesis, but unlike $Adar1^{-/-}$ mice, the defects in hematopoiesis are limited to the lymphoid compartment.

ADAR1's role in cell survival during development is not complemented by ADAR2 (13–15), and ADAR2 is not required for embryogenesis (135). $Adar2^{-/-}$ mice, although prone to seizures and shorterlived than wild-type mice, survive to adulthood (135). The most important role of ADAR2 in mice appears to be editing of GluR-B at the Q/R site (see Sections II.B.3 and II.C.3); the behavioral dysfunctions of $Adar2^{-/-}$ mice are rescued by replacing each wild-type GluR-B allele with a GluR-B allele encoding an edited RNA (135). The role of ADAR in embryonic development seems to be specific to mammals; deletion of the single ADAR allele in $Drosophila$ (136) or the adr-1 and adr-2 genes in C. elegans yields much less severe phenotypes (137). The mutant worms and flies are morphologically normal, although they display behavioral defects (136, 137).

3. Neurotransmitter Receptors

As demonstrated by ADAR knockout studies in mice, flies, and worms (135–137), RNA editing is important for proper functioning of the nervous system, and some of the best-characterized ADAR substrates are neurotransmitter receptor and channel mRNAs (see Section II.B.3). A-to-I editing of neurotransmitter receptor RNAs leads to amino acid substitutions that affect neurophysiology. The biologic effects of RNA editing on the AMPA glutamate receptor, serotonin 5-HT$_{2C}$ receptor, and the Kv1.1 potassium ion channel RNAs have been extensively characterized.

Ionotropic GluRs [reviewed in (89)] play a crucial role in the mediation of excitatory synaptic neurotransmission. In mammals, four main classes of ionotropic glutamate receptors (the AMPA, NMDA, kainate, and delta receptors) exist, which are subdivided into categories based on pharmacologic and structural properties. Each receptor is a multimeric assembly of four or five receptor subunits. There are four AMPA receptor subunits (GluR-A–D, also called GluR1–4), seven NMDA receptor subunits (NR1, NR2A-D, NR3, and NR3B), five kainate receptor subunits (KA-1, KA-2, and GluR5–7), and two delta receptor subunits. All GluR receptor subunits share a similar structure: an extracellular N-terminus, four hydrophobic transmembrane segments (M1–M4), and an intracellular C-terminus (89).

The glutamate receptors are a diverse group of receptors, and heterogeneity is achieved through a variety of mechanisms including multiple genes and alternative splicing. In addition, RNA editing by ADAR proteins also contributes to the variety of AMPA and kainate glutamate receptors (see Section II.B.3). Q/R editing of GluR RNA, which is performed primarily if not exclusively by ADAR2 (13, 15, 135), affects the calcium permeability of GluR channels: channels composed of the edited subunits of GluR-B [called "GluR-B(R)"] display a low-divalent permeability, whereas channels composed of unedited subunits of GluR-B [called "GluR-B(Q)"] exhibit a high-divalent permeability (138). In addition, Q/R editing also has been reported to affect assembly and subunit stoichiometry of AMPA GluRs; edited R subunits are unassembled and retained in the ER, whereas unedited Q subunits readily tetramerize and traffic to synapses (139). Editing at the R/G site, which is performed by both ADAR1 and ADAR2 (13, 15, 140), leads to faster recovery from receptor desensitization (94). The function of A-to-I editing at intronic "hotspots" in GluR pre-mRNA (98, 141) is currently unknown. The IFN-inducible ADAR1 enzyme efficiently edits the +60 intron 11 hotspot (140).

The 5-HT$_2$ serotonin receptors are a family of seven-transmembrane G-protein–coupled receptors, linked to phospholipase C with the production of inositol phosphates and diacylglycerol. In humans and rodents, the 5-HT$_2$ receptor family contains three receptor subtypes: 5-HT$_{2A}$R, 5-HT$_{2B}$R, and 5-HT$_{2C}$R (142). Transcripts encoding the 5-HT$_{2C}$R undergo site-selective editing at five different positions, termed A, B, C, D (95), and E (96, 143, 144). Editing at the five sites in the 5-HT$_{2C}$R RNA leads to changes in amino acids at three positions (amino acids 157, 159, and 161 in rat; 156, 158, and 160 in human) of the second extracellular loop of these seven-transmembrane receptors, and up to 14 different 5-HT$_{2C}$R protein isoforms can result from different combinations of edited or unedited adenosines at the five sites (*143*). Isoform expression patterns vary greatly in different regions of the brain (*25, 95, 96*) and between rats and humans (*144*), suggesting that the different isoforms have distinct biological functions. The edited VSV isoform (with valine, serine, and valine at positions 157, 159, and 161, respectively, generated by editing at sites ABCD) has a 10–15-fold less agonist potency than the unedited INI isoform (with isoleucine, asparagine, and isoleucine at positions 157, 159, and 161, respectively) (*95*). The difference in agonist potency between these two isoforms is caused by both decreased R–G coupling (*95*) and decreased agonist affinity (*143*). In addition to affecting agonist potency, editing of 5-HT$_{2C}$R RNA also affects constitutive G-protein coupling (*144*), defined as the activity in the absence of agonist (*145*). The VGV isoform, which is much more abundant in human brain than rat brain due to differences in E site editing between the two species, couples less efficiently to G-proteins than the unedited INI isoform, and although these two isoforms have the same maximal agonist activity, the basal level of activity of the VGV isoform was only 20% of the INI isoform (*144*). By silencing constitutive activity, efficient editing of the 5-HT$_{2C}$ receptor may function to reduce the "signal to noise ratio" at serotonergic synapses (*144*).

Studies have reported that editing of Kv1.1 RNA changes an isoleucine codon to a valine codon (*92*), and the resulting amino acid substitution leads to a channel with rapid recovery from fast inactivation (*146*). Cotransfection of Kv1.1 RNA with ADAR1 or ADAR2 expression constructs revealed that ADAR2 is the deaminase primarily responsible for editing Kv1.1 RNA. A role for mammalian ADAR1 in editing of potassium channels has not yet been identified. However, because both ADAR1 and ADAR2 edit at specific and distinct sites in GluR and 5-HT$_{2C}$R RNAs, it is conceivable that ADAR1 selectively edits other neurotransmitter RNAs at sites that cannot be edited by ADAR2.

4. Gene Silencing by RNA Interference

ADAR1 is implicated to play a role in RNA interference (RNAi), the posttranscriptional gene silencing pathway triggered by dsRNA [reviewed in (*147*)]. Gene silencing by RNAi occurs when long dsRNAs are cleaved into small interfering RNAs (siRNAs) by the enzyme Dicer. These siRNAs are then incorporated into the enzyme complex RISC (RNAi-induced silencing complex), which recognizes and cleaves messenger RNAs with homology to the siRNAs, thus inhibiting translation. Because RNAi depends on the presence of dsRNA, and because ADAR proteins are capable of binding to dsRNA and introducing multiple I-U mismatches [and ADARs were discovered as a dsRNA unwinding activity (55, 148)], it is possible that ADAR proteins interfere with the RNAi pathway. Some results obtained from cultured MEF cells and *C. elegans* are consistent with this notion (149–151).

In mammals, ADAR proteins have been demonstrated to limit the efficacy of RNAi. ADAR1 and ADAR2 were shown to bind to siRNAs, with the cytoplasmic p150 form of ADAR1 having the highest affinity (149). Short siRNAs (<36 bp) were not edited but still bound with high affinity. RNAi was significantly more potent in MEF cells deficient in ADAR1. Furthermore, overexpression of wild-type ADAR1 suppressed the RNAi effect, whereas overexpression of ADAR1 with three deleted RNA-binding domains did not. In addition, high doses of siRNAs have been found to induce mouse-enhanced RNAi gene and *Adar1* gene expression and reduce the efficiency of RNAi in cell culture and mice challenged by different doses of siRNAs (151). ADAR interferes with RNAi in cultured mammalian cells and perhaps whole animals, but the ADAR antagonism appears primarily to be the consequence of RNA binding rather than editing of the siRNAs (149).

In ADAR deficient worms, but not in wild-type worms, transgenes were able to induce gene silencing in somatic tissue through an RNAi pathway (150). In the wild-type worms, dsRNA was generated from the transgenes but was found to be edited, and it was proposed that this editing prevented initiation of the RNAi process (150). In worms, ADAR proteins appear to antagonize the RNAi silencing pathway through their editing activity. A chemotaxis dysfunction in worms deficient in ADAR proteins (137) is rescued by crossing ADAR deficient worms with worms that are defective in the RNAi response (152). A possible explanation for the rescue is that ADAR proteins edit RNAs needed for proper chemotaxis, and editing prevents these RNAs from entering the RNAi pathway. If not edited (i.e., in ADAR deficient worms), these genes are silenced; but if not edited and no RNAi occurs (i.e., in ADAR deficient and RNAi deficient worms), then these genes are not silenced and normal behavior results.

III. Protein Kinase Regulated by RNA

A. *PKR* Gene and Transcripts

1. GENE MULTIPLICITY AND CHROMOSOME ASSIGNMENT

Chromosome localization by FISH mapped the human *PKR* gene to a single locus, chromosome 2 band p21–22, and the mouse *Pkr* gene to chromosome 17 band E2 (153, 154). Comparison of nucleotide sequences determined for genomic and cDNA clones together with earlier Southern blot hybridization analyses is consistent with a single *Pkr* gene both in the mouse and human

(154–156). Two independently established lines of mice in which the *Pkr* gene has been disrupted are both viable. Mutant knockout mice with a targeted disruption of the N-terminal region of the protein that includes the AUG translation start and RNA-binding domain-RI (157) and mice with a disruption of the C-terminal catalytic domain (158) have been generated. Neither knockout genetically ablated for functional PKR is lethal, although the two $Pkr^{-/-}$ disruptions do show some phenotypic differences (157, 158). Although no consequences of loss of PKR on tumor suppression have been observed in either disruption, differences in antiviral responses, cytokine signaling responses in the interferon pathway, and tumor necrosis factor alpha (TNF-α)-induced apoptosis and the antiviral apoptosis in response to influenza virus are observed (130, 157, 158).

2. EXON–INTRON ORGANIZATION

As shown in Fig. 3, the human *PKR* gene spans about ∼50 kbp and includes 17 exons (32, 154). The mouse *Pkr* gene is smaller, about ∼30 kbp, although the exon–intron organization is highly conserved between the mouse and human genes (154, 155). Exons of the human *PKR* gene range in size from as small as 18 bp to as large as 840 bp, whereas the intron sizes vary from 78 bp to approximately 9.5 kbp (154). The difference in size between the mouse and human genes primarily is due to the larger size of some of the introns in the human gene (154). The cDNA for human PKR specifies a single long ORF (159). The translational initiation AUG codon for the predicted 551-amino acid PKR protein from human cells is present in exon 3 and the UAG termination codon for the ORF is found in exon 17, the 3′-most exon that is also the largest exon. The two copies of the dsRBD, designated RI and RII, are located in exons 4 and 6 within the N-terminal region of the protein (116). *Alu*-type DNA repeat elements have been identified in the 5′-flanking region upstream of exon 1 and also in the long 3′-proximal exon 17 that specifies the 3′-UTR of the PKR transcript (154, 155). The 5′-UTR of PKR mRNA is composed of exons 1 and 2 and a part of exon 3; exon 2 occurs in three alternative splice variants designated as α, β, and γ (160).

The mouse *Pkr* gene organization is similar to that of the human gene. Mouse *Pkr* consists of 16 exons and spans ∼30 kbp region on chromosome 17, band E2 (155). The exon sizes range from 35 bp to 750 bp, while the introns vary in size between 80 bp to ∼53 kbp (155). Exon 2 includes the translational start codon AUG that codes for the 515-amino acid PKR protein. The translational termination codon UAG is present in exon 16.

For the human *PKR* gene, two IFN-inducible transcripts of ∼2.5 and ∼6 kb are observed in cultured cells (161), whereas for the mouse ∼2.4-, ∼4.5-, and ∼6-kb transcripts are detected in cultured cells and tissues (22, 162, 163) by northern gel blot analysis. The sequences that have been reported

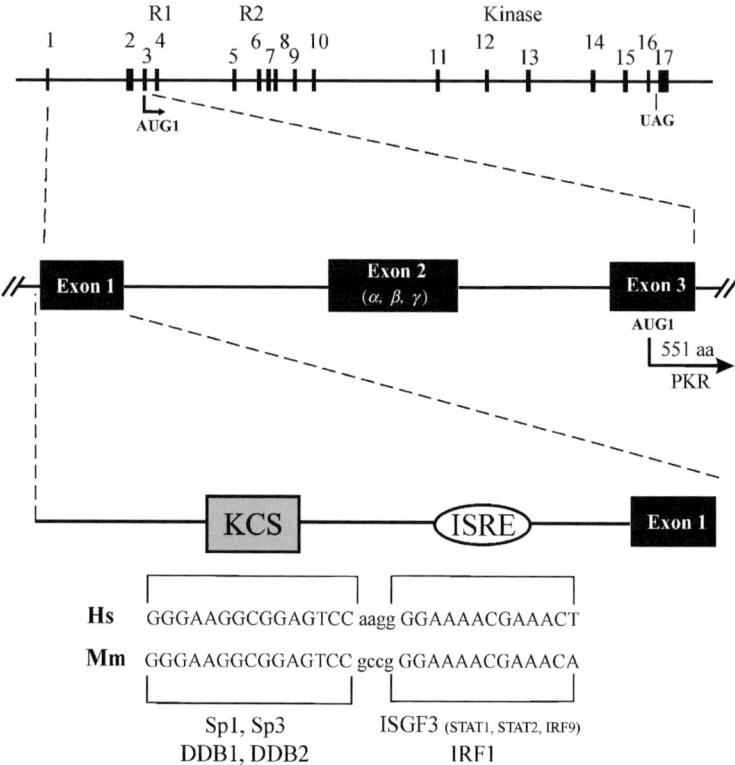

FIG. 3. Organization of the human *PKR* gene structure. Organization of introns and exons with regard to the structure of the *PKR* gene is represented in the top part of the figure. Exons are indicated by the numbers 1–17 and are represented by the filled boxes; introns and the 5'- and 3'-flanking regions are shown by the solid lines. The entire gene spans about ~50 kbp and has 17 exons that are present in the mature mRNA; there are three alternatively spliced forms of exon 2, 2α, 2β, and 2γ. The translation initiation site of the PKR protein (551 amino acids) is present in exon 3. The lower diagram depicts an enlargement of the IFN-inducible promoter region for the human and mouse genes. The sequences represent the 15-bp KCS element required for optimal basal and IFN-inducible transcriptional activity and the 13-bp ISRE element that confers inducibility; the proteins known to bind the elements are Sp1, Sp3, and DDB (at KCS), and ISGF3 and IRF1 (at ISRE). Human, Hs, *Homo sapiens*; mouse, Mm, *Mus musculus*.

for the human (*161, 164*) and mouse (*155, 161, 163*) PKR cDNAs correspond to the smallest (~2.4–2.5 kb) of the transcripts based on size and northern blot hybridization patterns. The molecular explanation for the origin of the larger RNA transcripts derived from the *PKR* gene remains unresolved, as does the functional significance of the differently sized RNA transcripts. Tissue-selective differences in the ratios of the three different-sized PKR transcripts are seen in the mouse model. The ~4.5-kb species is more abundant than the ~2.4- and ~6.0-kb RNAs in heart and lung tissues, whereas the ~2.4-kb mRNA is the predominant species in the testes and in cultured cells. Although multiple size classes of PKR transcripts are observed in human and mouse cells (*155, 159, 163, 164*), curiously only a single-sized transcript of ~7 kb is seen for porcine *PKR* (*165*).

3. TRANSCRIPTIONAL REGULATION

A single TATA-less promoter drives both basal and IFN-inducible transcription of the *PKR* gene (*32, 155, 166, 167*). The promoter region possesses a consensus 13-bp ISRE element that is responsible for the inducibility of PKR by type I interferons. The sequence is identical, with one exception, between the mouse and human genes. The sequence of the ISRE from the human *PKR* gene, 5'-GGAAAACGAAACT-3', differs from the mouse *Pkr* gene, which possesses an "A" instead of a "T" in the 3'-most position (*32, 155*). A novel 15-bp DNA element (5'-GGGAAGGCGGAGTCC-3') is present in both the human and mouse *PKR* promoters (Fig. 3). This element, designated KCS for kinase conserved sequence, is exactly conserved between the mouse and human *PKR* promoters in both sequence and position relative to the ISRE (*32*). Analysis of mutant promoter constructs with reporter genes, both deletion and substitution mutants, revealed that the KCS was required for basal as well as IFN-inducible transcriptional activity of the *PKR* promoter (*30, 32, 166, 168, 169*), and furthermore, that the KCS and ISRE likely function together as a unit (*30*). In addition to the KCS and ISRE, *cis*-acting sites for several known *trans*-acting protein factors are present in the *PKR* promoter region (*32, 155*). Among these are two especially important sites for Sp family factors, one within the KCS element and one upstream of the KCS element, that are necessary for optimal transcriptional activation of the *PKR* gene (*166, 167*).

Type I IFN-mediated transcriptional activation of the *PKR* gene is best characterized in the context of the JAK-STAT signaling pathway. Analysis of mouse embryo fibroblast (MEF) cell lines genetically deficient in the Jak1 kinase or the Stat1 transcription factor revealed that these two proteins of the JAK-STAT signal transduction pathway are required for type I IFN induction of the PKR protein but not for basal expression of PKR in MEF cells (*167*). The heterotrimeric interferon-stimulated gene factor 3 protein complex (ISGF3), composed of Stat1, Stat2, and IRF9 (also known as p48 or ISGF3γ) transcription factors (*170*), binds at the ISRE of the *PKR* promoter (*30, 166, 169, 171*). In addition to ISGF3, interferon regulatory factor 1 (IRF1) also binds at the ISRE of the *PKR* promoter (*166*). Pull-down assays with KCS element linked to Sepharose beads suggest that the Stat1 and Stat2 proteins also interact with proteins bound at the KCS element in the absence of the ISRE (*166*). Protein–protein interactions between Stat1 and Sp1 have been reported in the upregulation of intercellular adhesion molecule-1 gene by IFN-γ treatment (*172*). In hepatocytes, overexpression of IRF9, a component of ISGF3, mediates the IFN-α-mediated PKR activation (*173*), but neither IRF9 nor IRF1 detectably interacts with KCS-bound proteins (*166*). Sp3, along with components of ISGF3, form a higher molecular weight protein–DNA complex (iKIBP) detectable with KCS-ISRE DNA probes by electrophoretic mobility shift assays (EMSA); the iKIBP complex formation is IFN inducible (*166*).

Each position within the 15-bp KCS element has been mutated, substituting the nucleotide occurring in the wild-type sequence with the other three nucleotides, one position at a time, to generate a family of 45 mutants by site-directed mutagenesis. Analysis of these KCS mutants in reporter constructs with a wild-type ISRE revealed that the "Gs" at the 6, 7, and 9 positions of the KCS element (Fig. 3) are required for optimal *PKR* promoter activity, both basal and IFN inducible (*168*). Competitive EMSA analyses established the formation of specific protein-DNA complexes at the KCS element, designated as KBP for KCS-binding protein complex (*30, 169*). The composition of the KBP complex seen *in vitro* in cell-free nuclear extracts was established by multiple approaches, including antibody supershift analyses (*30, 168*), KCS-DNA pull-down assays (*166*), and ultimately by biochemical purification of the complexes and identification of the constituent proteins by mass spectrometry (*167, 171*). Computational analysis identified a Sp-binding site within the 5'-half of the KCS element (5'-GGGAAGG...). Supershift and pull-down assays revealed that two Sp-family members, Sp1 and Sp3, bound at the KCS element in an IFN-independent manner (*166, 168, 171*). Mass spectrometry and immunochemistry subsequently identified two subunits of DNA damage-binding protein, DDB1 and DDB2, as additional KCS-binding proteins (*171*).

Chromatin immunoprecipitation (ChIP) analyses importantly established that Stat1, Stat2, Sp1, Sp3, and DDB2 proteins all bind within the *PKR* promoter region *in vivo* (*166, 171*). As expected, the binding of Stat1 and Stat2, key components of ISGF3, is IFN dependent, but surprisingly the binding of Sp3 to the *PKR* promoter region *in vivo* also is IFN dependent when measured by ChIP analysis (*166*). By contrast, the binding *in vivo* of Sp1 and DDB2 measured by ChIP is not dependent upon prior IFN treatment (*171*). Depletion of DDB2 reduces KBP complex formation *in vitro* by EMSA (*171*). Although Stat1 is not required for basal transcription of *PKR*, the Sp family of transcription factors is necessary for optimal transcriptional activation of the *PKR* gene in the absence of IFN treatment (*167*).

B. PKR Protein

1. SEQUENCE AND FUNCTIONAL DOMAINS

The PKR protein conceptually can be viewed as a fusion between a regulatory subunit and a catalytic subunit to yield a single protein possessing different functional domains: an N-terminal regulatory domain with RNA-binding activity (dsRBD) that plays a role in dimerization and kinase activation, and a C-terminal catalytic domain responsible for the kinase activity (Fig. 4). The amino acid sequence information for PKR has been deduced from cDNA clones, several of which have been described. The PKR cDNA from human cell lines, isolated from libraries prepared using RNA from IFN-treated cells, is ~2.6 kbp in size (161, 164). cDNAs were isolated both from human Daudi (164) and human amnion U (161) cell lines. They possess a single long ORF of 1653 nt, sufficient to encode a 551-amino acid protein with a predicted molecular weight of 62 kDa (164). Exon 2 alternative splice variants (Fig. 3) all yield a 5′-UTR of less than 200 nt (160); the 3′-UTR is less than 800 nt (164). The sequence flanking the AUG codon (GAA<u>ATG</u>G) that initiates the 551 residue ORF has purines present at positions –3 and +4 and thus conforms to Kozak's consensus sequence for a strong translation initiation site (20, 174). Furthermore, ribosome pausing analysis directly identified this AUG as the major site of translation initiation within the deduced ORF (161). From the sequences of cDNA and genomic clones obtained for human PKR (154, 161, 164), it is likely that 3′-cleavage and subsequent polyadenylation of the PKR nascent transcript occur after the C(A) at nt 2373 that is positioned about 30 nt downstream from the hexanucleotide polyadenylation signal AUUAAA found at nucleotide position 2343–2348 (154).

Mouse PKR cDNAs isolated from mouse pre-B70Z/3 cells (163) and mouse FM3A cells (175) are about 2.3 kbp in size. Again, a single long ORF is found in the mouse cDNA that predicts a 517-amino acid protein in pre-B70Z/3 cells (163) but a slightly smaller protein, 515 amino acids, in mouse FM3A cells (175). In agreement with the latter 515 residue ORF, the genomic nucleotide sequence obtained from adult mouse liver DNA, when compared to that of the previously determined cDNA sequences, predicts a 515 amino acid mouse PKR protein (156). The cDNA for rat PKR also has been characterized, and the rat ORF is 514 amino acids (176). The deduced molecular weight from cDNA sequences gives a value of ~59 kDa for the mouse PKR protein, slightly smaller than that deduced for the human protein, ~62 kDa (156, 161). When analyzed by SDS-polyacrylamide gel electrophoresis, the apparent size of the phosphorylated form of PKR has long been known to differ between the mouse and human proteins, about ~67 kDa for the mouse protein and about ~69 kDa for the human protein (109, 120). No alternative splice variants within the coding region have been found for either mouse or

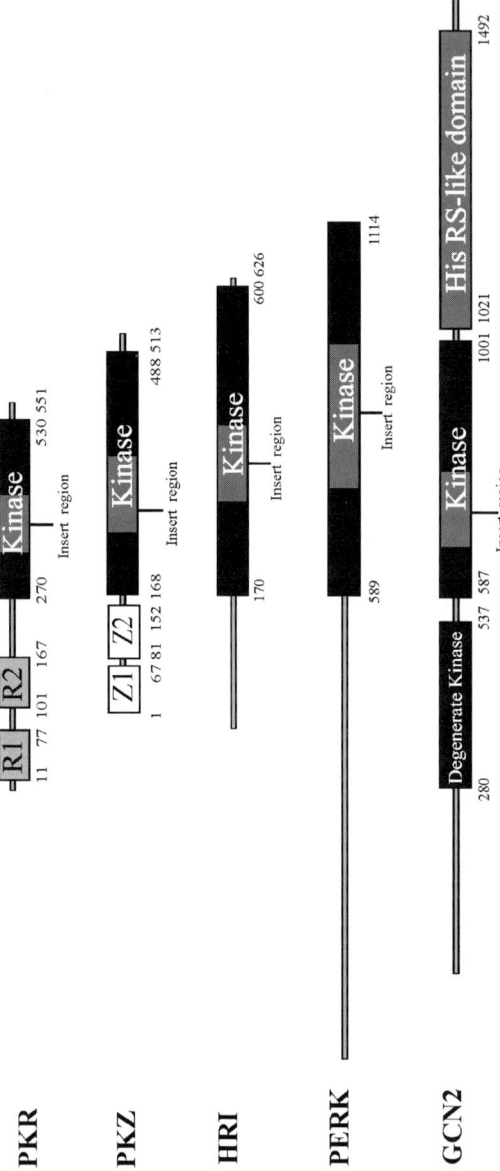

FIG. 4. Schematic diagrams of the eIF-2α protein kinase family members, including the RNA-dependent kinase PKR. The approximate location of the protein serine/threonine kinase catalytic domains (kinase) are shown for the four mammalian eIF-2α protein kinases that are stress responsive, and a fish PKZ homolog. Numbers immediately below each schematic refer to the first and last amino acid of each domain. PKR, the interferon-inducible protein kinase from human cells activated by pathogen infection (*161, 164*, accession number P19525); PKZ, the PKR homolog from fish cells, activated in response to stresses (*189*, accession number AAP49830); HRI, the hemin-regulated eIF-2α kinase from rabbit reticulocytes activated by hemin deficiency (*234*, accession number AAA31241); GCN2, the eIF-2α kinase from mouse embryo cells, regulated by serum conditions and metabolite availability (*237*, accession number CAB58363); PERK, the eIF-2α kinase from mouse fibroblast cells induced by ER stress as part of the unfolded protein response (*235*, accession number AAD03337). R1 and R2 indicate the two copies of the RNA-binding domains of PKR enzymes; Z1 and Z2 indicate the two Z-DNA-binding domains of PKZ in fish. The gray rectangle included within each kinase domain (black rectangle) represents the insert region between catalytic subdomains IV and V, a key feature of eIF-2α kinases.

human PKR. The codon phasing is conserved between PKR exons that specify the two copies of the RNA-binding domain (exons 3 and 5 for mouse, and exons 4 and 6 for human) as shown in Fig. 5. Such phasing would readily accommodate exon skipping involving one of the RNA-binding motif exons (*154*) with conservation of the ORF, but such exon skipping has not been reported. Strikingly, the codon phasing at the exon junctions is identical for the multiple exons that encode the dsRBD copies in ADAR and PKR, and conserved between mouse and human sequences for the two IFN-inducible genes (Fig. 5).

The N-terminal domain of PKR includes a repeated motif (dsRBD or R) that possesses dsRNA-binding activity as first identified by mutational and computational analyses of PKR, the core of which is about 20 amino acid residues (*116, 175, 177–180*) and is conserved among several proteins now known to be RNA-binding proteins, including human ADAR1, human ADAR2, *trans*-activating region (TAR) binding protein, *Drosophila* Staufen, *Xenopus* RBPA, and Dicer and Drosha of the RNA interference pathway

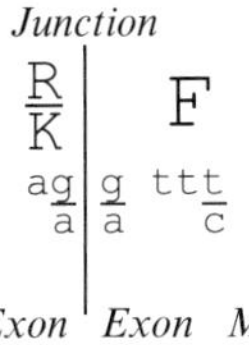

Gene	Exon	Exon	Motif	aa
Mm *ADAR1*	2	3	RI	61
Mm *ADAR1*	4	5	RII	46
Mm *ADAR1*	6	7	RIII	75
Hs *ADAR1*	2	3	RI	61
Hs *ADAR1*	4	5	RII	48
Hs *ADAR1*	6	7	RIII	75
Mm *PKR*	2	3	R1	40
Mm *PKR*	4	5	R2	40
Hs *PKR*	3	4	R1	40
Hs *PKR*	5	6	R2	42

FIG. 5. Codon phases and sizes of the exons containing the dsRBD of PKR and ADAR1. The codon phasing at the exon junctions is shown for the human *ADAR1* and *PKR* genes and the mouse *Adar1* and *Pkr* genes. The three RNA-binding R (dsRBD) domains of ADAR1 are found in exons 3 (RI), 5 (RII), and 7 (RIII) of the human and mouse genes. For PKR, the two R-motifs are found in exons 4 (RI) and 6 (RII) of the human gene, and exons 3 (R1) and 5 (R2) of the mouse gene. The sizes of the exons are shown in amino acids (aa). Human: Hs, *Homo sapiens*; mouse: Mm, *Mus musculus*.

(*18, 36, 44, 181, 182*). Although the N-terminal proximal copy of dsRBD is both necessary and sufficient for RNA-binding activity of PKR (*116, 177*), both dsRBD copies are required for optimal RNA-binding activity and for catalytic activity (*116, 177*).

The C-terminal half of PKR (amino acids 258–551 of the human protein) includes the 11 conserved catalytic subdomains characteristic of kinases and, importantly, an insert region is located between subdomains IV and V (*161, 164*). The organization of the regulatory and kinase subdomains of the PKR protein are remarkably conserved among the mammalian PKR proteins; the amino acid junction positions for 13 of the 15 protein coding exons are exactly conserved between the mouse and human proteins (*154*). Among the most important and widely used PKR mutants are those that alter the subdomain II sequence, including the invariant lysine residue at amino acid position 296 that, when substituted with arginine (K296R), greatly reduces PKR kinase catalytic activity (*183, 184*).

NMR and X-ray crystallography studies have shed important insights into the structural basis of PKR function. The solution structure of the ∼20-kDa N-terminal fragment of PKR that includes the two dsRBD/R domains has revealed a dumbbell shaped structure, consisting of two tandem linked dsRBDs (*185*). Although both dsRBDs contain an α-β-β-β-α-fold, only dsRBD2/RII has a specific interaction with C-terminal kinase domain, which might serve as a lock to keep kinase domain in a "closed" conformation in the latent PKR in the absence of dsRNA (*186*). The structure of full-length PKR protein has not yet been determined, but the crystal structures of the catalytic domain of PKR complexed with the eIF-2α substrate have been elucidated (*187*). The catalytic domain of PKR adopts a bilobal structure with the smaller N-terminal part of the catalytic domain (N lobe, human protein amino acids 258–369) and the larger C-terminal lobe (C lobe, amino acids 370–551) connected by a short hinge region (*187*). The bilobal structure of the PKR catalytic domain is typical of many protein kinases (*188*).

Fish and fish cell-lines express a novel PKR-like gene product (PKZ) that differs from the mammalian PKR proteins that have been characterized from multiple species including human, mouse, rat, and pig (Fig. 4). The fish PKR-like gene called Ca-PKR-like has been cloned and sequenced from UV-inactivated grass carp hemorrhage virus (GCHV) infected CAB cells (*189*) and from zebrafish (*190*). These PKR-like proteins show a low level of constitutive expression in normal fish cells, but are upregulated by virus infection or treatment with IFN or poly (rI):poly (rC). What is unique about the fish PKR-like PKZ kinase proteins is that they possess two copies of a Z-DNA-binding domain within their N-terminal region, instead of two dsRBDs characteristic of mammalian PKR proteins (*189, 190*). Hence, the fish proteins are designated PKZ rather than PKR. By circular dichroism analysis, the Zα domain of PKZ binds Z-DNA (*190*).

In addition to the fish PKZ (PKR-like) proteins, there are three other known proteins in the Zα family: the A-to-I RNA-editing enzyme ADAR1 (*18*), which was the first protein shown to bind Z-DNA (*46, 191*), the Z-DNA-binding protein (ZBP1/DLM-1)(*48*), and the poxvirus virulence factor E3L (*192*). All of these Z-DNA-binding proteins are relevant in the context of host–pathogen interactions. Why the fish PKZ and mammalian PKR proteins possess different regulatory domains at their N-terminus remains to be established, but undoubtedly is of functional and developmental importance to the organisms and their environments, including the pathogens that they encounter.

2. Subcellular Localization

In IFN-treated cells, PKR is present predominantly in the cytoplasm and associates with ribosomes (*130*). However, small amounts of PKR have also been detected in the nucleus by cell fractionation and immunofluorescence microscopy analyses (*161, 193*).

3. Activation of Kinase Activity

The latent PKR protein is monomeric and catalytically inactive (*130, 194*). Activation of the PKR protein involves dimerization and autophosphorylation, a process that occurs under a variety of physiologic conditions including viral infection (*3*) and cellular stress (*195*). Combined genetic and biochemical analyses have established that dimerization may be mediated by two biologically significant processes, RNA binding in virus-infected cells and protein interactions that may occur even in the absence of RNA binding (*109, 196–203*).

Activators of PKR may be grouped broadly into three categories: (1) RNA, either double-stranded or single-stranded with appropriate higher-ordered structure, including both synthetic and natural RNAs such as $(rI)_n$-$(rC)_n$, aptamers, and reovirus genome dsRNA and derived s1 transcript (*109, 120, 180, 204*); (2), cellular proteins, including the PKR activating protein (PACT) from human cells (*201*) and its murine homolog RAX (*202*); (3) chemical agents, including polyanions, such as dextransulfate, and heparin oligosaccharides with eight or more sugar residues (*203, 205*). The most broadly studied activators of PKR are RNAs with double-stranded character (*44, 130, 206*). Undistorted A-form dsRNA has its sequence-rich information buried in the major groove, and no sequence specificity has been observed in interactions between PKR and dsRNA *per se* (*109, 207*). Certain highly structured single-stranded viral RNA species also activate PKR, for example, HIV TAR RNA, reovirus s1 mRNA, and HDV RNA (*3, 204, 208*). The structural basis of the selectivity in the activation of PKR by some RNAs and not others remains an important question. Synthetic aptamer RNAs selected using the N-terminal RNA-binding region of PKR from a library of $\sim 10^{14}$ RNA sequences that contained a randomized region of 50 nt gave both activators and inhibitors of PKR (180) and illustrated the role of tandem A-G mismatches and noncontiguous helices in affecting the structures. Whereas PKR interacts with as little as 11-bp of dsRNA, kinase activation is found with dsRNA duplexes of ~30–50 bp length and is optimal with RNAs >50 bp (180, 209). This activation becomes an important practical issue when chemically synthesized dsRNAs are used to selectively silence gene expression in transfected cells, but 21-bp siRNAs do yield selective silencing without causing the activation of PKR and subsequent global inhibition of translation (206, 210, McAllister and Samuel, unpublished observations).

PACT and RAX, cellular protein activators of PKR from human and murine cells, respectively, were identified using a yeast two-hybrid screen (201, 202). PACT heterodimerizes with PKR through its dsRBDs similar to PKR's dsRBDs and activates PKR independent of dsRNA (201). PACT and RAX belong to the family of dsRNA-binding proteins and share 98% identity in their amino acid sequences (201, 211). In response to diverse stress signals, PACT is phosphorylated, which increases its affinity for association with PKR (211). PKR is activated through direct protein–protein interaction with PACT by a process that can occur in the absence of dsRNA (201). The PACT homologue RAX, which was cloned from an interleukin (IL)-3 dependent cell line, regulates PKR activation in response to IL-3 deprivation and stress signals. The phosphorylation on Ser18 of RAX appears necessary to mediate activation of PKR and subsequent kinase activity (212). Although the physiologic relevance is unclear, it is known that PKR binds to heparin and can be activated by heparin *in vitro* and *in vivo* (203, 205). The two heparin-binding domains of PKR were identified that are nonoverlapping with the two dsRBDs (213): the N-terminal heparin-binding domain (ATD) and the C-terminal heparin-binding domain (CTD) lie between amino acids 278–318 and 412–479, respectively. Substitution mutations within these two domains cause a loss of heparin binding and heparin-mediated activation of PKR (213), just as substitution mutations within the dsRBD domains of PKR cause a loss of dsRNA binding and dsRNA-mediated activation of PKR (4, 43, 130, 206, 210).

The mechanism by which dsRNA binding to PKR mediates dimerization and activation is not resolved (194), but in part involves a conformational change of the protein (186, 187, 197, 204, 209, 214). Among the initial evidence consistent with the notion that dsRNA binding mediates a change in PKR structure was the observations that binding of ATP to PKR is dsRNA dependent (204), and the subsequent finding that dsRNA-mediated dimerization promotes critical phosphorylation events in the activation loop of the kinase (215). Intermolecular association of PKR can occur in the absence of dsRNA-binding activity (216), and heterologous dimerization domains can functionally substitute for the dsRNA-binding domains of PKR (214).

Mutational studies of PKR (197), coupled with the crystal structures of two complexes of PKR and eIF-2α (187), further illustrate the importance of the dimerization of PKR mediated by phosphorylation for activation of the kinase.

The autophosphorylation of PKR may be either intramolecular (217, 218) or intermolecular (219–222). Multiple sites of phosphorylation on PKR have been identified, predominantly serine residues but also including threonine residues (120, 218, 223), but no tyrosine sites were found (163, 223). By phosphopeptide analysis, there appear to be four to five phosphorylation sites per PKR molecule. Three phosphorylation sites were identified in the third basic region, Thr258, Ser242, and Thr255 (224). Mutation of Thr258 resulted in reduced kinase activity, which was further exacerbated following mutation of Ser242 and Thr255 (224). Two additional phosphorylation sites, Thr446 and Thr451, which lie in the kinase activation loop and between subdomains VII–VIII, were identified (225). These two sites are required *in vivo* and *in vitro* for high-level kinase activity: the Thr451Ala mutant, which was inactive, and the Thr446Ala mutant both impaired autophosphorylation and reduced PKR toxicity in a yeast growth assay (225).

4. Substrates

Following activation, PKR can phosphorylate at least five protein substrates in addition to its own autophosphorylation that occurs during the activation process. The PKR substrates include: the α subunit of protein synthesis initiation factor 2, eIF-2α (120); the transcription factor inhibitor IκB (226, 227); the Tat protein of HIV (228); the 90-kDa NFAT protein (229); and the M-phase specific dsRNA binding phosphoprotein MPP4 (230). Among the PKR substrates, eIF-2α is the best-characterized, in both structural and functional terms. PKR catalyzes the phosphorylation of eIF-2α on Ser51 (120, 231). Ser51 phosphorylation of eIF-2α leads to an inhibition of translation by impairing the eIF-2B catalyzed guanine nucleotide exchange reaction (109, 232). A variety of physiologic changes, including IFN treatment and virus infection (3, 109, 130), cause the phosphorylation state of eIF-2α to increase and mRNA translation initiation to subsequently decrease.

It is now recognized, however, that PKR is only one member of a family of four related protein kinases that share the ability to phosphorylate the α subunit of eIF-2 on precisely the same site, Ser51 (Fig. 4). These eIF-2α kinases respond to different stress stimuli (109, 233). The stresses include pathogen infection in the case of the RNA-dependent kinase (PKR) and presumably also the Z-DNA-dependent (PKZ) kinase (109, 130); hemin-deficiency and heavy metals in the case of the heme-regulated inhibitor (HRI) kinase (234); the PKR-like ER kinase (PERK) activated in response to endoplasmic reticulum stress as an unfolded protein response pathway regulator (235, 236), and the kinase that responds to nutrient deprivation (GCN2) as a sensor of amino-acid levels through two his-tRNA-like domains within its C-terminus (237, 238). All of these eIF-2α kinases share a high identity in their kinase catalytic domain, and there is an insert region between their subdomains IV and V (Fig. 4), which distinguishes them from other serine/threonine kinases (239). In contrast, the similarity among the N-terminal regions of the different eIF-2α kinases is comparatively low, consistent with their responsiveness to different physiologic stress signals.

The fact that these four stress responsive kinases, PKR and PKZ, HRI, PERK, and GCN2 all phosphorylate the same site on eIF-2α illustrates the universality and importance of this modification of the protein synthesizing machinery as a mechanism of translational control. Mutational analysis indicates at least two of these kinases, PKR and GCN2, and eIF-2B, recognize overlapping surfaces on eIF-2α(240). In the case of PKR, the crystal structure of the catalytic domain reveals that the catalytic N-lobe mediates dimerization, and that eIF-2α binds to the C-lobe catalytic cleft of the kinase in a manner that induces local unfolding of the Ser51 phosphorylation site (187). This data suggests an ordered and allosteric process of PKR activation in which RNA-mediated catalytic domain dimerization triggers Thr446 autophosphorylation and subsequent eIF-2 substrate docking and phosphorylation (197).

5. Antagonists

Both RNA and protein antagonists of PKR function have been identified and characterized (130). At high concentration, synthetic and natural dsRNA and highly structured viral single-stranded RNAs including adenovirus VAI RNA, EBV EBER RNA, and HIV TAR RNA all antagonize PKR autophosphorylation (3, 109, 207, 241). Among the naturally occurring antagonists, the adenovirus VAI RNA, a pol III gene product synthesized in large amounts at late times after infection, was the first characterized inhibitor of PKR and the antiviral response (113). Adenovirus deletion mutants that do not express the VAI RNA show sensitivity to IFN, whereas wild-type adenovirus is relatively resistant to the antiviral actions of type I IFNs (113, 241). Numerous protein antagonists of PKR function also have been identified (4, 130), including both cellular proteins (P58^{IPK}, PP1) and several viral proteins (Fig. 6).

The cellular protein P58^{IPK} binds to PKR in a region spanning both of the PKR regulatory and kinase domains, via its TPR6 domain, to inhibit PKR activity (242, 243). P58^{IPK} is a member of the tetratricopeptide repeat (TRP) and J-domain protein families, and the TPR6 domain and the DnaJ-domain are required in its function as an inhibitor of PKR (242, 244, 245). In normal cells, P58^{IPK} is inactive due to the binding of its negative regulator hsp40, and is activated after influenza virus infection (246). In response to viral infection or other environmental stresses, P58^{IPK} binds to the chaperone, hsp70/Hsc70, leads to dissociation from hsp40 and downregulation of PKR (247). Another

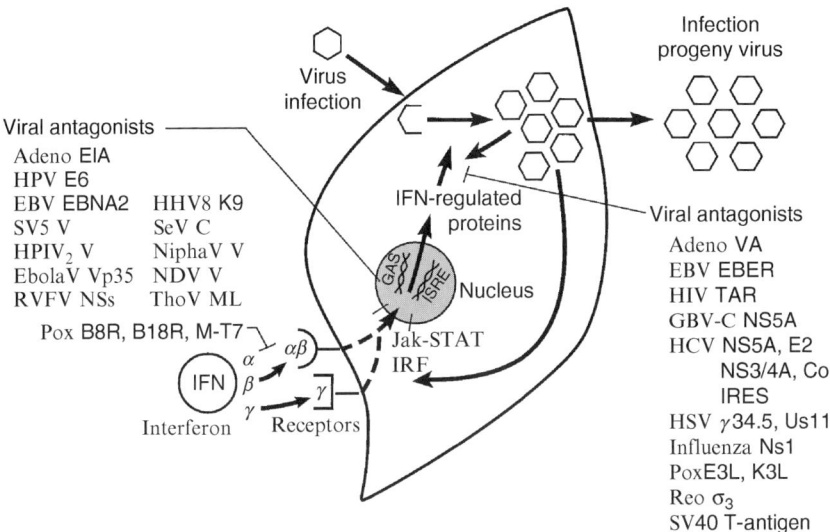

FIG. 6. Viral gene products that antagonize the interferon-mediated induction and the biochemical function of ADAR1 and PKR. Virus-encoded proteins that antagonize the signal transduction and transcriptional activation pathway leading to IFN-stimulated gene expression are listed at the *left* side of the diagram. The *right* side of the diagram lists viral gene products, both RNAs and proteins, that antagonize the biochemical activity of IFN-induced proteins including ADAR1 and PKR. (Adapted from (*130*) with permission.)

negative regulator of P58^{IPK}, P52^{rIPK}, possesses a region of homology to hsp90 (*248*), which may downregulate P58^{IPK} in response to different stress signals. Hsp90 chaperone complex has also been reported to facilitate the folding and/or maturation of PKR, and also as a repressor of PKR subsequently (*249*). The catalytic subunit of type 1 protein phosphatase (PP1), PP1α or PP1c, reduces dsRNA-mediated autoactivation of PKR and inhibits PKR phosphorylation activities (*250*). The PP1c inhibitor protein directly interacts with PKR via a PP1c-binding consensus motif, (R/K)(V/I/L)X(F/W/Y) present within the N-terminal regulatory domain of PKR (*250*). Coexpression of bacteriophage λ protein phosphatase along with the PKR cDNA expression plasmid in *E. coli* permitted production of an unphosphorylated, dsRNA responsive GST-tagged PKR protein (*251*). Finally, agents, such as phorbol ester, are reported to downregulate PKR in IFN-treated mouse fibroblast cells at the posttranslational level (*252*).

An effective mechanism by which viruses inhibit the host antiviral response is either to impair the action of an IFN-induced protein, such as PKR, or to impair the IFN-mediated signal transduction process (Fig. 6). Several different viral proteins are reported to antagonize PKR function including HCV NS5A and E2, HSV γ34.5 and US11, influenza virus NS1, reovirus s3, SV40 T-antigen, and vaccinia virus E3L and K3L (*130, 253*). The poxviruses, including vaccinia virus have contributed immensely to the elucidation of the mechanisms by which viral proteins may antagonize the antiviral actions of interferon. Two vaccinia virus proteins, K3L and E3L, act to antagonize PKR function through fundamentally different mechanisms as revealed from combined biochemical and biological studies with recombinantly expressed viral proteins or mutant viruses deficient in the function of the viral genes (*130, 253*). E3L is a dsRNA binding protein and a Z-DNA-binding protein (*130*). There appear to be two mechanisms by which E3L antagonizes PKR activity, one by sequestering the activator RNA (*254*) and the other by interacting with the substrate-binding region of PKR (*255*). Deletion mutants of vaccinia virus deficient in E3L expression are sensitive to the antiviral effects of IFN; virus mutants that express dsRNA binding-deficient E3L proteins display IFN sensitivity, whereas E3L mutants that retain dsRNA-binding activity complement the deletion and reverse the IFN-sensitive phenotype (*118, 129, 256, 257*). The vaccinia virus K3L protein is a virus-encoded homolog of eIF-2α that acts as a pseudosubstrate to block phosphorylation of eIF-2α, both competitively and noncompetitively (*118, 258–261*). In herpes simplex virus (HSV)-infected cells, the PP1 phosphatase is recruited by the HSV γ34.5 protein, leading to the dephosphorylation of eIF-2α and virus resistance to IFN (*262, 263*), although this antagonism is not sufficient for efficient viral replication (*264*). The RNA-binding protein Us11 of HSV also antagonizes PKR activation (*265, 266*).

IFN-α inhibits HCV replicons in Huh-6 and Huh-7 liver cells, but the biochemical mechanism of the inhibition has not yet been resolved. IFN-γ inhibits HCV replicons in Huh-7 cells, but not in Huh-6 cells, even though unrelated viruses are inhibited (*267*). Two HCV-encoded proteins—the nonstructural NS5A protein and the envelope glycoprotein E2—are reported to antagonize PKR, although the physiologic significance of these findings in the context of the host response to infectious HCV is not yet fully understood (*268, 269*). The RIG-I protein that has a role in TLR (Toll-like receptor)-independent, virus and dsRNA-dependent induction of IFN also is an important target for the NS3/4A protease of HCV (*270*). The IRES, an RNA structural element found in the 5′-UTR of the HCV genomic RNA that is involved in translation initiation, also is reported to antagonize PKR activation and block the PKR-mediated translational inhibition (*271*). With the development of a system for the efficient growth of infectious HCV in cultured cells, it now should be possible to more fully assess the significance of the interactions between *PKR* and *HCV* gene products, singularly and in combination, in IFN-treated as compared to untreated cells (*272, 273*). GB virus C/ hepatitis G virus (GBV-C) is a human virus closely related to HCV (*274, 275*). Sequence polymorphism present in the NS5A protein of GBV-C differentially affects PKR function (276).

C. Biological Activities of PKR

1. ANTIVIRAL ACTIVITY

PKR plays multiple roles in the host response to viral infection. Inhibition of virus replication by PKR-mediated events can be an important component of the host antiviral response, but this antiviral activity frequently is countered by virus-encoded gene products that antagonize PKR function. The replication of a broad range of RNA and DNA viruses (Table I) is blocked by a PKR-dependent antiviral response based on studies that involve either the overexpression of recombinant PKR in cultured cells, the use of cells and mice genetically deficient in the production of PKR, or the use of antagonists of PKR function (4, 130). For example, the replication of EMCV (159), HIV (277, 278), vaccinia virus (279) and VSV (280) is reduced by overexpression of the cDNA encoding catalytically active wild-type PKR but not by overexpression of a mutant PKR that lacks kinase activity. Use of a vaccinia virus recombinant (VVr) system engineered for IPTG-induced controlled expression of PKR permitted the analysis of the role of PKR in inhibiting vaccinia virus and VSV replication (281). Expression of an HIV-1 chimeric genome expressing wild-type PKR from the Nef ORF, but not the K296R mutant deficient in kinase activity, inhibited HIV protein production in a human T-cell line (277). However, in hematopoietic CD 34^+ stem cells, lentivirus-mediated transduction of PKR inhibited HIV-1 replication by only about twofold (282).

Mice lacking PKR are susceptible to VSV replication and lethality following intranasal infection, and also display increased susceptibility to mouse-adapted influenza virus infection (283). $Pkr^{-/-}$ mice also are not resistant to genital HSV-2 infection following transfection with the IFN-α transgene, in contrast to wild-type mice (284). Furthermore, MEF cells derived from $Pkr^{-/-}$ animals (158) are highly susceptible to VSV infection, as opposed to wild-type cells that express PKR, when treated with IFN-α (280, 285). Studies using various knockout cell lines also suggested that activation of PERK, the endoplasmic reticulum-resident protein kinase, can lead to eIF-2α phosphorylation in VSV infected MEF cells (286). PKR signaling occurs downstream of PERK. The effect of IFN on VSV replication clearly depends on the type of IFN as well as the kind of host cell line used. Inhibition of VSV replication by IFN-β is not mediated by PKR in neuronal cells (287), unlike IFN-γ mediated inhibition of VSV in neurons (288, 289). Dengue virus retains its sensitivity to both IFN-α and β in the absence of PKR (290).

IFN-γ mediated inhibition of coxsackie virus (CV) infection of human pancreatic cells is mediated by a PKR-signaling pathway (291). Classic swine fever virus replication in monocytes also is impaired by a lipopolysaccharide (LPS)-mediated PKR activation response (292). Finally, HSV-mediated NF-κB activation occurs through a PKR-dependent signaling response (293), which then activates a pro-apoptotic NF-κB-mediated pathway (294). In reovirus-infected primary cardiac monocyte cultures, PKR prevents reovirus-induced viral myocarditis by mediating IFN-β production (295). The NSs protein of bunyamwera virus (BUNV) inhibits the dsRNA-dependent IFN induction in the host, and hence the subsequent IFN-inducible gene expression needed to establish an antiviral response (296).

Several viruses of paramyxovirus family have evolved to interfere with the Stat protein function during the IFN-mediated signal transduction and transcriptional activation process (297). For example, the V protein of simian virus 5 (SV5) is involved in the ubiquitin-mediated proteasomal degradation of the Stat1 protein in human cells, subsequently impairing the IFN signaling cascade (298). Degradation of Stat1 mediated by the V protein of Newcastle disease virus also inhibits the IFN signaling pathway (299). Similarly, the V protein of human parainfluenza virus 2 (HPIV2) mediates the degradation of the Stat2 protein (300). The large subunit of the DNA damage-binding protein dimer, DDB1, has been identified in the ubiquitin ligase complexes that are part of the SV5 and HPIV2 V protein-mediated Stat inactivation responses (301). Curiously, the DDB1 protein also has been identified as part of the protein complex that binds at the KCS element of the *PKR* promoter, which is involved both in basal and IFN-inducible transcriptional activation of PKR (166, 167, 171).

2. SIGNAL TRANSDUCTION

PKR is implicated as an important mediator in several signaling pathways, including the NF-κB activation pathway and the p38 mitogen-activated kinase (MAPK) pathway, in response to a plethora of stress signals including proinflammatory cytokines, antigen stimulation of T and B cells, bacterial LPS, viral infection, UV irradiation, ionizing radiation, and oxidative stress. NF-κB is a centrally important transcriptional regulator of numerous genes involved in a wide range of biological responses such as inflammation, innate and adaptive immunity, and apoptosis (302, 303). In mammals, the NF-κB family of transcription factors includes several members, p65(RelA), RelB, c-Rel, p50/p105 (NF-κB1), and p52/p100 (NF-κB2), which are present in unstimulated cells in a latent form, in some instances bound to its inhibitor, IκB. NF-κB is activated in response to physiological and pathological stimuli (304). DsRNA triggers NF-κB activation through two distinct pathways in which PKR appears to act as an important factor.

In one pathway, dsRNA is recognized by TLR-3 on the cell surface, activating NF-κB through an IL-1 receptor-associated kinase-independent pathway (305, 306). PKR was identified as a component of the TAK1 complex and is indispensable for the TLR-3-mediated NF-κB activation (306). In the other NF-κB activation pathway, dsRNA is bound by cytoplasmic PKR, subsequently activating NF-κB directly through phosphorylation of IκB or indirectly by inducing IκB kinase activation (226, 307). Here, PKR serves as a signal transducer by directly interacting with IκB kinase (IKK) and NF-κB-inducing kinase (NIK) (308–310). IKK, an effector kinase that phosphorylates IκB (311, 312), is a large (~700- to 900-kDa multimeric complex) composed of two catalytic subunits, IKKα(IKK1) and IKKβ(IKK2), and a regulatory subunit, IKKγ (NEMO, NF-κB essential modulator). Upon phosphorylation of the IKK complex, IκB separates from the NF-κB–IκB complex, thereby releasing NF-κB that translocates into the nucleus. Whether PKR kinase catalytic activity is required for the activation of IKK is not yet fully clear (309, 313, 314). By using vaccinia virus recombinants, it was concluded that the interaction of PKR with IKK complex is not sufficient for NF-κB activation, but rather PKR catalytic activity is required (313). By contrast, others found that both wild-type PKR and the K296R catalytic mutant can activate the IKK complex, suggesting that PKR catalytic activity is not an obligate requirement (309). Furthermore, neither the PKR N-terminal dsRBD nor the C-terminal kinase domain were needed in this PKR-mediated NF-κB activation pathway as revealed by studies in NIH 3T3 stable cell lines expressing different PKR mutants (314). Thus, the precise role that PKR plays in the NF-κB activation pathway, as a catalytically active kinase or as a scaffold protein, remains to be defined. TNF-receptor associated factor (TRAF) family proteins have been identified as key downstream signal transducer, through interaction with PKR, an association that is required for PKR mediated-NF-κB activation pathway (315). PKR associates with multiple TRAF family members including TRAF2, TRAF3, and TRAF5 (315, 316).

The mammalian mitogen-activated protein kinase (MAPK) family members include $MAPK^{erk1/2}$, $MAPK^{p38}$, $MAPK^{jnk}$, $MAPK^{erk3/4}$, and $MAPK^{erk5}$. The MAPKs are activated through a sequential activation pathway comprising three kinases, a MAPK kinase kinase (MAPKKK), MAPK kinase (MAPKK), and MAPK (317). PKR has been found to mediate the activation of p38 and JNK MAPKs by specific proinflammatory stress stimuli (318). In the PKR-null mouse embryonic fibroblast cells, the activation of MAPK is abrogated, which can be rescued by reconstitution with functional human PKR (318). PKR has been shown to regulate p38 MAPK activation in response to dsRNA stimulation through interaction with and activation of MKK6, a member of MAPKK (319). An inhibitor of calmodulin-dependent kinase II (CaMKII) (KN-93) attenuated MKK3/6 and p38 as well as PKR and eIF-2α phosphorylation. Although CaM-KII was not affected by inhibition of PKR with 2-aminopurine, phosphorylation of MKK3/6 and p38 as well as eIF-2α were significantly reduced (320).

Finally, PKR-independent pathways also clearly contribute to the activation of NF-κB and MAPK, as activation of both are observed in *Pkr* null cells (*321, 322*). In *Pkr*$^{-/-}$ MEFs treated with dsRNA or infected with EMCV, activation of NF-κB is observed (*322*), and PKR-independent MAPK activation is observed in EMCV-infected *Pkr*$^{-/-}$ macrophages (*321*).

3. CELL GROWTH AND DEATH

Although PKR is well established as a component of the IFN-inducible antiviral response, PKR also has been implicated as an important regulator of cell growth and differentiation and programmed cell death (PCD) (*323, 324*). PKR has been shown to suppress tumor growth both *in vivo* and *in vitro*. Alterations in components of the PKR pathway that lead to a reduction of eIF-2α phosphorylation mediate the transformation of cells in culture and the formation of tumors in experimental animal models. Stable transformants of NIH 3T3 overexpressing dominant-negative PKR mutant forms, such as K296R/P (*325*), inhibitor protein P58^{IPK}, TAR RNA binding protein TRBP or mutated eIF-2α with the Ser51Ala substitution, display a transformed phenotype and are highly tumorigenic when injected into nude mice (*277, 325–328*). However, no evidence of tumor suppressor activity of PKR was found in two independent mouse mutants devoid of functional PKR by targeted gene disruption (*157, 158*). TRBP, which has several functions including inhibition of PKR activity, has been found to interact with Dicer and function in RNA silencing (*329*). This raises the possibility of interplay between RNAi and interferon-PKR pathways.

Apoptosis, or PCD, is a controlled biochemical process with characteristics of cellular shrinkage, membrane blebbing, and chromatin condensation initiated in response to a broad range of stimuli, including virus infection (*330*). PKR has been shown to have a proapoptotic role. PKR mediates apoptosis induced by a variety of stimuli (*323*), including dsRNA (*331–335*), virus infection (*336–340*), bacterial LPS (*332, 341*), serum starvation (*333*), TNF-α (*332, 333, 342, 343*), and UV radiation (*344*). HeLa cells were found to undergo apoptosis rapidly upon infection with recombinant vaccinia virus expressing active PKR, or a virus mutant deleted of the E3L gene whose product inhibits PKR, but not with a variant virus that expresses catalytically inactive PKR (*345*). Multiple subsequent studies utilizing *Pkr*$^{-/-}$ MEF cells (*332*), promonocytic U937 cells carrying a PKR antisense expression vector (*343*), NIH 3T3 cells expressing the PKR inhibitor P58^{IPK}, and HeLa cells with knocked-down PKR by RNAi (*335*) have provided considerable evidence that PKR is a pro-apoptotic factor. However, it also has been found that overexpression of PKR in inducible NIH3T3 PKR wild-type cells sequentially activated NF-κB and eIF-2α, first temporally promoting cell survival and delaying cell death mediated by NF-κB, and then inducing apoptosis through phosphorylation of eIF-2α (*340*). These findings suggest that PKR possesses both an anti- and proapoptotic activity thereby acting as a double-edge sword in the regulation of apoptosis (*340*).

Although the precise mechanism of cell apoptosis mediated by PKR remains unclear, several pathways have been implicated in the process. The best-characterized substrate of PKR, eIF-2α may be involved in the apoptosis induction, by inhibiting protein synthesis (*323*). Expression of a Ser51Ala mutant eIF-2α, which cannot be phosphorylated by PKR, induced transformation of NIH 3T3 cells, protecting them from serum deprivation, or TNF-α (*327, 333*); this mutant can also rescue apoptosis in HeLa cells induced by overexpression of PKR (*339*). Consistently, apoptosis was not observed in the monkey BSC-40 cells infected with recombinant vaccine virus expressing an eIF-2α variant (Ser51Ala) that cannot be phosphorylated (*339*). Apoptosis of COS-1 cells did occur after transfection with a different eIF-2α mutant, Ser51Asp, which mimics the phosphorylated form of the factor (*333*). It was proposed that phosphorylated eIF-2α induce apoptosis, by allowing preferential translation or by inhibiting the translation of the short-lived antiapoptotic proteins (*346*).

Activated NF-κB has been found to be involved in the PKR-mediated apoptosis pathway. The Rel/NF-κB family of transcription factors can both induce and inhibit the apoptotic response to a variety of stimuli (*303*). Coexpression of PKR with a repressor form of IκBα (Ser32,36Ala) or the inhibition of IκBα degradation abrogates apoptosis (*339*). Activated NF-κB induces expression of death-promoting transcription factors including p53 (*344*) as well as other death-promoting genes, Fas (*347*), Fas ligand (*348, 349*), and IRF1 (*350*). But, the mechanistically important target(s) of the PKR-mediated activation of NF-κB involved in apoptosis signaling remain unclear. Other proteins activated by PKR, such as p38 MAPK (*351*), Stat1 (*352*), and the TRAF family proteins (*315*), could be involved in the PKR-mediated apoptotic pathway.

IV. Possible Roles of ADAR1 and PKR in Human Genetic and Infectious Diseases

If the ADAR1 and PKR proteins do not perform their biochemical functions properly, or if they are improperly regulated, one might anticipate that pathological conditions could result. ADAR mutations and abnormal editing have been implicated in several diseases, and disruption of PKR and ADAR1 function by HCV-encoded proteins may contribute to reducing the response rate to IFN antiviral therapy.

A. ADAR1 and Human Diseases

Dyschromatosis symmetrica hereditaria, or DSH, is an autosomal dominant pigmentary dermatological disorder that causes asymptomatic skin lesions on the backs of the hands and feet. DSH has been studied, in particular, in patients in Japan and China, although the disorder occurs in families all over the world. The disease locus for DSH has been mapped to chromosome 1q21.3 (353–355), the locus that contains the *ADAR1* gene (11), and multiple *ADAR1* missense, nonsense, frameshift, and splicing mutations have been identified in DSH patients (354, 356–362). The effects of these *ADAR1* mutations on deaminase activity have not yet been studied, and the mechanism by which heterozygous *ADAR1* mutations cause the pigmentary disorder is not known. However, the correlation between DSH and *ADAR1* suggests that RNA editing by ADAR1 plays a role in development of epidermal layers or perhaps in protection against UV-induced apoptosis.

Misregulated editing of neurotransmitter receptor RNAs has also been observed in several mental and cognitive disorders, including schizophrenia, Alzheimer's disease, Huntington's disease, and depression. When serotonin 5-HT$_{2C}$R editing in the frontal cortex of schizophrenic patients was examined, an overall decrease in editing at all five 5-HT$_{2C}$R sites was seen (363). Reduction in editing at only the B site, a site that is edited primarily by ADAR1 (14, 15, 25), reached statistical significance. Also found was an increase in the number of 5-HT$_{2C}$R clones that were unedited (the INI isotype; see Section II.C.3); this form of 5-HT$_{2C}$R encodes a receptor with increased G-protein coupling and agonist potency, suggesting schizophrenics have enhanced 5-HT$_{2C}$R activity. However, in another patient group, differences in editing efficiencies of the 5-HT$_{2C}$R were not found when prefrontal cortex tissue from schizophrenic patients or patients that suffered from major depressive disorder were examined (364). Rather, a statistically significant difference in editing at the 5-HT$_{2C}$R A-site, also a site edited primarily by ADAR1 (14, 15, 25), was seen in patients who had committed suicide (364). Depression associated with IFN-α therapy might also be caused by altered editing of 5-HT$_{2C}$ receptors: IFN treatment of glioblastoma cell lines caused increased editing at the A, B, and C sites and significantly altered the expression of the VSI isoform (365). Finally, decreased editing of GluR-B at the Q/R site has been detected in the prefrontal cortex of Alzheimer's disease and schizophrenic subjects and in the striatum of Huntington's disease patients (366) and also in individual motor neurons from patients with amylotrophic lateral sclerosis (367).

B. PKR and Human Diseases

PKR has an important role in the control of cell proliferation and death, and changes in PKR levels or activity have been associated with many different kinds of diseases. PKR has been associated with breast carcinomas (368, 369); both breast carcinoma and several breast carcinoma-derived cell lines have a higher level of PKR than normal breast cells (368, 369), despite the inverse relationship of PKR and proliferative activity in many human malignancies (370–374). Although the PKR can bind dsRNA, kinase activity is minimal in the carcinoma cells.

Alteration of PKR level or activity is also associated with some blood diseases. Nonrandom chromosomal abnormalities of the human *PKR* gene locus, 2p21–22, is associated with myelodysplastic syndromes, malignant lymphomas, and acute myeloid leukemias (375). Two cell lines, human T-cell leukemia (Jurkat) and murine lymphocytic leukemia (L1210), express truncated versions of PKR (376, 377). Results revealed that PKR activity is abolished due to a soluble inhibitor present in samples from B-cell chronic lymphocytic leukemia patients, a disease of defective apoptosis, suggesting that PKR may play a role in the mutation and maintenance of leukemia (378). Inappropriate activation of PKR in bone marrow has been implicated to play a role in a different blood disease, Fanconi anemia (FA), which is characterized by bone marrow failure and leukemia (379, 380). Finally, PKR also has been associated with the pathogenesis of some neuronal degenerations such as Alzheimer's (381, 382) and Huntington's diseases (383).

Acknowledgments

We thank the many investigators within the ADAR1 and PKR fields for their basic and clinical research contributions that collectively made this chapter possible, and we apologize to those whose work may not have been cited. Work from our laboratory was supported in part by National Institutes of Health Research Grants AI-12520 and AI-20611 from the National Institute of Allergy and Infectious Diseases.

References

1. Isaacs, A., and Lindenmann, J. (1957). Virus interference. I. The interferon. *Proc. R. Soc. Lond. B. Biol. Sci.* **147**, 258–267.
2. Nagano, Y., and Kojima, Y. (1958). Inhibition of vaccinia infection by a liquid factor in tissues infected by homologous virus. *C. R. Seances. Soc. Biol. Fil.* **152**, 1627–1629.
3. Samuel, C. E. (1991). Antiviral actions of interferon. Interferon-regulated cellular proteins and their surprisingly selective antiviral activities. *Virology* **183**, 1–11.

4. Sen, G. C. (2001). Viruses and interferons. *Annu. Rev. Microbiol.* **55**, 255–281.
5. Stark, G. R., Kerr, I. M., Williams, B. R., Silverman, R. H., and Schreiber, R. D. (1998). How cells respond to interferons. *Annu. Rev. Biochem.* **67**, 227–264.
6. Pestka, S., Langer, J. A., Zoon, K. C., and Samuel, C. E. (1987). Interferons and their actions. *Annu. Rev. Biochem.* **56**, 727–777.
7. George, C. X., Wagner, M. V., and Samuel, C. E. (2005). Expression of interferon-inducible RNA adenosine deaminase ADAR1 during pathogen infection and mouse embryo development involves tissue-selective promoter utilization and alternative splicing. *J. Biol. Chem.* **280**, 15020–15028.
8. George, C. X., and Samuel, C. E. (1999). Characterization of the 5'-flanking region of the human RNA-specific adenosine deaminase *ADAR1* gene and identification of an interferon-inducible *ADAR1* promoter. *Gene* **229**, 203–213.
9. Liu, Y., George, C. X., Patterson, J. B., and Samuel, C. E. (1997). Functionally distinct double-stranded RNA-binding domains associated with alternative splice site variants of the interferon-inducible double-stranded RNA-specific adenosine deaminase. *J. Biol. Chem.* **272**, 4419–4428.
10. Wang, Y., Zeng, Y., Murray, J. M., and Nishikura, K. (1995). Genomic organization and chromosomal location of the human dsRNA adenosine deaminase gene: The enzyme for glutamate-activated ion channel RNA editing. *J. Mol. Biol.* **254**, 184–195.
11. Weier, H. U. G., George, C. X., Greulich, K. M., and Samuel, C. E. (1995). The interferon-inducible, double-stranded RNA-specific adenosine deaminase gene (DSRAD) maps to human chromosome 1q21.1–21.2. *Genomics* **30**, 372–375.
12. Weier, H. U. G., George, C. X., Lersch, R. A., Breitweser, S., Cheng, J. F., and Samuel, C. E. (2000). Assignment of the RNA-specific adenosine deaminase gene (*Adar*) to mouse chromosome 3f2 by *in situ* hybridization. *Cytogenet. Cell Genet.* **89**, 214–215.
13. Wang, Q., Khillan, J., Gadue, P., and Nishikura, K. (2000). Requirement of the RNA editing deaminase *Adar1* gene for embryonic erythropoiesis. *Science* **290**, 1765–1768.
14. Hartner, J. C., Schmittwolf, C., Kispert, A., Muller, A. M., Higuchi, M., and Seeburg, P. H. (2004). Liver disintegration in the mouse embryo caused by deficiency in the RNA-editing enzyme ADAR1. *J. Biol. Chem.* **279**, 4894–4902.
15. Wang, Q., Miyakoda, M., Yang, W., Khillan, J., Stachura, D. L., Weiss, M. J., and Nishikura, K. (2004). Stress-induced apoptosis associated with null mutation of *Adar1* RNA editing deaminase gene. *J. Biol. Chem.* **279**, 4952–4961.
16. Kawakubo, K., and Samuel, C. E. (2000). Human RNA-specific adenosine deaminase (*ADAR1*) gene specifies transcripts that initiate from a constitutively active alternative promoter. *Gene* **258**, 165–172.
17. George, C. X., and Samuel, C. E. (1999). Human RNA-specific adenosine deaminase ADAR1 transcripts possess alternative exon 1 structures that initiate from different promoters, one constitutively active and the other interferon inducible. *Proc. Natl. Acad. Sci. USA* **96**, 4621–4626.
18. Patterson, J. B., and Samuel, C. E. (1995). Expression and regulation by interferon of a double-stranded-RNA-specific adenosine deaminase from human cells: Evidence for two forms of the deaminase. *Mol. Cell. Biol.* **15**, 5376–5388.
19. Kim, U., Wang, Y., Sanford, T., Zeng, Y., and Nishikura, K. (1994). Molecular cloning of cDNA for double-stranded-RNA adenosine-deaminase, a candidate enzyme for nuclear-RNA editing. *Proc. Natl. Acad. Sci. USA* **91**, 11457–11461.
20. Kozak, M. (1989). The scanning model for translation: An update. *J. Cell Biol.* **108**, 229–241.
21. Brawerman, G. (1989). mRNA decay: Finding the right targets. *Cell* **57**, 9–10.

22. Shtrichman, R., Heithoff, D. M., Mahan, M. J., and Samuel, C. E. (2002). Tissue selectivity of interferon-stimulated gene expression in mice infected with dam(+) versus dam(−) *Salmonella enterica* Serovar typhimurium strains. *Infect. Immun.* **70,** 5579–5588.
23. Yang, J. H., Luo, X., Nie, Y., Su, Y., Zhao, Q., Kabir, K., Zhang, D., and Rabinovici, R. (2003). Widespread inosine-containing mRNA in lymphocytes regulated by ADAR1 in response to inflammation. *Immunology* **109,** 15–23.
24. Yang, J. H., Nie, Y., Zhao, Q., Su, Y., Pypaert, M., Su, H., and Rabinovici, R. (2003). Intracellular localization of differentially regulated RNA-specific adenosine deaminase isoforms in inflammation. *J. Biol. Chem.* **278,** 45833–45842.
25. Liu, Y., Emeson, R. B., and Samuel, C. E. (1999). Serotonin-2c receptor pre-mRNA editing in rat brain and *in vitro* by splice site variants of the interferon-inducible double-stranded RNA-specific adenosine deaminase ADAR1. *J. Biol. Chem.* **274,** 18351–18358.
26. Liu, Y., and Samuel, C. E. (1996). Mechanism of interferon action: Functionally distinct RNA-binding and catalytic domains in the interferon-inducible, double-stranded RNA-specific adenosine deaminase. *J. Virol.* **70,** 1961–1968.
27. O'Connell, M. A., Krause, S., Higuchi, M., Hsuan, J. J., Totty, N. F., Jenny, A., and Keller, W. (1995). Cloning of cDNAs encoding mammalian double-stranded RNA-specific adenosine deaminase. *Mol. Cell. Biol.* **15,** 1389–1397.
28. Patterson, J. B., Thomis, D. C., Hans, S. L., and Samuel, C. E. (1995). Mechanism of interferon action-double-stranded RNA-specific adenosine-deaminase from human-cells is inducible by alpha-interferon and gamma-interferon. *Virology* **210,** 508–511.
29. Darnell, J. E., Jr. (1997). Stats and gene regulation. *Science* **277,** 1630–1635.
30. Ward, S. V., Markle, D., Das, S., and Samuel, C. E. (2002). The promoter-proximal KCS element of the PKR kinase gene enhances transcription irrespective of orientation and position relative to the ISRE element and is functionally distinct from the KCS-like element of the ADAR deaminase promoter. *J. Interferon Cytokine Res.* **22,** 891–898.
31. Markle, D., Das, S., Ward, S. V., and Samuel, C. E. (2003). Functional analysis of the KCS-like element of the interferon-inducible RNA-specific adenosine deaminase ADAR1 promoter. *Gene* **304,** 143–149.
32. Kuhen, K. L., and Samuel, C. E. (1997). Isolation of the interferon-inducible RNA-dependent protein kinase PKR promoter and identification of a novel DNA element within the 5′-flanking region of human and mouse *Pkr* genes. *Virology* **227,** 119–130.
33. Gallo, A., Keegan, L. P., Ring, G. M., and O'Connell, M. A. (2003). An ADAR that edits transcripts encoding ion channel subunits functions as a dimer. *EMBO J.* **22,** 3421–3430.
34. Jaikaran, D. C., Collins, C. H., and MacMillan, A. M. (2002). Adenosine to inosine editing by ADAR2 requires formation of a ternary complex on the GluR-B R/G site. *J. Biol. Chem.* **277,** 37624–37629.
35. Cho, D. S., Yang, W., Lee, J. T., Shiekhattar, R., Murray, J. M., and Nishikura, K. (2003). Requirement of dimerization for RNA editing activity of adenosine deaminases acting on RNA. *J. Biol. Chem.* **278,** 17093–17102.
36. Melcher, T., Maas, S., Herb, A., Sprengel, R., Seeburg, P. H., and Higuchi, M. (1996). A mammalian RNA editing enzyme. *Nature* **379,** 460–464.
37. Lai, F., Drakas, R., and Nishikura, K. (1995). Mutagenic analysis of double-stranded-RNA adenosine-deaminase, a candidate enzyme for RNA editing of glutamate-gated ion-channel transcripts. *J. Biol. Chem.* **270,** 17098–17105.
38. Scott, J. (1995). A place in the world for RNA editing. *Cell* **81,** 833–836.
39. Maas, S., Melcher, T., Herb, A., Seeburg, P. H., Keller, W., Krause, S., Higuchi, M., and O'Connell, M. A. (1996). Structural requirements for RNA editing in glutamate receptor

pre-mRNAs by recombinant double-stranded RNA adenosine deaminase. *J. Biol. Chem.* **271**, 12221–12226.
40. Wilson, D. K., Rudolph, F. B., and Quiocho, F. A. (1991). Atomic structure of adenosine deaminase complexed with a transition-state analog: Understanding catalysis and immunodeficiency mutations. *Science* **252**, 1278–1284.
41. Betts, L., Xiang, S., Short, S. A., Wolfenden, R., and Carter, C. W., Jr. (1994). Cytidine deaminase. The 2.3 Å crystal structure of an enzyme: Transition-state analog complex. *J. Mol. Biol.* **235**, 635–656.
42. Macbeth, M. R., Schubert, H. L., Vandemark, A. P., Lingam, A. T., Hill, C. P., and Bass, B. L. (2005). Inositol hexakisphosphate is bound in the ADAR2 core and required for RNA editing. *Science* **309**, 1534–1539.
43. McCormack, S. J., Ortega, L. G., Doohan, J. P., and Samuel, C. E. (1994). Mechanism of interferon action motif I of the interferon-induced, RNA-dependent protein kinase (PKR) is sufficient to mediate RNA-binding activity. *Virology* **198**, 92–99.
44. Fierro-Monti, I., and Mathews, M. B. (2000). Proteins binding to duplexed RNA: One motif, multiple functions. *Trends Biochem. Sci.* **25**, 241–246.
45. Liu, Y., Lei, M., and Samuel, C. E. (2000). Chimeric double-stranded RNA-specific adenosine deaminase ADAR1 proteins reveal functional selectivity of double-stranded RNA-binding domains from ADAR1 and protein kinase PKR. *Proc. Natl. Acad. Sci. USA* **97**, 12541–12546.
46. Herbert, A., Alfken, J., Kim, Y. G., Mian, I. S., Nishikura, K., and Rich, A. (1997). A Z-DNA binding domain present in the human editing enzyme, double-stranded RNA adenosine deaminase. *Proc. Natl. Acad. Sci. USA* **94**, 8421–8426.
47. Schwartz, T., Rould, M. A., Lowenhaupt, K., Herbert, A., and Rich, A. (1999). Crystal structure of the Z domain of the human editing enzyme ADAR1 bound to left-handed Z-DNA. *Science* **284**, 1841–1845.
48. Schwartz, T., Behlke, J., Lowenhaupt, K., Heinemann, U., and Rich, A. (2001). Structure of the DLM-1-Z-DNA complex reveals a conserved family of Z-DNA-binding proteins. *Nat. Struct. Biol.* **8**, 761–765.
49. Herbert, A., and Rich, A. (1999). Left-handed Z-DNA: Structure and function. *Genetica* **106**, 37–47.
50. Kim, Y. G., Muralinath, M., Brandt, T., Pearcy, M., Hauns, K., Lowenhaupt, K., Jacobs, B. L., and Rich, A. (2003). A role for Z-DNA binding in vaccinia virus pathogenesis. *Proc. Natl. Acad. Sci. USA* **100**, 6974–6979.
51. Herbert, A., and Rich, A. (2001). The role of binding domains for dsRNA and Z-DNA in the *in vivo* editing of minimal substrates by ADAR1. *Proc. Natl. Acad. Sci. USA* **98**, 12132–12137.
52. Koeris, M., Funke, L., Shrestha, J., Rich, A., and Maas, S. (2005). Modulation of ADAR1 editing activity by Z-RNA *in vitro*. *Nucleic Acids Res.* **33**, 5362–5370.
53. Eckmann, C. R., and Jantsch, M. F. (1999). The RNA-editing enzyme ADAR1 is localized to the nascent ribonucleoprotein matrix on *Xenopus lampbrush* chromosomes but specifically associates with an atypical loop. *J. Cell Biol.* **144**, 603–615.
54. Desterro, J. M. P., Keegan, L. P., Jaffray, E., Hay, R. T., O'Connell, M. A., and Carmo-Fonseca, M. (2005). SUMO-1 modification alters ADAR1 editing activity. *Mol. Biol. Cell* **16**, 5115–5126.
55. Bass, B. L., and Weintraub, H. (1987). A developmentally regulated activity that unwinds RNA duplexes. *Cell* **48**, 607–613.
56. Wagner, R. W., Yoo, C., Wrabetz, L., Kamholz, J., Buchhalter, J., Hassan, N. F., Khalili, K., Kim, S. U., Perussia, B., McMorris, F. A., and Nishikura, K. (1990). Double-stranded RNA unwinding and modifying activity is detected ubiquitously in primary tissues and cell lines. *Mol. Cell. Biol.* **10**, 5586–5590.

57. Kim, U., Garner, T. L., Sanford, T., Speicher, D., Murray, J. M., and Nishikura, K. (1994). Purification and characterization of double-stranded RNA adenosine deaminase from bovine nuclear extracts. *J. Biol. Chem.* **269,** 13480–13489.
58. Hough, R. F., and Bass, B. L. (1997). Analysis of xenopus dsRNA adenosine deaminase cDNAs reveals similarities to DNA methyltransferases. *RNA* **3,** 356–370.
59. Eckmann, C. R., Neuteufl, A., Pfaffstetter, L., and Jantsch, M. F. (2001). The human but not the xenopus RNA-editing enzyme ADAR1 has an atypical nuclear localization signal and displays the characteristics of a shuttling protein. *Mol. Biol. Cell* **12,** 1911–1924.
60. Strehblow, A., Hallegger, M., and Jantsch, M. F. (2002). Nucleocytoplasmic distribution of human RNA-editing enzyme ADAR1 is modulated by double-stranded RNA-binding domains, a leucine-rich export signal, and a putative dimerization domain. *Mol. Biol. Cell* **13,** 3822–3835.
61. Nie, Y., Zhao, Q., Su, Y., and Yang, J. H. (2004). Subcellular distribution of ADAR1 isoforms is synergistically determined by three nuclear discrimination signals and a regulatory motif. *J. Biol. Chem.* **279,** 13249–13255.
62. Poulsen, H., Nilsson, J., Damgaard, C. K., Egebjerg, J., and Kjems, J. (2001). Crm1 mediates the export of ADAR1 through a nuclear export signal within the Z-DNA binding domain. *Mol. Cell. Biol.* **21,** 7862–7871.
63. Desterro, J. M., Keegan, L. P., Lafarga, M., Berciano, M. T., O'Connell, M., and Carmo-Fonseca, M. (2003). Dynamic association of RNA-editing enzymes with the nucleolus. *J. Cell Sci.* **116,** 1805–1818.
64. Kudo, N., Matsumori, N., Taoka, H., Fujiwara, D., Schreiner, E. P., Wolff, B., Yoshida, M., and Horinouchi, S. (1999). Leptomycin B inactivates CRM1/exportin 1 by covalent modification at a cysteine residue in the central conserved region. *Proc. Natl. Acad. Sci. USA* **96,** 9112–9117.
65. Nishi, K., Yoshida, M., Fujiwara, D., Nishikawa, M., Horinouchi, S., and Beppu, T. (1994). Leptomycin B targets a regulatory cascade of crm1, a fission yeast nuclear protein, involved in control of higher order chromosome structure and gene expression. *J. Biol. Chem.* **269,** 6320–6324.
66. Wang, Q., Zhang, Z., Blackwell, K., and Carmichael, G. G. (2005). Vigilins bind to promiscuously A-to-I-edited RNAs and are involved in the formation of heterochromatin. *Curr. Biol.* **15,** 384–391.
67. Nie, Y., Ding, L., Kao, P. N., Braun, R., and Yang, J. H. (2005). ADAR1 interacts with NF90 through double-stranded RNA and regulates NF90-mediated gene expression independently of RNA editing. *Mol. Cell. Biol.* **25,** 6956–6963.
68. Cattaneo, R. (1994). Biased (A–>I) hypermutation of animal RNA virus genomes. *Curr. Opin. Genet. Dev.* **4,** 895–900.
69. Cattaneo, R., Schmid, A., Eschle, D., Baczko, K., Meulen, V., and Billeter, M. A. (1988). Biased hypermutation and other genetic changes in defective measles viruses in human brain infections. *Cell* **55,** 255–265.
70. Murphy, D. G., Dimock, K., and Kang, C. Y. (1991). Numerous transitions in human parainfluenza virus 3 RNA recovered from persistently infected cells. *Virology* **181,** 760–763.
71. O'Hara, P., Nichol, S., Horodyski, F., and Holland, J. (1984). Vesicular stomatitis virus defective interfering particles can contain extensive genomic sequence rearrangements and base substitutions. *Cell* **36,** 915–924.
72. Knipe, D. M., Howley, P. M., Griffin, D. E., Martin, M., Roizman, B., and Straus, S. E. (2001). "Fields Virology." Lippincott, Williams & Wilkins, Philadelphia, PA.
73. Cattaneo, R., and Billeter, M. A. (1992). Mutations and A/I hypermutations in measles virus persistent infections. *Curr. Top. Microbiol. Immunol.* **176,** 63–74.

74. Cattaneo, R., Schmid, A., Spielhofer, P., Kaelin, K., Baczko, K., Meulen, V., Pardowitz, J., Flanagan, S., Rima, B. K., and Udem, S. A. (1989). Mutated and hypermutated genes of persistent measles viruses which caused lethal human brain diseases. *Virology* **173**, 415–425.
75. Baczko, K., Lampe, J., Liebert, U. G., Brinckmann, U., Meulen, V., Pardowitz, I., Budka, H., Cosby, S. L., Isserte, S., and Rima, B. K. (1993). Clonal expansion of hypermutated measles virus in a SSPE brain. *Virology* **197**, 188–195.
76. Kumar, M., and Carmichael, G. G. (1997). Nuclear antisense RNA induces extensive adenosine modifications and nuclear retention of target transcripts. *Proc. Natl. Acad. Sci. USA* **94**, 3542–3547.
77. Lei, M., Liu, Y., and Samuel, C. E. (1998). Adenovirus VAI RNA antagonizes the RNA-editing activity of the ADAR adenosine deaminase. *Virology* **245**, 188–196.
78. Taylor, D. R., Puig, M., Darnell, M. E. R., Mihalik, K., and Feinstone, S. M. (2005). New antiviral pathway that mediates hepatitis C virus replicon interferon sensitivity through ADAR1. *J. Virol.* **79**, 6291–6298.
79. Lai, M. M. C. (1995). The molecular biology of hepatitis delta virus. *Annu. Rev. Biochem.* **64**, 259–286.
80. Taylor, J. M. (2003). Replication of human hepatitis delta virus: Recent developments. *Trends Microbiol.* **11**, 185–190.
81. Kuo, M. Y. P., Chao, M., and Taylor, J. (1989). Initiation of replication of the human hepatitis-delta virus genome from cloned DNA: Role of delta-antigen. *J. Virol.* **63**, 1945–1950.
82. Chang, F. L., Chen, P. J., Tu, S. J., Wang, C. J., and Chen, D. S. (1991). The large form of hepatitis-delta antigen is crucial for assembly of hepatitis-delta virus. *Proc. Natl. Acad. Sci. USA* **88**, 8490–8494.
83. Luo, G. X., Chao, M., Hsieh, S. Y., Sureau, C., Nishikura, K., and Taylor, J. (1990). A specific base transition occurs on replicating hepatitis delta virus-RNA. *J. Virol.* **64**, 1021–1027.
84. Chao, M., Hsieh, S. Y., and Taylor, J. (1990). Role of 2 forms of hepatitis delta virus-antigen: Evidence for a mechanism of self-limiting genome replication. *J. Virol.* **64**, 5066–5069.
85. Glenn, J. S., and White, J. M. (1991). Transdominant inhibition of human hepatitis-delta virus genome replication. *J. Virol.* **65**, 2357–2361.
86. Jayan, G. C., and Casey, J. L. (2005). Effects of conserved RNA secondary structures on hepatitis delta virus genotype I RNA editing, replication, and virus production. *J. Virol.* **79**, 11187–11193.
87. Wong, S. K., and Lazinski, D. W. (2002). Replicating hepatitis delta virus RNA is edited in the nucleus by the small form of ADAR1. *Proc. Natl. Acad. Sci. USA* **99**, 15118–15123.
88. Jayan, G. C., and Casey, J. L. (2002). Inhibition of hepatitis delta virus RNA editing by short inhibitory RNA-mediated knockdown of ADAR1 but not ADAR2 expression. *J. Virol.* **76**, 12399–12404.
89. Mayer, M. L. (2005). Glutamate receptor ion channels. *Curr. Opin. Neurobiol.* **15**, 282–288.
90. Seeburg, P. H., and Hartner, J. (2003). Regulation of ion channel/neurotransmitter receptor function by RNA editing. *Curr. Opin. Neurobiol.* **13**, 279–283.
91. Seeburg, P. H. (2002). A-to-I editing: New and old sites, functions and speculations. *Neuron* **35**, 17–20.
92. Hoopengardner, B., Bhalla, T., Staber, C., and Reenan, R. (2003). Nervous system targets of RNA editing identified by comparative genomics. *Science* **301**, 832–836.
93. Raitskin, O., Cho, D. S., Sperling, J., Nishikura, K., and Sperling, R. (2001). RNA editing activity is associated with splicing factors in lnRNP particles: The nuclear pre-mRNA processing machinery. *Proc. Natl. Acad. Sci. USA* **98**, 6571–6576.
94. Lomeli, H., Mosbacher, J., Melcher, T., Hoger, T., Geiger, J. R., Kuner, T., Monyer, H., Higuchi, M., Bach, A., and Seeburg, P. H. (1994). Control of kinetic properties of ampa receptor channels by nuclear RNA editing. *Science* **266**, 1709–1713.

95. Burns, C. M., Chu, H., Rueter, S. M., Hutchinson, L. K., Canton, H., SandersBush, E., and Emeson, R. B. (1997). Regulation of serotonin-2C receptor G-protein coupling by RNA editing. *Nature* **387,** 303–308.
96. Wang, Q., O'Brien, P. J., Chen, C.-X., Cho, D.-S. C., Murray, J. M., and Nishikura, K. (2000). Altered G protein-coupling functions of RNA editing isoform and splicing variant serotonin2c receptors. *J. Neurochem.* **74,** 1290–1300.
97. Herb, A., Higuchi, M., Sprengel, R., and Seeburg, P. H. (1996). Q/R site editing in kainate receptor GluR5 and GluR6 pre-mRNAs requires distant intronic sequences. *Proc. Natl. Acad. Sci. USA* **93,** 1875–1880.
98. Higuchi, M., Single, F. N., Kohler, M., Sommer, B., Sprengel, R., and Seeburg, P. H. (1993). RNA editing of ampa receptor subunit GluR-B: A base-paired intron-exon structure determines position and efficiency. *Cell* **75,** 1361–1370.
99. Ma, J., Qian, R., Rausa, F. M., 3rd, and Colley, K. J. (1997). Two naturally occurring alpha2,6-sialyltransferase forms with a single amino acid change in the catalytic domain differ in their catalytic activity and proteolytic processing. *J. Biol. Chem.* **272,** 672–679.
100. Feng, H. C., Bhave, M., and Fairclough, R. J. (2000). Regulation of oxytocin receptor gene expression in sheep: Tissue specificity, multiple transcripts and mRNA editing. *J. Reprod. Fertil.* **120,** 187–200.
101. Tanoue, A., Koshimizu, T. A., Tsuchiya, M., Ishii, K., Osawa, M., Saeki, M., and Tsujimoto, G. (2002). Two novel transcripts for human endothelin B receptor produced by RNA editing/alternative splicing from a single gene. *J. Biol. Chem.* **277,** 33205–33212.
102. Levanon, E. Y., Hallegger, M., Kinar, Y., Shemesh, R., Djinovic-Carugo, K., Rechavi, G., Jantsch, M. F., and Eisenberg, E. (2005). Evolutionarily conserved human targets of adenosine to inosine RNA editing. *Nucleic Acids Res.* **33,** 1162–1168.
103. Levanon, E. Y., Eisenberg, E., Yelin, R., Nemzer, S., Hallegger, M., Shemesh, R., Fligelman, Z. Y., Shoshan, A., Pollock, S. R., Sztybel, D., Olshansky, M., Rechavi, G. *et al.* (2004). Systematic identification of abundant A-to-I editing sites in the human transcriptome. *Nat. Biotechnol.* **22,** 1001–1005.
104. Kim, D. D., Kim, T. T., Walsh, T., Kobayashi, Y., Matise, T. C., Buyske, S., and Gabriel, A. (2004). Widespread RNA editing of embedded Alu elements in the human transcriptome. *Genome Res.* **14,** 1719–1725.
105. Athanasiadis, A., Rich, A., and Maas, S. (2004). Widespread A-to-I RNA editing of alu-containing mRNAs in the human transcriptome. *PLoS Biol.* **2,** e391.
106. Morse, D. P., and Bass, B. L. (1999). Long RNA hairpins that contain inosine are present in *Caenorhabditis elegans* poly(A)+ RNA. *Proc. Natl. Acad. Sci. USA* **96,** 6048–6053.
107. Clutterbuck, D. R., Leroy, A., O'Connell, M. A., and Semple, C. A. (2005). A bioinformatic screen for novel A-I RNA editing sites reveals recoding editing in BC10. *Bioinformatics* **21,** 2590–2595.
108. Liu, Y., Wolff, K. C., Jacobs, B. L., and Samuel, C. E. (2001). Vaccinia virus E3L interferon resistance protein inhibits the interferon-induced adenosine deaminase A-to-I editing activity. *Virology* **289,** 378–387.
109. Samuel, C. E. (1993). The eIF-2 alpha protein kinases, regulators of translation in eukaryotes from yeasts to humans. *J. Biol. Chem.* **268,** 7603–7606.
110. Schneider, R. J., Weinberger, C., and Shenk, T. (1984). Adenovirus VAI RNA facilitates the initiation of translation in virus-infected cells. *Cell* **37,** 291–298.
111. Ma, Y., and Mathews, M. B. (1996). Structure, function, and evolution of adenovirus-associated RNA: A phylogenetic approach. *J. Virol.* **70,** 5083–5099.
112. Ma, Y., and Mathews, M. B. (1996). Secondary and tertiary structure in the central domain of adenovirus type 2 VA RNA1. *RNA* **2,** 937–951.

113. Kitajewski, J., Schneider, R. J., Safer, B., Munemitsu, S. M., Samuel, C. E., Thimmappaya, B., and Shenk, T. (1986). Adenovirus VAI RNA antagonizes the antiviral action of interferon by preventing activation of the interferon-induced eIF-2-alpha kinase. *Cell* **45**, 195–200.
114. O'Malley, R. P., Mariano, T. M., Siekierka, J., and Mathews, M. B. (1986). A mechanism for the control of protein-synthesis by adenovirus VA RNA1. *Cell* **44**, 391–400.
115. McCormack, S. J., and Samuel, C. E. (1995). Mechanism of interferon action: RNA-binding activity of full-length and R-domain forms of the RNA-dependent protein kinase PKR—determination of KD values for VAI and TAR RNAs. *Virology* **206**, 511–519.
116. McCormack, S. J., Thomis, D. C., and Samuel, C. E. (1992). Mechanism of interferon action: Identification of a RNA binding domain within the N-terminal region of the human RNA-dependent P1/eIF-2 alpha protein kinase. *Virology* **188**, 47–56.
117. Brandt, T. A., and Jacobs, B. L. (2001). Both carboxy- and amino-terminal domains of the vaccinia virus interferon resistance gene, E3L, are required for pathogenesis in a mouse model. *J. Virol.* **75**, 850–856.
118. Shors, S. T., Beattie, E., Paoletti, E., Tartaglia, J., and Jacobs, B. L. (1998). Role of the vaccinia virus E3L and K3L gene products in rescue of VSV and EMCV from the effects of IFN-alpha. *J. Interferon Cytokine Res.* **18**, 721–729.
119. Shors, T., Kibler, K. V., Perkins, K. B., Seider-Wulff, R., Banaszak, M. P., and Jacobs, B. L. (1997). Complementation of vaccinia virus deleted of the E3L gene by mutants of E3L. *Virology* **239**, 269–276.
120. Samuel, C. E. (1979). Mechanism of interferon action: Phosphorylation of protein synthesis initiation factor eIF-2 in interferon-treated human cells by a ribosome-associated kinase processing site specificity similar to hemin-regulated rabbit reticulocyte kinase. *Proc. Natl. Acad. Sci. USA* **76**, 600–604.
121. Chang, H. W., Watson, J. C., and Jacobs, B. L. (1992). The E3L gene of vaccinia virus encodes an inhibitor of the interferon-induced, double-stranded RNA-dependent protein-kinase. *Proc. Natl. Acad. Sci. USA* **89**, 4825–4829.
122. Rebouillat, D., and Hovanessian, A. G. (1999). The human 2′,5′-oligoadenylate synthetase family: Interferon-induced proteins with unique enzymatic properties. *J. Interferon Cytokine Res.* **19**, 295–308.
123. Rivas, C., Gil, J., Melkova, Z., Esteban, M., and Diaz-Guerra, M. (1998). Vaccinia virus E3L protein is an inhibitor of the interferon-induced 2–5A synthetase enzyme. *Virology* **243**, 406–414.
124. Herbert, A., Alfken, J., Kim, Y. G., Mian, I. S., Nishikura, K., and Rich, A. (1997). A Z-DNA binding domain present in the human editing enzyme, double-stranded RNA adenosine deaminase. *Proc. Natl. Acad. Sci. USA* **94**, 8421–8426.
125. Chang, H. W., and Jacobs, B. L. (1993). Identification of a conserved motif that is necessary for binding of the vaccinia virus E3L gene-products to double-stranded-RNA. *Virology* **194**, 537–547.
126. Yuwen, H., Cox, J. H., Yewdell, J. W., Bennink, J. R., and Moss, B. (1993). Nuclear localization of a double-stranded RNA-binding protein encoded by the vaccinia virus E3L gene. *Virology* **195**, 732–744.
127. Chang, H. W., Uribe, L. H., and Jacobs, B. L. (1995). Rescue of vaccinia virus lacking the E3L gene by mutants of E3L. *J. Virol.* **69**, 6605–6608.
128. Scadden, A. D., and Smith, C. W. (1997). A ribonuclease specific for inosine-containing RNA: A potential role in antiviral defence? *EMBO J.* **16**, 2140–2149.
129. Brandt, T., Heck, M. C., Vijaysri, S., Jentarra, G. M., Cameron, J. M., and Jacobs, B. L. (2005). The N-terminal domain of the vaccinia virus E3L-protein is required for neurovirulence, but not induction of a protective immune response. *Virology* **333**, 263–270.
130. Samuel, C. E. (2001). Antiviral actions of interferons. *Clin. Microbiol. Rev.* **14**, 778–809.

131. Jayan, G. C., and Casey, J. L. (2002). Increased RNA editing and inhibition of hepatitis delta virus replication by high-level expression of ADAR1 and ADAR2. *J. Virol.* **76,** 3819–3827.
132. Sato, S., Cornillez-Ty, C., and Lazinski, D. W. (2004). By inhibiting replication, the large hepatitis delta antigen can indirectly regulate amber/W editing and its own expression. *J. Virol.* **78,** 8120–8134.
133. Hartwig, D., Schoeneich, L., Greeve, J., Schutte, C., Dorn, I., Kirchner, H., and Hennig, H. (2004). Interferon-alpha stimulation of liver cells enhances hepatitis delta virus RNA editing in early infection. *J. Hepatol.* **41,** 667–672.
134. Rodig, S. J., Meraz, M. A., White, J. M., Lampe, P. A., Riley, J. K., Arthur, C. D., King, K. L., Sheehan, K. C. F., Yin, L., Pennica, D., Johnson, E. M., and Schreiber, R. D. (1998). Disruption of the Jak1 gene demonstrates obligatory and nonredundant roles of the Jaks in cytokine-induced biologic responses. *Cell* **93,** 373–383.
135. Higuchi, M., Stefan, M., Single, F. N., Hartner, J., Rozov, A., Burnashev, N., Feldmeyer, D., Sprengel, R., and Seeburg, P. H. (2000). Point mutation in an ampa receptor gene rescues lethality in mice deficient in the RNA-editing enzyme ADAR2. *Nature* **406,** 78–81.
136. Palladino, M. J., Keegan, L. P., O'Connell, M. A., and Reenan, R. A. (2000). A-to-I pre-mRNA editing in Drosophila is primarily involved in adult nervous system function and integrity. *Cell* **102,** 437–449.
137. Tonkin, L. A., Saccomanno, L., Morse, D. P., Brodigan, T., Krause, M., and Bass, B. L. (2002). RNA editing by ADARs is important for normal behavior in *Caenorhabditis elegans*. *EMBO J.* **21,** 6025–6035.
138. Burnashev, N., Monyer, H., Seeburg, P. H., and Sakmann, B. (1992). Divalent ion permeability of AMPA receptor channels is dominated by the edited form of a single subunit. *Neuron* **8,** 189–198.
139. Greger, I. H., Khatri, L., Kong, X., and Ziff, E. B. (2003). AMPA receptor tetramerization is mediated by Q/R editing. *Neuron* **40,** 763–774.
140. Liu, Y., and Samuel, C. E. (1999). Editing of glutamate receptor subunit B pre-mRNA by splice-site variants of interferon-inducible double-stranded RNA-specific adenosine deaminase ADAR1. *J. Biol. Chem.* **274,** 5070–5077.
141. Rueter, S. M., Burns, C. M., Coode, S. A., Mookherjee, P., and Emeson, R. B. (1995). Glutamate receptor RNA editing *in vitro* by enzymatic conversion of adenosine to inosine. *Science* **267,** 1491–1494.
142. Baxter, G., Kennett, G., Blaney, F., and Blackburn, T. (1995). 5-HT2 receptor subtypes: A family re-united? *Trends Pharmacol. Sci.* **16,** 105–110.
143. Fitzgerald, L. W., Iyer, G., Conklin, D. S., Krause, C. M., Marshall, A., Patterson, J. P., Tran, D. P., Jonak, G. J., and Hartig, P. R. (1999). Messenger RNA editing of the human serotonin 5-HT2C receptor. *Neuropsychopharmacology* **21,** 82S–90S.
144. Niswender, C. M., Copeland, S. C., Herrick-Davis, K., Emeson, R. B., and Sanders-Bush, E. (1999). RNA editing of the human serotonin 5-hydroxytryptamine 2c receptor silences constitutive activity. *J. Biol. Chem.* **274,** 9472–9478.
145. Lefkowitz, R. J., Cotecchia, S., Samama, P., and Costa, T. (1993). Constitutive activity of receptors coupled to guanine nucleotide regulatory proteins. *Trends Pharmacol. Sci.* **14,** 303–307.
146. Bhalla, T., Rosenthal, J. J. C., Holmgren, M., and Reenan, R. (2004). Control of human potassium channel inactivation by editing of a small mRNA hairpin. *Nat. Struct. Mol. Biol.* **11,** 950–956.
147. Meister, G., and Tuschl, T. (2004). Mechanisms of gene silencing by double-stranded RNA. *Nature* **431,** 343–349.
148. Rebagliati, M. R., and Melton, D. A. (1987). Antisense RNA injections in fertilized frog eggs reveal an RNA duplex unwinding activity. *Cell* **48,** 599–605.

149. Yang, W. D., Wang, Q. D., Howell, K. L., Lee, J. T., Cho, D. S. C., Murray, J. M., and Nishikura, K. (2005). ADAR1 RNA deaminase limits short interfering RNA efficacy in mammalian cells. *J. Biol. Chem.* **280**, 3946–3953.
150. Knight, S. W., and Bass, B. L. (2002). The role of RNA editing by ADARs in RNAi. *Mol. Cell* **10**, 809–817.
151. Hong, J., Qian, Z., Shen, S., Min, T., Tan, C., Xu, J., Zhao, Y., and Huang, W. (2005). High doses of siRNAs induce ERI-1 and ADAR-1 gene expression and reduce the efficiency of RNA interference in the mouse. *Biochem. J.* **390**, 675–679.
152. Tonkin, L. A., and Bass, B. L. (2003). Mutations in RNAi rescue aberrant chemotaxis of ADAR mutants. *Science* **302**, 1725.
153. Barber, G. N., Edelhoff, S., Katze, M. G., and Disteche, C. M. (1993). Chromosomal assignment of the interferon-inducible double-stranded RNA-dependent protein kinase (PRKR) to human chromosome 2p21-p22 and mouse chromosome 17 e2. *Genomics* **16**, 765–767.
154. Kuhen, K. L., Shen, X., Carlisle, E. R., Richardson, A. L., Weier, H. U., Tanaka, H., and Samuel, C. E. (1996). Structural organization of the human gene (*PKR*) encoding an interferon-inducible RNA-dependent protein kinase (PKR) and differences from its mouse homolog. *Genomics* **36**, 197–201.
155. Tanaka, H., and Samuel, C. E. (1994). Mechanism of interferon action: Structure of the mouse PKR gene encoding the interferon-inducible RNA-dependent protein kinase. *Proc. Natl. Acad. Sci. USA* **91**, 7995–7999.
156. Tanaka, H., and Samuel, C. E. (1995). Sequence of the murine interferon-inducible RNA-dependent protein kinase (PKR) deduced from genomic clones. *Gene* **153**, 283–284.
157. Yang, Y. L., Reis, L. F., Pavlovic, J., Aguzzi, A., Schafer, R., Kumar, A., Williams, B. R., Aguet, M., and Weissmann, C. (1995). Deficient signaling in mice devoid of double-stranded RNA-dependent protein kinase. *EMBO J.* **14**, 6095–6106.
158. Abraham, N., Stojdl, D. F., Duncan, P. I., Methot, N., Ishii, T., Dube, M., Vanderhyden, B. C., Atkins, H. L., Gray, D. A., McBurney, M. W., Koromilas, A. E., Brown, E. G. *et al.* (1999). Characterization of transgenic mice with targeted disruption of the catalytic domain of the double-stranded RNA-dependent protein kinase, PKR. *J. Biol. Chem.* **274**, 5953–5962.
159. Meurs, E. F., Watanabe, Y., Kadereit, S., Barber, G. N., Katze, M. G., Chong, K., Williams, B. R., and Hovanessian, A. G. (1992). Constitutive expression of human double-stranded RNA-activated p68 kinase in murine cells mediates phosphorylation of eukaryotic initiation factor 2 and partial resistance to encephalomyocarditis virus growth. *J. Virol.* **66**, 5805–5814.
160. Kawakubo, K., Kuhen, K. L., Vessey, J. W., George, C. X., and Samuel, C. E. (1999). Alternative splice variants of the human PKR protein kinase possessing different 5'-untranslated regions: Expression in untreated and interferon-treated cells and translational activity. *Virology* **264**, 106–114.
161. Thomis, D. C., Doohan, J. P., and Samuel, C. E. (1992). Mechanism of interferon action: cDNA structure, expression, and regulation of the interferon-induced, RNA-dependent P1/eIF-2 alpha protein kinase from human cells. *Virology* **188**, 33–46.
162. Tanaka, H., and Samuel, C. E. (2000). Mouse interferon-inducible RNA-dependent protein kinase *Pkr* gene: Cloning and sequence of the 5'-flanking region and functional identification of the minimal inducible promoter. *Gene* **246**, 373–382.
163. Icely, P. L., Gros, P., Bergeron, J. J., Devault, A., Afar, D. E., and Bell, J. C. (1991). TIK, a novel serine/threonine kinase, is recognized by antibodies directed against phosphotyrosine. *J. Biol. Chem.* **266**, 16073–16077.
164. Meurs, E., Chong, K., Galabru, J., Thomas, N. S., Kerr, I. M., Williams, B. R., and Hovanessian, A. G. (1990). Molecular cloning and characterization of the human double-stranded RNA-activated protein kinase induced by interferon. *Cell* **62**, 379–390.

165. Asano, A., Kon, Y., and Agui, T. (2004). The mRNA regulation of porcine double-stranded RNA-activated protein kinase gene. *J. Vet. Med. Sci.* **66**, 1523–1528.
166. Ward, S. V., and Samuel, C. E. (2003). The PKR kinase promoter binds both Sp1 and Sp3, but only Sp3 functions as part of the interferon-inducible complex with ISGF-3 proteins. *Virology* **313**, 553–566.
167. Das, S., Ward, S. V., Tacke, R. S., Suske, G., and Samuel, C. E. (2006). Activation of the RNA-dependent protein kinase PKR promoter in the absence of interferon is dependent upon Sp proteins. *J. Biol. Chem.* **281**, 3244–3253.
168. Kuhen, K. L., Vessey, J. W., and Samuel, C. E. (1998). Mechanism of interferon action: Identification of essential positions within the novel 15-base-pair KCS element required for transcriptional activation of the RNA-dependent protein kinase PKR gene. *J. Virol.* **72**, 9934–9939.
169. Kuhen, K. L., and Samuel, C. E. (1999). Mechanism of interferon action: Functional characterization of positive and negative regulatory domains that modulate transcriptional activation of the human RNA-dependent protein kinase PKR promoter. *Virology* **254**, 182–195.
170. Darnell, J. E., Jr., Kerr, I. M., and Stark, G. R. (1994). Jak-STAT pathways and transcriptional activation in response to IFNs and other extracellular signaling proteins. *Science* **264**, 1415–1421.
171. Das, S., Ward, S. V., Markle, D., and Samuel, C. E. (2004). DNA damage-binding proteins and heterogeneous nuclear ribonucleoprotein A1 function as constitutive KCS element components of the interferon-inducible RNA-dependent protein kinase promoter. *J. Biol. Chem.* **279**, 7313–7321.
172. Look, D. C., Pelletier, M. R., Tidwell, R. M., Roswit, W. T., and Holtzman, M. J. (1995). STAT1 depends on transcriptional synergy with Sp1. *J. Biol. Chem.* **270**, 30264–30267.
173. Tamada, Y., Nakao, K., Nagayama, Y., Nakata, K., Ichikawa, T., Kawamata, Y., Ishikawa, H., Hamasaki, K., Eguchi, K., and Ishii, N. (2002). P48 overexpression enhances interferon-mediated expression and activity of double-stranded RNA-dependent protein kinase in human hepatoma cells. *J. Hepatol.* **37**, 493–499.
174. Kozak, M. (1986). Regulation of protein synthesis in virus-infected animal cells. *Adv. Virus Res.* **31**, 229–292.
175. Feng, G. S., Chong, K., Kumar, A., and Williams, B. R. (1992). Identification of double-stranded RNA-binding domains in the interferon-induced double-stranded RNA-activated p68 kinase. *Proc. Natl. Acad. Sci. USA* **89**, 5447–5451.
176. Mellor, H., Flowers, K. M., Kimball, S. R., and Jefferson, L. S. (1994). Cloning and characterization of a cDNA encoding rat PKR, the double-stranded RNA-dependent eukaryotic initiation factor-2 kinase. *Biochim. Biophys. Acta* **1219**, 693–696.
177. Green, S. R., and Mathews, M. B. (1992). Two RNA-binding motifs in the double-stranded RNA-activated protein kinase, DAI. *Genes Dev.* **6**, 2478–2490.
178. Katze, M. G., Wambach, M., Wong, M. L., Garfinkel, M., Meurs, E., Chong, K., Williams, B. R., Hovanessian, A. G., and Barber, G. N. (1991). Functional expression and RNA binding analysis of the interferon-induced, double-stranded RNA-activated, 68,000-mr protein kinase in a cell-free system. *Mol. Cell. Biol.* **11**, 5497–5505.
179. Patel, R. C., Stanton, P., and Sen, G. C. (1994). Role of the amino-terminal residues of the interferon-induced protein kinase in its activation by double-stranded RNA and heparin. *J. Biol. Chem.* **269**, 18593–18598.
180. Bevilacqua, P. C., George, C. X., Samuel, C. E., and Cech, T. R. (1998). Binding of the protein kinase PKR to RNAs with secondary structure defects: Role of the tandem A-G mismatch and noncontiguous helixes. *Biochemistry (Mosc)* **37**, 6303–6316.
181. Tian, B., and Mathews, M. B. (2003). Phylogenetics and functions of the double-stranded RNA-binding motif: A genomic survey. *Prog. Nucleic Acid Res. Mol. Biol.* **74**, 123–158.

182. Hammond, S. M. (2005). Dicing and slicing: The core machinery of the RNA interference pathway. *FEBS Lett.* **579**, 5822–5829.
183. Barber, G. N., Tomita, J., Hovanessian, A. G., Meurs, E., and Katze, M. G. (1991). Functional expression and characterization of the interferon-induced double-stranded RNA activated p68 protein kinase from *Escherichia coli*. *Biochemistry (Mosc).* **30**, 10356–10361.
184. Thomis, D. C., and Samuel, C. E. (1992). Mechanism of interferon action: Autoregulation of RNA-dependent P1/eIF-2 alpha protein kinase (PKR) expression in transfected mammalian cells. *Proc. Natl. Acad. Sci. USA* **89**, 10837–10841.
185. Nanduri, S., Carpick, B. W., Yang, Y., Williams, B. R., and Qin, J. (1998). Structure of the double-stranded RNA-binding domain of the protein kinase PKR reveals the molecular basis of its dsRNA-mediated activation. *EMBO J.* **17**, 5458–5465.
186. Nanduri, S., Rahman, F., Williams, B. R., and Qin, J. (2000). A dynamically tuned double-stranded RNA binding mechanism for the activation of antiviral kinase PKR. *EMBO J.* **19**, 5567–5574.
187. Dar, A. C., Dever, T. E., and Sicheri, F. (2005). Higher-order substrate recognition of eIF2alpha by the RNA-dependent protein kinase PKR. *Cell* **122**, 887–900.
188. Huse, M., and Kuriyan, J. (2002). The conformational plasticity of protein kinases. *Cell* **109**, 275–282.
189. Hu, C. Y., Zhang, Y. B., Huang, G. P., Zhang, Q. Y., and Gui, J. F. (2004). Molecular cloning and characterisation of a fish PKR-like gene from cultured CAB cells induced by UV-inactivated virus. *Fish Shellfish Immunol.* **17**, 353–366.
190. Rothenburg, S., Deigendesch, N., Dittmar, K., Koch-Nolte, F., Haag, F., Lowenhaupt, K., and Rich, A. (2005). A PKR-like eukaryotic initiation factor 2alpha kinase from zebrafish contains Z-DNA binding domains instead of dsRNA binding domains. *Proc. Natl. Acad. Sci. USA* **102**, 1602–1607.
191. Liu, Y., Herbert, A., Rich, A., and Samuel, C. E. (1998). Double-stranded RNA-specific adenosine deaminase: Nucleic acid binding properties. *Methods* **15**, 199–205.
192. Ha, S. C., Lokanath, N. K., Quyen, D., Wu, C. A., Lowenhaupt, K., Rich, A., Kim, Y. G., and Kim, K. K. (2004). A poxvirus protein forms a complex with left-handed Z-DNA: Crystal structure of a yatapoxvirus zalpha bound to DNA. *Proc. Natl. Acad. Sci. USA* **101**, 14367–14372.
193. Jimenez-Garcia, L. F., Green, S. R., Mathews, M. B., and Spector, D. L. (1993). Organization of the double-stranded RNA-activated protein kinase DAI and virus-associated VA RNAi in adenovirus-2-infected Hela cells. *J. Cell Sci.* **106**(Pt. 1), 11–22.
194. Taylor, S. S., Haste, N. M., and Ghosh, G. (2005). PKR and eIF2alpha: Integration of kinase dimerization, activation, and substrate docking. *Cell* **122**, 823–825.
195. Chu, W. M., Ballard, R., Carpick, B. W., Williams, B. R., and Schmid, C. W. (1998). Potential Alu function: Regulation of the activity of double-stranded RNA-activated kinase PKR. *Mol. Cell. Biol.* **18**, 58–68.
196. Sen, G. C., and Lengyel, P. (1992). The interferon system. A bird's eye view of its biochemistry. *J. Biol. Chem.* **267**, 5017–5020.
197. Dey, M., Cao, C., Dar, A. C., Tamura, T., Ozato, K., Sicheri, F., and Dever, T. E. (2005). Mechanistic link between PKR dimerization, autophosphorylation, and eIF2alpha substrate recognition. *Cell* **122**, 901–913.
198. Katze, M. G. (1995). Regulation of the interferon-induced PKR: Can viruses cope? *Trends Microbiol.* **3**, 75–78.
199. Samuel, C. E. (1992). Role of the RNA-dependent protein kinase in the regulated expression of genes in transfected cells. *Pharmacol. Ther.* **54**, 307–317.
200. Sonenberg, N. (1990). Measures and countermeasures in the modulation of initiation factor activities by viruses. *New Biol.* **2**, 402–409.

201. Patel, R. C., and Sen, G. C. (1998). PACT, a protein activator of the interferon-induced protein kinase, PKR. *EMBO J.* **17**, 4379–4390.
202. Ito, T., Yang, M., and May, W. S. (1999). Rax a cellular activator for double-stranded RNA-dependent protein kinase during stress signaling. *J. Biol. Chem.* **274**, 15427–15432.
203. George, C. X., Thomis, D. C., McCormack, S. J., Svahn, C. M., and Samuel, C. E. (1996). Characterization of the heparin-mediated activation of PKR, the interferon-inducible RNA-dependent protein kinase. *Virology* **221**, 180–188.
204. Bischoff, J. R., and Samuel, C. E. (1989). Mechanism of interferon action. Activation of the human P1/eIF-2 alpha protein kinase by individual reovirus s-class mRNAs: s1 mRNA is a potent activator relative to s4 mRNA. *Virology* **172**, 106–115.
205. Hovanessian, A. G., and Galabru, J. (1987). The double-stranded RNA-dependent protein kinase is also activated by heparin. *Eur. J. Biochem.* **167**, 467–473.
206. Wang, Q., and Carmichael, G. G. (2004). Effects of length and location on the cellular response to double-stranded RNA. *Microbiol. Mol. Biol. Rev.* **68**, 432–452, table of contents.
207. Clemens, M. J., and Elia, A. (1997). The double-stranded RNA-dependent protein kinase PKR: Structure and function. *J. Interferon Cytokine Res.* **17**, 503–524.
208. Circle, D. A., Neel, O. D., Robertson, H. D., Clarke, P. A., and Mathews, M. B. (1997). Surprising specificity of PKR binding to delta agent genomic RNA. *RNA* **3**, 438–448.
209. Manche, L., Green, S. R., Schmedt, C., and Mathews, M. B. (1992). Interactions between double-stranded RNA regulators and the protein kinase DAI. *Mol. Cell. Biol.* **12**, 5238–5248.
210. Tian, B., Bevilacqua, P. C., Diegelman-Parente, A., and Mathews, M. B. (2004). The double-stranded-RNA-binding motif: Interference and much more. *Nat. Rev. Mol. Cell Biol.* **5**, 1013–1023.
211. Patel, C. V., Handy, I., Goldsmith, T., and Patel, R. C. (2000). PACT, a stress-modulated cellular activator of interferon-induced double-stranded RNA-activated protein kinase, PKR. *J. Biol. Chem.* **275**, 37993–37998.
212. Bennett, R. L., Blalock, W. L., and May, W. S. (2004). Serine 18 phosphorylation of RAX, the PKR activator, is required for PKR activation and consequent translation inhibition. *J. Biol. Chem.* **279**, 42687–42693.
213. Fasciano, S., Hutchins, B., Handy, I., and Patel, R. C. (2005). Identification of the heparin-binding domains of the interferon-induced protein kinase, PKR. *FEBS J.* **272**, 1425–1439.
214. Ung, T. L., Cao, C., Lu, J., Ozato, K., and Dever, T. E. (2001). Heterologous dimerization domains functionally substitute for the double-stranded RNA binding domains of the kinase PKR. *EMBO J.* **20**, 3728–3737.
215. Zhang, F., Romano, P. R., Nagamura-Inoue, T., Tian, B., Dever, T. E., Mathews, M. B., Ozato, K., and Hinnebusch, A. G. (2001). Binding of double-stranded RNA to protein kinase PKR is required for dimerization and promotes critical autophosphorylation events in the activation loop. *J. Biol. Chem.* **276**, 24946–24958.
216. Ortega, L. G., McCotter, M. D., Henry, G. L., McCormack, S. J., Thomis, D. C., and Samuel, C. E. (1996). Mechanism of interferon action. Biochemical and genetic evidence for the intermolecular association of the RNA-dependent protein kinase PKR from human cells. *Virology* **215**, 31–39.
217. Berry, M. J., Knutson, G. S., Lasky, S. R., Munemitsu, S. M., and Samuel, C. E. (1985). Mechanism of interferon action. Purification and substrate specificities of the double-stranded RNA-dependent protein kinase from untreated and interferon-treated mouse fibroblasts. *J. Biol. Chem.* **260**, 11240–11247.
218. Galabru, J., and Hovanessian, A. (1987). Autophosphorylation of the protein kinase dependent on double-stranded RNA. *J. Biol. Chem.* **262**, 15538–15544.
219. Kostura, M., and Mathews, M. B. (1989). Purification and activation of the double-stranded RNA-dependent eIF-2 kinase DAI. *Mol. Cell. Biol.* **9**, 1576–1586.

220. Romano, P. R., Green, S. R., Barber, G. N., Mathews, M. B., and Hinnebusch, A. G. (1995). Structural requirements for double-stranded RNA binding, dimerization, and activation of the human eIF-2 alpha kinase DAI in *Saccharomyces cerevisiae*. *Mol. Cell. Biol.* **15,** 365–378.
221. Thomis, D. C., and Samuel, C. E. (1993). Mechanism of interferon action: Evidence for intermolecular autophosphorylation and autoactivation of the interferon-induced, RNA-dependent protein kinase PKR. *J. Virol.* **67,** 7695–7700.
222. Thomis, D. C., and Samuel, C. E. (1995). Mechanism of interferon action: Characterization of the intermolecular autophosphorylation of PKR, the interferon-inducible, RNA-dependent protein kinase. *J. Virol.* **69,** 5195–5198.
223. Lasky, S. R., Jacobs, B. L., and Samuel, C. E. (1982). Mechanism of interferon action. Characterization of sites of phosphorylation in the interferon-induced phosphoprotein P1 from mouse fibroblasts: Evidence for two forms of P1. *J. Biol. Chem.* **257,** 11087–11093.
224. Taylor, D. R., Lee, S. B., Romano, P. R., Marshak, D. R., Hinnebusch, A. G., Esteban, M., and Mathews, M. B. (1996). Autophosphorylation sites participate in the activation of the double-stranded-RNA-activated protein kinase PKR. *Mol. Cell. Biol.* **16,** 6295–6302.
225. Romano, P. R., Garcia-Barrio, M. T., Zhang, X., Wang, Q., Taylor, D. R., Zhang, F., Herring, C., Mathews, M. B., Qin, J., and Hinnebusch, A. G. (1998). Autophosphorylation in the activation loop is required for full kinase activity *in vivo* of human and yeast eukaryotic initiation factor 2alpha kinases PKR and GCN2. *Mol. Cell. Biol.* **18,** 2282–2297.
226. Kumar, A., Haque, J., Lacoste, J., Hiscott, J., and Williams, B. R. (1994). Double-stranded RNA-dependent protein kinase activates transcription factor NF-kappaB by phosphorylating IkappaB. *Proc. Natl. Acad. Sci. USA* **91,** 6288–6292.
227. Offermann, M. K., Zimring, J., Mellits, K. H., Hagan, M. K., Shaw, R., Medford, R. M., Mathews, M. B., Goodbourn, S., and Jagus, R. (1995). Activation of the double-stranded-RNA-activated protein kinase and induction of vascular cell adhesion molecule-1 by Poly (I). Poly (C) in endothelial cells. *Eur. J. Biochem.* **232,** 28–36.
228. McMillan, N. A., Chun, R. F., Siderovski, D. P., Galabru, J., Toone, W. M., Samuel, C. E., Mak, T. W., Hovanessian, A. G., Jeang, K. T., and Williams, B. R. (1995). HIV-1 Tat directly interacts with the interferon-induced, double-stranded RNA-dependent kinase, PKR. *Virology* **213,** 413–424.
229. Langland, J. O., Kao, P. N., and Jacobs, B. L. (1999). Nuclear factor-90 of activated T-cells: A double-stranded RNA-binding protein and substrate for the double-stranded RNA-dependent protein kinase, PKR. *Biochemistry (Mosc)* **38,** 6361–6368.
230. Patel, R. C., Vestal, D. J., Xu, Z., Bandyopadhyay, S., Guo, W., Erme, S. M., Williams, B. R., and Sen, G. C. (1999). DRBP76, a double-stranded RNA-binding nuclear protein, is phosphorylated by the interferon-induced protein kinase, PKR. *J. Biol. Chem.* **274,** 20432–20437.
231. Pathak, V. K., Schindler, D., and Hershey, J. W. (1988). Generation of a mutant form of protein synthesis initiation factor eIF-2 lacking the site of phosphorylation by eIF-2 kinases. *Mol. Cell. Biol.* **8,** 993–995.
232. Gale, M., Jr., Tan, S. L., and Katze, M. G. (2000). Translational control of viral gene expression in eukaryotes. *Microbiol. Mol. Biol. Rev.* **64,** 239–280.
233. Proud, C. G. (2005). EIF2 and the control of cell physiology. *Semin. Cell Dev. Biol.* **16,** 3–12.
234. Chen, J. J., Throop, M. S., Gehrke, L., Kuo, I., Pal, J. K., Brodsky, M., and London, I. M. (1991). Cloning of the cDNA of the heme-regulated eukaryotic initiation factor 2 alpha (eIF-2 alpha) kinase of rabbit reticulocytes: Homology to yeast GCN2 protein kinase and human double-stranded-RNA-dependent eIF-2 alpha kinase. *Proc. Natl. Acad. Sci. USA* **88,** 7729–7733.
235. Harding, H. P., Zhang, Y., and Ron, D. (1999). Protein translation and folding are coupled by an endoplasmic-reticulum-resident kinase. *Nature* **397,** 271–274.

236. Shi, Y., Vattem, K. M., Sood, R., An, J., Liang, J., Stramm, L., and Wek, R. C. (1998). Identification and characterization of pancreatic eukaryotic initiation factor 2 alpha-subunit kinase, PEK, involved in translational control. *Mol. Cell. Biol.* **18,** 7499–7509.
237. Berlanga, J. J., Santoyo, J., and Haro, C. (1999). Characterization of a mammalian homolog of the GCN2 eukaryotic initiation factor 2alpha kinase. *Eur. J. Biochem.* **265,** 754–762.
238. Clemens, M. J. (2001). Initiation factor eIF2 alpha phosphorylation in stress responses and apoptosis. *Prog. Mol. Subcell. Biol.* **27,** 57–89.
239. Hanks, S. K., and Hunter, T. (1995). Protein kinases 6. The eukaryotic protein kinase superfamily: Kinase (catalytic) domain structure and classification. *FASEB J.* **9,** 576–596.
240. Dey, M., Trieselmann, B., Locke, E. G., Lu, J., Cao, C., Dar, A. C., Krishnamoorthy, T., Dong, J., Sicheri, F., and Dever, T. E. (2005). PKR and GCN2 kinases and guanine nucleotide exchange factor eukaryotic translation initiation factor 2b (eIF2b) recognize overlapping surfaces on eIF2alpha. *Mol. Cell. Biol.* **25,** 3063–3075.
241. Mathews, M. B., and Shenk, T. (1991). Adenovirus virus-associated RNA and translation control. *J. Virol.* **65,** 5657–5662.
242. Lee, T. G., Tang, N., Thompson, S., Miller, J., and Katze, M. G. (1994). The 58,000-dalton cellular inhibitor of the interferon-induced double-stranded RNA-activated protein kinase (PKR) is a member of the tetratricopeptide repeat family of proteins. *Mol. Cell. Biol.* **14,** 2331–2342.
243. Tang, N. M., Ho, C. Y., and Katze, M. G. (1996). The 58-kDa cellular inhibitor of the double stranded RNA-dependent protein kinase requires the tetratricopeptide repeat 6 and DNAj motifs to stimulate protein synthesis *in vivo*. *J. Biol. Chem.* **271,** 28660–28666.
244. Lee, T. G., Tomita, J., Hovanessian, A. G., and Katze, M. G. (1990). Purification and partial characterization of a cellular inhibitor of the interferon-induced protein kinase of mr 68,000 from influenza virus-infected cells. *Proc. Natl. Acad. Sci. USA* **87,** 6208–6212.
245. Lee, T. G., Tomita, J., Hovanessian, A. G., and Katze, M. G. (1992). Characterization and regulation of the 58,000-dalton cellular inhibitor of the interferon-induced, dsRNA-activated protein kinase. *J. Biol. Chem.* **267,** 14238–14243.
246. Melville, M. W., Hansen, W. J., Freeman, B. C., Welch, W. J., and Katze, M. G. (1997). The molecular chaperone hsp40 regulates the activity of P58^{IPK}, the cellular inhibitor of PKR. *Proc. Natl. Acad. Sci. USA* **94,** 97–102.
247. Melville, M. W., Tan, S. L., Wambach, M., Song, J., Morimoto, R. I., and Katze, M. G. (1999). The cellular inhibitor of the PKR protein kinase, P58^{IPK}, is an influenza virus-activated co-chaperone that modulates heat shock protein 70 activity. *J. Biol. Chem.* **274,** 3797–3803.
248. Gale, M., Jr., Blakely, C. M., Hopkins, D. A., Melville, M. W., Wambach, M., Romano, P. R., and Katze, M. G. (1998). Regulation of interferon-induced protein kinase PKR: Modulation of P58^{IPK} inhibitory function by a novel protein, P52^{rIPK}. *Mol. Cell. Biol.* **18,** 859–871.
249. Donze, O., Abbas-Terki, T., and Picard, D. (2001). The Hsp90 chaperone complex is both a facilitator and a repressor of the dsRNA-dependent kinase PKR. *EMBO J.* **20,** 3771–3780.
250. Tan, S. L., Tareen, S. U., Melville, M. W., Blakely, C. M., and Katze, M. G. (2002). The direct binding of the catalytic subunit of protein phosphatase 1 to the PKR protein kinase is necessary but not sufficient for inactivation and disruption of enzyme dimer formation. *J. Biol. Chem.* **277,** 36109–36117.
251. Matsui, T., Tanihara, K., and Date, T. (2001). Expression of unphosphorylated form of human double-stranded RNA-activated protein kinase in *Escherichia coli*. *Biochem. Biophys. Res. Commun.* **284,** 798–807.
252. Zhou, Y., Chase, B. I., Whitmore, M., Williams, B. R., and Zhou, A. (2005). Double-stranded RNA-dependent protein kinase (PKR) is downregulated by phorbol ester. *FEBS J.* **272,** 1568–1576.

253. Haller, O., Kochs, G., and Weber, F. (2005). The interferon response circuit: Induction and suppression by pathogenic viruses. *Virology* **344,** 119–130.
254. Shors, T., and Jacobs, B. L. (1997). Complementation of deletion of the vaccinia virus E3L gene by the *Escherichia coli* RNase III gene. *Virology* **227,** 77–87.
255. Sharp, T. V., Moonan, F., Romashko, A., Joshi, B., Barber, G. N., and Jagus, R. (1998). The vaccinia virus E3L gene product interacts with both the regulatory and the substrate binding regions of PKR: Implications for PKR autoregulation. *Virology* **250,** 302–315.
256. Langland, J. O., and Jacobs, B. L. (2004). Inhibition of PKR by vaccinia virus: Role of the N- and C-terminal domains of E3L. *Virology* **324,** 419–429.
257. Vijaysri, S., Talasela, L., Mercer, A. A., McInnes, C. J., Jacobs, B. L., and Langland, J. O. (2003). The orf virus E3L homologue is able to complement deletion of the vaccinia virus E3L gene *in vitro* but not *in vivo*. *Virology* **314,** 305–314.
258. Sharp, T. V., Witzel, J. E., and Jagus, R. (1997). Homologous regions of the alpha subunit of eukaryotic translational initiation factor 2 (eIF2alpha) and the vaccinia virus K3L gene product interact with the same domain within the dsRNA-activated protein kinase (PKR). *Eur. J. Biochem.* **250,** 85–91.
259. Dar, A. C., and Sicheri, F. (2002). X-ray crystal structure and functional analysis of vaccinia virus K3L reveals molecular determinants for PKR subversion and substrate recognition. *Mol. Cell* **10,** 295–305.
260. Kawagishi-Kobayashi, M., Silverman, J. B., Ung, T. L., and Dever, T. E. (1997). Regulation of the protein kinase PKR by the vaccinia virus pseudosubstrate inhibitor K3L is dependent on residues conserved between the K3L protein and the PKR substrate eIF2alpha. *Mol. Cell. Biol.* **17,** 4146–4158.
261. Carroll, K., Elroy-Stein, O., Moss, B., and Jagus, R. (1993). Recombinant vaccinia virus K3L gene product prevents activation of double-stranded RNA-dependent, initiation factor 2 alpha-specific protein kinase. *J. Biol. Chem.* **268,** 12837–12842.
262. He, B., Gross, M., and Roizman, B. (1997). The gamma(1)34.5 protein of herpes simplex virus 1 complexes with protein phosphatase 1alpha to dephosphorylate the alpha subunit of the eukaryotic translation initiation factor 2 and preclude the shutoff of protein synthesis by double-stranded RNA-activated protein kinase. *Proc. Natl. Acad. Sci. USA* **94,** 843–848.
263. Herbert, A. M., Bagg, J., Walker, D. M., Davies, K. J., and Westmoreland, D. (1995). Seroepidemiology of herpes virus infections among dental personnel. *J. Dent.* **23,** 339–342.
264. Cheng, G., Yang, K., and He, B. (2003). Dephosphorylation of eIF-2alpha mediated by the gamma(1)34.5 protein of herpes simplex virus type 1 is required for viral response to interferon but is not sufficient for efficient viral replication. *J. Virol.* **77,** 10154–10161.
265. Cassady, K. A., Gross, M., and Roizman, B. (1998). The herpes simplex virus us11 protein effectively compensates for the gamma1(34.5) gene if present before activation of protein kinase R by precluding its phosphorylation and that of the alpha subunit of eukaryotic translation initiation factor 2. *J. Virol.* **72,** 8620–8626.
266. Poppers, J., Mulvey, M., Khoo, D., and Mohr, I. (2000). Inhibition of PKR activation by the proline-rich RNA binding domain of the herpes simplex virus type 1 US11 protein. *J. Virol.* **74,** 11215–11221.
267. Windisch, M. P., Frese, M., Kaul, A., Trippler, M., Lohmann, V., and Bartenschlager, R. (2005). Dissecting the interferon-induced inhibition of hepatitis C virus replication by using a novel host cell line. *J. Virol.* **79,** 13778–13793.
268. Gale, M., Jr., Tan, S. L., Wambach, M., and Katze, M. G. (1996). Interaction of the interferon-induced PKR protein kinase with inhibitory proteins P58^{IPK} and vaccinia virus K3L is mediated by unique domains: Implications for kinase regulation. *Mol. Cell. Biol.* **16,** 4172–4181.

269. Taylor, D. R., Shi, S. T., Romano, P. R., Barber, G. N., and Lai, M. M. (1999). Inhibition of the interferon-inducible protein kinase PKR by HCV E2 protein. *Science* **285,** 107–110.
270. Breiman, A., Grandvaux, N., Lin, R., Ottone, C., Akira, S., Yoneyama, M., Fujita, T., Hiscott, J., and Meurs, E. F. (2005). Inhibition of RIG-I-dependent signaling to the interferon pathway during hepatitis C virus expression and restoration of signaling by IKKepsilon. *J. Virol.* **79,** 3969–3978.
271. Vyas, J., Elia, A., and Clemens, M. J. (2003). Inhibition of the protein kinase PKR by the internal ribosome entry site of hepatitis C virus genomic RNA. *RNA* **9,** 858–870.
272. Wakita, T., Pietschmann, T., Kato, T., Date, T., Miyamoto, M., Zhao, Z., Murthy, K., Habermann, A., Krausslich, H. G., Mizokami, M., Bartenschlager, R., and Liang, T. J. (2005). Production of infectious hepatitis C virus in tissue culture from a cloned viral genome. *Nat. Med.* **11,** 791–796.
273. Lindenbach, B. D., Evans, M. J., Syder, A. J., Wolk, B., Tellinghuisen, T. L., Liu, C. C., Maruyama, T., Hynes, R. O., Burton, D. R., McKeating, J. A., and Rice, C. M. (2005). Complete replication of hepatitis C virus in cell culture. *Science* **309,** 623–626.
274. Simons, J. N., Leary, T. P., Dawson, G. J., Pilot-Matias, T. J., Muerhoff, A. S., Schlauder, G. G., Desai, S. M., and Mushahwar, I. K. (1995). Isolation of novel virus-like sequences associated with human hepatitis. *Nat. Med.* **1,** 564–569.
275. Linnen, J., Wages, J., Jr., Zhang-Keck, Z. Y., Fry, K. E., Krawczynski, K. Z., Alter, H., Koonin, E., Gallagher, M., Alter, M., Hadziyannis, S., Karayiannis, P., Fung, K. *et al.* (1996). Molecular cloning and disease association of hepatitis G virus: A transfusion-transmissible agent. *Science* **271,** 505–508.
276. Xiang, J., Martinez-Smith, C., Gale, M., Jr., Chang, Q., Labrecque, D. R., Schmidt, W. N., and Stapleton, J. T. (2005). GB virus type C NS5A sequence polymorphisms: Association with interferon susceptibility and inhibition of PKR-mediated eIF2alpha phosphorylation. *J. Interferon Cytokine Res.* **25,** 261–270.
277. Benkirane, M., Neuveut, C., Chun, R. F., Smith, S. M., Samuel, C. E., Gatignol, A., and Jeang, K. T. (1997). Oncogenic potential of TAR RNA binding protein TRBP and its regulatory interaction with RNA-dependent protein kinase PKR. *EMBO J.* **16,** 611–624.
278. Muto, N. F., Martinand-Mari, C., Adelson, M. E., and Suhadolnik, R. J. (1999). Inhibition of replication of reactivated human immunodeficiency virus type 1 (HIV-1) in latently infected U1 cells transduced with an HIV-1 long terminal repeat-driven PKR cDNA construct. *J. Virol.* **73,** 9021–9028.
279. Lee, S. B., and Esteban, M. (1993). The interferon-induced double-stranded RNA-activated human p68 protein kinase inhibits the replication of vaccinia virus. *Virology* **193,** 1037–1041.
280. Stojdl, D. F., Abraham, N., Knowles, S., Marius, R., Brascy, A., Lichty, B. D., Brown, E. G., Sonenberg, N., and Bell, J. C. (2000). The murine double-stranded RNA-dependent protein kinase PKR is required for resistance to vesicular stomatitis virus. *J. Virol.* **74,** 9580–9585.
281. Gil, J., and Esteban, M. (2004). Vaccinia virus recombinants as a model system to analyze interferon-induced pathways. *J. Interferon Cytokine Res.* **24,** 637–646.
282. Dimitrova, D. I., Yang, X., Reichenbach, N. L., Karakasidis, S., Sutton, R. E., Henderson, E. E., Rogers, T. J., and Suhadolnik, R. J. (2005). Lentivirus-mediated transduction of PKR into CD34(+) hematopoietic stem cells inhibits HIV-1 replication in differentiated t cell progeny. *J. Interferon Cytokine Res.* **25,** 345–360.
283. Balachandran, S., Roberts, P. C., Brown, L. E., Truong, H., Pattnaik, A. K., Archer, D. R., and Barber, G. N. (2000). Essential role for the dsRNA-dependent protein kinase PKR in innate immunity to viral infection. *Immunity* **13,** 129–141.
284. Carr, D. J., Tomanek, L., Silverman, R. H., Campbell, I. L., and Williams, B. R. (2005). RNA-dependent protein kinase is required for alpha-1 interferon transgene-induced resistance to genital herpes simplex virus type 2. *J. Virol.* **79,** 9341–9345.

285. Durbin, R. K., Mertz, S. E., Koromilas, A. E., and Durbin, J. E. (2002). PKR protection against intranasal vesicular stomatitis virus infection is mouse strain dependent. *Viral Immunol.* **15,** 41–51.
286. Baltzis, D., Qu, L. K., Papadopoulou, S., Blais, J. D., Bell, J. C., Sonenberg, N., and Koromilas, A. E. (2004). Resistance to vesicular stomatitis virus infection requires a functional cross talk between the eukaryotic translation initiation factor 2 alpha kinases PERK and PKR. *J. Virol.* **78,** 12747–12761.
287. Trottier, M. D., Jr., Palian, B. M., and Reiss, C. (2005). Vsv replication in neurons is inhibited by type I IFN at multiple stages of infection. *Virology* **333,** 215–225.
288. Chesler, D. A., Dodard, C., Lee, G. Y., Levy, D. E., and Reiss, C. S. (2004). Interferon-gamma-induced inhibition of neuronal vesicular stomatitis virus infection is STAT1 dependent. *J. Neurovirol.* **10,** 57–63.
289. Komatsu, T., Bi, Z., and Reiss, C. S. (1996). Interferon-gamma induced type I nitric oxide synthase activity inhibits viral replication in neurons. *J. Neuroimmunol.* **68,** 101–108.
290. Diamond, M. S., and Harris, E. (2001). Interferon inhibits dengue virus infection by preventing translation of viral RNA through a PKR-independent mechanism. *Virology* **289,** 297–311.
291. Flodstrom-Tullberg, M., Hultcrantz, M., Stotland, A., Maday, A., Tsai, D., Fine, C., Williams, B., Silverman, R., and Sarvetnick, N. (2005). RNase l and double-stranded RNA-dependent protein kinase exert complementary roles in islet cell defense during coxsackievirus infection. *J. Immunol.* **174,** 1171–1177.
292. Knoetig, S. M., McCullough, K. C., and Summerfield, A. (2002). Lipopolysaccharide-induced impairment of classical swine fever virus infection in monocytic cells is sensitive to 2-aminopurine. *Antiviral Res.* **53,** 75–81.
293. Taddeo, B., Luo, T. R., Zhang, W., and Roizman, B. (2003). Activation of NF-kappaB in cells productively infected with HSV-1 depends on activated protein kinase R and plays no apparent role in blocking apoptosis. *Proc. Natl. Acad. Sci. USA* **100,** 12408–12413.
294. Taddeo, B., Zhang, W., Lakeman, F., and Roizman, B. (2004). Cells lacking NF-kappaB or in which NF-kappaB is not activated vary with respect to ability to sustain herpes simplex virus 1 replication and are not susceptible to apoptosis induced by a replication-incompetent mutant virus. *J. Virol.* **78,** 11615–11621.
295. Stewart, M. J., Blum, M. A., and Sherry, B. (2003). PKR's protective role in viral myocarditis. *Virology* **314,** 92–100.
296. Streitenfeld, H., Boyd, A., Fazakerley, J. K., Bridgen, A., Elliott, R. M., and Weber, F. (2003). Activation of PKR by bunyamwera virus is independent of the viral interferon antagonist NSs. *J. Virol.* **77,** 5507–5511.
297. Goodbourn, S., Didcock, L., and Randall, R. E. (2000). Interferons: Cell signalling, immune modulation, antiviral response and virus countermeasures. *J. Gen. Virol.* **81,** 2341–2364.
298. Didcock, L., Young, D. F., Goodbourn, S., and Randall, R. E. (1999). The v protein of simian virus 5 inhibits interferon signalling by targeting STAT1 for proteasome-mediated degradation. *J. Virol.* **73,** 9928–9933.
299. Huang, Z., Krishnamurthy, S., Panda, A., and Samal, S. K. (2003). Newcastle disease virus V protein is associated with viral pathogenesis and functions as an alpha interferon antagonist. *J. Virol.* **77,** 8676–8685.
300. Andrejeva, J., Young, D. F., Goodbourn, S., and Randall, R. E. (2002). Degradation of STAT1 and STAT2 by the V proteins of simian virus 5 and human parainfluenza virus type 2, respectively: Consequences for virus replication in the presence of alpha/beta and gamma interferons. *J. Virol.* **76,** 2159–2167.
301. Ulane, C. M., and Horvath, C. M. (2002). Paramyxoviruses SV5 and HPIV2 assemble STAT protein ubiquitin ligase complexes from cellular components. *Virology* **304,** 160–166.

302. Karin, M., and Lin, A. (2002). NF-kappaB at the crossroads of life and death. *Nat. Immunol.* **3**, 221–227.
303. Kucharczak, J., Simmons, M. J., Fan, Y., and Gelinas, C. (2003). To be, or not to be: NF-kappaB is the answer—role of Rel/NF-kappaB in the regulation of apoptosis. *Oncogene* **22**, 8961–8982.
304. Bonizzi, G., and Karin, M. (2004). The two NF-kappaB activation pathways and their role in innate and adaptive immunity. *Trends Immunol.* **25**, 280–288.
305. Alexopoulou, L., Holt, A. C., Medzhitov, R., and Flavell, R. A. (2001). Recognition of double-stranded RNA and activation of NF-kappaB by toll-like receptor 3. *Nature* **413**, 732–738.
306. Jiang, Z., Zamanian-Daryoush, M., Nie, H., Silva, A. M., Williams, B. R., and Li, X. (2003). Poly(I-C)-induced toll-like receptor 3 (TLR3)-mediated activation of NFkappa B and map kinase is through an interleukin-1 receptor-associated kinase (IRAK)-independent pathway employing the signaling components TLR3-TRAF6-TAK1-TAB2-PKR. *J. Biol. Chem.* **278**, 16713–16719.
307. Chu, W. M., Ostertag, D., Li, Z. W., Chang, L., Chen, Y., Hu, Y., Williams, B., Perrault, J., and Karin, M. (1999). JNK2 and IKKbeta are required for activating the innate response to viral infection. *Immunity* **11**, 721–731.
308. Zamanian-Daryoush, M., Mogensen, T. H., DiDonato, J. A., and Williams, B. R. (2000). NF-kappaB activation by double-stranded-RNA-activated protein kinase (PKR) is mediated through NF-kappaB-inducing kinase and IkappaBkinase. *Mol. Cell. Biol.* **20**, 1278–1290.
309. Bonnet, M. C., Weil, R., Dam, E., Hovanessian, A. G., and Meurs, E. F. (2000). PKR stimulates NF-kappaB irrespective of its kinase function by interacting with the IkappaBkinase complex. *Mol. Cell. Biol.* **20**, 4532–4542.
310. Gil, J., Alcami, J., and Esteban, M. (2000). Activation of NF-kappa B by the dsRNA-dependent protein kinase, PKR involves the I kappa B kinase complex. *Oncogene* **19**, 1369–1378.
311. DiDonato, J. A., Hayakawa, M., Rothwarf, D. M., Zandi, E., and Karin, M. (1997). A cytokine-responsive I kappa B kinase that activates the transcription factor NF-kappaB. *Nature* **388**, 548–554.
312. Regnier, C. H., Song, H. Y., Gao, X., Goeddel, D. V., Cao, Z., and Rothe, M. (1997). Identification and characterization of an IkappaBkinase. *Cell* **90**, 373–383.
313. Gil, J., Rullas, J., Garcia, M. A., Alcami, J., and Esteban, M. (2001). The catalytic activity of dsRNA-dependent protein kinase, PKR, is required for NF-kappaB activation. *Oncogene* **20**, 385–394.
314. Ishii, T., Kwon, H., Hiscott, J., Mosialos, G., and Koromilas, A. E. (2001). Activation of the I kappa Balpha kinase (IKK) complex by double stranded RNA-binding defective and catalytic inactive mutants of the interferon-inducible protein kinase PKR. *Oncogene* **20**, 1900–1912.
315. Gil, J., Garcia, M. A., Gomez-Puertas, P., Guerra, S., Rullas, J., Nakano, H., Alcami, J., and Esteban, M. (2004). TRAF family proteins link PKR with NF-kappa B activation. *Mol. Cell. Biol.* **24**, 4502–4512.
316. Oganesyan, G., Saha, S. K., Guo, B., He, J. Q., Shahangian, A., Zarnegar, B., Perry, A., and Cheng, G. (2006). Critical role of TRAF3 in the toll-like receptor-dependent and -independent antiviral response. *Nature* **439**, 208–211.
317. Widmann, C., Gibson, S., Jarpe, M. B., and Johnson, G. L. (1999). Mitogen-activated protein kinase: Conservation of a three-kinase module from yeast to human. *Physiol. Rev.* **79**, 143–180.
318. Goh, K. C., deVeer, M. J., and Williams, B. R. (2000). The protein kinase PKR is required for p38 MAPK activation and the innate immune response to bacterial endotoxin. *EMBO J.* **19**, 4292–4297.

319. Silva, A. M., Whitmore, M., Xu, Z., Jiang, Z., Li, X., and Williams, B. R. (2004). Protein kinase R (PKR) interacts with and activates mitogen-activated protein kinase kinase 6 (MKK6) in response to double-stranded RNA stimulation. *J. Biol. Chem.* **279**, 37670–37676.
320. Gardner, O. S., Shiau, C. W., Chen, C. S., and Graves, L. M. (2005). Peroxisome proliferator-activated receptor gamma-independent activation of p38 MAPK by thiazolidinediones involves calcium/calmodulin-dependent protein kinase II and protein kinase R: Correlation with endoplasmic reticulum stress. *J. Biol. Chem.* **280**, 10109–10118.
321. Moran, J. M., Moxley, M. A., Buller, R. M., and Corbett, J. A. (2005). Encephalomyocarditis virus induces PKR-independent mitogen-activated protein kinase activation in macrophages. *J. Virol.* **79**, 10226–10236.
322. Iordanov, M. S., Wong, J., Bell, J. C., and Magun, B. E. (2001). Activation of NF-kappaB by double-stranded RNA (dsRNA) in the absence of protein kinase R and RNase l demonstrates the existence of two separate dsRNA-triggered antiviral programs. *Mol. Cell. Biol.* **21**, 61–72.
323. Barber, G. N. (2005). The dsRNA-dependent protein kinase, PKR and cell death. *Cell Death Differ.* **12**, 563–570.
324. Samuel, C. E., Kuhen, K. L., George, C. X., Ortega, L. G., Rende-Fournier, R., and Tanaka, H. (1997). The PKR protein kinase—an interferon-inducible regulator of cell growth and differentiation. *Int. J. Hematol.* **65**, 227–237.
325. Meurs, E. F., Galabru, J., Barber, G. N., Katze, M. G., and Hovanessian, A. G. (1993). Tumor suppressor function of the interferon-induced double-stranded RNA-activated protein kinase. *Proc. Natl. Acad. Sci. USA* **90**, 232–236.
326. Koromilas, A. E., Roy, S., Barber, G. N., Katze, M. G., and Sonenberg, N. (1992). Malignant transformation by a mutant of the IFN-inducible dsRNA-dependent protein kinase. *Science* **257**, 1685–1689.
327. Donze, O., Jagus, R., Koromilas, A. E., Hershey, J. W., and Sonenberg, N. (1995). Abrogation of translation initiation factor eIF-2 phosphorylation causes malignant transformation of NIH 3T3 cells. *EMBO J.* **14**, 3828–3834.
328. Barber, G. N., Wambach, M., Thompson, S., Jagus, R., and Katze, M. G. (1995). Mutants of the RNA-dependent protein kinase (PKR) lacking double-stranded RNA binding domain I can act as transdominant inhibitors and induce malignant transformation. *Mol. Cell. Biol.* **15**, 3138–3146.
329. Haase, A. D., Jaskiewicz, L., Zhang, H., Laine, S., Sack, R., Gatignol, A., and Filipowicz, W. (2005). TRBP, a regulator of cellular PKR and HIV-1 virus expression, interacts with dicer and functions in RNA silencing. *EMBO Rep.* **6**, 961–967.
330. Barber, G. N. (2001). Host defense, viruses and apoptosis. *Cell Death Differ.* **8**, 113–126.
331. Kibler, K. V., Shors, T., Perkins, K. B., Zeman, C. C., Banaszak, M. P., Biesterfeldt, J., Langland, J. O., and Jacobs, B. L. (1997). Double-stranded RNA is a trigger for apoptosis in vaccinia virus-infected cells. *J. Virol.* **71**, 1992–2003.
332. Der, S. D., Yang, Y. L., Weissmann, C., and Williams, B. R. (1997). A double-stranded RNA-activated protein kinase-dependent pathway mediating stress-induced apoptosis. *Proc. Natl. Acad. Sci. USA* **94**, 3279–3283.
333. Srivastava, S. P., Kumar, K. U., and Kaufman, R. J. (1998). Phosphorylation of eukaryotic translation initiation factor 2 mediates apoptosis in response to activation of the double-stranded RNA-dependent protein kinase. *J. Biol. Chem.* **273**, 2416–2423.
334. Balachandran, S., Kim, C. N., Yeh, W. C., Mak, T. W., Bhalla, K., and Barber, G. N. (1998). Activation of the dsRNA-dependent protein kinase, PKR, induces apoptosis through FADD-mediated death signaling. *EMBO J.* **17**, 6888–6902.
335. Matsumoto, S., Miyagishi, M., Akashi, H., Nagai, R., and Taira, K. (2005). Analysis of double-stranded RNA-induced apoptosis pathways using interferon-response noninducible small interfering RNA expression vector library. *J. Biol. Chem.* **280**, 25687–25696.

336. Wada, N., Matsumura, M., Ohba, Y., Kobayashi, N., Takizawa, T., and Nakanishi, Y. (1995). Transcription stimulation of the Fas-encoding gene by nuclear factor for interleukin-6 expression upon influenza virus infection. *J. Biol. Chem.* **270,** 18007–18012.
337. Takizawa, T., Ohashi, K., and Nakanishi, Y. (1996). Possible involvement of double-stranded RNA-activated protein kinase in cell death by influenza virus infection. *J. Virol.* **70,** 8128–8132.
338. Yeung, M. C., Chang, D. L., Camantigue, R. E., and Lau, A. S. (1999). Inhibitory role of the host apoptogenic gene PKR in the establishment of persistent infection by encephalomyocarditis virus in U937 cells. *Proc. Natl. Acad. Sci. USA* **96,** 11860–11865.
339. Gil, J., Alcami, J., and Esteban, M. (1999). Induction of apoptosis by double-stranded-RNA-dependent protein kinase (PKR) involves the alpha subunit of eukaryotic translation initiation factor 2 and NF-kappaB. *Mol. Cell. Biol.* **19,** 4653–4663.
340. Donze, O., Deng, J., Curran, J., Sladek, R., Picard, D., and Sonenberg, N. (2004). The protein kinase PKR: A molecular clock that sequentially activates survival and death programs. *EMBO J.* **23,** 564–571.
341. Hsu, L. C., Park, J. M., Zhang, K., Luo, J. L., Maeda, S., Kaufman, R. J., Eckmann, L., Guiney, D. G., and Karin, M. (2004). The protein kinase PKR is required for macrophage apoptosis after activation of toll-like receptor 4. *Nature* **428,** 341–345.
342. Saelens, X., Kalai, M., and Vandenabeele, P. (2001). Translation inhibition in apoptosis: Caspase-dependent PKR activation and eIF2-alpha phosphorylation. *J. Biol. Chem.* **276,** 41620–41628.
343. Yeung, M. C., Liu, J., and Lau, A. S. (1996). An essential role for the interferon-inducible, double-stranded RNA-activated protein kinase PKR in the tumor necrosis factor-induced apoptosis in U937 cells. *Proc. Natl. Acad. Sci. USA* **93,** 12451–12455.
344. Cuddihy, A. R., Li, S., Tam, N. W., Wong, A. H., Taya, Y., Abraham, N., Bell, J. C., and Koromilas, A. E. (1999). Double-stranded-RNA-activated protein kinase PKR enhances transcriptional activation by tumor suppressor P53. *Mol. Cell. Biol.* **19,** 2475–2484.
345. Lee, S. B., and Esteban, M. (1994). The interferon-induced double-stranded RNA-activated protein kinase induces apoptosis. *Virology* **199,** 491–496.
346. Kaufman, R. J. (1999). Double-stranded RNA-activated protein kinase mediates virus-induced apoptosis: A new role for an old actor. *Proc. Natl. Acad. Sci. USA* **96,** 11693–11695.
347. Behrmann, I., Walczak, H., and Krammer, P. H. (1994). Structure of the human APO-1 gene. *Eur. J. Immunol.* **24,** 3057–3062.
348. Takahashi, T., Tanaka, M., Inazawa, J., Abe, T., Suda, T., and Nagata, S. (1994). Human fas ligand: Gene structure, chromosomal location and species specificity. *Int. Immunol.* **6,** 1567–1574.
349. Kasibhatla, S., Brunner, T., Genestier, L., Echeverri, F., Mahboubi, A., and Green, D. R. (1998). DNA damaging agents induce expression of fas ligand and subsequent apoptosis in T lymphocytes via the activation of NF-kappa B and AP-1. *Mol. Cell* **1,** 543–551.
350. Kumar, A., Yang, Y. L., Flati, V., Der, S., Kadereit, S., Deb, A., Haque, J., Reis, L., Weissmann, C., and Williams, B. R. (1997). Deficient cytokine signaling in mouse embryo fibroblasts with a targeted deletion in the PKR gene: Role of IRF-1 and NF-kappaB. *EMBO J.* **16,** 406–416.
351. Xia, Z., Dickens, M., Raingeaud, J., Davis, R. J., and Greenberg, M. E. (1995). Opposing effects of ERK and JNK-p38 MAP kinases on apoptosis. *Science* **270,** 1326–1331.
352. Wong, A. H., Tam, N. W., Yang, Y. L., Cuddihy, A. R., Li, S., Kirchhoff, S., Hauser, H., Decker, T., and Koromilas, A. E. (1997). Physical association between STAT1 and the interferon-inducible protein kinase PKR and implications for interferon and double-stranded RNA signaling pathways. *EMBO J.* **16,** 1291–1304.
353. He, P. P., He, C. D., Cui, Y., Yang, S., Xu, H. H., Li, M., Yuan, W. T., Gao, M., Liang, Y. H., Li, C. R., Xu, S. J., Chen, S. J. *et al.* (2004). Refined localization of dyschromatosis symmetrica

hereditaria gene to a 9.4-cm region at 1q21–22 and a literature review of 136 cases reported in China. *Br. J. Dermatol.* **150,** 633–639.
354. Miyamura, Y., Suzuki, T., Kono, M., Inagaki, K., Ito, S., Suzuki, N., and Tomita, Y. (2003). Mutations of the RNA-specific adenosine deaminase gene (DSRAD) are involved in dyschromatosis symmetrica hereditaria. *Am. J. Hum. Genet.* **73,** 693–699.
355. Zhang, X. J., Gao, M., Li, M., Li, C. R., Cui, Y., He, P. P., Xu, S. J., Xiong, X. Y., Wang, Z. X., Yuan, W. T., Yang, S., and Huang, W. (2003). Identification of a locus for dyschromatosis symmetrica hereditaria at chromosome 1q11–1q21. *J. Invest. Dermatol.* **120,** 776–780.
356. Cui, Y., Wang, J., Yang, S., Gao, M., Chen, J. J., Yan, K. L., Xiao, F. L., Huang, W., and Zhang, X. J. (2005). Identification of a novel mutation in the DSRAD gene in a chinese pedigree with dyschromatosis symmetrica hereditaria. *Arch. Dermatol. Res* **296,** 543–545.
357. Gao, M., Wang, P. G., Yang, S., Hu, X. L., Zhang, K. Y., Zhu, Y. G., Ren, Y. Q., Du, W. H., Zhang, G. L., Cui, Y., Chen, J. J., Yan, K. L. *et al.* (2005). Two frameshift mutations in the RNA-specific adenosine deaminase gene associated with dyschromatosis symmetrica hereditaria. *Arch. Dermatol.* **141,** 193–196.
358. Li, C. R., Li, M., Ma, H. J., Luo, D., Yang, L. J., Wang, D. G., Zhu, X. H., Yue, X. Z., Chen, W. Q., and Zhu, W. Y. (2005). A new arginine substitution mutation of DSRAD gene in a chinese family with dyschromatosis symmetrica hereditaria. *J. Dermatol. Sci.* **37,** 95–99.
359. Li, M., Jiang, Y. X., Liu, J. B., Yang, S., He, P. P., Gao, M., Wei, S. C., Yan, K. L., Huang, W., and Zhang, X. J. (2004). A novel mutation of the DSRAD gene in a chinese family with dyschromatosis symmetrica hereditaria. *Clin. Exp. Dermatol.* **29,** 533–535.
360. Liu, Q., Liu, W., Jiang, L., Sun, M., Ao, Y., Zhao, X., Song, Y., Luo, Y., Lo, W. H., and Zhang, X. (2004). Novel mutations of the RNA-specific adenosine deaminase gene (DSRAD) in chinese families with dyschromatosis symmetrica hereditaria. *J. Invest. Dermatol.* **122,** 896–899.
361. Suzuki, N., Suzuki, T., Inagaki, K., Ito, S., Kono, M., Fukai, K., Takama, H., Sato, K., Ishikawa, O., Abe, M., Shimizu, H., Kawai, M. *et al.* (2005). Mutation analysis of the *ADAR1* gene in dyschromatosis symmetrica hereditaria and genetic differentiation from both dyschromatosis universalis hereditaria and acropigmentatio reticularis. *J. Invest. Dermatol.* **124,** 1186–1192.
362. Zhang, X.-J., He, P.-P., Li, M., He, C.-D., Yan, K.-L., Cui, Y., Yang, S., Zhang, K.-Y., Gao, M., Chen, J.-J., Li, C.-R., Jin, L. *et al.* (2004). Seven novel mutations of the *ADAR* gene in chinese families and sporadic patients with dyschromatosis symmetrica hereditaria (DSH). *Hum. Mutat.* **23,** 629–630.
363. Sodhi, M. S., Burnet, P. W., Makoff, A. J., Kerwin, R. W., and Harrison, P. J. (2001). RNA editing of the 5-HT(2C) receptor is reduced in schizophrenia. *Mol. Psychiatry* **6,** 373–379.
364. Niswender, C. M., Herrick-Davis, K., Dilley, G. E., Meltzer, H. Y., Overholser, J. C., Stockmeier, C. A., Emeson, R. B., and Sanders-Bush, E. (2001). RNA editing of the human serotonin 5-HT2C receptor: Alterations in suicide and implications for serotonergic pharmacotherapy. *Neuropsychopharmacology* **24,** 478–491.
365. Yang, W., Wang, Q., Kanes, S. J., Murray, J. M., and Nishikura, K. (2004). Altered RNA editing of serotonin 5-HT2C receptor induced by interferon: Implications for depression associated with cytokine therapy. *Brain Res. Mol. Brain Res.* **124,** 70–78.
366. Akbarian, S., Smith, M. A., and Jones, E. G. (1995). Editing for an ampa receptor subunit RNA in prefrontal cortex and striatum in alzheimer's disease, Huntington's disease and schizophrenia. *Brain Res.* **699,** 297–304.
367. Kwak, S., and Kawahara, Y. (2005). Deficient RNA editing of GluR2 and neuronal death in amyotropic lateral sclerosis. *J. Mol. Med.* **83,** 110–120.
368. Haines, G. K., Cajulis, R., Hayden, R., Duda, R., Talamonti, M., and Radosevich, J. A. (1996). Expression of the double-stranded RNA-dependent protein kinase (p68) in human breast tissues. *Tumour Biol.* **17,** 5–12.

369. Savinova, O., Joshi, B., and Jagus, R. (1999). Abnormal levels and minimal activity of the dsRNA-activated protein kinase, PKR, in breast carcinoma cells. *Int. J. Biochem. Cell Biol.* **31,** 175–189.
370. Haines, G. K., 3rd, Becker, S., Ghadge, G., Kies, M., Pelzer, H., and Radosevich, J. A. (1993). Expression of the double-stranded RNA-dependent protein kinase (p68) in squamous cell carcinoma of the head and neck region. *Arch. Otolaryngol. Head Neck Surg.* **119,** 1142–1147.
371. Haines, G. K., Ghadge, G., Thimmappaya, B., and Radosevich, J. A. (1992). Expression of the protein kinase p-68 recognized by the monoclonal antibody TJ4C4 in human lung neoplasms. *Virchows Arch. B. Cell Pathol. Incl. Mol. Pathol.* **62,** 151–158.
372. Haines, G. K., 3rd, Panos, R. J., Bak, P. M., Brown, T., Zielinski, M., Leyland, J., and Radosevich, J. A. (1998). Interferon-responsive protein kinase (p68) and proliferating cell nuclear antigen are inversely distributed in head and neck squamous cell carcinoma. *Tumour Biol.* **19,** 52–59.
373. Singh, C., Haines, G. K., Talamonti, M. S., and Radosevich, J. A. (1995). Expression of p68 in human colon cancer. *Tumour Biol.* **16,** 281–289.
374. Vojdani, A., Ghoneum, M., Choppa, P. C., Magtoto, L., and Lapp, C. W. (1997). Elevated apoptotic cell population in patients with chronic fatigue syndrome: The pivotal role of protein kinase RNA. *J. Intern. Med.* **242,** 465–478.
375. Mitelman, F., Kaneko, Y., and Trent, J. M. (1990). Report of the committee on chromosome changes in neoplasia. *Cytogenet. Cell Genet.* **55,** 358–386.
376. Abraham, N., Jaramillo, M. L., Duncan, P. I., Methot, N., Icely, P. L., Stojdl, D. F., Barber, G. N., and Bell, J. C. (1998). The murine PKR tumor suppressor gene is rearranged in a lymphocytic leukemia. *Exp. Cell Res.* **244,** 394–404.
377. Li, S., and Koromilas, A. E. (2001). Dominant negative function by an alternatively spliced form of the interferon-inducible protein kinase PKR. *J. Biol. Chem.* **276,** 13881–13890.
378. Hii, S. I., Hardy, L., Crough, T., Payne, E. J., Grimmett, K., Gill, D., and McMillan, N. A. (2004). Loss of PKR activity in chronic lymphocytic leukemia. *Int. J. Cancer* **109,** 329–335.
379. Pang, Q., Keeble, W., Diaz, J., Christianson, T. A., Fagerlie, S., Rathbun, K., Faulkner, G. R., O'Dwyer, M., and Bagby, G. C., Jr. (2001). Role of double-stranded RNA-dependent protein kinase in mediating hypersensitivity of Fanconi anemia complementation group C cells to interferon gamma, tumor necrosis factor-alpha, and double-stranded RNA. *Blood* **97,** 1644–1652.
380. Zhang, X., Li, J., Sejas, D. P., Rathbun, K. R., Bagby, G. C., and Pang, Q. (2004). The Fanconi anemia proteins functionally interact with the protein kinase regulated by RNA (PKR). *J. Biol. Chem.* **279,** 43910–43919.
381. Chang, R. C., Suen, K. C., Ma, C. H., Elyaman, W., Ng, H. K., and Hugon, J. (2002). Involvement of double-stranded RNA-dependent protein kinase and phosphorylation of eukaryotic initiation factor-2alpha in neuronal degeneration. *J. Neurochem.* **83,** 1215–1225.
382. Onuki, R., Bando, Y., Suyama, E., Katayama, T., Kawasaki, H., Baba, T., Tohyama, M., and Taira, K. (2004). An RNA-dependent protein kinase is involved in tunicamycin-induced apoptosis and Alzheimer's disease. *EMBO J.* **23,** 959–968.
383. Peel, A. L., Rao, R. V., Cottrell, B. A., Hayden, M. R., Ellerby, L. M., and Bredesen, D. E. (2001). Double-stranded RNA-dependent protein kinase, PKR, binds preferentially to Huntington's disease (HD) transcripts and is activated in HD tissue. *Hum. Mol. Genet.* **10,** 1531–1538.
384. Lai, F., Chen, C. X., Carter, K. C., and Nishikura, K. (1997). Editing of glutamate receptor B subunit ion channel RNAs by four alternatively spliced DRADA2 double-stranded RNA adenosine deaminases. *Mol. Cell. Biol.* **17,** 2413–2424.

385. Chen, C. X., Cho, D. S., Wang, Q., Lai, F., Carter, K. C., and Nishikura, K. (2000). A third member of the RNA-specific adenosine deaminase gene family, ADAR3, contains both single- and double-stranded RNA binding domains. *RNA* **6,** 755–767.
386. Rueter, S. M., Dawson, R. T., and Emeson, R. B. (1999). Regulation of alternative splicing by RNA editing. *Nature* **399,** 75–80.
387. Slavov, D., and Gardiner, K. (2002). Phylogenetic comparison of the pre-mRNA adenosine deaminase *ADAR2* genes and transcripts: Conservation and diversity in editing site sequence and alternative splicing patterns. *Gene* **299,** 83–94.
388. Kawahara, Y., Ito, K., Ito, M., Tsuji, S., and Kwak, S. (2005). Novel splice variants of human ADAR2 mRNA: Skipping of the exon encoding the dsRNA-binding domains, and multiple c-terminal splice sites. *Gene* **363,** 193–201.
389. Guidotti, L. G., Morris, A., Mendez, H., Koch, R., Silverman, R. H., Williams, B. R., and Chisari, F. V. (2002). Interferon-regulated pathways that control hepatitis B virus replication in transgenic mice. *J. Virol.* **76,** 2617–2621.
390. Al-Khatib, K., Williams, B. R., Silverman, R. H., Halford, W. P., and Carr, D. J. (2002). Absence of PKR attenuates the anti-HSV-1 activity of an adenoviral vector expressing murine IFN-beta. *J. Interferon Cytokine Res.* **22,** 861–871.

Establishment and Regulation of Chromatin Domains: Mechanistic Insights from Studies of Hemoglobin Synthesis

EMERY H. BRESNICK,
KIRBY D. JOHNSON,
SHIN-IL KIM, AND
HOGUNE IM

Department of Pharmacology,
University of Wisconsin Medical School,
383 Medical Sciences Center,
Madison, Wisconsin 53706

I. Introduction	436
II. β-Globin Locus as a Model System to Dissect Chromatin Domain Regulation	437
A. Murine and Human β-Globin Loci	438
B. β-Globin Locus Control Region Structure/Function	438
C. Models of Long-Range Transcriptional Activation	439
III. Protein Components of the Endogenous β-Globin Chromatin Domain	441
A. NF-E2 and Other Maf Response Element Binding Proteins	442
B. Hematopoietic GATA Proteins	443
C. Erythroid Kruppel-Like Factor	447
D. Chromatin Remodeling Components of the β-Globin Chromatin Domain	447
IV. Developmentally Dynamic Histone Modification Pattern of the Murine β-Globin Locus: Molecular Determinants and Functional Implications	449
A. Erythroid Cell-Specific Histone Modification Pattern	449
B. Histone Acetyltransferase vs Histone Deacetylase Actions	450
V. Multistep Model of β-Globin Chromatin Domain Activation	450
A. Dissecting Molecular Determinants of Chromatin Domain Assembly and Regulation via Genetic Complementation Analysis	450
B. Coregulator Interactions	452
C. Segregating Individual Reaction Steps	453
VI. A Role for the LCR in Establishing and Regulating the Erythroid Cell-Specific β-Globin Chromatin Domain	454
VII. General Insights into Chromatin Domain Establishment and Regulation	455
A. Multiple Distant Transcriptional Elements as Common Components of Chromatin Domains	455
B. Discontinuous Histone Modification Patterns	456

C. Exquisite Specificity of *Cis*-Element Utilization in Chromatin........... 457
D. Chromosome Context-Dependent *Trans*-Acting Factor Function....... 458
VIII. Concluding Remarks 458
References 459

I. Introduction

Although the organization of DNA into chromatin and the resulting functional consequences have been studied for decades, only recently has a consensus emerged regarding the importance of chromatin for regulating nuclear processes. DNA replication, transcription, recombination, and repair are dynamically controlled by multiprotein chromatin remodeling complexes that regulate nucleosome structure and positioning, as well as chromatin modifying complexes that posttranslationally modify histone and nonhistone components of chromatin. Not surprisingly, DNA sequences that nucleate the assembly of transcriptional complexes, such as enhancers and locus control regions (LCRs), function in the establishment and regulation of specific chromatin structures as an intrinsic part of the transcription process. Despite extensive progress in solidifying the functional importance of chromatin organization and defining key mediators of chromatin-based regulatory mechanisms, much remains to be learned about how cell- and tissue-specific chromatin domains are established and regulated. Throughout this chapter, "chromatin domain" will refer to a broad chromosomal region from approximately several thousand to several hundred thousand base pairs in which a gene or gene cluster resides, rather than the classically defined ~100-kb topologically constrained chromosomal loop identified by microscopy.

DNA regulatory elements or motifs as small as four nucleotides (nt) are scattered throughout chromatin domains intermixed with identical motifs that are inaccessible to regulatory factors. Taken together with the combinatorial nature of histone modifications, the involvement of multiple large heteromeric protein complexes, and diverse cell signaling inputs, analysis of how chromatin domains are established and regulated poses serious challenges. *In vitro* transcription systems utilizing simple chromatin templates represent a powerful means of dissecting biochemical functions of a defined set of regulatory factors. However, analyses of entire chromatin domains *in vitro* have not been described. Even if domain-wide templates could be reconstituted, validating the biological relevance of such templates will be difficult if not impossible, as multiple components of the native structure of domains are not known. Alternatively, molecular genetic approaches have been used to analyze chromatin domain regulation in cultured cells and in mice. Of particular note are studies of the β-globin locus, which contains a cluster of genes encoding the β-globin subunits of the hemoglobin tetramer. This chapter summarizes progress on dissecting mechanisms underlying the establishment and regulation of the erythroid cell-specific β-globin chromatin domain, while emphasizing general principles that have emerged from this work.

II. β-Globin Locus as a Model System to Dissect Chromatin Domain Regulation

Efforts to understand mechanisms governing cell type-specific transcription were greatly facilitated by the identification of large numbers of sequence-specific DNA-binding proteins that act as classical transcription factors termed *trans*-acting factors (*1*). However, as wrapping DNA around the surface of a nucleosome (*2, 3*) and higher order chromatin folding (*4*) can severely restrict *trans*-acting factor access to DNA (*5*), chromatin organization is a crucial determinant of transcriptional regulation. Studies on the erythroid cell-specific *β-globin* genes have yielded major progress in defining contributions of *trans*-acting factor- and chromatin-based mechanisms to cell type-specific transcription. For example, studies of the chicken *β-globin* genes in primary erythroid cells isolated from chicken embryos led to key discoveries, including the identification of one of the first cell type-specific transcription factors, the erythroid-specific protein GATA-1 (*6, 7*), the description of broad histone acetylation throughout a chromatin domain (*8*), and the identification of an insulator element that segregates a chromatin domain from neighboring domains (*9*).

Surprisingly, despite the sequence conservation of *β-globin* locus coding regions, the organization of the mammalian and chicken β-globin loci differ greatly. Extensive noncoding sequences of mammalian β-globin loci, including a plethora of prospective *cis*-elements, are not conserved with respect to the chicken locus (*10*). By contrast to mammalian red blood cells, chicken red blood cells do not undergo enucleation (*11*). Thus, despite the merits of providing large numbers of primary chicken erythroid cells, it is unclear whether all key mechanisms regulating the chicken *β-globin* genes can be extrapolated to the mammalian *β-globin* genes or if the mammalian genes are regulated via unique mechanisms not operational in chickens. This chapter shall review predominantly mechanistic studies conducted in mammalian systems, including genetic complementation analysis of *trans*-acting factor function in biologically relevant murine erythroid cell lines and analysis of *cis*-element function via targeted deletions in mice.

A. Murine and Human β-Globin Loci

The murine β-globin locus consists of four functional genes, Ey, βH1, βmajor, and βminor (Fig. 1), which are differentially expressed during red blood cell development or erythropoiesis (12, 13). The differential expression during erythropoiesis, termed hemoglobin switching, which is poorly understood, will not be the subject of detailed discussion in this chapter. Ey and βH1 are active during embryonic erythropoiesis, whereas βmajor and βminor are active in adult erythroid cells. Although it was thought that Ey and βH1 are coexpressed in embryonic or primitive erythroid cells, meticulous analysis revealed a "maturational switch" in which βH1 is expressed prior to Ey during mouse embryogenesis (13). The human locus contains five functional genes expressed in the order of their chromosomal arrangement (ε-Gγ-Aγ-δ-β) with ε expressed in the embryo, Gγ and Aγ during fetal development, and δ and β in the adult (Fig. 1) (14).

Upstream of the murine and human β-globin genes reside four erythroid-specific DNaseI hypersensitive sites (HS1-HS4) (15–17) collectively termed the LCR (18, 19). The LCR is defined by its ability to confer copy number-dependent and position-independent expression to linked β-globin transgenes (19). Additional HSs exist upstream of the LCR (20, 21). Human HS5 (22, 23) appears to be analogous to the chromatin insulator chicken β-globin HS4, but unlike chicken HS4 that is active in diverse contexts, HS5-insulating activity has been reported to be developmental stage specific (24). Despite this potential functional conservation, as noted earlier, sequence conservation between noncoding sequences of the mammalian and loci is exceptionally low.

B. β-Globin Locus Control Region Structure/Function

Considerable excitement arose from the discovery that linking the LCR to a transgene counteracts chromosome position effects that create highly variable transgene expression, elevates transgene expression, and induces broad

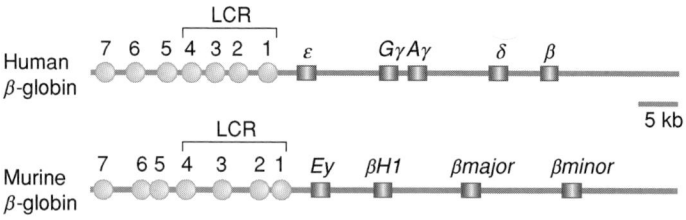

FIG. 1. Organization of the human and murine β-globin loci. The human locus contains β-globin genes expressed during embryogenesis (ε), fetal development (Gγ and Aγ), and the adult (δ and β). The mouse locus contains genes expressed during embryogenesis (Ey and βH1) and in the adult (βmajor and βminor). Multiple DNaseI hypersensitive sites are depicted as spheres upstream of the β-globin genes. HS1–HS4 are designated as the locus control region (LCR).

chromatin accessibility (18, 19, 25). Based on the unusually strong erythroid cell-specific enhancer activity of HS2 in transient and stable transfection assays (26–29) and the unique activity of HS3 to confer copy number-dependent expression to a linked transgene in mice (30), it was thought that the individual HSs have qualitatively distinct functions. Thus, the LCR "holocomplex" would require all HSs to appropriately activate the β-globin locus (30). This concept was further supported by stable transfection studies demonstrating synergism among the individual HSs of the LCR (31, 32). However, studies of LCR function in the context of the endogenous β-globin locus (33–39) yielded results inconsistent with conclusions derived from transgenic and transfection experiments.

Analysis of a mutation that deleted the LCR and upstream sequences in a patient with severe hispanic β-thalassemia yielded the first evidence that the LCR mediates high-level transcription of the endogenous β-globin locus (40). This mutation correlated with locus-wide loss of general DNaseI sensitivity, a parameter indicative of chromatin accessibility. Importantly, targeted deletion of the endogenous murine LCR revealed that the four HSs of the LCR are crucial for conferring high-level transcription of the β-globin genes (36–39), confirming results from transgenic studies. However, despite numerous studies in cultured cells and transgenic mice demonstrating that individual HSs confer unique and essential functions, the targeted deletions of individual HSs failed to provide evidence for such functions. By contrast, the knockout results support a mechanism in which the HSs function additively (39), and therefore similarly, at the endogenous locus. In addition, despite the large reduction in β-globin transcription upon loss of the LCR, developmental switching of the β-globin genes was not impaired.

Unlike results obtained from targeted deletion studies with the endogenous murine β-globin locus, deletions of individual HSs at the human β-globin locus within a yeast artificial chromosome (β-YAC) in transgenic mice can strongly inhibit β-globin gene transcription (41–45). Thus, either the mouse or human loci are subject to fundamentally distinct regulatory mechanisms, the β-YACs integrated at ectopic chromosomal sites are aberrantly sensitive to locus deletions, or deletions within the endogenous murine β-globin locus trigger molecular changes that compensate for deletion of uniquely important HSs. Although it is unlikely that the murine and human loci have fundamentally distinct regulatory mechanisms, major additional work needs to be conducted to assess these possibilities.

C. Models of Long-Range Transcriptional Activation

Given the essential role for the LCR in conferring high-level transcription to the β-globin genes over a long distance on a chromosome, considerable efforts have been directed toward determining how the daunting array of *cis*-elements comprising the LCR communicates with the promoters. Perhaps the simplest model involves "looping" (*46*) in which one or more LCR-associated components physically contact promoter-bound factors, and this contact is essential for or enhances RNA polymerase II (Pol II) recruitment and/or function. Evidence supporting looping at the β-globin locus has emerged from application of the chromosome conformation capture (3C) assay (*47–49*), which allows one to assess the relative proximity of regions separated in three-dimensional space (*50*). A modified RNA fluorescence *in situ* hybridization method (RNA tagging and recovery of associated proteins) was also used to demonstrate proximity of the LCR to downstream regions of the β-globin locus (*51*).

Tracking and linking models have also been proposed to explain LCR function. Tracking assumes that factors, such as Pol II, are recruited far upstream of the *β-globin* genes, for example, at the LCR or at sequences upstream of the LCR, and then processively migrate along the chromosome to a downstream promoter (*52*). The linking model incorporates features of looping and tracking models. LCR-bound factors become physically linked to factors at the promoters via a continuous chain of protein–protein interactions along the chromosome (*53*). These models are not mutually exclusive, as tracking can occur on looped templates, and it would not be surprising if an extensive network of protein–protein interactions links regulatory sequences at the β-globin locus (*54–56*).

The discovery that a large amount of Pol II occupies the LCR HSs (*57*) has important implications for mechanisms underlying LCR function. Pol II occupies the central ∼200–300 base pair "core" regions of the HSs, but not sequences between the HSs, at least not at levels comparable to the HS cores (*58*). In principle, Pol II might be recruited to the HSs, or elongating Pol II might stall at these regions. However, blockade of transcriptional elongation does not redistribute LCR-bound Pol II (*58*), suggesting that Pol II is recruited to the HSs.

β-globin genes are repressed in CB3 murine erythroleukemia cells lacking the erythroid cell- and megakaryocyte-specific p45 subunit of the nuclear factor-erythroid 2 (NF-E2) transcription factor. The p45/NF-E2 locus is disrupted in CB3 cells by retroviral insertion. Importantly, p45/NF-E2 expression reactivates the endogenous *βmajor* gene (*59–63*), providing strong evidence that NF-E2 is a crucial determinant of adult β-globin transcription. Pol II occupies the LCR, but not the *βmajor* promoter, in CB3 cells (*57*). p45/NF-E2 expression in these cells induces Pol II recruitment to the promoter concomitant with *βmajor* transcriptional activation, suggesting a long-range polymerase transfer (LPT) mechanism in which the LCR-bound Pol II is transferred to the promoter in a p45/NF-E2-dependent manner. LPT is also supported by *in vitro* studies demonstrating p45/NF-E2-mediated Pol II transfer on simple

DNA templates under defined conditions (64) and by results in diverse systems (65).

Given the Pol II occupancy at the LCR complex, an obvious question is whether LCR-bound Pol II must be transcriptionally competent or actively engaged in transcription to function, especially considering the explosion of work documenting rampant intergenic transcription throughout genomes (66). Alternatively, is transfer the sole or predominant function? LCR-bound Pol II is phosphorylated on serine 5 of the Pol II large subunit carboxy-terminal domain (57), and serine 5 phosphorylation is linked to transition of a stalled to an elongating Pol II (67, 68). Consistent with the phosphorylation data, Pol II initiates transcription from LCR HSs in cell-free assays and in cells (52, 69), and intergenic transcripts are detected within the chromatin domain (70). Major work is required, however, to determine if the resulting transcripts and/or processed derivatives have regulatory potential, Pol II processivity is linked to chromatin modification, or transcription reflects promiscuous activity of a small fraction of LCR-bound Pol II.

Results have highlighted the functional importance of dynamic changes in the subnuclear localization of transcription factors and even entire chromosomal segments (71–76). Cell biological studies have provided evidence for the existence of transcription and replication "factories," which are highly enriched in transcription and replication factors, respectively (77–81). The LCR mediates extrusion of the β-globin locus from a subnuclear region or chromosomal territory in proerythroblasts prior to β-globin gene activation (82). Subnuclear organization of the β-globin chromatin domain has emerged as an important parameter in the establishment and regulation of the β-globin chromatin domain, but this remains a relatively unexplored frontier.

Although the identification of DNA regulatory elements represents an important first step in dissecting transcription mechanisms, multiple members of a transcription factor family often act through an identical *cis*-element in a chromosome context-specific manner to elicit distinct biological functions. It is therefore essential to identify and analyze the function of endogenous factors that occupy and function through regulatory motifs of the β-globin chromatin domain.

III. Protein Components of the Endogenous β-Globin Chromatin Domain

A key question in the field has been whether the LCR binds unique components that do not occupy and function at promoters and classical enhancers, or if LCR, enhancer, and promoter binding factors are identical.

Sequence conservation analyses have been useful for pinpointing potential regulatory *cis*-elements within the β-globin chromatin domain (*83, 84*). Coupled with traditional gel shift and DNaseI footprinting data, numerous erythroid-specific and broadly expressed DNA-binding activities have been identified. (*85–87*). The LCR-binding proteins identified are the same factors that function through promoters and enhancers.

A. NF-E2 and Other Maf Response Element Binding Proteins

Early mechanistic studies on LCR function focused on understanding the particularly strong erythroid cell-specific enhancer activity of HS2 (*26–28, 88, 89*). Other HSs of the LCR were considerably less effective in conferring enhancer activity in transient and stable transfection assays (*29, 90*). A tandem *cis*-element mediating the enhancer activity binds activator protein-1 (AP-1) (*27*), a common transcription factor that translates diverse cell signals into gene expression changes (*91*). However, since HS2 enhancer activity is erythroid cell specific, and multiple components of AP-1 dimeric complexes are broadly expressed, it seemed likely that a cell type-restricted factor confers the enhancer activity. Such a factor p45/NF-E2 was purified, cloned, and shown to be expressed in erythroid and megakaryocytic cells (*92, 93*). p45/NF-E2 heterodimerizes with p18, a member of the small Maf family of basic leucine zipper transcription factors (*94*); "NF-E2" denotes the heterodimer. As small Maf proteins can also assemble homodimers lacking p45/NF-E2, which bind the same *cis*-element as p45/NF-E2-Maf heterodimers, the DNA sequence motif has been termed a Maf responsive element (MARE) (*95*). Maf homodimers, which can mediate transcriptional repression, occupy HS2 at the repressed β-globin locus in cells lacking p45/NF-E2 (*96*).

As noted earlier, p45/NF-E2 expression reactivates the repressed β*major* gene in CB3 cells (*59–63*), providing strong evidence that p45/NF-E2 is a crucial determinant of adult β-globin transcription in this system. Chromatin immunoprecipitation (ChIP) analysis revealed p45/NF-E2 occupancy at HS2, at other HSs of the LCR, and to a lesser extent at the β*major* promoter, providing strong evidence that p45/NF-E2 functions directly at the β-globin locus (*57, 96–99*). p45/NF-E2 occupancy at spatially segregated regions of the locus is differentially regulated. While p45/NF-E2 occupies HS2, independent of the hematopoietic transcription factor GATA-1 (discussed later in detail), GATA-1 promotes p45/NF-E2 occupancy at HS1 and HS3 (*99, 100*).

p45/NF-E2 occupancy at the LCR in CB3 cells is associated with increases in certain epigenetic marks, including histone H3 acetylation of lysines 9 and 14, and H3 methylation at lysine 4 (H3-dimeK4), at the β*major* promoter and open reading frame (ORF) (*63, 101*). Histone acetylation increases factor access to nucleosomal-binding sites (3, 102), and a modest increase in histone acetylation counteracts higher-order chromatin folding (4). H3-dimeK4 can mark "active" chromatin regions (101, 103, 104), exclude the repressive nucleosome remodeling and deacetylase (NuRD) complex (105), and promote binding of the WD40 repeat protein WDR5 (106).

Little or no Pol II resides at the βmajor promoter in CB3 cells, and p45/NF-E2 expression rescues high-level Pol II occupancy at the promoter (57, 100). p45/NF-E2-mediated rescue of βmajor transcription in CB3 cells requires the sumoylation site lysine 368 (107). Acetylation of the basic domain of the p45/NF-E2 heterodimeric partner MafG enhances NF-E2 DNA-binding activity (108). Moreover, as mitogen-activated protein kinase signaling increases p45/NF-E2-mediated transactivation in the context of stably integrated chromatin templates (109, 110), NF-E2 activity is dynamically regulated in cells.

The biological relevance of the evidence implicating p45/NF-E2 in β-globin transcriptional regulation has been questioned as targeted deletion of p45/NF-E2 in mice is not associated with a pronounced defect in β-globin gene expression or erythropoiesis (111, 112). The p45/NF-E2 knockout results in severe disruption of megakaryopoiesis, and p45/NF-E2 has since been shown to have important functions in the megakaryocyte lineage (113). Thus, in the context of β-globin transcriptional activation, either other factors function redundantly with p45/NF-E2, p45/NF-E2 loss results in compensatory mechanisms that abrogate the p45/NF-E2 requirement, or the p45/NF-E2 requirement is unique to CB3 cells and possibly other erythroleukemia cells. As p45/NF-E2 occupies the LCR in erythroid cell lines and primary erythroid cells (57, 99, 101, 114), and based on CB3 genetic complementation results (59–62), NF-E2 likely has an important function at the endogenous β-globin locus. Although several potential compensatory factors exist, including NF-E2-related factors 1 and 2 (Nrf1 and Nrf2) (115, 116), Bach1 and Bach2 (117, 118), and AP-1 components (91), efforts to establish whether these factors compensate have been inconclusive (119).

B. Hematopoietic GATA Proteins

By contrast to the uncertain role of NF-E2 in β-globin transcriptional regulation, the hematopoietic zinc finger transcription factor GATA-1, one of three hematopoietic GATA factors, has been definitively shown to directly regulate transcription of the β-globin genes (86, 120, 121). GATA-1 was discovered as a DNA-binding activity that interacts with GATA motifs [(A/T)GATA(A/G)] at promoters and an enhancer of the chicken β-globin genes (7, 122, 123). The GATA-1 carboxy-terminal zinc finger binds GATA motifs (124, 125), while the amino-terminal zinc finger stabilizes DNA binding (126–128) and engages in protein–protein interactions with the nine zinc finger coregulator Friend of GATA-1 (FOG-1) (129, 130). FOG-1 increases

GATA-1 chromatin occupancy at certain chromatin target sites (*131, 132*) and is required for "GATA switches" (*121*) in which GATA-1 displaces GATA-2 from chromatin (*121, 131*). GATA-2 is expressed prior to GATA-1 during hematopoiesis and is essential for the function of hematopoietic stem cells and multipotent hematopoietic precursor cells. Thus, it has been proposed that GATA switches at *GATA-2* target genes are essential for hematopoiesis to proceed (*121, 133*). Besides FOG-1 binding, GATA-1 forms distinct multiprotein complexes (*55, 56*) that mediate activation or repression of target genes in a context-dependent manner.

The targeted deletion of *GATA-1* definitively established the crucial role of GATA-1 in erythropoiesis and β-globin transcription (*134, 135*). To circumvent the complexities associated with analyzing biochemical mechanisms *in vivo*, a simple adult or definitive erythroid cell system was developed that allows for facile genetic complementation analysis. *GATA-1* was disrupted via homologous recombination in murine embryonic stem (ES) cells, and an immortalized proerythroblast-like cell line (G1E) was derived (*136*). Activation of an estrogen receptor ligand-binding domain fusion to GATA-1 (ER-GATA-1) stably expressed in G1E cells induces high-level transcription of the adult *β-globin* genes and erythroid maturation (*137, 138*). Importantly, gene profiling analysis indicates that ER-GATA-1-mediated G1E cell maturation recapitulates a normal window of erythropoiesis (*138*). ER-GATA-1 activation increases Pol II occupancy at the LCR and induces Pol II recruitment to the *βmajor* promoter (*100*). As p45/NF-E2 is also required for Pol II recruitment to the promoter, p45/NF-E2 levels are unchanged in GATA-1-null cells, and GATA-1 levels are unchanged in p45/NF-E2-null CB3 cells, Pol II recruitment requires both GATA-1 and p45/NF-E2 (*57, 100*).

Based on the simplicity and abundance (>200 at the β-globin chromatin domain) of the GATA motif, understanding how GATA-1 instigates specific molecular actions, such as Pol II recruitment, at the β-globin domain is complex. ER-GATA-1 and endogenous GATA-1 occupy a small subset of the GATA motifs at the endogenous murine β-globin locus, including motifs at HS1, HS2, HS3, HS4, and the *βmajor* promoter (Fig. 2B) (*99, 100*). Importantly, ER-GATA-1 expression at levels comparable to endogenous GATA-1 recapitulates the GATA-1 occupancy pattern (*100*), further justifying the biological relevance of G1E-ER-GATA-1 cells. Although chromatin accessibility and DNA sequence conservation are important parameters for predicting the probability of whether a given motif will be occupied in chromatin, initial studies suggest that neither of these parameters yield high accuracy predictions of GATA-1 occupancy (*99*).

All ER-GATA-1-occupied sites at the β-globin locus and additional loci are also occupied by endogenous GATA-2 and FOG-1 when stably expressed ER-GATA-1 is inactive in G1E cells (Fig. 2C) (*99*). Although the biological functions of GATA-2 at these sites are unknown, it is tempting to speculate that

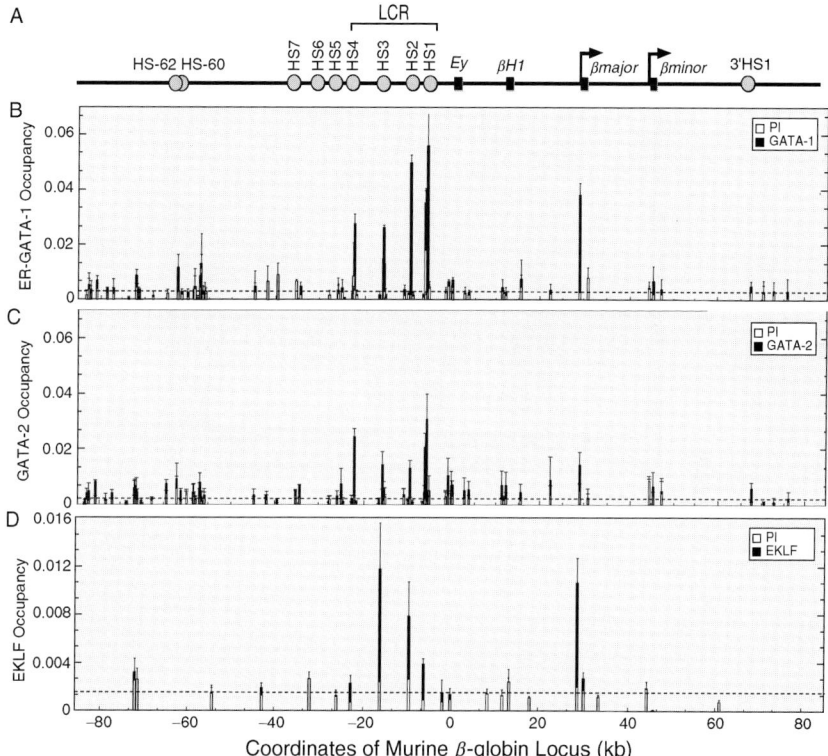

FIG. 2. Chromatin domain-wide profiles of ER-GATA-1, GATA-2, and EKLF occupancy. (A) Organization of the murine β-globin locus. HSs are depicted as spheres. (B) Quantitative ChIP analysis of ER-GATA-1 occupancy at conserved (A/T)GATA(A/G) (WGATAR), WGATA, and GATAR motifs. Note that ER-GATA-1 was expressed at levels comparable to or less than endogenous mouse erythroleukemia (MEL) cell GATA-1. (C) Quantitative ChIP analysis of endogenous GATA-2 occupancy at conserved WGATAR, WGATA, and GATAR motifs, (D) Quantitative ChIP analysis of endogenous EKLF occupancy at conserved CACCC motifs [modified from (99) PI, preimmune].

GATA-2 and/or a GATA-2/FOG-1 complex establishes key features of the active-chromatin structure at genes in which GATA-1 occupancy confers transcriptional activation. As establishment of H3-dimeK4 at the LCR is GATA-1 and NF-E2 independent (99, 101), it will be important to test whether GATA-2 mediates establishment of H3-dimeK4. In addition, occupancy of a target site by GATA-2, followed by FOG-1 recruitment, or by a preformed GATA-2-FOG-1 complex, might facilitate subsequent GATA-1 occupancy (131).

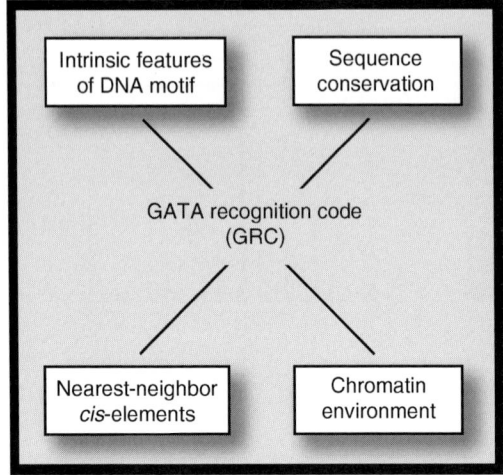

FIG. 3. GATA recognition code. The model assumes that the determinants of GATA factor chromatin occupancy are multifactorial, including the depicted parameters and potentially additional parameters.

Nevertheless, the highly selective GATA motif utilization by GATA-1 suggests the existence of a GATA recognition code (GRC) in which multiple parameters, such as intrinsic features of GATA motifs, nearest-neighbor *cis*-elements, and the chromatin microenvironment, collectively determine whether GATA motifs are competent for factor occupancy in chromatin (Fig. 3). Intrinsic sequence differences between GATA motifs, for example, WGATAR vs WGATA and GATAR appear to be important determinants of chromatin occupancy (99). Sequence conservation is an important parameter, although initial studies indicate that multispecies sequence comparisons do not predict chromatin occupancy with high fidelity (99, 139, 140). It is easy to envision how nearest-neighbor *cis*-elements influence chromatin occupancy. If a *cis*-element in the neighborhood of a GATA motif mediates recruitment of chromatin modifying/remodeling enzymes, chromatin structure changes mediated by such enzymes might dramatically affect GATA factor access to the GATA motif. Chromatin environment is almost certainly a crucial parameter, and this parameter can be subdivided into more specific categories, including spatial relationship between motif and nucleosome position, linker histone occupancy, epigenetic marks, and higher order structures. Although extensive evidence supports a direct role for GATA-1 in activating β-globin transcription, considerable work is required to elucidate the GRC and post-chromatin occupancy mechanisms that dictate context-dependent transcriptional outputs.

C. Erythroid Kruppel-Like Factor

Analogous to GATA-1, which was purified and cloned from erythroid cells based on its erythroid cell-specific DNA-binding activity (6, 7), erythroid Krupple-like factor (EKLF) was discovered by differential cloning of erythroid cell-enriched cDNAs (141). EKLF and Krupple-like factor (KLF) family members bind CACCC motifs (142, 143), and such motifs are implicated in β-globin promoter transcriptional regulation (144–146). The targeted deletion of EKLF in mice provides strong evidence that EKLF is a crucial regulator of adult β-globin transcription (147–149). EKLF mediates late stages of erythroid differentiation, but is not essential for adult erythroid progenitor expansion (150). Recent work indicates that EKLF has important functions in the primitive erythroid linage (151). Given the crucial role of EKLF in adult β-globin transcriptional activation and in β-globin gene switching (152–156), it is of obvious importance to consider how EKLF interfaces with GATA-1, NF-E2, and other regulatory factors in the establishment and regulation of the β-globin chromatin domain.

ChIP analysis revealed EKLF occupancy at HS2, HS3, and the βmajor promoter and weakly at HS1 (99), supporting a direct role in adult β-globin transcriptional activation. Analogous to the stringent specificity of GATA motif utilization, the majority of CACCC motifs are not occupied by endogenous EKLF (Fig. 2D) (99). GATA-1 and EKLF interact physically and functionally (157, 158), raising the questions of whether these factors co-occupy or bind chromatin sites independently. ER-GATA-1 and EKLF occupancy correlate at HS2 and HS3 but not HS4 and HS1 (99). Thus, ER-GATA-1 and EKLF do not appear to co-occupy all chromatin target sites.

EKLF interacts with cyclic AMP response element-binding protein (CREB)-binding protein (CBP), the CBP paralog p300 (159), and the BRG1 subunit of the SWI/SNF chromatin remodeling complex (159–161), suggesting that EKLF functions in part via chromatin modification and remodeling. Consistent with this notion, targeted deletion of EKLF abrogates DNaseI hypersensitivity at HS3 and the β-globin promoter (162), and EKLF mediates chromatin remodeling in vitro (160).

D. Chromatin Remodeling Components of the β-Globin Chromatin Domain

As the SWI/SNF complex (163, 164) is required for EKLF-mediated activation of the human β-globin promoter in vitro (160, 165), EKLF function at the endogenous locus might require SWI/SNF activity. Based on a physical interaction with BRG1 (*165*), it was suggested that EKLF recruits SWI/SNF to chromatin. Surprisingly, in uninduced G1E-ER-GATA-1 cells, in which ER-GATA-1 is largely inactive, BRG1 occupies HS4, HS3, HS2, *βmajor* promoter, and *βmajor* intron 2 (*99*). ER-GATA-1 activity is therefore not required for BRG1 recruitment to the locus. As ER-GATA-1 activation elevates EKLF levels via a transcriptional mechanism (*99*), maximal EKLF levels are also not required for BRG1 recruitment. ER-GATA-1 activation uniquely increases BRG1 occupancy at HS3, suggesting a requirement for higher SWI/SNF levels at HS3, but not at other sites.

Whereas the ChIP studies demonstrating BRG-1 occupancy at the locus are compelling, functional analyses are required to dissect whether occupancy translates into altered β-globin transcription. A hypomorphic allele of BRG-1 was isolated via an ethylnitrosourea mutagenesis screen in mice (*166*). The mutation resides within the BRG1 ATPase domain, but the mutant retains ATPase activity. The *BRG1* mutant is competent to form a stable SWI/SNF complex and occupy endogenous target genes. Although the mutant occupies the LCR, β-globin transcription is severely reduced in mouse embryos expressing the mutant. Accordingly, BRG-1 function is crucial for β-globin transcriptional activation, but the underlying mechanisms are not understood.

DNA-binding studies with naked DNA templates provided evidence for SWI/SNF interactions at a pyrimidine-rich element between the human fetal and adult *β-globin* genes (*167, 168*). Deletion of the sequence-mediating binding *in vitro* delays fetal to adult *β-globin* gene switching in transgenic mice (*167*). The purified "PYR" complex contains Ikaros, a transcription factor initially implicated in regulating lymphopoiesis (*169*), but was shown to function in multiple hematopoietic lineages (*170*). The purification of erythroid cell factors binding the HS2 MARE motif resulted in the identification of SWI/SNF in a complex with heterogeneous nuclear ribonucleoprotein C1/C2 and the NuRD complex (*171*).

The following questions have emerged from the studies described earlier. Which factors mediate GATA-1-independent BRG-1 recruitment? Why is additional BRG-1 required at HS3? Is BRG-1 functional without ER-GATA-1 activity? Is BRG-1 function induced post-recruitment? Does BRG-1 function at the β-globin locus reflect a canonical activity to remodel and move nucleosomes (*163*), or are novel mechanisms involved? These issues need to be considered in the context of how the multiple regulators of β-globin transcription function through the many prospective regulatory elements of the β-globin locus to confer erythroid cell-specific and developmental stage-specific expression. Such an integrative analysis requires knowledge of how the endogenous β-globin locus assembles into an erythroid cell-specific chromatin structure that selectively restricts or facilitates factor access to key regulatory regions.

IV. Developmentally Dynamic Histone Modification Pattern of the Murine β-Globin Locus: Molecular Determinants and Functional Implications

Through the use of ChIP (172), and ChIP coupled with genomic microarrays (ChIP-chip) (173, 174), histone modification patterns across broad chromosomal regions can be readily determined. Such patterns pinpoint key regulatory regions that often reside within islands of enriched epigenetic marks (175). Furthermore, the histone modification pattern of an endogenous locus provides a biologically relevant readout to monitor how dynamic regulatory processes, such as cell signaling and developmental changes in factor levels/activities, converge upon sites within a chromatin domain.

A. Erythroid Cell-Specific Histone Modification Pattern

The β-globin locus exhibits a developmentally dynamic, erythroid-specific histone modification pattern (20, 101, 114, 176, 177). In adult erythroid cells, the LCR and the adult β-globin genes reside within chromatin containing high levels of acetylated histones H3 and H4 and H3-dimeK4. These histone modifications are erythroid cell-specific as histone acetylation is not enriched in fetal brain and in nonerythroid cell lines. The histone modification pattern of mouse erythroleukemia (MEL) cells resembles that of the locus in E13.5 fetal liver (114), validating the MEL cell system for analyzing determinants and functional consequences of the pattern.

By contrast to the adult β-globin genes, the embryonic β-globin genes (Ey and βH1) reside in a chromatin domain containing low levels of acetylated H3, acetylated H4, and H3-dimeK4 (20, 101, 114, 176). A limited analysis of histone acetylation within the central subdomain containing the embryonic Ey and βH1 genes and the βmajor promoter in mouse yolk sac containing primitive erythroid cells provided evidence for broadly enriched acetylation in primitive erythroid cells (114). In human K562 erythroleukemia cells expressing both the embryonic ε- and the fetal Gγ- and Aγ-globin genes, the β-globin locus is broadly enriched in histone acetylation and H3-dimeK4 (177). Hemin-induced erythroid maturation of K562 cells is accompanied by an approximate two-fold increase in embryonic ε-globin transcription and two-fold increase in histone acetylation throughout the locus with little or no change in H3-dimeK4 (177). As the evidence supporting a developmentally dynamic histone modification pattern accrued from studies in multiple cell systems differing in developmental stage, further studies are required using a single system in which multipotent hematopoietic precursors differentiate into erythroid cells.

B. Histone Acetyltransferase vs Histone Deacetylase Actions

Histone deacetylase (HDAC) inhibitors can reactivate the repressed human fetal β-*globin* gene transcription in patients and in certain cell systems (*178–182*). However, it is unclear whether this therapeutically beneficial activity involves a direct regulation of the histone modification pattern of the β-globin locus.

Since the repressed murine embryonic β-*globin* genes reside within a broad region of low-level acetylation in MEL cells (*114*), the HDAC inhibitors butyrate and trichostatin A (TSA) were used to determine if suppressing HDAC activity establishes broad hyperacetylation within the central subdomain (*183*). The low-level acetylation might result from an overt defect in histone acetyltransferase (HAT) recruitment to this region, persistent HDAC action at this region, or both mechanisms. Butyrate or TSA, with or without the DNA methylase inhibitor 5-azacytidine that also induces fetal hemoglobin (*184*), restored high-level H4, but not H3 acetylation, at the central subdomain. Restoration of H4 acetylation, however, was insufficient to reactivate the repressed genes. Thus, maintenance of low-level H4 acetylation at the central subdomain involves continuous HDAC function at this region. H3 acetylation is not restored upon HDAC inhibition indicating that HATs are unable to access this region and/or HDAC inhibitor-insensitive HDACs maintain low-level acetylation.

Many questions regarding the role of HATs and HDACs in establishing and regulating the β-globin chromatin domain remain unanswered. For example, what specific HDACs are involved? How do HDACs uniquely access and function at the central subdomain without dominantly functioning at the LCR and the adult β-*globin* genes? What mechanisms oppose HAT access and/or function at the central subdomain? In this regard, tractable, biologically relevant systems are needed that allow for detailed mechanistic analysis. The innovative development of immortalized lines lacking GATA-1 (*136*) and NF-E2 (*59*) has provided an entrée into this problem.

V. Multistep Model of β-Globin Chromatin Domain Activation

A. Dissecting Molecular Determinants of Chromatin Domain Assembly and Regulation via Genetic Complementation Analysis

Genetic complementation analyses in G1E cells led to the development of a multistep model to explain how GATA-1 contributes to the establishment of the active β-globin chromatin domain (Fig. 4). GATA-1 deficiency in G1E cells

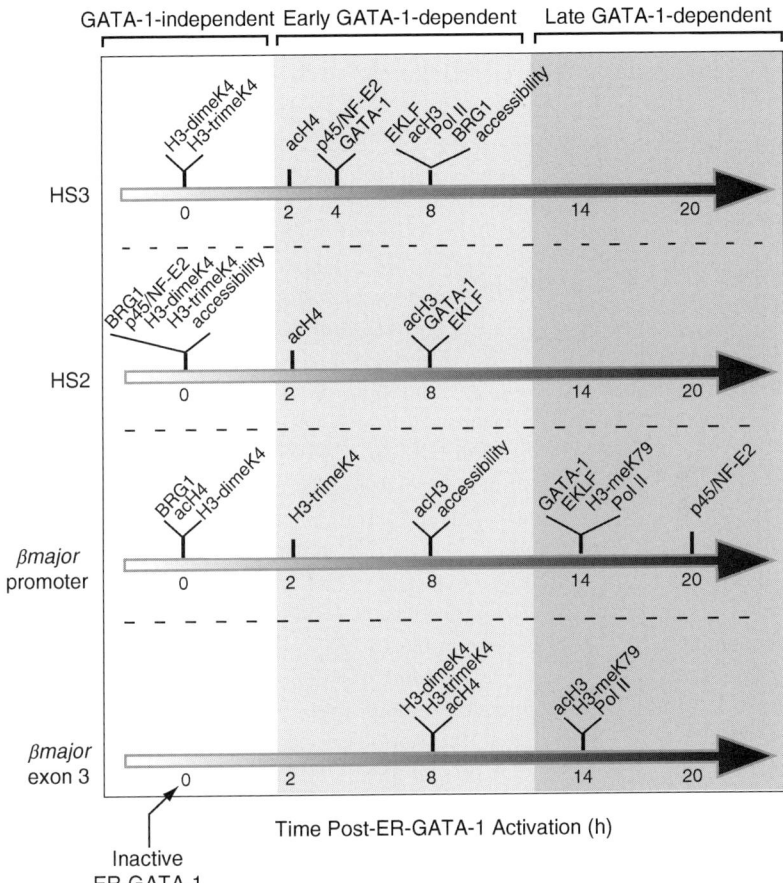

FIG. 4. Multistep model of β-globin locus activation. The model [modified from (99)] depicts molecular events measured at the endogenous murine β-globin locus various times after tamoxifen mediated activation of ER-GATA-1 stably expressed in G1E cells. Molecular events are indicated at the time in which their values equal 50 ± 20% of the maximal value achieved during the kinetic analysis. Three phases in the establishment of the active β-globin domain are indicated at the top. In the GATA-1-independent phase, certain molecular events have already occurred, indicating that other cellular factors mediate their establishment. Note that ER-GATA-1 occupancy at the LCR precedes ER-GATA-1 occupancy at the promoter. A central aspect of the model is that differential utilization of GATA motifs within the chromatin domain underlies the multistep activation mechanism.

is associated with modestly reduced histone H3 and H4 acetylation at the βmajor promoter, βmajor ORF, and the LCR. Acetylation is rescued upon activation of stably expressed ER-GATA-1 (99, 101, 185). p45/NF-E2 deficiency in CB3 cells is associated with modestly reduced H3 acetylation, and a lesser reduction in H4 acetylation, at the *βmajor* promoter and ORF. Stable expression of p45/NF-E2 rescues acetylation (57, 63, 101). p45/NF-E2 deficiency also results in reduced H3-dimeK4 at the *βmajor* promoter and ORF, but not at the LCR, and again p45/NF-E2 expression rescues H3-dimeK4 (101). Reduced H3-dimeK4 is not simply a consequence of transcriptional repression, since *βmajor* transcription is abrogated in G1E cells, and H3-dimeK4 levels at the promoter in these cells are comparable to that of GATA-1-expressing cells (99, 101). Neither GATA-1 nor p45/NF-E2 deficiency influences the highly enriched, erythroid cell-specific H3-dimeK4 at the LCR.

The results described in the previous paragraph indicate that considerable erythroid cell-specific acetylation and H3-dimeK4 persist at the locus in cells lacking GATA-1 or p45/NF-E2. Although analyses of epigenetic marks have not been conducted in cells lacking both GATA-1 and NF-E2, additional factors appear to be required for establishing and/or maintaining the erythroid cell-specific histone modification pattern. Defining such factors has potential to reveal previously unidentified regulators of *β*-globin transcription.

Another epigenetic mark linked to transcriptional activation, H3-meK79 (186), is regulated by both GATA-1 and NF-E2 (176). H3-meK79 is catalyzed by the disruptor of telomeric silencing (DOT1) methyltransferase (186). In *Saccharomyces cerevisiae*, H3-meK79 has been shown to exclude silent information regulator (SIR) repressive complexes (186), thereby, maintaining chromosomal regions as active chromatin. Stable expression of p45/NF-E2 in CB3 cells rescues H3-meK79 at the *βmajor* promoter and ORF. Similarly, during the later stages of ER-GATA-1-mediated transcriptional activation in the G1E system, H3-meK79 levels increase at the *βmajor* promoter and ORF (99). Of particular interest is the finding that H3-meK79 levels are low at the LCR, despite other epigenetic marks associated with active chromatin being similarly high at the LCR and the adult *β-globin* promoters and genes (99, 176).

B. Coregulator Interactions

The specific HAT(s) and histone methyltransferases that mediate the effects of p45/NF-E2 and GATA-1 on the histone modification pattern are unknown, although p45/NF-E2, GATA-1, and EKLF physically and functionally interact with the HATs CBP/p300 (159, 187, 188). CBP/p300 occupies the LCR and the *βmajor* promoter (63, 185). In addition to catalyzing histone acetylation, CBP/p300 acetylates nonhistone proteins, including GATA-1 (189), EKLF (159), and p45/NF-E2 (108). CBP/p300 interacts with other HATs, including the p300/CBP-associated factor (PCAF) complex that catalyzes histone and nonhistone protein acetylation (190, 191).

As discussed earlier, FOG-1 is an essential GATA-1 coregulator that mediates transcriptional activation and repression (86). FOG-1 is crucial for assembly of the active *β*-globin chromatin domain, as GATA-1 access to certain chromatin sites is facilitated by FOG-1 (*131, 132*), and FOG-1 is required for GATA-1-mediated looping that brings the LCR in proximity of the adult β-globin genes (*49*). It will be intriguing if future work reveals direct links between FOG-1 and other coactivators that mediate β-globin chromatin domain activation.

The amino-terminal activation domain of p45/NF-E2 contains two PPXY sequences that bind WW domains *in vitro* and are required for activation of endogenous *βmajor* transcriptional activation (*62*). Although the PPXY motifs are not required for p45/NF-E2 to activate transcription in a transient transfection assay and to occupy HS2 at the endogenous locus, a mutant lacking the PPXY motifs is unable to induce H3 acetylation, H3-dimeK4, and Pol II recruitment (*63*). Multiple WW domain proteins, including YAP (*192, 193*) and TAZ (*194*), are transcriptional coactivators, but their involvement in mediating p45/NF-E2 function is unclear. Members of the homologous to E6-associated protein carboxy-terminus (HECT) domain ubiquitin ligase family contain WW domains that mediate substrate interactions (*195*). WW domains from the WWP1 ubiquitin ligase bind the p45/NF-E2 PPXY motifs (*62, 196*). The related proteins Itch and E6-associated protein (E6-AP) have been reported to be NF-E2 (*197*) and nuclear receptor coactivators (*198*), respectively.

C. Segregating Individual Reaction Steps

Considerable progress has been made in determining the order-of-events in ER-GATA-1-mediated establishment of the active murine β-globin chromatin domain. ER-GATA-1 activation in G1E cells with a maximally effective concentration of estradiol or tamoxifen for varying times, or by titrating ER-GATA-1 activity with a range of agonist concentrations allows one to dissect the temporal relationships among multiple ER-GATA-1-instigated reaction steps. Such experiments revealed that ER-GATA-1 occupies LCR sites prior to occupancy at the *βmajor* promoter, and the LCR is preferentially occupied at a lower level of ER-GATA-1 activity (*99*). As the LCR and *βmajor* promoter contain GATA motifs indistinguishable from those that bind GATA-1 with high affinity *in vitro*, the results can be explained by an apparently higher affinity of the GATA motifs in the LCR chromatin.

The intriguing sequential occupancy of spatially segregated regions of the locus raised the possibility that ER-GATA-1 instigates certain regulatory events first at the LCR, followed by events constituting a late phase in activation. These studies led to a three-phase model: (i) GATA-1-independent establishment of certain epigenetic features and factor occupancy at the locus; (ii) GATA-1-dependent factor loading and epigenetic modifications at the LCR; and (iii) GATA-1-dependent Pol II recruitment to the promoter, activator loading on the promoter, and epigenetic modifications at the promoter and ORF (Fig. 4) (*99*). This mechanism, in which each phase consists of multiple individual reactions, provides a strong foundation for efforts to delineate the kinetics of additional regulatory steps and to molecularly modulate individual steps within each phase. Further studies are expected to reveal whether individual reactions within a single phase and those spanning multiple phases are interlinked or proceed independently.

VI. A Role for the LCR in Establishing and Regulating the Erythroid Cell-Specific β-Globin Chromatin Domain

Given the crucial role of the LCR in conferring high-level transcription of the β-globin genes (199, 200), loss of general DNaseI sensitivity locus-wide that accompanies the LCR deletion in hispanic β-thalassemia (40), and the hallmark LCR activity to counteract chromosome position effects (19), it was assumed that the LCR has a major role in establishing the active chromatin structure of the β-globin locus. Surprisingly, however, targeted deletions of HS1–HS4, and also the individual HSs in mice do not support this assumption (36). Loss of the LCR does not abrogate the locus-wide general DNaseI sensitivity (39), does not have an obvious effect on H4 acetylation at the adult β-globin promoters and ORF, and results in a modest (201) or no (202) reduction in H3 acetylation at the adult β-globin promoters and ORF. Furthermore, although the LCR was believed to be required for *trans*-acting factor access to the promoter, GATA-1 and p45/NF-E2 occupancy was detected at the *βmajor* promoter in the LCR-deleted allele (49, 96).

Since GATA-1 and NF-E2 are crucial for Pol II recruitment to the adult β-globin promoters and both factors occupy the LCR (57, 100), it is attractive to propose that LCR occupancy is required for Pol II recruitment to the promoters. However, at least 50% of the Pol II present on a wild-type allele occupies the *βmajor* promoter of the LCR-deleted allele (203). Since GATA-1 and p45/NF-E2 can be detected at the promoter without the LCR, promoter-bound GATA-1 and p45/NF-E2 might recruit Pol II to the promoter. Since transcription of the LCR-deleted allele is <5% of the wild-type allele, at least one activity of the LCR is to stimulate transcription post-Pol II recruitment, and this appears to involve phosphorylation of the Pol II carboxy-terminal domain (203). For transcriptional regulatory elements that function via chromatin modification and remodeling, it would not be surprising if such elements influence multiple steps in transcription, including preinitiation complex assembly, promoter clearance, and elongation.

If the LCR is not essential for establishing the erythroid cell-specific histone modification pattern and for recruiting factors and Pol II to the β-globin locus, how are these molecular steps regulated? Importantly, Pol II is lower at the promoter of the LCR-deleted vs the wild-type allele (203), consistent with an important role of the LCR in recruiting maximal levels of Pol II. Regarding the histone modification pattern, the LCR resides within an island of highly enriched epigenetic marks (20, 101, 114), and deletion of the LCR obviously eliminates the potential for these sequences to assemble an erythroid cell-specific chromatin structure. The apparent LCR-independent establishment of enriched histone acetylation at the adult β-globin genes and promoters does not rule out an important function for the LCR in chromatin remodeling/modification, as this function might be primarily manifested at and immediately surrounding the LCR.

Studies have only begun to investigate the impact of the LCR on the locus-wide epigenetic marks, and the number of such marks known has increased almost exponentially (204). Extensive semi-quantitative analyses are required to assess factor occupancy at the promoters with and without the LCR at distinct developmental stages and with LCR-deleted erythroid cells derived from human ES cells. One can envision how the LCR deletion might be associated with chromosomal changes that allow for promiscuous transcription factor and Pol II access to the promoter, but such changes have not yet been identified. HS1–HS4 are, therefore, insufficient to establish certain erythroid cell-specific molecular hallmarks of the β-globin locus, indicating that additional *cis*-elements remain to be identified, or new models are required to understand how the LCR confers long-range transactivation.

VII. General Insights into Chromatin Domain Establishment and Regulation

Although the β-globin locus is used as a model to elucidate mechanisms underlying cell type-specific and developmentally regulated transcription, it is possible that mechanisms regulating the β-globin genes are unique and not applicable to large numbers of mammalian genes. Clearly, this is not the case, as multiple general principles have emerged from this work.

A. Multiple Distant Transcriptional Elements as Common Components of Chromatin Domains

An obvious discovery of general importance that emerged from β-globin chromatin domain studies is the development of the concept of an LCR. The demonstration that multiple HSs of the LCR confer high-level transcription over a long distance on the chromosome, and the development of the concept that LCRs function via protecting chromatin domains from chromosomal position effects have far-reaching implications (18, 19). Although the mechanistic conclusion regarding the capacity of the LCR to establish and maintain active chromatin structure has been questioned as discussed earlier, certainly, the LCR has this activity when linked to transgenes. Major additional work is required to investigate the relationship between LCR activities in the context of transgenes vs the endogenous locus. An ever-increasing number of genes have been shown to be controlled via long-range regulatory elements consisting of multiple cell type-specific HSs (85, 205) resembling the β-globin LCR, and conceivably such elements would fulfill the classical definition of an LCR upon rigorous analysis in transgenic mice.

B. Discontinuous Histone Modification Patterns

Based on the early work on chromatin domain structure/function that correlated histone acetylation with active chromatin (206), and subsequent studies demonstrating that the chicken β-globin locus in erythroid cells is broadly enriched in acetylated histones (207), a paradigm emerged in which active chromatin domains were believed to be broadly enriched in histone acetylation. However, detailed studies of the histone modification pattern at the murine β-globin locus demonstrating a discontinuous and developmentally dynamic histone modification pattern are inconsistent with this generalization (20, 114).

Enhancers and LCRs often reside in islands of enriched histone modifications associated with active chromatin, resembling chromatin structure at promoters and ORFs; intergenic sites lacking regulatory elements and repressed genes within active domains can be devoid of such epigenetic marks. Analogous to the murine β-globin chromatin domain in adult erythroid cells, the *Gata-2* domain is characterized by a developmentally dynamic histone modification pattern in which cell type-specific epigenetic marks are enriched at upstream HSs, the promoter, and ORF (133, 139). The endothelial cell-specific human *Notch4* chromatin domain exhibits an endothelial cell-specific histone modification pattern in which epigenetic marks associated with active chromatin are surprisingly low throughout most of the ORF (208). However, the promoter and the 5'-portion of the ORF are enriched in acetylated histones H3 and H4 and H3-dimeK4. Thus, considerable opportunity exists for regulatory factors to sculpt unique histone modification patterns, which elicit functional consequences via regulation of local chromatin structure, factor access, and higher order chromatin organization.

Not only is murine β-globin locus chromatin characterized by a discontinuous pattern of histone acetylation in adult erythroid cells, but other epigenetic marks including H3-dimeK4 are also nonuniformly distributed throughout the domain (*101*). H3 and H4 acetylation and H3-dimeK4 colocalize within the domain, suggesting that establishment of these histone modifications at such regions might be intricately linked. In this regard, multiple scenarios have been described in which establishment of a given histone modification is linked to that of another (*204*). Studies at the murine β-globin chromatin domain reveal that histone acetylation, but not H3-dimeK4, at the *βmajor* promoter is induced by GATA-1 (*101*). Furthermore, p45/NF-E2 preferentially induces histone H3 vs H4 acetylation at the promoter (*57, 101*). Clearly, certain components of the composite histone modification pattern are differentially established and regulated, despite the spatial concordance of these components. Nevertheless, existing data can be viewed as being the tip-of-the-iceberg vis-à-vis dissecting interrelationships between individual components of the histone modification pattern.

C. Exquisite Specificity of *Cis*-Element Utilization in Chromatin

Naked DNA-binding assays, including site-selection assays that isolate preferentially bound oligonucleotides, are routinely used to define the DNA-binding specificity of *trans*-acting factors (*172*), and it is assumed that this specificity can be extrapolated to chromatin occupancy in cells. Combined with evolutionary conservation analysis of prospective binding motifs, promising chromosome-wide predictions of CREB (*209*) and Nuclear Factor κ-B (*210*) occupancy and function have been reported. However, direct analyses of chromatin occupancy at the endogenous β-globin chromatin domain and other domains by ChIP revealed GATA-1 and GATA-2 occupancy at only a small subset of GATA motifs (*99, 133, 139, 140*). GATA factor chromatin occupancy sites in cells does not appear to be subject to binary regulation in which sites are fully accessible or inaccessible. Accessible sites can have distinct apparent affinities (e.g., LCR HSs are occupied by relatively low ER-GATA-1 activity, and *βmajor* promoter occupancy requires high ER-GATA-1 activity) (*99*). Initial studies of EKLF chromatin occupancy in cells reveal occupancy at only a small subset of CACCC motifs (*99*). While *cis*-element sequence and conservation have utility for predicting chromatin occupancy for certain factors, it would not be surprising if chromatin occupancy rules for certain factors are multifactorial. It is attractive to consider that novel rules beyond *cis*-element sequence and conservation specify chromatin occupancy, and elucidating such rules is expected to have exceptionally broad biological implications. In principle, such studies will be greatly facilitated via ChIP-chip methodology, but it will be crucial to validate genome-wide ChIP-chip data sets, as the false negative and positive rates might be considerable.

D. Chromosome Context-Dependent *Trans*-Acting Factor Function

Despite the GATA-1 occupancy of multiple regions within a chromatin domain, GATA-1 instigates unique molecular actions at spatially segregated sites. This finding highlights the context-dependent nature of *trans*-acting factor function to establish and regulate chromatin domains. This concept of spatially segregated factors eliciting spatially distinct functions is also nicely illustrated by the finding that p45/NF-E2 occupies the β-globin LCR and β*major* promoter, but only establishes H3-dimeK4 at the promoter and ORF (*101*). Similarly, GATA-1 induces H3-meK79 at the β*major* promoter and ORF but not at GATA-1-occupied regions within the LCR (99). Moreover, despite p45/NF-E2 occupancy at both the LCR and the β*major* promoter, p45/NF-E2 is crucial for Pol II recruitment to the promoter, but not at the LCR (57). These findings highlight the context-dependent nature of the molecular steps instigated by *trans*-acting factors that establish and regulate chromatin domains. Undoubtedly, this important area is ripe for extensive investigation.

VIII. Concluding Remarks

Regardless of the exact details of how the LCR confers long-range transactivation, distal and proximal regulatory elements of the murine β-globin chromatin domain are brought into proximity of each other via looping (47–49), similar to results from studies in simpler biological systems (46, 50). While it is easy to conceptualize how such loops form with a limited number of proteins bound to a small enhancer and nearby promoter, how can reproducible structures be assembled at a ~100-kb chromatin domain, given the apparent plethora of nonhistone factors, histone modifications, and histone variants within such a broad region? Is the looped structure stabilized via combinatorial protein–protein interactions involving many components or a small defined set of regulatory proteins? The demonstrations that single-factor deficiencies, either EKLF (*48*) or GATA-1 (*49*), abrogate looping suggest that looping is a delicate process in which individual proteins promote loop assembly and disassembly.

Perhaps we are overestimating the complexity of chromatin domain regulation, as many *in vitro* protein interactors might not be functionally important in cells. Furthermore, despite the presence of a specific and developmentally dynamic histone modification pattern, nucleosomal filament and higher-order chromatin structure might not be precise with respect to the integral protein–protein interactions and three-dimensional geometry of the chromatin domain. It remains a possibility that our knowledge of rules governing the establishment and regulation of chromatin domains is analogous to early world explorers who devised first-generation navigational maps based on the fallacious premise that the world is flat! Sorting out these big-picture issues necessitates extensive additional experimentation using diverse approaches and systems, and the merits of incorporating systems biology-level integrative analysis (211) need to be carefully considered. Further work on the compelling problem of how cell type-specific chromatin domains are established and regulated in normal cells and how the underlying mechanisms go awry in disease states promises to be highly rewarding and to yield a continuous stream of focused and serendipitous discoveries for many years to come.

Acknowledgments

Work on chromatin domains in the Bresnick laboratory is supported by NIH Grants DK50107, DK55700, and DK68634. Kirby Johnson is supported by an NIH K01 award (DK069488). Hogune Im is a Predoctoral Fellow of the American Heart Association.

References

1. Kadonaga, J. T. (2004). Regulation of RNA polymerase II transcription by sequence-specific DNA binding factors. *Cell* **116,** 247–257.
2. Archer, T. K., Cordingley, M. G., Wolford, R. G., and Hager, G. L. (1991). Transcription factor access is mediated by accurately positioned nucleosomes on the mouse mammary tumor virus promoter. *Mol. Cell. Biol.* **11,** 688–698.
3. Lee, D. Y., Hayes, J. J., Pruss, D., and Wolffe, A. P. (1993). A positive role for histone acetylation in transcription factor access to nucleosomal DNA. *Cell* **72,** 73–84.
4. Tse, C., Sera, T., Wolffe, A. P., and Hansen, J. C. (1998). Disruption of higher-order folding by core histone acetylation dramatically enhances transcription of nucleosomal arrays by RNA polymerase III. *Mol. Cell Biol.* **18,** 4629–4638.
5. Hager, G. L., Archer, T. K., Fragoso, G., Bresnick, E. H., Tsukagoshi, Y., John, S., and Smith, C. L. (1993). Influence of chromatin structure on the binding of transcription factors to DNA. *Cold Spring Harb. Symp. Quant. Biol.* **58,** 63–71.
6. Evans, T., and Felsenfeld, G. (1989). The erythroid-specific transcription factor Eryf1: A new finger protein. *Cell* **58,** 877–885.
7. Tsai, S. F., Martin, D. I., Zon, L. I., D'Andrea, A. D., Wong, G. G., and Orkin, S. H. (1989). Cloning of cDNA for the major DNA-binding protein of the erythroid lineage through expression in mammalian cells. *Nature* **339,** 446–451.
8. Hebbes, T. R., Thorne, A. W., and Crane-Robinson, C. (1988). A direct link between core histone acetylation and transcriptionally active chromatin. *EMBO J.* **7,** 1395–1402.
9. Chung, J. H., Whiteley, M., and Felsenfeld, G. (1993). A 5′ element of the chicken beta-globin domain serves as an insulator in human erythroid cells and protects against position effect in Drosophila. *Cell* **74,** 505–514.

10. Hardison, R. (1998). Hemoglobins from bacteria to man: Evolution of different patterns of gene expression. *J. Exp. Biol.* **201**, 1099–1117.
11. Kingsley, P. D., Malik, J., Fantauzzo, K. A., and Palis, J. (2004). Yolk sacderived primitive erythroblasts enucleate during mammalian embryogenesis. *Blood* **104**, 19–25.
12. Whitelaw, E., Tsai, S. F., Hogben, P., and Orkin, S. H. (1990). Regulated expression of globin chains and the erythroid transcription factor GATA-1 during erythropoiesis in the developing mouse. *Mol. Cell. Biol.* **10**, 6596–6606.
13. Kingsley, P. D., Malik, J., Emerson, R. L., Bushnell, T. P., McGrath, K. E., Bloedorn, L. A., Bulger, M., and Palis, J. (2006). "Maturational" globin switching in primary primitive erythroid cells. *Blood* **107**(4), 1665–1672.
14. Stamatoyannopoulos, G. (1991). Human hemoglobin switching. *Science* **252**, 383.
15. Tuan, D., and London, I. M. (1984). Mapping of DNase I-hypersensitive sites in the upstream DNA of human embryonic epsilon-globin gene in K562 leukemia cells. *Proc. Natl. Acad. Sci. USA* **81**, 2718–2722.
16. Tuan, D., Solomon, W., Li, Q., and London, I. M. (1985). The "beta-like-globin" gene domain in human erythroid cells. *Proc. Natl. Acad. Sci. USA* **82**, 6384–6388.
17. Forrester, W. C., Thompson, C., Elder, J. T., and Groudine, M. (1986). A developmentally stable chromatin structure in the human beta-globin gene cluster. *Proc. Natl. Acad. Sci. USA* **83**, 1359–1363.
18. Forrester, W. C., Takegawa, S., Papayannopoulou, T., Stamatoyannopoulos, G., and Groudine, M. (1987). Evidence for a locus activation region: The formation of developmentally stable hypersensitive sites in globin-expressing hybrids. *Nucleic Acids Res.* **15**, 10159–10177.
19. Grosveld, F., van Assendelft, G. B., Greaves, D. R., and Kollias, G. (1987). Position-independent, high-level expression of the human beta-globin gene in transgenic mice. *Cell* **51**, 975–985.
20. Bulger, M., Schubeler, D., Bender, M. A., Hamilton, J., Farrell, C. M., Hardison, R. C., and Groudine, M. (2003). A complex chromatin landscape revealed by patterns of nuclease sensitivity and histone modification within the mouse beta-globin locus. *Mol. Cell. Biol.* **23**, 5234–5244.
21. Li, Q., Zhang, M., Duan, Z., and Stamatoyannopoulos, G. (1999). Structural analysis and mapping of DNase I hypersensitivity of HS5 of the beta-globin locus control region. *Genomics* **61**, 183–193.
22. Li, Q., and Stamatoyannopoulos, G. (1994). Hypersensitive site 5 of the human beta locus control region functions as a chromatin insulator. *Blood* **84**, 1399–1401.
23. Tanimoto, K., Sugiura, A., Omori, A., Felsenfeld, G., Engel, J. D., and Fukamizu, A. (2003). Human beta-globin locus control region HS5 contains CTCF- and developmental stage-dependent enhancer-blocking activity in erythroid cells. *Mol. Cell. Biol.* **23**, 8946–8952.
24. Wai, A. W., Gillemans, N., Raguz-Bolognesi, S., Pruzina, S., Zafarana, G., Meijer, D., Philipsen, S., and Grosveld, F. (2003). HS5 of the human beta-globin locus control region: A developmental stage-specific border in erythroid cells. *EMBO J.* **22**, 4489–4500.
25. Talbot, D., Collis, P., Antoniou, M., Vidal, M., Grosveld, F., and Greaves, D. R. (1989). A dominant control region from the human beta-globin locus conferring integration site-independent gene expression. *Nature* **338**, 352–355.
26. Ney, P. A., Sorrentino, B. P., Lowrey, C. H., and Nienhuis, A. W. (1990). Inducibility of the HS II enhancer depends on binding of an erythroid specific nuclear protein. *Nucleic Acids Res.* **18**, 6011–6017.
27. Ney, P. A., Sorrentino, B. P., McDonagh, K. T., and Nienhuis, A. W. (1990). Tandem AP-1-binding sites within the human beta-globin dominant control region function as an inducible enhancer in erythroid cells. *Genes Dev.* **4**, 993–1006.
28. Talbot, D., and Grosveld, F. (1991). The 5'HS2 of the globin locus control region enhances transcription through the interaction of a multimeric complex binding at two functionally distinct NF-E2 binding sites. *EMBO J.* **10**, 1391–1398.

29. Moon, A. M., and Ley, T. J. (1991). Functional properties of the beta-globin locus control region in K562 erythroleukemia cells. *Blood* **77**, 2272–2284.
30. Ellis, J., Tan-Un, K. C., Harper, A., Michalovich, D., Yannoutsos, N., Philipsen, S., and Grosveld, F. (1996). A dominant chromatin-opening activity in 5′ hypersensitive site 3 of the human beta-globin locus control region. *EMBO J.* **15**, 562–568.
31. Bresnick, E. H., and Tze, L. (1997). Synergism between hypersensitive sites confers long-range gene activation by the beta-globin locus control region. *Proc. Natl. Acad. Sci. USA* **94**, 4566–4571.
32. Jackson, J. D., Petrykowska, H., Philipsen, S., Miller, W., and Hardison, R. (1996). Role of DNA sequences outside the cores of DNase hypersensitive sites (HSs) in functions of the beta-globin locus control region. Domain opening and synergism between HS2 and HS3. *J. Biol. Chem.* **271**, 11871–11878.
33. Fiering, S., Epner, E., Robinson, K., Zhuang, Y., Telling, A., Hu, M., Martin, D. I., Enver, T., Ley, T. J., and Groudine, M. (1995). Targeted deletion of 5′HS2 of the murine beta-globin LCR reveals that it is not essential for proper regulation of the beta-globin locus. *Genes Dev.* **9**, 2203–2213.
34. Hug, B. A., Wesselschmidt, R. L., Fiering, S., Bender, M. A., Epner, E., Groudine, M., and Ley, T. J. (1996). Analysis of mice containing a targeted deletion of beta-globin locus control region 5′ hypersensitive site 3. *Mol. Cell. Biol.* **16**, 2906–2912.
35. Bender, M. A., Reik, A., Close, J., Telling, A., Epner, E., Fiering, S., Hardison, R., and Groudine, M. (1998). Description and targeted deletion of 5′ hypersensitive site 5 and 6 of the mouse beta-globin locus control region. *Blood* **92**, 4394–4403.
36. Epner, E., Reik, A., Cimbora, D., Telling, A., Bender, M. A., Fiering, S., Enver, T., Martin, M., Kennedy, M., Keller, G., and Groudine, M. (1998). The beta-globin LCR is not necessary for an open chromatin structure or developmentally regulated transcription of the native mouse beta-globin locus. *Mol. Cell* **2**, 447–455.
37. Reik, A., Telling, A., Zitnik, G., Cimbora, D., Epner, E., and Groudine, M. (1998). The locus control region is necessary for gene expression in the human beta-globin locus but not the maintenance of an open chromatin structure in erythroid cells. *Mol. Cell. Biol.* **18**, 5992–6000.
38. Bender, M. A., Bulger, M., Close, J., and Groudine, M. (2000). Beta-globin gene switching and DNase I sensitivity of the endogenous beta-globin locus in mice do not require the locus control region. *Mol. Cell* **5**, 387–393.
39. Bender, M. A., Roach, J. N., Halow, J., Close, J., Alami, R., Bouhassira, E. E., Groudine, M., and Fiering, S. N. (2001). Targeted deletion of 5′HS1 and 5′HS4 of the beta-globin locus control region reveals additive activity of the DNaseI hypersensitive sites. *Blood* **98**, 2022–2027.
40. Forrester, W. C., Epner, E., Driscoll, M. C., Enver, T., Brice, M., Papayannopoulou, T., and Groudine, M. (1990). A deletion of the human beta-globin locus activation region causes a major alteration in chromatin structure and replication across the entire beta-globin locus. *Genes Dev.* **4**, 1637–1649.
41. Navas, P. A., Peterson, K. R., Li, Q., Skarpidi, E., Rohde, A., Shaw, S. E., Clegg, C. H., Asano, H., and Stamatoyannopoulos, G. (1998). Developmental specificity of the interaction between the locus control region and embryonic or fetal globin genes in transgenic mice with an HS3 core deletion. *Mol. Cell. Biol.* **18**, 4188–4196.
42. Navas, P. A., Peterson, K. R., Li, Q., McArthur, M., and Stamatoyannopoulos, G. (2001). The 5′HS4 core element of the human beta-globin locus control region is required for high-level globin gene expression in definitive but not in primitive erythropoiesis. *J. Mol. Biol.* **312**, 17–26.
43. Peterson, K. R., Clegg, C. H., Navas, P. A., Norton, E. J., Kimbrough, T. G., and Stamatoyannopoulos, G. (1996). Effect of deletion of 5′HS3 or 5′HS2 of the human

beta-globin locus control region on the developmental regulation of globin gene expression in beta-globin locus yeast artificial chromosome transgenic mice. *Proc. Natl. Acad. Sci. USA* **93,** 6605–6609.
44. Bungert, J., Dave, U., Lim, K. C., Lieuw, K. H., Shavit, J. A., Liu, Q., and Engel, J. D. (1995). Synergistic regulation of human beta-globin gene switching by locus control region elements HS3 and HS4. *Genes Dev.* **9,** 3083–3096.
45. Bungert, J., Tanimoto, K., Patel, S., Liu, Q., Fear, M., and Engel, J. D. (1999). Hypersensitive site 2 specifies a unique function within the human beta-globin locus control region to stimulate globin gene transcription. *Mol. Cell. Biol.* **19,** 3062–3072.
46. Schleif, R. (1992). DNA looping. *Annu. Rev. Biochem.* **61,** 199–223.
47. Tolhuis, B., Palstra, R. J., Splinter, E., Grosveld, F., and de Laat, W. (2002). Looping and interaction between hypersensitive sites in the active beta-globin locus. *Mol. Cell* **10,** 1453–1475.
48. Drissen, R., Palstra, R. J., Gillemans, N., Splinter, E., Grosveld, F., Philipsen, S., and de Laat, W. (2004). The active spatial organization of the beta-globin locus requires the transcription factor EKLF. *Genes Dev.* **18,** 2485–2490.
49. Vakoc, C. R., Letting, D. L., Gheldof, N., Sawado, T., Bender, M. A., Groudine, M., Weiss, J., Dekker, J., and Blobel, G. A. (2005). Proximity among distant regulatory elements at the beta globin locus requires GATA-1 and FOG-1. *Mol. Cell* **17,** 453–462.
50. Dekker, J., Rippe, K., Dekker, M., and Kleckner, N. (2002). Capturing chromosome conformation. *Science* **295,** 1306–1311.
51. Carter, D., Chakalova, L., Osborne, C. S., Dai, Y. F., and Fraser, P. (2002). Long-range chromatin regulatory interactions *in vivo. Nat. Genet.* **32,** 623–626.
52. Ling, J., Ainol, L., Zhang, L., Yu, X., Pi, W., and Tuan, D. (2004). HS2 enhancer function is blocked by a transcriptional terminator inserted between the enhancer and the promoter. *J. Biol. Chem.* **279,** 51704–51713.
53. Bulger, M., and Groudine, M. (1999). Looping versus linking: Toward a model for long-distance gene activation. *Genes Dev.* **13,** 2465–2477.
54. Brand, M., Ranish, J. A., Kummer, N. T., Hamilton, J., Igarashi, K., Francastel, C., Chi, T. H., Crabtree, R., Aebersold, R., and Groudine, M. (2003). Dynamic changes in transcription factor complexes during erythroid differentiation revealed by quantitative proteomics. *Nat. Struct. Mol. Biol.* **11,** 73–80.
55. Hong, W., Nakazawa, M., Chen, Y. Y., Kori, R., Vakoc, C. R., Rakowski, C., and Blobel, G. A. (2005). FOG-1 recruits the NuRD repressor complex to mediate transcriptional repression by GATA-1. *EMBO J.* **24,** 67–78.
56. Rodriquez, P., Bonte, E., Krijgsveld, J., Kolodziej, K. E., Guyot, B., Heck, A. J., Vyas, P., de Boer, F., Grosveld, F., and Strouboulis, J. (2005). GATA-1 forms distinct activating and repressive complexes in erythroid cells. *EMBO J.* **24,** 2354–2366.
57. Johnson, K. D., Christensen, H. M., Zhao, B., and Bresnick, E. H. (2001). Distinct mechanisms control RNA polymerase II recruitment to a tissue-specific locus control region and a downstream promoter. *Mol. Cell* **8,** 465–471.
58. Johnson, K. D., Grass, J. A., Im, H., Park, C., Choi, K., and Bresnick, E. H. (2003). Highly restricted localization of RNA polymerase II to the hypersensitive site cores of a tissue-specific locus control region. *Mol. Cell. Biol.* **23,** 6468–6493.
59. Lu, S. J., Rowan, S., Bani, M. R., and Ben-David, Y. (1994). Retroviral integration within the Fli-2 locus results in inactivation of the erythroid transcription factor NF-E2 in Friend erythroleukemias: Evidence that NF-E2 is essential for globin expression. *Proc. Natl. Acad. Sci. USA* **91,** 8398–8402.
60. Kotkow, K. J., and Orkin, S. H. (1995). Dependence of globin gene expression in mouse erythroleukemia cells on the NF-E2 heterodimer. *Mol. Cell. Biol.* **15,** 4640–4647.

61. Bean, T. L., and Ney, P. A. (1997). Multiple regions of p45 NF-E2 are required for beta-globin gene expression in erythroid cells. *Nucleic Acids Res.* **25**, 2509–2515.
62. Mosser, E. A., Kasanov, J. D., Forsberg, E. C., Kay, B. K., Ney, P. A., and Bresnick, E. H. (1998). Physical and functional interactions between the transactivation domain of the hematopoietic transcription factor NF-E2 and WW domains. *Biochemistry* **37**, 13686–13695.
63. Kiekhaefer, C. M., Grass, J. D., and Bresnick, E. H. (2004). WW domain binding motif within the activation domain of the hematopoietic activator p45/NF-E2 is essential for establishment of a tissue-specific histone modification pattern. *J. Biol. Chem.* **279**, 7456–7461.
64. Vieira, K. F., Levings, P. P., Hill, M. A., Cruselle, V. J., Kang, S. H., Engel, J. D., and Bungert, J. (2004). Recruitment of transcription complexes to the beta-globin gene locus *in vivo* and *in vitro*. *J. Biol. Chem.* **279**, 50350–50357.
65. Szutorisz, H., Dillon, N., and Tora, L. (2005). The role of enhancers as centres for general transcription factor recruitment. *Trends Biochem. Sci.* **30**, 593–599.
66. Cheng, J., Kapranov, P., Drenkow, J., Dike, S., Brubaker, S., Patel, S., Long, J., Stern, D., Tammana, G., Helt, G., Sementchenko, V., Piccolboni, A. *et al.* (2005). Transcriptional maps of 10 human chromosomes at 5-nucleotide resolution. *Science* **308**, 1149–1154.
67. Komarnitsky, P., Cho, E. J., and Buratowski, S. (2000). Different phosphorylated forms of RNA polymerase II and associated mRNA processing factors during transcription. *Genes Dev.* **14**, 2452–2460.
68. Lin, P. S., Marshall, N. F., and Dahmus, M. E. (2002). CTD phosphatase: Role in RNA polymerase II cycling and the regulation of transcript elongation. *Prog. Nucleic Acid Res. Mol. Biol.* **72**, 333–365.
69. Leach, K. M., Nightingale, K., Igarashi, K., Levings, P. P., Engel, J. D., Becker, P. B., and Bungert, J. (2001). Reconstitution of human beta-globin locus control region hypersensitive sites in the absence of chromatin assembly. *Mol. Cell. Biol.* **21**, 2629–2640.
70. Gribnau, J., Diderich, K., Pruzina, S., Calzolari, R., and Fraser, P. (2000). Intergenic transcription and developmental remodeling of chromatin subdomains in the human beta-globin locus. *Mol. Cell* **5**, 377–386.
71. Francastel, C., Walters, M. C., Groudine, M., and Martin, D. I. (1999). A functional enhancer suppresses silencing of a transgene and prevents its localization close to centrometric heterochromatin. *Cell* **99**, 259–269.
72. Kosak, S. T., Skok, J. A., Medina, K. L., Riblet, R., Le Beau, M. M., Fisher, A. G., and Singh, H. (2002). Subnuclear compartmentalization of immunoglobulin loci during lymphocyte development. *Science* **296**, 158–162.
73. Kosak, S. T., and Groudine, M. (2004). Form follows function: The genomic organization of cellular differentiation. *Genes Dev.* **18**, 1371–1384.
74. Misteli, T. (2004). Spatial positioning: A new dimension in genome function. *Cell* **119**, 153–156.
75. Nagaich, A. K., Rayasam, G. V., Martinez, E. D., Becker, M., Qiu, Y., Johnson, T. A., Elbi, C., Fletcher, S., John, S., and Hager, G. L. (2004). Subnuclear targeting and gene targeting by steroid receptors. *Ann. N. Y. Acad. Sci.* **1024**, 213–220.
76. Zaidi, S. K., Javed, A., Choi, J.-E., Pratap, J., Javed, A., Montecino, M., Stein, J. L., van Wijnen, J. B., Lian, J. B., and Stein, G. S. (2005). The dynamic organization of gene-regulatory networks in nuclear microenvironments. *EMBO Rep.* **6**, 128–133.
77. Chakalova, L., Debrand, E., Mitchell, J. A., Osborne, C. S., and Fraser, P. (2005). Replication and transcription: Shaping the landscape of the genome. *Nat. Rev. Genet.* **6**, 669–677.
78. Cook, P. R. (2002). Predicting three-dimensional genome structure from transcriptional activity. *Nat. Genet.* **32**, 347–352.
79. Swedlow, J. R., Agard, D. A., and Sedat, J. W. (1993). Chromosome structure inside the nucleus. *Curr. Opin. Cell Biol.* **5**, 412–416.

80. van Driel, R., Fransz, P. F., and Verschure, P. J. (2003). The eukaryotic genome: A system regulated at different hierarchical levels. *J. Cell Sci.* **116**, 4067–4075.
81. Spector, D. L. (2003). The dynamics of chromosome organization and gene regulation. *Annu. Rev. Biochem.* **72**, 573–608.
82. Ragoczy, T., Telling, A., Sawado, T., Groudine, M., and Kosak, S. T. (2003). A genetic analysis of chromosome territory looping: Diverse roles for distal regulatory elements. *Chromosome Res.* **11**, 513–525.
83. Shelton, D. A., Stegman, L., Hardison, R., Miller, W., Bock, J. H., Slightom, J. L., Goodman, M., and Gumucio, D. L. (1997). Phylogenetic footprinting of hypersensitive site 3 of the beta-globin locus control region. *Blood* **89**, 3457–3469.
84. Hardison, R., Slightom, J. L., Gumucio, D. L., Goodman, M., Stojanovic, N., and Miller, W. (1997). Locus control regions of mammalian beta-globin gene clusters: Combining phylogenetic analyses and experimental results to gain functional insights. *Gene* **205**, 73–94.
85. Li, Q., Peterson, K. R., Fang, X. R., and Stamatoyannopoulos, G. (2002). Locus control regions. *Blood* **100**, 3077–3086.
86. Cantor, A. B., and Orkin, S. H. (2002). Transcriptional regulation of erythropoiesis: An affair involving multiple partners. *Oncogene* **21**, 3368–3376.
87. Johnson, K. D., and Bresnick, E. H. (2002). Dissecting long-range transcriptional mechanisms by chromatin immunoprecipitation. *METHODS* **26**, 27–36.
88. Tuan, D. Y., Solomon, W. B., London, I. M., and Lee, D. P. (1989). An erythroid-specific, developmental-stage-independent enhancer far upstream of the human "beta-like globin" genes. *Proc. Natl. Acad. Sci. USA* **86**, 2554–2558.
89. Moon, A. M., and Ley, T. J. (1990). Conservation of the primary structure, organization, and function of the human and mouse beta-globin locus-activating regions. *Proc. Natl. Acad. Sci. USA* **87**, 7693–7697.
90. Fraser, P., Hurst, J., Collis, P., and Grosveld, F. (1990). DNaseI hypersensitive sites 1, 2 and 3 of the human beta-globin dominant control region direct position-independent expression. *Nucleic Acids Res.* **18**, 3503–3508.
91. Eferl, R., and Wagner, E. F. (2003). AP1: A double-edged sword in tumorigenesis. *Nat. Rev. Cancer* **3**, 859–868.
92. Andrews, N. C., Erdjument-Bromage, H., Davidson, M. B., Tempst, P., and Orkin, S. H. (1993). Erythroid transcription factor NF-E2 is a haematopoietic-specific basic-leucine zipper protein. *Nature* **362**, 722–728.
93. Ney, P. A., Andrews, N. C., Jane, S. M., Safer, B., Purucker, M. E., Weremowicz, S., Morton, S. C., Goff, S. C., Orkin, S. H., and Nienhuis, A. W. (1993). Purification of the human NF-E2 complex: cDNA cloning of the hematopoietic cell-specific subunit and evidence for an associated partner. *Mol. Cell. Biol.* **13**, 5604–5612.
94. Andrews, N. C., Kotkow, K. J., Ney, P. A., Erdjument-Bromage, H., Tempst, P., and Orkin, S. H. (1993). The ubiquitous subunit of erythroid transcription factor NF-E2 is a small basic-leucine zipper protein related to the v-maf oncogene. *Proc. Natl. Acad. Sci. USA* **90**, 11488–11492.
95. Motohashi, H., Shavit, J. A., Igarashi, K., Yamamoto, M., and Engel, J. D. (1997). The world according to Maf. *Nucleic Acids Res.* **25**, 2953–2959.
96. Sawado, T., Igarashi, K., and Groudine, M. (2001). Activation of beta-major globin gene transcription is associated with recruitment of NF-E2 to the beta-globin LCR and gene promoter. *Proc. Natl. Acad. Sci. USA* **98**, 10226–10231.
97. Daftari, P., Gavva, N. R., and Shen, C. K. (1999). Distinction between AP1 and NF-E2 factor-binding at specific chromatin regions in mammalian cells. *Oncogene* **18**, 5482–5486.
98. Forsberg, E. C., Downs, K. M., and Bresnick, E. H. (2000). Direct interaction of NF-E2 with hypersensitive site 2 of the beta-globin locus control region in living cells. *Blood* **96**, 334–339.

99. Im, H., Grass, J. A., Johnson, K. D., Kim, S.-I., Boyer, M. E., Imbalzano, A. N., Bieker, J. J., and Bresnick, E. H. (2005). Chromatin domain activation via GATA-1 utilization of a small subset of dispersed GATA motifs within a broad chromosomal region. *Proc. Natl. Acad. Sci. USA* **102**, 17065–17070.
100. Johnson, K. D., Grass, J. D., Boyer, M. E., Kiekhaefer, C. M., Blobel, G. A., Weiss, M. J., and Bresnick, E. H. (2002). Cooperative activities of hematopoietic regulators recruit RNA polymerase II to a tissue-specific chromatin domain. *Proc. Natl. Acad. Sci. USA* **99**, 11760–11765.
101. Kiekhaefer, C. M., Grass, J. A., Johnson, K. D., Boyer, M. E., and Bresnick, E. H. (2002). Hematopoietic activators establish an overlapping pattern of histone acetylation and methylation within a tissue-specific chromatin domain. *Proc. Natl. Acad. Sci. USA* **99**, 14309–14314.
102. Vettese-Dadey, M., Grant, P. A., Hebbes, T. R., Crane-Robinson, C., Allis, C. D., and Workman, J. L. (1996). Acetylation of histone H4 plays a primary role in enhancing transcription factor binding to nucleosomal DNA *in vitro*. *EMBO J.* **15**, 2508–2518.
103. Litt, M. D., Simpson, M., Gaszner, M., Allis, C. D., and Felsenfeld, G. (2001). Correlation between histone lysine methylation and developmental changes at the chicken {beta}-globin locus. *Science* **9**, 9.
104. Litt, M. D., Simpson, M., Recillas-Targa, F., Prioleau, M. N., and Felsenfeld, G. (2001). Transitions in histone acetylation reveal boundaries of three separately regulated neighboring loci. *EMBO J.* **20**, 2224–2235.
105. Zegerman, P., Canas, B., Pappin, D., and Kouzarides, T. (2002). Histone H3 lysine 4 methylation disrupts binding of nucleosome remodeling and deacetylase (NuRD) repressor complex. *J. Biol. Chem.* **277**, 11621–11624.
106. Wysocka, J., Swigut, T., Milne, T. A., Dou, Y., Zhang, X., Burlingame, A. L., Roeder, R. G., Brivanlou, A. H., and Allis, C. D. (2005). WDR5 associates with histone H3 methylated at K4 and is essential for H3 K4 methylation and vertebrate development. *Cell* **121**, 859–872.
107. Shyu, Y. C., Lee, T. L., Ting, C. Y., Wen, S. C., Hsieh, L. J., Li, Y. C., Hwang, J. L., Lin, C. C., and Shen, C. K. (2005). Sumoylation of p45/NF-E2: Nuclear positioning and transcriptional activation of the mammalian beta-like globin locus. *Mol. Cell. Biol.* **25**, 10365–10378.
108. Hung, H. L., Kim, A. Y., Hong, W., Rakowski, C., and Blobel, G. A. (2001). Stimulation of NF-E2 DNA binding by CREB-binding protein (CBP)-mediated acetylation. *J. Biol. Chem.* **276**, 10715–10721.
109. Versaw, W. K., Blank, V., Andrews, N. M., and Bresnick, E. H. (1998). Mitogen-activated protein kinases enhance long-range activation by the beta-globin locus control region. *Proc. Natl. Acad. Sci. USA* **95**, 8756–8760.
110. Forsberg, E. C., Zaboikina, T. N., Versaw, W. K., Ahn, N. G., and Bresnick, E. H. (1999). Enhancement of beta-globin locus control region-mediated transactivation by mitogen-activated protein kinases through stochastic and graded mechanisms. *Mol. Cell. Biol.* **19**, 5565–5575.
111. Shivdasani, R. A., Rosenblatt, M. F., Zucker-Franklin, D., Jackson, C. W., Hunt, P., Saris, C. J., and Orkin, S. H. (1995). Transcription factor NF-E2 is required for platelet formation independent of the actions of thrombopoietin/MGDF in megakaryocyte development. *Cell* **81**, 695–704.
112. Shivdasani, R. A., and Orkin, S. H. (1995). Erythropoiesis and globin gene expression in mice lacking the transcription factor NF-E2. *Proc. Natl. Acad. Sci. USA* **92**, 8690–8694.
113. Schulze, H., and Shivdasani, R. A. (2005). Mechanisms of thrombopoiesis. *J. Thromb. Haemost.* **3**, 1717–1724.
114. Forsberg, E. C., Downs, K. M., Christensen, H. M., Im, H., Nuzzi, P. A., and Bresnick, E. H. (2000). Developmentally dynamic histone acetylation pattern of a tissue-specific chromatin domain. *Proc. Natl. Acad. Sci. USA* **97**, 14494–14499.

115. Chan, J. Y., Han, X. L., and Kan, Y.-W. (1993). Cloning of Nrf1, an NF-E2-related transcription factor, by genetic selection in yeast. *Proc. Natl. Acad. Sci. USA* **90,** 11371–11375.
116. Chan, K., Lu, R., Chang, J. C., and Kan, Y. W. (1996). NRF2, a member of the NFE2 family of transcription factors, is not essential for murine erythropoiesis, growth, and development. *Proc. Natl. Acad. Sci. USA* **93,** 13943–13948.
117. Oyake, T., Itoh, K., and Motohashi, H. (1996). Bach proteins belong to a novel family of BTB-basic leucine zipper transcription factors that interact with MafK and regulate transcription through the NF-E2 site. *Mol. Cell. Biol.* **16,** 6083–6095.
118. Igarashi, K., Hoshino, H., Muto, A., Suwabe, N., Nishikawa, S., Nakauchi, H., and Yamamoto, M. (1998). Multivalent DNA binding complex generated by small Maf and Bach1 as a possible biochemical basis for beta-globin locus control region complex. *J. Biol. Chem.* **273,** 11783–11790.
119. Martin, F., van Deursen, J. M., Shivdasani, R. A., Jackson, C. W., Troutman, A. G., and Ney, P. A. (1998). Erythroid maturation and globin gene expression in mice with combined deficiency of NF-E2 and Nrf-2. *Blood* **91,** 3459–3466.
120. Orkin, S. H. (1992). GATA-binding transcription factors in hematopoietic cells. *Blood* **80,** 575–581.
121. Bresnick, E. H., Martowicz, M. L., Pal, S., and Johnson, K. D. (2005). Developmental control via GATA factor interplay at chromatin domains. *J. Cell. Physiol.* **205,** 1–9.
122. Reitman, M., and Felsenfeld, G. (1988). Mutational analysis of the chicken beta-globin enhancer reveals two positive-acting domains. *Proc. Natl. Acad. Sci. USA* **85,** 6267–6271.
123. Evans, T., Reitman, M., and Felsenfeld, G. (1988). An erythrocyte-specific DNA-binding factor recognizes a regulatory sequence common to all chicken globin genes. *Proc. Natl. Acad. Sci. USA* **85,** 5976–5980.
124. Merika, M., and Orkin, S. H. (1993). DNA-binding specificity of GATA family transcription factors. *Mol. Cell. Biol.* **13,** 3999–4010.
125. Ko, L. J., and Engel, J. D. (1993). DNA-binding specificities of the GATA transcription factor family. *Mol. Cell. Biol.* **13,** 4011–4022.
126. Martin, D. I., and Orkin, S. H. (1990). Transcriptional activation and DNA binding by the erythroid factor GF-1/NF-E1/Eryf 1. *Genes Dev.* **4,** 1886–1898.
127. Trainor, C. D., Omichinski, J. G., Vandergon, T. L., Gronenborn, A. M., Clore, G. M., and Felsenfeld, G. (1996). A palindromic regulatory site within vertebrate GATA-1 promoters requires both zinc fingers of the GATA-1 DNA-binding domain for high-affinity interaction. *Mol. Cell. Biol.* **16,** 2238–2247.
128. Newton, A., MacKay, J. P., and Crossley, M. (2001). The N-terminal zinc finger of the erythroid transcription factor GATA-1 binds GATC motifs in DNA. *J. Biol. Chem.* **276,** 35794–35801.
129. Tsang, A. P., Visvader, J. E., Turner, C. A., Fujuwara, Y., Yu, C., Weiss, M. J., Crossley, M., and Orkin, S. H. (1997). FOG, a multitype zinc finger protein as a cofactor for transcription factor GATA-1 in erythroid and megakaryocytic differentiation. *Cell* **90,** 109–119.
130. Crispino, J. D., Lodish, M. B., MacKay, J. P., and Orkin, S. H. (1999). Use of altered specificity mutants to probe a specific protein-protein interaction in differentiation: The GATA-1: FOG complex. *Mol. Cell* **3,** 219–228.
131. Pal, S., Cantor, A. B., Johnson, K. D., Moran, T., Boyer, M. E., Orkin, S. H., and Bresnick, E. H. (2004). Coregulator-dependent facilitation of chromatin occupancy by GATA-1. *Proc. Natl. Acad. Sci. USA* **101,** 980–985.
132. Letting, D. L., Chen, Y. Y., Rakowski, C., Reedy, S., and Blobel, G. A. (2004). Context-dependent regulation of GATA-1 by friend of GATA-1. *Proc. Natl. Acad. Sci. USA* **101,** 476–481.
133. Grass, J. A., Boyer, M. E., Paul, S., Wu, J., Weiss, M. J., and Bresnick, E. H. (2003). GATA-1-dependent transcriptional repression of GATA-2 via disruption of positive autoregulation and domain-wide chromatin remodeling. *Proc. Natl. Acad. Sci. USA* **100,** 8811–8816.

134. Pevny, L., Simon, M. C., Robertson, E., Klein, W. H., Tsai, S. F., D'Agati, V., Orkin, S. H., and Costantini, F. (1991). Erythroid differentiation in chimaeric mice blocked by a targeted mutation in the gene for transcription factor GATA-1. *Nature* **349**, 257–260.
135. Fujiwara, Y., Browne, C. P., Cunniff, K., Goff, S. C., and Orkin, S. H. (1996). Arrested development of embryonic red cell precursors in mouse embryos lacking transcription factor GATA-1. *Proc. Natl. Acad. Sci. USA* **93**, 12355–12358.
136. Weiss, M. J., Yu, C., and Orkin, S. H. (1997). Erythroid-cell-specific properties of transcription factor GATA-1 revealed by phenotypic rescue of a gene-targeted cell line. *Mol. Cell. Biol.* **17**, 1642–1651.
137. Gregory, T., Yu, C., Ma, A., Orkin, S. H., Blobel, G. A., and Weiss, M. J. (1999). GATA-1 and erythropoietin cooperate to promoter erythroid cell survival by regulating bcl-xl expression. *Blood* **94**, 87–96.
138. Welch, J. J., Watts, J. A., Vakoc, C. R., Yao, Y., Wang, H., Hardison, R. C., Blobel, G. A., Chodosh, L. A., and Weiss, M. J. (2004). Global regulation of erythroid gene expression by transcription factor GATA-1. *Blood* **104**, 3136–3147.
139. Martowicz, M. L., Grass, J. A., Boyer, M. E., Guend, H., and Bresnick, E. H. (2005). Dynamic GATA factor interplay at a multi-component regulatory region of the GATA-2 locus. *J. Biol. Chem.* **280**, 1724–1732.
140. Pal, S., Nemeth, M. J., Bodine, D. M., Miller, J. L., Svaren, J., Thein, S. L., Lowry, P. J., and Bresnick, E. H. (2004). Neurokinin-B transcription in erythroid cells: Direct activation by the hematopoietic transcription factor GATA-1. *J. Biol. Chem.* **279**, 31348–31356.
141. Miller, I. J., and Bieker, J. J. (1993). A novel, erythroid cell-specific murine transcription factor that binds to the CACCC element and is related to the Kruppel family of nuclear proteins. *Mol. Cell. Biol.* **13**, 2776–2786.
142. Bieker, J. J. (2001). Kruppel-like factors: Three fingers in many pies. *J. Biol. Chem.* **276**, 34355–34358.
143. Basu, P., Sargent, T. G., Redmond, L. C., Aisenberg, J. C., Kransdorf, E. P., Wang, S. Z., Ginder, G. D., and Lloyd, J. A. (2004). Evolutionary conservation of KLF transcription factors and functional conservation of human gamma-globin gene regulation in chicken. *Genomics* **84**, 311–319.
144. Orkin, S. H., Kazazian, H. H., Antonarakis, S. E., Goff, S. C., Boehm, C. D., Sexton, J. P., Waber, P. G., and Giardina, P. J. (1982). Linkage of beta-thalassaemia mutations and beta-globin gene polymorphisms with DNA polymorphisms in human beta-globin gene cluster. *Nature* **296**, 627–631.
145. Orkin, S. H., Antonarakis, S. E., and Kazazian, H. H. (1984). Base substitution at position −88 in a beta-thalassemic globin gene. Further evidence for the role of distal promoter element ACACCC. *J. Biol. Chem.* **867**, 9–8681.
146. Myers, R. M., Tilly, K., and Maniatis, T. (1986). Fine structure genetic analysis of a beta-globin promoter. *Science* **232**, 613–618.
147. Perkins, A. C., Sharpe, A. H., and Orkin, S. H. (1995). Lethal beta-thalassaemia in mice lacking the erythroid CACCC- transcription factor EKLF. *Nature* **375**, 318–322.
148. Nuez, B., Michalovich, D., Bygrave, A., Ploemacher, R., and Grosveld, F. (1995). Defective haematopoiesis in fetal liver resulting from inactivation of the EKLF gene. *Nature* **375**, 316–318.
149. Milot, E., Strouboulis, J., Trimborn, T., Wijgerde, M., de Boer, E., Langeveld, A., Tan-Un, K., Vergeer, N., Yannoutsos, N., Grosveld, F., and Fraser, P. (1996). Heterochromatin effects on the frequency and duration of LCR-mediated gene transcription. *Cell* **87**, 105–114.
150. Drissen, R., von Lindern, M., Kolbus, A., Driegen, S., Steinlein, P., Beug, H., Grosveld, F., and Philipsen, S. (2005). The erythroid phenotype of EKLF-null mice: Defects in hemoglobin metabolism and membrane stability. *Mol. Cell. Biol.* **25**, 5205–5214.

151. Hodge, D., Coghill, E., Keys, J., Maguire, T., Hartmann, B., McDowall, A., Weiss, M., Grimmond, S., and Perkins, A. (2006). A global role for EKLF in definitive and primitive erythropoiesis. *Blood* **107**(8), 3359–3370.
152. Donze, D., Townes, T. M., and Bieker, J. J. (1995). Role of erythroid Kruppel-like factor in human gamma- to beta-globin gene switching. *J. Biol. Chem.* **270**, 1955–1959.
153. Perkins, A. C., Gaensler, K. M., and Orkin, S. H. (1996). Silencing of human fetal globin expression is impaired in the absence of the adult beta-globin gene activator protein EKLF. *Proc. Natl. Acad. Sci. USA* **93**, 12267–12271.
154. Wijgerde, M., Gribnau, J., Trimborn, T., Nuez, B., Philipsen, S., Grosveld, F., and Fraser, P. (1996). The role of EKLF in human beta-globin gene competition. *Genes Dev.* **10**, 2894–2902.
155. Asano, H., and Stamatoyannopoulos, G. (1998). Activation of beta-globin promoter by erythroid Kruppel-like factor. *Mol. Cell. Biol.* **18**, 102–109.
156. Gillemans, N., Tewari, R., Lindeboom, F., Rottier, R., de Wit, T., Wijgerde, M., Grosveld, F., and Philipsen, S. (1998). Altered DNA-binding specificity mutants of EKLF and Sp1 show that EKLF is an activator of the beta-globin locus control region *in vivo*. *Genes Dev.* **12**, 2863–2873.
157. Gregory, R. C., Taxman, D. J., Seshasayee, D., Kensinger, M. H., Bieker, J. J., and Wojchowski, D. M. (1996). Functional interaction of GATA1 with erythroid Kruppel-like factor and Sp1 at defined erythroid promoters. *Blood* **87**, 1793–1801.
158. Merika, M., and Orkin, S. H. (1995). Functional synergy and physical interactions of the erythroid transcription factor GATA-1 with the Kruppel family proteins Sp1 and EKLF. *Mol. Cell. Biol.* **15**, 2437–2447.
159. Zhang, W., Kadam, S., Emerson, B. M., and Bieker, J. J. (2001). Site-specific acetylation by p300 or CREB binding protein regulates erythroid Kruppel-like factor transcriptional activity via its interaction with the SWI-SNF complex. *Mol. Cell. Biol.* **21**, 2413–2422.
160. Armstrong, J. A., Bieker, J. J., and Emerson, B. M. (1998). A SWI/SNF-related chromatin remodeling complex, E-RC1, is required for tissue-specific transcriptional regulation by EKLF *in vitro*. *Cell* **95**, 93–104.
161. Brown, R. C., Pattison, S., van Ree, J., Coghill, E., Perkins, A., Jane, S. M., and Cunningham, J. M. (2002). Distinct domains of erythroid Kruppel-like factor modulate chromatin remodeling and transactivation at the endogenous beta-globin gene promoter. *Mol. Cell. Biol.* **22**, 161–170.
162. Tewari, R., Gillemans, N., Wijgerde, M., Nuez, B., von Lindern, M., Grosveld, F., and Philipsen, S. (1998). Erythroid Kruppel-like factor (EKLF) is active in primitive and definitive erythroid cells and is required for the function of 5′HS3 of the beta-globin locus control region. *EMBO J.* **17**, 2334–2341.
163. Imbalzano, A. N., and Xiao, H. (2004). Functional properties of ATP-dependent chromatin remodeling enzymes. *Adv. Protein Chem.* **67**, 157–179.
164. Wang, W. (2003). The SWI/SNF family of ATP-dependent chromatin remodelers: Similar mechanisms for diverse functions. *Curr. Top. Microbiol. Immunol.* **274**, 143–169.
165. Kadam, S., McAlpine, G. S., Phelan, M. L., Kingston, R. E., Jone, K. A., and Emerson, B. M. (2000). Functional selectivity of recombinant mammalian SWI/SNF subunits. *Genes Dev.* **14**, 2441–2451.
166. Bultman, S. J., Gebuhr, T. C., and Magnuson, T. (2005). A Brg 1 mutation that uncouples ATPase activity from chromatin remodeling reveals an essential role for SWI/SNF complexes in beta-globin expression and erythroid development. *Genes Dev.* **19**, 2849–2861.
167. O'Neill, D., Yang, J., Erdjument-Bromage, H., Bornschlegel, K., Tempst, P., and Bank, A. (1999). Tissue-specific and developmental stage-specific DNA binding by a mammalian SWI/SNF complex associated with human fetal-to-adult globin gene switching. *Proc. Natl. Acad. Sci. USA* **96**, 349–354.

168. O'Neill, D. W., Schoetz, S. S., Lopez, R. A., Castle, M., Rabinowitz, L., Shor, E., Krawchuk, M. G., Goll, M. G., Renz, M., Seelig, H. P., Han, S., Seong, R. H. *et al.* (2000). An ikaros-containing chromatin-remodeling complex in adult-type erythroid cells. *Mol. Cell. Biol.* **20,** 7572–7582.
169. Georgopoulos, K., Bigby, M., Wang, J. H., Molnar, A., Wu, P., Winandy, S., and Sharpe, A. (1994). The lkaros gene is required for the development of all lymphoid lineages. *Cell* **79,** 143–156.
170. Lopez, R. A., Schoetz, S., DeAngelis, K., O'Neill, D., and Bank, A. (2002). Multiple hematopoietic defects and delayed globin switching in lkaros null mice. *Proc. Natl. Acad. Sci. USA* **99,** 602–607.
171. Mahajan, M. C., Narkikar, G. J., Boyapaty, G., Kingston, R. E., and Weissman, S. M. (2005). Heterogeneous nuclear ribonucleoprotein C1/C2. MeCP1, and SWI/SNF form a chromatin remodeling complex at the beta-globin locus control region. *Proc. Natl. Acad. Sci. USA* **102,** 15012–15017.
172. Im, H., Grass, J. A., Johnson, K. D., Boyer, M. E., Wu, J., and Bresnick, E. H. (2004). Measurement of protein-DNA interactions *in vivo* by chromatin immunoprecipitation. *Methods Mol. Biol.* **284,** 129–146.
173. Weinmann, A. S., Yan, P. S., Oberley, M. J., Huang, T. H., and Farnham, P. J. (2002). Isolating human transcription factor targets by coupling chromatin immunoprecipitation and CpG island microarray analysis. *Genes Dev.* **16,** 235–244.
174. Ren, B., Robert, F., Wyrick, J. J., Aparicio, O., Jennings, E. G., Simon, I., Zeitlinger, J., Schreiber, N., Hannett, N., Kanin, E., Volkert, T. L., Wilson, C. J. *et al.* (2000). Genome-wide location and function of DNA binding proteins. *Science* **290,** 2306–2309.
175. Forsberg, E. C., and Bresnick, E. H. (2001). Histone acetylation beyond promoters: Long-range acetylation patterns in the chromatin world. *Bioessays* **23,** 820–830.
176. Im, H., Park, C., Feng, Q., Johnson, K. D., Kiekhaefer, C. M., Choi, K., Zhang, Y., and Bresnick, E. H. (2003). Dynamic regulation of histone H3 methylated at lysine 79 within a tissue-specific chromatin domain. *J. Biol. Chem.* **278,** 18346–18352.
177. Kim, A., and Dean, A. (2004). Developmental stage differences in chromatin subdomains of the beta-globin locus. *Proc. Natl. Acad. Sci. USA* **101,** 7028–7033.
178. Perrine, S. P., Rudolph, A., Faller, D. V., Roman, C., Cohen, R. A., Chen, S. J., and Kan, Y. W. (1988). Butyrate infusions in the ovine fetus delay the biologic clock for globin gene switching. *Proc. Natl. Acad. Sci. USA* **85,** 8540–8552.
179. Liakopoulou, E., Blau, C. A., Li, Q., Josephson, B., Wolf, J. A., Fournarakis, B., Raisys, V., Dover, T., Papayannopoulou, T., and Stamatoyannopoulos, G. (1995). Stimulation of fetal hemoglobin production by short chain fatty acids *Blood* **86,** 3227–3235.
180. McCaffrey, P. G., Newsome, D. A., Fibach, E., Yoshida, M., and Su, M. S. (1997). Induction of gamma-globin by histone deacetylase inhibitors. *Blood* **90,** 2075–2083.
181. Cao, H., Satamatoyannopoulos, G., and Jung, M. (2004). Induction of human gamma globin gene expression by histone deacetylase inhibitors. *Blood* **103,** 701–709.
182. Atweh, G. F., Sutton, M., Nassif, I., Boosalis, M., Dover, G. J., Wallenstein, S., Wright, E., McMahon, G., Stamatoyannopoulos, G., Faller, D. V., and Perrine, S. P. (1999). Sustained induction of fetal hemoglobin by pulse butyrate therapy in sickle cell disease. *Blood* **93,** 1790–1797.
183. Im, H., Grass, J. A., Christensen, H. M., Perkins, A., and Bresnick, E. H. (2002). Histone deacetylase-dependent establishment and maintenance of broad low-level histone acetylation within a tissue-specific chromatin domain. *Biochem.* **41,** 15152–15160.
184. Ley, T. J., DeSimone, J., Noguchi, C. T., Turner, P. H., Schechter, A. N., Heller, P., and Nienhuis, A. W. (1983). 5-Azacytidine increases gamma-globin synthesis and reduces the proportion of dense cells in patients with sickle cell anemia. *Blood* **62,** 370–380.

185. Letting, D. L., Rakowski, C., Weiss, M. J., and Blobel, G. A. (2003). Formation of a tissue-specific histone acetylation pattern by the hematopoietic transcription factor GATA-1. *Mol. Cell. Biol.* **23**, 1334–1340.
186. van Leeuwen, F., Gafken, P. R., and Gottschling, D. E. (2002). Dot1p modulates silencing in yeast by methylation of the nucleosome core. *Cell* **109**, 745–756.
187. Blobel, G. A., Nakajima, T., Eckner, R., Montminy, M., and Orkin, S. H. (1998). CREB-binding protein cooperates with transcription factor GATA-1 and is required for erythroid differentiation. *Proc. Natl. Acad. Sci. USA* **95**, 2061–2066.
188. Forsberg, E. C., Johnson, K., Zaboikina, T. N., Mosser, E. A., and Bresnick, E. H. (1999). Requirement of an E1A-sensitive coactivator for long-range transactivation by the beta-globin locus control region. *J. Biol. Chem.* **274**, 26850–26859.
189. Hung, H. L., Lau, J., Kim, A. Y., Weiss, M. J., and Blobel, G. A. (1999). CREB-binding protein acetylates hematopoietic transcription factor GATA-1 at functionally important sites. *Mol. Cell. Biol.* **19**, 3496–3505.
190. Yang, X. J., Ogryzko, V. V., Nishikawa, J., Howard, B. H., and Nakatani, Y. (1996). A p300/CBP-associated factor that competes with the adenoviral oncoprotein E1A. *Nature* **382**, 319–324.
191. Ogryzko, V. V., Kotani, T., Zhang, X., Schlitz, R. L., Howard, T., Yang, X. J., Howard, B. H., Qin, J., and Nakatani, Y. (1998). Histone-like TAFs within the PCAF histone acetylase complex. *Cell* **94**, 35–44.
192. Yagi, R., Chen, L. F., Shigesada, K., Murakami, Y., and Ito, Y. (1999). A WW domain-containing yes-associated protein (YAP) is a novel transcriptional coactivator. *EMBO J.* **18**, 2551–2562.
193. Komuro, A., Nagai, M., Navin, N. E., and Sudol, M. (2003). WW domain-containing protein YAP associates with ErbB-4 and acts as a cotranscriptional activator for the carboxyl-terminal fragment of ErbB-4 that translocates to the nucleus. *J. Biol. Chem.* **278**, 33334–33341.
194. Kanai, F., Marignani, P. A., Sarbassova, D., Yagi, R., Hall, R. A., Donowiwtz, M., Hisaminato, T., Fujiwara, T., Ito, Y., Cantley, L. C., and Yaffe, M. B. (2000). TAZ: A novel transcriptional coactivator regulated by interactions with 14-3-3 and PDZ domain proteins. *EMBO J.* **19**, 6778–6791.
195. Staub, O., Dho, S., Henry, P., Correa, J., Ishikawa, T., McGlade, J., and Rotin, D. (1996). WW domains of Nedd4 bind to the proline-rich PY motifs in the epithelial Na+ channel deleted in Liddle's syndrome. *EMBO J.* **15**, 2371–2380.
196. Huang, K., Johnson, K. D., Petcherski, A. G., Vandergon, T., Mosser, E. A., Copeland, N. G., Jenkins, J., Kimble, J., and Bresnick, E. H. (2000). A HECT domain ubiquitin ligase closely related to the mammalian protein WWP1 is essential for *Caenorhabditis elegans* embryogenesis. *Gene* **252**, 137–145.
197. Chen, X., Wen, S., Fukuda, M. N., Gavva, N. R., Hsu, D., Akamo, T. O., Yang-Feng, T., and Shen, C. K. (2001). Human Itch is a coregulator of the hematopoietic transcription factor NF-E2. *Genomics* **73**, 238–241.
198. Nawaz, Z., Lonard, D. M., Smith, C. L., Lev-Lehman, E., Tsai, S. Y., Tsai, M. J., and O'Malley, B. W. (1999). The Angelman syndrome-associated protein, E6-AP, is a coactivator for the nuclear hormone receptor superfamily. *Mol. Cell. Biol.* **19**, 1182–1189.
199. Bender, M. A., Bulger, M., Close, J., and Groudine, M. (2000). Beta-globin gene switching and DNaseI sensitivity of the endogenous beta-globin locus in mice do not require the locus control region. *Mol. Cell* **5**, 387–393.
200. Epner, E., Reik, A., Cimbora, D., Telling, A., Bender, M. A., Fiering, S., Enver, T., Martin, M., Kennedy, M., Keller, G., and Groudine, M. (1998). The beta-globin LCR is not

necessary for an open chromatin structure or developmentally regulated transcription of the native mouse beta-globin locus. *Mol. Cell* **2,** 447–455.
201. Schubeler, D., Francastel, C., Cimbora, D. M., Reik, A., Martin, D. I., and Groudine, M. (2000). Nuclear localization and histone acetylation: A pathway for chromatin opening and transcriptional activation of the human beta-globin locus. *Genes Dev.* **14,** 940–950.
202. Schubeler, D., Groudine, M., and Bender, M. A. (2001). The murine beta-globin locus control region regulates the rate of transcription but not the hyperacetylation of histones at the active genes. *Proc. Natl. Acad. Sci. USA* **98,** 11432–11437.
203. Sawado, T., Halow, J., Bender, M. A., and Groudine, M. (2003). The beta-globin locus control region (LCR) functions primarily by enhancing the transition from transcription initiation to elongation. *Genes Dev.* **17,** 1009–1118.
204. Fischle, W., Wang, Y., and Allis, C. D. (2003). Histone and chromatin crosstalk. *Curr. Opin. Cell Biol.* **15,** 172–183.
205. Sproul, D., Gilbert, N., and Bickmore, W. A. (2005). The role of chromatin structure in regulating the expression of clustered genes. *Nat. Rev. Genet.* **6,** 775–781.
206. Allfrey, V., Faulkner, R. M., and Mirsky, A. E. (1964). Acetylation and methylation of histones and their possible role in the regulation of RNA synthesis. *Proc. Natl. Acad. Sci. USA* **51,** 786–794.
207. Hebbes, T. R., Clayton, A. L., Thorne, A. W., and Crane-Robinson, C. (1994). Core histone hyperacetylation co-maps with generalized DNase I sensitivity in the chicken beta-globin chromosomal domain. *EMBO J.* **13,** 1823–1830.
208. Wu, J., Iwata, F., Grass, J. A., Osborne, C. S., Elnitski, L., Fraser, P., Ohneda, O., Yamamoto, M., and Bresnick, E. H. (2005). Molecular determinants of NOTCH4 transcription in vascular endothelium. *Mol. Cell. Biol.* **25,** 1458–1474.
209. Conkright, M. D., Guzman, E., Flechner, L., Su, A. I., Hogenesch, J. B., and Montminy, M. (2003). Genome-wide analysis of CREB target genes reveals a core promoter requirement for cAMP responsiveness. *Mol. Cell* **11,** 1101–1108.
210. Udalova, I. A., Mott, R., Field, D., and Kwiatkowski, D. (2002). Quantitative prediction of NF-kappa B DNA-protein interaction. *Proc. Natl. Acad. Sci. USA* **99,** 8167–8172.
211. Minokawa, T., Wikramanayake, A. H., and Davidson, E. H. (2005). Cis-regulatory inputs of the *wnt8* gene in the sea urchin endomesoderm network. *Dev. Biol.* **288,** 545–558.

Detecting the Unusual: Natural Killer Cells

> Armin Volz and Britta Radeloff
>
> Institut für Immungenetik
> Charité-Universitätsmedizin Berlin,
> Campus Virchow-Klinikum,
> Spanndauer Damm 130,
> 14050 Berlin, Germany

I. Introduction	474
II. Discovery and Characterization of Natural Killer Cells	474
A. Definition and Characterization of Killer Cells	474
B. Heterogeneity of Natural Killer Cells	477
C. Target Specificity of Natural Killer Cells	477
D. Target Recognition	478
III. Natural Killer Cell Receptors	479
A. Receptors Belonging to the Immunoglobulin Superfamily	479
B. Receptors Belonging to the C-Type Lectin Family	490
C. Adaptor Molecules	497
D. Organization and Evolution of Receptor Complexes	498
IV. Expression of NK-Receptor Genes	501
A. Specificity of NK Cells	501
B. Complex Phenotypes	502
C. Single-Cell Analysis	502
D. Promoter Analysis	512
V. Natural Killer Cells and Disease	514
A. Tumor Immune Evasion	514
B. Pathogen Immune Evasion	516
C. Autoimmunity	517
VI. Concluding Remarks	518
References	519

Abbreviations: AML, acute myeloid leukemia; CRD, carbohydrate-recognition domain; Cyp, cytoplasmic domain; Ig, domain of the immunoglobulin super family; HCMV, human cytomegalovirus; IFN-γ, interferon-γ; ITAM, immunoreceptor tyrosine-based activation motif; ITIM, immunoreceptor tyrosine-based inhibition motif; ITSM, immunoreceptor tyrosine-based switch motif; KIR, killer Ig-like receptors; LILR, leukocyte immunoglobulin-like receptors; LRC, leukocyte receptor complex; mAB, monoclonal antibody; MCMV, mouse cytomegalovirus; MHC, major histocompatibility complex; NCR, natural cytotoxicity triggering receptors; NK cell, natural killer cell; NKC, natural killer complex; TM, transmembrane domain.

I. Introduction

In the 1880s, when Emil von Behring and Shibasaburo Kitasato discovered that the serum of vaccinated individuals contained antibodies specific to the applied pathogen, they laid the foundation for modern immunology. For several decades, the underlying adaptive immune system and the respective target-specific agents, they had provided evidence for, were the major subjects immunologists worked on. In the early 1970s, it became apparent (*1–4*) that additional nonadaptive cytotoxic defense mechanisms must exist that are not only B- and T-cell independent but also too fast for an adaptive response. Since no target-specific educational process was needed to render the effector cells cytotoxic, the members of this new class of lymphocytes were named "natural killers" (NK) (*5, 6*). They mediate an early immune response by direct killing of malignantly transformed, infected, and otherwise stressed cells and by the production of proinflammatory cytokines such as interferon-γ (IFN-γ), tumor necrosis factor-α (TNF-α), and granulocyte–macrophage colony stimulating factor (GM-CSF) (*7*).

The main purpose of this chapter is to provide a detailed summary about human NK cells and their sophisticated competencies to discriminate between autologous healthy cells and autologous unhealthy cells that have been rendered "unusual" by malignant transformation, virus infection, or other means. In this context, heterogeneity of the NK cell population will be a major topic since this population heterogeneity (*8, 9*), and not the exceptional properties of a single NK cell, provides the basis of broad specificity (Fig. 1). This heterogeneity results mainly from a highly variable and at least in part stochastic expression of numerous inhibitory and eventually also activating receptors at the cell surface and it is generally agreed that the surface expression of these receptors and the subsequent signal integration process are part of a regulatory network. The fine tuning of this network, probably by a nontarget-specific educational process, provides not only the basis for self-tolerance of NK cells to normal tissues but allows also fast and sensitive detection of unusual properties of diseased cells, leading subsequently to their elimination.

II. Discovery and Characterization of Natural Killer Cells

A. Definition and Characterization of Killer Cells

In the early 1970s, in addition to the known B and T cells, a third group of lymphocytes was described. This new group comprises about 5–15% of peripheral blood lymphocytes and provides cytotoxic reactivity against syngeneic, allogeneic, and xenogeneic cells immediately after the first contact. Most of the

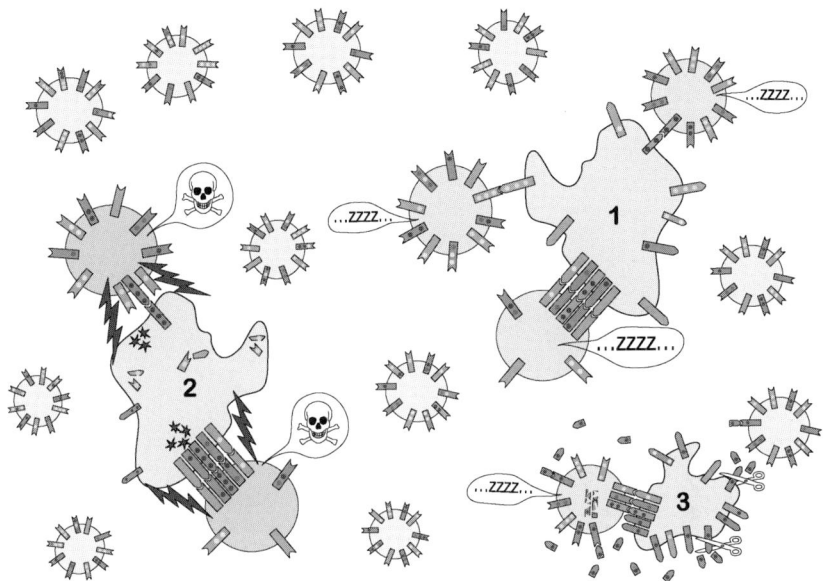

Fig. 1. Principles of NK cell recognition. The figure shows the principle of NK cell recognition. NK cells are shown as round shaped cells in red (cytolytic NK cell), in green (inhibited NK cell) and in gray (without target contact). Target cells are shown in yellow. Receptors on NK cells and corresponding ligands are color coded. Inhibitory receptors are red and activating receptors are green. The variety of receptors is demonstrated by differently colored dots. Target cell 1 is a healthy autologous cell not susceptible to NK cell lysis. Target cell 2 is a cell infected with a virus that leads to MHC antigen degradation and consequently to NK cell-mediated cytolysis. Target cell 3 is a tumor cell that evades NK cytolysis by shedding of MIC proteins. (See Color Insert.)

targets were derived from virus-induced tumors (1, 2, 5, 6, 10, 11) (Fig. 2), but cytotoxicity could also be demonstrated for sheep red blood cells (Fig. 3) (3). Their cytotoxic function did not depend on antibodies, complement, or other serum components. As classified by surface markers, they were neither T nor B lymphocytes and adaptation to antigens or clonal expansion seemed to be irrelevant for target recognition and cytotoxicity. In contrast to educated cytotoxic T lymphocytes they were called *"natural"* killer cells (6).

Initially, due to lack of specific markers, NK cells were isolated by serum albumin (3) and Percoll gradients (12) or by depletion of adherent cells using carbonyl iron, glass beads, or nylon fibers (5, 13, 14). The respective cell isolates contained large granular lymphocytes (LGL) and exhibited, in addition to antibody-dependent cell-mediated cytotoxicity, most of the antileukemia NK activity of the source population (14).

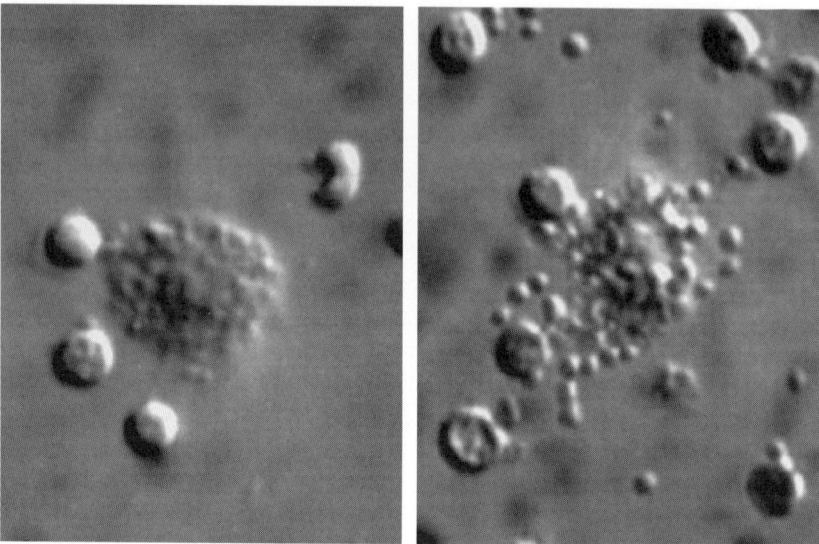

Fig. 2. Large granular lymphocytes detecting a tumor cell. Differential interference contrast micrograph of mouse P815 plasmocytoma cells (the large structured cell in the middle of the left picture). Cytotoxic cells (small cells) recognize the tumor (left) and induce apoptosis, which is visible as membrane blebbing (right) (kindly supplied by Dr. Thornthwaite).

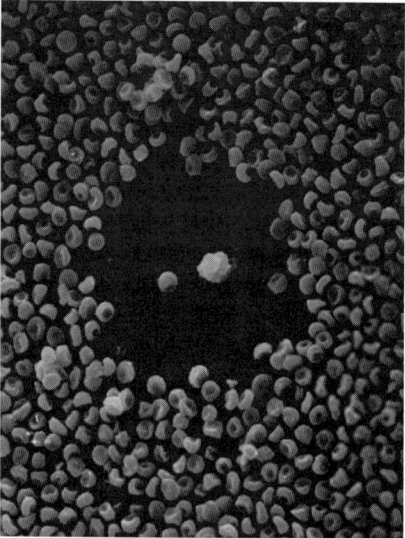

Fig. 3. Complement independent plaque forming cells. The scanning electron micrograph of a plaque cytogram assay shows a large granular lymphocyte that has destroyed sheep red blood cells in its surrounding leaving an erythrocyte free area around the cytotoxic cell (kindly supplied by Dr. Thornthwaite).

B. Heterogeneity of Natural Killer Cells

Competitive killing assays with a variety of target and LGL cells revealed that quite a number of different structures are involved in the NK reaction (15, 16). The functional relevance of this heterogeneity could be shown in single-cell assays with Percoll gradient purified LGL (12). Dependent on the target-cell lines, the LGL-binding frequency varied widely between 10% and 54%, while target-cell line mixtures led to binding percentages surpassing the highest single-cell line. However, only 32–86% of the bound LGLs showed lytic activity without previous activation. Interferon treatment elevated cytotoxicity and binding levels for some LGL target combinations significantly, but with other combinations reactivity was reduced (12). In addition to this functional heterogeneity, staining of single donor NK cells with a selection of mAB revealed also heterogeneity regarding surface protein expression (17). Analysis of over 200 LGL clones did not only reinforce the high heterogeneity in target-cell specificity but led also to the impression that density of some of the analyzed surface markers varies substantially upon environmental changes (17). Secretion of cytokines did also point to heterogeneity. Upon challenge with different stimulating substances, like lipopolisaccharide, phytohemagglutinin, or complete viruses, subsets of LGLs were found to secrete cytokine cocktails unique for the respective substance (17). According to these data, it was concluded that rodent and human LGL belong to a distinct lymphocyte lineage, which is functional without idiotypic surface receptors or clonal selection but features an extensive level of heterogeneity (8, 18).

C. Target Specificity of Natural Killer Cells

A wide range of tumor cell lines was found to be susceptible to NK cell-mediated lysis and natural cytotoxicity was, therefore, initially considered to operate without a particular structure on the target-cell surface. However, the concluded nonspecific or nonselective activity was incompatible with two findings: not all tumor cell lines were sensitive for NK cell-mediated killing (14) and cytotoxicity depended on the source of the NK cells. Since reactivity against MHC negative cells was strong and antibodies masking MHC antigens did not reduce cytotoxicity, NK cells were defined in 1983 as non-MHC restricted (19), although H-2 haplotype dependency had been demonstrated (20). The key for this puzzling selectivity had been published about 10 years before but was not related to NK activity: the F1 hybrid resistance (4). This F1-host antiparental-bone marrow graft reaction had been shown to operate very fast (9–48 hr after transplantation), was independent of acute irradiation and was not detectable before the 3rd week of age in mice. This fits perfectly to NK cells that are permanently under arms, do not proliferate, which leads to radiation resistance and do not emerge before the 3rd week after birth, with maximal activity at ~10–12 weeks (15). Hybrid resistance was linked to parental heterozygosity of loci within the major histocompatibility complex (MHC) and was functional only if parental cells possess "transplantation antigens" not shared by F1 hybrid mice (4).

Although NK cells had meanwhile been confirmed to be the effector cells of F1 hybrid resistance (21, 22), the "recessive" segregation of this transplantation antigen seemed to be incompatible with the dominant segregation known for MHC phenotypes. This discrepancy could be resolved by the missing self-model (23, 24) postulating that one function of NK cells is to detect and eliminate cells that lack or express reduced levels of self MHC class I antigens. Validity of this model was initially demonstrated with NK cells of mice transgenic for an additional MHC class I gene. These NK cells rejected grafts of tumor cells (25) and bone marrow cells (26) of the parent nontransgenic strains lacking the respective gene product.

D. Target Recognition

Shortly after verification of the target molecules, an inhibitory receptor, Ly49 [new nomenclature: killer cell lectin-like receptor subfamily A, member 1 (Klra1)], was described in mice (27). It belongs to the family of C-type lectins, is expressed by a subset of NK and T cells and mediates an inhibitory signal upon interaction with MHC class I molecules of the appropriate haplotype (28). The respective gene was located on mouse chromosome 6 within a large gene complex, the "natural killer" complex, which turned out to contain many further NK receptors of the C-type lectin family (29) (see also Section III.D.2). Many of them belong also to subfamily A, are expressed clonally (30) and interact haplotype specific with distinct classic MHC class I molecules. In addition, members of other subfamilies specific, for example, for the nonclassic MHC class I molecule Qa-1 (Section III.B.3) (31), or MHC-like proteins (Rae1 and H60) (32, 33) are also present. In humans, KLRA1 seems to be a pseudogene (34) and haplotype-specific surveillance of classic MHC class I molecules is mediated by KIR (35–37). Although these genes evolved completely independently, they were found to convey a surprising high similarity regarding target specificity, expression pattern, and signal transduction (38) (Section III.A.1).

MHC-specific inhibitory receptors could mostly explain specificity of F1 hybrid resistance (39) and NK reactivity against tumor cell lines. However, it was expected that inhibitory NK receptors would need an activating counterpart responsible for the proper induction of cytotoxicity and cytokine release (40). Such a receptor, Fcgr3, was known to be expressed by NK cells (41) and mediates antibody-dependent cell-mediated cytotoxicity (42). Further activating receptors known at that time are encoded also in the NK complex (43–45) and are described in detail, together with numerous additional activating and inhibitory receptors that were found later on (in section III).

Activating and inhibitory receptors are coexpressed on NK cells and for several receptors clonal distribution has been described (38, 45–47). These results explained the heterogeneity of NK cell populations (Section II.B) and suggested a distinct specificity for each single cell (47). The resulting immense killer cell repertoire is necessary and effective to discriminate a plethora of differently differentiated healthy cells from cells with numerous distinct aberrations due to neoplastic transformation, pathogen infection, or other stressors. However, the mechanism that conveys tolerance to normal self, probably by balancing activating and inhibitory activities is unknown but clearly effective, because no autoimmune state that results from NK cell dysfunction has been described (48).

III. Natural Killer Cell Receptors

Numerous receptors are expressed on the surface of NK cells. They provide the interface for the interaction with other cells and are, therefore, the basis for immune surveillance. In this chapter, we describe the receptors that are responsible for discrimination between autologous healthy cells and autologous damaged cells, the targets of NK cell attack.

A. Receptors Belonging to the Immunoglobulin Superfamily

Proteins of the immunoglobulin (Ig) superfamily contain so-called Ig domains, which all share a characteristic protein fold. Based on size and sequence, V-type immunoglobulin domains are similar to the variable regions of antibody molecules, while C2-type Ig domains are similar to the constant regions. The extracellular domain of the receptors described in this chapter are composed of Ig domains. They are type I integral membrane proteins (C-terminus within the cytoplasm) that confer target specificity to NK cells but many of them may be found also on other cells of the immune system. They belong to several subfamilies that are described in the next sections. The genes of these subfamilies are mostly clustered, whereas most of them have been localized to the LRC (cf. Section III.D.2) on human chromosome 19 (49, 50).

1. KILLER CELL IG-LIKE RECEPTORS

The first target specificity mediating human NK-cell receptors were described in 1990 (35, 36) as surface molecules that inhibit cytolytic activity when cross-linked by antibodies. The family members were therefore designated killer inhibitory receptors (51, 52). However, it became clear later on that the KIR family encompasses also activating receptors (53) (Table I). To resolve this inconsistency, the family members were renamed to killer cell Ig-like receptors.

TABLE I
KILLER CELL IG-LIKE RECEPTORS

KIR	Functionality	Protein domains	Ligand	Alias
2DL1	Inhibitory	2 Ig C2 (D1 + D2), TM, Cyp 2 ITIM	HLA-C$_{Lys80}$	nkat1, cl-42, CD158A, p58.1
2DL2	Inhibitory	2 Ig C2 (D1 + D2), TM, Cyp 2 ITIM	HLA-C$_{Asn80}$	nkat6, cl-43, CD158B1
2DL3	Inhibitory	2 Ig C2 (D1 + D2), TM, Cyp 2 ITIM	HLA-C$_{Asn80}$	nkat2a, cl-6, CD158B1, p58
2DL4	Inhibitory + activating	2 Ig C2 (D0 + D2), TM$_{Arg}$, Cyp 1 ITIM	HLA-G	KIR103AS, CD158D, p49
2DL5A	Inhibitory	2 Ig C2 (D0 + D2), TM, Cyp 2 ITIM	?	CD158F
2DL5B	Inhibitory	2 Ig C2 (D0 + D2), TM, Cyp 2 ITIM	?	KIR2DL5.2/3/4
2DS1	Activating	2 Ig C2 (D1 + D2), TM$_{Lys}$, Cyp$_{short}$	HLA-C$_{Lys80}$	EB6actl, CD158H
2DS2	Activating	2 Ig C2 (D1 + D2), TM$_{Lys}$, Cyp$_{short}$	HLA-C ?	nkat5, CD158J, cl-49
2DS3	Activating	2 Ig C2 (D1 + D2), TM$_{Lys}$, Cyp$_{short}$	HLA-C ?	nkat7
2DS4	Activating	2 Ig C2 (D1 + D2), TM$_{Lys}$, Cyp$_{short}$	HLA-C ?	nkat8, CD158I, cl-39
2DS5	Activating	2 Ig C2 (D1 + D2), TM$_{Lys}$, Cyp$_{short}$	HLA-C ?	nkat9, CD158G
3DL1	Inhibitory	3 Ig C2 (D0 + D1 + D2), TM, Cyp 2 ITIM	HLA-B$_{Bw4}$	nkat3, NKB1, cl-2, cl-11, NKB1B, AMB11, CD158E1/2
3DL2	Inhibitory	3 Ig C2 (D0 + D1 + D2), TM, Cyp 2 ITIM	HLA-A3, -A11	nkat4, cl-5, CD158K
3DL3	Inhibitory	3 Ig C2 (D0 + D1 + D2), TM, Cyp 2 ITIM	?	KIRC1, KIR3DL7, KIR44, CD158Z
3DS1	Activating	3 Ig C2 (D0 + D1 + D2), TM$_{Lys}$, Cyp$_{short}$	HLA-B$_{Bw4}$	nkat10
2DP1	Pseudogene	—	—	KIRZ, KIRY, KIR15, KIR2DL6
3DP1	Pseudogene	—	—	KIRX, KIR48, CD158C, KIR2DS6

Inhibitory KIRs interact with the α1 and α2 domains of the ubiquitously expressed HLA class I molecules (Table I) and position 7 and 8 of the bound peptide (54). They become, upon ligand binding, tyrosine phosphorylated at their ITIMs, which are part of the cytoplasmic domain. Phosphorylated ITIMs recruit and activate tyrosine phosphatases like SHP-1 and SHP-2, resulting in the abrogation of NK cell effector functions (55, 56). Accordingly, decreased or absent expression of MHC class I molecules reduces or abolishes inhibition and results, in agreement with the missing self-hypothesis (23, 24), in triggering of cytotoxicity and cytokine release (57). Sufficient surface expression of HLA class I antigens, therefore, protects healthy autologous cells from lysis.

Activating KIR have lost most of the cytoplasmic domain and consequently also the inhibitory motives. Association to adaptor molecules (cf. Section III.C) that contribute immunoreceptor tyrosine-based activation motifs compensates for the missing functional elements. To allow discrimination of inhibitory from activating as well as two from three domain variants, these features are reflected in the gene designations: the inhibitory killer cell Ig-like receptor number 1 with 3 Ig domains and long cytoplasmic tail results in *KIR3DL1*, while the activating killer cell Ig-like receptor number 1 with 2 Ig domains and short cytoplasmic tail results in *KIR2DS1*.

Seven of the 14 human *KIR* genes described are inhibitory, six are activating, and one is bifunctional (58). In contrast to many other genomic regions, the gene content of the KIR cluster of different individuals is not conserved (58–60) but highly variable. KIR haplotypes have, therefore, been divided into two functionally distinct groups: the group A haplotypes contain only one activating *KIR* gene (*KIR2DS4*) and in many individuals even this last activating receptor has lost its function by a 22-bp deletion (61, 62). The type B haplotypes contain up to five genes coding for activating receptors (*KIR2DS1–5* and *KIR3DS1*) and additional inhibitory *KIR* genes (*KIR2DL5A/B*) might be present. Therefore, they are much more complicated and numerous different B haplotypes have been described (62–64). This unusual bimodal distribution into haplotypes that are rich or poor in activating *KIRs*, points to selective pressure exerted by two different opposing forces (cf. Section III.D.3). A special case is the HLA-G-specific *KIR2DL4*. It was shown to be present in all haplotypes and is expressed by most NK cells. Unlike other members of the KIR family, *KIR2DL4* is localized predominantly in endosomes and mediates endocytosis of its ligand, soluble HLA-G. Furthermore, engagement of *KIR2DL4* by soluble ligand activates a proinflammatory/proangiogenic response, consistent with a role in promoting vascularization during early pregnancy (65). Additionally, *KIR2DL4* seems to transduce also inhibitory signals (66).

Specificity of the inhibitory KIRs is well investigated (summarized in Table I) (52, 53, 55–67). Receptors with three Ig domains (KIR3DL) interact with HLA-A and -B antigens but however, only with a subset of alleles (67–69).

A significant proportion of KIR haplotypes even lacks any apparent HLA-B self-ligand leading to a considerable amount of individuals that lack protective HLA/KIR3DL combinations (63, 70). The major inhibitory KIR ligand in humans is, therefore, HLA-C and most haplotypes or at least genotypes encode inhibitory KIRs that confer tolerance to HLA-C expressing healthy autologous cells.

According to their ability to interact with distinct two domain KIRs, HLA-C alleles were divided into two broad groups: C1 with asparagine at position 80 and C2 with lysine at the respective position (71). Group C1 interactions with their cognate receptors KIR2DL2 and -3 are weaker than HLA-C2/KIR2DL1 interactions (72, 73). This leads, together with a high degree of *KIR* gene polymorphism, to different inhibitory potentials of distinct KIR (74–76). These differences seem to be associated with susceptibility or resistance to disease (cf. Section V).

Function and target molecules of activating KIRs are still obscure. Since no phenotype could be assigned to the relatively frequent group A homozygotes that are completely deficient of activating KIR (62), it might be concluded that activating KIR are dispensable in normal life. However, phylogenetic analysis revealed that activating KIR have been developed independently by different species (77), which points to significant evolutionary pressure. Such pressure might be imposed by pathogens, and an activating killer receptor (*Klra8*) is not only specific for the MCMV protein m157 but also expression of *KLRA8* confers resistance to MCMV infection.

Due to highest homology of the ligand-binding domains of some inhibitory-activating receptor pairs, which results in some cases from an allelic relation, the ligands of activating KIR were initially expected to be also HLA class I molecules. However, mutational and structural analysis of KIR2DS1 and KIR2DS2 indicate that HLA-C is not their cognate ligand (72, 78, 79). On the other hand, a chimera of the extracellular domain of KIR2DS4 with an inhibitory cytoplasmic tail was capable to inhibit lyses of HLA-Cw4 positive cells (80) and demonstrates that interaction of activating KIR with HLA class I molecules might result at least in a weak signal. In agreement with this assumption, correlative studies indicate that activating KIR eventually combined with weak HLA-C1/*KIR2DL* interactions assist in fighting HIV (81) but might also promote autoimmune disease (82).

The high number of *KIR* genes, the varying gene content of different KIR haplotypes and a high degree of polymorphism within each gene (76) combined with the high variability of HLA class I ligands leads to a very complex situation regarding NK cell specificity. On the single NK cell level, the situation is even more complex, since KIR (except K2DL4) and other receptors are clonally expressed (35, 38, 45–47). Hence, NK cells show numerous different receptor expression patterns (9) and consequently different specificities. The distribution of these phenotypes seems to be defined genetically, since the KIR repertoire normalizes usually to a donor-specific pattern about 3 months after transplantation (*83, 84*). These donor-specific phenotypes explain the MHC-based variety of NK specificities (*38*) and play a major role in the outcome of hematopoietic stem cell transplantation (*85*) (cf. Section V). They are also responsible for the F1 hybrid resistance, which initiated NK cell research (cf. Section II.C).

2. Leukocyte Immunoglobulin-Like Receptors

LILRs have simultaneously been discovered by several groups (*86–88*), which led to a couple of different names summarized in Table II. Five activating, five inhibitory, and one soluble LILR have been described, all encoded within the leukocyte receptor cluster (*49*). They are mainly expressed on myelomonocytic cells (*87*), where they are thought to influence other immune receptors (*89*). Most cell types of the innate as well as the adaptive immune system express at least one member of the LILR family (Table II). For most LILRs no ligands are known (Table II).

The inhibitory receptors LILRB1, -B2, and -B4 are expressed on a minor subset of NK cells (*86, 90*). LILRB1 and -B2, interact with a broad spectrum of HLA class I molecules (Table II) (*91*). However, the best endogenous ligand seems to be HLA-G (*92*). It results in upregulation of LILRB1, -B2, -B4, and *KIR2DL4* (*93*) in antigen-presenting cells, NK cells, and T cells and seem to protect, in combination with KLRD1/KLRC1, the fetus against the attack of decidual NK cells (*94, 95*). The rolling of NK cells prior to extravasation is also disturbed by HLA-G, an effect that is in part mediated by LILRB1. In contrast to T cell receptors (TCRs) and KIRs, HLA-specific LILRs do not interact with the polymorphic $\alpha 1$ and $\alpha 2$ domains of HLA molecules but with the conserved $\alpha 3$ domain and $\beta 2$-microglobulin (*96–98*). This explains not only the broad specificity of LILRB1 and -B2 but also the competition with CD8 for the same binding site (*92*). It seems to be relevant for the induction of peripheral tolerance (*99*).

LILRB1 and to an $\sim$1000-fold lower extend also LILRB2 are very strong receptors for UL18 (*86*). This protein is produced by HCMV and resembles HLA Class I antigens (*100*). It binds also endogenous peptides and associates with $\beta 2$-microglobulin (*98*). UL18 does not interact with TCR and can therefore not provoke a cytotoxic T cell response. UL18 may, therefore, act as a surrogate ligand for LILRB1 to prevent NK cell lysis of CMV-infected cells (*101, 102*), which are susceptible to NK cell attacks due to virus-induced downregulation of HLA class I expression (*103*). However, cytotoxicity of polyclonal NK cells was rather increased than diminished by UL18 expression (*104*). An explanation may be that LILRB1 is expressed only by a minor subset of NK cells while the major targets are monocytic and dendritic cells.

TABLE II
Leukocyte Immunoglobulin-Like Receptors

Gene	Functionality	Protein domains	Ligand	Alias	Predominant expression
LILRA1	Activating	4 Ig C2, TM$_{Arg}$, Cyp$_{short}$	HLA-B27	LIR-6, CD85i	Monocyte
LILRA2	Activating	4 Ig C2, TM$_{Arg}$, Cyp$_{short}$	HLA-interaction predicted	LIR-7, ILT1, CD85H	Monocyte, eosinophil, B-cell, NK-cell
LILRA3	Soluble	4 Ig C2	HLA-interaction predicted	LIR-4, HM43, ILT6, HM31, CD85e	Monocyte, B-cell, macrophage
LILRA4	Activating	4 Ig C2, TM$_{Arg}$, Cyp$_{short}$	?	ILT7, CD85g	Monocyte, dendritic cell
LILRA5	Activating	2 Ig C2, TM$_{Arg}$, Cyp$_{short}$	?	ILT11, LIR9, CD85f	Monocyte, neutrophil
LILRA6	Activating	4 Ig C2, TM$_{Arg}$, Cyp$_{short}$	?	ILT8, CD85b	
LILRB1	Inhibitory	4 Ig C2, TM, Cyp 4 ITIM	HLA-A, B, C, E, F, G, UL18	LIR-1, ILT2, MIR-7, CD85j	Monocyte, dendritic cell, eosinophil, B cell, NK cell, T cell
LILRB2	Inhibitory	4 Ig C2, TM, Cyp 3 ITIM	HLA-A, B, C, F, G, UL18	LIR-2, ILT4, MIR-10, CD85d	Monocyte, dendritic cell, eosinophil, B-cell, NK-cell
LILRB3	Inhibitory	4 Ig C2, TM, Cyp 4 ITIM	?	LIR-3, HL9, ILT5, CD85a	Monocyte, dendritic cell, eosinophil, B cell
LILRB4	Inhibitory	2 Ig C2, TM, Cyp 3 ITIM	?	LIR-5, ILT3, HM18, CD85k	Monocyte, mast cell, macrophage, NK cell, dendritic cell, B cell
LILRB5	Inhibitory	4 Ig C2, TM, Cyp 2 ITIM	?	LIR-8, CD85c	NK cells
LILRP1	Pseudogene	4 Ig C2 exons	–	–	–
LILRP2	Pseudogene	4 Ig C2 exons	–	–	–

For more details regarding receptor specificity and expression profiling, please refer to Ref. (88).

In total, we can conclude, that the clonal expression of a subset of LILR receptors on NK cells contributes significantly to their broad specificity, mainly by protecting healthy autologous cells from cytolysis.

3. Low-Affinity Receptor for the Fc Fragment of Immunoglobulin Gamma Variant III B

The low-affinity IgG receptor is expressed by a subset of NK cells (*105, 106*). It associates for signal transduction and cell-surface expression with the adaptor proteins FcεRIγ (*107*) or CD3ζ (*105, 108*) (Table V). It interacts with antibody aggregates, is responsible for antibody dependent-cell-mediated cytotoxicity of NK cells (*109*) and tumor therapies with mAB are mainly mediated by the low-affinity receptor for the Fc fragment of IgG variant III B (FCGR3B) (*110*). Unfortunately, the affinity is relatively low and normal serum levels of IgG effectively compete for binding of therapeutic antibodies (*109*). Since some alleles of CD16 have even lower affinities to IgG, very high concentrations are needed for some individuals (*110*). Low-fucose IgG was demonstrated to show higher affinity to FCGR3B and induces strong antibody-dependent cell-mediated cytotoxicity even with low-IgG concentrations (*111*) and might lead to better clinical efficacy. A selective depletion of $CD16^+$ NK cells was found during HIV infection (*112*).

4. Natural Cytotoxicity Triggering Receptors

NCR are cell-surface glycoproteins that activate NK cells after ligand interaction (*7*). Three NCRs were discovered in search of monoclonal anti-NK cell antibodies that are able to invoke NK-mediated killing of Fc receptor positive tumor cells (*113–115*).

(a) NCR1 was found to be expressed on all NK cells, including $CD56^{bright}$ $CD16^-$ and immature NK cells (*113*), but the expression level varies largely and allows subdivision of NK cells into $NCR1^{bright}$ and $NCR1^{dull}$ subgroups (*7*), whereas $NCR1^{bright}$ NK cells revealed low levels of KLRC2 and KIR2DS (*7*). In some donors, however, the whole spectrum of surface expression was observed. In such NK cells, the cytolytic activity correlates mostly with the level of NCR1 expression (*116*). This observation might provide a link to the missing antitumor reactivity of the immune system in AML patients (*117*). Their NK cells show reduced levels of NCR and the tumor cells are almost devoid of the unknown endogenous NCR ligands. NCR1 and NCR3 expression is also regulated by the immune modulators prolactin and cortisol. Prolactin dramatically increases NK-mediated killing of the K562 cell line, whereas cortisol abolishes this activity (*118*).

NCR1 orthologs of mouse and rat display ~60% homology and show an expression pattern comparable to the human NCR1 (119, 120). In double-knockout mice, lacking the adaptor molecules CD3ζ and FcεR1γ, functional studies revealed that these adaptors are essential for NCR signal transduction (cf. Section III.C and Table V). Their absence reduced cytolytic activity of NK cells against most target-cells profoundly.

Interaction of hemagglutinin of influenza virus and hemagglutinin-neuraminidase of parainfluenza virus with NCR1 and NCR2 was demonstrated (121, 122). Both receptors bind to sialic acids, which are probably part of the single N-linked or two O-linked oligosaccharides attached to the NCRs. Since several virus families use sialic acid as a target for virus entry into host cells, this might be a general strategy for NK cell recognition of a substantial subset of viruses.

(b) NCR2 is an activating receptor that promotes cytotoxicity and cytokine release upon ligand binding via the adaptor molecule DAP12 (cf. Section III.C and Table V) (115). An additional ITIM like sequence predicted in the NCR2 cytoplasmic domain is not functional (123). NCR2 expression is absent on fresh peripheral blood NK cells but may be induced by IL-2 in culture (115). Since only a minor subset of γδ T lymphocytes is also positive (124), expression of NCR2 may be used as specific marker for activated NK cells (7). In such an analysis with HIV patients, NCR2 was found on a substantial percentage of NKs (125) and promotes cytotoxicity against NCR2 ligand (NCR2L) positive $CD4^+$ T cells. Unfortunately, NCR2L is induced by a part of the HIV-1 envelope protein gp41 resulting in a progressive $CD4^+$ T cell depletion by NK cells, which is correlated to the increase of the virus load (125). As mentioned, influenza and related viruses may also be recognized by NCR2 (see III.A.4.a) (122). A murine homolog for NCR2 was not found.

(c) NCR3 is, as NCR1, expressed on all NK cells, including $CD56^{bright}$ $CD16^-$ and immature NK cells and is the major triggering receptor for cytotoxicity against some tumors (114). The human NCR3 gene has six differentially spliced transcripts, which differ in their Ig domains (V type or C type) and each of the extracellular domains can be linked to one of three different intracellular domains (126). Cross-linking of NCR3 with a mAB results in cytotoxicity and cytokine production (7). Probably due to the low affinity of NCR3, no ligand has been identified, but experiments with soluble NCR3 clearly demonstrated, that cellular ligands must exist (127). In the analyzed mouse strains (*Mus musculus*), NCR3 was found to be a pseudogene (128).

Cytotoxicity against many tumor targets depends on NCR interaction. Blocking of an individual NCR resulted in most instances in a reduced cytotoxicity (114, 115, 129, 130). However, masking of all NCRs simultaneously abrogated cytolytic NK activity almost completely (114, 129). Although different adaptor molecules are involved, NCRs activate the same NK cell effector functions. Simultaneous interaction of more than one receptor with its specific cellular ligands leads, therefore, to synergistic signal amplification (131). However, activating signals are virtually abrogated at a very early state in the signal cascade by KLRAC1/KLRAD engagement (131) and probably also by other inhibitory NK receptors.

NCR are not clonally expressed (114, 115, 130), but surface expression seems to be highly variable (116). Since NCR surface density is directly correlated to the degree of NCR mediated cytolytic attack against target cells, it might be concluded that NK cell specificity is extensively affected by the NCR expression level (116).

5. Natural Killer Receptor 2B4

Natural killer receptor 2B4 (CD244) (2B4, Table III) is a member of the signaling lymphocytic activation molecule (SLAM) subfamily and belongs to the CD2 receptor family (132). It is encoded in the CD2 cluster and is expressed in almost all NK cells and many further lymphoid and myeloid lineages (Table III) (133, 134). CD244 was initially described in mice (135) as a receptor that mediates non-MHC restricted killing (136) and activation of murine $\gamma\delta$ T cells. (137). The initial interpretation that mouse CD244 is an activating receptor was due to redirected killing experiments (135), in which addition of CD244-specific antibodies had enhanced cytotoxicity. Later on it became clear that the antibody had masked inhibition by CD244, while activation was mediated by other receptors (138). However, in humans, CD244 seems to be an activating receptor (139, 140). The four-immunoreceptor tyrosine-based switch motifs (ITSM) that are present in the cytoplasmic domain of CD244 recruit the SH2 domain protein 1A (SH2D1A, SAP) which signals via the SRC tyrosine kinase FYN. Together with the "linker for activation of T cells" (LAT), which is also associated with CD244 (141, 142), FYN activates the effector cell via SLAM phosphorylation (140). However, in X-linked lymphoproliferative disease the signal transduction cascade is interrupted. Patients are devoid of functional SH2D1A resulting in effector cells that cannot be activated by CD244 (143). In addition, the four ITSMs recruit SHP1 and mediate inhibitory signals (143). In consequence, cells expressing the CD244 ligand do not only fail to induce cytotoxicity but also inhibit NK cell effector functions.

CD48, the ligand of CD244 (133, 144), is expressed by all nucleated hematopoietic cells and by human endothelial cells (145). It was originally

TABLE III
FURTHER RECEPTORS OF THE IMMUNOGLOBULIN SUPERFAMILY

Gene	Functionality	Protein domains	Ligand	Alias	Predominant expression
NCR1	Activating	2 Ig C2, TM$_{Arg}$, Cyp$_{short}$	?	NKp46	All NK cells, immature NK cells
NCR2	Activating	1 Ig V, TM$_{Lys}$, Cyp 1 ITIM	?	NKp44, LY95	Activated NK cells, subset of γδT cells
NCR3	Activating	1 Ig V/C2, TM$_{Arg}$, Cyp$_{short}$	?	NKp30, 1C7	All NK cells, immature NK cells
FCGR3B	Activating	2 Ig, TM$_{Arg}$, Cyp	IgG	CD16b	NK cells
CD244	Activating	1 Ig V, 1 Ig C2, TM, Cyp 4 ITSM	CD48	NKR2B4, NAIL, Nmrk, SLAMF4	All NK cells, subset of T cells, monocytes, mast cells, granulocytes
LAIR	Inhibitory	1 Ig C2, TM, Cyp 2 ITIM	?	—	Most peripheral blood mononuclear leukocytes
CEACAM1	Inhibitory	1 Ig V, 3 Ig C2, TM, Cyp 2 ITIM	CEACAM1, CEACAM5	BGP1, CD66A	Ubiquitous
SIGLEC7	Inhibitory	1 Ig V, 2 Ig C2, TM, Cyp 2 ITIM	Sialic acid	p75/AIRM1, QA79	NK cells, monocytes, cytotoxic T cells
SIGLEC9	Inhibitory	1 Ig V, 2 Ig C2, TM, Cyp 1 ITIM+1 ITSM	Sialic acid	—	50% of NK cells, monocytes, neutrophils, bone marrow, placenta, spleen, and fetal liver

identified as a protein that is induced by infection of B cells with Epstein-Barr virus (EBV) (*146*). Since NK cells of X-linked lymphoproliferative disease patients are inactivated by EBV-induced CD48, the early NK cell response to EBV infection is missing. Consequently, the infected cells are not eliminated and the virus propagates fulminantly (*147*), which results in a high mortality. Since Epstein–Barr virus infects at least 95% of people in the United States (*48*), it was suggested that EBV is responsible for the development of the activating function of CD244 [it seems to be inhibitory in rodents (*48*)].

6. OTHER RECEPTORS

Several other receptors containing domains of the immunoglobulin super family have been detected on NK cells and other leukocytes. However, most of these receptors are not well characterized yet. They will be shortly introduced in the later sections.

The leukocyte-associated immunoglobulin-like receptor 1 (LAIR1) was identified as receptor that inhibits cytotoxicity even in the presence of strong positive signals (*148*). Further analysis revealed that LAIR1 and the signal mediating phosphatase SHP1 constitutively associated (*149*). The ligand has not been discovered yet, but seems to be no MHC molecule (*148*). There is another LAIR gene in the leucocyte receptor complex (*49*), but it lacks TM and cytoplasmic tail and is, therefore, suspected to be no receptor.

The carcinoembryonic antigen related cell-adhesion molecule 1 (CEACAM1) belongs to a large and multifunctional family (*150*) and is the only family member that is expressed in NK cells (*151*). CEACAM1 binds in a homophilic interaction to CEACAM1 of another cell (*152*) or to CEACAM5 and has antiproliferative properties in carcinomas (*153, 154*). In NK cells, homophilic interactions of CEACAM1 inhibit cytolysis independently of MHC class I recognition (*155*) and support NK self-tolerance in TAP2-deficient patients (*156*). In this context, it is noteworthy that in healthy donors only 12% of NK cells express CEACAM1 in contrast to 80% in TAP2-deficient patients (*156*). Additionally, soluble CEACAM1 in serum is reduced in patients, which increases inhibition of cytolysis in CEACAM1 positive NK cells (*157*). Due to these findings, it was suggested that CEACAM1 is another receptor that mediates self-tolerance to NK cells (*48*).

Sialic acid-binding immunoglobulin-like lectins are inhibitory receptors specific for variants of sialic acid (*158*). SIGLEC7 and -9 are expressed on NK cells (Table III). Since SIGLEC7 interaction with ganglioside GD3 (*159*) inhibits cytolysis of NK cells (*160*), and GD3 is expressed by cells of the central nervous system (*161*), it might confer NK cell tolerance to this tissue. However, the exact ligand specificity and function of the SIGLECs is unknown.

B. Receptors Belonging to the C-Type Lectin Family

Natural killer cell lectin-like receptors are clustered in the NKC (cf. Section III.D.2) (*162*) on human chromosome 12 and belong to the C-type (calcium dependent) lectin family. They are type II integral membrane proteins (N-terminus within the cytoplasm) with a single transmembrane domain (*163*). According to their homology, they are divided into several subclasses.

1. Subfamily A (KLRA1)

Killer cell lectin-like receptors of the family A were initially discovered in mice (*27*). An mAB (SW5E6) specific for Klra3 revealed that they are clonally expressed by NK cells (*164*) and a subset of T cells (*27*). By masking Klra3 with SW5E6 *in vivo*, a correlation between Klra3 function and NK cell specificity in hybrid resistance was demonstrated (*164, 165*). In-between, 23 KLRA members were discovered in mice with a broad spectrum of MHC allelic specificity [reviewed in (*166*)]. However, humans have only a single gene of that family (*34*) and premature stop codons probably render it nonfunctional (*167*).

2. Subfamily B (KLRB)

The first killer cell lectin-like receptor subfamily B member was discovered in rat (*168*). Subsequently, five mouse Klrb family members were described (*48, 169*), whereas Klrb1c is expressed by all mouse NK cells of certain inbred strains and is, therefore, used as major lineage specific marker (*170*). Cross-linking of Klrb1c on the surface of rodent NK cells triggers NK cell-mediated cytotoxicity, secretion of cytokines, and IP3 generation (*168, 171*) but other family members (Klrb1b + d) transduce inhibitory signals (*172*). The ligands of Klrb1b, d (Clec2d/Clr-b) and f (Clec2i/Clr-g) have been discovered recently (*173*). Clec2d seems to be the major NK cell inhibitor in mice without MHC expression (*48*) and its expression on hematopoietic tumors is often reduced dramatically (*173*). It might, therefore, be another example for the missing self-hypothesis.

KLRB1, the only human member of this family, is a disulfide-linked homodimer that is present on the majority (73–97%) of $CD3^-/CD56^+$ NK cells as well as a subset (16–47%) of T cells (*45*). KLRB1 positive as well as negative phenotypes seem to be stable in cultured NK cells. In contrast to the rodent Klrb family members, no signal trancducing motives could be demonstrated for KLRB1, and anti-KLRB1 antibodies do not enable freshly isolated peripheral blood NK cells to lyse NK-resistant FcR-bearing targets in a mAb redirected cytotoxicity assay (*45*). However, KLRB1 might still be an activating receptor, since cytolysis of the murine P815 line by some NK clones could be eliminated by masking of the receptor and women with implantation failures exhibited a significantly increased KLRB1 expression in CD56$^+$/CD3$^+$ NKT cells (*174*).

3. SUBFAMILY C (KLRC)

The members of the killer cell lectin-like receptor subfamily C were initially discovered by the search for genes expressed preferentially by NK cells (*175*). One clone of the thus defined NK group 2 (NKG2) hybridized to a number of related transcripts (KLRC1, C2, and K1) (*44*). Several of these have been proven functional, while the status of KLRC4 is unclear. Their properties will be discussed in the later sections.

(a) KLRC1 forms disulfide bound heterodimers with KLRD1 (CD94) (*43, 176, 177*) and contains ITIM motives in the cytoplasmic domain (Table IV). Search for KLRC1 ligands revealed that transfection of a broad spectrum of HLA class I molecules into cytolysis susceptible target cells abolished KLRC1-dependent cytotoxicity (*43, 178–180*). Since an anti-KLRD1 antibody was able to reverse inhibition, it was suggested that KLRC1/KLRD1 complexes interact with a conserved part of most HLA class 1 antigens (*53*) including the non-classic HLA-E antigen. In this context it is of note that HLA-E reaches the cell surface only, if suitable peptides are available (*181*) and such peptides are part of the leader sequences of many HLA heavy chains. However, only peptides derived from HLA molecules previously found to "interact" with KLRC were presented by HLA-E and abolished cytotoxicity (*182, 183*), while leader peptides of "noninteracting" HLA antigens or an artificially disabled HLA-G leader (*95*) did not reduce cytotoxicity. In addition, HLA-E tetramers were used to demonstrate that this HLA antigen interacts only with receptors of the KLRC family (*184, 185*).
In summary, this means that HLA-E may inhibit NK cells only if other HLA molecules are translated into the endoplasmic reticulum and supply thereby leader peptides for HLA-E stabilization. Hence, a single receptor complex achieves surveillance of the normal biosynthesis of many different HLA class I allotypes and confers by this means tolerance to healthy cells.

(b) KLRC2 is an activating homolog of KLRC1, but the affinity for HLA-E peptide complexes is usually lower (*186*). Cells expressing both receptors are, therefore, inactivated by ligand interaction (*187*). A significant exception to this pattern seems to be the HLA-G–derived leader peptide that triggered cytotoxicity very efficiently against such cells (*187*). KLRC2 forms also heterodimers with KLRD1 (*188*), but needs to associate with TYOBP for signal transduction (Table V) (*189*). It is a potent activating receptor in redirected killing assays and is expressed

TABLE IV
NATURAL KILLER CELL RECEPTORS OF THE C-TYPE LECTIN FAMILY

Gene	Functionality	Domains	Known ligand	Alias	Predominant expression
KLRA1	Pseudogene	–	–	Ly49, LY49L	IL2-activated NK cells
KLRB1	Unclear	CRD, TM, Cyp	–	CD161, hNKRP1A, CLEC5B	Subset of NK cells, and T cells
KLRC1	Inhibitory	CRD, TM, Cyp 2 ITIM	HLA-E	NKG2A, NKG2B, CD159A	Subset of NK cells, and T cells
KLRC2	Activating	CRD, TM $_{Lys}$, Cyp	HLA-E	NKG2C	Subset of NK cells, and T cells
KLRC3	Activating	CRD, TM $_{Lys}$, Cyp	HLA-E	NKG2E, NKG2H	Subset of NK cells
KLRC4	Pseudogene	CRD$_{truncated}$, TM $_{Lys}$, Cyp ITIM	–	NKG2F	?
KLRD1	Partner in KLRC heterodimer	CRD, TM, Cyp	HLA-E	CD94	Subset of NK cells, and T cells
KLRG1	Inhibitory	CRD, TM, Cyp ITIM	?	MAFA, 2F1, MAFAL	Subset of NK cells, activated T cells, mast cells and basophils
KLRK1	Activating	CRD, TM $_{Arg}$, Cyp	MICA, MICB, ULBPs, RAETs	NKG2D, KLR, D12S2489E	NK cells, $\alpha\beta$T cells, $\gamma\delta$T cells

TABLE V
Adaptor Molecules of Activating NK Receptors

Adaptor	Official name	NK receptor	AS in TM	Alias
HCST	Hematopoietic cell signal transducer	KLRK1	Asp	DAP10, KAP10, DNAX, PIK3AP
TYROBP	TYRO protein tyrosine kinase–binding protein	KIR2DS1-5, KIR3DS1, KLRC2, -3, -4/KLRD1, Klrk1-s$_{mouse}$, NCR2	Asp	DAP12, PLOSL, KARAP
CD3ζ	CD3Z antigen, zeta polypeptide	NCR1, NCR3, CD16	Asp	CD3H, CD3Q
FcεR1γ	High-affinity Fc fragment of IgE receptor I gamma	KIR2DL4, LILRA1-6, NCR1, CD16	Asp	

on subpopulations of NK and on cytotoxic T-cells (190, 191). It gives rise to the NK-like activity of some CTLs in vitro (188, 191), but its physiological role is still enigmatic, particularly after a study, which demonstrated that about 4% of healthy donors were homozygous for a deletion mutant of KLRC2 (192, 193).

(c) KLRC3 seems to be another activating homolog of KLRC1 (95% homology in extracellular domain) (194) but its affinity to HLA-E peptide complexes is almost identical to KLRC1 (195). The two differently spliced variants identified show the same specificity (195). Both variants are expressed as heterodimers with KLRD1 and signal transduction is probably mediated by TYRO protein tyrosine kinase-binding protein (TYROBP) (Table V) (189). They are clonally expressed, whereas the combined expression level is lower than KLRC1 but significantly higher than KLRC2 (196).

(d) KLRC4 is the last family member (197). The truncated carbohydrate-recognition domain does not support association with KLRD1, which prevents surface expression (198). It contains an ITIM like motive in the cytoplasmic domains and associates with TYROBP, but its function is, due to its intracellular retention, questionable. Exons of KLRC3 and KLRK1 overlap to some extend.

Based on the initial experiments, only permissive and nonpermissive HLA heavy chain leaders were differentiated for KLRC1/KLRD1 functionality. However, a study demonstrated a broad span of affinities for KLRCx/KLRD1/HLA-E$_{\text{leader peptide}}$ combinations (195) that might result in very complex activation/inhibition patterns. In this context, it is not surprising anymore that the spectrum of class Ia HLA antigens monitored indirectly by KLRC1/C2/C3 and directly by KIRs is only partially overlapping, pointing out that both systems play a complementary role in the expression surveillance of most HLA class I molecules. This process is known to be altered in certain virus-infected or tumor cells (190). Especially the HLA expression linked immune evasion strategies of the HCMV are well investigated (103). The leader peptide providing HLA-C and -G antigens, for example, are not only spared from the virus-induced HLA class I degradation process (199), but also a virus-encoded leader peptide for HLA-E is supported and may consequently lead to NK cell inhibition (200).

KLRC1, C2, and C3 are all clonally expressed (196, 201, 202), whereas in healthy donors, almost all KLRD1$^+$ cells (45–65% of CD3$^-$CD56$^+$ cells) are positive with a so-called KLRC1 specific antibody. In HIV infected people, however, the percentage of KLRC1$^+$ NK cells is dramatically reduced, especially in patients with high-virus load (46) and KLRC2 is elevated accordingly. These KLRC2$^+$ cells were found to express reduced amounts of other activating receptors (NCR) but the expression of inhibitory KIR was significantly elevated pointing to a balancing mechanism operating in NK cells.

4. Subfamily D (KLRD)

KLRD1 forms disulfide-linked heterodimers with most members of the KLRC family (188). Expression was found to be sometimes also KLRC independent and in myeloma patients, nonfunctional KLRD1 molecules could be demonstrated in cytotoxic T cells (203). Structural analysis suggests that the C-terminal half of the HLA-E bound peptide is contacted by KLRD1 (195) and might explain that many substitutions have similar effects on the affinity of the different KLRCx/KLRD1 heterodimers. Expression was demonstrated to be clonal by surface staining with KLRD1-specific antibodies (~50% of NK cells) (46). However, this might be an underestimation, since KLRD1 reaches the surface only if bound to a KLRC family member.

5. Subfamily G (KLRG)

KLRG1 belongs to the superfamily of C-type lectins and was initially described in rat. The extracellular domain of the subsequently identified human KLRG1 shows marked homology to other KLR family members (204). In mice, an ITIM like motive in the cytoplasmic domain of Klrg1 inhibits NK cell functions (205) and expression seems to be modulated by MHC molecules (206). Rat Klrg1 binds to α-D-mannose residues and seems to mediate cell adhesion, but cognate ligands have not been described (207). The human KLRG1 contains also an ITIM and is expressed at least by $CD3^-CD56^{dim}$ NK cells but not by decidual NK cells (204).

6. Subfamily K (KLRK)

KLRK1 was, together with members of the KLRC family, identified in a screen for genes that are expressed preferentially by human NK cells (44) (Section III.B.3). Like the other KLR receptors, it belongs to the C-type lectin superfamily but the homology to subfamily C members is very limited (208). KLRK1 is expressed as homodimer in almost all NK cells as well as in most cytotoxic T cells. IL15 further enhances expression of KLRK1 (209). For proper surface expression, association with the adaptor protein HCST is necessary (208) (cf. Section III.C.2). The ligands of KLRK1 are the MHC class-I chain-related proteins A and B (MICA/MICB) (210), the UL16-binding proteins (ULBP1-3) (211–212), and the retinoic acid early transcripts (RAET1E–RAET1N). All the ligands are generally absent on healthy cells or expressed at very low levels (213). However, in malignantly transformed or otherwise stressed cells, their surface expression is markedly upregulated (213–215). The protozoans *Cryptosporidium*, for example, that are the major cause of diarrheal illness worldwide, induce MICA and MICB expression in human ileal tissue (209). The interaction of KLRK1 with these ligands results in stimulation of cytotoxicity (210, 211, 216) against the infected ileal cells and leads, together with other parts of the immune system, finally to the clearance of the pathogen. In addition to cytotoxicity, the interaction of KLRK1 with multivalent ligands like the ULBPs leads to the production and release of several cytokines in NK cells, including IFN-γ, tumor-necrosis factor (TNF), lymphotoxin and granulocyte–macrophage colony-stimulating factor (GM-CSF), as well as chemokines such as CCL4 (macrophage inflammatory protein 1β, MIP1β) and CCL1 (211, 217).

The domain structure of MIC proteins is similar to MHC molecules but they bind neither peptides nor β2microglobulin and seem to be regulated by a heat shock protein like promoter (218). Their high degree of polymorphism (211) results in varying affinities to KLRK1 and might result from coevolution with pathogens (219). Numerous primary tumors, like early bone marrow myeloma cells (220) or cell lines with tumor origin, express MIC proteins (213, 214). Most of them are, therefore, susceptible to cytolysis by NKs. Later stage tumors have survived cytotoxicity and are negative for KLRK1 ligands (220). Such MIC negative targets may be rendered susceptible for NK cytolysis by MICA transfection (210).

ULBPs and RAET1s belong to another protein family. Their extra cellular domain is only distantly related to the α1 and α2 domains of HLA molecules and the α3 domain is missing. Only RAET1E contains a TM, all others are glycosylphosphatidylinositol-anchored (211–212, 221, 222). The family members show considerable sequence diversity, ranging as low as $\sim$35% amino acid identity for some protein pairs (212). Significant amounts of RAET1 RNA were detected in various healthy cells (211, 212), but surface expression was very low or could not be demonstrated (223). On tumor cells, however, considerable ULDP surface expression has been demonstrated regularly, and correlates with cytotoxicity of NK cells (224). Other tumors, like AML, show typically no expression of KLRK1 ligands and consequently impaired immunogenicity. After cytokine induced differentiation, they express not only ULBP1 but also NCR ligands and become susceptible to NK-mediated lysis (117). It might, therefore, be concluded that the ULBP surface expression is not only regulated on the transcriptional level but also by a further transcription independent regulatory mechanism (223).

Although KLRK1 binds many significantly different ligands (225–227), the ligand-binding affinities are generally higher than for other NK receptors (218). An explanation for this unusual structural feature may be that the KLRK1 amino-acid residues that dominate binding and the respective contact residues of the ligands are mainly conserved for the different ligand receptor pairs (225). The higher strength of the interaction might be appropriate since cognate ligands of KLRK1 are absent on healthy cells and should always be considered if expressed, while, for example, ligands of inhibitory receptors are always present and should not hamper NK cells by long-lasting cell contacts. Whether the signal delivered by ULBPs and MICs is sufficient to cause activation of NK cells *in vivo* may depend on the relative numbers of NK cell-activating and inhibitory receptors involved in target recognition and on the avidity of receptor engagement (228).

C. Adaptor Molecules

1. TYRO Protein Tyrosine Kinase-Binding Protein

TYROBP is an adaptor protein that associates with numerous activating receptors of NK (Table V) (229) and myeloid cells [reviewed in (230)]. The negatively charged aspartic acid in the transmembrane domain (TM) mediates binding to the lysine in the TM of the activating receptors (231). In addition, other residues might play a crucial role, since strict receptor/adaptor specificity has been observed (232, 233) and amino acid exchanges in the TM lead to loss of function (234). TYROBP is expressed as disulfide-linked homodimer and contains a single immunoreceptor tyrosine-based activation motif (ITAM) in the cytoplasmic domain (229). Upon receptor engagement, the ITAM becomes tyrosine phosphorylated and binds the cytoplasmic protein tyrosine kinases SYK and ZAP70 (231, 235). This interaction results in intracellular calcium mobilization and subsequent cellular activation (235).

NK development in Tyrobp knockout (236) and knock in loss-of-function mice (237) is normal and their repertoire of inhibitory MHC class I receptors is intact, but natural cytotoxicity toward tumor cell targets is restricted. Additionally, dendritic cells are impaired in Tyrobp loss-of-function mutants (237). In humans, the recessive disease polycystic lipomembranous osteodysplasia with sclerosing leukoencephalopathy (PLOSL) results from a TYROBP loss-of-function-mutation (238). However, patient NK cells were normal regarding number and cytotoxicity against K562 and no increased frequency of malignancies or susceptibility to viral infection was observed. It was, therefore, suggested that the major function of TYROBP is exerted in cells of the monocyte/macrophage lineage (238). In agreement with a substantial number of humans lacking TYROBP dependent activating KIR it would, therefore, be of minor relevance in NK cells (77).

2. Hematopoietic Cell Signal Transducer

HCST is an adaptor protein with homology to TYROBP. It forms disulfide-linked homodimers, associates via an aspartic acid in the TM with KLRK1 (208) and is essential for KLRK1 surface expression in NK cells (239). Signal transduction of HCST is ITAM independent (Syk-family independent) but allows fully featured cytotoxicity. It recruits via an Tyr-X-X-Met SH2 domain-binding site the p85 subunit of the phosphatidyl inositol 3-kinase (PI3K) and the antiapoptotic kinase AKT (240). NK-cell responses of DAP10-deficient mice against KLRK1 ligand positive target cells are only partially impaired, and it was demonstrated at least in mice that a shorter variant of Klrk1 (Klra1s) is able to use the DAP12 adaptor for signal transduction (240).

3. High-Affinity Receptor I for Fc Fragments of IgE, Gamma Polypeptide

FcεR1γ is an adaptor protein with high homology to CD3ζ (241). It forms disulfide-linked homodimers as well as heterodimers with CD3ζ (242) and associates with several activating NK (Table V) (107) and immunoglobulin Fc receptors (106). FcεR1γ is, at least in some cases, essential for surface expression of the linked receptor molecules (105) and contains an ITAM in the cytoplasmic domain. For signal transduction it preferentially recruits the protein tyrosine kinase Syk (243).

4. CD3Z Antigen, Zeta Polypeptide

CD3ζ is an adaptor protein, which was initially discovered as component of the (TCR) complex (244), forms disulfide-linked homodimers and interacts with the cytoskeleton (245). In NK cells, it forms also heterodimers with FcεR1γ and associates with activating NK receptors (Table V). Upon target recognition, activating signals are transduced via an ITAM in the cytoplasmic domain. CD3ζ expression is reduced in T and NK cells of colorectal carcinoma patients. This reduction of CD3ζ levels correlates with tumor progression and might be an immune evasion strategy of the tumor (246–247).

D. Organization and Evolution of Receptor Complexes

Most of the NK-cell receptors are clustered in two large complexes (Fig. 4). The LRC on human chromosome 19q13.4 (248, 249) contains most of the NK-cell receptors with an immunoglobulin superfamily fold. An extended LRC has been defined (250) because further receptor genes (SIGLECs) belonging to the Ig superfamily have been localized close to the LRC. The second chromosomal region, the NK complex (NKC) is located on human chromosome 12p12-p13.1 (162) and encompasses mainly receptors of the C-type lectin family.

1. Leucocyte Receptor Complex

The receptor genes located within the LRC belong to different gene families and have been grouped by phylogeny, gene organization, and structure into clusters (251). The cluster with the highest variability within the LRC is the relatively young KIR region (Fig. 4) (50). It arose during primate development by probably still ongoing gene duplication and deletion events (60, 252) and

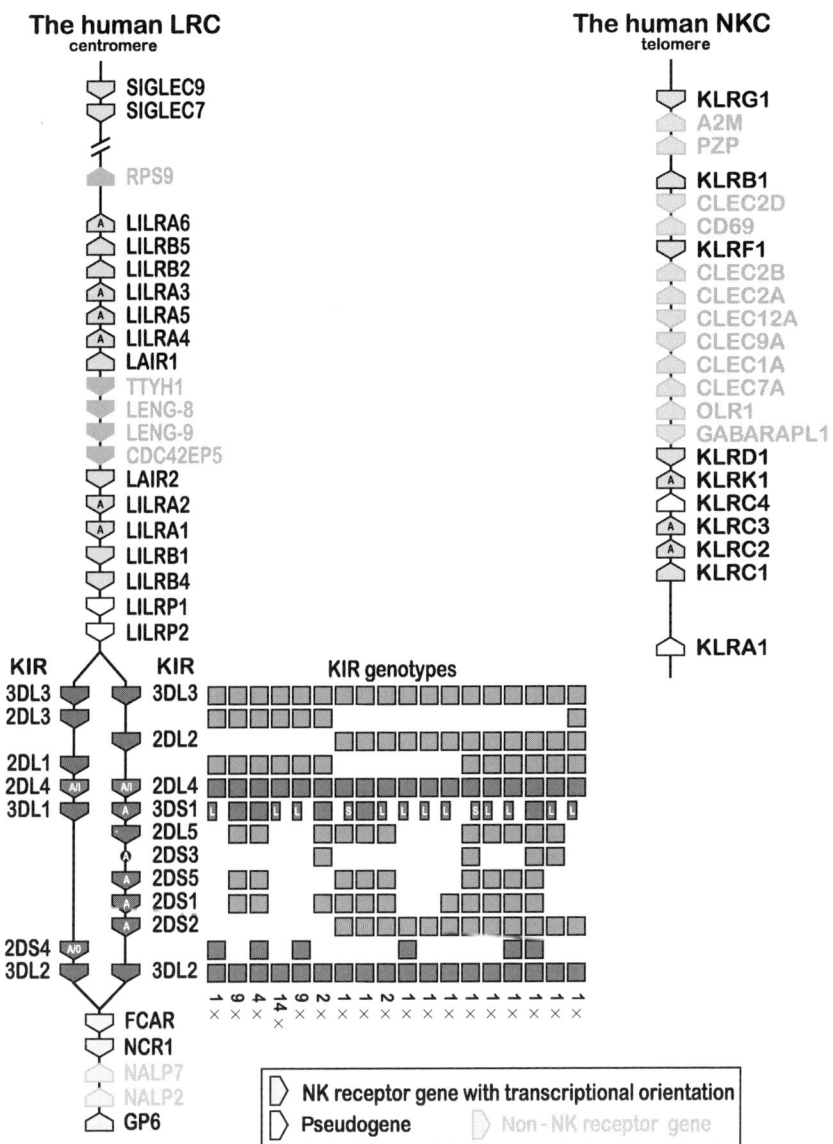

Fig. 4. Organization of gene clusters containing natural killer cell receptors. The human genomic organization of the LRC and the NKC is shown. The linear organization is derived from the genomic sequence (49, 50, 163). The two sequenced KIR haplotypes are shown in different branches. In order to indicate that many more haplotypes exist, 18 genotypes (59) are shown together with their observed frequency. Activating receptors (A) and possible null alleles (0) are indicated.

subsequent diversification by minor mutations led to an extremely polymorphic region (76). In the current human population, many haplotypes with different gene content exist (Fig. 4) (58, 59, 62–64, 253), and very limited similarity between the LRCs of different primates was observed (254–257). Evidence for selective processes was obtained from single nucleotide polymorphism haplotypes and allele frequencies (258). Defense against pathogens was suggested to be the driving force for the development of this immense diversity, but the proof of this hypothesis is missing (cf. Section III.D.3).

The LILR cluster displays a much lower diversity. Apart from some haplotyes missing a functional *LILRA3* gene, the gene content is not variable (50). LILRs are also less polymorphic than KIRs (251) and an analysis of the repetitive elements suggested the contention of a greater maturity (50). We tried to disentangle the ancestry of the individual LILR loci (259), but comparison of the human LILR region with the region of our closest relative, the chimpanzee, revealed that evolution of the LILR region was probably more complex than expected (255, 260).

In summary, the LRC is a highly dynamic region, supplying the genetic basis for many NK receptors. It is the hotbed of new receptor variants, especially in the KIR region, were the tandem duplicated genes simplify exon shuffling as well as birth and death of receptor genes (253, 257), for example, by unequal crossing over.

2. Natural Killer Complex

The NKC contains the genes of many C2-type lectin-like receptors involved in NK function (163). These loci are also organized into clusters with high homology between genes. The comparison of the NKC of different species revealed less variability compared to the LRC (261), whereas different copy numbers of the respective genes are the main difference. In rodents, for example, the Klra genes are heavily amplified and exert a KIR-like function, while in humans KLRA1 seems to be a pseudogene (167). The NKC is less variable than the LRC and evolutionarily much older (163, 261).

3. Evolution

The evolution of the NK cell function is obviously very complex and goes far beyond the scope of this chapter. However, some insights regarding the evolution of killer cell receptors help to understand the specificity of NK cells and shall be mentioned here.

There are several lines of evidence that the primordial genes giving rise to the IgSF and the C2-lectin-like NK-receptor families encoded inhibitory receptors (77). After several rounds of duplication and diversification of such an ancient gene, several mutations have probably accumulated in one of the respective copies. The resulting positively charged residue in the TM allowed the interaction with an adaptor molecule (cf. Section III.C), which led, together with the subsequent loss of the inhibitory motives in the cytoplasmic domain, to an activating receptor. Other mutations have probably changed the ligand specificity (78). Such transformations occurred not only independently in several mammalian species for the same gene family but also within the structurally different, but functionally similar, Klra genes of rodents (77). For the chicken Ig-like receptor (CHIR) family, we could demonstrate that the conversion of inhibitory into activating variants happened even twice (262). For KIRs, this transformation dates back 13.5–18 million years ago and especially the activating variants evolved extraordinarily fast since that time (256).

The independent development of these activating receptors by many different species points to strong positive selection and involvement in pathogen defense was suggested as driving force (77). Altogether, the following hypothesis arose: viruses downregulate MHC expression to evade adaptive immune responses (263). To avoid detection by NK cells due to insufficient MHC class I expression, virus-encoded MHC-like molecules are expressed that bind to inhibitory receptors on NK cells but not to TCR. The host counteracted by evolving activating receptors that bind to the viral MHC-like proteins but not to host MHC molecules (77, 264). This theory is supported by the missing H2 downregulation dependent host response against MCMV (265, 266). The MCMV-encoded protein m157 suppresses the immune reaction by interaction with the inhibitory receptor Klra9. However, MCMV-resistant mouse strains reactivate immune defenses by expression of the m157-specific activating receptor Klra8 (166).

Another example might be CD244. In mice, it seems to be an inhibitory receptor, but in humans it is activating (48). This activating functionality is essential for the defense against HCMV (cf. Section III.A.5), and since HCMV infects most humans, CD244 seems to be under very strong selective pressure. However, it is unclear whether these examples are just exceptions or the rule and it will be necessary to find specific ligands for further activating Klra and KIR receptors to solve this problem finally.

IV. Expression of NK-Receptor Genes

A. Specificity of NK Cells

All receptors described in section III were found to be expressed by NK-cells and in many cases also by other cells of the immune system like B-cells, T cells, or phagocytes. Cell-surface expression was analyzed using receptor-specific antibodies and fluorescence activated cell sorting (FACS) in many studies. Some of the receptors, like CD244 (*267*), NCR1 (*129*), NCR3 (*114*), and SIGLEC7 (*268*), were found on virtually all NK cells in these studies, others, like NCR2 (*115*) or KLRC2 (*46*), were detected only on IL-2–activated NK cells or in HIV infected patients, respectively.

A different situation was found for KIR, KLRC1, LILRB1, and CD16. These receptors are used only by a fraction of NK cells (*46–47, 269, 270*) and a stochastic process, which was described for murine NK receptors (*KLRA*) (*271*), seems to be responsible for the expression of these receptors (see later) also in humans. Because of this process, each NK cell exhibits its own unique receptor mixture. In addition, clonal distribution of receptors is donor dependent (*9*). The resulting KIR expression patterns were found to be determined epigenetically and they are stable over time (*272*). Since specificity of the cytotoxic effector function of NK-cells is directly dependent on the presence/absence of distinct NK receptors on the NK-cell surface (*47*), the stochastic occurrence of the respective receptors leads to an enormous repertoire of different NK-cell specificities that allow detection of multitudinous different aberrations found in infected or malignantly transformed cells.

B. Complex Phenotypes

The high number of receptor families, most of them with several highly similar family members, makes FACS analysis not suitable to disentangle such complex receptor phenotypes. The examination of gene expression patterns, although sometimes not reflecting the whole picture on the protein level, provides the specificity to analyze numerous transcripts of different receptors in parallel. For such experiments, sensitivity is a major issue, since NK receptors are not expressed abundantly on the relatively small and inactive peripheral blood NK cells. Additionally, several independent receptor-specific reactions need to be conducted with the very limiting amount of RNA provided by a single NK cell. Previously, a preamplification techniques called *three primed end amplification* (TPEA) (*273*) was used to analyze the expression of *KIR* and *NKG2A* (*274*) in single NK cells. Unfortunately, we had not only to realize that the study is flawed due to inadequate receptor-specific primers (*275*) but also figured out that the published variants of the TPEA design (*273, 276*) do not lead to proper preamplification.

C. Single-Cell Analysis

1. Sensitivity Considerations

A mammalian cell contains on average between 20 and 40 pg total RNA (*277, 278*), which is equivalent to about 0.5–1.0 pg mRNA or 10^5–10^6 mRNA molecules. Handling such small RNA amounts is already challenging, but an uneven distribution of different RNA species (*279*) complicates the situation further. While several hundred transcripts of housekeepers may be present (280), mRNAs of genes with regulatory functions like, for example, receptors are usually underrepresented and for mRNAs encoding transcription factors, less than one copy per cell has been found regularly (281). If such rare transcripts of several different genes shall be analyzed in a single cell, the whole RNA content of the initial cell must be preamplified (282). Described methods like TPEA (273), Poly-A-PCR (283), global single-cell reverse transcription-PCR (GSC RT-PCR) (284), and SMART [Clontech (285)] address these problems and allow direct amplification of RNA from a single cell. Some suffer from weak amplification (<20-fold), sophisticated handling, or potential loss of sample material. Other techniques are timeconsuming or need expensive equipment (286, 287).

2. Adaptor-Mediated Transcript Amplification

As delineated in the earlier section, we expected very low copy numbers for the NK-receptor transcripts. To avoid sample loss, which is inherent to any manipulation, we decided to take advantage of the simplicity of TPEA (273, 276), which is very similar to the Microarray Target Amplification Kit (Roche, Basel, Switzerland). However, we realized relatively soon that amplification is very weak, since the proposed second strand primers hamper the amplification instead of promoting it (275). Hence, we changed the design from a universal second strand adaptor primer (SSAP) that theoretically fits all cDNAs, to a gene-specific variant (Fig. 5) and immediately obtained the expected exponential amplification. Subsequently, we verified by control reactions, for example, by omission of individual primers and by sequencing of reaction products, that the reaction follows the theoretical path.

The modified method was named adaptor-mediated transcript amplification (AmTA), since the ends of all double-stranded cDNAs are adapted to each other. This is achieved by first and second strand adaptor primers (FSAP and SSAP in Fig. 5) that are composed of tag sequences (5′ end), which are identical for all first and second strand primers, respectively, and a specific part (poly-T or a part of the gene selected) at the 3′ end. After first and second strand synthesis, a single tag primer pair (TFP and TRP in Fig. 5) is used to promote exponential amplification. In total, AmTA facilitates reverse transcription and up to 10^6-fold preamplification in a single tube and excludes, thereby, sample loss by tube switching or other manipulations. As a result, it is particularly suitable for the analysis of rare transcripts in single cells but has also been used for the quantification of rare splicing variants of HLA-G (291).

However, the number of different transcripts that may be analyzed after such a preamplification is limited, since a second strand adaptor primer needs to be added for each gene or gene family of interest. On the other hand, this

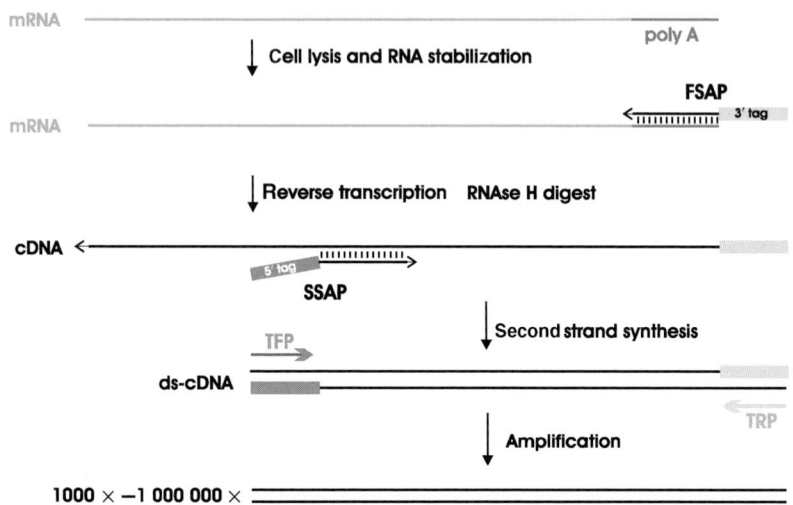

FIG. 5. Schematic representation of AmTA. Primers are indicated as vertical lines with arrowheads pointing to the 3' end. FSAP: first-strand adaptor primer; SSAP: second-strand adaptor primer; TFP: tag-specific forward primer; TRP: tag-specific reverse primer.

approach offers the possibility to adapt product length and amplification efficiencies of the targets and allows, therefore, semiquantitative analysis.

3. AmTA Reliability and Sensitivity

Using single-cell equivalents of NK-cell line YT (289) RNA, the method was optimized (buffer composition, reverse transcriptases, polymerases) to yield reliable results for housekeeper and NK-receptor transcripts. In control experiments with up to 50 cells per reaction, but without reverse transcriptase, it was ensured that genomic DNA is not subject to preamplification. Sensitivity and reproducibility was examined by the preamplification of serial dilutions of YT-cell lysates covering a range between 5 and 1/100 cell equivalents.

The quantification results for the housekeeper glyceraldehyde-3-phosphate dehydrogenase (GAPDH) are summarized in Fig. 6A. The diagram shows up to a 1/20 cell equivalent a good correlation between RNA input and the corresponding results, but the correlation between dilution and the averaged readout is not linear but close to exponential (mind the logarithmic scaling). Additionally, the dilution series containing 5–1/6 cell equivalents does not fit the dilution series containing 1/10–1/100 cell equivalents. Since it was reproducibly obtained in two independent experiments each with three replicates, we concluded that a systematic error inherent to the generation of the

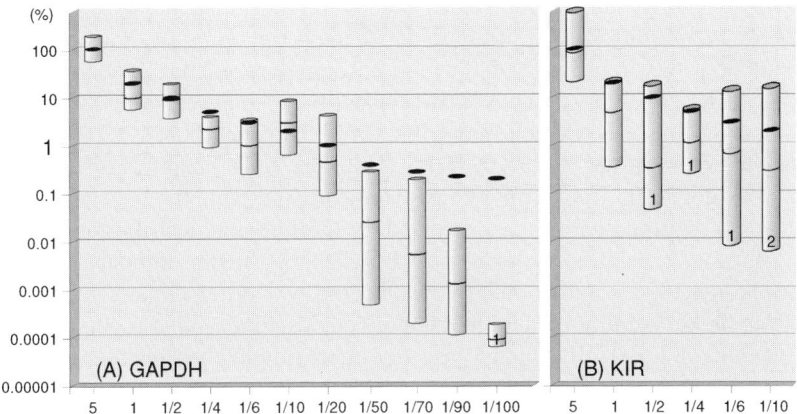

Fig. 6. Sensitivity of AmTA. Cell lysate corresponding to the number or fraction of cells given at the bottom was first subjected to AmTA. 1/80 of the AmTA product was quantified by real-time PCR with primers specific for the house keeper GAPDH (A) or the *KIR* gene family (B). The resulting concentrations are given as percentage of the mean five-cell expression value in percent. Maximum and minimum values of five parallel experiments correspond to the top/bottom of the columns, respectively. Mean values are given as black rings and expected values calculated on the basis of the mean five-cell results are indicated as black ovals. Numbers in the columns correspond to negative PCR reactions.

dilution series must be responsible. The first 5 dilution steps are made by direct transfer of 5, 1, 0.5, 0.25, and 0.17 μl aliquots, from the undiluted solution containing lysed YT cells (1 cell μl^{-1}) but subsequent dilution steps (1/10–1/100) were done by transferring 5, 2.5, 1, 0.71, 0.56, and 0.5 μl from an intermediate 1:50 dilution (3 μl initial cell lysate + 147 μl lysis buffer). We hypothesized, therefore, that pipetting, especially of very small volumes, leads to a significant loss of RNA whereas lower RNA concentrations intensify the problem.

Although not relevant for the preamplification of total single cells, we tried to quantify this phenomenon to learn more about the reliability of the method and about obstacles associated with ultrasmall RNA amounts. Two cells were lysed in cup A, half of the volume was transferred to cup B, and after a 1-min incubation to allow adsorption, transferred with a new tip completely to cup C. After preamplification with AmTA, the GAPDH content was measured by real time PCR. A loss of material between 45% and 75% of the initial content was found in the C cups, while only a single B cup was positive. This shows clearly that manipulating such small RNA amounts leads to significant loss of material.

Another distinct feature of the dilution series is the dramatic increase of the dispersion of values for higher dilutions. Since probabilistic effects build up significantly if the theoretical target copy number in an aliquot approaches one, we were interested in the number of transcripts present in a single cell. About 300–700 transcripts of the house keeper beta-actin were found per kidney cell (290) and ∼250 GAPDH transcripts in mouse-cell lines and tissues (279) as well as the NK-cell line HNK (275). In conclusion, a 1/100 sample contains theoretically ∼2–3 GAPDH transcripts. Such small transcript numbers are, due to statistical spread, not evenly distributed in the huge amount of solvent. The probabilities are expressed by the binomial distribution (Fig. 7) and show, in agreement to our results (Fig. 6A), that under optimal conditions one of ten 1/100 aliquots should be empty. However, other aliquots might, with reasonable probability, contain much more copies, for example, 7 copies with a probability of 8%. Unfortunately, this does not completely explain a variability of three orders of magnitude. We suggest that in addition to the statistical spread, the PCR process might suffer under reduced efficiency, for example, due to adsorption of one or several of the few target molecules to the wall,

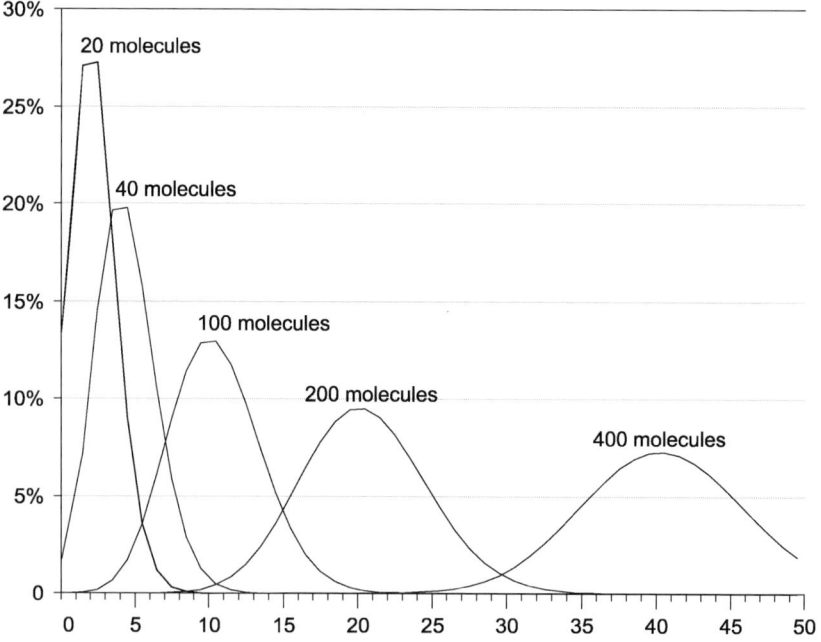

FIG. 7. Binomial distribution. Binomial distribution delineating the probable transcript content of aliquots of RNA solutions. Blotted are the probabilities (ordinate) of getting the number of molecules indicated at the bottom if one-tenth of the whole volume is taken. Different curves correspond to different initial numbers of transcripts in the total volume.

which desorb later in the reaction. The resulting delay in PCR initiation would lead to an underestimation of the transcript concentration.

The scattering of the KIR-specific reaction data and the respective manipulation dependent loss of transcripts is comparable to the results of factor 5–10 further diluted GAPDH reactions. Assuming that the adsorption is saturable but not sequence dependent, relative loss of material should depend on total RNA concentration and not on the concentration of a specific transcript, which fits precisely the situation found and is in line with only ~40 KIR transcripts per NK cell (275). The number of negative KIR-specific reactions is slightly higher than expected form the binomial distribution but the number of replications is obviously not sufficient to do proper statistics. In this context, the disturbing effects become disproportionally prominent only if the numbers of transcripts approach a dilution in which only a few thousand mRNA molecules are present per reaction. Additionally, increased spread and loss of material seem to be mainly due to the additional manipulations inherent to the dilution process. Since ~500,000 mRNAs and huge amounts of rRNA are present in an undivided single cell and manipulations, like pipetting the target RNA or tube switching, are completely omitted in single-cell AmTA, we conclude that this method is particularly suitable for the semiquantitative analysis of rare NK-receptor transcripts in single NK cells.

4. NK Lysate Preamplification and Housekeeper Expression

Using AmTA, an analysis of NK-cell receptor expression was conducted. The preliminary data will allow a first impression on the variability of NK cells: 28 single cells of the NK-cell line YT (289) were picked with a micropipette under microscopic control and transferred into lysis buffer (273). Following cell lysis, the single-cell material as well as isolated total RNA (~1 cell aliquot) of YT (control) was subjected to AmTA as described (291) but with second strand adaptor primers for NK receptors (Table VI). Subsequent reactions specific for the housekeeper transcripts of GAPDH, β2-microglobulin (B2M), β-actin (ACTB), ribosomal protein L5 (RPL5), and ribosomal protein S9 (RPS9) were positive for the total RNA control and the 26 YT-cell reactions, but negative for two YT cells, indicating that the single-cell transfer failed in two cases. Housekeeper cell-to-cell variability is, apart from two outliers, moderate (~25%) and the expression levels of four genes are similar. Only RPS9 revealed a factor ~10 lower expression level and higher variability (not shown). In the main, the arithmetic mean values calculated for the 24 YT cells matched the housekeeper values of the respective YT control RNA relatively closely (Fig. 8 ACTB). Subsequently, NK cells of a female Caucasian donor were purified with the Dynal NK cell negative isolation kit and 100 of these

TABLE VI
Primers Used for AmTA and Single Cell Analysis

Gene	Gene-specific adaptor primer	Gene-specific forward primer	Gene-specific reverse primer
B2M	CTGACTCTATCTAATGCTCC CTGCTTGCTTGCTTTT	AGCTCTAGGAGGGCTGGCAACT	TTCCCCCAAATTCTAAGCAG
ACTB	CTGACTCTATCTAATGCTCC ACGAAACTACCTTCAACTC	CGTGGACATCCGTAAAGACC	ACATCTGCTGGAAGGTGGAC
G3PDH	CTGCATCTATCTAATGCTCC GCATCCTGGGCTACACT	CGACCACTTTGTCAAGCTCA	AGGGGTCTACATGGCAACTG
CD16	CTGACTCTATCTAATGCTCC GGAAGGAAAGCGCA	TGCAGGGACTGTAAAACCACCT	TGCTTTATTGGAACCAAGAAATGTTGC
KIR	CTGACTCTATCTAATGCTCC AGGAYTCTGATGAACAAGACC	AGGGAGACAACAGCCCCTGTC	TTCCTCAGTGTGATTGCAGCC
LAIR1	CTGACTCTATCTAATGCTCC AACCTCCAGTGACCCCAGA	TATCTGCCCTGCTGACCCTAAA	GAAGGCAGAGAGTGGGTCCG
SIGLEC7	CTGACTCTATCTAATGCTCC GGGGACCCACAGACCAA	CCTCAGGGGAGGAAAGAGAG	TTGTTGCTGCCTTCTTCTCCT
LILRA	CTGACTCTATCTAATGCTCC TYHNHRDHACCAGGCTG	GGAGAATCTCATCCGCATGG	GGCTTCTCTGGCTGTGCTG
LILRB	CTGACTCTATCTAATGCTCC TYHNHRDHACCAGGCTG	GCTGCTGCATCTGAAGCC	TTCCTGGGATGGAGGAGG

KLRC1	CTGACTCTATCTAATGCTCC TTTGTGTCGGGCTCAT	TAGAATAGTGGTTGCCAATGTCTG	CTGGATAGCTTTATTGAAGTGTCG
KLRC2	CTGACTCTATCTAATGCTCC TTTGTGTCGGGCTCAT	TAGAATAGTGGTTGCCAATGTCTC	CTGGATAGCTTTATTGAAGTGTCA
KLRK1	CTGACTCTATCTAATGCTCC TCATGCTGCCACTTTT	ATCTTAAAGGCATTATTCTCCAGCCT	TCAGAGGCAAAATTCCTTCTTAACTG
KLRD1	CTGACTCTATCTAATGCTCC AAAATTTAACTCACACTGCCC	GGAGCATAGTGGCAAGATCATA	AGCCTGGGCAACATAGC
NCR1	CTGACTCTATCTAATGCTCC CCTGGCCTTTCTAGTCC	TCAGCAGGAAGAGGACTAGAGA	ACACCAGCTTTCAGATCCAC
NCR2	CTGACTCTATCTAATGCTCC GGGGACATATGGTGGA	GAGCTCAGGAGCCTGGATAC	AAAGTGTGTTCATCATCATCATCGC
NCR3	CTGACTCTATCTAATGCTCC AAAGGTCCAAGAGGGC	ACATCTGCTTCCCCCAGTCC	CTTCCACCCACTCTGGGGTC
CD244	CTGACTCTATCTAATGCTCC TTCATTAATTCAGCCTTCCAG	ACCACAGCCCTTCCTTCAATAG	GTCATCCACTGTGCCAATTCC
FCERIG	CTGACTCTATCTAATGCTCC TTTGGCTTCTTGGTTCTTCC	CATATGCCTGCATGCCATTAAC	GGCAGTTTTATTGGGGTTCTC
TYROBP	CTGACTCTATCTAATGCTCC CAGCGACCCGAAACA	GAGACCGAGTCGCCCTTATC	GGGTCTGTATCGCGTAG

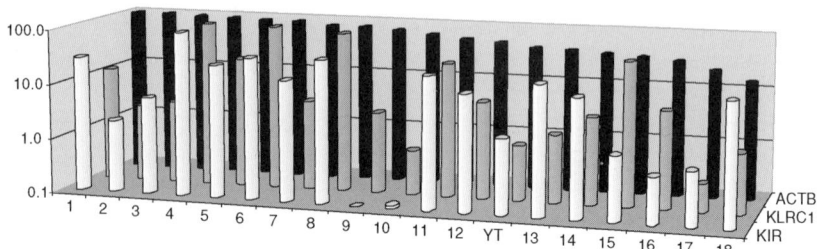

FIG. 8. Single-cell expression analysis of three genes in 18 YT cells. The presented 18 cells are depicted at the bottom; YT is a single cell aliquot of RNA isolated from $\sim 10^7$ YT cells. Gene designations are depicted at the right. The cell with the maximal value was set to 100% for each gene analyzed.

cells were applied to AmTA. Housekeepers were quantified to select the cells with the most uniform distribution.

5. Receptor Expression in the Cell Line YT

For each selected preamplified cDNA, 16 gene or gene-family-specific reactions were performed (Table VII). In contrast to the housekeeper genes, variability increases dramatically and values cover three orders of magnitude for many receptors (Fig. 8). Nevertheless, all selected YT cells were found to express KLRC1 and a member of the KIR family. All other analyzed transcripts were detected only in varying subfractions of cells or not at all (summarized in the last line of Table VII). We suggest that the complete absence of LILRs or LAIR1 in YT is due to clonal nonexpression in the NK cell that gave rise to this cell line. The variability for most of the other genes, however, must have different reasons. We favor a regulation mechanism that does not depend on the irreversible hypomethylation of the promoter region (cf. Section IV.D). It might enable NK cells to fine tune tolerance in the periphery. It can, however, not be excluded completely that methodological problems like, alternatively spliced variants missing the analyzed region, mutations located at the 3′ end of primer-binding sites, a still insufficient sensitivity for extremely rare transcripts, or a combination thereof, lead to an overestimation of variability.

The missing expression of two receptors is worth mentioning separately. NCR2 could not be detected in YT, although cells were cultured with interleukin 2, which usually results in induction of NCR2 expression. The complete absence of the adaptor protein TYROBP in YT cells is also noteworthy, since activating KIR receptors are only functional in combination with this protein (cf. Section III.C.1).

TABLE VII
NK-Receptor Expression in Single Cells of a Female Caucasian Donor

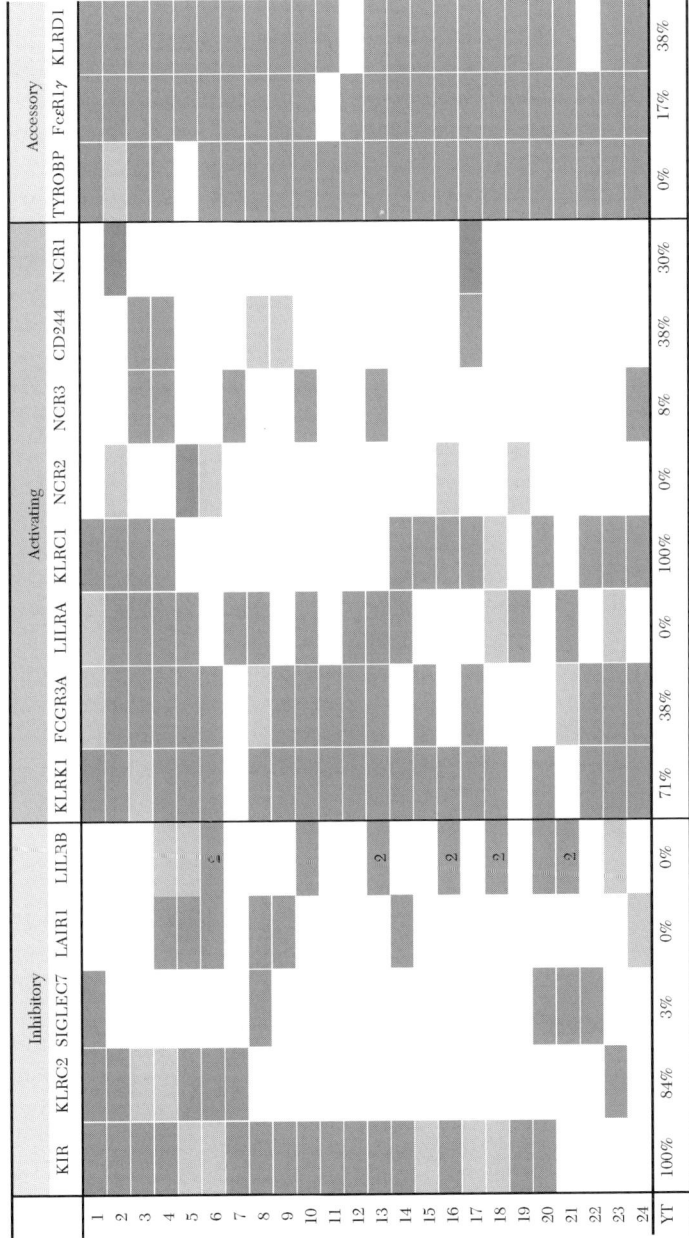

Displayed are the results of 24 single NK cells purified from peripheral blood with the Dynal NK cell negative isolation kit. The line at the bottom summarizes the frequency of receptor expression in single cells of the cell line YT. Dark gray, strong positive; light gray, expression less than 2% compared to the cell with the maximal value of this analysis; white, negative; 2, two variants visible on control gels. Please note, that the female donor has a KIR B-hyplotype containing also activating *KIR* genes (285).

6. Natural Killer Phenotypes *In Vivo*

The results obtained for freshly isolated NK cells are closer to the frequencies described in the literature in most cases. The 24 cells with the highest values of β-actin (ACTB) are summarized in Table VII. Although the *LILRA*, *LILRB*, and *KIR* gene families were analyzed as a whole, no cells with matching expression patterns could be observed. This approves impressively that a healthy donor is equipped with a broad spectrum of NK cells with different phenotypes.

As expected, almost all cells are positive for KLRK1, KLRD1, and the adaptor molecules. However, some cells are KIR negative. These cells are CD4/CD8 negative (not shown) and the results regarding the other receptors are compatible with NK cells. Additionally, a further analysis with cells of a different donor resulted in a significantly higher frequency of KIR negative cells (not shown). As discussed in the previous section, it might not be excluded completely that shortcomings of the applied methodology results in false negatives, but our results at least indicate that KIR negative NK cells might exist in significant amounts.

The expression frequency of NCR1, NCR3, and CD244 is much lower than described in the literature (*113, 114, 133, 134*). On the other hand, NCR1 expression levels might vary widely (*116*), it was found to be downregulated in tumor patients (*117, 292*) and it is, together with NCR3, regulated hormonally (*118*). In the case of NCR3, extensive alternative splicing adds a further level of complexity (*126*) and for CD244 it has been shown that expression might be reduced to less than 25% of cells in HIV patients (*293*). These examples show that the expression frequency of these receptors is not necessarily close to 100% in all individuals but they do not explain the discrepancy for the actual analysis of a healthy donor. They, therefore, need to be revisited.

D. Promoter Analysis

One of the major factors responsible for the regulation of protein expression is transcriptional control and NK-cell receptors seem to be no exception. However, in most cases, differentiated cells with the same functional properties express mainly the same genes. That this is different for NK-cell receptors was initially demonstrated in mice. The MHC-specific Klra receptors are expressed clonally (*294*) together with the Klrc family (*295*). The selection of active genes is stochastic, but they are activated in a chronologically ordered fashion during differentiation (*295*). The underlying mechanism, a probabilistical transcriptional switch, was described for the Klra genes (*271*). The functionality is summarized in Fig. 9 and explains the allele-specific expression pattern.

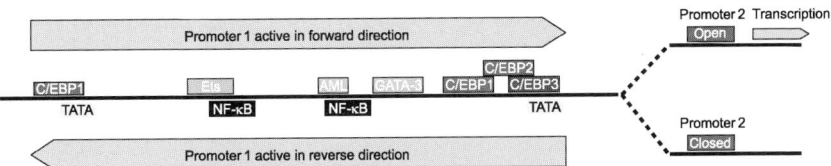

Fig. 9. Probabilistical transcriptional switch of KLRA receptors in mice. Early in differentiation of NK cells, transcription factors are present only for promoter 1. Transcription factors bind and lead to the assembly of a transcriptional complex driving transcription into one direction, which excludes the assembly of a complex pointing into the reverse direction. The probability that the complex assembles in a given orientation depends on availability of transcription factors and their specificity to the precise sequence of the respective binding sites. Transcription from promoter 1 in forward direction irreversibly opens promoter 2, while transcription in reverse direction does not. Later in differentiation, promoter 1 is switched of and promoter 2 becomes active. Promoter 2 is functional only, if it was "opened" previously.

In humans, KIR are responsible for the surveillance of classic MHC class I expression. Most of them (except *KIR2DL4* and *KIR3DL3*) show not only a very similar functionality, if compared to Klra, but also the same clonal expression pattern. A prediction of transcription factor-binding sites, based on the sequence similarity of the promoter regions between different *KIR* genes, revealed the presence of sequence motives similar to the probabilistical transcriptional switch of mice (288). In addition, mice transgenic for a part of the human LRC expressed KIR and LILR similar to humans regarding cell types and frequency (296). This very high degree of similarity regarding functional as well as regulatory properties must be based on convergent evolution, since the receptors belong to completely different structural families and evolved independently in man and mice.

Methylation of a small CpG island surrounding the transcriptional start sites of most *KIR* genes was found to determine mostly whether a KIR gene is expressed or not in finally differentiated NK cells (272, 297). It is also responsible for the monoallelic expression of KIR (297), while histone modifications are less relevant for promoter activity (298). The methylation pattern and consequently also the clonal expression of KIR is stable in cell lines. The expression level as well as the clonal expression pattern of KIR is also stable over time in healthy donors (9, 299). Comparison of the KIR expression pattern before and after HLA-matched stem cell transplantation revealed in many cases that the donor pattern is stable even under these circumstances (299). This is in agreement with the fact that repertoire differences were found to be dominated by the KIR genotype in sibling pairs and only minor effects are associated with the HLA haplotype of the environment (299). Polymorphisms in the promoter region that could be responsible for KIR repertoire differences have been demonstrated (300).

In conclusion, the on/off state defined by methylation is the major switch that decides whether a KIR might be expressed or not in a given NK cell. The level of KIR expression, however, seems to be determined by classic promoter functionality (*301*) based on subtle differences between KIR promoter sequences (*302*). In the KIR expressing subset of T cells, the transcriptional control mechanism seems to be distinct (*303*).

V. Natural Killer Cells and Disease

Many viruses, including HIV and HCMV, reduce MHC class I surface expression to avoid T-cell dependent cytotoxicity (*103, 304, 305*). Due to reduced MHC expression, NK cells would be expected to attack the affected cells and eliminate the problem. A similar situation is found for malignantly transformed cells (*306, 307*). However, many of the respective viruses have managed somehow to avoid effective NK-cell elimination and a reduction of systemic NK-cell reactivity has been demonstrated at least for some tumors (*215, 246, 308–310*). The underlying mechanisms become recognizable by experimental advances and are expected to provide starting points for new therapies (*103, 311, 312*). They will be discussed shortly in the later sections.

A. Tumor Immune Evasion

1. Soluble MHC Class I-Related Chains A and B

In healthy individuals, the MHC class I–related chains A and B (MICA/MICB) molecules are not surface expressed by most cells. Only cells of the gastrointestinal epithelium are an exception (*213*). In malignantly transformed and pathogen-affected cells, however, the MIC proteins are expressed frequently (*214, 313–315*). Their promoter contains heat shock elements similar to those of the HSP70 genes, their expression is linked to the DNA damage pathway (*316*) and a pathogen-dependent mechanisms of induction has also been described (*317*). It is, therefore, generally agreed that the surface expression of the MIC proteins signals cellular stress to the immune system (*210*). NK cells, $\gamma\delta$ T cells and $\alpha\beta$ CD8$^+$ T cells detect the MIC proteins with their KLRK1 receptors and raise defenses against such stressed cells (*318*). In some tumors, like glioblastoma, the strong expression of MHC class I antigens more than counterbalances this activating signal. Nevertheless, transfection experiments revealed that only MICA negative glioblastoma cells were able to grow progressively (*319*). Other tumors release a soluble form of the MIC (sMIC) proteins by proteolytic shedding (*320, 321*). These soluble forms promote KLRK1 internalization and degradation rendering effector cells inactive

(313, 322–325) and contribute, therefore, to tumor immune evasion (323, 326). Similar properties were attributed to the ULBP (313, 324).

Despite their limited functionality in tumor elimination *in vivo*, MIC proteins signal cell stress and might, therefore, be used for diagnosis or medical intervention. sMICA levels, for example, correlate in many cases well with tumor progression and were, therefore, suggested as prognostic markers for malignancies (313, 321, 323). Retinoic acid, which upregulates MICA expression, was considered as stimulant for innate immunity against human hepatocellular carcinoma (315) and antibody/MICA chimera were used to sensitize MICA⁻ tumor cells to specific lysis by NK cells (327).

2. Transforming Growth Factor-Beta

Many tumors secrete the potent immunosuppressive molecule transforming growth factor-beta (TGF-β1). It has been demonstrated that soluble TGF-β1 downregulates the expression of KLRK1 and suppresses by this means NK-dependent lysis of tumor cells and cytokine secretion (328). Additionally, KLRK1 expression might be reduced by membrane bound TGF-β1 on regulatory T cells (329). In NK cells, the effect was specific for KLRK1, since KLRC1/KLRD1, FCGR3, CD44, and CD244 expression was unchanged (328). In a subset of human T lymphocytes, however, the expression of inhibitory KLRC1/KLRD1 receptors was increased (330). The reduction of KLRK1 was inversely correlated with the level of active TGF-β1 in serum, which in turn was found to increase with tumor progression (331–332). The previously described inhibition of expansion, cytotoxicity, and cytokine production of purified NK cells *in vitro* (333, 334) by TGF-β1 is, therefore, at least in part mediated by loss of KLRK1 expression.

3. HLA-G

HLA-G is an almost monomorphic nonclassic MHC class I antigen. Its expression is mostly restricted to the feto-maternal interface during pregnancy. HLA-G has been shown to inhibit NK cell-mediated lysis of the semiallogeneic fetal cells and influence cytokine release of NK cells (335). This tolerance seems to be mediated directly by HLA-G-specific inhibitory receptors on NK cells and by dentritic cells that become tolerogenic by LILRB2/HLA-G interaction (336). Alternatively spliced variants of HLA-G, some of which are soluble, have also been found (337). These soluble forms were suggested to induce T cell apoptosis (338), but it is still a matter of debate whether significant amounts of these molecules are produced by the trophoblast (291, 339). It has been demonstrated that HLA-G is also expressed by breast, colon, and lung cancer and in tumor-infiltrating lymphocytes. Consequently, it was supposed to promote tumor immune evation (340, 341). Additionally, HLA-G inhibits rolling of activated human NK cells in an experimental system and might reduce extravasation of leukocytes in tumors (342). However, the mode of HLA-G induction as well as its role in tumor immune evasion remains, due to insufficient reliable data, speculative.

4. ACTIVATION INDUCED NK CELL DEATH

A study, demonstrated that cytolysis of tumor cells *in vitro* resulted in apoptosis of the attacking NK cells. This effect could be blocked with antibodies masking NCR1–3 or induced by cross-linking the NCRs (343). The degree of target-cell lysis was directly proportional to apoptosis in a redirected-killing assay and FCGR3B cross-linking led to the same phenomenon (344). Further experiments demonstrated that in NK cells, the NCR-mediated activating signals induced Fas ligand (FasL) expression and release, which subsequently forced the NK cells to commit suicide (343). This effect could be abolished by Cyclosporin A that suspends FasL expression (345). Since NCR1–3 and FCGR3B belong to the main receptors that induce NK-cell cytotoxicity (cf. Section III.A.4) and CD2 induced activation of NK cells leads also to NK-cell death (346), it might be concluded that strong cytolytic activity results not only in target destruction but also limits NK cell activity. This seems to be an intrinsic regulatory mechanism to avoid NK-cell autoreactivity but might be used also for tumor immune escape (343).

5. CD3ζ

Autoimmune disease, infectious disease, and tumors were shown to result in downregulation of the adaptor molecule CD3ζ (347). Since CD3ζ is essential for normal function of T cells as well as NK cells, lymphocytes of the adaptive and the innate immunity are impaired. Because this phenomenon is common to many different pathologies, it can be suggested that it is the normal mechanisms providing tolerance after an excessive and potentially hazardous inflammatory immune response (347).

Reduction of CD3ζ expression may be mediated by limiting amounts of extracellular L-arginine (348). Such a reduction might be caused by the stimulation of macrophages with TH2 cytokines. Tumors, however, seem to have the same potential (349) leading to low levels of L-arginine within the tumor. This could initially impair only infiltrating lymphocytes (350, 351) and might lead in later stages to systemical anergy of lymphocytes in cancer patients (246, 352). However, there might be other mechanisms for CD3ζ downregulation but data are limiting.

B. Pathogen Immune Evasion

Strategies of pathogens to avoid NK cell-mediated elimination are less well studied. Best analyzed is the HCMV that is directly after infection fend of by NK cells (353). It has developed many strategies to avoid elimination, including inhibition of antigen presentation, avoidance of complement-mediated lysis, and impairment of macrophages (354). To inactivate NK cells and cells of the monocytic lineage, HCMV expresses UL18, a protein that mimics MHC class I expression and interacts with inhibitory NK receptors (100). Further inhibitory signals are mediated by the leader peptide of UL40, which is presented by HLA-E and inhibits NK cells by KLRC1/KLRD1 interaction (cf. Section III. B.3). The described mechanisms lead to inhibitory signals, but HCMV does also reduce stress-induced activating signals by KLRK1 ligands (326). It produces UL16 that binds to MICB and ULBP1 and -2 (211, 217), which delays their surface expression (355) and renders HCMV-infected cells less susceptible to NK cytotoxicity (356). Further details are summarized in a review (357).

Other viruses use different strategies: HIV, for example, downregulates only HLA-A and -B. The remaining HLA-C and -E molecules on infected cells interact with inhibitory receptors and protect from NK cell lysis (358). However, many NK cells seem to be dysfunctional in HIV patients by other means. There are, for example, unusual subsets with low levels of activating receptors expanded in viremic individuals (359) and CD244 expression is significantly reduced (293). On the other hand, epidemiological studies revealed that the combination of the activating KIR3DS1 and selected HLA-Bw4 alleles is correlated with slower AIDS progression (81).

In cells infected by influenza and other viruses, the affinity of *KIR2DL1* and LILRB1 to their MHC class I ligands is increased and only activated NK-cells seem to overcome this enhanced engagement of inhibitory receptors and show killing (360). Other evasion strategies concerning NK-cell immunity seem to exist, but their mechanisms are just emerging (357).

C. Autoimmunity

Combinations of certain HLA/KIR genotypes have been correlated with susceptibility to autoimmune diseases. Since KIR are expressed by NK cells, subsets of $\gamma\delta^+$ T cells, and subsets of memory and effector $\alpha\beta+$ T cells, it is unclear which lineage is responsible for this malfunction of the immune system. The knowledge regarding the molecular mechanisms is, therefore, very limited. One hypothesis makes an excess of activating KIR responsible for disease development (361). This excess is in part due to the activating *KIR* genes found in B-haplotypes (cf. Section III.D.1), but disease association is found only if the excessive activating potential is not counterbalanced by inhibitory HLA/KIR combinations. Examples are psoriatic arthritis (362) and Type I diabetes (363). Psoriasis vulgaris (364) and idiopathic bronchiectasis (365) might also belong to this group. The rationale of this hypothesis is that defense against pathogens and eventually also malignancy (366) was dominant during evolution, despite an associated risk of developing autoimmunity (361).

The reverse case with an excess of NK inhibition was also associated with disease. The remodeling of spiral arteries in maternal decidua is at least in part induced by NK-cell activity. If fetal cells express HLA-C molecules that match inhibitory receptors on maternal NK cells, remodeling may be incomplete leading to high blood pressure during pregnancy (preeclampsia) (367). Excess of NK cell inhibition should also result from the interaction of HLA-B27 homodimers with inhibitory KIR in patients with spondyloarthropathies and rheumatoid arthritis (368, 369). However, the respective NK cells were found to be in an activated state. This might be due to cytokines released by $CD4^+CD28^-$ T cells upon interaction of their anomalously expressed KLRK1 with stress-induced MIC ligands on synoviocytes (370).

The autoimmune potential of NK receptors is obviously disadvantageous under normal circumstances. However, it might also be used for medical treatment, which is based on hybrid resistance found initially for mice (cf. Section II.C). Donor NK cells that are not tolerant to host lymphocytes eliminate remaining tumor cells after stem cell transplantation and improve prognosis of patients (84–85, 371–373). Missing tolerance results mainly from a lag of KIR ligands on the tumor cells of the host, but activating KIR are also beneficial if tumor cells express matching HLA antigens (372).

VI. Concluding Remarks

The innate immunity is the first line of defense after a primary infection. It limits or eradicates pathogens before T and B cells can mount efficient responses and prevents by these means clinical manifestations of disease in many cases. NK cells and phagocytes can enter and defend a tissue almost as soon as it becomes infected, since they are permanently under arms and do not require pathogen-specific clonal expansion. Over a long time, it was unclear how NK cells differentiate between autologous healthy cells and cells that are aberrant due to pathogen infection or malignant transformation.

Meanwhile, the molecular mechanisms become more and more evident: NK cells need an activating signal that is in most cases provided by the aberrant cell itself. The signal is generated if such a cell is stressed, whereas heatshock, viral infection, or DNA damage are some examples that have been demonstrated yet. The signal is provided to NK cells by the expression of surface proteins that are detected by activating receptors. To avoid autoimmunity due to minor aberrant expression of these cell stress-induced markers, NK cells check the target-cell integrity with inhibitory receptors prior to elimination. Most of these inhibitory receptors survey proper HLA expression, but other surface markers like sialic acids are also relevant. To avoid that inhibitory signals are always dominant, most inhibitory receptors are expressed clonally, which results in a broad spectrum of different specificities that should contain always the necessary combination. This double check preceding NK activation, combined with the induction of NK-cell apoptosis after strong cytolytic activity, ensures that no autoreactive NK cells develop. However, the existence of functional NK cells in HLA negative individuals makes an additional educational process likely.

In conclusion, NK cells are really detecting the "unusual," whereas effector functions are only applied if "expected" molecules are absent and "unexpected" molecules are present on a potential target cell.

Acknowledgments

This study was supported by the Deutsche Forschungsgemeinschaft (Vo715/4-1) and the Sonnenfeldstiftung (Berlin). The cell line YT was a kind gift of Dr. J. Yodoi to Dr. B. Uchanska-Ziegler. We thank Dr. Barbara Uchanska-Ziegler and Angelika Zank for providing cultured YT cells, Dr. Thornthwaite for supplying photographs, and Dr. Cordula Petter for proofreading.

References

1. Oldham, R. K., and Herberman, R. B. (1973). Evaluation of cell-mediated cytotoxic reactivity against tumor associated antigens with 125I-iododeoxyuridine labeled target cells. *J. Immunol.* **111,** 862–871.
2. Rosenberg, E. B., Herberman, R. B., Levine, P. H., Halterman, R. H., McCoy, J. L., and Wunderlich, J. R. (1972). Lymphocyte cytotoxicity reactions to leukemia-associated antigens in identical twins. *Int. J. Cancer* **9,** 648–658.
3. Thornthwaite, J. T., and Leif, R. C. (1974). The plaque cytogram assay. I. Light and scanning electron microscopy of immunocompetent cells. *J. Immunol.* **113,** 1897–1908.
4. Cudkowicz, G., and Bennett, M. (1971). Peculiar immunobiology of bone marrow allografts. II. Rejection of parental grafts by resistant F 1 hybrid mice. *J. Exp. Med.* **134,** 1513–1528.
5. Kiessling, R., Klein, E., Pross, H., and Wigzell, H. (1975). "Natural" killer cells in the mouse. II. Cytotoxic cells with specificity for mouse Moloney leukemia cells. Characteristics of the killer cell. *Eur. J. Immunol.* **5,** 117–121.
6. Kiessling, R., Klein, E., and Wigzell, H. (1975). "Natural" killer cells in the mouse. I. Cytotoxic cells with specificity for mouse Moloney leukemia cells. Specificity and distribution according to genotype. *Eur. J. Immunol.* **5,** 112–117.
7. Moretta, A., Bottino, C., Vitale, M., Pende, D., Cantoni, C., Mingari, M. C., Biassoni, R., and Moretta, L. (2001). Activating receptors and coreceptors involved in human natural killer cell-mediated cytolysis. *Annu. Rev. Immunol.* **19,** 197–223.
8. Kimber, I., and Moore, M. (1985). Mechanism and regulation of natural cytotoxicity. Mini-review on cancer research. *Exp. Cell Biol.* **53,** 69–84.

9. Valiante, N. M., Uhrberg, M., Shilling, H. G., Lienert-Weidenbach, K., Arnett, K. L., D'Andrea, A., Phillips, J. H., Lanier, L. L., and Parham, P. (1997). Functionally and structurally distinct NK cell receptor repertoires in the peripheral blood of two human donors. *Immunity* **7,** 739–751.
10. Nunn, M. E., Herberman, R. B., and Holden, H. T. (1977). Natural cell-mediated cytotoxicity in mice against non-lymphoid tumor cells and some normal cells. *Int. J. Cancer* **20,** 381–387.
11. Herberman, R. B., Nunn, M. E., and Lavrin, D. H. (1975). Natural cytotoxic reactivity of mouse lymphoid cells against syngeneic acid allogeneic tumors. I. Distribution of reactivity and specificity. *Int. J. Cancer* **16,** 216–229.
12. Timonen, T., Ortaldo, J. R., and Herberman, R. B. (1982). Analysis by a single cell cytotoxicity assay of natural killer (NK) cells frequencies among human large granular lymphocytes and of the effects of interferon on their activity. *J. Immunol.* **128,** 2514–2521.
13. Herberman, R. B., Nunn, M. E., Holden, H. T., and Lavrin, D. H. (1975). Natural cytotoxic reactivity of mouse lymphoid cells against syngeneic and allogeneic tumors. II. Characterization of effector cells. *Int. J. Cancer* **16,** 230–239.
14. Kiessling, R., Petranyi, G., Karre, K., Jondal, M., Tracey, D., and Wigzell, H. (1976). Killer cells: A functional comparison between natural, immune T-cell and antibody-dependent *in vitro* systems. *J. Exp. Med.* **143,** 772–780.
15. Zöller, M., and Matzku, S. (1980). Characterization of natural cytotoxicity *in vitro* in a spontaneous rat tumor model. *J. Immunol.* **124,** 1683–1690.
16. Callewaert, D. M., Kaplan, J., Johnson, D. F., and Peterson, W. D., Jr. (1979). Spontaneous cytotoxicity of cultured human cell lines mediated by normal peripheral blood lymphocytes. II. Specificity for target antigens. *Cell. Immunol.* **42,** 103–112.
17. Ortaldo, J. R., and Herberman, R. B. (1984). Heterogeneity of natural killer cells. *Annu. Rev. Immunol.* **2,** 359–394.
18. Phillips, W. H., Ortaldo, J. R., and Herberman, R. B. (1980). Selective depletion of human natural killer cells on monolayers of target cells. *J. Immunol.* **125,** 2322–2327.
19. Koren, H. S., and Herberman, R. B. (1983). Natural killing: Present and future (summary of the workshop on natural killer cells). *J. Natl. Cancer Inst.* **70,** 785.
20. Kiessling, R., Petranyi, G., Klein, G., and Wigzell, H. (1976). Non-T-cell resistance against a mouse Moloney lymphoma. *Int. J. Cancer* **17,** 275–281.
21. Kiessling, R., Hochman, P. S., Haller, O., Shearer, G. M., Wigzell, H., and Cudkowicz, G. (1977). Evidence for a similar or common mechanism for natural killer cell activity and resistance to hemopoietic grafts. *Eur. J. Immunol.* **7,** 655–663.
22. Kiessling, R., Petranyi, G., Klein, G., and Wigzel, H. (1975). Genetic variation of *in vitro* cytolytic activity and *in vivo* rejection potential of non-immunized semi-syngeneic mice against a mouse lymphoma line. *Int. J. Cancer* **15,** 933–940.
23. Kärre, K., Ljunggren, H. G., Piontek, G., and Kiessling, R. (1986). Selective rejection of H-2-deficient lymphoma variants suggests alternative immune defence strategy. *Nature* **319,** 675–678.
24. Ljunggren, H. G., and Karre, K. (1990). In search of the "missing self": MHC molecules and NK cell recognition*Immunol. Today* **11,** 237–244.
25. Höglund, P., Ljunggren, H. G., Ohlen, C., Ahrlund-Richter, L., Scangos, G., Bieberich, C., Jay, G., Klein, G., and Karre, K. (1988). Natural resistance against lymphoma grafts conveyed by H-2Dd transgene to C57BL mice. *J. Exp. Med.* **168,** 1469–1474.
26. Öhlen, C., Kling, G., Hoglund, P., Hansson, M., Scangos, G., Bieberich, C., Jay, G., and Karre, K. (1989). Prevention of allogeneic bone marrow graft rejection by H-2 transgene in donor mice. *Science* **246,** 666–668.
27. Chan, P. Y., and Takei, F. (1989). Molecular cloning and characterization of a novel murine T cell surface antigen, YE1/48. *J. Immunol.* **142,** 1727–1736.

28. Karlhofer, F. M., Ribaudo, R. K., and Yokoyama, W. M. (1992). MHC class I alloantigen specificity of Ly-49+ IL-2-activated natural killer cells. *Nature* **358**, 66–70.
29. Yokoyama, W. M., Daniels, B. F., Seaman, W. E., Hunziker, R., Margulies, D. H., and Smith, H. R. (1995). A family of murine NK cell receptors specific for target cell MHC class I molecules. *Semin. Immunol.* **7**, 89–101.
30. Held, W., Dorfman, J. R., Wu, M. F., and Raulet, D. H. (1996). Major histocompatibility complex class I-dependent skewing of the natural killer cell Ly49 receptor repertoire. *Eur. J. Immunol.* **26**, 2286–2292.
31. Vance, R. E., Kraft, J. R., Altman, J. D., Jensen, P. E., and Raulet, D. H. (1998). Mouse CD94/NKG2A is a natural killer cell receptor for the nonclassical major histocompatibility complex (MHC) class I molecule Qa-1(b). *J. Exp. Med.* **188**, 1841–1848.
32. Nomura, M., Zou, Z., Joh, T., Takihara, Y., Matsuda, Y., and Shimada, K. (1996). Genomic structures and characterization of Rae1 family members encoding GPI-anchored cell surface proteins and expressed predominantly in embryonic mouse brain. *J. Biochem. (Tokyo)* **120**, 987–995.
33. Malarkannan, S., Shih, P. P., Eden, P. A., Horng, T., Zuberi, A. R., Christianson, G., Roopenian, D., and Shastri, N. (1998). The molecular and functional characterization of a dominant minor H antigen, H60. *J. Immunol.* **161**, 3501–3509.
34. Westgaard, I. H., Berg, S. F., Orstavik, S., Fossum, S., and Dissen, E. (1998). Identification of a human member of the Ly-49 multigene family. *Eur. J. Immunol.* **28**, 1839–1846.
35. Moretta, A., Tambussi, G., Bottino, C., Tripodi, G., Merli, A., Ciccone, E., Pantaleo, G., and Moretta, L. (1990). A novel surface antigen expressed by a subset of human CD3-CD16+ natural killer cells. Role in cell activation and regulation of cytolytic function. *J. Exp. Med.* **171**, 695–714.
36. Moretta, A., Bottino, C., Pende, D., Tripodi, G., Tambussi, G., Viale, O., Orengo, A., Barbaresi, M., Merli, A., and Ciccone, E. (1990). Identification of four subsets of human CD3-CD16+ natural killer (NK) cells by the expression of clonally distributed functional surface molecules: Correlation between subset assignment of NK clones and ability to mediate specific alloantigen recognition. *J. Exp. Med.* **172**, 1589–1598.
37. Moretta, A., Vitale, M., Bottino, C., Orengo, A. M., Morelli, L., Augugliaro, R., Barbaresi, M., Ciccone, E., and Moretta, L. (1993). P58 molecules as putative receptors for major histocompatibility complex (MHC) class I molecules in human natural killer (NK) cells. Anti-p58 antibodies reconstitute lysis of MHC class I-protected cells in NK clones displaying different specificities. *J. Exp. Med.* **178**, 597–604.
38. Moretta, L., Ciccone, E., Poggi, A., Mingari, M. C., and Moretta, A. (1994). Ontogeny, specific functions and receptors of human natural killer cells. *Immunol. Lett.* **40**, 83–88.
39. Yu, Y. Y., George, T., Dorfman, J. R., Roland, J., Kumar, V., and Bennett, M. (1996). The role of Ly49A and 5E6(Ly49C) molecules in hybrid resistance mediated by murine natural killer cells against normal T cell blasts. *Immunity* **4**, 67–76.
40. Raulet, D. H., and Held, W. (1995). Natural killer cell receptors: The offs and ons of NK cell recognition. *Cell* **82**, 697–700.
41. Simmons, D., and Seed, B. (1988). The Fc gamma receptor of natural killer cells is a phospholipid-linked membrane protein. *Nature* **333**, 568–570.
42. Timonen, T., Ortaldo, J. R., and Herberman, R. B. (1981). Characteristics of human large granular lymphocytes and relationship to natural killer and K cells. *J. Exp. Med.* **153**, 569–582.
43. Lazetic, S., Chang, C., Houchins, J. P., Lanier, L. L., and Phillips, J. H. (1996). Human natural killer cell receptors involved in MHC class I recognition are disulfide-linked heterodimers of CD94 and NKG2 subunits. *J. Immunol.* **157**, 4741–4745.

44. Houchins, J. P., Yabe, T., McSherry, C., and Bach, F. H. (1991). DNA sequence analysis of NKG2, a family of related cDNA clones encoding type II integral membrane proteins on human natural killer cells. *J. Exp. Med.* **173**, 1017–1020.
45. Lanier, L. L., Chang, C., and Phillips, J. H. (1994). Human NKR-P1A. A disulfide-linked homodimer of the C-type lectin superfamily expressed by a subset of NK and T lymphocytes. *J. Immunol.* **153**, 2417–2428.
46. Mela, C. M., Burton, C. T., Imami, N., Nelson, M., Steel, A., Gazzard, B. G., Gotch, F. M., and Goodier, M. R. (2005). Switch from inhibitory to activating NKG2 receptor expression in HIV-1 infection: Lack of reversion with highly active antiretroviral therapy. *AIDS* **19**, 1761–1769.
47. Draghi, M., Yawata, N., Gleimer, M., Yawata, M., Valiante, N. M., and Parham, P. (2005). Single-cell analysis of the human NK cell response to missing self and its inhibition by HLA class I. *Blood* **105**, 2028–2035.
48. Kumar, V., and McNerney, M. E. (2005). A new self: MHC-class-I-independent natural-killer-cell self-tolerance. *Nat. Rev. Immunol.* **5**, 363–374.
49. Wende, H., Volz, A., and Ziegler, A. (2000). Extensive gene duplications and a large inversion characterize the human leukocyte receptor cluster. *Immunogenetics* **51**, 703–713.
50. Wilson, M. J., Torkar, M., Haude, A., Milne, S., Jones, T., Sheer, D., Beck, S., and Trowsdale, J. (2000). Plasticity in the organization and sequences of human KIR/ILT gene families. *Proc. Natl. Acad. Sci. USA* **97**, 4778–4783.
51. Long, E. O., Colonna, M., and Lanier, L. L. (1996). Inhibitory MHC class I receptors on NK and T cells: A standard nomenclature. *Immunol. Today* **17**, 100.
52. Döhring, C., Scheidegger, D., Samaridis, J., Cella, M., and Colonna, M. (1996). A human killer inhibitory receptor specific for HLA-A1,2. *J. Immunol.* **156**, 3098–3101.
53. Moretta, A., Bottino, C., Vitale, M., Pende, D., Biassoni, R., Mingari, M. C., and Moretta, L. (1996). Receptors for HLA class-I molecules in human natural killer cells. *Annu. Rev. Immunol.* **14**, 619–648.
54. Rajagopalan, S., and Long, E. O. (1997). The direct binding of a p58 killer cell inhibitory receptor to human histocompatibility leukocyte antigen (HLA)-Cw4 exhibits peptide selectivity. *J. Exp. Med.* **185**, 1523–1528.
55. Lanier, L. L. (1998). NK cell receptors. *Annu. Rev. Immunol.* **16**, 359–393.
56. Long, E. O. (1999). Regulation of immune responses through inhibitory receptors. *Annu. Rev. Immunol.* **17**, 875–904.
57. Moretta, A., Biassoni, R., Bottino, C., Pende, D., Vitale, M., Poggi, A., Mingari, M. C., and Moretta, L. (1997). Major histocompatibility complex class I-specific receptors on human natural killer and T lymphocytes. *Immunol. Rev.* **155**, 105–117.
58. Uhrberg, M., Valiante, N. M., Shum, B. P., Shilling, H. G., Lienert-Weidenbach, K., Corliss, B., Tyan, D., Lanier, L. L., and Parham, P. (1997). Human diversity in killer cell inhibitory receptor genes. *Immunity* **7**, 753–763.
59. Hsu, K. C., Liu, X. R., Selvakumar, A., Mickelson, E., O'Reilly, R. J., and Dupont, B. (2002). Killer Ig-like receptor haplotype analysis by gene content: Evidence for genomic diversity with a minimum of six basic framework haplotypes, each with multiple subsets. *J. Immunol.* **169**, 5118–5129.
60. Yawata, M., Yawata, N., Abi-Rached, L., and Parham, P. (2002). Variation within the human killer cell immunoglobulin-like receptor (KIR) gene family. *Crit. Rev. Immunol.* **22**, 463–482.
61. Maxwell, L. D., Wallace, A., Middleton, D., and Curran, M. D. (2002). A common KIR2DS4 deletion variant in the human that predicts a soluble KIR molecule analogous to the KIR1D molecule observed in the rhesus monkey. *Tissue Antigens* **60**, 254–258.

62. Yawata, M., Yawata, N., McQueen, K. L., Cheng, N. W., Guethlein, L. A., Rajalingam, R., Shilling, H. G., and Parham, P. (2002). Predominance of group A KIR haplotypes in Japanese associated with diverse NK cell repertoires of KIR expression. *Immunogenetics* **54,** 543–550.
63. Uhrberg, M., Parham, P., and Wernet, P. (2002). Definition of gene content for nine common group B haplotypes of the Caucasoid population: KIR haplotypes contain between seven and eleven KIR genes. *Immunogenetics* **54,** 221–229.
64. Norman, P. J., Stephens, H. A., Verity, D. H., Chandanayingyong, D., and Vaughan, R. W. (2001). Distribution of natural killer cell immunoglobulin-like receptor sequences in three ethnic groups. *Immunogenetics* **52,** 195–205.
65. Rajagopalan, S., Bryceson, Y. T., Kuppusamy, S. P., Geraghty, D. E., Meer, A. V., Joosten, I., and Long, E. O. (2005). Activation of NK cells by an endocytosed receptor for soluble HLA-G. *PLoS Biol.* **4,** e9.
66. Kikuchi-Maki, A., Catina, T. L., and Campbell, K. S. (2005). Cutting edge: KIR2DL4 transduces signals into human NK cells through association with the Fc receptor gamma protein. *J. Immunol.* **174,** 3859–3863.
67. Cella, M., Longo, A., Ferrara, G. B., Strominger, J. L., and Colonna, M. (1994). NK3-specific natural killer cells are selectively inhibited by Bw4-positive HLA alleles with isoleucine 80. *J. Exp. Med.* **180,** 1235–1242.
68. Vitale, M., Sivori, S., Pende, D., Augugliaro, R., Di Donato, C., Amoroso, A., Malnati, M., Bottino, C., Moretta, L., and Moretta, A. (1996). Physical and functional independency of p70 and p58 natural killer (NK) cell receptors for HLA class I: Their role in the definition of different groups of alloreactive NK cell clones. *Proc. Natl. Acad. Sci. USA* **93,** 1453–1457.
69. Pende, D., Biassoni, R., Cantoni, C., Verdiani, S., Falco, M., Di Donato, C., Accame, L., Bottino, C., Moretta, A., and Moretta, L. (1996). The natural killer cell receptor specific for HLA-A allotypes: A novel member of the p58/p70 family of inhibitory receptors that is characterized by three immunoglobulin-like domains and is expressed as a 140-kD disulphide-linked dimer. *J. Exp. Med.* **184,** 505–518.
70. Grau, R., Lang, K. S., Wernet, D., Lang, P., Niethammer, D., Pusch, C. M., and Handgretinger, R. (2004). Cytotoxic activity of natural killer cells lacking killer-inhibitory receptors for self-HLA class I molecules against autologous hematopoietic stem cells in healthy individuals. *Exp. Mol. Pathol.* **76,** 90–98.
71. Colonna, M., Borsellino, G., Falco, M., Ferrara, G. B., and Strominger, J. L. (1993). HLA-C is the inhibitory ligand that determines dominant resistance to lysis by NK1- and NK2-specific natural killer cells. *Proc. Natl. Acad. Sci. USA* **90,** 12000–12004.
72. Winter, C. C., Gumperz, J. E., Parham, P., Long, E. O., and Wagtmann, N. (1998). Direct binding and functional transfer of NK cell inhibitory receptors reveal novel patterns of HLA-C allotype recognition. *J. Immunol.* **161,** 571–577.
73. Boyington, J. C., and Sun, P. D. (2002). A structural perspective on MHC class I recognition by killer cell immunoglobulin-like receptors. *Mol. Immunol.* **38,** 1007–1021.
74. Carr, W. H., Pando, M. J., and Parham, P. (2005). KIR3DL1 polymorphisms that affect NK cell inhibition by HLA-Bw4 ligand. *J. Immunol.* **175,** 5222–5229.
75. Gardiner, C. M., Guethlein, L. A., Shilling, H. G., Pando, M., Carr, W. H., Rajalingam, R., Vilches, C., and Parham, P. (2001). Different NK cell surface phenotypes defined by the DX9 antibody are due to KIR3DL1 gene polymorphism. *J. Immunol.* **166,** 2992–3001.
76. Shilling, H. G., Guethlein, L. A., Cheng, N. W., Gardiner, C. M., Rodriguez, R., Tyan, D., and Parham, P. (2002). Allelic polymorphism synergizes with variable gene content to individualize human KIR genotype. *J. Immunol.* **168,** 2307–2315.

77. Abi-Rached, L., and Parham, P. (2005). Natural selection drives recurrent formation of activating killer cell immunoglobulin-like receptor and Ly49 from inhibitory homologues. *J. Exp. Med.* **201,** 1319–1332.
78. Biassoni, R., Pessino, A., Malaspina, A., Cantoni, C., Bottino, C., Sivori, S., Moretta, L., and Moretta, A. (1997). Role of amino acid position 70 in the binding affinity of p50.1 and p58.1 receptors for HLA-Cw4 molecules. *Eur. J. Immunol.* **27,** 3095–3099.
79. Vales-Gomez, M., Reyburn, H. T., Erskine, R. A., and Strominger, J. (1998). Differential binding to HLA-C of p50-activating and p58-inhibitory natural killer cell receptors. *Proc. Natl. Acad. Sci. USA* **95,** 14326–14331.
80. Katz, G., Markel, G., Mizrahi, S., Arnon, T. I., and Mandelboim, O. (2001). Recognition of HLA-Cw4 but not HLA-Cw6 by the NK cell receptor killer cell Ig-like receptor two-domain short tail number 4. *J. Immunol.* **166,** 7260–7267.
81. Martin, M. P., Gao, X., Lee, J. H., Nelson, G. W., Detels, R., Goedert, J. J., Buchbinder, S., Hoots, K., Vlahov, D., Trowsdale, J., Wilson, M., O'Brien, S. J. *et al.* (2002). Epistatic interaction between KIR3DS1 and HLA-B delays the progression to AIDS. *Nat. Genet.* **31,** 429–434.
82. Martin, M. P., Nelson, G., Lee, J. H., Pellett, F., Gao, X., Wade, J., Wilson, M. J., Trowsdale, J., Gladman, D., and Carrington, M. (2002). Cutting edge: Susceptibility to psoriatic arthritis: Influence of activating killer Ig-like receptor genes in the absence of specific HLA-C alleles. *J. Immunol.* **169,** 2818–2822.
83. Shilling, H. G., McQueen, K. L., Cheng, N. W., Shizuru, J. A., Negrin, R. S., and Parham, P. (2003). Reconstitution of NK cell receptor repertoire following HLA-matched hematopoietic cell transplantation. *Blood* **101,** 3730–3740.
84. Leung, W., Iyengar, R., Turner, V., Lang, P., Bader, P., Conn, P., Niethammer, D., and Handgretinger, R. (2004). Determinants of antileukemia effects of allogeneic NK cells. *J. Immunol.* **172,** 644–650.
85. Dupont, B., and Hsu, K. C. (2004). Inhibitory killer Ig-like receptor genes and human leukocyte antigen class I ligands in haematopoietic stem cell transplantation. *Curr. Opin. Immunol.* **16,** 634–643.
86. Cosman, D., Fanger, N., Borges, L., Kubin, M., Chin, W., Peterson, L., and Hsu, M. L. (1997). A novel immunoglobulin superfamily receptor for cellular and viral MHC class I molecules. *Immunity* **7,** 273–282.
87. Colonna, M., Navarro, F., Bellon, T., Llano, M., Garcia, P., Samaridis, J., Angman, L., Cella, M., and Lopez-Botet, M. (1997). A common inhibitory receptor for major histocompatibility complex class I molecules on human lymphoid and myelomonocytic cells. *J. Exp. Med.* **186,** 1809–1818.
88. Wagtmann, N., Rojo, S., Eichler, E., Mohrenweiser, H., and Long, E. O. (1997). A new human gene complex encoding the killer cell inhibitory receptors and related monocyte/macrophage receptors. *Curr. Biol.* **7,** 615–618.
89. Brown, D., Trowsdale, J., and Allen, R. (2004). The LILR family: Modulators of innate and adaptive immune pathways in health and disease. *Tissue Antigens* **64,** 215–225.
90. Borges, L., Hsu, M. L., Fanger, N., Kubin, M., and Cosman, D. (1997). A family of human lymphoid and myeloid Ig-like receptors, some of which bind to MHC class I molecules. *J. Immunol.* **159,** 5192–5196.
91. Vitale, M., Castriconi, R., Parolini, S., Pende, D., Hsu, M. L., Moretta, L., Cosman, D., and Moretta, A. (1999). The leukocyte Ig-like receptor (LIR)-1 for the cytomegalovirus UL18 protein displays a broad specificity for different HLA class I alleles: analysis of LIR-1 + NK cell clones. *Int. Immunol.* **11,** 29–35.
92. Shiroishi, M., Tsumoto, K., Amano, K., Shirakihara, Y., Colonna, M., Braud, V. M., Allan, D. S., Makadzange, A., Rowland-Jones, S., Willcox, B., Jones, E. Y., van der Merwe, P. A. *et al.* (2003). Human inhibitory receptors Ig-like transcript 2 (ILT2) and ILT4 compete with CD8

for MHC class I binding and bind preferentially to HLA-G. *Proc. Natl. Acad. Sci. USA* **100**, 8856–8861.
93. Lemaoult, J., Zafaranloo, K., Le Danff, C., and Carosella, E. D. (2005). HLA-G up-regulates ILT2, ILT3, ILT4, and KIR2DL4 in antigen presenting cells, NK cells, and T cells. *FASEB J.* **19**, 662–664.
94. Biassoni, R., Bottino, C., Millo, R., Moretta, L., and Moretta, A. (1999). Natural killer cell-mediated recognition of human trophoblast. *Semin. Cancer Biol.* **9**, 13–18.
95. Navarro, F., Llano, M., Bellon, T., Colonna, M., Geraghty, D. E., and Lopez-Botet, M. (1999). The ILT2(LIR1) and CD94/NKG2A NK cell receptors respectively recognize HLA-G1 and HLA-E molecules co-expressed on target cells. *Eur. J. Immunol.* **29**, 277–283.
96. Willcox, B. E., Thomas, L. M., and Bjorkman, P. J. (2003). Crystal structure of HLA-A2 bound to LIR-1, a host and viral major histocompatibility complex receptor. *Nat. Immunol.* **4**, 913–919.
97. Chapman, T. L., and Bjorkman, P. J. (1998). Characterization of a murine cytomegalovirus class I major histocompatibility complex (MHC) homolog: Comparison to MHC molecules and to the human cytomegalovirus MHC homolog. *J. Virol.* **72**, 460–466.
98. Chapman, T. L., Heikeman, A. P., and Bjorkman, P. J. (1999). The inhibitory receptor LIR-1 uses a common binding interaction to recognize class I MHC molecules and the viral homolog UL18. *Immunity* **11**, 603–613.
99. Chang, C. C., Ciubotariu, R., Manavalan, J. S., Yuan, J., Colovai, A. I., Piazza, F., Lederman, S., Colonna, M., Cortesini, R., Dalla-Favera, R., and Suciu-Foca, N. (2002). Tolerization of dendritic cells by T(S) cells: The crucial role of inhibitory receptors ILT3 and ILT4. *Nat. Immunol.* **3**, 237–243.
100. Fahnestock, M. L., Johnson, J. L., Feldman, R. M., Neveu, J. M., Lane, W. S., and Bjorkman, P. J. (1995). The MHC class I homolog encoded by human cytomegalovirus binds endogenous peptides. *Immunity* **3**, 583–590.
101. Farrell, H. E., Vally, H., Lynch, D. M., Fleming, P., Shellam, G. R., Scalzo, A. A., and Davis-Poynter, N. J. (1997). Inhibition of natural killer cells by a cytomegalovirus MHC class I homologue *in vivo*. *Nature* **386**, 510–514.
102. Cretney, E., Degli-Esposti, M. A., Densley, E. H., Farrell, H. E., Davis-Poynter, N. J., and Smyth, M. J. (1999). m144, a murine cytomegalovirus (MCMV)-encoded major histocompatibility complex class I homologue, confers tumor resistance to natural killer cell-mediated rejection. *J. Exp. Med.* **190**, 435–444.
103. Tortorella, D., Gewurz, B. E., Furman, M. H., Schust, D. J., and Ploegh, H. L. (2000). Viral subversion of the immune system. *Annu. Rev. Immunol.* **18**, 861–926.
104. Leong, C. C., Chapman, T. L., Bjorkman, P. J., Formankova, D., Mocarski, E. S., Phillips, J. H., and Lanier, L. L. (1998). Modulation of natural killer cell cytotoxicity in human cytomegalovirus infection: The role of endogenous class I major histocompatibility complex and a viral class I homolog. *J. Exp. Med.* **187**, 1681–1687.
105. Lanier, L. L., Yu, G., and Phillips, J. H. (1991). Analysis of Fc gamma RIII (CD16) membrane expression and association with CD3 zeta and Fc epsilon RI-gamma by site-directed mutation. *J. Immunol.* **146**, 1571–1576.
106. Daeron, M. (1997). Fc receptor biology. *Annu. Rev. Immunol.* **15**, 203–234.
107. Wirthmueller, U., Kurosaki, T., Murakami, M. S., and Ravetch, J. V. (1992). Signal transduction by Fc gamma RIII (CD16) is mediated through the gamma chain. *J. Exp. Med.* **175**, 1381–1390.
108. Lanier, L. L., Yu, G., and Phillips, J. H. (1989). Co-association of CD3 zeta with a receptor (CD16) for IgG Fc on human natural killer cells. *Nature* **342**, 803–805.
109. Preithner, S., Elm, S., Lippold, S., Locher, M., Wolf, A., Silva, A. J., Baeuerle, P. A., and Prang, N. S. (2005). High concentrations of therapeutic IgG1 antibodies are needed to compensate for inhibition of antibody-dependent cellular cytotoxicity by excess endogenous immunoglobulin G. *Mol. Immunol.* **43**, 1183–1193.

110. Bowles, J. A., and Weiner, G. J. (2005). CD16 polymorphisms and NK activation induced by monoclonal antibody-coated target cells. *J. Immunol. Methods* **304**, 88–99.
111. Niwa, R., Sakurada, M., Kobayashi, Y., Uehara, A., Matsushima, K., Ueda, R., Nakamura, K., and Shitara, K. (2005). Enhanced natural killer cell binding and activation by low-fucose IgG1 antibody results in potent antibody-dependent cellular cytotoxicity induction at lower antigen density. *Clin. Cancer Res.* **11**, 2327–2336.
112. Lucia, B., Jennings, C., Cauda, R., Ortona, L., and Landay, A. L. (1995). Evidence of a selective depletion of a CD16+ CD56+ CD8+ natural killer cell subset during HIV infection. *Cytometry* **22**, 10–15.
113. Sivori, S., Vitale, M., Morelli, L., Sanseverino, L., Augugliaro, R., Bottino, C., Moretta, L., and Moretta, A. (1997). p46, a novel natural killer cell-specific surface molecule that mediates cell activation. *J. Exp. Med.* **186**, 1129–1136.
114. Pende, D., Parolini, S., Pessino, A., Sivori, S., Augugliaro, R., Morelli, L., Marcenaro, E., Accame, L., Malaspina, A., Biassoni, R., Bottino, C., Moretta, L. *et al.* (1999). Identification and molecular characterization of NKp30, a novel triggering receptor involved in natural cytotoxicity mediated by human natural killer cells. *J. Exp. Med.* **190**, 1505–1516.
115. Vitale, M., Bottino, C., Sivori, S., Sanseverino, L., Castriconi, R., Marcenaro, E., Augugliaro, R., Moretta, L., and Moretta, A. (1998). NKp44, a novel triggering surface molecule specifically expressed by activated natural killer cells, is involved in non-major histocompatibility complex-restricted tumor cell lysis. *J. Exp. Med.* **187**, 2065–2072.
116. Sivori, S., Pende, D., Bottino, C., Marcenaro, E., Pessino, A., Biassoni, R., Moretta, L., and Moretta, A. (1999). NKp46 is the major triggering receptor involved in the natural cytotoxicity of fresh or cultured human NK cells. Correlation between surface density of NKp46 and natural cytotoxicity against autologous, allogeneic or xenogeneic target cells. *Eur. J. Immunol.* **29**, 1656–1666.
117. Nowbakht, P., Ionescu, M. C., Rohner, A., Kalberer, C. P., Rossy, E., Mori, L., Cosman, D., De Libero, G., and Wodnar-Filipowicz, A. (2005). Ligands for natural killer cell-activating receptors are expressed upon the maturation of normal myelomonocytic cells but at low levels in acute myeloid leukemias. *Blood* **105**, 3615–3622.
118. Mavoungou, E., Bouyou-Akotet, M. K., and Kremsner, P. G. (2005). Effects of prolactin and cortisol on natural killer (NK) cell surface expression and function of human natural cytotoxicity receptors (NKp46, NKp44 and NKp30). *Clin. Exp. Immunol.* **139**, 287–296.
119. Biassoni, R., Pessino, A., Bottino, C., Pende, D., Moretta, L., and Moretta, A. (1999). The murine homologue of the human NKp46, a triggering receptor involved in the induction of natural cytotoxicity. *Eur. J. Immunol.* **29**, 1014–1020.
120. Falco, M., Cantoni, C., Bottino, C., Moretta, A., and Biassoni, R. (1999). Identification of the rat homologue of the human NKp46 triggering receptor. *Immunol. Lett.* **68**, 411–414.
121. Mandelboim, O., Lieberman, N., Lev, M., Paul, L., Arnon, T. I., Bushkin, Y., Davis, D. M., Strominger, J. L., Yewdell, J. W., and Porgador, A. (2001). Recognition of haemagglutinins on virus-infected cells by NKp46 activates lysis by human NK cells. *Nature* **409**, 1055–1060.
122. Arnon, T. I., Lev, M., Katz, G., Chernobrov, Y., Porgador, A., and Mandelboim, O. (2001). Recognition of viral hemagglutinins by NKp44 but not by NKp30. *Eur. J. Immunol.* **31**, 2680–2689.
123. Campbell, K. S., Yusa, S., Kikuchi-Maki, A., and Catina, T. L. (2004). NKp44 triggers NK cell activation through DAP12 association that is not influenced by a putative cytoplasmic inhibitory sequence. *J. Immunol.* **172**, 899–906.
124. Cantoni, C., Bottino, C., Vitale, M., Pessino, A., Augugliaro, R., Malaspina, A., Parolini, S., Moretta, L., Moretta, A., and Biassoni, R. (1999). NKp44, a triggering receptor involved in tumor cell lysis by activated human natural killer cells, is a novel member of the immunoglobulin superfamily. *J. Exp. Med.* **189**, 787–796.

125. Vieillard, V., Strominger, J. L., and Debre, P. (2005). NK cytotoxicity against CD4+ T cells during HIV-1 infection: A gp41 peptide induces the expression of an NKp44 ligand. *Proc. Natl. Acad. Sci. USA* **102**, 10981–10986.
126. Neville, M. J., and Campbell, R. D. (1999). A new member of the Ig superfamily and a V-ATPase G subunit are among the predicted products of novel genes close to the TNF locus in the human MHC. *J. Immunol.* **162**, 4745–4754.
127. Warren, H. S., Jones, A. L., Freeman, C., Bettadapura, J., and Parish, C. R. (2005). Evidence that the cellular ligand for the human NK cell activation receptor NKp30 is not a heparan sulfate glycosaminoglycan. *J. Immunol.* **175**, 207–212.
128. Hollyoake, M., Campbell, R. D., and Aguado, B. (2005). NKp30 (NCR3) is a pseudogene in 12 inbred and wild mouse strains, but an expressed gene in *Mus caroli*. *Mol. Biol. Evol.* **22**, 1661–1672.
129. Moretta, A., Biassoni, R., Bottino, C., Mingari, M. C., and Moretta, L. (2000). Natural cytotoxicity receptors that trigger human NK-cell-mediated cytolysis. *Immunol. Today* **21**, 228–234.
130. Pessino, A., Sivori, S., Bottino, C., Malaspina, A., Morelli, L., Moretta, L., Biassoni, R., and Moretta, A. (1998). Molecular cloning of NKp46: A novel member of the immunoglobulin superfamily involved in triggering of natural cytotoxicity. *J. Exp. Med.* **188**, 953–960.
131. Augugliaro, R., Parolini, S., Castriconi, R., Marcenaro, E., Cantoni, C., Nanni, M., Moretta, L., Moretta, A., and Bottino, C. (2003). Selective cross-talk among natural cytotoxicity receptors in human natural killer cells. *Eur. J. Immunol.* **33**, 1235–1241.
132. Mathew, P. A., Garni-Wagner, B. A., Land, K., Takashima, A., Stoneman, E., Bennett, M., and Kumar, V. (1993). Cloning and characterization of the 2B4 gene encoding a molecule associated with non-MHC-restricted killing mediated by activated natural killer cells and T cells. *J. Immunol.* **151**, 5328–5337.
133. Kubin, M. Z., Parshley, D. L., Din, W., Waugh, J. Y., Davis-Smith, T., Smith, C. A., Macduff, B. M., Armitage, R. J., Chin, W., Cassiano, L., Borges, L., Petersen, M. *et al.* (1999). Molecular cloning and biological characterization of NK cell activation-inducing ligand, a counterstructure for CD48. *Eur. J. Immunol.* **29**, 3466–3477.
134. Boles, K. S., Nakajima, H., Colonna, M., Chuang, S. S., Stepp, S. E., Bennett, M., Kumar, V., and Mathew, P. A. (1999). Molecular characterization of a novel human natural killer cell receptor homologous to mouse 2B4. *Tissue Antigens* **54**, 27–34.
135. Moretta, A., Bottino, C., Tripodi, G., Vitale, M., Pende, D., Morelli, L., Augugliaro, R., Barbaresi, M., Ciccone, E., and Millo, R. (1992). Novel surface molecules involved in human NK cell activation and triggering of the lytic machinery. *Int. J. Cancer Suppl.* **7**, 6–10.
136. Garni-Wagner, B. A., Purohit, A., Mathew, P. A., Bennett, M., and Kumar, V. (1993). A novel function associated molecule related to non-MHC-restricted cytotoxicity mediated by activated natural killer cells and T cells. *J. Immunol.* **151**, 60–70.
137. Schuhmachers, G., Ariizumi, K., Mathew, P. A., Bennett, M., Kumar, V., and Takashima, A. (1995). Activation of murine epidermal gamma delta T cells through surface 2B4. *Eur. J. Immunol.* **25**, 1117–1120.
138. Sivori, S., Parolini, S., Falco, M., Marcenaro, E., Biassoni, R., Bottino, C., Moretta, L., and Moretta, A. (2000). 2B4 functions as a co-receptor in human NK cell activation. *Eur. J. Immunol.* **30**, 787–793.
139. Valiante, N. M., and Trinchieri, G. (1993). Identification of a novel signal transduction surface molecule on human cytotoxic lymphocytes. *J. Exp. Med.* **178**, 1397–1406.
140. Tangye, S. G., Cherwinski, H., Lanier, L. L., and Phillips, J. H. (2000). 2B4-mediated activation of human natural killer cells. *Mol. Immunol.* **37**, 493–501.
141. Bottino, C., Augugliaro, R., Castriconi, R., Nanni, M., Biassoni, R., Moretta, L., and Moretta, A. (2000). Analysis of the molecular mechanism involved in 2B4-mediated NK cell activation:

Evidence that human 2B4 is physically and functionally associated with the linker for activation of T cells. *Eur. J. Immunol.* **30**, 3718–3722.
142. Jevremovic, D., Billadeau, D. D., Schoon, R. A., Dick, C. J., Irvin, B. J., Zhang, W., Samelson, L. E., Abraham, R. T., and Leibson, P. J. (1999). Cutting edge: A role for the adaptor protein LAT in human NK cell-mediated cytotoxicity. *J. Immunol.* **162**, 2453–2456.
143. Parolini, S., Bottino, C., Falco, M., Augugliaro, R., Giliani, S., Franceschini, R., Ochs, H. D., Wolf, H., Bonnefoy, J. Y., Biassoni, R., Moretta, L., Notarangelo, L. D. *et al.* (2000). X-linked lymphoproliferative disease. 2B4 molecules displaying inhibitory rather than activating function are responsible for the inability of natural killer cells to kill Epstein-Barr virus-infected cells. *J. Exp. Med.* **192**, 337–346.
144. Nakajima, H., Cella, M., Langen, H., Friedlein, A., and Colonna, M. (1999). Activating interactions in human NK cell recognition: The role of 2B4-CD48. *Eur. J. Immunol.* **29**, 1676–1683.
145. Korinek, V., Stefanova, I., Angelisova, P., Hilgert, I., and Horejsi, V. (1991). The human leucocyte antigen CD48 (MEM-102) is closely related to the activation marker Blast-1. *Immunogenetics* **33**, 108–112.
146. Yokoyama, S., Staunton, D., Fisher, R., Amiot, M., Fortin, J. J., and Thorley-Lawson, D. A. (1991). Expression of the Blast-1 activation/adhesion molecule and its identification as CD48. *J. Immunol.* **146**, 2192–2200.
147. Engel, P., Eck, M. J., and Terhorst, C. (2003). The SAP and SLAM families in immune responses and X-linked lymphoproliferative disease. *Nat. Rev. Immunol.* **3**, 813–821.
148. Meyaard, L., Adema, G. J., Chang, C., Woollatt, E., Sutherland, G. R., Lanier, L. L., and Phillips, J. H. (1997). LAIR-1, a novel inhibitory receptor expressed on human mononuclear leukocytes. *Immunity* **7**, 283–290.
149. Sathish, J. G., Johnson, K. G., Fuller, K. J., LeRoy, F. G., Meyaard, L., Sims, M. J., and Matthews, R. J. (2001). Constitutive association of SHP-1 with leukocyte-associated Ig-like receptor-1 in human T cells. *J. Immunol.* **166**, 1763–1770.
150. Hammarstrom, S. (1999). The carcinoembryonic antigen (CEA) family: Structures, suggested functions and expression in normal and malignant tissues. *Semin. Cancer Biol.* **9**, 67–81.
151. Moller, M. J., Kammerer, R., Grunert, F., and von Kleist, S. (1996). Biliary glycoprotein (BGP) expression on T cells and on a natural-killer-cell sub-population. *Int. J. Cancer* **65**, 740–745.
152. Markel, G., Gruda, R., Achdout, H., Katz, G., Nechama, M., Blumberg, R. S., Kammerer, R., Zimmermann, W., and Mandelboim, O. (2004). The critical role of residues 43R and 44Q of carcinoembryonic antigen cell adhesion molecules-1 in the protection from killing by human NK cells. *J. Immunol.* **173**, 3732–3739.
153. Beauchemin, N., Kunath, T., Robitaille, J., Chow, B., Turbide, C., Daniels, E., and Veillette, A. (1997). Association of biliary glycoprotein with protein tyrosine phosphatase SHP-1 in malignant colon epithelial cells. *Oncogene* **14**, 783–790.
154. Busch, C., Hanssen, T. A., Wagener, C., and O'Brink, B. (2002). Down-regulation of CEA-CAM1 in human prostate cancer: Correlation with loss of cell polarity, increased proliferation rate, and Gleason grade 3 to 4 transition. *Hum. Pathol.* **33**, 290–298.
155. Markel, G., Lieberman, N., Katz, G., Arnon, T. I., Lotem, M., Drize, O., Blumberg, R. S., Bar-Haim, E., Mader, R., Eisenbach, L., and Mandelboim, O. (2002). CD66a interactions between human melanoma and NK cells: A novel class I MHC-independent inhibitory mechanism of cytotoxicity. *J. Immunol.* **168**, 2803–2810.
156. Markel, G., Mussaffi, H., Ling, K. L., Salio, M., Gadola, S., Steuer, G., Blau, H., Achdout, H., de Miguel, M., Gonen-Gross, T., Hanna, J., Arnon, T. I. *et al.* (2004). The mechanisms controlling NK cell autoreactivity in TAP2-deficient patients. *Blood* **103**, 1770–1778.

157. Markel, G., Achdout, H., Katz, G., Ling, K. L., Salio, M., Gruda, R., Gazit, R., Mizrahi, S., Hanna, J., Gonen-Gross, T., Arnon, T. I., Lieberman, N. *et al.* (2004). Biological function of the soluble CEACAM1 protein and implications in TAP2-deficient patients. *Eur. J. Immunol.* **34,** 2138–2148.
158. Crocker, P. R., and Varki, A. (2001). Siglecs, sialic acids and innate immunity. *Trends Immunol.* **22,** 337–342.
159. Yamaji, T., Teranishi, T., Alphey, M. S., Crocker, P. R., and Hashimoto, Y. (2002). A small region of the natural killer cell receptor, Siglec-7, is responsible for its preferred binding to alpha 2,8-disialyl and branched alpha 2,6-sialyl residues. A comparison with Siglec-9. *J. Biol. Chem.* **277,** 6324–6332.
160. Nicoll, G., Avril, T., Lock, K., Furukawa, K., Bovin, N., and Crocker, P. R. (2003). Ganglioside GD3 expression on target cells can modulate NK cell cytotoxicity via siglec-7-dependent and -independent mechanisms. *Eur. J. Immunol.* **33,** 1642–1648.
161. Urmacher, C., Cordon-Cardo, C., and Houghton, A. N. (1989). Tissue distribution of GD3 ganglioside detected by mouse monoclonal antibody R24. *Am. J. Dermatopathol.* **11,** 577–581.
162. Renedo, M., Arce, I., Rodriguez, A., Carretero, M., Lanier, L. L., Lopez-Botet, M., and Fernandez-Ruiz, E. (1997). The human natural killer gene complex is located on chromosome 12p12–p13. *Immunogenetics* **46,** 307–311.
163. Yokoyama, W. M., and Plougastel, B. F. (2003). Immune functions encoded by the natural killer gene complex. *Nat. Rev. Immunol.* **3,** 304–316.
164. Sentman, C. L., Hackett, J., Jr., Kumar, V., and Bennett, M. (1989). Identification of a subset of murine natural killer cells that mediates rejection of Hh-1d but not Hh-1b bone marrow grafts. *J. Exp. Med.* **170,** 191–202.
165. Yokoyama, W. M. (1995). Hybrid resistance and the Ly-49 family of natural killer cell receptors. *J. Exp. Med.* **182,** 273–277.
166. Dimasi, N., and Biassoni, R. (2005). Structural and functional aspects of the Ly49 natural killer cell receptors. *Immunol. Cell Biol.* **83,** 1–8.
167. Barten, R., and Trowsdale, J. (1999). The human Ly-49L gene. *Immunogenetics* **49,** 731–734.
168. Chambers, W. H., Vujanovic, N. L., DeLeo, A. B., Olszowy, M. W., Herberman, R. B., and Hiserodt, J. C. (1989). Monoclonal antibody to a triggering structure expressed on rat natural killer cells and adherent lymphokine-activated killer cells. *J. Exp. Med.* **169,** 1373–1389.
169. Giorda, R., and Trucco, M. (1991). Mouse NKR-P1. A family of genes selectively coexpressed in adherent lymphokine-activated killer cells. *J. Immunol.* **147,** 1701–1708.
170. Ryan, J. C., Turck, J., Niemi, E. C., Yokoyama, W. M., and Seaman, W. E. (1992). Molecular cloning of the NK1.1 antigen, a member of the NKR-P1 family of natural killer cell activation molecules. *J. Immunol.* **149,** 1631–1635.
171. Ryan, J. C., Niemi, E. C., Goldfien, R. D., Hiserodt, J. C., and Seaman, W. E. (1991). NKR-P1, an activating molecule on rat natural killer cells, stimulates phosphoinositide turnover and a rise in intracellular calcium. *J. Immunol.* **147,** 3244–3250.
172. Carlyle, J. R., Martin, A., Mehra, A., Attisano, L., Tsui, F. W., and Zuniga-Pflucker, J. C. (1999). Mouse NKR-P1B, a novel NK1.1 antigen with inhibitory function. *J. Immunol.* **162,** 5917–5923.
173. Iizuka, K., Naidenko, O. V., Plougastel, B. F., Fremont, D. H., and Yokoyama, W. M. (2003). Genetically linked C-type lectin-related ligands for the NKRP1 family of natural killer cell receptors. *Nat. Immunol.* **4,** 801–807.
174. Ntrivalas, E. I., Bowser, C. R., Kwak-Kim, J., Beaman, K. D., and Gilman-Sachs, A. (2005). Expression of killer immunoglobulin-like receptors on peripheral blood NK cell subsets of women with recurrent spontaneous abortions or implantation failures. *Am. J. Reprod. Immunol.* **53,** 215–221.

175. Houchins, J. P., Yabe, T., McSherry, C., Miyokawa, N., and Bach, F. H. (1990). Isolation and characterization of NK cell or NK/T cell-specific cDNA clones. *J. Mol. Cell Immunol.* **4,** 295–304.
176. Carretero, M., Cantoni, C., Bellon, T., Bottino, C., Biassoni, R., Rodriguez, A., Perez-Villar, J. J., Moretta, L., Moretta, A., and Lopez-Botet, M. (1997). The CD94 and NKG2-A C-type lectins covalently assemble to form a natural killer cell inhibitory receptor for HLA class I molecules. *Eur. J. Immunol.* **27,** 563–567.
177. Sivori, S., Vitale, M., Bottino, C., Marcenaro, E., Sanseverino, L., Parolini, S., Moretta, L., and Moretta, A. (1996). CD94 functions as a natural killer cell inhibitory receptor for different HLA class I alleles: Identification of the inhibitory form of CD94 by the use of novel monoclonal antibodies. *Eur. J. Immunol.* **26,** 2487–2492.
178. Phillips, J. H., Chang, C., Mattson, J., Gumperz, J. E., Parham, P., and Lanier, L. L. (1996). CD94 and a novel associated protein (94AP) form a NK cell receptor involved in the recognition of HLA-A, HLA-B, and HLA-C allotypes. *Immunity* **5,** 163–172.
179. Perez-Villar, J. J., Melero, I., Navarro, F., Carretero, M., Bellon, T., Llano, M., Colonna, M., Geraghty, D. E., and Lopez-Botet, M. (1997). The CD94/NKG2-A inhibitory receptor complex is involved in natural killer cell-mediated recognition of cells expressing HLA-G1. *J. Immunol.* **158,** 5736–5743.
180. Moretta, A., Vitale, M., Sivori, S., Bottino, C., Morelli, L., Augugliaro, R., Barbaresi, M., Pende, D., Ciccone, E., Lopez-Botet, M., and Moretta, L. (1994). Human natural killer cell receptors for HLA-class I molecules. Evidence that the Kp43 (CD94) molecule functions as receptor for HLA-B alleles. *J. Exp. Med.* **180,** 545–555.
181. Lee, N., Llano, M., Carretero, M., Ishitani, A., Navarro, F., Lopez-Botet, M., and Geraghty, D. E. (1998). HLA-E is a major ligand for the natural killer inhibitory receptor CD94/NKG2A. *Proc. Natl. Acad. Sci. USA* **95,** 5199–5204.
182. Borrego, F., Ulbrecht, M., Weiss, E. H., Coligan, J. E., and Brooks, A. G. (1998). Recognition of human histocompatibility leukocyte antigen (HLA)-E complexed with HLA class I signal sequence-derived peptides by CD94/NKG2 confers protection from natural killer cell-mediated lysis. *J. Exp. Med.* **187,** 813–818.
183. Braud, V., Jones, E. Y., and McMichael, A. (1997). The human major histocompatibility complex class Ib molecule HLA-E binds signal sequence-derived peptides with primary anchor residues at positions 2 and 9. *Eur. J. Immunol.* **27,** 1164–1169.
184. Braud, V. M., Allan, D. S., O'Callaghan, C. A., Soderstrom, K., D'Andrea, A., Ogg, G. S., Lazetic, S., Young, N. T., Bell, J. I., Phillips, J. H., Lanier, L. L., and McMichael, A. J. (1998). HLA-E binds to natural killer cell receptors CD94/NKG2A, B and C. *Nature* **391,** 795–799.
185. Brooks, A. G., Borrego, F., Posch. P. E., Patamawenu, A., Scorzelli, C. J., Ulbrecht, M., Weiss, E. H., and Coligan, J. E. (1999). Specific recognition of HLA-E, but not classical, HLA class I molecules by soluble CD94/NKG2A and NK cells. *J. Immunol.* **162,** 305–313.
186. Vales-Gomez, M., Reyburn, H. T., Erskine, R. A., Lopez-Botet, M., and Strominger, J. L. (1999). Kinetics and peptide dependency of the binding of the inhibitory NK receptor CD94/NKG2-A and the activating receptor CD94/NKG2-C to HLA-E. *EMBO J.* **18,** 4250–4260.
187. Llano, M., Lee, N., Navarro, F., Garcia, P., Albar, J. P., Geraghty, D. E., and Lopez-Botet, M. (1998). HLA-E-bound peptides influence recognition by inhibitory and triggering CD94/NKG2 receptors: Preferential response to an HLA-G-derived nonamer. *Eur. J. Immunol.* **28,** 2854–2863.
188. Cantoni, C., Biassoni, R., Pende, D., Sivori, S., Accame, L., Pareti, L., Semenzato, G., Moretta, L., Moretta, A., and Bottino, C. (1998). The activating form of CD94 receptor complex: CD94 covalently associates with the Kp39 protein that represents the product of the NKG2-C gene. *Eur. J. Immunol.* **28,** 327–338.

189. Lanier, L. L., Corliss, B., Wu, J., and Phillips, J. H. (1998). Association of DAP12 with activating CD94/NKG2C NK cell receptors. *Immunity* **8**, 693–701.
190. Lopez-Botet, M., Llano, M., Navarro, F., and Bellon, T. (2000). NK cell recognition of non-classical HLA class I molecules. *Semin. Immunol.* **12**, 109–119.
191. Guma, M., Busch, L. K., Salazar-Fontana, L. I., Bellosillo, B., Morte, C., Garcia, P., and Lopez-Botet, M. (2005). The CD94/NKG2C killer lectin-like receptor constitutes an alternative activation pathway for a subset of CD8+ T cells. *Eur. J. Immunol.* **35**, 2071–2080.
192. Hikami, K., Tsuchiya, N., Yabe, T., and Tokunaga, K. (2003). Variations of human killer cell lectin-like receptors: Common occurrence of NKG2-C deletion in the general population. *Genes Immun.* **4**, 160–167.
193. Miyashita, R., Tsuchiya, N., Hikami, K., Kuroki, K., Fukazawa, T., Bijl, M., Kallenberg, C. G., Hashimoto, H., Yabe, T., and Tokunaga, K. (2004). Molecular genetic analyses of human NKG2C (KLRC2) gene deletion. *Int. Immunol.* **16**, 163–168.
194. Adamkiewicz, T. V., McSherry, C., Bach, F. H., and Houchins, J. P. (1994). Natural killer lectin-like receptors have divergent carboxy-termini, distinct from C-type lectins. *Immunogenetics* **39**, 218.
195. Kaiser, B. K., Barahmand-Pour, F., Paulsene, W., Medley, S., Geraghty, D. E., and Strong, R. K. (2005). Interactions between NKG2x immunoreceptors and HLA-E ligands display overlapping affinities and thermodynamics. *J. Immunol.* **174**, 2878–2884.
196. Brostjan, C., Bellon, T., Sobanov, Y., Lopez-Botet, M., and Hofer, E. (2002). Differential expression of inhibitory and activating CD94/NKG2 receptors on NK cell clones. *J. Immunol. Methods* **264**, 109–119.
197. Glienke, J., Sobanov, Y., Brostjan, C., Steffens, C., Nguyen, C., Lehrach, H., Hofer, E., and Francis, F. (1998). The genomic organization of NKG2C, E, F, and D receptor genes in the human natural killer gene complex. *Immunogenetics* **48**, 163–173.
198. Kim, D. K., Kabat, J., Borrego, F., Sanni, T. B., You, C. H., and Coligan, J. E. (2004). Human NKG2F is expressed and can associate with DAP12. *Mol. Immunol.* **41**, 53–62.
199. Schust, D. J., Tortorella, D., Seebach, J., Phan, C., and Ploegh, H. L. (1998). Trophoblast class I major histocompatibility complex (MHC) products are resistant to rapid degradation imposed by the human cytomegalovirus (HCMV) gene products US2 and US11. *J. Exp. Med.* **188**, 497–503.
200. Tomasec, P., Braud, V. M., Rickardsm, C., Powell, M. B., McSharry, B. P., Gadola, S., Cerundolo, V., Borysiewicz, L. K., McMichael, A. J., and Wilkinson, G. W. (2000). Surface expression of HLA-E, an inhibitor of natural killer cells, enhanced by human cytomegalovirus gpUL40. *Science* **287**, 1031.
201. Lopez-Botet, M., Carretero, M., Bellon, T., Perez-Villar, J. J., Llano, M., and Navarro, F. (1998). The CD94/NKG2 C-type lectin receptor complex. *Curr. Top. Microbiol. Immunol.* **230**, 41–52.
202. Yabe, T., McSherry, C., Bach, F. H., Fisch, P., Schall, R. P., Sondel, P. M., and Houchins, J. P. (1993). A multigene family on human chromosome 12 encodes natural killer-cell lectins. *Immunogenetics* **37**, 455–460.
203. Besostri, B., Beggiato, E., Bianchi, A., Mariani, S., Coscia, M., Peola, S., Foglietta, M., Boccadoro, M., Pileri, A., Moretta, L., and Massaia, M. (2000). Increased expression of non-functional killer inhibitory receptor CD94 in CD8+ cells of myeloma patients. *Br. J. Haematol.* **109**, 46–53.
204. Butcher, S., Arney, K. L., and Cook, G. P. (1998). MAFA-L, an ITIM-containing receptor encoded by the human NK cell gene complex and expressed by basophils and NK cells. *Eur. J. Immunol.* **28**, 3755–3762.
205. Robbins, S. H., Nguyen, K. B., Takahashi, N., Mikayama, T., Biron, C. A., and Brossay, L. (2002). Cutting edge: Inhibitory functions of the killer cell lectin-like receptor G1 molecule during the activation of mouse NK cells. *J. Immunol.* **168**, 2585–2589.

206. Corral, L., Hanke, T., Vance, R. E., Cado, D., and Raulet, D. H. (2000). NK cell expression of the killer cell lectin-like receptor G1 (KLRG1), the mouse homolog of MAFA, is modulated by MHC class I molecules. *Eur. J. Immunol.* **30**, 920–930.
207. Abramson, J., Xu, R., and Pecht, I. (2002). An unusual inhibitory receptor – the mast cell function-associated antigen (MAFA). *Mol. Immunol.* **38**, 1307–1313.
208. Wu, J., Song, Y., Bakker, A. B., Bauer, S., Spies, T., Lanier, L. L., and Phillips, J. H. (1999). An activating immunoreceptor complex formed by NKG2D and DAP10. *Science* **285**, 730–732.
209. Dann, S. M., Wang, H. C., Gambarin, K. J., Actor, J. K., Robinson, P., Lewis, D. E., Caillat-Zucman, S., and White, A. C., Jr. (2005). Interleukin-15 activates human natural killer cells to clear the intestinal protozoan cryptosporidium. *J. Infect. Dis.* **192**, 1294–1302.
210. Bauer, S., Groh, V., Wu, J., Steinle, A., Phillips, J. H., Lanier, L. L., and Spies, T. (1999). Activation of NK cells and T cells by NKG2D, a receptor for stress-inducible MICA. *Science* **285**, 727–729.
211. Cosman, D., Mullberg, J., Sutherland, C. L., Chin, W., Armitage, R., Fanslow, W., Kubin, M., and Chalupny, N. J. (2001). ULBPs, novel MHC class I-related molecules, bind to CMV glycoprotein UL16 and stimulate NK cytotoxicity through the NKG2D receptor. *Immunity* **14**, 123–133.
212. Radosavljevic, M., Cuillerier, B., Wilson, M. J., Clement, O., Wicker, S., Gilfillan, S., Beck, S., Trowsdale, J., and Bahram, S. (2002). A cluster of ten novel MHC class I related genes on human chromosome 6q24.2-q25.3. *Genomics* **79**, 114–123.
213. Groh, V., Bahram, S., Bauer, S., Herman, A., Beauchamp, M., and Spies, T. (1996). Cell stress-regulated human major histocompatibility complex class I gene expressed in gastrointestinal epithelium. *Proc. Natl. Acad. Sci. USA* **93**, 12445–12450.
214. Groh, V., Rhinehart, R., Secrist, H., Bauer, S., Grabstein, K. H., and Spies, T. (1999). Broad tumor-associated expression and recognition by tumor-derived gamma delta T cells of MICA and MICB. *Proc. Natl. Acad. Sci. USA* **96**, 6879–6884.
215. Jinushi, M., Takehara, T., Tatsumi, T., Hiramatsu, N., Sakamori, R., Yamaguchi, S., and Hayashi, N. (2005). Impairment of natural killer cell and dendritic cell functions by the soluble form of MHC class I-related chain A in advanced human hepatocellular carcinomas. *J. Hepatol.* **43**, 1013–1020.
216. Pende, D., Cantoni, C., Rivera, P., Vitale, M., Castriconi, R., Marcenaro, S., Nanni, M., Biassoni, R., Bottino, C., Moretta, A., and Moretta, L. (2001). Role of NKG2D in tumor cell lysis mediated by human NK cells: Cooperation with natural cytotoxicity receptors and capability of recognizing tumors of nonepithelial origin. *Eur. J. Immunol.* **31**, 1076–1086.
217. Kubin, M., Cassiano, L., Chalupny, J., Chin, W., Cosman, D., Fanslow, W., Mullberg, J., Rousseau, A. M., Ulrich, D., and Armitage, R. (2001). ULBP1, 2, 3: Novel MHC class I-related molecules that bind to human cytomegalovirus glycoprotein UL16, activate NK cells. *Eur. J. Immunol.* **31**, 1428–1437.
218. Li, P., Morris, D. L., Willcox, B. E., Steinle, A., Spies, T., and Strong, R. K. (2001). Complex structure of the activating immunoreceptor NKG2D and its MHC class I-like ligand MICA. *Nat. Immunol.* **2**, 443–451.
219. Steinle, A., Li, P., Morris, D. L., Groh, V., Lanier, L. L., Strong, R. K., and Spies, T. (2001). Interactions of human NKG2D with its ligands MICA, MICB, and homologs of the mouse RAE-1 protein family. *Immunogenetics* **53**, 279–287.
220. Carbone, E., Neri, P., Mesuraca, M., Fulciniti, M. T., Otsuki, T., Pende, D., Groh, V., Spies, T., Pollio, G., Cosman, D., Catalano, L., Tassone, P. *et al.* (2005). HLA class I, NKG2D, and natural cytotoxicity receptors regulate multiple myeloma cell recognition by natural killer cells. *Blood* **105**, 251–258.
221. Chalupny, J., Cosman, D., Mullberg, J., Chin, W., Cassiano, L., Means, G., Derry, J., Russell, C., Armitage, R., and Sutherland, C. (2000). Soluble forms of the novel, MHC class I-related

molecules, ULBP1 and ULBP2, bind to, and functionally activate NK cells. *FASEB J.* **14**, A1018.
222. Cerwenka, A., and Lanier, L. L. (2001). Natural killer cells, viruses and cancer. *Nat. Rev. Immunol.* **1**, 41–49.
223. Diefenbach, A., Hsia, J. K., Hsiung, M. Y., and Raulet, D. H. (2003). A novel ligand for the NKG2D receptor activates NK cells and macrophages and induces tumor immunity. *Eur. J. Immunol.* **33**, 381–391.
224. Pende, D., Rivera, P., Marcenaro, S., Chang, C. C., Biassoni, R., Conte, R., Kubin, M., Cosman, D., Ferrone, S., Moretta, L., and Moretta, A. (2002). Major histocompatibility complex class I-related chain A and UL16-binding protein expression on tumor cell lines of different histotypes: Analysis of tumor susceptibility to NKG2D-dependent natural killer cell cytotoxicity. *Cancer Res.* **62**, 6178–6186.
225. McFarland, B. J., Kortemme, T., Yu, S. F., Baker, D., and Strong, R. K. (2003). Symmetry recognizing asymmetry: Analysis of the interactions between the C-type lectin-like immunoreceptor NKG2D and MHC class I-like ligands. *Structure (Camb)* **11**, 411–422.
226. Strong, R. K. (2002). Asymmetric ligand recognition by the activating natural killer cell receptor NKG2D, a symmetric homodimer. *Mol. Immunol.* **38**, 1029–1037.
227. Radaev, S., Rostro, B., Brooks, A. G., Colonna, M., and Sun, P. D. (2001). Conformational plasticity revealed by the cocrystal structure of NKG2D and its class I MHC-like ligand ULBP3. *Immunity* **15**, 1039–1049.
228. Raulet, D. H. (2003). Roles of the NKG2D immunoreceptor and its ligands. *Nat. Rev. Immunol.* **3**, 781–790.
229. Olcese, L., Cambiaggi, A., Semenzato, G., Bottino, C., Moretta, A., and Vivier, E. (1997). Human killer cell activatory receptors for MHC class I molecules are included in a multimeric complex expressed by natural killer cells. *J. Immunol.* **158**, 5083–5086.
230. Tomasello, E., and Vivier, E. (2005). KARAP/DAP12/TYROBP: Three names and a multiplicity of biological functions. *Eur. J. Immunol.* **35**, 1670–1677.
231. Lanier, L. L., Corliss, B. C., Wu, J., Leong, C., and Phillips, J. H. (1998). Immunoreceptor DAP12 bearing a tyrosine-based activation motif is involved in activating NK cells. *Nature* **391**, 703–707.
232. Rosenberg, S. A., and Lotze, M. T. (1986). Cancer immunotherapy using interleukin-2 and interleukin-2-activated lymphocytes. *Annu. Rev. Immunol.* **4**, 681–709.
233. Lanier, L. L. (2000). Turning on natural killer cells. *J. Exp. Med.* **191**, 1259–1262.
234. Bottino, C., Falco, M., Sivori, S., Moretta, L., Moretta, A., and Biassoni, R. (2000). Identification and molecular characterization of a natural mutant of the p50.2/KIR2DS2 activating NK receptor that fails to mediate NK cell triggering. *Eur. J. Immunol.* **30**, 3569–3574.
235. McVicar, D. W., Taylor, L. S., Gosselin, P., Willette-Brown, J., Mikhael, A. I., Geahlen, R. L., Nakamura, M. C., Linnemeyer, P., Seaman, W. E., Anderson, S. K., Ortaldo, J. R., and Mason, L. H. (1998). DAP12-mediated signal transduction in natural killer cells. A dominant role for the Syk protein-tyrosine kinase. *J. Biol. Chem.* **273**, 32934–32942.
236. Bakker, A. B., Hoek, R. M., Cerwenka, A., Blom, B., Lucian, L., McNeil, T., Murray, R., Phillips, L. H., Sedgwick, J. D., and Lanier, L. L. (2000). DAP12-deficient mice fail to develop autoimmunity due to impaired antigen priming. *Immunity* **13**, 345–353.
237. Tomasello, E., Desmoulins, P. O., Chemin, K., Guia, S., Cremer, H., Ortaldo, J., Love, P., Kaiserlian, D., and Vivier, E. (2000). Combined natural killer cell and dendritic cell functional deficiency in KARAP/DAP12 loss-of-function mutant mice. *Immunity* **13**, 355–364.
238. Paloneva, J., Kestila, M., Wu, J., Salminen, A., Bohling, T., Ruotsalainen, V., Hakola, P., Bakker, A. B., Phillips, J. H., Pekkarinen, P., Lanier, L. L., Timonen, T. *et al.* (2000). Loss-of-function mutations in TYROBP (DAP12) result in a presenile dementia with bone cysts. *Nat. Genet.* **25**, 357–361.

239. Wu, J., Cherwinski, H., Spies, T., Phillips, J. H., and Lanier, L. L. (2000). DAP10 and DAP12 form distinct, but functionally cooperative, receptor complexes in natural killer cells. *J. Exp. Med.* **192**, 1059–1068.
240. Billadeau, D. D., Upshaw, J. L., Schoon, R. A., Dick, C. J., and Leibson, P. J. (2003). NKG2D-DAP10 triggers human NK cell-mediated killing via a Syk-independent regulatory pathway. *Nat. Immunol.* **4**, 557–564.
241. Kuster, H., Thompson, H., and Kinet, J. P. (1990). Characterization and expression of the gene for the human Fc receptor gamma subunit. Definition of a new gene family. *J. Biol. Chem.* **265**, 6448–6452.
242. Orloff, D. G., Ra, C. S., Frank, S. J., Klausner, R. D., and Kinet, J. P. (1990). Family of disulphide-linked dimers containing the zeta and eta chains of the T-cell receptor and the gamma chain of Fc receptors. *Nature* **347**, 189–191.
243. Taylor, N., Jahn, T., Smith, S., Lamkin, T., Uribe, L., Liu, Y., Durden, D. L., and Weinberg, K. (1997). Differential activation of the tyrosine kinases ZAP-70 and Syk after Fc gamma RI stimulation. *Blood* **89**, 388–396.
244. Weissman, A. M., Samelson, L. E., and Klausner, R. D. (1986). A new subunit of the human T-cell antigen receptor complex. *Nature* **324**, 480–482.
245. Caplan, S., Zeliger, S., Wang, L., and Baniyash, M. (1995). Cell-surface-expressed T-cell antigen-receptor zeta chain is associated with the cytoskeleton. *Proc. Natl. Acad. Sci. USA* **92**, 4768–4772.
246. Matsuda, M., Petersson, M., Lenkei, R., Taupin, J. L., Magnusson, I., Mellstedt, H., Anderson, P., and Kiessling, R. (1995). Alterations in the signal-transducing molecules of T cells and NK cells in colorectal tumor-infiltrating, gut mucosal and peripheral lymphocytes: Correlation with the stage of the disease. *Int. J. Cancer* **61**, 765–772.
247. Tsan, M. F. (2005). Toll-like receptors, inflammation and cancer. *Semin. Cancer Biol.* **16**, 32–37.
248. Wende, H., Colonna, M., Ziegler, A., and Volz, A. (1999). Organization of the leukocyte receptor cluster (LRC) on human chromosome 19q13.4. *Mamm. Genome* **10**, 154–160.
249. Torkar, M., Haude, A., Milne, S., Beck, S., Trowsdale, J., and Wilson, M. J. (2000). Arrangement of the ILT gene cluster: A common null allele of the ILT6 gene results from a 6.7-kbp deletion. *Eur. J. Immunol.* **30**, 3655–3662.
250. Trowsdale, J., Barten, R., Haude, A., Stewart, C. A., Beck, S., and Wilson, M. J. (2001). The genomic context of natural killer receptor extended gene families. *Immunol. Rev.* **181**, 20–38.
251. Martin, A. M., Kulski, J. K., Witt, C., Pontarotti, P., and Christiansen, F. T. (2002). Leukocyte Ig-like receptor complex (LRC) in mice and men. *Trends Immunol.* **23**, 81–88.
252. Martin, A. M., Freitas, E. M., Witt, C. S., and Christiansen, F. T. (2000). The genomic organization and evolution of the natural killer immunoglobulin-like receptor (KIR) gene cluster. *Immunogenetics* **51**, 268–280.
253. Martin, A. M., Kulski, J. K., Gaudieri, S., Witt, C. S., Freitas, E. M., Trowsdale, J., and Christiansen, F. T. (2004). Comparative genomic analysis, diversity and evolution of two KIR haplotypes A and B. *Gene* **335**, 121–131.
254. Kelley, J., and Trowsdale, J. (2005). Features of MHC and NK gene clusters. *Transpl. Immunol.* **14**, 129–134.
255. Sambrook, J. G., Bashirova, A., Palmer, S., Sims, S., Trowsdale, J., Abi-Rached, L., Parham, P., Carrington, M., and Beck, S. (2005). Single haplotype analysis demonstrates rapid evolution of the killer immunoglobulin-like receptor (KIR) loci in primates. *Genome Res.* **15**, 25–35.
256. Khakoo, S. I., Rajalingam, R., Shum, B. P., Weidenbach, K., Flodin, L., Muir, D. G., Canavez, F., Cooper, S. L., Valiante, N. M., Lanier, L. L., and Parham, P. (2000). Rapid evolution of NK cell receptor systems demonstrated by comparison of chimpanzees and humans. *Immunity* **12**, 687–698.

257. Rajalingam, R., Parham, P., and Abi-Rached, L. (2004). Domain shuffling has been the main mechanism forming new hominoid killer cell Ig-like receptors. *J. Immunol.* **172**, 356–369.
258. Norman, P. J., Cook, M. A., Carey, B. S., Carrington, C. V., Verity, D. H., Hameed, K., Ramdath, D. D., Chandanayingyong, D., Leppert, M., Stephens, H. A., and Vaughan, R. W. (2004). SNP haplotypes and allele frequencies show evidence for disruptive and balancing selection in the human leukocyte receptor complex. *Immunogenetics* **56**, 225–237.
259. Volz, A., Wende, H., Laun, K., and Ziegler, A. (2001). Genesis of the ILT/LIR/MIR clusters within the human leukocyte receptor complex. *Immunol. Rev.* **181**, 39–51.
260. Canavez, F., Young, N. T., Guethlein, L. A., Rajalingam, R., Khakoo, S. I., Shum, B. P., and Parham, P. (2001). Comparison of chimpanzee and human leukocyte Ig-like receptor genes reveals framework and rapidly evolving genes. *J. Immunol.* **167**, 5786–5794.
261. Kelley, J., Walter, L., and Trowsdale, J. (2005). Comparative genomics of natural killer cell receptor gene clusters. *PLoS Genet.* **1**, e27.
262. Laun, K., Coggill, P., Palmer, S., Sims, S., Ning, Z., Ragoussis, J., Volpi, E., Wilson, N., Beck, S., Ziegler, A., and Volz, A. (2006). The leukocyte receptor complex in chicken is characterized by massive expansion and diversification of immunoglobulin-like loci. *PLoS Genet.* in press.
263. Ambagala, A. P., Solheim, J. C., and Srikumaran, S. (2005). Viral interference with MHC class I antigen presentation pathway: The battle continues. *Vet. Immunol. Immunopathol.* **107**, 1–15.
264. Arase, H., and Lanier, L. L. (2004). Specific recognition of virus-infected cells by paired NK receptors. *Rev. Med. Virol.* **14**, 83–93.
265. Arase, H., Mocarski, E. S., Campbell, A. E., Hill, A. B., and Lanier, L. L. (2002). Direct recognition of cytomegalovirus by activating and inhibitory NK cell receptors. *Science* **296**, 1323–1326.
266. Smith, H. R., Heusel, J. W., Mehta, I. K., Kim, S., Dorner, B. G., Naidenko, O. V., Iizuka, K., Furukawa, H., Beckman, D. L., Pingel, J. T., Scalzo, A. A., Fremont, D. H. *et al.* (2002). Recognition of a virus-encoded ligand by a natural killer cell activation receptor. *Proc. Natl. Acad. Sci. USA* **99**, 8826–8831.
267. Boles, K. S., Stepp, S. E., Bennett, M., Kumar, V., and Mathew, P. A. (2001). 2B4 (CD244) and CS1: Novel members of the CD2 subset of the immunoglobulin superfamily molecules expressed on natural killer cells and other leukocytes. *Immunol. Rev.* **181**, 234–249.
268. Falco, M., Biassoni, R., Bottino, C., Vitale, M., Sivori, S., Augugliaro, R., Moretta, L., and Moretta, A. (1999). Identification and molecular cloning of p75/AIRM1, a novel member of the sialoadhesin family that functions as an inhibitory receptor in human natural killer cells. *J. Exp. Med.* **190**, 793–802.
269. Penack, O., Gentilini, C., Fischer, L., Asemissen, A. M., Scheibenbogen, C., Thiel, E., and Uharek, L. (2005). CD56dimCD16neg cells are responsible for natural cytotoxicity against tumor targets. *Leukemia* **19**, 835–840.
270. Loza, M. J., and Perussia, B. (2004). The IL-12 signature: NK cell terminal CD56+high stage and effector functions. *J. Immunol.* **172**, 88–96.
271. Saleh, A., Davies, G. E., Pascal, V., Wright, P. W., Hodge, D. L., Cho, E. H., Lockett, S. J., Abshari, M., and Anderson, S. K. (2004). Identification of probabilistic transcriptional switches in the Ly49 gene cluster: A eukaryotic mechanism for selective gene activation. *Immunity* **21**, 55–66.
272. Santourlidis, S., Trompeter, H. I., Weinhold, S., Eisermann, B., Meyer, K. L., Wernet, P., and Uhrberg, M. (2002). Crucial role of DNA methylation in determination of clonally distributed killer cell Ig-like receptor expression patterns in NK cells. *J. Immunol.* **169**, 4253–4261.
273. Dixon, A. K., Richardson, P. J., Lee, K., Carter, N. P., and Freeman, T. C. (1998). Expression profiling of single cells using 3 prime end amplification (TPEA) PCR. *Nucleic Acids Res.* **26**, 4426–4431.

274. Husain, Z., Alper, C. A., Yunis, E. J., and Dubey, D. P. (2002). Complex expression of natural killer receptor genes in single natural killer cells. *Immunology* **106**, 373–380.
275. Radeloff, B., Laun, K., Zirra, M., and Volz, A. (2003). Killer immunoglobulin-like receptor expression on single cells: A cautionary note. *Immunology* **110**, 421–426.
276. Heams, T., and Kupiec, J. J. (2003). Modified 3′-end amplification PCR for gene expression analysis in single cells. *Biotechniques* **34**, 712–714, 716.
277. Roozemond, R. C. (1976). Ultramicrochemical determination of nucleic acids in individual cells using the Zeiss UMSP-I microspectrophotometer. Application to isolated rat hepatocytes of different ploidy classes. *Histochem. J.* **8**, 625–638.
278. Uemura, E. (1980). Age-related changes in neuronal RNA content in rhesus monkeys (Macaca mulatta). *Brain Res. Bull.* **5**, 117–119.
279. Carter, M. G., Sharov, A. A., VanBuren, V., Dudekula, D. B., Carmack, C. E., Nelson, C., and Ko, M. S. (2005). Transcript copy number estimation using a mouse whole-genome oligonucleotide microarray. *Genome Biol.* **6**, R61.
280. Hastie, N. D., and Bishop, J. O. (1976). The expression of three abundance classes of messenger RNA in mouse tissues. *Cell* **9**, 761–774.
281. Holland, M. J. (2002). Transcript abundance in yeast varies over six orders of magnitude. *J. Biol. Chem.* **277**, 14363–14366.
282. Kawasaki, E. S. (2004). Microarrays and the gene expression profile of a single cell. *Ann. N. Y. Acad. Sci.* **1020**, 92–100.
283. Brady, G., Billia, F., Knox, J., Hoang, T., Kirsch, I. R., Voura, E. B., Hawley, R. G., Cumming, R., Buchwald, M., and Siminovitch, K. (1995). Analysis of gene expression in a complex differentiation hierarchy by global amplification of cDNA from single cells. *Curr. Biol.* **5**, 909–922.
284. Brail, L. H., Jang, A., Billia, F., Iscove, N. N., Klamut, H. J., and Hill, R. P. (1999). Gene expression in individual cells: Analysis using global single cell reverse transcription polymerase chain reaction (GSC RT-PCR). *Mutat. Res.* **406**, 45–54.
285. Gustincich, S., Contini, M., Gariboldi, M., Puopolo, M., Kadota, K., Bono, H., LeMieux, J., Walsh, P., Carninci, P., Hayashizaki, Y., Okazaki, Y., and Raviola, E. (2004). Gene discovery in genetically labeled single dopaminergic neurons of the retina. *Proc. Natl. Acad. Sci. USA* **101**, 5069–5074.
286. Nitin, N., Santangelo, P. J., Kim, G., Nie, S., and Bao, G. (2004). Peptide-linked molecular beacons for efficient delivery and rapid mRNA detection in living cells. *Nucleic Acids Res.* **32**, e58.
287. Santangelo, P. J., Nix, B., Tsourkas, A., and Bao, G. (2004). Dual FRET molecular beacons for mRNA detection in living cells. *Nucleic Acids Res.* **32**, e57.
288. Radeloff, B., Nagler, L., Zirra, M., Ziegler, A., and Volz, A. (2005). Specific amplification of cDNA ends (SPACE): A new tool for the analysis of rare transcripts and its application for the promoter analysis of killer cell receptor genes. *DNA Seq.* **16**, 44–52.
289. Yodoi, J., Teshigawara, K., Nikaido, T., Fukui, K., Noma, T., Honjo, T., Takigawa, M., Sasaki, M., Minato, N., and Tsudo, M. (1985). TCGF (IL 2)-receptor inducing factor(s). I. Regulation of IL 2 receptor on a natural killer-like cell line (YT cells). *J. Immunol.* **134**, 1623–1630.
290. Femino, A. M., Fay, F. S., Fogarty, K., and Singer, R. H. (1998). Visualization of single RNA transcripts in situ. *Science* **280**, 585–590.
291. Blaschitz, A., Juch, H., Volz, A., Hutter, H., Daxboeck, C., Desoye, G., and Dohr, G. (2005). The soluble pool of HLA-G produced by human trophoblasts does not include detectable levels of the intron 4-containing HLA-G5 and HLA-G6 isoforms. *Mol. Hum. Reprod.* **11**, 699–710.
292. Romero, A. I., Thoren, F. B., Brune, M., and Hellstrand, K. (2006). NKp46 and NKG2D receptor expression in NK cells with CD56 and CD56 phenotype: Regulation by histamine and reactive oxygen species. *Br. J. Haematol.* **132**, 91–98.

293. Ostrowski, S. R., Ullum, H., Pedersen, B. K., Gerstoft, J., and Katzenstein, T. L. (2005). 2B4 expression on natural killer cells increases in HIV-1 infected patients followed prospectively during highly active antiretroviral therapy. *Clin. Exp. Immunol.* **141**, 526–533.
294. Kubota, A., Kubota, S., Lohwasser, S., Mager, D. L., and Takei, F. (1999). Diversity of NK cell receptor repertoire in adult and neonatal mice. *J. Immunol.* **163**, 212–216.
295. Williams, N. S., Kubota, A., Bennett, M., Kumar, V., and Takei, F. (2000). Clonal analysis of NK cell development from bone marrow progenitors *in vitro*: Orderly acquisition of receptor gene expression. *Eur. J. Immunol.* **30**, 2074–2082.
296. Belkin, D., Torkar, M., Chang, C., Barten, R., Tolaini, M., Haude, A., Allen, R., Wilson, M. J., Kioussis, D., and Trowsdale, J. (2003). Killer cell Ig-like receptor and leukocyte Ig-like receptor transgenic mice exhibit tissue- and cell-specific transgene expression. *J. Immunol.* **171**, 3056–3063.
297. Chan, H. W., Kurago, Z. B., Stewart, C. A., Wilson, M. J., Martin, M. P., Mace, B. E., Carrington, M., Trowsdale, J., and Lutz, C. T. (2003). DNA methylation maintains allele-specific KIR gene expression in human natural killer cells. *J. Exp. Med.* **197**, 245–255.
298. Chan, H. W., Miller, J. S., Moore, M. B., and Lutz, C. T. (2005). Epigenetic control of highly homologous killer Ig-like receptor gene alleles. *J. Immunol.* **175**, 5966–5974.
299. Shilling, H. G., Young, N., Guethlein, L. A., Cheng, N. W., Gardiner, C. M., Tyan, D., and Parham, P. (2002). Genetic control of human NK cell repertoire. *J. Immunol.* **169**, 239–247.
300. Vilches, C., Gardiner, C. M., and Parham, P. (2000). Gene structure and promoter variation of expressed and nonexpressed variants of the KIR2DL5 gene. *J. Immunol.* **165**, 6416–6421.
301. Trompeter, H. I., Gomez-Lozano, N., Santourlidis, S., Eisermann, B., Wernet, P., Vilches, C., and Uhrberg, M. (2005). Three structurally and functionally divergent kinds of promoters regulate expression of clonally distributed killer cell Ig-like receptors (KIR), of KIR2DL4, and of KIR3DL3. *J. Immunol.* **174**, 4135–4143.
302. Van Bergen, J., Stewart, C. A., van den Elsen, P. J., and Trowsdale, J. (2005). Structural and functional differences between the promoters of independently expressed killer cell Ig-like receptors. *Eur. J. Immunol.* **35**, 2191–2199.
303. Xu, J., Vallejo, A. N., Jiang, Y., Weyand, C. M., and Goronzy, J. J. (2005). Distinct transcriptional control mechanisms of killer immunoglobulin-like receptors in natural killer (NK) and in T cells. *J. Biol. Chem.* **280**, 24277–24285.
304. Scott-Algara, D., and Paul, P. (2002). NK cells and HIV infection: Lessons from other viruses. *Curr. Mol. Med.* **2**, 757–768.
305. Schwartz, O., Marechal, V., Le Gall, S., Lemonnier, F., and Heard, J. M. (1996). Endocytosis of major histocompatibility complex class I molecules is induced by the HIV-1 Nef protein. *Nat. Med.* **2**, 338–342.
306. Garrido, F., Ruiz-Cabello, F., Cabrera, T., Perez-Villar, J. J., Lopez-Botet, M., Duggan-Keen, M., and Stern, P. L. (1997). Implications for immunosurveillance of altered HLA class I phenotypes in human tumours. *Immunol. Today* **18**, 89–95.
307. Garrido, F. (1996). HLA and cancer. *Tissue Antigens* **47**, 361–363.
308. Fujimiya, Y., Bakke, A., Chang, W. C., Linker-Israeli, M., Udis, B., Horwitz, D., and Pattengale, P. K. (1986). Natural killer-cell immunodeficiency in patients with chronic myelogenous leukemia. I. Analysis of the defect using the monoclonal antibodies HNK-1 (LEU-7) and B73.1. *Int. J. Cancer* **37**, 639–649.
309. Costello, R. T., Sivori, S., Marcenaro, E., Lafage-Pochitaloff, M., Mozziconacci, M. J., Reviron, D., Gastaut, J. A., Pende, D., Olive, D., and Moretta, A. (2002). Defective expression and function of natural killer cell-triggering receptors in patients with acute myeloid leukemia. *Blood* **99**, 3661–3667.
310. Miller, J. S. (2002). Biology of natural killer cells in cancer and infection. *Cancer Invest.* **20**, 405–419.

311. Hsu, K. C., and Dupont, B. (2005). Natural killer cell receptors: Regulating innate immune responses to hematologic malignancy. *Semin. Hematol.* **42,** 91–103.
312. Tanaka, J., Asaka, M., and Imamura, M. (2005). Potential role of natural killer cell receptor-expressing cells in immunotherapy for leukemia. *Int. J. Hematol.* **81,** 6–12.
313. Salih, H. R., Antropius, H., Gieseke, F., Lutz, S. Z., Kanz, L., Rammensee, H. G., and Steinle, A. (2003). Functional expression and release of ligands for the activating immunoreceptor NKG2D in leukemia. *Blood* **102,** 1389–1396.
314. Vetter, C. S., Groh, V., Spies, S. P., thor, T., Brocker, E. B., and Becker, J. C. (2002). Expression of stress-induced MHC class I related chain molecules on human melanoma. *J. Invest. Dermatol.* **118,** 600–605.
315. Jinushi, M., Takehara, T., Tatsumi, T., Kanto, T., Groh, V., Spies, T., Kimura, R., Miyagi, T., Mochizuki, K., Sasaki, Y., and Hayashi, N. (2003). Expression and role of MICA and MICB in human hepatocellular carcinomas and their regulation by retinoic acid. *Int. J. Cancer* **104,** 354–361.
316. Gasser, S., Orsulic, S., Brown, E. J., and Raulet, D. H. (2005). The DNA damage pathway regulates innate immune system ligands of the NKG2D receptor. *Nature* **436,** 1186–1190.
317. Tieng, V., Le Bouguenec, C., Bertheau, M. L., du, P., Desreumaux, P., Janin, A., Charron, D., and Toubert, A. (2002). Binding of *Escherichia coli* adhesin AfaE to CD55 triggers cell-surface expression of the MHC class I-related molecule MICA. *Proc. Natl. Acad. Sci. USA* **99,** 2977–2982.
318. Cerwenka, A., Baron, J. L., and Lanier, L. L. (2001). Ectopic expression of retinoic acid early inducible-1 gene (RAE-1) permits natural killer cell-mediated rejection of a MHC class I-bearing tumor *in vivo*. *Proc. Natl. Acad. Sci. USA* **98,** 11521–11526.
319. Friese, M. A., Platten, M., Lutz, S. Z., Naumann, U., Aulwurm, S., Bischof, F., Buhring, H. J., Dichgans, J., Rammensee, H. G., Steinle, A., and Weller, M. (2003). MICA/NKG2D-mediated immunogene therapy of experimental gliomas. *Cancer Res.* **63,** 8996–9006.
320. Salih, H. R., Rammensee, H. G., and Steinle, A. (2002). Cutting edge: Down-regulation of MICA on human tumors by proteolytic shedding. *J. Immunol.* **169,** 4098–4102.
321. Wu, J. D., Higgins, L. M., Steinle, A., Cosman, D., Haugk, K., and Plymate, S. R. (2004). Prevalent expression of the immunostimulatory MHC class I chain-related molecule is counteracted by shedding in prostate cancer. *J. Clin. Invest.* **114,** 560–568.
322. Groh, V., Wu, J., Yee, C., and Spies, T. (2002). Tumour-derived soluble MIC ligands impair expression of NKG2D and T-cell activation. *Nature* **419,** 734–738.
323. Oppenheim, D. E., Roberts, S. J., Clarke, S. L., Filler, R., Lewis, J. M., Tigelaar, R. E., Girardi, M., and Hayday, A. C. (2005). Sustained localized expression of ligand for the activating NKG2D receptor impairs natural cytotoxicity *in vivo* and reduces tumor immunosurveillance. *Nat. Immunol.* **6,** 928–937.
324. Raffaghello, L., Prigione, I., Airoldi, I., Camoriano, M., Levreri, I., Gambini, C., Pende, D., Steinle, A., Ferrone, S., and Pistoia, V. (2004). Downregulation and/or release of NKG2D ligands as immune evasion strategy of human neuroblastoma. *Neoplasia* **6,** 558–568.
325. Doubrovina, E. S., Doubrovin, M. M., Vider, E., Sisson, R. B., O'Reilly, R. J., Dupont, B., and Vyas, Y. M. (2003). Evasion from NK cell immunity by MHC class I chain-related molecules expressing colon adenocarcinoma. *J. Immunol.* **171,** 6891–6899.
326. Collins, R. W. (2004). Human MHC class I chain related (MIC) genes: Their biological function and relevance to disease and transplantation. *Eur. J. Immunogenet.* **31,** 105–114.
327. Germain, C., Larbouret, C., Cesson, V., Donda, A., Held, W., Mach, J. P., Pelegrin, A., and Robert, B. (2005). MHC class I-related chain a conjugated to antitumor antibodies can sensitize tumor cells to specific lysis by natural killer cells. *Clin. Cancer Res.* **11,** 7516–7522.

328. Lee, J. C., Lee, K. M., Kim, D. W., and Heo, D. S. (2004). Elevated TGF-beta1 secretion and down-modulation of NKG2D underlies impaired NK cytotoxicity in cancer patients. *J. Immunol.* **172,** 7335–7340.
329. Ghiringhelli, F., Menard, C., Terme, M., Flament, C., Taieb, J., Chaput, N., Puig, P. E., Novault, S., Escudier, B., Vivier, E., Lecesne, A., Robert, C. *et al.* (2005). CD4+CD25+ regulatory T cells inhibit natural killer cell functions in a transforming growth factor-(beta)-dependent manner. *J. Exp. Med.* **202,** 1075–1085.
330. Bertone, S., Schiavetti, F., Bellomo, R., Vitale, C., Ponte, M., Moretta, L., and Mingari, M. C. (1999). Transforming growth factor-beta-induced expression of CD94/NKG2A inhibitory receptors in human T lymphocytes. *Eur. J. Immunol.* **29,** 23–29.
331. Hasegawa, Y., Takanashi, S., Kanehira, Y., Tsushima, T., Imai, T., and Okumura, K. (2001). Transforming growth factor-beta1 level correlates with angiogenesis, tumor progression, and prognosis in patients with nonsmall cell lung carcinoma. *Cancer* **91,** 964–971.
332. Kong, F. M., Washington, M. K., Jirtle, R. L., and Anscher, M. S. (1996). Plasma transforming growth factor-beta 1 reflects disease status in patients with lung cancer after radiotherapy: A possible tumor marker. *Lung Cancer* **16,** 47–59.
333. Rook, A. H., Kehrl, J. H., Wakefield, L. M., Roberts, A. B., Sporn, M. B., Burlington, D. B., Lane, H. C., and Fauci, A. S. (1986). Effects of transforming growth factor beta on the functions of natural killer cells: Depressed cytolytic activity and blunting of interferon responsiveness. *J. Immunol.* **136,** 3916–3920.
334. Bellone, G., Aste-Amezaga, M., Trinchieri, G., and Rodeck, U. (1995). Regulation of NK cell functions by TGF-beta 1. *J. Immunol.* **155,** 1066–1073.
335. Le Bouteiller, P., and Blaschitz, A. (1999). The functionality of HLA-G is emerging. *Immunol. Rev.* **167,** 233–244.
336. Ristich, V., Liang, S., Zhang, W., Wu, J., and Horuzsko, A. (2005). Tolerization of dendritic cells by HLA-G. *Eur. J. Immunol.* **35,** 1133–1142.
337. Hunt, J. S., Petroff, M. G., McIntire, R. H., and Ober, C. (2005). HLA-G and immune tolerance in pregnancy. *FASEB J.* **19,** 681–693.
338. Fournel, S., Aguerre-Girr, M., Huc, X., Lenfant, F., Alam, A., Toubert, A., Bensussan, A., and Le Bouteiller, P. (2000). Cutting edge: Soluble HLA-G1 triggers CD95/CD95 ligand-mediated apoptosis in activated CD8+ cells by interacting with CD8. *J. Immunol.* **164,** 6100–6104.
339. Lemaoult, J., Rouas-Freiss, N., and Carosella, E. D. (2005). HLA-G5 expression by trophoblast cells: The facts. *Mol. Hum. Reprod.* **11,** 719–722.
340. Urosevic, M., Trojan, A., and Dummer, R. (2002). HLA-G and its KIR ligands in cancer – another enigma yet to be solved? *J. Pathol.* **196,** 252–253.
341. Lefebvre, S., Antoine, M., Uzan, S., McMaster, M., Dausset, J., Carosella, E. D., and Paul, P. (2002). Specific activation of the non-classical class I histocompatibility HLA-G antigen and expression of the ILT2 inhibitory receptor in human breast cancer. *J. Pathol.* **196,** 266–274.
342. Forte, P., Pazmany, L., Matter-Reissmann, U. B., Stussi, G., Schneider, M. K., and Seebach, J. D. (2001). HLA-G inhibits rolling adhesion of activated human NK cells on porcine endothelial cells. *J. Immunol.* **167,** 6002–6008.
343. Poggi, A., Massaro, A. M., Negrini, S., Contini, P., and Zocchi, M. R. (2005). Tumor-induced apoptosis of human IL-2-activated NK cells: Role of natural cytotoxicity receptors. *J. Immunol.* **174,** 2653–2660.
344. Trinchieri, G. (1989). Biology of natural killer cells. *Adv. Immunol.* **47,** 187–376.
345. Furuke, K., Shiraishi, M., Mostowski, H. S., and Bloom, E. T. (1999). Fas ligand induction in human NK cells is regulated by redox through a calcineurin-nuclear factors of activated T cell-dependent pathway. *J. Immunol.* **162,** 1988–1993.

346. Ida, H., and Anderson, P. (1998). Activation-induced NK cell death triggered by CD2 stimulation. *Eur. J. Immunol.* **28**, 1292–1300.
347. Baniyash, M. (2004). TCR zeta-chain downregulation: Curtailing an excessive inflammatory immune response. *Nat. Rev. Immunol.* **4**, 675–687.
348. Rodriguez, P. C., Zea, A. H., DeSalvo, J., Culotta, K. S., Zabaleta, J., Quiceno, D. G., Ochoa, J. B., and Ochoa, A. C. (2003). L-Arginine consumption by macrophages modulates the expression of CD3 zeta chain in T lymphocytes. *J. Immunol.* **171**, 1232–1239.
349. Wu, C. W., Chi, C. W., Lin, E. C., Lui, W. Y., P'eng, F. K., and Wang, S. R. (1994). Serum arginase level in patients with gastric cancer. *J. Clin. Gastroenterol.* **18**, 84–85.
350. Finke, J. H., Zea, A. H., Stanley, J., Longo, D. L., Mizoguchi, H., Tubbs, R. R., Wiltrout, R. H., O'Shea, J. J., Kudoh, S., and Klein, E. (1993). Loss of T-cell receptor zeta chain and p56lck in T-cells infiltrating human renal cell carcinoma. *Cancer Res.* **53**, 5613–5616.
351. Ungefroren, H., Voss, M., Bernstorff, W. V., Schmid, A., Kremer, B., and Kalthoff, H. (1999). Immunological escape mechanisms in pancreatic carcinoma. *Ann. N. Y. Acad. Sci.* **880**, 243–251.
352. Frydecka, I., Kaczmarek, P., Bocko, D., Kosmaczewska, A., Morilla, R., and Catovsky, D. (1999). Expression of signal-transducing zeta chain in peripheral blood T cells and natural killer cells in patients with Hodgkin's disease in different phases of the disease. *Leuk Lymphoma* **35**, 545–554.
353. Biron, C. A., Nguyen, K. B., Pien, G. C., Cousens, L. P., and Salazar-Mather, T. P. (1999). Natural killer cells in antiviral defense: Function and regulation by innate cytokines. *Annu. Rev. Immunol.* **17**, 189–220.
354. Hengel, H., Brune, W., and Koszinowski, U. H. (1998). Immune evasion by cytomegalovirus –survival strategies of a highly adapted opportunist. *Trends Microbiol.* **6**, 190–197.
355. Rolle, A., Mousavi-Jazi, M., Eriksson, M., Odeberg, J., Soderberg-Naucler, C., Cosman, D., Karre, K., and Cerboni, C. (2003). Effects of human cytomegalovirus infection on ligands for the activating NKG2D receptor of NK cells: Up-regulation of UL16-binding protein (ULBP) 1 and ULBP2 is counteracted by the viral UL16 protein. *J. Immunol.* **171**, 902–908.
356. Vales-Gomez, M., Browne, H., and Reyburn, H. T. (2003). Expression of the UL16 glycoprotein of human cytomegalovirus protects the virus-infected cell from attack by natural killer cells. *BMC Immunol.* **4**, 4.
357. Lodoen, M. B., and Lanier, L. L. (2005). Viral modulation of NK cell immunity. *Nat. Rev. Microbiol.* **3**, 59–69.
358. Cohen, G. B., Gandhi, R. T., Davis, D. M., Mandelboim, O., Chen, B. K., Strominger, J. L., and Baltimore, D. (1999). The selective downregulation of class I major histocompatibility complex proteins by HIV-1 protects HIV-infected cells from NK cells. *Immunity* **10**, 661–671.
359. Mavilio, D., Lombardo, G., Benjamin, J., Kim, D., Follman, D., Marcenaro, E., O'Shea, M. A., Kinter, A., Kovacs, C., Moretta, A., and Fauci, A. S. (2005). Characterization of CD56-/CD16+ natural killer (NK) cells: A highly dysfunctional NK subset expanded in HIV-infected viremic individuals. *Proc. Natl. Acad. Sci. USA* **102**, 2886–2891.
360. Achdout, H., Arnon, T. I., Markel, G., Gonen-Gross, T., Katz, G., Lieberman, N., Gazit, R., Joseph, A., Kedar, E., and Mandelboim, O. (2003). Enhanced recognition of human NK receptors after influenza virus infection. *J. Immunol.* **171**, 915–923.
361. Rajagopalan, S., and Long, E. O. (2005). Understanding how combinations of HLA and KIR genes influence disease. *J. Exp. Med.* **201**, 1025–1029.
362. Nelson, G. W., Martin, M. P., Gladman, D., Wade, J., Trowsdale, J., and Carrington, M. (2004). Cutting edge: Heterozygote advantage in autoimmune disease: Hierarchy of protection/susceptibility conferred by HLA and killer Ig-like receptor combinations in psoriatic arthritis. *J. Immunol.* **173**, 4273–4276.

363. van der Slik, A. R., Koeleman, B. P., Verduijn, W., Bruining, G. J., Roep, B. O., and Giphart, M. J. (2003). KIR in type 1 diabetes: Disparate distribution of activating and inhibitory natural killer cell receptors in patients versus HLA-matched control subjects. *Diabetes* **52**, 2639–2642.
364. Luszczek, W., Manczak, M., Cislo, M., Nockowski, P., Wisniewski, A., Jasek, M., and Kusnierczyk, P. (2004). Gene for the activating natural killer cell receptor, KIR2DS1, is associated with susceptibility to psoriasis vulgaris. *Hum. Immunol.* **65**, 758–766.
365. Boyton, R. J., Smith, J., Ward, R., Jones, M., Ozerovitch, L., Wilson, R., Rose, M., Trowsdale, J., and Altmann, D. M. (2005). HLA-C and killer cell immunoglobulin-like receptor (KIR) genes in idiopathic bronchiectasis. *Am. J. Respir. Crit. Care Med.* **173**, 327–333.
366. Naumova, E., Mihaylova, A., Stoitchkov, K., Ivanova, M., Quin, L., and Toneva, M. (2005). Genetic polymorphism of NK receptors and their ligands in melanoma patients: Prevalence of inhibitory over activating signals. *Cancer Immunol. Immunother.* **54**, 172–178.
367. Hiby, S. E., Walker, J. J., O'shaughnessy, K. M., Redman, C. W., Carrington, M., Trowsdale, J., and Moffett, A. (2004). Combinations of maternal KIR and fetal HLA-C genes influence the risk of preeclampsia and reproductive success. *J. Exp. Med.* **200**, 957–965.
368. Kollnberger, S., Bird, L., Sun, M. Y., Retiere, C., Braud, V. M., McMichael, A., and Bowness, P. (2002). Cell-surface expression and immune receptor recognition of HLA-B27 homodimers. *Arthritis Rheum.* **46**, 2972–2982.
369. Chan, A. T., Kollnberger, S. D., Wedderburn, L. R., and Bowness, P. (2005). Expansion and enhanced survival of natural killer cells expressing the killer immunoglobulin-like receptor KIR3DL2 in spondylarthritis. *Arthritis Rheum.* **52**, 3586–3595.
370. Groh, V., Bruhl, A., El Gabalawy, H., Nelson, J. L., and Spies, T. (2003). Stimulation of T cell autoreactivity by anomalous expression of NKG2D and its MIC ligands in rheumatoid arthritis. *Proc. Natl. Acad. Sci. USA.* **100**, 9452–9457.
371. Cook, M. A., Milligan, D. W., Fegan, C. D., Darbyshire, P. J., Mahendra, P., Craddock, C. F., Moss, P. A., and Briggs, D. C. (2004). The impact of donor KIR and patient HLA-C genotypes on outcome following HLA-identical sibling hematopoietic stem cell transplantation for myeloid leukemia. *Blood* **103**, 1521–1526.
372. Verheyden, S., Schots, R., Duquet, W., and Demanet, C. (2005). A defined donor activating natural killer cell receptor genotype protects against leukemic relapse after related HLA-identical hematopoietic stem cell transplantation. *Leukemia* **19**, 1446–1451.
373. Bishara, A., De Santis, D., Witt, C. C., Brautbar, C., Christiansen, F. T., Or, R., Nagler, A., and Slavin, S. (2004). The beneficial role of inhibitory KIR genes of HLA class I NK epitopes in haploidentically mismatched stem cell allografts may be masked by residual donor-alloreactive T cells causing GVHD. *Tissue Antigens* **63**, 204–211.

Index

A

Acinetobacter sp. strain ADP1 MutS, inactivation of, 13
Adaptor-mediated transcript amplification (AmTA), 503–504
 primers used for, 508–509
 reliability and sensitivity, 504–507
Adaptor proteins and NK receptors, 493
 CD3Z antigen, zeta polypeptide, 498
 hematopoietic cell signal transducer (HCST), 497–498
 high-affinity receptor I for Fc fragments of IgE gamma polypeptide, 498
 tyro protein tyrosine kinase-binding protein (TYROBP), 497
ADAR1 gene, 370
 homozygosity for, 371
Adar2$^{-/-}$ mice, behavioral dysfunctions of, 387
ADAR1 protein(s)
 antagonists, 383–384
 anti-HDV activity of, 385–386
 biological activities of
  antiviral activity, 385–386
  apoptosis and embryogenesis, 386–387
  gene silencing by RNA interference, 389–390
 catalytic domain of, 377
 and human diseases, 409–410
 and PKR in human genetic and infectious diseases
  human diseases, 409–410
  PKR and human diseases, 410
 sequence and functional domains, 375–379
 subcellular localization, 379–380
 substrates, 380–383
Adenosine deaminase acting on RNA (ADAR1), 362
 ADAR1 gene and transcripts, 370
  exon-intron organization, 371–374
  gene multiplicity and chromosome assignment, 370–371
  transcriptional regulation, 374–375
 biological activities of
  antiviral activity, 384–386
  apoptosis and embryogenesis, 386–387
  gene silencing by RNA interference, 389–390
Adenovirus VAI RNA, 383
"Amber/W", 381
AmTA. *See* Adaptor-mediated transcript amplification
Angiogenesis and HDC/histamine-mediated cancer development, 254–255
Antimycobacterial action of D3, role of macrophage reactive oxygen intermediates (ROI) in, 72
Antisense oligonucleotide (ASO) therapy, 313–314
ASO therapy. *See* Antisense oligonucleotide (ASO) therapy
ATPase activity, 9
ATP$_\gamma$S, 23
ATP hydrolysis, 21, 26
ATP to *E. coli* MutS, binding of, 24–25
Autoimmunity in NK cells and receptors, 517–518

B

Bacillus Calmette-Guerin (BCG), role of cd14/cr3 in phagocytosis of, 76
Bacillus subtilis, 13
Bacteriophage repressors, 156–157
Barley yellow dwarf virus (BYDV), RNA–RNA interactions in, 340–343
BCG, CD14-dependent phagocytosis of, 76–77
Binding of MutS, 23
Breast carcinomas and PKR, 410

Buckle, in double helical DNA, 150
Bunyamwera virus (BUNV), NSs protein of, 405

C

Caenorhabditis elegans T20 H4.4, 377
Calladine's rules, DNA structure, 151
Camptothecins
  Chk1 activation pathway by, 210–211
  Chk2 activation pathway by, 211–212
  for induction and stabilization of Top1 cleavage complex, 184, 185–186, 187
  pharmacological properties of, 187
  Top1 cleavage complexes by, 188
Campylobacter jejuni, 32
Cancer
  eIF2B activity in mRNA translation and, 282–283
  and HDC. *See* Cancer development and HDC
  localization of PTPs in loss/gain region of, 300
  prevention, role of selenium-containing proteins in, 99
  PTP genes mediated in, 303
  PTPs mutations in, 300, 302–303
Cancer development and HDC
  angiogenesis role in HDC/histamine-mediated cancer development, 254–255
  HDC expression in different cancers, 251–254
  role of histamine/HDC in development of gastric cancer, 256–258, 259
  role of inflammation induced by histamine/HDC for cancer development, 255–256
Canonical translation initiation factors, 351–352
Capindependent translational enhancer (3' CITE), 340–344
Carcinoembryonic antigen related cell-adhesion molecule 1 (CEACAM1), 475
Carcinogens, and Top1 cleavage complex, 184, 187
Cation-dependent 434 repressor gene regulatory activity, *in vivo* studies of, 169–170
CD48, 487, 489
CD244. *See* NK receptor 2B4

Cd14/cr3, role of, in phagocytosis of bacillus Calmette-Guerin (BCG), 76
CDNA for rat PKR, 396
CD3ζ antigen. *See* CD3Z antigen
CD3Z antigen, zeta polypeptide, 493, 498
  tumor evasion and, 516
CEACAM1. *See* Carcinoembryonic antigen related cell-adhesion molecule 1
Cell death phosphatases, 309
Cell-specific β-globin chromatin domain, role for LCR in establishing and regulating the erythroid, 454–455
Cell tropism, 351
Chk1 activation pathway by camptothecin, 210–211
Chk 2 activation pathway by camptothecins, 211–212
Chromatin
  cis-element utilization in, exquisite specificity of, 457
  domain establishment and regulation
    chromosome context-dependent trans-acting factor function, 458
    discontinuous histone modification patterns, 456–457
    exquisite specificity of cis-element utilization in chromatin, 457
    multiple distant transcriptional elements ascommon components of chromatin domains, 455–456
  domains, multiple distant transcriptional elements ascommon components of, 455–456
  immunoprecipitation (ChIP) analysis, 442
  remodeling components of β-globin chromatin domain, 447–448
Chromatin immunoprecipitation (ChIP), 394
Chromosome context-dependent trans-acting factor function, 458
Circularization of mRNA during translation, 337–339
Cisplatin, 27
CITE. *See* Capindependent translational enhancer
Closed-loop model for mRNA circularization during translation, 337–339
C-myc gene, camptothecin effect on, 189
Colon cancers, histamine/HDC in, 251–252
Computer modeling of HDC active site and catalytic mechanism, 245–246

INDEX

ConA treatment, 237
Conditional knockout mouse models, See tRNA$^{[Ser]Sec}$ in selenoprotein biosynthesis, 127–128
Cos-7 cells, 249, 251
Coxsackie virus (CV) infection of human pancreatic cells, inhibition of, 405
CpG, methylation of, 513–514
CPT-11, 187
CRP, 34
C-type lectin family, NK cell receptors from, 492–493
   subfamily A (KLRA1), 490
   subfamily B (KLRB), 490–491
   subfamily C (KLRC), 491, 494–495
   subfamily D (KLRD), 495
   subfamily G (KLRG), 495
   subfamily K (KLRK), 495–497
Cytohesin, regulation by rho and pi 3-kinase and requirement for, 63–66

D

DACH. See Diaminocyclohexane
Dam methylase, 33
Dam methyltransferase, 7
Dam MTase mutants of *S. enterica*, 34
Dam mutants, methylation deficient *E. coli*, 27
DAR. See Differentiation-associated repair
DAZ. See Deleted in AZoospermia (DAZ)-family proteins
DAZ gene, 344
Dcm recognition sites, 5
Dcp1. See Dipeptidyl carboxypeptidase
DDC. See L-DOPA decarboxylase
Decitabine, 315
Deinococcus radiodurans, 5
Deleted in AZoospermia (DAZ)-family proteins, protein-protein interactions in, 354–356
Dephosphorylation of proteins, 297–298
D(GATC), MutH on undermethylation of, 10
Diabetes, eIF2B activity in mRNA translation and, 277–280
Diaminocyclohexane (DACH), 27
Dideoxynucleotides, 8
Differentiation-associated repair (DAR), 2–3
1α,25-Dihydroxyvitamin D_3 and PI 3-Kinase, 66–67
   D_3-induced antimicrobial activity against salmonella enterica, 74–75
   D_3-induced monocyte antimycobacterial activity regulation, 71–74
   VDR-PI 3-kinase signaling complex regulates myeloid cell differentiation, 68–71
   vitamin D_3 action and signaling, 66–68
Dipeptidyl carboxypeptidase (Dcp1), 328
Discontinuous histone modification patterns, 456–457
DNA
   biosynthetic errors, repair of, 7
   damagebinding protein
     DDB1, 386
     DDB2, 386
   in double helical
     buckle, 150
     helical rise, 149
     helical twist, 149
     lateral displacement, 149
     propeller twist, 150
     roll angle, 149
     tilt, 149
   flexibility of and indirect readout, 147
   free energy (ΔG) twisting (torsion) of, 161
   helix, average structure of, 146
   lesions, types of, 2
   methylation, 34
   mismatch repair, 3–6
     methylation-independent, 28–31
     methyl-directed, 7–15
     pathways, common, 6
     proteins, application of, 35–37
     proteins, molecular structure of, 15–22
     proteins and processes, applications of, 35–37
     vs bacterial virulence, 31–35
   polymerase, proofreading activity of, 4
   recognition mechanism, indirect readout as, 145–146
   regulatory elements in, 436
   repair in, 2
   repair system
     definition, 2
     in human copying system, 2
   replication, fidelity of, 4
   sequence–dependent alterations, in protein–DNA complex, 147
   sequences
     binding of proteins to specific, 144

DNA (cont.)
  DNA structure and indirect readout of, by 434 repressor, 159–162
  structure, extrinsic forces and
    divalent cations and major groove, 154
    monovalent cations and minor groove, 153–154
  structure(s)
    and indirect readout of DNA sequence by 434 repressor, 159–162
    instrinsic forces and, 151–152
    intrinsic forces and, 150–151
    in 434 repressor–DNA complexes, polymorphism of, 160
DNA-binding assays, naked, 457
DNA-damaging agents, 3
  on mismatch repair, effect of, 27–28
DNA methylation and *HDC* gene regulation, 234–235
DNA methyltransferases (DNMT), 315
DNA-scanning process, 25–26
DNA sequence–dependent structural polymorphisms, structural and physicochemical basis for
  DNA conformation displays sequence-dependent polymorphism, 148–150
  extrinsic forces and DNA structure, 152–153
    divalent cations and major groove, 154
    monovalent cations and minor groove, 153–154
  instrinsic forces and DNA structure, 150–152
  sequence-dependent polymorphisms and indirect readout, 150
DNMT. *See* DNA methyltransferases
L-DOPA decarboxylase (DDC), 245
  flexible loop structure in, 246–247
Drosophila twine gene, 356
DSH. *See* Dyschromatosis symmetrica hereditaria
DsRBD of PKR and ADAR1, exons containing, 396
Dyschromatosis symmetrica hereditaria (DSH), 409

E

E. coli. *See Escherichia coli*
EFsec, 114–115

EIF-2α phosphorylation
  eIF2B GEF activity regulation by, 272–274
  translational control, 370
EIF-2α protein kinase family members, diagrams of, 395
EIF(2B). *See* Eukaryotic initiation factor 2B
EIF2B mutations in VWM, 285–286
EIF4E, 334
  interaction between PABP and, 337–338
  overexpression, in cancer, 282
EIF4E-binding proteins (4E-BP), hyperphosphorylation of, 281
EIF2ε phosphorylation, eIF2B GEF activity regulation by, 274
EIF4G, 334
  interaction between PABP and, 337–338
EIF2-GDP binary complex, 272
EKLF abrogates DNaseI, deletion of, 447
Endogenous β-globin chromatin domain, protein components of, 441–442
  chromatin remodeling components of, 447–448
  erythroid Kruppel-like factor, 447
  hematopoietic GATA proteins, 443–447
  NF-E2 and other maf response element binding proteins, 442–443
Endogenous DNA lesions, and Top1 cleavage complex, 184, 187
Endonucleases, for Top1-DNA damage repair, 206, 208–209
Endothelin B receptor, 382
Enterochromaffinlike (ECL) cells, gastrin stimulation of, 236
Epigenetic regulation
  of *HDC* gene expression, 234–235
  of *PTP* genes, 303–304
Epigenetic therapy and PTP, 315–316
ERK. *See* Extracellular signal-regulated kinases
Erythroid Kruppel-like factor, 447
Escherichia coli, 1–2
  dam mutants, 28
  methyl-directed DNA mismatch repair in, 12
  mismatch repair system of, 4
  MutH, 15
  MutH, MutL, and MutS proteins, structures of, 16
  MutL, 18

INDEX

MutS in, 8
mutS mutation, 29
   phenotype of, 4
   very short patch mismatch repair system (VSPMMR) of, 5
Escherichia coli cytidine deaminase, 377
Eukaryotic cellular mRNAs, translation in classical, 337–339
Eukaryotic initiation factor (eIF)2B, 272
   eIF2B GEF activity, regulation of
      regulation by pyridine dinucleotides, 274
      role of eIF2α phosphorylation, 272–274
      role of eIF2ε phosphorylation, 274
   role in mRNA translation alterations under various pathophysiological and physiological conditions
      cancer, 282–283
      diabetes, 277–280
      resistance exercise, 280–282
      sepsis, 282–285
      VWM disease, 285–287
   subunit composition and function, 275–277
Eukaryotic Sec tRNA$^{[Ser]Sec}$
   evolution of, 104–109
   isoforms, amounts and distributions of, 104
   tRNA and gene structures, 99–102
   Um34 Sec tRNA$^{[ser]sec}$ methylase, 103–104
   Um34 synthesis, 102–103
Extracellular signal-regulated kinases (ERK)
   map kinase pathway signaling and *HDC* gene regulation, 236
   phosphorylation of, 236
Extrinsic forces and DNA structure, 152–153
   divalent cations and major groove, 154
   monovalent cations and minor groove, 153–154
Ey, 438

F

FcεR1γ. *See* High-affinity receptor I for Fc fragments of IgE gamma polypeptide
Fcgr3 receptor, 478
F1 hybrid resistance and NK cells, 477–478
Flaviviruses, 336
Free energy (ΔG) twisting (torsion) of DNA, 161

G

GacA, 14
GacS, 14
GAPDH. *See* Glyceraldehyde-3-phosphate dehydrogenase
Gastric cancer
   cells, 236
   and histamine/HDC, 256–258, 259
Gastrin, 233
   for *HDC* gene expression, 238–239
   signaling pathway by, for *HDC* gene regulation, 236
   stimulation of ECL cells by, 236
GATA
   proteins, hematopoietic, 443–447
   recognition code (GRC), 446
GATA-1, targeted deletion of, 444
GB virus C/hepatitis G virus (GBV-C), 404
Genes, physicochemical constitution of, 2
Gene therapy and PTP, 312–314
Genomic sequences, *Helicobacter pylori*, 13
β-globins
   genes, 450
      coding, 437
      transcription of, 439
   locus
      activation, multistep model of, 451
   locus, system to dissect chromatin domain regulation, 437
      control region structure/function, 438–439
      models of long-range transcriptional activation, 439–441
   murine and human β-globin loci, 438
β-globin chromatin domain
   activation, multistep model of
      coregulator interactions, 452–453
      dissecting molecular determinants of chromatin domain assembly and regulation, 450–452
      segregating individual reaction steps, 453–454
   chromatin remodeling components of, 447–448
"GluR-B(Q)", 388
"GluR-B(R)", 388
Glyceraldehyde-3-phosphate dehydrogenase (GAPDH), and AmTA, 504–507
GPCR. *See* G-protein–coupled receptors
G-protein, 272

G-protein–coupled receptors (GPCRs), 53
Gram-negative bacteria, 9
GRC. *See* GATA recognition code

H

H274, 242
βH1, 438
HAT. *See* Histone acetyltransferases
HBV helper virus, 381
HCST. *See* Hematopoietic cell signal transducer
HDAC inhibitors butyrate and trichostatin A (TSA), 450
HDAg. *See* Hepatitis delta antigen
HDAg-S, *trans*-dominant inhibitors of, 385
HDC. *See* L-Histidine decarboxylase
HDC gene(s), 233
  expression
    epigenetic regulation of, 234–235
    signal transduction pathways in, 235–238
    transcription factors and *cis*-elements involved in, 238–239
  HDC processing
    and degradation, 242–243
    and enzyme activity, 242
    and intracellular localization, 243–244
    regulation of, 244–245
Helical rise, in double helical DNA, 149
Helical twist, in double helical DNA, 149
Helicobacter pylori, 32
  analysis, of complete genomic sequences, 13
  infection, 233, 256
  and *HDC* gene regulation, 237–238
Hematopoietic cell signal transducer (HCST), 493, 497–498
Hematopoietic lineages and *HDC* gene regulation, 237–238
Hemimethylated DNA molecules, 9
Heminregulated inhibitor (HRI), 273
Hemophilus influenzae, 14, 32–33
  dam, 34
  MutS, 25
Hepatic α2,6-sialyltransferase transcript, 382
Hepatitis delta antigen (HDAg), 381
HexAB systems, 4
HexB, 18
HFIR. *See* Homology-facilitated illegitimate recombination

High-affinity receptor I for Fc fragments of IgE gamma polypeptide (FcεR1γ), 493, 498
His606, 19
Histamine, functions of, 232–233
L-Histidine decarboxylase (HDC), 232
  in cancer development, role of
    angiogenesis role in HDC/histamine-mediated cancer development, 254–255
    HDC expression in different cancers, 251–254
    role of histamine/HDC in development of gastric cancer, 256–258, 259
    role of inflammation induced by histamine/HDC for cancer development, 255–256
  future aspects of, 249, 251
  posttranscriptional regulation of, 239
    computer modeling of HDC active site and catalytic mechanism, 245–246
    flexible loop domain and structural compaction of the dimer, 246–249
    HDC is translated as a 74-kDa protein undergoing posttranslational processing, 240–241
    HDC processing. *See* HDC processing
    regulation of HDC processing, 244–245
  transcriptional regulation of, 233
    epigenetic regulation of *HDC* gene expression, 234–235
    signal transduction pathways involved in *HDC* gene expression, 235–238
Histone acetyltransferases (HAT), 234–235, 450
Histone deacetylases (HDAC), 234–235
  inhibitors, 315–316
Histone mRNA, 344–350
Histone protein(s)
  and gene expression, 234
  RNA-protein interactions in synthesis of, 344–350
HIV and NK cells, 517
HLA-G and tumor evasion, 515–516
HMSH proteins to ATP, binding of, 24–25
Homology-facilitated illegitimate recombination (HFIR), 14
H1 receptors in colon cancer, 252
5-HT$_2$ serotonin receptors, 388
Human *ADAR1* cDNA, 371

INDEX

549

Human *ADAR1* gene, 370–371
  structure, organization of, 372
Human ADAR proteins, diagrams of, 376
Human β-globin loci, 438
Human cytomegalovirus (HCMV), 483, 501, 516–517
Human genetic and infectious diseases, ADAR1 and PKR in
  ADAR1 and human diseases, 409–410
  PKR and human diseases, 410
Human parainfluenza virus 2 (HPIV2), V protein of, 405
Human *PKR* gene
  exons of, 391
  organization of, 391
Hycamtin, 187
Hydrolysis-dependent translocation model, 24, 25
Hypergastrinemia, 257–258
Hypermutation in contingency loci, acquisition of, 32
HZPF100. *See* Zinc-finger protein

I

ICAM-1. *See* Intercellular adhesion molecule-1
IGF-1 levels in sepsis, 284
IL-6, role of HDC/histamine in regulation of, 256
Immunoglobulin family, NK cell receptors from, 488
  killer cell Ig-like receptors, 479, 480, 481–483
  leukocyte immunoglobulin-like receptors (LILR), 483, 484, 485
  low-affinity receptor for Fc fragment IgG variant III B, 485
  natural cytotoxicity triggering receptors, 485–487
  natural killer receptor 2B4, 487, 489
  other receptors, 489
Indirect readout by bacteriophage repressor proteins
  bacteriophage developmental decisions, 154–156
  DNA structure and indirect readout of DNA sequence by 434 repressor, 159–162

evidence for indirect readout of DNA sequence, 157–159
insights from structural studies
  role of arginine 43, 165–166
  role of solvent-mediated protein–DNA contacts, 166–169
lambdoid phage repressor structure, 156–157
role of extrinsic and intrinsic forces in indirect readout by 434 repressor, 162–165
Indirect readout for DNA recognition
  mechanism, 145–146
  sequence-dependent polymorphisms and, 150
  strategies
    altering the flexibility of DNA, 147
    average structure of DNA helix, 146
    modulating the conformation of protein–DNA complex, 147–148
Infant rat model, with *H. influenzae* strain R3549, 34
Inflammation by histamine/HDC and cancer development, 255–256
Initiation factors, for mRNA translation, 266
  See also Eukaryotic initiation factor (eIF)2B
Instrinsic forces and DNA structure, 151–152
Insulin and eIF2B GEF activity
  in diabetes, 279–280
  in resistance exercise, 281
  in sepsis, 283–284
Integration host factor (IHF), *Escherichia coli* in, 146
Intercellular adhesion molecule-1(ICAM-1), 61–62
  LPS-induced monocyte adherence to, 66
Interferon(s)
  as antiviral agents, 369
  regulating expression of genes, 370
  regulatory factor 1 (IRF1), 393
Interferon-stimulated gene factor 3 (ISGF3), 393
Internal ribosome entry segment/site (IRES), 349–353
Intrinsic forces and DNA structure, 150–151
IRES. *See* Internal ribosome entry segment/site

IRF1. *See* Interferon(s), regulatory factor 1
ISGF3. *See* Interferon-stimulated gene factor 3

J

$Jak1^{-/-}$ mice, phenotype of, 387

K

K308, 246
KCS. *See* Kinase conserved sequence
Killer cell Ig-like receptors (KIR), 479, 480, 481–483, 488
Killer cell lectin-like receptor subfamily A, member 1 (KLRA1), 478, 490, 492
Killer cell lectin-like receptor subfamily B (KLRB), 490–491, 492
Killer cell lectin-like receptor subfamily C (KLRC), 491, 492, 494–495
Killer cell lectin-like receptor subfamily D (KLRD), 492, 495
Killer cell lectin-like receptor subfamily G (KLRG), 492, 495
Killer cell lectin-like receptor subfamily K (KLRK), 492, 495–497
Kinase conserved sequence (KCS), 374
KIR. *See* Killer cell Ig-like receptors
KIR genes, 481–482
Kissing-loop structure in RNA–RNA interactions, 340–344
KLF4, *HDC* gene regulation by, 238–239
KLRA1. *See* Killer cell lectin-like receptor subfamily A
KLRB1. *See* Killer cell lectin-like receptor subfamily B
KLRC. *See* Killer cell lectin-like receptor subfamily C
KLRD1. *See* Killer cell lectin-like receptor subfamily D
KLRG1. *See* Killer cell lectin-like receptor subfamily G
KLRK1. *See* Killer cell lectin-like receptor subfamily K
Knockout animal models, drawbacks of, 305
Kruppel-like factor, erythroid, 447

Kunjin virus, 336
Kv1.1 RNA, editing of, 389

L

Lactococcus lactis, 29
LAIR1. *See* Leukocyte-associated immunoglobulin-like receptor 1
Large granular lymphocytes (LGL), 475, 477
Lateral displacement, in double helical DNA, 149
LC20, N-terminus of, 18
Leucocyte receptor complex (LRC), of NK-cell receptors, 498, 500
Leukocyte-associated immunoglobulin-like receptor 1 (LAIR1), 489
Leukocyte immunoglobulin-like receptors (LILR), 483, 484, 485, 488
LILR. *See* Leukocyte immunoglobulin-like receptors
Lipopolysaccharides (LPS), 54
Lipopolysaccharide and phosphatidylinositol 3-kinase (PI 3-kinase), 54–55
  bimodal regulation of IRAK-1 dependent on CD14/TLR4, CR3
  involvement of CD14/TLR4, CR3, and PI 3-kinase in IRAK-1 degradation, 60–61
  IRAK–1 degradation and tolerance to LPS, 60
  dependent activation of PKC-ζ in LPS-treated human monocytes for LPS-induced activation of PKC-ζ, 58–59
  in monocytes, 58
  LPS-induced adherence is regulated by Rho and PI 3-kinase
  monocyte adherence induced by LPS involves CD14 and LFA-1, 61–63
  regulation by RHO and PI 3-kinase and requirement for cytohesin, 63–66
  LPS-induced association and coordinate activation of $p^{53/56lyn}$ and PI 3-Kinase
  bacterial LPS induces CD14-dependent activation of PI 3-kinase in monocytes, 56–57
  LPS activates $P^{53/56lyn}$, 57–58
  LPS signaling pathway and LPS receptors, 55–56

INDEX

LN40, C-terminus of, 18
LN40–AMPPNP complex, crystal structure of, 19
Long-patch mismatch DNA repair (LPMMR), 5
Low-affinity IgG receptor, 485, 488
LPMMR. See Long-patch mismatch DNA repair
LPS-induced activation of PKC-ζ, attenuation of, 59
LRC. See Leucocyte receptor complex
L30 ribosomal protein, 114–115
Luteoviridae family, RNA–RNA interactions in, 340–341
Luteovirus, RNA–RNA interactions in, 340, 343
Ly49, 478
Lys116, 17
Lys610, 19
Lysis-lysogeny switch of temperate lambdoid bacteriophages, 156
Lys613 (P–2), 19

M

M. BOVIS BCG, CR3-mediated phagocytosis of, 77–79
M. tuberculosis H37Rv, 71–72
βmajor, 438, 442
malM567, 28
malM582, 28
malM594, 28
Mammalian class I_A PI 3-Ks, 53
Mammalian mitogen-activated protein kinase (MAPK), members of, 406–407
Mammalian Sec tRNA$^{[Ser]Sec}$, 101
Mammalian topoisomerase families, 180–183
Mast cells, HDC gene expression in, 237
MEF. See Mouse embryo fibroblast
Melanoma cells, 252
Methionyl-tRNA$_i$ (met-tRNA$_i$), 266
Methylation of PTP genes, 303
Methyl-directed DNA mismatch repair, 7–15
in E. coli, 12
Methyl-directed pathway, bidirectional excision capability of, 23
N-Methyl-N'-nitro-N-nitrosoguanidine (MNNG), 27
MHC class I-related chains A and B and tumor evasion, 514–515

MIC proteins, soluble form of (sMIC), 514–515
βminor, 438
Mismatched nucleotides and the excision machinery, communication between, 22–27
MNNG. See N-methyl-N'-nitro-N-nitrosoguanidine
Molecular switch model, 24, 25
Mononuclear phagocytes, 52
Mouse Adar1 gene structure, 373
Mouse embryo fibroblast (MEF), 393
Mre11/Rad50/Xrs2 complex (MRX) endonuclease, 209
mRNAs for eukaryotic cells, 337–339
mRNA translation
  protein–protein interactions in
    DAZ-family proteins, 354–356
    NSP3 of rotaviruses, 356–358
  5'-3' RNA–RNA interactions in, 339–344
Murine adenosine deaminase, X-ray crystal structures of, 377
Murine β-globin locus, 438
  coordinates of, 445
Murine β-globin locus, dynamic histone modification pattern of, 450
  erythroid cell-specific histone modification pattern, 449
Murine loci, 438
Murine Tenr, 377
Muscle protein synthesis, 280–281
Mus81/Mms4 endonuclease, 209
'Mut,' 7
MutH
  comparison of crystal structure of, 15
  homologues, 9
MutHLSU systems, 4
MutK encoding, 14
MutL, 18, 25
  N-terminal domain (NTD) of, 17
MutL, 5
MutL–MutH complex, molecular model of, 19–20
MutL NTD, 19
MutS
  binding of, 23
  dimerization of, 21
  DNA-binding activity of, 21
  homodimers, 26

MutS (cont.)
  protein, 4
  proteins function as homodimers, 21
MutS, 5
  in E. coli, 8
  gene
    from Pseudomonas putida, 14
    of S. typhimurium, 29
MutS–DNA complexes, 8
MutS homologues (MSH) genes, 31
MutS-mismatch complex, 10
MutS–MutL, sliding clamp complex, interaction of, 25
MutS–MutL–heteroduplex ternary complexes, 26
Mycobacteria, 32
  PI 3-kinase and phagocytosis of
    cross-talk between CD14 and CR3: regulation by PI 3-kinase and cytohesin-1, 76–79
    receptors and signaling, 75–76
    rescue of phagosome maturation by vitamin D3 Is PI 3-kinase dependent, 79–80
    receptors and signaling of phagocytosis and, 75–76
Mycobacterium smegmatis, 32
Myeloid leukemia cells, 67

N

Natural cytotoxicity triggering receptors (NCR), 488
  NCR1, 485–486
  NCR2, 476
  NCR3, 486–487
  and tumor invasion, 516
Natural killer cells
  definition and characterization of, 474–476
  heterogeneity of, 477
  principles of recognition of, 475
  target recognition, 478–479
  target specificity of, 477–478
Natural killer cell receptors
  adaptor molecules
    CD3Z antigen, zeta polypeptide, 498
    hematopoietic cell signal transducer (HCST), 497–498
    high-affinity receptor I for Fc fragments of IgE gamma polypeptide, 498
    tyro protein tyrosine kinase-binding protein (TYROBP), 497
  from C-type lectin family, 492–493
    subfamily A (KLRA1), 490
    subfamily B (KLRB), 490–491
    subfamily C (KLRC), 491, 494–495
    subfamily D (KLRD), 495
    subfamily G (KLRG), 495
    subfamily K (KLRK), 495–497
  from immunoglobulin superfamily, 488
    killer cell Ig-like receptors, 479, 480, 481–483
    leukocyte immunoglobulin-like receptors (LILR), 483, 484, 485
    low-affinity receptor for Fc fragment IgG variant III B, 485
    natural cytotoxicity triggering receptors, 485–487
    natural killer receptor 2B4, 487, 489
    other receptors, 489
  organization and evolution of receptor complexes
    evolution, 500–501
    leukocyte receptor complex (LRC), 498, 500
    natural killer complex (NKC), 500
  organization of gene clusters containing, 499
Natural killer cells and disease
  autoimmunity, 517–518
  pathogen immune evasion, 516–517
  tumor immune evasion
    CD3ζ, 516
    HLA-G, 515–516
    NCR, 516
    soluble MHC class I-related chains A and B, 514–515
    transforming growth factor-beta, 515
Natural killer complex (NKC), 500
Natural killer phenotypes in vivo, 512
NCR. See Natural cytotoxicity triggering receptors
Necrovirus, 340–341
Neisseria meningitides, 32
Neurotransmitter receptor transcript substrates, RNA editing of, 381–382
NF-E2 binding proteins, 442–443
NKC. See Natural killer complex
NK cells. See Natural killer cells

INDEX

NK lysate preamplification and housekeeper expression, 507, 508–509, 510
NK receptor 2B4 (CD244), 487, 488, 489
NK-receptor genes, expression of
  complex phenotypes, 502
  promoter analysis, 502–514
  single-cell analysis
    adaptor-mediated transcript amplification, 503–504
    AmTA reliability and sensitivity, 504–507
    natural killer phenotypes *in vivo*, 512
    NK lysate preamplification and housekeeper expression, 507, 508–509, 510
    receptor expression in the cell line YT, 510
    sensitivity considerations, 502–503
    specifity of NK cells, 501–502
Nonreceptor-type protein tyrosine phosphatases (NRPTP), 299
NRPTP. *See* Nonreceptor-type protein tyrosine phosphatases
NSP3 of rotaviruses, protein-protein interactions, 356–358
N-terminal ATPase, 18
N-terminal DNA-binding domain of 438 repressor, structure of, 158
N-terminal domain (NTD) of MutL, 17
Nucleotides, hydrolysis-independent diffusion for several, 25

O

Open reading frames (ORFs), 2
434 O_R region, 155
Oxaliplatin, 27
Oxytocin receptor, 382

P

P110, 365. *See also* ADAR1 protein(s)
P150, 365. *See also* ADAR1 protein(s)
P. aeruginosa mutS gene, disruption of, 32
Pancreatic tumors, HDC expression in, 253
Paramyxovirus family, viruses of, 405
Pathogen immune evasion and NK cells, 516–517
PDK. *See* Phosphoinositide-dependent kinases

Pharmacological properties of camptothecins, 187
Phase variation, bacterial pathogens, 31–32
Phenotypic characteristics attributable to the function of *PTP* genes, 306–308
Phosphatidylinositol 3-kinase (PI 3-kinase), 62
  antisense S-oligo to the p110 subunit of, 69
  classification and structure of, 53
  signalling, lipid products as downstream mediators of, 53–54
Phosphatidylinositol 3-kinase (PI 3-kinase) and 1α,25-Dihydroxyvitamin D_3, 66–67
  D_3-induced antimicrobial activity against salmonella enterica, 74–75
  D_3-induced monocyte antimycobacterial activity regulation
    phagocyte oxidative burst, 72
    resistance to *M. tuberculosis*, 71–72
    translocation of phagocyte oxidase components to the plasma membrane, 72–74
  VDR-PI 3-kinase signaling complex regulates myeloid cell differentiation
    D3-induced CD14 and CD11B expression is PI 3-kinase dependent, 68
    induction of CD14, 68–70
    vitamin D_3 action and signaling, 66–68
Phosphatidylinositol 3-kinase (PI 3-kinase) and lipopolysaccharide, 54–55
  bimodal regulation of IRAK-1 dependent on CD14/TLR4, CR3, 60–61
  dependent activation of PKC-ζ n LPS-treated human monocytes, 58–60
  LPS-induced adherence is regulated by Rho and PI 3-kinase, 61–66
  LPS-induced association and coordinate activation of $p^{53/56lyn}$ and PI 3-kinase, 56–58
  LPS signaling pathway and lps receptors, 55–56
Phosphatidylinositol 3-kinase (PI 3-kinase) and phagocytosis of mycobacteria
  cross-talk between CD14 and CR3
    CD14-dependent phagocytosis of BCG, 76
    CR3-mediated phagocytosis of *M. bovis* BCG, 77–79

Phosphatidylinositol 3-kinase
(PI 3-kinase) and phagocytosis of
mycobacteria (*cont.*)
role of CD14/CR3 in phagocytosis of
BCG, 76–77
phagocytosis of mycobacteria: receptors and
signaling, 75–76
rescue of phagosome maturation by vitamin
D3 Is PI 3-kinase dependent, 79–80
Phosphoinositide-dependent kinases
(PDK), 54
Phosphoinositide (P'tide) metabolites, 52
Picornavirus
IRES, 351
translation, RNA–protein interactions in,
350–353
P58^{IPK}, cellular protein, 402
PKC pathways and *HDC* gene
regulation, 236
PKR
activators of, 398–399
and ADAR1 in human genetic and infectious
diseases
ADAR1 and human diseases, 409–410
PKR and Human Diseases, 410
biological activities
antiviral activity, 404–405
cell growth and death, 407–408
signal transduction, 405–407
and breast carcinomas, 410
cDNAs, 392
gene and transcripts
exon–intron organization, 391–392
gene multiplicity and chromosome
assignment, 390
transcriptional regulation, 392–394
proteins
activation of kinase activity, 398–400
antagonists, 401–404
sequence and functional
domains, 394–398
subcellular localization, 398
substrates, 400–401
PKR gene, 370
P45/NF-E2 expression, 442
Polarity effect, 332
Poliovirus (PV), 336–337
Poly(A)-binding protein (PABP), 334
eIF4E, eIF4G, and, 337–338

Posphosery-tRNA$^{[Ser]Sec}$ kinase, 110–111
Posttranscriptional regulation of HDC, 239
computer modeling of HDC active site and
catalytic mechanism, 245–246
flexible loop domain and structural
compaction of the dimer, 246–249
HDC is translated as a 74-kDa protein
undergoing posttranslational
processing, 240–241
HDC processing, 242–244
regulation of HDC processing, 244–245
Potassium channels, mammalian ADAR1 in
editing of, 389
Pridoxal phosphate (PLP), 232
Prokaryotic cellular mRNAs, transcription and
translation in, 332, 333
Prokaryotic DNA repair proteins, phylogenetic
analysis of, 31
Propeller twist, in double helical DNA, 150
Protein dephosphorylation, 297–298
Protein–DNA complex, conformation
modulation of, 147–148
Protein kinase regulated by RNA. *See* PKR
Protein phosphorylation, 297–298
Protein-protein interactions, in mRNA
translation
DAZ-family proteins, 354–356
NSP3 of rotaviruses, 356–358
Protein tyrosine kinases (PTK), 298
Protein tyrosine phosphatases (PTP), 298
as drug targets for cancer, 311
epigenetic therapy, 315–316
gene therapy, 312–314
small molecule inhibitors, 314
tumor suppressor and oncogenic properties
of, 317–318
Pseudocircularization of RNA
molecules, 332–337
Pseudomonas aeruginosa, 32
Pseudomonas putida, *mutS* gene from, 14
Pseudomonas sp., phase variation in, 14
PTEN, mutations in, 283
PTP. *See* Protein tyrosine phosphatases
PTP genes
genetic and epigenetic alterations of, 299
epigenetic regulation of PTPs, 303–304
PTPs localized to regions of loss/gain in
human cancers, 300, 301
PTPs mutated in cancers, 300, 302–303

INDEX

transformation-related phenotypes
 attributable to, 304–305, 306–308,
 309–311
PTPN2 overexpression, 310
PTPN7 overexpression, 311
PTPRF gene, 300, 302
PTPRG gene, 300, 302, 309
PTPRH expression, 305
PTPROt expression, 310
PTPRT gene, 300, 302
Pyridine dinucleotides, eIF2B GEF activity
 regulation by, 274
Pyrococcus furiosus, 13

R

Rabbit reticulocyte lysate (RRL), 272–273
Rad27 endonuclease, 209
RAET. *See* Retinoic acid early transcripts
Rat ADAR1 cDNA, 373–374
Rat PKR, cDNA for, 396
Reactive oxygen intermediates (ROI), role of
 macrophage in, antimycobacterial action
 of D3, 72
Receptor-type protein tyrosine phosphatases
 (RPTP), 299
434 repressor, 145
 binding site, role of N2-NH2 group and
  number of base pair hydrogen bonds
  of, 163
 examination of, 165
Repressor-DNA contacts, noncontacted base
 sequence effects on, 167
Resistance exercise, eIF2B activity in mRNA
 translation and, 280–282
Retinoic acid early transcripts
 (RAET), 495–496
Reversible cleavage complexes of Top1, 188
RNA, adenosine deaminase acting on
 ADAR1 gene and transcripts
  exon–intron organization, 371–374
  gene multiplicity and chromosome
   assignment, 370–371
  transcriptional regulation, 374–375
 ADAR1 proteins
  antagonists, 383–384
  sequence and functional
   domains, 375–379

subcellular localization, 379–380
substrates, 380–383
biological activities of ADAR1
 antiviral activity, 384–386
 apoptosis and embryogenesis, 386–387
 gene silencing by RNA interference,
  389–390
 neurotransmitter receptors, 387–389
RNA, protein kinase regulated by
 biological activities
  antiviral activity, 404–405
  cell growth and death, 407–408
  signal transduction, 405–407
 PKR gene and transcripts
  exon–intron organization, 391–392
  gene multiplicity and chromosome
   assignment, 390
  transcriptional regulation, 392–394
 PKR proteins
  activation of kinase activity, 398–400
  antagonists, 401–404
  sequence and functional domains,
   394–398
  subcellular localization, 398
  substrates, 400–401
RNAbinding activity (dsRBD), N-terminal
 regulatory domain with, 394
RNA-editing enzyme, 370
RNA interference (RNAi), 309
 role of ADAR1 in, 389
RNA interference (RNAi) technology, 103
RNA-protein interactions, in mRNA translation
 case of histone protein synthesis,
  344–350
 case of picornavirus translation,
  350–353
RNA-recognition motif (RRM), 344
5′-3′ RNA–RNA interactions, in mRNA
 translation, 339–344
RNA-specific C-6 adenosine deaminase. *See*
 Adenosine deaminase acting on RNA
RNA viruses, 334–336
Roll angle, in double helical DNA, 149
Rotaviruses, protein–protein interactions
 in, 356–358
Rotavirus X protein associated with NSP3
 (RoXaNI), 357
RoXaNI. *See* Rotavirus X protein associated
 with NSP3

RPTP. *See* Receptor-type protein tyrosine phosphatases
RPTPη gene, adenovirus-mediated expression of, 313

S

Saccharomyces cerevisiae HRA400 gene products, 377
Salmonella enterica, 32
  dam MTase mutants of, 34
Salmonella flagella antigens, 31
Salmonella infection, 374
Salmonella serovars, 34
Salmonella typhimurium, 34, 375
  dam mutants, 35
  in gastrointestinal tract, 34
  mutS gene of, 29
SBP2, 114–115
Sec. *See* Selenocysteine
Sec codon, UGA as, 131–133
Sec insertion into protein
  EFsec, SBP2, the L30 Ribosomal Protein, 114–115
  SECIS elements, 112–114
  UGA, 116
Sec tRNA$^{[Ser]Sec}$, 99
  evolution of, 104–109
  isoforms, amounts and distributions of, 104
  roles of, 130–131
  in selenoprotein biosynthesis, mouse models for elucidating the role of, 123
    conditional knockout mouse models, 127–128
    roles of Sec tRNA$^{[Ser]Sec}$, 130–131
    transgenic/conditional knockout mouse models, 129–130
    transgenic/knockout mouse models, 128–129
    transgenic mouse models, 124–126
  structure of eukaryotic, 100–101
Selenium, 98
  and selenoprotein hierarchy, 116–120
Selenium-containing proteins, in cancer prevention, 99
Selenocysteine
  biosynthesis, 109
  posphosery-tRNA$^{[Ser]Sec}$ kinase and Sec synthases, 110–111

  selenophosphate synthetase, 111
  seryl-tRNA synthetase, 110
  UGA codes for, 99
Selenophosphate synthetase, 111
Selenoprotein(s), 98
  identity, function, and targeted removal
    identity of Sec UGA codons, 120–121
    selenoproteome, 122
    targeted removal, 122–123
Sepsis, eIF2B activity in mRNA translation and, 282–285
Sequence-dependent polymorphisms and indirect readout, 150
Ser51Ala mutant eIF-2a, expression of, 408
Serratia marcescens dam mutants, 35
Seryl-tRNA synthetase, 110
Sexual recombination, 2
Short excision repair tracts, 5
Sialic acid-binding immunoglobulin-like lectins (SIGLEC), 489
SIGLEC. *See* Sialic acid-binding immunoglobulin-like lectins
Signaling lymphocytic activation molecule (SLAM), 487
Signal transduction pathways in *HDC* gene expression, 235–238
Single-cell analysis of NK-receptor genes
  adaptor-mediated transcript amplification, 503–504
  AmTA reliability and sensitivity, 504–507
  natural killer phenotypes *in vivo*, 512
  NK lysate preamplification and housekeeper expression, 507, 510
  primers used for, 508–509
  receptor expression in the cell line YT, 510
  sensitivity considerations, 502–503
Single-strand DNA binding protein (SSB), 4
SL. *See* Stem-loop structure
SLBP. *See* Stem-loop binding protein
SLBP interacting protein (SLIP1), 348–349
"Sleeping Beauty" system, 313
Small interference RNA (siRNA), 309
Small molecule inhibitors of PTP, 314–315
SSB. *See* Single-strand DNA binding protein
Stable gene silencing of isoforms of PI 3-kinase and perspectives
  perspectives, 84
  silencing of p110α subunit of PI 3-kinase

INDEX

monocyte adherence induced by D_3, 81–83
stable silencing of P110a PI 3-kinase subunit, 80–81
Stable silencing of P110α PI 3-kinase subunit in human monocytic cell lines, 80–81
Staphylococcus aureus, 32
Static transactivation model, 24, 25
Stem-loop binding protein (SLBP), 344–350
Stem-loop structure (SL), 341
  in histone protein synthesis, 344–350
Stomach peptide hormones, 236
Streptococcus enterica serovar *typhimurium* lacking Dam methyl transferase (MTase), mutants of, 34
Streptococcus pneumoniae, 4, 28, 32
  chromosome, *mal* region of, 28
  mismatch repair systems of, 4
  mutants, 29
"Suicide complexes", 188

T

TATA binding protein (TBP), 145
TBP. *See* TATA Binding Protein
Tdp1, 195
TE. *See* 3′ translational enhancer
Temperate lambdoid bacteriophages, lysis-lysogeny switch of, 156
TGF-β1. *See* Transforming growth factor-beta
Therapeutic genes, 312
Thermus aquaticus MutS dimer, 21
Thermus thermophilus, 13
  MutS protein, 22
Thrombopoietin, 232
Tilt, in double helical DNA, 149
Time-post-ER-GATA-1 activation, 451
Tip60 protein, 235
Tissue tropism, 351
TNFBP. *See* Tumor necrosis factor (TNF)-binding protein
Tomato bushy stunt virus (TBSV), RNA–RNA interactions in, 340–343
Tombusviridae family, RNA-RNA interactions in, 340–341
Top1. *See* Topoisomerase I

Top1-associated DNA damage, 188–189
  checkpoint activation
    Chk1 activation by camptothecin, 210–211
    Chk2 activation by camptothecin, 211–212
    Chk1 vs Chk2 activation by camptothecin and, 212–213
  repair, 189, 191
    reversal of Top1-DNA covalent complexes by 5′-end religation, 190
    Top1 excision by endonucleases, 203, 208–209
    Top1 excision by tyrosyl-DNA-phosphodiesterase, 190, 195, 203, 206
Top1–DNA covalent complexes by 5′-end religation, reversal of, 190
Top1 excision, for Top1-DNA damage repair
  by endonucleases, 206, 208–209
  by tyrosyl-DNA-phosphodiesterase, 190, 195, 203, 206
Topoisomerase I (Top1), 175. *See also* Topoisomerase cleavage complexes
  functions and catalytic mechanisms, 180–183
Topoisomerase cleavage complexes, 181–182
  conversion of, into DNA damage, 188–189
  induction and stabilization of, 184, 185–186, 187
Top1 poisons
  genetic alterations conferring hypersensitivity to
    in budding yeast, 196–200
    in fission yeast, 204–205
  genetic alterations sensitizing mammalian cells to, 193–194
Transcriptional regulation of HDC, 233
  epigenetic regulation of *HDC* gene expression, 234–235
  signal transduction pathways involved in *HDC* gene expression, 235–238
  transcription factors and *cis*-elements involved in *HDC* gene expression, 238–239
Transcription factors for *HDC* gene expression, 238–239
Transforming growth factor-beta (TGF-β1) and tumor evasion, 515
Transgenic/conditional knockout mouse models, Sec tRNA[Ser]Sec in selenoprotein biosynthesis, 129–130

Transgenic mouse models, Sec tRNA[Ser]Sec in selenoprotein biosynthesis, 124–126
3' Translational enhancer (TE), 340–344
Trichostatin A (TSA), HDAC inhibitors butyrate and, 450
TRNA and gene structures, eukaryotic Sec tRNA[Ser]Sec, 99–103
Trp repressor, 145
Tumor cell lines and NK cells, 477
Tumor immune evasion and NK cells
  CD3ζ, 516
  HLA-G, 515–516
  NCR, 516
  soluble MHC class I-related chains A and B, 514–515
  transforming growth factor-beta, 515
Tumor necrosis factor (TNF)-binding protein (TNFBP), administration of, 284–285
Tumor suppressor genes, 299
Tumor suppressor protein phosphatase, mutations in, 283
Type II diabetes mellitus, 280
Type 1-protein phosphatase (PP1), catalytic subunit of, 402–403
TYROBP. *See* Tyro protein tyrosine kinase-binding proteins
Tyro protein tyrosine kinase-binding proteins (TYROBP), 493, 497
Tyrosyl-DNA-phosphodiesterase, for Top1-DNA damage repair, 190, 195, 203, 206

U

UGA, 116
  codes for Sec, 99
UL18, 483
UL16-binding proteins (ULBP), 495–497
ULBP. *See* UL16-binding proteins

Um34
  Sec tRNA[ser]sec methylase, 103–104
  synthesis, 102–103
Untranslated region (UTR), 340–344
5'-untranslated region (UTR), sequences, 233–234
UTR. *See* Untranslated region

V

Vaccinia virus
  E3L protein, 383–384
  proteins, 403
  recombinant (VVr) system, use of, for PKR, 404
Vanishing white matter (VWM) disease, eIF2B activity in mRNA translation and, 285–287
VDRE. *See* Vitamin D response element
Very short patch mismatch repair system (VSPMMR)
  of *E. coli*, 5
VGV isoform, in human brain, 389
Vibrio cholerae, 14
Vitamin D response element (VDRE), 67
VSV replication, effect of IFN on, 405
VWM disease. *See* Vanishing white matter disease

Y

Yersinia pseudotuberculosis, 34
YT cell line, NK receptor expression in, 510

Z

Zinc-finger protein (hZPF100), 345

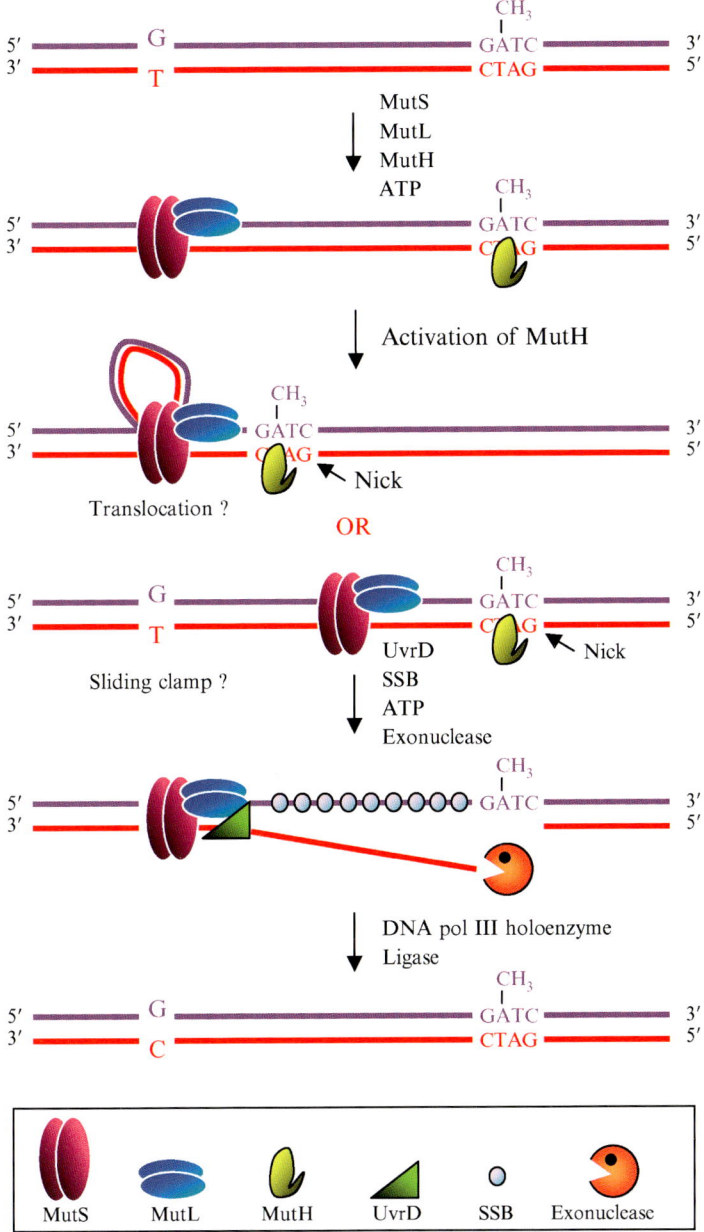

JOSEPH ET AL., FIG. 2. Schematic representation of methyl-directed DNA mismatch repair in *E. coli*. Only the 5'-3' exonuclease activity is shown.

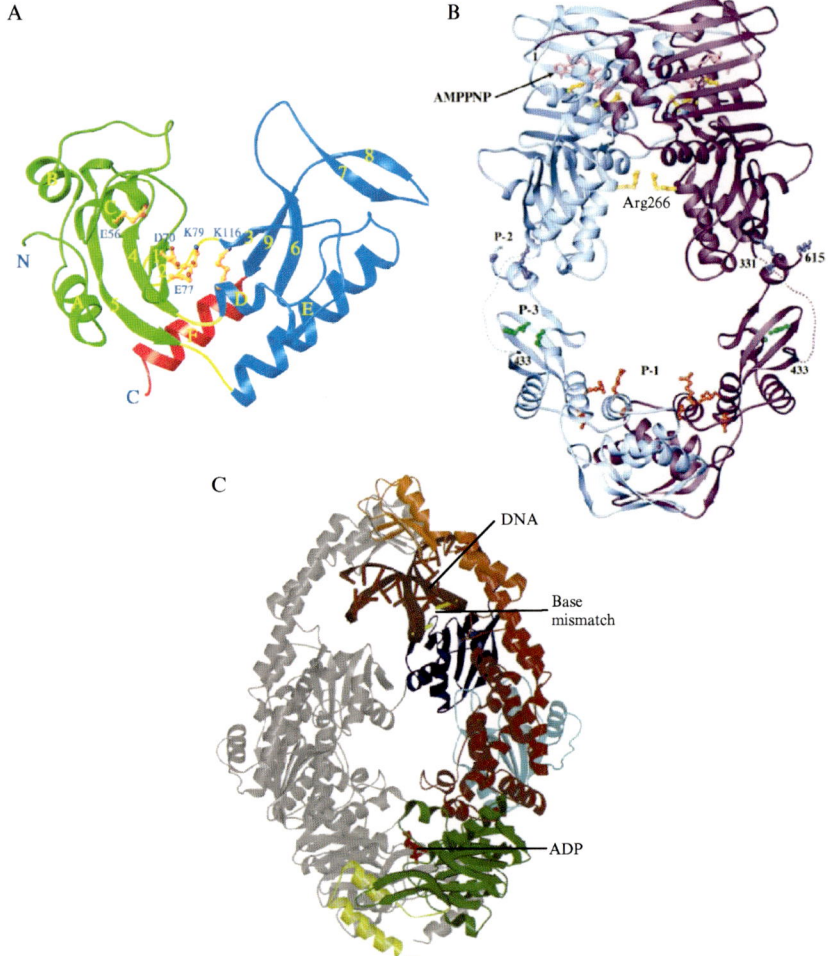

JOSEPH ET AL., FIG. 3. The structures of *E. coli* MutH, MutL, and MutS proteins. (A) Ribbon diagram of MutH. The N-terminal subdomain is colored green and the C-terminal subdomain blue, the C-terminal "lever" red, and linker peptides between the two subdomains are yellow. The five active site residues are shown. (B) A ribbon diagram of full-length MutL (LN40 and LC20 placed to share a common dyad axis). The C-terminus of LN40 (331) and N-terminus of LC20 (433) are linked by a dotted line. AMPPNP (pink), Asn33 (Mg^{2+} chelation), Glu29 (ATP hydrolysis), and Arg266 (DNA binding) shown in yellow. Clusters of positively charged residues (DNA binding) are represented as P-1 (red), P-2 (blue), and P-3 (green) patches. (C) MutS structure with mismatch-binding monomer shown in domains: N-terminal mismatch-recognition domain, dark blue; connector domain, light blue; core domain, red; clamp, orange; ATPase domain, green with red ADP; and helix-turn-helix domain, yellow. The other monomer is shown in gray. DNA is shown in red, with a yellow mismatch. [Reprinted by permission from Macmilan Publishers *EMBO J.* **23**, 4134–4145 (2004); **17**, 1526–1534 (1998); *Curr. Opin. Struct. Biol.* **11**, 47–52 (2001).]

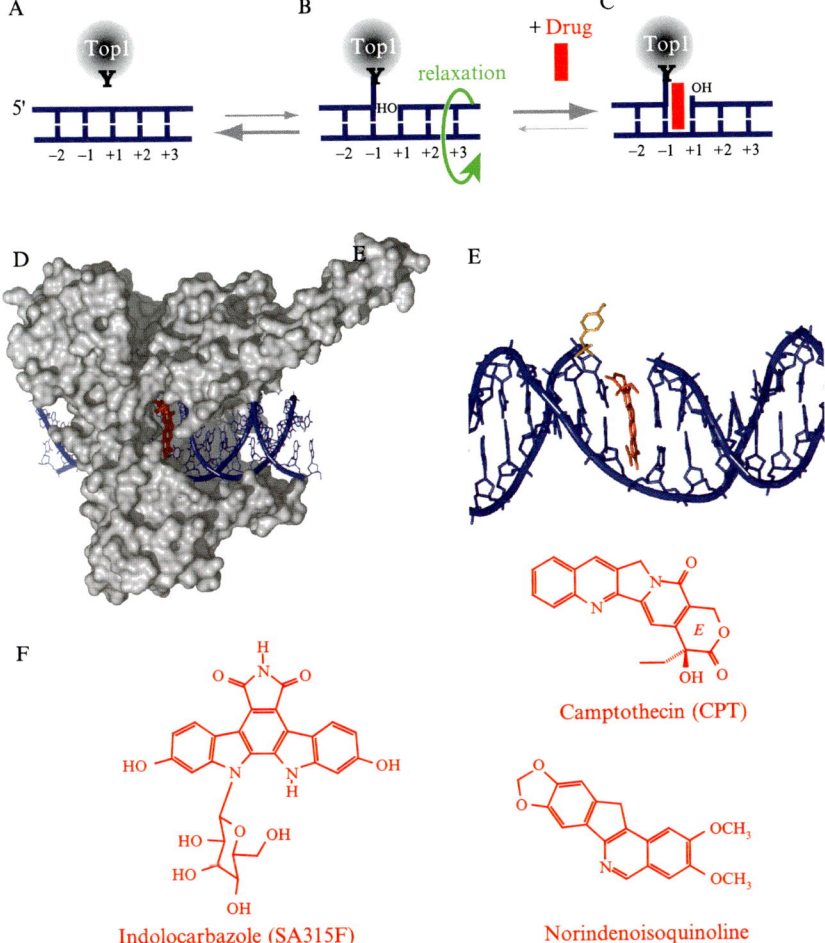

POMMIER ET AL., FIG. 3. Trapping of Top1 cleavage complexes by camptothecin and noncamptothecin inhibitors. (A) Under physiological conditions, Top1 is associated with chromatin in noncovalent complexes. (B) A small fraction of Top1 forms cleavage complexes that relax DNA supercoiling by controlled rotation of the cleaved strand around the intact strand (green curved arrow). (C) Anticancer drugs, such as those shown in panel F, reversibly trap the Top1 cleavage complex by inhibiting religation. (D) Crystal structure of camptothecin bound to the Top1–DNA cleavage complex [from (29)] showing "interfacial inhibition" (26, 27) of the Top1 cleavage complex by camptothecin. Interfacial inhibition also applies to noncamptothecin Top1 inhibitors shown in panel F (29, 30). (E) Same structure as in panel D. The Top1 has been removed except for the catalytic tyrosine (in orange). Camptothecin is shown intercalated between the base pairs flanking the Top1 cleavage site. (F) Structures of three Top1 inhibitors.

VOLZ AND RADELOFF, FIG. 1. Principles of NK cell recognition. The figure shows the principle of NK cell recognition. NK cells are shown as round shaped cells in red (cytolytic NK cell), in green (inhibited NK cell) and in gray (without target contact). Target cells are shown in yellow. Receptors on NK cells and corresponding ligands are color coded. Inhibitory receptors are red and activating receptors are green. The variety of receptors is demonstrated by differently colored dots. Target cell 1 is a healthy autologous cell not susceptible to NK cell lysis. Target cell 2 is a cell infected with a virus that leads to MHC antigen degradation and consequently to NK cell-mediated cytolysis. Target cell 3 is a tumor cell that evades NK cytolysis by shedding of MIC proteins.